Entropy in Image Analysis

Entropy in Image Analysis

Special Issue Editor

Amelia Carolina Sparavigna

MDPI • Basel • Beijing • Wuhan • Barcelona • Belgrade

Special Issue Editor
Amelia Carolina Sparavigna
Polytechnic University of Turin
Italy

Editorial Office
MDPI
St. Alban-Anlage 66
4052 Basel, Switzerland

This is a reprint of articles from the Special Issue published online in the open access journal *Entropy* (ISSN 1099-4300) from 2018 to 2019 (available at: https://www.mdpi.com/journal/entropy/special_issues/entropy_image_analysis)

For citation purposes, cite each article independently as indicated on the article page online and as indicated below:

LastName, A.A.; LastName, B.B.; LastName, C.C. Article Title. *Journal Name* **Year**, *Article Number*, Page Range.

ISBN 978-3-03921-092-3 (Pbk)
ISBN 978-3-03921-093-0 (PDF)

Contents

About the Special Issue Editor

Amelia Carolina Sparavigna (Dr.) is a physics researcher working mainly in the field of condensed matter physics and image processing. She graduated from the University of Torino in 1982 and obtained a Ph.D. in Physics at the Politecnico of Torino in 1990. Since 1993, she has carried out teaching and research activities at the Politecnico of Torino, as assistant and aggregate professor. Her scientific researches cover the fields of thermal transport and Boltzmann equation, liquid crystals, and the related image processing of polarized light microscopy. She has proposed new methods of image processing inspired by physical quantities such as the coherence length. Her recent works mainly concern the problem of the image segmentation. She is also interested in the history of physics and science. The papers that she has published in international journals are mainly on the topics of phonon thermal transport, the elastic theory of nematic liquid crystals, and the texture transitions of liquid crystals, investigated by means of image processing.

entropy

MDPI

Editorial

Entropy in Image Analysis

Amelia Carolina Sparavigna

Department of Applied Science and Technology, Polytechnic University of Turin, 10129 Turin, Italy; amelia.sparavigna@polito.it

Received: 13 May 2019; Accepted: 15 May 2019; Published: 17 May 2019

Keywords: image entropy; Shannon entropy; generalized entropies; image processing; image segmentation; medical imaging; remote sensing; security

---

Image analysis is playing a very essential role in numerous research areas in the fields of science and technology, ranging from medical imaging to the computer science of automatic vision. Being involved in several applications, which are mainly based on a constant innovation of technologies, image analysis always requires different approaches and new algorithms, including continual upgrades and improvements of existing methods. Accordingly, the range of the analyses in which it is engaged can be considered as wide as the prospective future technologies.

A challenge of image analysis is obtaining meaningful information by extracting specific features from images contained in large databases or during real-time acquisition. Among the problems requiring feature extraction are, to name a few, face detection and recognition, character recognition, and parametric determinations for augmented reality and other technologies. Another challenge is the secure encryption and decryption of multimedia data. These tasks demand highly sophisticated numerical and analytical methods.

The contributions to this Special Issue provide a good overview of the most important demands and solutions concerning the abovementioned extraction, encryption, and decryption of data. In all the contributions, entropy plays a pivotal role. In the following, the reader can find subjects and problems according to the order of their publication.

Lu et al. [1] consider a method for establishing an automatic and efficient image retrieval system. The proposed solution is an adaptive weighting method based on entropy and relevance feedback. Among the advantages of the proposed solution, an improved retrieval ability and accuracy of feature extraction are featured.

Zhu et al. investigate image data security in [2]. Image encryption is necessary to protect digital image transmission. The method proposed in the article is based on chaos and Secure Hash Algorithm 256 (SHA-256). Experimental results were used to check the algorithm, showing that it is safe and reliable.

Saqib and Kazmi [3] propose a solution to the problem of the retrieval and delivery of contents from audio-video repositories, in order to achieve faster browsing of collections. The compression of data is achieved by means of keyframes, which are representative frames of the salient features of the videos.

Karawia [4] reports an encryption algorithm for image data security to protect the transmission of multiple images. The algorithm is based on the combination of mixed image elements (MIES) and a two-dimensional economic map. Pure image elements (PIES) are used. The analysis of the experimental results verifies the proposed algorithm as efficient and secure.

A chaos-based image encryption scheme is the subject of an improved cryptanalysis proposed in [5] by Zhu et al. Their analysis integrates permutation, diffusion, and linear transformation processes. A color image encryption scheme is also given. Experimental results and a security analysis of the proposed cryptosystem are provided as well.

As pointed out in Yang et al. [6], distortions are usually introduced in images by their acquisition, and by their compression, transmission, and storage. An image quality assessment (IQA) method is therefore required. In their contribution, the authors propose an effective blind IQA approach for natural scenes and validate its performance.

Lin et al. [7] investigate a problem of medical analysis, concerning ultrasound entropy imaging. This imaging is compared with acoustic structure quantification (ASQ), a typical method for analyzing backscattered statistics. To illustrate this analysis, they describe a case study on the fat accumulation in the liver.

As stressed in Li et al. [8], it is not possible to capture all the details of a scene by means of a single exposure. Multi-exposure image fusion is required. In the algorithm proposed by the authors, the image texture entropy has its most relevant role in the adaptive selection of image patch sizes.

Image encryption returns in [9], where Huang and Ye propose an encryption algorithm based on a chaotic map. The two-dimensional chaotic map is the 2D Sine Logistic Modulation Map (2D-SLMM). The sophisticated use of keystream, time delay, and diffusion gives a high sensitivity to keys and plain images.

Today, the classification of hyperspectral images, those currently used for mapping the state of the Earth's surface, is fundamental. Consequently, approaches to characterize the quality of classified maps are required. Shadman Roodposhti et al. [10] discuss the uncertainty assessment of the emerging classification methods.

Mejia et al. [11] consider one of the fundamental tools of medical imaging. It is the imaging technique based on the reconstruction of positron emission tomography (PET) data. The authors propose a method that includes models of a priori structures to capture anatomical spatial dependencies of the PET images.

In this Special Issue, research devoted to the study of the surface quality of 3D printed objects is highlighted. An application of this is proposed by Fastowicz et al. [12]. The method is based on the analysis of the surface regularity during the printing process. In the case of the detection of low quality, some corrections can be made or the printing process aborted.

In Li et al. [13], a new approach to the registration of images is described. The method is based on Arimoto entropy with gradient distributions. The proposed approach provides a nonrigid alignment, based on an optimal solution of a cost function.

Miao et al. propose, in [14], a method for evaluating the anti-skid performance of asphalt pavement surfaces. Three-dimensional macro- and micro-textures of asphalt surfaces are detected. The method based on entropy is compared to the traditional macrotexture parameter Mean Texture Depth index.

Mello Román et al. [15] report a processing that improves the details of infrared images in [15]. The method aims to enhance contrast. At the same time, it preserves the natural appearance of images. A multiscale top-hat transform is used.

The encryption of images is also a hot topic of this Special Issue. Wen et al. [16] present another relevant work on this subject. Their paper illustrates a study of the image encryption algorithm based on DNA encoding and spatiotemporal chaos (IEA-DESC). It is shown that the IEA-DESC algorithm has some inherent security problems that need a careful check.

Nagy et al., in their article concerning the imaging of colonoscopy [17], propose a research with its framework in the methods based on the structural Rényi entropy. The aim of their work is to contribute to computer-aided diagnoses in finding colorectal polyps. The authors investigate characteristic curves that can be used to distinguish polyps and other structures in colonoscopy images.

Information entropy is involved in binary images and primality, as shown by an article in the Special Issue which deals with the hidden structure of prime numbers [18]. As demonstrated by the author, Emanuel Guariglia, the construction of binary images enables the generalization of numerical studies, which have indicated a fractal-like behavior of the prime-indexed primes (PIPs). PIPs are compared to Ramanujan primes to investigate their fractal-like behavior as well.

In Lang and Jia [19], the Kapur entropy for a color image segmentation is discussed. A new hybrid whale optimization algorithm (WOA), possessing a differential evolution (DE) as a local search strategy, is proposed to better balance the exploitation and exploration phases of optimization. Experimental results of the WOA-DE algorithm are proposed.

Li et al. [20] address image encryption by means of a method that integrates a hyperchaotic system, pixel-level Dynamic Filtering, DNA computing, and operations on 3D Latin Cubes—namely, a DFDLC image encryption. Experiments show that the proposed DFDLC encryption can achieve state-of-the-art results.

The problem of the multilevel thresholding segmentation of color images is considered in the work of Song et al. [21], according to a method based on a chaotic Electromagnetic Field Optimization (EFO) algorithm. The entropy involved in the method is fuzzy entropy. The EFO algorithm is a process inspired by the electromagnetic theory developed in physics.

The q-sigmoid functions, based on non-extensive Tsallis statistics, appear in [22]. Sergio Rodrigues et al. use them to enhance the regions of interest in digital images. The potential of q-sigmoid is demonstrated in the task of enhancing regions in ultrasound images, which are highly affected by speckle noise.

This Special Issue ends with a work devoted to an image processing method for person re-identification [23]. The method proposed by Ma et al. is based on a new deep hash learning, which is an improvement on the conventional method. Experiments show that the proposed method has comparable performances or outperforms other hashing methods.

As we have seen from the short descriptions of its contributions, this Special Issue shows that entropy in image analysis can have several variegated applications. However, applications of entropy are not limited to those described here. For this reason, the Guest Editor hopes that the readers, besides enjoying the present works, can receive positive hints from the reading and fruitful inspirations for future research and publications.

Acknowledgments: I express my thanks to the authors of the contributions of this Special Issue and to the journal *Entropy* and MDPI for their support during this work.

Conflicts of Interest: The author declares no conflict of interest.

References

1. Lu, X.; Wang, J.; Li, X.; Yang, M.; Zhang, X. An Adaptive Weight Method for Image Retrieval Based Multi-Feature Fusion. *Entropy* **2018**, *20*, 577. [CrossRef]
2. Zhu, S.; Zhu, C.; Wang, W. A New Image Encryption Algorithm Based on Chaos and Secure Hash SHA-256. *Entropy* **2018**, *20*, 716. [CrossRef]
3. Saqib, S.; Kazmi, S. Video Summarization for Sign Languages Using the Median of Entropy of Mean Frames Method. *Entropy* **2018**, *20*, 748. [CrossRef]
4. Karawia, A. Encryption Algorithm of Multiple-Image Using Mixed Image Elements and Two Dimensional Chaotic Economic Map. *Entropy* **2018**, *20*, 801. [CrossRef]
5. Zhu, C.; Wang, G.; Sun, K. Improved Cryptanalysis and Enhancements of an Image Encryption Scheme Using Combined 1D Chaotic Maps. *Entropy* **2018**, *20*, 843. [CrossRef]
6. Yang, X.; Li, F.; Zhang, W.; He, L. Blind Image Quality Assessment of Natural Scenes Based on Entropy Differences in the DCT Domain. *Entropy* **2018**, *20*, 885. [CrossRef]
7. Lin, Y.; Liao, Y.; Yeh, C.; Yang, K.; Tsui, P. Ultrasound Entropy Imaging of Nonalcoholic Fatty Liver Disease: Association with Metabolic Syndrome. *Entropy* **2018**, *20*, 893. [CrossRef]
8. Li, Y.; Sun, Y.; Zheng, M.; Huang, X.; Qi, G.; Hu, H.; Zhu, Z. A Novel Multi-Exposure Image Fusion Method Based on Adaptive Patch Structure. *Entropy* **2018**, *20*, 935. [CrossRef]
9. Huang, X.; Ye, G. An Image Encryption Algorithm Based on Time-Delay and Random Insertion. *Entropy* **2018**, *20*, 974. [CrossRef]
10. Shadman Roodposhti, M.; Aryal, J.; Lucieer, A.; Bryan, B. Uncertainty Assessment of Hyperspectral Image Classification: Deep Learning vs. Random Forest. *Entropy* **2019**, *21*, 78. [CrossRef]

11. Mejia, J.; Ochoa, A.; Mederos, B. Reconstruction of PET Images Using Cross-Entropy and Field of Experts. *Entropy* **2019**, *21*, 83. [CrossRef]
12. Fastowicz, J.; Grudziński, M.; Tecław, M.; Okarma, K. Objective 3D Printed Surface Quality Assessment Based on Entropy of Depth Maps. *Entropy* **2019**, *21*, 97. [CrossRef]
13. Li, B.; Shu, H.; Liu, Z.; Shao, Z.; Li, C.; Huang, M.; Huang, J. Nonrigid Medical Image Registration Using an Information Theoretic Measure Based on Arimoto Entropy with Gradient Distributions. *Entropy* **2019**, *21*, 189. [CrossRef]
14. Miao, Y.; Wu, J.; Hou, Y.; Wang, L.; Yu, W.; Wang, S. Study on Asphalt Pavement Surface Texture Degradation Using 3-D Image Processing Techniques and Entropy Theory. *Entropy* **2019**, *21*, 208. [CrossRef]
15. Mello Román, J.; Vázquez Noguera, J.; Legal-Ayala, H.; Pinto-Roa, D.; Gomez-Guerrero, S.; García Torres, M. Entropy and Contrast Enhancement of Infrared Thermal Images Using the Multiscale Top-Hat Transform. *Entropy* **2019**, *21*, 244. [CrossRef]
16. Wen, H.; Yu, S.; Lü, J. Breaking an Image Encryption Algorithm Based on DNA Encoding and Spatiotemporal Chaos. *Entropy* **2019**, *21*, 246. [CrossRef]
17. Nagy, S.; Sziová, B.; Pipek, J. On Structural Entropy and Spatial Filling Factor Analysis of Colonoscopy Pictures. *Entropy* **2019**, *21*, 256. [CrossRef]
18. Guariglia, E. Primality, Fractality, and Image Analysis. *Entropy* **2019**, *21*, 304. [CrossRef]
19. Lang, C.; Jia, H. Kapur's Entropy for Color Image Segmentation Based on a Hybrid Whale Optimization Algorithm. *Entropy* **2019**, *21*, 318. [CrossRef]
20. Li, T.; Shi, J.; Li, X.; Wu, J.; Pan, F. Image Encryption Based on Pixel-Level Diffusion with Dynamic Filtering and DNA-Level Permutation with 3D Latin Cubes. *Entropy* **2019**, *21*, 319. [CrossRef]
21. Song, S.; Jia, H.; Ma, J. A Chaotic Electromagnetic Field Optimization Algorithm Based on Fuzzy Entropy for Multilevel Thresholding Color Image Segmentation. *Entropy* **2019**, *21*, 398. [CrossRef]
22. Sergio Rodrigues, P.; Wachs-Lopes, G.; Morello Santos, R.; Coltri, E.; Antonio Giraldi, G. A q-Extension of Sigmoid Functions and the Application for Enhancement of Ultrasound Images. *Entropy* **2019**, *21*, 430. [CrossRef]
23. Ma, X.; Yu, C.; Chen, X.; Zhou, L. Large-Scale Person Re-Identification Based on Deep Hash Learning. *Entropy* **2019**, *21*, 449. [CrossRef]

entropy

MDPI

Article

An Adaptive Weight Method for Image Retrieval Based Multi-Feature Fusion

Xiaojun Lu, Jiaojuan Wang, Xiang Li, Mei Yang and Xiangde Zhang *

College of Sciences, Northeastern University, Shenyang 110819, China; luxiaojun@mail.neu.edu.cn (X.L.); 17640044931@163.com (J.W.); lxiang_1226@163.com (X.L.); yyangm1104@163.com (M.Y.)

* Correspondence: zhangxiangde@mail.neu.edu.cn; Tel.: +86-24-8368-7680

Received: 23 June 2018; Accepted: 31 July 2018; Published: 6 August 2018

Abstract: With the rapid development of information storage technology and the spread of the Internet, large capacity image databases that contain different contents in the images are generated. It becomes imperative to establish an automatic and efficient image retrieval system. This paper proposes a novel adaptive weighting method based on entropy theory and relevance feedback. Firstly, we obtain single feature trust by relevance feedback (supervised) or entropy (unsupervised). Then, we construct a transfer matrix based on trust. Finally, based on the transfer matrix, we get the weight of single feature through several iterations. It has three outstanding advantages: (1) The retrieval system combines the performance of multiple features and has better retrieval accuracy and generalization ability than single feature retrieval system; (2) In each query, the weight of a single feature is updated dynamically with the query image, which makes the retrieval system make full use of the performance of several single features; (3) The method can be applied in two cases: supervised and unsupervised. The experimental results show that our method significantly outperforms the previous approaches. The top 20 retrieval accuracy is 97.09%, 92.85%, and 94.42% on the dataset of Wang, UC Merced Land Use, and RSSCN7, respectively. The Mean Average Precision is 88.45% on the dataset of Holidays.

Keywords: image retrieval; multi-feature fusion; entropy; relevance feedback

1. Introduction

As an important carrier of information, it is significant to do efficient research with images [1–6]. Large-scale image retrieval has vast applications in many domains such as image analysis, search of image over internet, medical image retrieval, remote sensing, and video surveillance [7–24]. There are two common image retrieval systems: text-based image retrieval system and content-based image retrieval system. Text-based image retrieval system requires experienced experts to mark images, which is very expensive and time-consuming [7]. Content-based retrieval systems can be divided into two categories [8]. One is based on global features indexed with hashing strategies; another is local scale invariant features indexed by a vocabulary tree or a k-d tree. The two characteristics have pros and cons, and their performance complements each other [6,8]. In recent years, many excellent works focused on improving the accuracy and efficiency have been done [6]. A dynamically updating Adaptive Weights Allocation Algorithm (AWAA) which rationally allocates fusion weights proportional to their contributions to matching is proposed previously [7], which helps ours gain more complementary and helpful image information during feature fusion. In a previous paper [8], the authors improve reciprocal neighbor based graph fusion approach for feature fusion by the SVM prediction strategy, which increases the robustness of original graph fusion approach. In another past paper [9], the authors propose a graph-based query specific fusion approach where multiple retrieval sets are merged and are reranked by conducting a link analysis on a fused graph, which is capable of adaptively integrating the strengths of the retrieval methods using local or holistic features for different queries without any supervision. In a previous paper [10], the authors propose a simple yet effective late fusion method at score level by score curve and weighting different features in a query-adaptive manner. In another previous paper [11], the authors present a novel framework for color image retrieval through combining the ranking results of the different descriptors through various post-classification methods. In a past work [12], the authors propose robust discriminative extreme learning machine (RDELM), which enhances the discrimination capacity of ELM for RF. In a previous paper [13], the authors present a novel visual word integration of Scale Invariant Feature Transform (SIFT) and Speeded-Up Robust Features (SURF). The visual words integration of SIFT and SURF adds the robustness of both features to image retrieval. In another past work [14], an improved algorithm for center adjustment of RBFNNs and a novel algorithm for width determination have been proposed to optimize the efficiency of the Optimum Steepest Decent (OSD) algorithm, which achieves fast convergence speed, better and same network response in fewer train data. In a previous paper [15], an edge orientation difference histogram (EODH) descriptor and image retrieval system based on EODH and Color-SIFT was shown. In a previous paper [16], the authors investigate the late fusion of FREAK and SIFT to enhance the performance of image retrieval. In a previous paper [17], the authors propose to compress the CNN features using PCA and obtain a good performance. In a previous paper [18], the authors improve recent methods for large scale image search, which includes introducing a graph-structured quantizer and using binary.

Although the above methods have achieved good results, the performance of the retrieval system still has much room for improvement. In order to improve the performance of the retrieval system, it is an effective strategy to integrate multiple features for image retrieval [19–27]. Measurement level fusion is widely used, but how to determine the weight of each feature to improve the retrieval performance is still a very important problem [10,20,28]. In a previous paper [20], the author uses average global weight to fuse Color and Texture features for image retrieval. In a previous paper [9], the authors propose a graph-based query specific fusion approach without any supervision. In a previous paper [10], the author uses the area under the score curve of retrieval based on a single feature as the weight of the feature. The performances of different weight determination methods are different. The adaptive weights can achieve better retrieval performance than the global weights. In order to further improve the performance of the retrieval system, unlike previous weight determination methods, this paper proposes a new adaptive weight determination method based on relevance feedback and entropy theory to fuse multiple features. Our method has three outstanding advantages. (1) The retrieval system combines the performance of multiple features and has better retrieval accuracy and generalization ability than single feature retrieval system; (2) In each query, the weight of a single feature is updated dynamically with the query image, which makes the retrieval system make full use of the performance of several single features; (3) Unsupervised image retrieval means that there is no manual participation in the retrieval process. In an image search, no supervision is more popular than supervision. If we pursue higher retrieval accuracy, supervision is necessary. But from the perspective of user experience, unsupervised is better. It is worth mentioning that the method can be applied in two cases: supervised and unsupervised. Getting our method, firstly, we obtain single feature trust based on relevance feedback (supervised) or entropy (unsupervised); next, we construct a transfer matrix based on trust; finally, based on the transfer matrix, we get the weight of single feature through several iterations, which makes full use of single feature information of image and can achieve higher retrieval accuracy.

2. Related Work

For the image retrieval system integrating multi-features at measurement level, this paper mainly focus on how to determine the weight of each feature to improve the retrieval accuracy. In this section, we mainly introduce some work related to our method.

2.1. Framework

The main process of common system framework for image retrieval based on fusion of multiple features at the metric level is as follows [28–32]. Firstly, we extract several features of image and build benchmark image database. Then, when users enter images, we calculate the similarity between the query image and images of the database based on several features, separately. Finally, we get the comprehensive similarity measure by weighting several similarities and output retrieval results based on it.

2.2. The Ways to Determine Weight

A lot of work has been done to improve the performance of the retrieval system with multiple features [33,34]. At present, feature fusion is mainly carried out on three levels [8]: feature level, index level, and sorting level. The method proposed in this paper is applicable to the fusion of measurement level. Traditionally, there are two ways to determine the weight of feature, the global weight [11,20,32], and the adaptive weight [10,35], the pros/cons of each are listed in Table 1. The former is reciprocal of the number of features or decided by experienced experts, which leads the retrieval system to have poor generalization performance and low retrieval performance for different retrieval images. The latter is derived from retrieval feedback based on this feature, which is better than the global weight. However, in the sum or product fusion, the distinction between good features and bad features, is not obvious. If the weights of the bad features in the retrieval work are large, it will also reduce the retrieval performance to a certain extent. In order to clearly distinguish good features and bad features and the retrieval system can make full use of their performance to achieve better retrieval accuracy, a new adaptive weight retrieval system is proposed. Firstly, we obtain single feature trust based on relevance feedback (supervised) or entropy (unsupervised). Next, we construct a transfer matrix based on trust. Finally, based on the transfer matrix, we get the weight of single feature through several iterations, which makes full use of single feature information of image, and can achieve higher retrieval accuracy.

Table 1. Comparison of ways to determine weight.

Method	Pros	Cons
the global weight	short retrieval time	poor generalization performance/low retrieval performance
the adaptive weight	good generalization performance/excellent retrieval performance	long retrieval time

The common weighted fusion methods of measurement level are maximum fusion, multiplication fusion [10], and sum fusion [11,32]. The comprehensive metric obtained by maximum fusion is obtained from the feature with the maximum weight. The comprehensive metric obtained by multiplication fusion is the product of different weighted similarity measures. The comprehensive metric obtained by sum fusion is the adding of different weighted similarity measures. Specifically, K features labeled as are fused, q is a query image, $p_k \in \{p_1, p_2, \ldots, p_n\}$ is a target image of database $\Omega = \{p_1, p_2, \ldots, p_n\}$. Each method of fusion is shown as follows:

The maximum fusion:

$$sim(q) = \underset{D_i(q)}{\arg} \max\{{w_q}^{(i)} | i = 1, 2, \ldots\ldots, K\} \tag{1}$$

The multiplication fusion:

$$sim(q) = \prod_{i=1}^{K} w_q{}^{(i)} D_i(q),\ \{i = 1, 2, \ldots\ldots, K\} \tag{2}$$

The multiplication fusion:

$$sim(q) = \sum_{i=1}^{K} w_q{}^{(i)} D_i(q),\ \{i = 1, 2, \ldots\ldots, K\} \tag{3}$$

Here, q is a query image. K is the number of feature. $w_q{}^i$ is weight of $F_i \in \{F_1, F_2, \ldots, F_K\}$. $D_i(q) \in \{D_1(q), D_2(q), \ldots, D_K(q)\}$ is the similarity vector between the query image q and images of database $\Omega = \{p_1, p_2, \ldots, p_n\}$, which is calculated based on feature $F_i \in \{F_1, F_2, \ldots, F_K\}$. $sim(q)$ is Comprehensive similarity measure.

2.3. Relevance Feedback

The relevance feedback algorithm [34] is used to solve the semantic gap problem in content-based image retrieval, and the results obtained by relevance feedback are very similar to those of human [36,37]. The main steps of relevance feedback are as follows: first, the retrieval system provides primary retrieval results according to the retrieval keys provided by the user; then, the user determines which retrieval results are pleasant; finally, the system then provides new retrieval results according to the user's feedback. In this paper, we get the trust of single feature under the supervised condition through relevance feedback. Under the condition of supervision, this paper obtains the trust of single feature through relevance feedback.

3. Proposed Method

In this section, we will introduce our framework and adaptive weight strategy.

3.1. Our Framework

For a specific retrieval system, the weight of each feature is static in different queries. It causes low retrieval performance. In order to overcome the shortcoming, a new image retrieval system based on multi-feature is proposed. The basic framework of the retrieval system is shown in Figure 1.

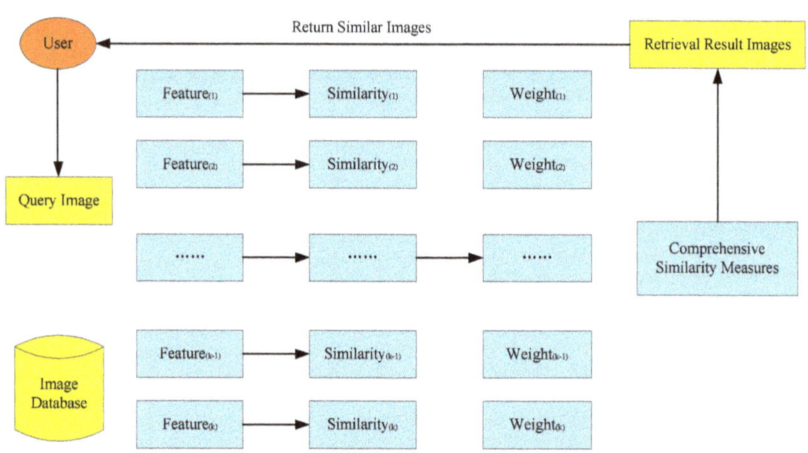

Figure 1. The proposed retrieval system framework.

In the database creation phase, firstly, we extract features separately; then, we calculate the entropy of different feature dimensions based on each feature; finally, we save features and entropies to get the image feature database. The original image database is a collection of large numbers of images. The established image feature database and the original image database are in a one-to-one correspondence, for example, the image 1.jpg is stored in the image database. The storage form of the image feature database is 1.jpg (image name), feature, and entropy. In this paper, what we call an image database is actually an image feature database.

In the image search phase, when users enter images, firstly, we calculate the similarity between the query image and the images of database based on each feature separately; then, we get the trust of a single feature; finally, we get the comprehensive similarity measure by weighting several measures and output retrieval results based on it.

Specifically, K features labeled as are fused, q is a query image, $p_k \in \{p_1, p_2, \ldots, p_n\}$ is a target image of database $\Omega = \{p_1, p_2, \ldots, p_n\}$. The proposed fusion method is as follows.

Firstly, considering that it will take a long time to calculate similarity measures using several features, we get binary feature as follows:

For each bit of feature $F_i \in \{F_1, F_2, \ldots, F_K\}$, we output binary codes $F_i \in \{F_1^{\bullet}, F_2, \ldots, F_K\}$ by:

$$ave(F_i) = \frac{\sum\limits_{j=1}^{m} F_i(c_j)}{m} \tag{4}$$

$$F_i(c_j) = \begin{cases} 1 & F_i(c_j) \geq ave(F_i) \\ 0 & F_i(c_j) < ave(F_i) \end{cases} \quad i \in \{1, 2, \ldots, K\} \tag{5}$$

Here, $ave(F_i)$ is the mean of feature $F_i \in \{F_1, F_2, \ldots, F_K\}$, m is the dimension of feature $F_i \in \{F_1, F_2, \ldots, F_K\}$, $F_i(c_j)$ is the j-th component of feature $F_i \in \{F_1, F_2, \ldots, F_K\}$.

Then, we calculate the distance between q and p, then normalize it:

$$d^i(k) = d^i(q, p_k) = \sum_{j=1}^{m} w_j |Fq^i(j) - Fp_k{}^i(j)| \qquad k \in \{1, 2, \ldots, n\}, i \in \{1, 2, \ldots, K\} \tag{6}$$

$$D_i(q) = 1 - \frac{1}{\sum\limits_{k=1}^{n} d^i(k)} (d^i(1), d^i(2), \ldots, d^i(n)) \quad (\mathrm{i} \in \{1, 2, \ldots, K\}) \tag{7}$$

Here, $D_i(q) \in \{D_1(q), D_2(q), \ldots, D_K(q)\}$ is the similarity vector between the query image q and images of database $\Omega = \{p_1, p_2, \ldots, p_n\}$, which is calculated based on feature $F_i \in \{F_1, F_2, \ldots, F_K\}$. n is the total number of images. Fq^i, $Fp_k{}^i$ respectively represent the feature $F_i \in \{F_1, F_2, \ldots, F_K\}$ of q and of $p_k \in \{p_1, p_2, \ldots, p_n\}$.

We calculate the comprehensive measure $sim(q)$ by fusing multiple features:

$$sim(q) = \sum_{i=1}^{K_1} \tilde{w}_q^{(i)} D_i(q) + \sum_{i=K_1+1}^{K} w_q{}^{(i)} D_i(q) \tag{8}$$

Here $\tilde{w}_q{}^{(i)} \in \left\{\tilde{w}_q{}^{(1)}, \tilde{w}_q{}^{(2)}, \ldots, \tilde{w}_q{}^{(K_1)}\right\}$, $w_q{}^{(i)} \in \{w_q{}^{(K_1+1)}, w_q{}^{(K_1+2)}, \ldots, w_q{}^{(K)}\}$, K_1 are the weight of a good feature, the weight of a bad feature, and the number of good features, respectively.

Finally, we sort the similarity $sim(q)$ and get the final search results.

3.2. Entropy of Feature

Information entropy is the expected value of the information contained in each message [38], represented as (n is the number of messages):

$$H_{(x)} = E(I_{(x)}) = \sum_{j=1}^{N} p(x)\log_2 p(x)^{-1} \tag{9}$$

Here, X is a random phenomenon. X contains N possibility. $p(x)$ is the probability of x. $H(X)$ is the nondeterminacy of the occurrence of X.

In our work, the entropy of j-th dimension feature is calculated as follows:

$$H_j = -\frac{1}{\log_2 n}\sum_{i=1}^{n}\frac{f_{ij}}{\sum\limits_{i=1}^{n} f_{ij}}\log_2\frac{\sum\limits_{i=1}^{n} f_{ij}}{f_{ij}} \quad , j \in \{1,2,\ldots,m\} \tag{10}$$

Here, N is the number of images in the database. M is the feature dimension. $f_{ij}, i \in \{1,2,\ldots,n\}$, $j \in \{1,2,\ldots,m\}$ is the j-th dimension feature of i-th image.

The weights of j-th dimension is calculated as follows:

$$w_j = \frac{e^{(1-H_j)}}{\sum\limits_{j=1}^{m} e^{(1-H_j)}}, j \in \{1,2,\ldots,m\} \tag{11}$$

Here, H_j is the entropy of j-th dimension feature. w_j is the weight of j-th dimension.

When all the values of feature are equal, the entropy H_j is 1. The weight of each feature component is equal to $\frac{1}{m}$.

3.3. Adaptive Weight Strategy

To overcome the problem of low retrieval performance caused by the weight determination method used with multiple feature fusion, this paper proposes a new method to obtain single feature weight. Our method can be applied to supervised learning and unsupervised learning. The specific methods are as follows:

Under the circumstances of supervision, the weight of a single feature is obtained based Relevance Feedback. $D_i(q) \in \{D_1(q), D_2(q), \ldots, D_K(q)\}$ is the similarity vector between the query image q and images of database, which is calculated based on feature $F_i \in \{F_1, F_2, \ldots, F_K\}$. We sort $D_i(q) \in \{D_1(q), D_2(q), \ldots, D_K(q)\}$ and return search results by it. The results are labeled as $a^i = \{{a_1}^i, {a_2}^i, \ldots\ldots, {a_t}^i\}$. Here, t represents the predefined number of returned images. The retrieved results are evaluated according to relevant feedback. The $pre_x, pre_y \in \{pre_1, pre_2, \ldots, pre_K\}$ as trust of single feature retrieval is calculated. That is to say, we rely on the feedback to evaluate the retrieval results, and then use the evaluation index on the dataset to calculate the retrieval performance that is the trust of the feature. For example, on the Wang dataset with the precision as the evaluation index, we search images based on $F_i \in \{F_1, F_2, \ldots, F_K\}$. If we find have h1 similar images in the h retrieval results by relevant feedback, we believe the trust of $F_i \in \{F_1, F_2, \ldots, F_K\}$ is h1/h. By several iterations, the weight of single feature is as follows: firstly, we structure the transfer matrix $H_{kk} = \{H(x,y)\}$, representing the performance preference among each feature. Note that the feature $F_x \in \{F_1, F_2, \ldots, F_K\}$ goes to feature $F_y \in \{F_1, F_2, \ldots, F_K\}$ with a bias of $H(x,y)$, the detailed construction process of $H_{KK} = \{H(x,y)\}$ is as follows:

$$\begin{aligned} & if \quad pre_y >= pre_x \\ & \qquad H(x,y) = e^{\alpha(pre_y - pre_x)} \qquad (\alpha \geq 1) \\ & else \\ & \qquad H(x,y) = |pre_y - pre_x| \end{aligned}$$

When the trust of $F_y \in \{F_1, F_2, \ldots, F_K\}$ is greater than $F_x \in \{F_1, F_2, \ldots, F_K\}$, in order to obtain better retrieval result, we believe that $F_x \in \{F_1, F_2, \ldots, F_K\}$ can be replaced by $F_y \in \{F_1, F_2, \ldots, F_K\}$. The replacement depends on the parameter α. The larger α is, the more the retrieval system depends on $F_y \in \{F_1, F_2, \ldots, F_K\}$. The $\alpha \geq 1$ is because $F_y \in \{F_1, F_2, \ldots, F_K\}$ is better than $F_x \in \{F_1, F_2, \ldots, F_K\}$, we need to get $e^{\alpha(pre_y - pre_x)} > |pre_y - pre_x|$, so that the weight of $F_y \in \{F_1, F_2, \ldots, F_K\}$ is larger and retrieval system relies more on $F_y \in \{F_1, F_2, \ldots, F_K\}$. When the trust of $F_y \in \{F_1, F_2, \ldots, F_K\}$ is equal to $F_x \in \{F_1, F_2, \ldots, F_K\}$, we believe that the $F_x \in \{F_1, F_2, \ldots, F_K\}$ can be replaced by $F_y \in \{F_1, F_2, \ldots, F_K\}$ the replacement bias $H(x, y)$ is 1. When the trust of $F_y \in \{F_1, F_2, \ldots, F_K\}$ is less than $F_x \in \{F_1, F_2, \ldots, F_K\}$, we think that $F_x \in \{F_1, F_2, \ldots, F_K\}$ can still be replaced by $F_y \in \{F_1, F_2, \ldots, F_K\}$, but the replacement bias $H(x, y)$ is relatively small. One benefit is that although retrieval performance based on some of the features of image retrieval is poor, we still believe that it is helpful for the retrieval task.

Then, the weight of a single feature is obtained by using the preference matrix. We initialize the weight w_1 to $w_1 = \left\{\frac{1}{K}, \frac{1}{K}, \ldots, \frac{1}{K}\right\}$. $w = \{w_{F_1}, w_{F_2}, \ldots, w_{F_K}\}$ is the weight of a single feature. The w_d is the newly acquired weights through iterations. The w_{d-1} is the weight of the previous iteration. We use the transfer matrix $H_{KK} = \{H(x, y)\}$ to iterate the weights based on formula 12.

$$\begin{bmatrix} w'_{F_1} \\ w'_{F_2} \\ \vdots \\ w'_{F_K} \end{bmatrix}_{(d)} = \gamma w'_{d-1} \begin{bmatrix} w'_{F_1} \\ w'_{F_2} \\ \vdots \\ w'_{F_K} \end{bmatrix}_{(d-1)} + (1-r) \begin{bmatrix} H(F_1, F_2) & \cdots & H(F_1, F_K) \\ H(F_2, F_1) & \cdots & H(F_2, F_K) \\ \vdots & & \vdots \\ H(F_K, F_1) & \cdots & H(F_K, F_K) \end{bmatrix} \begin{bmatrix} w'_{F_1} \\ w'_{F_2} \\ \vdots \\ w'_{F_K} \end{bmatrix}_{(d-1)}, (\gamma \in [0, 1]) \quad (12)$$

The w_d depends not only on the choice of features depending on the transfer matrix, but also on the w_{d-1} obtained from the previous calculation. The degree of dependence on the above two depends on the parameter γ. An obvious advantage of this voting mechanism is that it will not affect the final result because of a relatively poor decision. The process is as follows:

$$\begin{aligned} & w_d = \left\{\tfrac{1}{K}, \tfrac{1}{K}, \ldots, \tfrac{1}{K}\right\}, \qquad d = 1 \\ & repeat \\ & \qquad w'_d = \gamma w'_{d-1} + (1-r) H H w'_{d-1} \quad (\gamma \in [0, 1]) \\ & \qquad w_d = w_d / sum(w_d) \\ & \qquad d \leftarrow d + 1 \\ & Until \quad \|w_d - w_{d-1}\| < \varepsilon \quad (\varepsilon \geq 0) \\ & return \quad w_d \end{aligned}$$

- Good features and bad features

In our method, the weight of a single feature is different for different queries. In order to improve the retrieval accuracy, we hope that the features with better retrieval performance can have larger weight than those with poor retrieval performance. For this reason, we divide features into good features and bad features according to retrieval performance. We search image based on $F_y \in \{F_1, F_2, \ldots, F_K\}$ and $F_x \in \{F_1, F_2, \ldots, F_K\}$, respectively. If the retrieval performance of $F_y \in \{F_1, F_2, \ldots, F_K\}$ is better than $F_x \in \{F_1, F_2, \ldots, F_K\}$, we think that $F_y \in \{F_1, F_2, \ldots, F_K\}$ is a good feature and $F_x \in \{F_1, F_2, \ldots, F_K\}$ is a bad feature. Good features and bad features are specifically defined as follows:

$$\begin{aligned} & if \quad pre_y >= pre_x \\ & \qquad pre_y \in \{good\_feature\} \\ & else \\ & \qquad pre_x \in \{bad\_feature\} \end{aligned} \quad (13)$$

Here, $pre_y \in \{pre_1, pre_2, \ldots, pre_K\}$ is the retrieval performance of $F_y \in \{F_1, F_2, \ldots, F_K\}$, $pre_x \in \{pre_1, pre_2, \ldots, pre_K\}$ is the retrieval performance of $F_x \in \{F_1, F_2, \ldots, F_K\}$.

- Our method for unsupervised

Image retrieval based on the above adaptive weight strategy is a supervised retrieval process and users need to participate in the feedback of single feature trust. In the actual application process, users may prefer the automatic retrieval system. That is to say, unsupervised retrieval system without manual participation is more popular. Therefore, considering the advantages of unsupervised image retrieval, we further study this method and propose an adaptive weight method under unsupervised conditions. The unsupervised method is basically the same as the supervised method. The only difference is, in contrast to the supervised process, the weight of a single feature is obtained based entropy rather than relevant feedback.

First, the entropy of $D_i(q) = (d^{*i}(1), d^{*i}(2), \ldots\ldots, d^{*i}(n))$ is:

$$H_i = -\frac{1}{\log_2 n} \sum_{j=1}^{n} \frac{d^{*i}(j)}{\sum_{j=1}^{n} d^{*i}(j)} \log_2 \left(\frac{\sum_{j=1}^{n} d^{*i}(j)}{d^{*i}(j)} \right), i \in \{1, 2, \ldots, k\} \tag{14}$$

Here, $D_i(q) \in \{D_1(q), D_2(q), \ldots, D_K(q)\}$ is the similarity vector between the query image q and images of database, which is calculated based on feature $F_i \in \{F_1, F_2, \ldots, F_K\}$. n is the total number of images. $d^{*i}(j)$ is the similarity between the query image q and j-th image of database.

Then, the trust of $D_i(q) \in \{D_1(q), D_2(q), \ldots, D_K(q)\}$ is:

$$pre_i = H_i \tag{15}$$

Here, $D_i(q) \in \{D_1(q), D_2(q), \ldots, D_K(q)\}$ is the similarity vector between the query image q and images of database, which is calculated based on feature $F_i \in \{F_1, F_2, \ldots, F_K\}$. $pre_i \in \{pre_1, pre_2, \ldots, pre_K\}$ is the retrieval performance of $F_i \in \{F_1, F_2, \ldots, F_K\}$.

After gaining trust, the weight seeking process is the same as the supervised state.

4. Performance Evaluation

4.1. Features

The features we choose in this article are as follows:

- Color features. For each image, we compute 2000-dim HSV histogram (H, S, and V are 20, 10, and 10).
- CNN-feature1. The model we used to get CNN feature is VGG-16 [39]. We directly use pre-trained models to extract features from the fc7 layer as CNN features.
- CNN-feature2. The model we used to get CNN feature is AlexNet which is pre-trained by Simon, M., Rodner, E., Denzler, J., in their previous work [40]. We directly use the model to extract features from the fc7 layer as CNN features. The dimension of the feature is 4096.

The extraction methods of color feature, cnn-feature1, and cnn-feature2 belong to the results of the original papers and are well-known. So we did not retell it. However, the feature extraction code we adopted has been shared to the website at https://github.com/wangjiaojuan/An-adaptive-weight-method-for-image-retrieval-based-multi-feature-fusion.

4.2. Database and Evaluation Standard

- Wang (Corel 1K) [41]. That contains 1000 images that are divided into 10 categories. The precision of Top-r images is used as the evaluation standard of the retrieval system.
- Holidays [42]. That includes 1491 personal holiday pictures and is composed of 500 categories. mAp is used to evaluate the retrieval performance.

- UC Merced Land Use [43]. That contains 21 categories. Each category has 100 remote sensing images. Each image is taken as query in turn. The precision of Top-r images is used as the evaluation standard of the retrieval system.
- RSSCN7 [44]. That contains 2800 images which are divided into 7 categories. Each category has 400 images. Each image is taken as query in turn. The precision of Top-r images is used as the evaluation standard of the retrieval system.

The precision of Top-r images is calculated as follows:

$$precision = \frac{N_r}{r} \tag{16}$$

Here, N_r is the number of relevant images matching to the query image, r is the total number of results returned by the retrieval system.

The mAp is calculated as follows:

$$mAp = \frac{1}{|Q|}\sum_{i=1}^{|Q|}\frac{1}{RN_i}\sum_{j=1}^{RN_i}P({RS^j}_i) \tag{17}$$

Here, $|Q|$ is the number of query images, suppose $q_i \in Q$ is a retrieval image, RN_i is the total number of relevant images matching to q_i, ${RS^j}_i$ is ${RS^j}_i\_th$ similar image of query result and ${NR^j}_i$ is location information, $P({RS^j}_i)$ is the evaluation of retrieval results of q_i and is calculated as follows:

$$P({RS^j}_i) = \frac{{RS^j}_i}{{NR^j}_i} \tag{18}$$

4.3. Evaluation of the Effectiveness of Our Method

The main innovations of our method are as follows. (1) Based on entropy, we weigh features to improve the accuracy of similarity measurement; (2) Under the supervised condition, we obtain the single feature weight based on related feedback and fuse multi-feature at the measurement level to improve the retrieval precision; (3) Under the unsupervised condition, we obtain the single feature weight based on entropy and fuse multiple features at the measurement level to improve the retrieval precision. To verify the effectiveness of the method, we carried out experiments on Holidays, Wang, UC Merced Land Use, and RSSCN7.

We have done the following experiments. (1) Retrieve image based on CNN1-feature, Color feature, and CNN2-feature, respectively. At the same time, experiments are carried out under two conditions: entropy and no entropy; (2) under the state of supervision, retrieve image by fusing three different features which respectively uses relevance feedback and our method; (3) under the state of unsupervision, retrieve image by fusing three different features which respectively uses average global weights and our method. An implementation of the code is available at https://github.com/wangjiaojuan/An-adaptive-weight-method-for-image-retrieval-based-multi-feature-fusion.

4.3.1. Unsupervised

Under the unsupervised condition, in order to verify the effectiveness of the adaptive weight method proposed in this paper, we carried out experiments on Holidays, Wang, UC Merced Land Use, and RSSCN7 datasets. Table 2 shows a comparison of retrieval results based on AVGand OURS. On the Holidays dataset, our method is better than RF, and improves the retrieval precision by 5.12%. On the Wang dataset, our method improves the retrieval accuracy by 0.35% (Top 20), 0.47% (Top 30), and 0.58% (Top 50) compared with AVG. On the UC Merced Land Use dataset, our method improves the retrieval accuracy by 6.61% (Top 20), 9.33% (Top 30), and 12.59% (Top 50) compared with AVG. On the RSSCN7 dataset, our method improves the retrieval accuracy by 2.61% (Top 20), 3.14% (Top 30), and 3.84% (Top 50) compared with AVG.

Table 2. Comparison of retrieval results based on AVG and OURS under unsupervised conditions.

Database	Holidays	Wang (Top)			UC Merced Land Use (Top)			RSSCN7 (Top)		
		20	30	50	20	30	50	20	30	50
AVG	0.7872	0.9446	0.9274	0.8924	0.8468	0.7851	0.6866	0.8842	0.8611	0.8251
OURS	0.8384	0.9481	0.9321	0.8982	0.9129	0.8784	0.8125	0.9103	0.8925	0.8635

On Wang, UC Merced Land Use, RSSCN7, and Holidays, 50 images were randomly selected as query images, separately. We search similar images by our method. Figure 2 shows the change of weight with precision of each single feature. The abscissa is the features. From left to right, three points as 1 group, shows the precision and weights of each single feature of the same image retrieval. For example, in Figure 2a, the abscissa of 1–3 represents the three features of the first image in the 50 images selected from the Holidays. The blue line represents the weight, and the red line indicates the retrieval performance. We can see that the feature whose retrieval performance is excellent can obtain a relatively large weight by our method. That is to say, our method can make better use of good performance features, which is helpful to improve the retrieval performance.

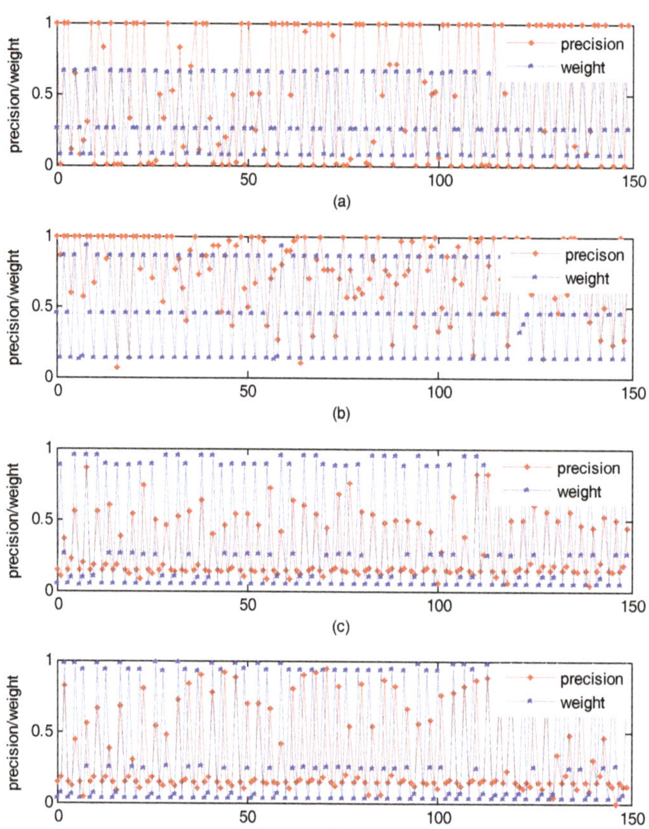

Figure 2. Under unsupervised condition, the change of weight obtained by our method with precision. (**a**) Experiment result on Holidays; (**b**) Experiment result on Wang; (**c**) Experiment result on UC Merced Land Use; (**d**) Experiment result on RSSCN7.

On Wang, UC Merced Land Use, and RSSCN7, one image was randomly selected as a query image and Top 10 retrieval results obtained by our method, respectively. On Holidays, one image was randomly selected as query image, respectively, and the Top 4 retrieval results obtained by our method. Figure 3 shows the retrieval results. The first image in the upper left corner is a query image that is labeled "query". The remaining images are the corresponding similar images that are labeled by a similarity measure such as 0.999. In accordance with similarity from large to small, we arrange retrieval results from left to right and from top to bottom.

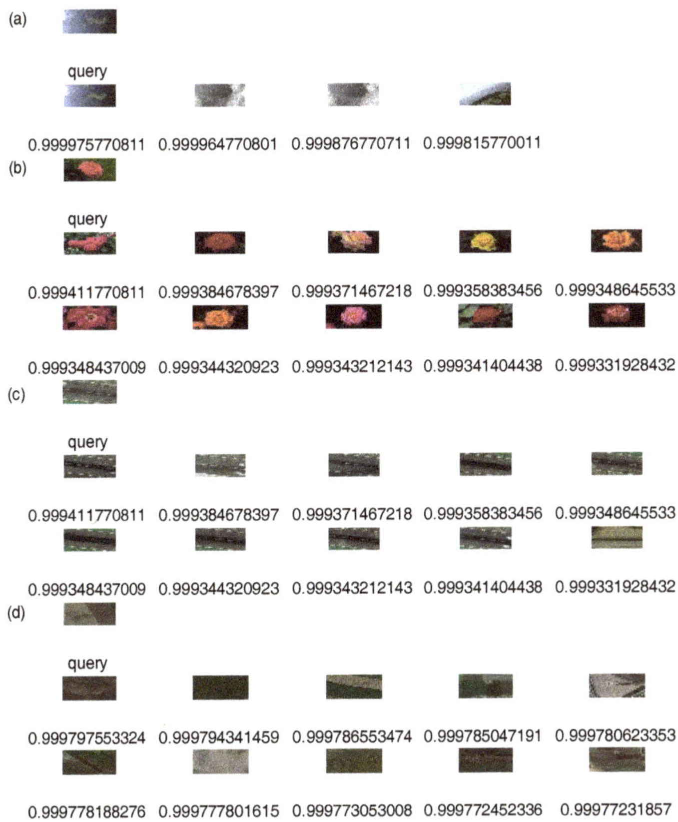

Figure 3. Under unsupervised condition, retrieval results were displayed. (**a**) Experiment result on Holidays; (**b**) Experiment result on Wang; (**c**) Experiment result on UC Merced Land Use; (**d**) Experiment result on RSSCN7.

4.3.2. Supervised

Under supervised conditions, in order to verify the effectiveness of the adaptive weight method proposed in this paper, we carried out experiments on Holidays, Wang, UC Merced Land Use, and RSSCN7 datasets. Table 3 shows a comparison of retrieval results based on RF and OURS. On the Holidays dataset, our method is better than RF to improve the retrieval precision by 0.26%. On the Wang dataset, our method improves the retrieval accuracy by 0.38% (Top 20), 0.38% (Top 30), and 0.34% (Top 50) compared with RF. On the UC Merced Land Use dataset, our method improves the retrieval accuracy by 0.38% (Top 20), 0.45% (Top 30), and 0.05% (Top 50) compared with RF. On the

RSSCN7 dataset, our method improves the retrieval accuracy by 0.84% (Top 20), 0.84% (Top 30), and 0.63% (Top 50) compared with RF.

Table 3. Comparison of retrieval results based on RF and OURS under supervised conditions.

Database	Holidays	Wang (Top)			UC Merced Land Use (Top)			RSSCN7 (Top)		
		20	30	50	20	30	50	20	30	50
RF	0.8819	0.9671	0.9539	0.9260	0.9247	0.8881	0.8250	0.9358	0.9191	0.8892
OURS	0.8845	0.9709	0.9577	0.9294	0.9285	0.8926	0.8255	0.9442	0.9275	0.8955

Similar to unsupervised state, on Wang, UC Merced Land Use, RSSCN7, and Holidays, 50 images were randomly selected as query images, separately. We search similar images by our method. Figure 4 shows the change of weight with precision of each single feature. The abscissa is the features. From left to right, three points as 1 group, shows the precision and weight of each single feature of same image retrieval. For example, in Figure 2a, the abscissa 1–3 represents the three features of the first image in the 50 images selected from the Holidays. The blue line represents the weight and the red line indicates the retrieval performance. We can see that the retrieval performance of feature got by relevance feedback is excellent, and can obtain a relatively large weight by our method. That is to say, our method can make better use of good performance features, which is helpful to improve the retrieval performance.

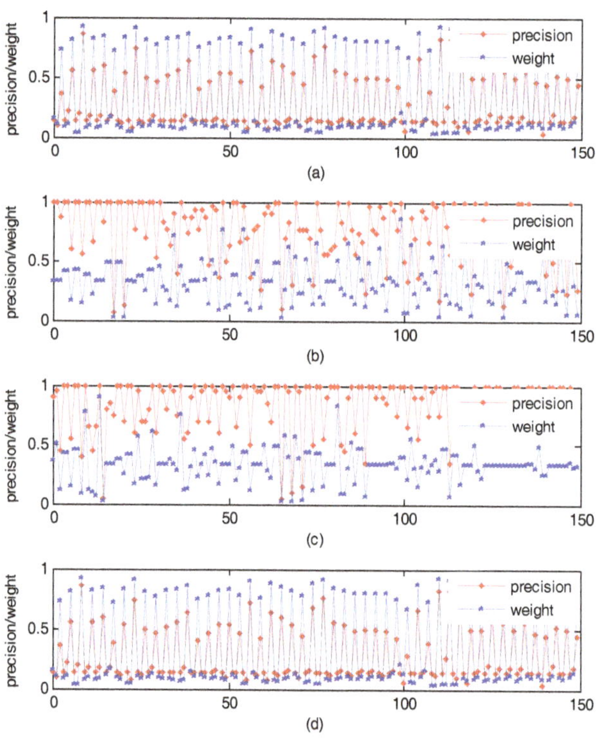

Figure 4. Under supervised condition, the change of weight that obtained by our method with precision. (**a**) Experiment result on Holidays; (**b**) Experiment result on Wang; (**c**) Experiment result on UC Merced Land Use; (**d**) Experiment result on RSSCN7.

Similar to unsupervised state, on Wang, UC Merced Land Use, RSSCN7, one image was randomly selected as query image, Top 10 retrieval results were obtained by through our method, respectively. On Holidays, one image was randomly selected as query image, respectively, Top 4 retrieval results obtained by our method. Figure 5 shows the retrieval results. The first image in the upper left corner is a query image that is labeled "query". The remaining images are the corresponding similar images that are labeled by similarity measure such as 0.999. In accordance with similarity from large to small, we arrange them from left to right and from top to bottom.

Figure 5. Under supervised condition, retrieval results were displayed. (**a**) Experiment result on Holidays; (**b**) Experiment result on Wang; (**c**) Experiment result on UC Merced Land Use; (**d**) Experiment result on RSSCN7.

4.4. Comparison with Others Methods

In order to illustrate the performance of supervised and unsupervised methods compared with existing methods. In Table 4, we show the comparison results on the Wang dataset (Top 20). Under the state of unsupervision, the precision of our method is 97.09%, which is about 26% higher than previous methods listed [13,14]. Compared with a previous paper [12], it increased by approximately 9.26%. Compared with a previous paper [15], it increased by 24.42%. Compared with a previous paper [16], it increased by about 22.29%. Under the state of unsupervision, the precision of our method is 94.81%, which is about 23.72% higher than [13,14]. Compared with a previous paper [12], it increased by about 6.98%. Compared with a previous paper [15], it increased by 22.14%. Compared with a previous paper [16], it increased by about 20.01%. From the results, we can see that the method has achieved good results both under supervision and unsupervision. As suggested in Section 3, the supervised method requires users to participate in the feedback of single feature trust, which may cause some users' aversion. The unsupervised method does not require users to participate in the selection of features, and directly outputs the retrieved images. The unsupervised method or supervised method is determined by the designer according to the actual use of the retrieval system. When we focus on user experience, we choose to be unsupervised. If we focus on higher retrieval accuracy, we choose to be supervised. After deciding whether to adopt supervised or unsupervised, the designer can make use of the corresponding solutions proposed in this paper to improve retrieval performance.

Table 5 shows the comparison results on the Holidays dataset. The map of our method is 88.45%. Compared with a previous paper [7], it increased by about 1.55%. Compared with a previous paper [8], it increased by 2.93%. Compared with a previous paper [9], it increased by about 3.81%. Compared with a previous paper [10], it increased by about 0.45%. Compared with [17], it increased by about 9.15%. Compared with a previous paper [18], it increased by about 3.65%.

(Note: To avoid misunderstanding, we do not use an abbreviation of each solution here, but the methods used in comparison are introduced in introduction.)

Table 4. Comparison with others methods on Wang.

Method	Ours		[11]	[12]	[13]	[14]	[15]	[16]
	Supervised	**Unsupervised**						
Africa	**87.70**	81.95	51.00	-	69.75	58.73	74.60	63.64
Beach	**99.35**	98.80	90.00	-	54.25	48.94	37.80	60.99
Buildings	**98.15**	97.25	58.00	-	63.95	53.74	53.90	68.21
Buses	**100.00**	100.00	78.00	-	89.65	95.81	96.70	92.75
Dinosaurs	**100.00**	100.00	100.00	-	98.7	98.36	99.00	100.00
Elephants	**99.10**	97.45	84.00	-	48.8	64.14	65.90	72.64
Flowers	99.95	99.45	**100.00**	-	92.3	85.64	91.20	91.54
Horses	**100.00**	100.00	100.00	-	89.45	80.31	86.90	80.06
Mountains	**98.00**	92.15	84.00	-	47.3	54.27	58.50	59.67
Food	**88.65**	81.05	38.00	-	70.9	63.14	62.20	58.56
Mean	**97.09**	**94.81**	78.3	87.83	70.58	70.31	72.67	74.80

Table 5. Comparison with others methods on Holidays.

Method	Ours	[7]	[8]	[9]	[10]	[17]	[18]
mAp	**88.45**	86.9	85.52	84.64	88.0	79.3	84.8

5. Discussion

Fusing multiple features can elevate the retrieval performance of retrieval system effectively. Meanwhile, in the process of multi-feature fusion, the proper single feature weight is helpful to further improve retrieval performance. This paper proposes a method to obtain single feature weights to fuse multiple features for image retrieval.

Retrieval results on daily scene datasets, which are Holidays and Wang, and remote sensing datasets, which are UC Merced Land Use and RSSCN7, show that compared with single feature and fusing multiple features by averaging global weights and relevance feedback, our method has better retrieval performance.

In the future work, there are two aspects of work that are worth doing. On the one hand, considering image retrieval based on multi-feature fusion increases the retrieval time; we will research how to improve the efficiency of retrieval. Many researches on image retrieval have been carried out on large-scale datasets, which may contain up to several million pictures, and it is very time-consuming to search for the images we need from the massive images. It is significant to improve the efficiency of retrieval. On the other hand, considering other forms of entropy have achieved good results in the image field [45,46], we will research other forms of entropy used in image retrieval. Meanwhile, considering the image decomposition and the classification of image patches has achieved outstanding results [47–50]. We can use the idea of image decomposition and the classification of image patches to extract better image description for retrieval system. It is significant to improve the performance of retrieval.

Author Contributions: X.L. (Xiaojun Lu), J.W. conceived and designed the experiments, performed the experiments and analyzed the data. J.W., X.L. (Xiang Li) and M.Y. wrote the manuscript. X.Z. refined expression of the article. All authors have read and approved the final version of the manuscript.

Funding: This work was supported by the National Natural Science Foundation of China (Grant No. 61703088).

Conflicts of Interest: The authors declare no conflicts of interest.

References

1. Dharani, T.; Aroquiaraj, I.L. A survey on content based image retrieval. In Proceedings of the 2013 International Conference on Pattern Recognition, Informatics and Mobile Engineering (PRIME), Salem, India, 21–22 February 2013; pp. 485–490.
2. Lu, X.; Yang, Y.; Zhang, W.; Wang, Q.; Wang, Y. Face Verification with Multi-Task and Multi-Scale Feature Fusion. *Entropy* **2017**, *19*, 228. [CrossRef]
3. Zhu, Z.; Zhao, Y. Multi-Graph Multi-Label Learning Based on Entropy. *Entropy* **2018**, *20*, 245. [CrossRef]
4. Al-Shamasneh, A.R.; Jalab, H.A.; Palaiahnakote, S.; Obaidellah, U.H.; Ibrahim, R.W.; El-Melegy, M.T. A New Local Fractional Entropy-Based Model for Kidney MRI Image Enhancement. *Entropy* **2018**, *20*, 344. [CrossRef]
5. Liu, R.; Zhao, Y.; Wei, S.; Zhu, Z.; Liao, L.; Qiu, S. Indexing of CNN features for large scale image search. *arXiv* **2015**, arXiv:1508.00217. [CrossRef]
6. Shi, X.; Guo, Z.; Zhang, D. Efficient Image Retrieval via Feature Fusion and Adaptive Weighting. In Proceedings of the Chinese Conference on Pattern Recognition, Singapore, 5–7 November 2016; Springer: Berlin, Germany, 2016; pp. 259–273.
7. Zhou, Y.; Zeng, D.; Zhang, S.; Tian, Q. Augmented feature fusion for image retrieval system. In Proceedings of the 5th ACM on International Conference on Multimedia Retrieval, Shanghai, China, 23–26 June 2015; ACM: New York, NY, USA, 2015; pp. 447–450.
8. Coltuc, D.; Datcu, M.; Coltuc, D. On the Use of Normalized Compression Distances for Image Similarity Detection. *Entropy* **2018**, *20*, 99. [CrossRef]
9. Zhang, S.; Yang, M.; Cour, T.; Yu, K.; Metaxas, D.N. *Query Specific Fusion for Image Retrieval*; Computer Vision–ECCV 2012; Springer: Berlin/Heidelberg, Germany, 2012; pp. 660–673.
10. Zheng, L.; Wang, S.; Tian, L.; He, F.; Liu, Z.; Tian, Q. Query-adaptive late fusion for image search and person re-identification. In Proceedings of the IEEE Conference on Computer Vision and Pattern Recognition, Boston, MA, USA, 7–12 June 2015; pp. 1741–1750.
11. Walia, E.; Pal, A. Fusion framework for effective color image retrieval. *J. Vis. Commun. Image Represent.* **2014**, *25*, 1335–1348. [CrossRef]
12. Liu, S.; Feng, L.; Liu, Y.; Wu, J.; Sun, M.; Wang, W. Robust discriminative extreme learning machine for relevance feedback in image retrieval. *Multidimens. Syst. Signal Process.* **2017**, *28*, 1071–1089. [CrossRef]
13. Ali, N.; Bajwa, K.B.; Sablatnig, R.; Chatzichristofis, S.A.; Iqbal, Z.; Rashid, M.; Habib, H.A. A novel image retrieval based on visual words integration of SIFT and SURF. *PLoS ONE* **2016**, *11*, e0157428. [CrossRef] [PubMed]
14. Montazer, G.A.; Giveki, D. An improved radial basis function neural network for object image retrieval. *Neurocomputing* **2015**, *168*, 221–233. [CrossRef]
15. Tian, X.; Jiao, L.; Liu, X.; Zhang, X. Feature integration of EODH and Color-SIFT: Application to image retrieval based on codebook. *Signal Process. Image Commun.* **2014**, *29*, 530–545. [CrossRef]
16. Ali, N.; Mazhar, D.A.; Iqbal, Z.; Ashraf, R.; Ahmed, J.; Khan, F.Z. Content-Based Image Retrieval Based on Late Fusion of Binary and Local Descriptors. *arXiv* **2017**, arXiv:1703.08492.

17. Babenko, A.; Slesarev, A.; Chigorin, A.; Lempitsky, V. Neural codes for image retrieval. In Proceedings of the European conference on computer vision, Zurich, Switzerland, 6–12 September 2014; Springer: Cham, Switzerland, 2014; pp. 584–599.
18. Jégou, H.; Douze, M.; Schmid, C. Improving bag-of-features for large scale image search. *Int. J. Comput. Vis.* **2010**, *87*, 316–336. [CrossRef]
19. Liu, P.; Guo, J.-M.; Chamnongthai, K.; Prasetyo, H. Fusion of color histogram and LBP-based features for texture image retrieval and classification. *Inf. Sci.* **2017**, *390*, 95–111. [CrossRef]
20. Kong, F.H. Image retrieval using both color and texture features. In Proceedings of the 2009 International Conference on Machine Learning and Cybernetics, Hebei, China, 12–15 July 2009; Volume 4, pp. 2228–2232.
21. Zheng, Y.; Huang, X.; Feng, S. An image matching algorithm based on combination of SIFT and the rotation invariant LBP. *J. Comput.-Aided Des. Comput. Graph.* **2010**, *22*, 286–292.
22. Yu, J.; Qin, Z.; Wan, T.; Zhang, X. Feature integration analysis of bag-of-features model for image retrieval. *Neurocomputing* **2013**, *120*, 355–364. [CrossRef]
23. Wang, X.; Han, T.X.; Yan, S. An HOG-LBP human detector with partial occlusion handling. In Proceedings of the 2009 IEEE 12th International Conference on Computer Vision, Kyoto, Japan, 29 September–2 October 2009; pp. 32–39.
24. Fagin, R.; Kumar, R.; Sivakumar, D. Efficient similarity search and classification via rank aggregation. In Proceedings of the 2003 ACM SIGMOD International Conference on Management of Data, San Diego, CA, USA, 10–12 June 2003; ACM: New York, NY, USA, 2003; pp. 301–312.
25. Terrades, O.R.; Valveny, E.; Tabbone, S. Optimal classifier fusion in a non-bayesian probabilistic framework. *IEEE Trans. Pattern Anal. Mach. Intell.* **2009**, *31*, 1630–1644. [CrossRef] [PubMed]
26. Li, Y.; Zhang, Y.; Tao, C.; Zhu, H. Content-based high-resolution remote sensing image retrieval via unsupervised feature learning and collaborative affinity metric fusion. *Remote Sens.* **2016**, *8*, 709. [CrossRef]
27. Mourão, A.; Martins, F.; Magalhães, J. Assisted query formulation for multimodal medical case-based retrieval. In Proceedings of the ACM SIGIR Workshop on Health Search & Discovery: Helping Users and Advancing Medicine, Dublin, Ireland, 28 July–1 August 2013.
28. De Herrera, A.G.S.; Schaer, R.; Markonis, D.; Müller, H. Comparing fusion techniques for the ImageCLEF 2013 medical case retrieval task. *Comput. Med. Imaging Graph.* **2015**, *39*, 46–54. [CrossRef] [PubMed]
29. Deng, J.; Berg, A.C.; Li, F.-F. Hierarchical semantic indexing for large scale image retrieval. In Proceedings of the IEEE Conference on Computer Vision and Pattern Recognition (CVPR), Colorado Springs, CO, USA, 20–25 June 2011; pp. 785–792.
30. Lin, K.; Yang, H.-F.; Hsiao, J.-H.; Chen, C.-S. Deep learning of binary hash codes for fast image retrieval. In Proceedings of the 2015 IEEE Conference on Computer Vision and Pattern Recognition Workshops (CVPRW), Boston, MA, USA, 7–12 June 2015; pp. 27–35.
31. Ahmad, J.; Sajjad, M.; Mehmood, I.; Rho, S.; Baik, S.W. Saliency-weighted graphs for efficient visual content description and their applications in real-time image retrieval systems. *J. Real Time Image Pr.* **2017**, *13*, 431–447. [CrossRef]
32. Yu, S.; Niu, D.; Zhao, X.; Liu, M. Color image retrieval based on the hypergraph and the fusion of two descriptors. In Proceedings of the 2017 10th International Congress on Image and Signal Processing, BioMedical Engineering and Informatics (CISP-BMEI), Shanghai, China, 14–16 October 2017; pp. 1–6.
33. Liu, Z.; Blasch, E.; John, V. Statistical comparison of image fusion algorithms: Recommendations. *Inf. Fusion* **2017**, *36*, 251–260. [CrossRef]
34. Zhou, X.S.; Huang, T.S. Relevance feedback in image retrieval: A comprehensive review. *Multimedia Syst.* **2003**, *8*, 536–544. [CrossRef]
35. Zhu, X.; Jing, X.Y.; Wu, F.; Wang, Y.; Zuo, W.; Zheng, W.S. Learning Heterogeneous Dictionary Pair with Feature Projection Matrix for Pedestrian Video Retrieval via Single Query Image. In Proceedings of the Thirty-First AAAI Conference on Artificial Intelligence, San Francisco, CA, USA, 4–9 February 2017; pp. 4341–4348.
36. Wang, X.; Wang, Z. A novel method for image retrieval based on structure elements' descriptor. *J. Vis. Commun. Image Represent.* **2013**, *24*, 63–74. [CrossRef]
37. Bian, W.; Tao, D. Biased discriminant euclidean embedding for content-based image retrieval. *IEEE Trans. Image Process.* **2010**, *19*, 545–554. [CrossRef] [PubMed]

38. Zheng, W.; Mo, S.; Duan, P.; Jin, X. An improved pagerank algorithm based on fuzzy C-means clustering and information entropy. In Proceedings of the 2017 3rd IEEE International Conference on Control Science and Systems Engineering (ICCSSE), Beijing, China, 17–19 August 2017; pp. 615–618.
39. Simonyan, K.; Zisserman, A. Very deep convolutional networks for large-scale image recognition. *arXiv* **2014**, arXiv:1409.1556.
40. Simon, M.; Rodner, E.; Denzler, J. Imagenet pre-trained models with batch normalization. *arXiv* **2016**, arXiv:1612.01452.
41. Hiremath, P.S.; Pujari, J. Content based image retrieval using color, texture and shape features. In Proceedings of the 15th International Conference on Advanced Computing and Communications (ADCOM 2007), Guwahati, India, 18–21 December 2007; pp. 780–784.
42. Jégou, H.; Douze, M.; Schmid, C. Hamming embedding and weak geometry consistency for large scale image search-extended version. In Proceedings of the 10th European Conference on Computer Vision, Marseille, France, 12–18 October 2008.
43. Yang, Y.; Newsam, S. Bag-of-visual-words and spatial extensions for land-use classification. In Proceedings of the 18th SIGSPATIAL International Conference on Advances in Geographic Information Systems, San Jose, CA, USA, 2–5 November 2010; ACM: New York, NY, USA, 2010; pp. 270–279.
44. Zou, Q.; Ni, L.; Zhang, T.; Wang, Q. Deep learning based feature selection for remote sensing scene classification. *IEEE Geosci. Remote Sens. Lett.* **2015**, *12*, 2321–2325. [CrossRef]
45. Ramírez-Reyes, A.; Hernández-Montoya, A.R.; Herrera-Corral, G.; Domínguez-Jiménez, I. Determining the entropic index q of Tsallis entropy in images through redundancy. *Entropy* **2016**, *18*, 299. [CrossRef]
46. Hao, D.; Li, Q.; Li, C. Digital Image Stabilization Method Based on Variational Mode Decomposition and Relative Entropy. *Entropy* **2017**, *19*, 623. [CrossRef]
47. Zhu, Z.; Yin, H.; Chai, Y.; Li, Y.; Qi, G. A novel multi-modality image fusion method based on image decomposition and sparse representation. *Inf. Sci.* **2018**, *432*, 516–529. [CrossRef]
48. Wang, K.; Qi, G.; Zhu, Z.; Chai, Y. A novel geometric dictionary construction approach for sparse representation based image fusion. *Entropy* **2017**, *19*, 306. [CrossRef]
49. Zhu, Z.; Qi, G.; Chai, Y.; Chen, Y. A novel multi-focus image fusion method based on stochastic coordinate coding and local density peaks clustering. *Future Internet* **2016**, *8*, 53. [CrossRef]
50. Fang, Q.; Li, H.; Luo, X.; Ding, L.; Rose, T.M.; An, W.; Yu, Y. A deep learning-based method for detecting non-certified work on construction sites. *Adv. Eng. Inform.* **2018**, *35*, 56–68. [CrossRef]

entropy

MDPI

Article

A New Image Encryption Algorithm Based on Chaos and Secure Hash SHA-256

Shuqin Zhu [1], **Congxu Zhu** [2,3,4,*] **and Wenhong Wang** [1]

1 School of Computer and Science, Liaocheng University, Liaocheng 252059, China; shuqinzhu2008@163.com (S.Z.); wangwenhong@lcu.edu.cn (W.W.)
2 School of Information Science and Engineering, Central South University, Changsha 410083, China
3 School of Physics and Electronics, Central South University, Changsha 410083, China
4 Guangxi Colleges and Universities Key Laboratory of Complex System Optimization and Big Data Processing, Yulin Normal University, Yulin 537000, China
* Correspondence: zhucx@csu.edu.cn; Tel.: +86-0731-8882-7601

Received: 23 August 2018; Accepted: 17 September 2018; Published: 19 September 2018

Abstract: In order to overcome the difficulty of key management in "one time pad" encryption schemes and also resist the attack of chosen plaintext, a new image encryption algorithm based on chaos and SHA-256 is proposed in this paper. The architecture of confusion and diffusion is adopted. Firstly, the surrounding of a plaintext image is surrounded by a sequence generated from the SHA-256 hash value of the plaintext to ensure that each encrypted result is different. Secondly, the image is scrambled according to the random sequence obtained by adding the disturbance term associated with the plaintext to the chaotic sequence. Third, the cyphertext (plaintext) feedback mechanism of the dynamic index in the diffusion stage is adopted, that is, the location index of the cyphertext (plaintext) used for feedback is dynamic. The above measures can ensure that the algorithm can resist chosen plaintext attacks and can overcome the difficulty of key management in "one time pad" encryption scheme. Also, experimental results such as key space analysis, key sensitivity analysis, differential analysis, histograms, information entropy, and correlation coefficients show that the image encryption algorithm is safe and reliable, and has high application potential.

Keywords: chaotic system; image encryption; permutation-diffusion; SHA-256 hash value; dynamic index

1. Introduction

In recent years, with the rapid development of computer technology, digital image processing technology has also rapidly developed and penetrated into all aspects of life, such as remote sensing, industrial detection, medicine, meteorology, communication, investigation, intelligent robots, etc. Therefore, image information has attracted widespread attention. Image data security is very important, especially in the special military, commercial and medical fields. Image encryption has become one of the ways to protect digital image transmission. However, the image data has the characteristics of large amounts of data, strong correlation and high redundancy, which lead to low encryption efficiency and low security, so the traditional encryption algorithms, such as Data Encryption Standard (DES) and Advanced Encryption Standard (AES), cannot meet the needs of image encryption [1]. Chaos has the characteristics of high sensitivity to the initial conditions and system parameters, no periodicity, pseudo randomness, ergodicity and chaotic sequences can be generated and regenerated accurately, so it is especially suitable for image encryption. Therefore, many image encryption algorithms have been put forward using chaotic system. In 1998, the American scholar Fridrich put forward the classical substitution-diffusion architecture for image encryption [2]. This structure subsequently has drawn world-wide concern, and nowadays, most of the image encryption schemes based on chaos adopt this structure and achieved satisfactory encryption effect, such as pixel-level scrambling approaches [3–5], enhanced diffusion schemes [6], improved hyper-chaotic sequences [7], linear hyperbolic chaotic system [8], and bit-level confusion methods [9–11]. However, only using low dimensional chaotic system to encrypt images cannot guarantee enough security. Some works on cryptanalysis [12–18] show that many chaos-based encryption schemes were insecure, and the main reason is that the encryption key has nothing to do with the plaintext. For examples, an image encryption algorithm with only one round diffusion operation is proposed in [19]. The algorithm has the advantages of easy implementation, low complexity and high sensitivity to cyphertext and plaintext, but Diab et al. [20] cryptanalyzed this algorithm and broke the algorithm with only one chosen plaintext. Akhavan et al. [21] cryptanalyzed an image encryption algorithm based on DNA encoding and the curve cryptography and found that the algorithm cannot resist chosen plaintext attacks. Using a skew tent chaotic map, Zhang [22] proposed a novel image encryption method, which adopted a cyphertext feedback mechanism to resist chosen plaintext attacks, but Zhu et al. [23] cracked the algorithm by applying a chosen plaintext combined with chosen cyphertext attack. Various plaintext-related key stream generation mechanisms have been proposed to improve the ability to resist chosen plaintext attacks [24–27]. In most of these algorithms, the SHA-256 hash value of image is used as the external key of the encryption system, so that the encryption keys of different images are different, so as to achieve the effect of "one time pad". Taking the scheme in [28] as an example, firstly, the initial values and parameters of the two-dimensional Logistic chaotic map are calculated from the SHA 256 hash of the original image and given values. Secondly, the initial values and system parameters of the chaotic system are updated by using the Hamming distance of the original image. So the generated random sequence is related to the plaintext image. This encryption method has the advantages of high sensitivity to plaintext and strong attack against plaintext. However, the decryption end needs not only the initial key which is not related to the plaintext, but also the key related to the plaintext. Therefore, decrypting different cyphertext requires different plaintext-related keys, which essentially makes the system work in OTP fashion and greatly increases the complexity for applications.

Concerned about the above issue, we propose to encrypt images based on permutation–diffusion framework using secure hash algorithm SHA-256. Two innovations are the main contributs of this work. Firstly, the hash value of the plaintext image is converted into the number in the range of [0, 255], which is added as the random number around the plaintext image, rather than as the external key of encryption system. This can resist chosen plaintext attacks, and does not need the hash value of the plaintext image in the decryption phase. Secondly, in the permutation and diffusion processes, the generation of random sequences is related to intermediate cyphertext. In this way, the key used to encrypt different images is the initial value of the chaotic system, but the generated key stream is different.

2. Preliminaries

2.1. Adding Surrounding Pixels

A hash function is any function that can be used to map data of arbitrary size to data of a fixed size. Here, we use SHA-256 to generate the 256-bit hash value V, which can be divided into 32 blocks with the same size of 8-bit, the i-th block $v_i \in [0, 255]$, $i = 1, 2, \ldots, 32$, so V can be expressed as $V = v_1, v_2, \ldots, v_{32}$. Suppose the size of the plain-image P is $m \times n$, obtain an integer k as:

$$k = fix(2(m+n+1)/32) + 1 \tag{1}$$

where, $fix(x)$ rounds the elements of x to the nearest integers towards zero. Then we generate a sequence H that has ($32k$) elements by:

$$H = repmat(V,\ [1,\ k]) \tag{2}$$

where, *repmat*(V, [1, k]) creates a large matrix H consisting of a $1 \times k$ tiling of copies of V, e.g., *repmat*([3, 6, 9], [1, 2]) = [3, 6, 9, 3, 6, 9]. Then, matrix RI of size $2 \times (n + 2)$ is formed by taking the first $2n + 4$ numbers of the sequence H, and the CI matrix of size $2 \times m$ is formed by taking the remaining $2m$ numbers of H. The elements of RI and CI have the same representation format as the pixels of P. For example, The SHA-256 hash value of the plaintext image "cameraman" of size 256×256 is the character string S, which is: S = "d6f35e24b1f70a68a37c9b8bfdcd91dc3977d7a98e67d453eb6f8003b6c6 9443".

According to the string S, we can get a sequence V of length 32. V = (214, 243, 94, 36, 177, 247, 10, 104, 163, 124, 155, 139, 253, 205, 145, 220, 57, 119, 215, 169, 142, 103, 212, 83, 235, 111, 128, 3, 182, 198, 148, 67). So, the sequence H of length 1028 can be obtained as H = (214, 243, 94, 36, 177, 247, 10, . . . , 214, 243). Similarly, matrices RI and CI are also obtained, as shown below:

$$RI = \begin{pmatrix} 214 & 243 & 94 & \ldots & 148 & 67 & 214 \\ 243 & 94 & 36 & \ldots & 67 & 214 & 243 \end{pmatrix}_{2\times(256+1)}$$

$$CI = \begin{pmatrix} 94 & 36 & 177 & \ldots & 67 & 214 & 243 \\ 94 & 36 & 177 & \ldots & 67 & 214 & 243 \end{pmatrix}_{2\times(256+1)}$$

RI and CI will surround the plaintext image. These values will affect all pixels after the confusion and diffusion operation. Figure 1 shows a numerical example of using RI and CI to add pixels to the image "cameraman". Figure 1b shows the result of the operation. It can be seen that the underscore is derived from RI and the value of bold is from CI.

156	159	...	152	152
160	154	...	155	153
...	...	...	...	...
121	126	...	130	113
121	126	...	130	113

(**a**)

<u>214</u>	<u>243</u>	<u>94</u>	<u>...</u>	<u>148</u>	<u>67</u>	<u>214</u>
94	156	159	...	152	152	**94**
36	160	154	...	155	153	**36**
...	...	...	...	...	...	**...**
214	121	126	...	130	113	**214**
243	121	126	...	130	113	**243**
<u>243</u>	<u>94</u>	<u>36</u>	<u>...</u>	<u>67</u>	<u>214</u>	<u>243</u>

(**b**)

Figure 1. An example of adding surrounding pixels. (**a**) plain-image P; (**b**) operation result.

2.2. Hyper-Chaotic System and Chebyshev Map

The scheme is based on a hyper-chaotic system and two Chebyshev maps. We will use a four dimensional hyper-chaotic system with five system parameters and four initial conditions [29], which can be modeled by Equation (3):

$$\begin{cases} dx/dt = a(y-x)+w \\ dy/dt = dx - xz + cy \\ dz/dt = xy - bz \\ dw/dt = yz + ew \end{cases} \tag{3}$$

where, a, b, c, d and e are parameters of the system. When a = 35, b = 3, c = 12, d = 7 and $e \in$ (0.085, 0.798), the system is hyper-chaotic and has two positive Lyapunov exponents, $LE1$ = 0.596, $LE2$ = 0.154. So the system is in a hyper-chaotic state. The system attractor curves are presented in Figure 2.

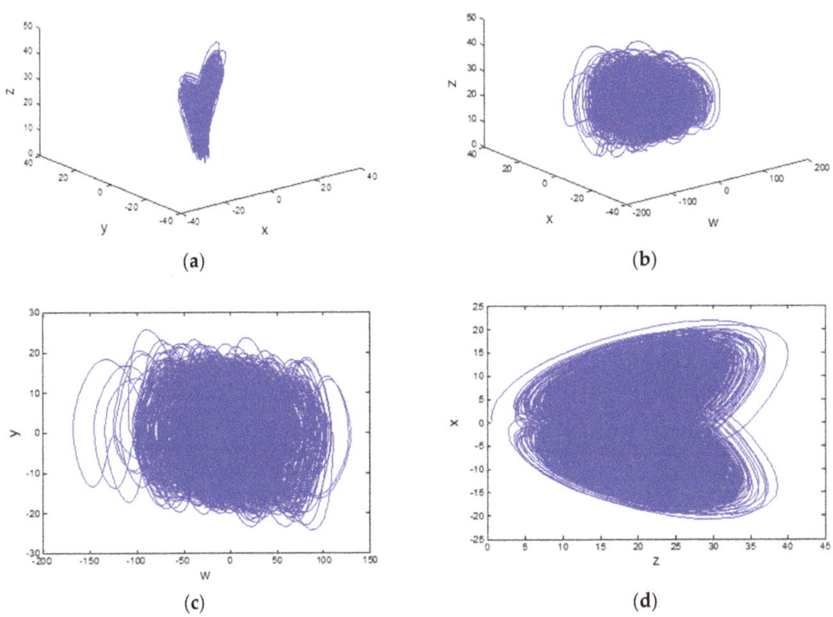

Figure 2. Hyper-chaotic attractor. (**a**) (x-y-z) plane; (**b**) (w-x-z) plane; (**c**) (w-y) plane; (**d**) (x-z) plane.

The two Chebyshev maps are modeled by Equation (4):

$$\begin{cases} u_1(i+1) = \cos(4 \times \arccos(u_1(i))) \\ u_2(i+1) = \cos(4 \times \arccos(u_2(i))) \end{cases} \tag{4}$$

where, $u_1(1)$ and $u_2(1)$ are initial values.

2.3. The Generation of Random Sequences of the Encryption System

The initial values of the chaotic system are given, then we iterate the hyper-chaotic system (1) to produce four sequences denoted as $X = [x(i)]$, $Y = [y(i)]$, $Z = [z(i)]$ and $W = [w(i)]$, respectively, where, $i = 1, 2, \ldots$ At the same time, $u_1(1)$ and $u_2(1)$ are given, the two Chebyshev maps from Equation (4) are iterated to generate two sequences denoted as U_1 and U_2, respectively. To further enhance the complexity of sequences, These six chaotic sequences X, Y, Z, W, U_1 and U_2 are transformed into three real value sequences D_1, D_2 and D_3 in the interval [0, 1] by the following Formulas (5)–(7), then transform three real value sequences D_1, D_2 and D_3 into three integer value sequences S, V and T by the following Formulas (8)–(10), so we get three sequences $S = \{s(1), s(2), \ldots, s(l)\}$, $V = \{v(1), v(2), \ldots, v(l)\}$, $T = \{t(1), t(2), \ldots, t(l)\}$, which will be used in the later encryption process, where, $s(i)$, $v(i)$ and $t(i) \in \{0, 1, \ldots, 255\}$, $i = 1,2, \ldots, l$:

$$D_1 = \cos^2((X + Y + Z)/3) \quad (5)$$

$$D_2 = \cos^2((W + U_1 + U_2)/3) \quad (6)$$

$$D_3 = \cos^2((X + Z + U_2)/3) \quad (7)$$

$$S = \text{mod}\left(\text{round}(D_1 \times 10^{15}), 256\right) \quad (8)$$

$$V = \text{mod}\left(\text{round}(D_2 \times 10^{15}), 256\right) \quad (9)$$

$$T = \text{mod}\left(\text{round}(D_3 \times 10^{15}), 256\right) \quad (10)$$

where, round(x) rounds x to the nearest integer, and mod(x, y) returns the remainder after x is divided by y. The sequence D_1 is used to scramble images, while D_2, S, V and T are used for image diffusion operation. Figure 3 is the numerical distribution curve of chaotic key sequence S, V and T. the abscissa represents 256 gray levels and the ordinate represents the frequency of each gray level. From Figure 3, it can be seen that the key flow S, V and T distribute evenly, and the pseudo-randomness is good.

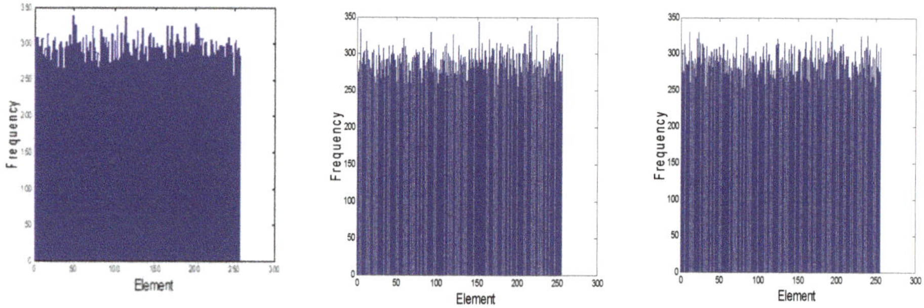

(**a**) (**b**) (**c**)

Figure 3. Histogram of three CPRNG sequences. (**a**) The histogram of sequence S; (**b**)The histogram of sequence V; (**c**)The histogram of sequence T.

2.4. Statistical Test Analysis of the Three CPRNG Sequences S, V and T

In order to measure randomness of the three CPRNG sequences S, V and T, we use the NIST SP800–22 statistical test suite (Rev1a, Information Technology Laboratory, Computer Security Resource Center, Gaithersburg, MD, USA), which consists of 15 statistical tests. Each test result is converted to a p-value for judgement, and when applying the NIST test suite, a significance level $\alpha = 0.01$ is chosen for testing. If the p-value $\geq \alpha$, then the test sequence is considered to be pseudo-random.

Setting different initial conditions of chaotic system, and using systems (3) and (4) as well as Equations (5)–(10), 1000 sequences S, 1000 sequences V and 1000 sequences T are generated, respectively. The parameters used in the test are set as: $a = 35$, $b = 3$, $c = 12$, $d = 7$, $e = 0.1583$, $x(0) = 0.398$, $y(0) = 0.45$, $z(0) = 0.78$, $w(0) = 0.98$, $u_1(1) = 0.58$ and $u2(1)$ varies from 0.0005 to 0.9995 with a variable step size of 0.0001. Hence, 1000 sequences of $\{S, V, T\}$ can be generated. The length of each integer sequence is 125,000 and each integer has 8 bits. Then three decimal integer sequences are turned into three binary sequences by converting each decimal number into an 8-bit binary number and connecting them together. Therefore, each binary sequence has the length of 1,000,000 bits (125,000 $\times$ 8 = 100,000). Unlike the bit sequence generation method introduced in the related literature [30], the method of generating bit sequences in our scheme can be demonstrated by the following simple example. Suppose the decimal integer sequence S has three 8-bit integers, S = [23, 106, 149], where, $23 = (0001\ 0111)_2$,

106 = $(0110\ 1010)_2$, 149 = $(1001\ 0101)_2$. Then the binary sequence S' corresponding to the decimal integer sequences S has the following form: S' = [0 0 0 1 0 1 1 1 0 1 1 0 1 0 1 0 1 0 0 1 0 1 0 1]. 15 statistical items (some items include two sub indicators) were tested by using the NIST SP800-22 suite, and the results from all statistical tests are given in Table 1. From Table 1, we can see that all the p-values from all 1000 sequences are greater than the significance level $\alpha = 0.01$, indicating that the tests meet the requirements of SP800-22 randomness, and the pass rate is also in acceptable range. Compared with the results of relevant literature [30], the overall result is not very different. However, the linear complexity index of our scheme is obviously better than that of reference [30], but the Rank index is slightly worse than that of reference [30].

Table 1. NIST SP800-22 standard test of pseudo-random sequence S', V' and T'.

Statistical Test Name	S'		V'		T'	
	Pass Rate	p-Value	Pass Rate	p-Value	Pass Rate	p-Value
Frequency(monobit)	99.5%	0.9346	99.3%	0.4058	99.4%	0.4708
Block Frequency	99.2%	0.8068	99.1%	0.6079	99.0%	0.5485
The Run Test	99.5%	0.4088	99.6%	0.4317	99.5%	0.5493
Longest Run of Ones	98.6%	0.1481	98.8%	0.4555	98.6%	0.4419
Rank	98.5%	0.0465	98.3%	0.0467	98.1%	0.0103
DFT Spectral	99.3%	0.9537	99.1%	0.5365	99.3%	0.6539
Non-Overlapping Templates	99.1%	0.6163	99.0%	0.5348	98.8%	0.4807
Overlapping Templates	98.8%	0.7597	98.6%	0.5331	98.4%	0.6420
Universal Statistical Test	98.5%	0.5825	98.3%	0.4624	98.2%	0.4171
Linear Complexity	98.9%	0.2215	98.7%	0.4642	98.5%	0.4936
Serial Test 1	99.1%	0.3358	98.9%	0.2421	98.7%	0.2602
Serial Test 2	99.2%	0.2046	99.4%	0.4207	99.3%	0.2315
Approximate Entropy	98.8%	0.7522	98.6%	0.6033	98.8%	0.4784
Cumulative Sums (forward)	99.6%	0.4752	99.8%	0.8023	99.7%	0.8163
Cumulative Sums (Reverse)	99.4%	0.8898	99.2%	0.6596	99.3%	0.8101
Random Excursions	98.7%	0.1599	98.8%	0.1713	98.6%	0.1314
Random Excursions Variant	98.9%	0.3226	98.4%	0.1564	98.6%	0.0942

3. Architecture of the Proposed Cryptosystem

In this paper, we use the classical permutation-diffusion image encryption structure. During the permutation process, we use the permutation sequence generated by the chaotic system to shuffle the pixels. However, the permutation does not change the pixel value, but makes the statistical relationship between cyphertext and key complicated, so that the opponent cannot infer the key statistics from the statistical relationship between cyphertext. Diffusion means that each bit of the plaintext affects many bits of the cyphertext, or that each bit of the cyphertext is affected by many bits of the plaintext, thus enhancing the sensitivity of the cyphertext.

3.1. Encryption Algorithm

The encryption process consists of three stages. Firstly, generating key streams by using the hyper-chaotic system and adding surrounding pixels to the plaintext image. Secondly, performing the permutation process. Thirdly, performing the diffusion process. The architecture of the encryption process is shown in Figure 4, and the operation procedures are described as follows:

Step 1: Assume that the size of the plaintext image is $m \times n$, adding surrounding pixels to the plaintext image matrix $P_{m\times n}$ According to the method described in Section 2.1 to get image matrix $P'_{(m+2)\times(n+2)}$. The matrix $P'_{(m+2)\times(n+2)}$ is converted to a one dimensional vector $P_0 = \{p_0(1), p_0(2), \ldots, p_0(l)\}$, where $l = (m+2) \times (n+2)$.

Step 2: Produce the required chaotic sequences D_1, D_2, S, V and T of length l for encryption according to the method described in Section 2.3.

Step 3: Permuting P_0 obtained in step 1 according to Equations (11) and (12). In order to make the scrambling sequence related to plaintext to prevent the chosen plaintext attack, a disturbance term g associated with the plaintext is added according to Equation (10) when the scrambling sequence h is generated, where $g = sum(P_0)/(256 \times l)$, Therefore, the scrambling sequence $h = \{h(1), h(2), \ldots, h(l)\}$ is different when encrypting different plaintext images. In Equation (11), *floor*(x) rounds x to the nearest integers towards minus infinity:

$$h(i) = i + \mathrm{mod}[floor(D_1(i) \times g \times 10^{14}), l - i], i = 1, 2, 3, \ldots, l. \tag{11}$$

$$\begin{cases} temp = p_0(i) \\ p_0(i) = p_0(h(i)) \quad , i = 1, 2, 3, \ldots, l. \\ p_0(h(i)) = temp \end{cases} \tag{12}$$

Step 4: Perform confusion and diffusion. Encrypt the first element in p_0 by Equation (13):

$$c(1) = \mathrm{mod}((p_0(1) + s(1), 256) \oplus \mathrm{mod}((t(1) + v(1)), 256). \tag{13}$$

Step 5: Set $i = 2, 3, \ldots, l$, calculate the dynamic indexes kt_1 and kt_2 by Equations (14) and (15), which are used for encrypting the i-th element in p_0. Obviously, $kt_1(i) \in [1, i-1]$, $kt_2(i) \in [i+1, l]$:

$$kt_1(i) = floor(s(i)/256 \times (i-1)) + 1, \tag{14}$$

$$kt_2(i) = floor(v(i)/256 \times (l - i - 1)) + i + 1. \tag{15}$$

Step 6: Encrypt the i-th element according to the following Equations (16)–(18):

$$tt(i) = \mathrm{mod}(floor(D_2(i) \times c(i-1)) \times 10^4, 256), i = 1, 2, 3, \ldots, l-1. \tag{16}$$

$$c(i) = \mathrm{mod}((p_0(i) + c(kt_1(i))), 256) \oplus \mathrm{mod}((tt(i) + p_0(kt_2(i))), 256), i = 1, 2, \ldots, l-1. \tag{17}$$

$$c(l) = \mathrm{mod}((p_0(l) + c(kt_1(l))), 256) \oplus tt(l) \tag{18}$$

From Equation (16), for different plain images, the sequence $[tt(i)]$ will be different, that will lead to the different i-th encrypted value.

Step 7: The final cyphertext sequence $CC = [cc(1), cc(2), \ldots, cc(l)]$ is obtained by using Equation (19). Transform the diffused vector CC into the $m \times n$ matrix, then the cypher image is obtained:

$$cc(i) = c(i) \oplus t(i) \tag{19}$$

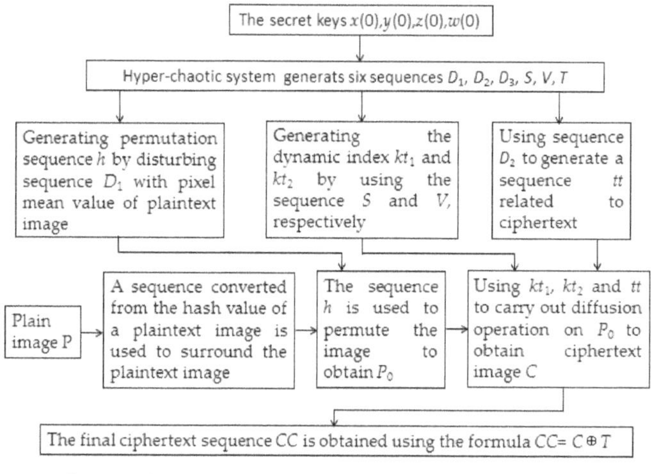

Figure 4. The architecture of the proposed encryption algorithm.

3.2. Decryption Algorithm

The decryption process is the process of transforming cyphertext into plaintext, and the reverse process of encryption. The decryption process is described as follows:

Step 1: Produce the required chaotic sequences D_1, D_2, S, V and T of length l for decryption according to the method described in Section 2.3 and calculate the dynamic indexes kt_1 and kt_2 according to Equations (14) and (15).

Step 2: The cyphertext image is translated into a one dimensional vector $CC = [cc(1), cc(2), \ldots, cc(l)]$. The intermediate cyphertext C is obtained by:

$$c(i) = cc(i) \oplus t(i) \tag{20}$$

Step 3: Calculate the sequence tt according to Equation (16) and decrypt the last element in p_0 by:

$$p0(l) = \mathrm{mod}(c(l) \oplus tt(l) - c(kt_1(l)), 256) \tag{21}$$

Step 4: In the opposite direction, we decrypt the plaintext pixel $P_0(l-1), P_0(l-2), \ldots, P_0(2)$ by Equation (22). Finally, the pixel P_0 (1) is decrypted as:

$$p_0(i) = \mathrm{mod}(c(i) \oplus \mathrm{mod}(tt(i) + p_0(kt_2(i)), 256) - c(kt_1(i)), 256), i = l-1, l-2, l-3, \ldots, 2. \tag{22}$$

$$p_0(1) = \mathrm{mod}(c(1) \oplus \mathrm{mod}(t(1) + v(1)), 256) - s(1), 256) \tag{23}$$

Step 5: Perform inverse permutation. Because the sum of pixel values before and after scrambling remains unchanged, the g value can be calculated by the sequence P_0 decrypted in Step 4, Thus, the sequence $H = \{h(1), h(2), \ldots, h(l)\}$ can be obtained by Equation (11). It should be noted that this process is reversed in the direction of encryption, from the last pixel to the first pixel, that is:

$$\begin{cases} temp = p_0(i) \\ p_0(i) = p_0(h(i)) \\ p_0(h(i)) = temp \end{cases}, i = l-1, l-2, l-3, \ldots, 1. \tag{24}$$

Finally, the decrypted sequence P_0 is transformed into a matrix P' of size $(m + 2) \times (n + 2)$. Discarding the first row, the last row, and the first column and the last column of the matrix P', and we can obtain a matrix P of size $m \times n$. P is the recovered plaintext image.

3.3. Application of the Algorithm for Color Images

A color image is composed of three main components, i.e., R, G and B. The hash values of R, G and B matrices are computed respectively, and then the hash values are transformed into sequences according to the method of Section 2.1. Then adding surrounding pixels to R, G and B by using the sequences to obtain three new matrices R', G' and B', respectively. Then R', G' and B' are encrypted in parallel and similar to the encryption of gray level image. Decryption process of matrixes R, G and B is also similar to the proposed decryption process in Section 3.2.

3.4. The Advantages in the New Encryption Scheme

(1) The method of surrounding pixels generated by the SHA-256 hash value of the plaintext image is adopted, which can enhance the ability of the encryption system to resist chosen plaintext attacks. In general, selecting an image of all the same pixel values to chosen plaintext attack, which can eliminate the global scrambling effect. But in the new encryption algorithms, even encrypt an image of all the same pixel values, because the first step is to add surrounding pixels to the image, then the image is not an image of all the same pixel values. On the other hand, the hash value of the image is not needed in decryption, which reduces the difficulty of key management.

(2) In the permutation process, by adding a perturbation g (g = sum(P_0)/(256 × l)) to the chaotic sequence D_1, the permutation sequence h is generated by Equation (10). Therefore, h is related to plaintext, which can resist the chosen plaintext attack. At the same time, g is not part of the decryption key, which reduces the difficulty of key management.
(3) From Equation (16), it is known that the sequence tt is related to the transition cyphertext c, so the sequence tt is different when encrypting different images, which further strengthens the ability of the encryption system to resist chosen plaintext attack.
(4) From the cyphertext feedback mechanism of Equation (17), It can be seen that our encryption algorithm is sensitive to plaintext.

4. Simulation Results

In this paper, the standard 256 × 256 image of "cameraman" is used as the input image. Matlab 2014a (MathWorks, Natick, MA, USA) is utilized to simulate the encryption and decryption operations and set parameters ($x(0)$, $y(0)$, $z(0)$, $w(0)$) = (0.398, 0.456, 0.784, 0.982). The continuous hyper-chaotic system (3) was solved by the ode45 solver of Matlab. The time step used in this algorithm is the adaptive variable step size instead of fixed step size. Figure 5a is the scrambled image, Figure 5b is the encrypted image and the decrypted image is shown in Figure 5c. It can be seen that the cyphertext image is a chaotic image, and has nothing to do with the original image. Therefore, the encryption effect of the algorithm is good.

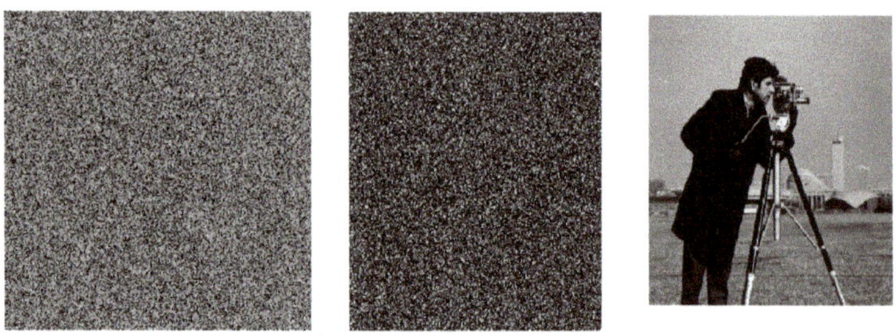

(a) (b) (c)

Figure 5. Experimental results. (**a**) scrambled image; (**b**) cyphertext image; (**c**) decrypted plaintext image.

5. Security Analysis

In this section, we will discuss the security analysis of the proposed encryption scheme with the traditional 8-bit gray image as an example.

5.1. Key Space

In cryptography, the larger the key space, the stronger the ability to resist brute force attacks. In the proposed cryptosystem, the keys are the initial value $x(0)$, $y(0)$, $z(0)$, $w(0)$ of the chaotic system (3), the chaotic system parameter e and the initial values $u_1(1)$, $u_2(1)$ of the two Chebyshev maps (4). The precision of $x(0)$, $y(0)$, $z(0)$, $w(0)$, $u_1(1)$ and $u_2(1)$ is 10^{-15}, while the precision of the parameter e is 10^{-12} for $e \in (0.085, 0.798)$, so the key space size will be $(10^{15})^6 \times 10^{12} = 10^{102} \approx 2^{339}$. In Reference [1], Li pointed out that the effective key space of the image encryption system should be greater than 2^{100} in order to prevent brute force attacks, so the key space of our algorithm is sufficiently large to resist against brute-force attacks.

Table 2 lists the key space of several similar algorithms. By comparison, the key space of this algorithm is better than most algorithms' key space. The size of the key space depends not only on the number of keys, but also on the number of possible values for each key. The problem of numerical chaotic systems is that the finite precision of the machines (e.g., computers) leads to performance degradation [31–34], such as: the key space is reduced, some weak keys appear, and the randomness of the sequence is reduced. In order to identify and avoid weak keys, we need to calculate the Lyapunov exponents of chaotic systems, or plot the phase space trajectories of the system.

Table 2. Key space comparisons.

Encryption Algorithm	Key Space
Proposed scheme	2^{339}
Reference [24]	2^{149}
Reference [25]	2^{256}
Reference [31]	2^{299}
Reference [32]	2^{375}
Reference [33]	$>2^{128}$
Reference [34]	2^{357}

5.2. *Key Sensitivity*

The key sensitivity can be evaluated in two aspects: First, the cyphertext image will be completely different when encrypting the same plaintext image with slightly different keys, which is measured by the change rate t of the cyphertext image. Second, no information about the plaintext image is available in the image decrypted from the wrong key, even though there is a very small difference between the wrong key and the correct key.

The specific method of calculating the change rate t of cyphertext image is as follows: first, the change of the key is kh, and the other parameters remain unchanged. For example, when calculating the sensitivity of the key a, the cyphertext C_1 is obtained by encrypting the plaintext image with the key a. In the same way, the cyphertext image C_2 and the cyphertext image C_3 are obtained by encrypting the plaintext image with the key $a + kh$ and the key $a - kh$, respectively. We can calculate the pixel difference rate t_1 between C_1 and C_2 and the pixel difference rate t_2 between C_1 and C_3, respectively. Then, the change rate of cyphertext image $t = (t_1 + t_2)/2$ is obtained. The sensitivity of each key is calculated by this method, as shown in Table 3, where the change kh is 10^{-15} for each key $x(0)$, $y(0)$, $z(0)$, $w(0)$. The calculated results show that the new algorithm is very sensitive to the initial secret key values.

The sensitivity test results of Table 3 confirmed that the sensitivity of the proposed algorithm to the keys $x(0)$, $y(0)$, $z(0)$, $w(0)$ is very high, and can reach more than 10^{-15}. In order to evaluate the sensitivity of the key in the second aspects, we select the following error keys *Key*1, *Key*2, *Key*3 and *Key*4 to decrypt the original cyphertext image, and the decryption result is shown in Figure 6.

$$Key1 = \{x(0), y(0), z(0), w(0)\} = (0.398 + 10^{-15}, 0.456, 0.784, 0.982),$$
$$Key2 = \{x(0), y(0), z(0), w(0)\} = (0.398, 0.456 + 10^{-15}, 0.784, 0.982),$$
$$Key3 = \{x(0), y(0), z(0), w(0)\} = (0.398, 0.456, 0.784 + 10^{-15}, 0.982),$$
$$Key4 = \{x(0), y(0), z(0), w(0)\} = (0.398, 0.456, 0.784, 0.982 + 10^{-15}).$$

Table 3. Sensitivity tests for each initial secret key value.

Keys	Change Rate of Cyphertext Image t
$x(0)$	0.9963
$y(0)$	0.9964
$z(0)$	0.9976
$w(0)$	0.9975

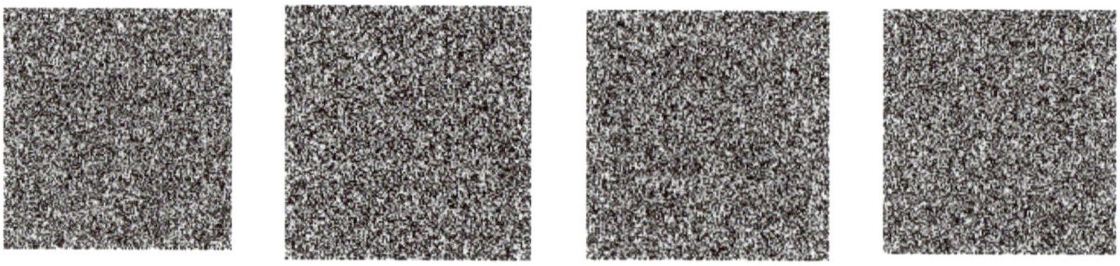

(a) (b) (c) (d)

Figure 6. The image decrypted by the wrong keys. (**a**) decryption result of key *Key*1; (**b**) decryption result of *Key*2; (**c**) decryption result of *Key*3; (**d**) decryption result of *Key*4.

5.3. Plaintext Sensitivity

The sensitivity of the algorithm to plaintext means that a small change in plaintext will cause a huge change in the corresponding cyphertext. This is one of the criteria for cryptographic security analysis. From the encryption process, we can see that the algorithm is sensitive to plaintext: first, if the image has a little difference, the hash value will be completely different. Therefore, the two matrices *RI* and *CI* generated in Section 2.1 and the disturbance term g will be different, so the scrambled image will be different and will lead to the great change of pseudo-random sequence *tt*. Experimentally, *NPCR* (number of pixels change rate) and *UACI* (unified average changing intensity) are used to measure the degree of sensitivity of image encryption algorithms to plaintext. *NPCR* and *UACI* respectively represent the percentage and change degree of the number of pixels in the encrypted image after the pseudo-random change of the gray value of a pixel in the original image. The formulas for the calculation of *NPCR* and *UACI* are as follows:

$$NPCR = \frac{1}{M \times N}\sum_{i=1}^{M}\sum_{j=1}^{N} D(i,j) \times 100\%, \tag{25}$$

$$UACI = \left(\sum_{i=1}^{M}\sum_{j=1}^{N}\left(\frac{|x(i,j) - x'(i,j)|}{255 \times M \times N}\right)\right) \times 100\%, \tag{26}$$

$$D(i,j) = \begin{cases} 1, \ if \ x(i,j) \neq x'(i,j) \\ 0, \ if \ x(i,j) = x'(i,j) \end{cases}. \tag{27}$$

where, $M \times N$ is the size of the image. $x(i,j)$ represent the pixel in a coordinate (i,j) of the cyphertext image corresponding to the original plaintext image, and $x'(i,j)$ represent the pixel in a coordinate (i,j) of the cyphertext image corresponding to the changed plaintext image. For 256 bit grayscale images, the expected values of *NPCR* and *UACI* are 99.6094% and 33.4635%, respectively. In this paper, four classical images ("cameraman", "pepper", "rice" and "autumn") are selected to be tested. In each image, the pixel values of randomly selected 200 pixels are changed, and their maximum, minimum, and average values of *NPCR* are listed in the Table 4. In order to show the superiority of our encryption algorithm, the maximum and minimum values of *NPCR* and *UACI* for each image calculated by the encryption algorithm of Referemce [35] are shown in Table 5.

Table 4. *NPCR* and *UACI* test results of slight change of plaintext in our algorithm.

Images		Rice	Autumn	Pepper	Cameraman
NPCR%	Max	99.8943	99.7932	99.9012	99.7821
	Min	99.5426	99.4213	99.3809	99.4608
	Average	99.6062	99.6115	99.5956	99.5697
UACI%	Max	33.5698	33.7754	33.8712	33.6590
	Min	33.3216	33.5500	33.4919	33.1958
	Average	33.4419	33.6319	33.5418	33.3618

Table 5. *NPCR* and *UACI* test results of slight change of plaintext in [35].

Images		Rice	Autumn	Pepper	Cameraman
NPCR%	Max	99.8812	99.6623	99.8719	99.8864
	Min	99.4961	99.5512	99.5698	99.5091
	Average	99.6006	99.6098	99.5796	99.5692
UACI%	Max	33.5612	33.6067	33.8523	33.7019
	Min	33.3187	33.5602	33.4967	33.2195
	Average	33.4297	33.5897	33.5154	33.3478

As shown in Tables 4 and 5, the average values of *NPCR* and *UACI* of the four images of the new algorithm are higher than the average of *NPCR* and *UACI* in Reference [31], thus proving that our algorithm has better performance in resisting differential attacks.

5.4. Statistical Analysis

Statistical analysis mainly includes: histogram analysis, chi-square test, adjacent pixel correlation analysis and information entropy analysis. In this part we will evaluate the algorithm from above aspects.

5.4.1. Statistical Histogram Analysis

Gray histogram is a function of gray level, which reflects the distribution of gray level in the image and describes the number of pixels of each gray level in the image, but does not contain the position information of these pixels in the image. Figure 7 are the histograms of the Pepper image and the corresponding cipher images. In the histogram, the horizontal axis denotes the gray level, and the vertical axis denotes the pixel number of each gray level. It can be seen that the probability distribution of plaintext image histogram presents a single peak distribution, while the corresponding cyphertext image histogram probability distribution is close to the equal probability distribution, so the cyphertext image is a pseudo-random image.

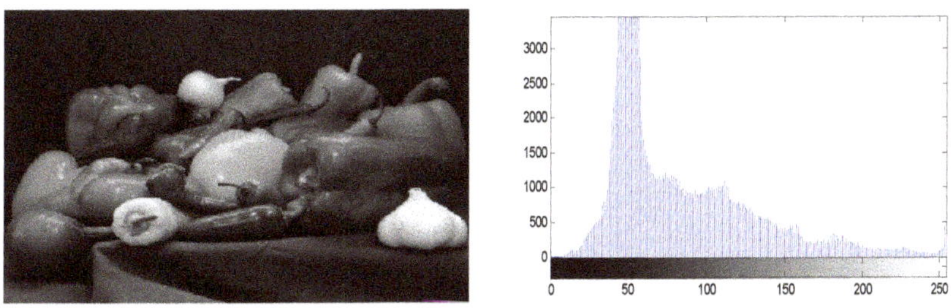

(a) (b)

Figure 7. *Cont.*

(c)

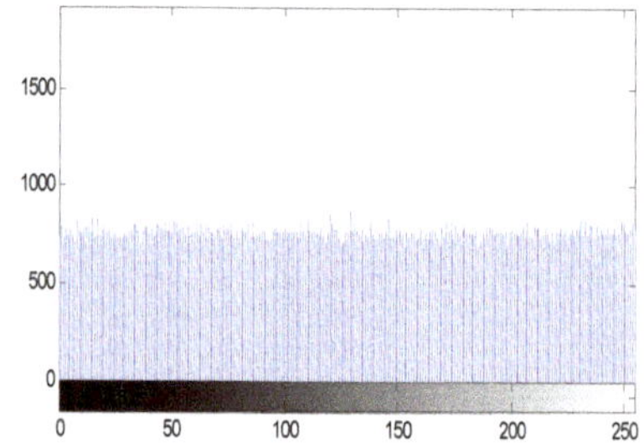

(d)

Figure 7. Histogram analysis. (**a**) The plaintext image of "pepper"; (**b**) the histogram of the plaintext image; (**c**) the cyphertext image; (**d**) the histogram of the cyphertext image.

5.4.2. Chi-Square Test

Figure 7d shows that the histogram of the cyphertext image is uniformly distributed, which can resist statistical attacks and can be proved by the chi-square test [36], which is described by the following expression:

$$x^2 = \sum_{k=1}^{256} \frac{(v_k - e)^2}{e} \tag{28}$$

where, v_k is the actual frequency of each gray level, and e is the expected frequency of each gray level. For different sizes of images, e is different, for example, e is 256 for the image cameraman of size 256×256. however, e is 768 for the image pepper of size 384×512. The smaller the chi square value, the better the uniformity of cyphertext images. For the confidence level a = 0.05, if the chi square value does not exceed 295.25, it is considered to pass the test. In this paper, the cyphertext image is generated by changing one bit of the ordinary image. The process is repeated 30 times. for confidence level = 0.05, and the average results for the four images "cameraman", "pepper", "rice" and "autumn" are demonstrated in Table 6.

Table 6. Chi-test results of 30 encrypted images under confidence level is 0.05.

Test Images	x^2 of Plain Image	x^2 of Cypherimage in [33]	x^2 of Cypherimage in Our Algorithm
cameraman	16,711,680	288.9823 < 295.25	285.3125 < 295.25
pepper	50,135,040	269.3387 < 295.25	260.3421 < 295.25
rice	96,312	284.2387 < 295.25	278.6172 < 295.25
autumn	18,122,850	289.9832 < 295.25	288.5792 < 295.25

It can be seen that the chi-square value of the histogram of the cyphertext images are less than 295.5, which means that the histogram of the cyphertext image has passed Chi-square test for confidence level =0.05 and better than in Kulsoom [37].

5.4.3. Information Entropy

The entropy of an image is expressed as the average number of bits of the set of gray levels of an image. It also describes the average amount of information of an image source. The greater the entropy, the more confusing the information provided by the image. For discrete two-dimensional images, the formula of information entropy E is shown in Equation (29):

$$E = -\sum_{i=0}^{n} p_i \log_2(p_i) \tag{29}$$

where, p_i is the probability of the occurrence of gray value i. When the probability distribution of cyphertext is equal probability distribution, that is, the probability of each value of [0, 255] is $1/256$, the maximum entropy is 8 bits. The information entropy of four cyphertext images of "rice", "cameraman", "autumn" and "pepper" is listed in the second column of Table 7, At the same time, the information entropy of the cyphertext image obtained by the other algorithms in [38–42] are listed in columns 3–7 of Table 7. It can be seen that the information entropy of the encrypted image of the four images is very close to 8 bits and our algorithm has greater superiority.

Table 7. Entropy of cyphertext images.

Images	This Paper	Ref. [34]	Ref. [39]	Ref. [40]	Ref. [41]	Ref. [42]
Rice (256 × 256)	7.9973	7.9864	7.9936	7.9643	7.9875	7.9968
cameraman (256 × 256)	7.9989	7.9763	7.9952	7.9867	7.9946	7.9865
autumn (206 × 345)	7.9968	7.9564	7.9962	7.9698	7.9864	7.9972
pepper (512 × 512)	7.9992	7.9819	7.9983	7.9949	7.9896	7.9993

5.4.4. Pixel Correlation Analysis

In a natural image, there is a high correlation between each pixel and its adjacent pixels, which means that there is a small difference in the gray value in the larger area of the image. One of the goals of encrypted image is to reduce the correlation between adjacent pixels, and the smaller the correlation, the better the encryption effect, the higher the security. Correlation mainly includes the correlation between horizontal pixels, vertical pixels and diagonal pixels. Firstly, 4000 pixels are selected randomly from the "cameraman" image and the corresponding cyphertext image as the base points, and 4000 pairs of adjacent pixels are collected along the horizontal, vertical and diagonal directions respectively, and the correlation distribution maps in these three directions are drawn, as shown in Figure 8. It can be seen that there is a strong correlation between adjacent pixels of plaintext image, showing a linear relationship, and for cyphertext image, this correlation is greatly weakened, showing a strong randomness. This indicates that the image encryption effect is good and the security is high.

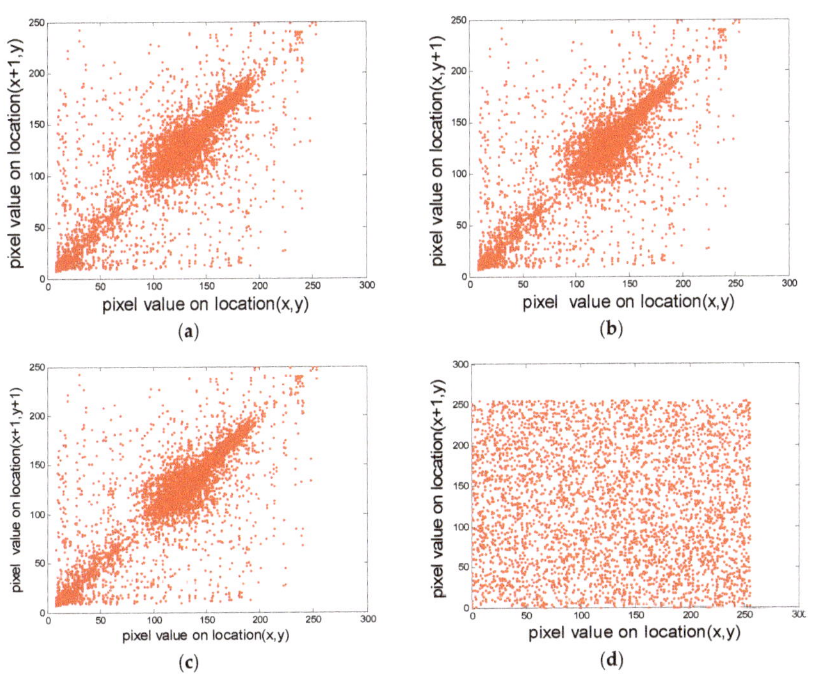

Figure 8. *Cont.*

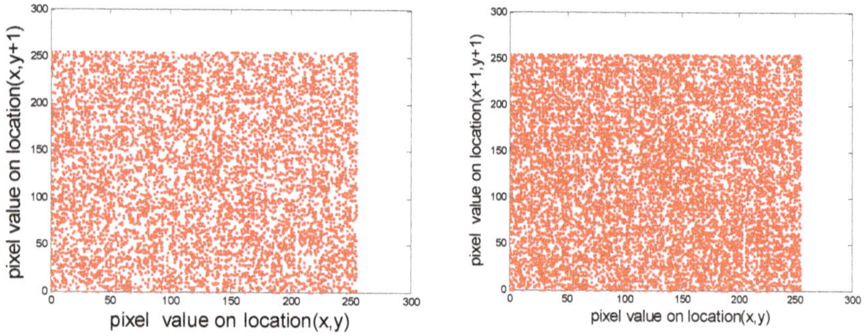

Figure 8. The correlation plots of the cameraman image and the corresponding ciphered image of cameraman. (**a**) Horizontal correlation of the cameraman image; (**b**) Vertical correlation of the cameraman image; (**c**) Diagonal correlation of the cameraman image; (**d**) Horizontal correlation of the cyphered image of cameraman; (**e**) Vertical correlation of the cyphered image of cameraman; (**f**) diagonal correlation of the cyphered image of cameraman.

In order to further quantify the linear correlation between the adjacent pixels, the correlation coefficients can be calculated as:

$$xc = \frac{n \sum\limits_{i=1}^{n} x_i y_i - \sum\limits_{i=1}^{n} x_i \sum\limits_{i=1}^{n} y_i}{\sqrt{n \sum\limits_{i=1}^{n} {x_i}^2 - \left(\sum\limits_{i=1}^{n} x_i\right)^2} \sqrt{n \sum\limits_{i=1}^{n} {y_i}^2 - \left(\sum\limits_{i=1}^{n} y_i\right)^2}} \tag{30}$$

where, x_i and y_i represent the gray values of two adjacent pixels, respectively, and n represents the number of pixel pairs selected. The correlation coefficients of adjacent pixels of the original image and the cyphertext image are shown in Table 8. It can be seen that the absolute values of the correlation coefficients between adjacent pixels in three directions of the plaintext image are very close to 1, while the absolute values of the correlation coefficients between adjacent pixels in each direction of the corresponding cyphertext image are close to 0, and the correlation is weakened.

Table 8. Correlation coefficients between adjacent elements of original image and encrypted image.

Images	Horizontal	Vertical	Diagonal
Plaintext "Rice"	0.9427	0.9263	0.8994
Cyphertext "Rice"	−0.0046	0.0287	−0.0361
Plaintext "Cameraman"	0.9588	0.9360	0.9095
Cypher "Cameraman"	−0.0017	−0.0279	0.0047
Plaintext "autumn"	0.9675	0.9845	0.9821
Cyphertext "autumn"	−0.0087	0.0142	0.0098
Plaintext "pepper"	0.9894	0.9931	0.9847
Cyphertext "pepper"	−0.0055	−0.0194	−0.0295

5.5. Computational Speed Analysis

Finally, the time complexity of the algorithm for encryption/decryption is evaluated. Several images for different sizes have been considered and the time complexity is given. The time complexity analysis is achieved on zn Intel(R) Pentium(R) Dual Core processor CPU (2.3 GHz with 2 GB RAM) personal computer. The algorithm is developed in Matlab R2014a and compiled by 7.14 on Windows 7 Home Premium edition. The results are shown in Table 9. This shows that our algorithm is faster than most of the algorithms in [39,40,43,44], but it is just slightly slower than that in [40].

In general, the encryption algorithm based on the spatial domain is faster than the algorithm based on the image frequency domain [45].

Table 9. Comparison of encryption or dcryption time (EDT) of 8-bit gray level images for different image sizes.

Image Size ($M \times N$)	Ref. [39]	Ref. [40]	Ref. [43]	Ref. [44]	Proposed System
64×64	0.07	0.03	0.19	0.61	0.02
128×128	0.19	0.08	0.29	2.17	0.06
256×256	0.46	0.18	6.01	7.73	0.22
512×512	1.88	0.97	35.59	31.59	0.85
1024×1024	3.62	2.94	253.88	169.21	3.11

6. Conclusions

A new image encryption scheme is proposed, which includes three main components: adding pixels around the image, pixel scrambling and pixel diffusion. Firstly, the hash value of the plaintext image is converted into a pseudo-random sequence, then adding pseudo-random sequences to the surrounding area of plaintext images. The pseudo-random sequences used in the permutation and diffusion process are related to the plaintext image, which can resist chosen plaintext attack. Different from previous algorithms, our algorithm transforms the hash value of the plaintext image into a pseudo-random sequence, which takes the pseudo-random sequence as part of the encrypted image, while the previous algorithm takes the hash value of the plaintext image as a part of the key. In our encryption algorithm, the key of the encryption system is only the initial value of the chaotic system, which reduces the difficulty of key management. In addition, the algorithm also has the following advantages, which can be demonstrated by theoretical analysis and experimental results: the key space is large, the cyphertext is very sensitive to plaintext and keys, the distribution of pixels in encrypted image is uniform, the correlation between adjacent pixels of cyphertext is very low, and the information entropy of cyphertext images is close to the ideal value of 8, Therefore, the proposed algorithm has good application prospects in secure image communication and storage applications.

Author Contributions: Conceptualization, S.Z. and C.Z.; Methodology, S.Z.; Software, C.Z.; Validation, S.Z., C.Z. and W.W.; Formal Analysis, S.Z.; Investigation, C.Z.; Resources, W.W.; Data Curation, W.W.; Writing—Original Draft Preparation, S.Z.; Writing—Review & Editing, C.Z.; Visualization, W.W.; Supervision, C.Z.; Project Administration, S.Z.; Funding Acquisition, S.Z.

Funding: This research was funded by [the National Natural Science Foundation of China] grant number [61472451]; [the Open Project of Guangxi Colleges and Universities Key Laboratory of Complex System Optimization and Big Data Processing] grant number [2016CSOBDP0103]; [the Shan Dong Province Nature Science Foundation] grant number [ZR2017MEM019]; [The Science Research Fund of Liaocheng University] grant number [318011606]; [Scientific Research Projects of Universities in Shan Dong Province] grant number [J18KA336].

Acknowledgments: The authors are thankful to the reviewers for their comments and suggestions to improve the quality of the manuscript.

Conflicts of Interest: The authors declare no conflict of interest.

References

1. Alvarez, G.; Li, S. Some basic cryptographic requirements for chaos-based cryptosystems. *Int. J. Bifurc. Chaos* **2006**, *16*, 2129–2151. [CrossRef]
2. Fridrich, J. Symmetric ciphers based on two-dimensional chaotic maps. *Int. J. Bifurc. Chaos* **1998**, *8*, 1259–1284. [CrossRef]
3. Hsiao, H.-I.; Lee, J. Color image encryption using chaotic nonlinear adaptive filter. *Signal Process.* **2015**, *117*, 281–309. [CrossRef]
4. Mirzaei, O.; Yaghoobi, M.; Irani, H. A new image encryption method: Parallel sub-image encryption with hyper chaos. *Nonlinear Dyn.* **2012**, *67*, 557–566. [CrossRef]

5. Wu, Y.; Zhou, Y.; Agaian, S.; Noonan, J.P. A symmetric image cipher using wave perturbations. *Signal Process.* **2014**, *102*, 122–131. [CrossRef]
6. Patidar, V.; Pareek, N.K.; Purohit, G.; Sud, K.K. Modified substitution-diffusion image cipher using chaotic standard and logistic maps. *Commun. Nonlinear Sci. Numer. Simul.* **2010**, *15*, 2755–2765. [CrossRef]
7. Zhu, C. A novel image encryption scheme based on improved hyperchaotic sequences. *Opt. Commun.* **2012**, *285*, 29–37. [CrossRef]
8. Zhang, Y.; Xiao, D.; Shu, Y.; Li, J. A novel image encryption scheme based on a linear hyperbolic chaotic system of partial differential equations. *Signal Process. Image Commun.* **2013**, *28*, 292–300. [CrossRef]
9. Zhang, W.; Wong, K.-W.; Yu, H.; Zhu, Z.-L. A symmetric color image encryption algorithm using the intrinsic features of bit distributions. *Commun. Nonlinear Sci. Numer. Simul.* **2013**, *18*, 584–600. [CrossRef]
10. Zhang, Y.; Xiao, D. An image encryption scheme based on rotation matrix bit-level permutation and block diffusion. *Commun. Nonlinear Sci. Numer. Simul.* **2014**, *19*, 74–82. [CrossRef]
11. Chai, X. An image encryption algorithm based on bit level brownian motion and new chaotic systems. *Multimed. Tools Appl.* **2017**, *76*, 1159–1175. [CrossRef]
12. Zhu, C.-X.; Sun, K.-H. Cryptanalysis and improvement of a class of hyperchaos based image encryption algorithms. *Acta Phys. Sin.* **2012**, *61*, 120503. [CrossRef]
13. Ozkaynak, F.; Yavuz, S. Analysis and improvement of a novel image fusion encryption algorithm based on DNA sequence operation and hyper-chaotic system. *Nonlinear Dyn.* **2014**, *78*, 1311–1320. [CrossRef]
14. Zhu, C.; Xu, S.; Hu, Y.; Sun, K. Breaking a novel image encryption scheme based on brownian motion and pwlcm chaotic system. *Nonlinear Dyn.* **2015**, *79*, 1511–1518. [CrossRef]
15. Yap, W.-S.; Phan, R.C.W.; Yau, W.-C.; Heng, S.-H. Cryptanalysis of a new image alternate encryption algorithm based on chaotic map. *Nonlinear Dyn.* **2015**, *80*, 1483–1491. [CrossRef]
16. Chen, L.; Ma, B.; Zhao, X.; Wang, S. Differential cryptanalysis of a novel image encryption algorithm based on chaos and line map. *Nonlinear Dyn.* **2016**, *87*, 1797–1807. [CrossRef]
17. Li, C.; Lin, D.; Lu, J. Cryptanalyzing an image-scrambling encryption algorithm of pixel bits. *IEEE Multimed.* **2017**, *24*, 64–71. [CrossRef]
18. Zhu, C.; Sun, K. Cryptanalyzing and improving a novel color image encryption algorithm using RT-enhanced chaotic tent maps. *IEEE Access* **2018**, *6*, 18759–18770. [CrossRef]
19. Norouzi, B.; Mirzakuchaki, S.; Seyedzadeh, S.; Mosavi, M.R. A simple, sensitive and secure image encryption algorithm based on hyper-chaotic system with only one round diffusion process. *Multimed. Tools Appl.* **2014**, *71*, 1469–1497. [CrossRef]
20. Diab, H.; El-semary, A.M. Secure image cryptosystem with unique key streams via hyper-chaotic system. *Signal Process.* **2018**, *142*, 53–68. [CrossRef]
21. Akhavan, A.; Samsudin, A.; Akhshani, A. Cryptanalysis of an image encryption algorithm based on DNA encoding. *Opt. Laser Technol.* **2017**, *95*, 94–99. [CrossRef]
22. Zhang, G.; Liu, Q. A novel image encryption method based on total shuffling scheme. *Opt. Commun.* **2011**, *284*, 2775–2780. [CrossRef]
23. Zhu, C.; Liao, C.; Deng, X. Breaking and improving an image encryption scheme based on total shuffling scheme. *Nonlinear Dyn.* **2013**, *71*, 25–34. [CrossRef]
24. Wang, X.; Zhu, X.; Wu, X.; Zhang, Y. Image encryption algorithm based on multiple mixed hash functions and cyclic shift. *Opt. Lasers Eng.* **2017**, *107*, 370–379. [CrossRef]
25. Guesmi, R.; Farah, M.A.B.; Kachouri, A.; Samet, M. A novel chaos-based image encryption using DNA sequence operation and secure hash algorithm sha-2. *Nonlinear Dyn.* **2016**, *83*, 1123–1136. [CrossRef]
26. Liu, H.; Wang, X. Color image encryption based on one-time keys and robust chaotic maps. *Comput. Math. Appl.* **2010**, *59*, 3320–3327. [CrossRef]
27. Huang, L.; Cai, S.; Xiao, M.; Xiong, X. A simple chaotic map-based image encryption system using both plaintext related permutation and diffusion. *Entropy* **2018**, *20*, 535. [CrossRef]
28. Chai, X.; Chen, Y.; Broyde, L. A novel chaos-based image encryption algorithm using DNA sequence operations. *Opt. Lasers Eng.* **2017**, *88*, 197–213. [CrossRef]
29. Li, Y.; Liu, X.; Chen, G.; Liao, X. A new hyperchaotic lorenz-type system: Generation, analysis, and implementation. *Int. J. Circuit Theory Appl.* **2011**, *39*, 865–879. [CrossRef]

30. Stoyanov, B.P. Pseudo-random bit generator based on chebyshev map. In *Application of Mathematics in Technical and Natural Sciences*; Todorov, M.D., Ed.; American Institute of Physics: College Park, MD, USA, 2013; Volume 1561, pp. 369–372.
31. Li, S.; Chen, G.; Mou, X. On the dynamical degradation of digital piecewise linear chaotic maps. *Int. J. Bifurc. Chaos* **2005**, *15*, 3119–3151. [CrossRef]
32. Li, S.; Chen, G.; Wong, K.-W.; Mou, X.; Cai, Y. Baptista-type chaotic cryptosystems: Problems and countermeasures. *Phys. Lett. A* **2004**, *332*, 368–375. [CrossRef]
33. Curiac, D.-I.; Volosencu, C. Chaotic trajectory design for monitoring an arbitrary number of specified locations using points of interest. *Math. Probl. Eng.* **2012**, *2012*, 1–18. [CrossRef]
34. Curiac, D.I.; Iercan, D.; Dranga, O.; Dragan, F.; Banias, O. Chaos-Based Cryptography: End of the Road? In Proceedings of the International Conference on Emerging Security Information, System and Technologies, Valencia, Spain, 14–20 October 2007; pp. 71–76.
35. Wang, X.-Y.; Gu, S.-X.; Zhang, Y.-Q. Novel image encryption algorithm based on cycle shift and chaotic system. *Opt. Lasers Eng.* **2015**, *68*, 126–134. [CrossRef]
36. Khanzadi, H.; Eshghi, M.; Borujeni, S.E. Image encryption using random bit sequence based on chaotic maps. *Arab. J. Sci. Eng.* **2014**, *39*, 1039–1047. [CrossRef]
37. Kulsoom, A.; Xiao, D.; Aqeel Ur, R.; Abbas, S.A. An efficient and noise resistive selective image encryption scheme for gray images based on chaotic maps and DNA complementary rules. *Multimed. Tools Appl.* **2016**, *75*, 1–23. [CrossRef]
38. Wang, X.-Y.; Zhang, Y.-Q.; Bao, X.-M. A novel chaotic image encryption scheme using DNA sequence operations. *Opt. Lasers Eng.* **2015**, *73*, 53–61. [CrossRef]
39. Stoyanov, B.; Kordov, K. Image encryption using chebyshev map and rotation equation. *Entropy* **2015**, *17*, 2117–2139. [CrossRef]
40. Stoyanov, B.; Kordov, K. Novel image encryption scheme based on chebyshev polynomial and duffing map. *Sci. World J.* **2014**, *2014*, 283639. [CrossRef] [PubMed]
41. Seyedzade, S.M.; Mirzakuchaki, S.; Atani, R.E. A novel image encryption algorithm based on hash function. In Proceedings of the Iranian Conference on Machine Vision and Image Processing, Isfahan, Iran, 27–28 October 2011; pp. 1–6.
42. Chai, X.-L.; Gan, Z.-H.; Yuan, K.; Lu, Y.; Chen, Y.-R. An image encryption scheme based on three-dimensional brownian motion and chaotic system. *Chin. Phys. B* **2017**, *26*, 020504. [CrossRef]
43. Wang, X.; Zhang, J. An image scrambling encryption using chaos-controlled poker shuffle operation. In Proceedings of the International Symposium on Biometrics and Security Technologies, Islamabad, Pakistan, 23–24 April 2008; pp. 41–46.
44. Rehman, A.U.; Liao, X.; Kulsoom, A.; Abbas, S.A. Selective encryption for gray images based on chaos and DNA complementary rules. *Multimed. Tools Appl.* **2015**, *74*, 4655–4677. [CrossRef]
45. Ramadan, N.; Ahmed, H.H.; El-khamy, S.E.; Abd El-Samie, F.E. Permutation-substitution image encryption scheme based on a modified chaotic map in transform domain. *J. Central South Univ.* **2017**, *24*, 2049–2057. [CrossRef]

entropy

MDPI

Article

Video Summarization for Sign Languages Using the Median of Entropy of Mean Frames Method

Shazia Saqib *,† and Syed Asad Raza Kazmi

Department of Computer Science, Government College University, Lahore 54000, Pakistan; arkazmi@gcu.edu.pk
* Correspondence: shaziasaqib@lgu.edu.pk; Tel.: +92-321-499-3631
† Current address: Lahore Garrison University, Lahore 54000, Pakistan.

Received: 4 September 2018; Accepted: 27 September 2018; Published: 29 September 2018

Abstract: Multimedia information requires large repositories of audio-video data. Retrieval and delivery of video content is a very time-consuming process and is a great challenge for researchers. An efficient approach for faster browsing of large video collections and more efficient content indexing and access is video summarization. Compression of data through extraction of keyframes is a solution to these challenges. A keyframe is a representative frame of the salient features of the video. The output frames must represent the original video in temporal order. The proposed research presents a method of keyframe extraction using the mean of consecutive k frames of video data. A sliding window of size $k/2$ is employed to select the frame that matches the median entropy value of the sliding window. This is called the Median of Entropy of Mean Frames (MME) method. MME is mean-based keyframes selection using the median of the entropy of the sliding window. The method was tested for more than 500 videos of sign language gestures and showed satisfactory results.

Keywords: entropy; keyframes; Shannon's entropy; sign languages; video summarization; video skimming

1. Introduction

Gesture recognition is a giant leap toward the touch-free interface. The information conveyed through gestures is either in the form of static gestures or in the form of continuous gestures [1]. The continuous gestures are represented by videos [2]. A video itself cannot be recognized. A video needs to be summarized for analysis of its content. Video summarization is used to prepare a reduced size of the video in the form of frames that can be used for indexing or content analysis. This research aims at a keyframe extraction technique that can, in turn, be used for object recognition and information retrieval. Every video can be converted into frames. A keyframe refers to the image frame that represents the maximum information contained in a group of frames [3]. The keyframe defines the starting and ending points of any transition. The position of the keyframe tells us about the timing of any event. Combining all keyframes results in the abstract of the particular video. The idea of keyframe usage is very powerful as it saves a great deal of processing time and requires less storage. Figure 1 shows a few frames at a time; orange frames are the frames with mean values. Keyframes are basically the representative frames of a video. Using an appropriate technique, keyframes can be located among all frames of the video. These frames represent the video content and thus reduce the amount of storage and processing needed.

Figure 1. Video converted to frames and mean of k frames.

The selection of "correct" keyframe is based on the application as well as the personal "definition" of what the summary should represent. Figure 2 shows the mean frames in a sliding window whose median of entropy is being calculated. The size of the sliding window is chosen such that it has an odd number of elements.

Researchers have described keyframe extraction into either "sequence-based approaches" or "cluster-based approaches" [4]. The first type of approaches uses the temporal information and visual features to identify the keyframes. Consecutive frames are compared and the variation in consecutive frames is estimated. When a substantial change in the frame is detected, that frame is selected as the keyframe. Cluster-based approaches divide the video stream into shots. The frames that represent the shot are chosen as candidate keyframes. The clustering process should maintain the temporal order of the frames [4].

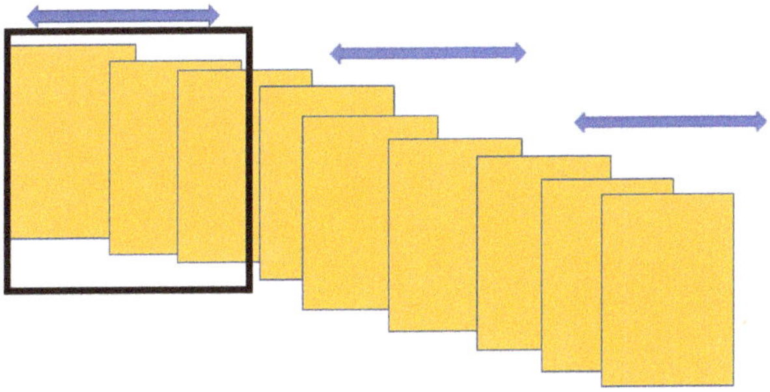

Figure 2. Keyframe selection using entropy measure through the sliding window of size $k/2$.

The process of selecting keyframes passes through video information analysis, meaningful clip selection, and output generation. For a good summary of video information, we must determine salient features, the descriptors in the visual component, the audio component if any, and the textual components such as closed captions. A shot can change by a "CUT", which is a sudden change between two adjacent frames, or a "FADE", which occurs by a steady change in brightness. Another is "DISSOLVE", which is similar to FADE but is sandwiched between two shots. One scene gets dimmer and the incoming scene gets brighter, and the 2nd shot finally replaces the first one [4]. All the methods of video summarization are grouped into the following:

1.1. Static Video Summarization

The video is sampled either uniformly or randomly. The complete video is divided into frames. Out of these frames, one or more will be representative of the content of the video, helping in generating video summaries [4].

1.2. Methods Based on Clustering Techniques

These techniques combine similar frames/shots. Some features are then extracted from this group of frames. Based on this, one or more frames are extracted from the cluster. Different features such as luminance, color histogram, a motion vector, and k-means clustering are used in making the decision for keyframe selection [4].

1.3. Dynamic Video Summarization

This is also called video skimming, which is actually a summary video of all the important scenes from an input stream. It forms an abstraction of the video. Singular Value Decomposition (SVD), and motion model and semantic analysis, are the few techniques that are used for dynamic video summarization [4].

The rest of the paper is organized as follows: Section 2 covers related work, Section 3 shows the algorithm and experimental work. Section 4 elaborates the results of the experiment, and Section 5 concludes and suggests future work.

2. Related Work

A great deal of work has been done on video summarization. Sheena and Narayanan used the histogram of consecutive frames. In this method, the threshold difference of histograms is calculated to find keyframe from video data from the KTH action database. Their algorithm is good both in terms of fidelity value as well as compression ratio [3]. Khattabi et al. analyzed the static and dynamic methods of producing video summaries [4]. Tsai et al. have related transmitted information and image noise, they investigated the effect of noise on blurring. They further analyzed the use of smoothing filters for improving the noise and blurring, their results gave reasonable performance in medical imaging [5]. Fauvet et al. used the computation of the dominant image motion and the geometrical properties that result in a change in a frame in the considered shot. They improved their own technique at computational cost using an energy function. They tested their technique on sports videos and obtained satisfactory results [6]. Vasconcelos et al. presented a technique for characterization and analysis of video data. They used Bayesian architecture to analyze the content of videos on a semantic basis [7]. Mikolajczyk et al. compared the detection rate with the false positive rate. They used differential invariants, steerable filters, Scale Invariant Feature Transform (SIFT) descriptors, moment invariants, complex filters, and cross-correlation. Their research shows that SIFT descriptors yielded the best results. Steerable filters also proved to be a good choice [8]. Sebastian et al. proposed a technique that divides the frames of the video into blocks. They used the mean, variance, skew, and kurtosis histogram of every block and compared them with the corresponding blocks of the next frame. They selected the frame with the highest mean as the keyframe. The method is based on the color distribution [9]. Supriya Kamoji et al. captured the motion in a video to find the keyframes. To analyze this motion, block matching techniques based on Diamond Search and Three Step Search were compared. The comparison process is on the varied nature of videos. The summarization factor was increased at the cost of precision during the summarization process [10]. Mentzelopoulos et al. compared all of the current keyframe extraction algorithms. They proposed the use of Entropy-Difference for spatial frame segmentation [11].

Cahuina et al. proposed a technique based on local descriptors for semantic video summarization and tested the technique on 100 videos. Their technique achieved a recognition level of 99%. They used color information with local descriptors to produce video summaries [12]. Shi et al. proposed a key frame extraction method for video copyright protection. Their technique is based on the difference of frames using features such as color and structure. For final results, optimization is done on a number of keyframes that have been selected [13]. Zhao et al. proposed the use of local motion features extracted from their neighborhood. Their method uses a hierarchical spatial pyramid structure giving very good results over standard benchmark datasets [14]. Hasebe et al. proposed a new method to find the keyframes for input videos. The technique works in the wavelet transform domain. As a first step, shot boundaries are sorted out so that initial keyframes may be defined. Secondly, feature vectors are grouped into clusters for these selected frames. The results are tested on the basis of processing speed and precision rates [15]. Mahmoud et al. have suggested the use of VGRAPH that uses color as well as texture features. The video is divided into shots based on color features. The technique uses a nearest neighbor graph using textural features [16].

Ciocca et al. proposed an algorithm based on the difference between two consecutive frames of a video sequence and used the visual content changes. They used a color histogram, wavelet statistics, and an edge direction histogram. Similarity measures are determined and combined with the frame difference. The method even detects very minor changes. The proposed method dynamically selects a variable number of keyframes from different shots [17]. Ejaz et al. combined the features of Red Green Blue (RGB) color channels, histograms, and moments to find the keyframes. The technique is adaptive as it combines current and old iterations. The summaries produced by these techniques are as good as those created by humans [18]. Rajendra et al. reviewed previous work on content-based information processing for multimedia data. They focused on how to browse andhow to add new features, learning, effective computing semantic queries, high-performance indexing, and evaluation techniques [19]. Girgensohn et al. designed an algorithm to find keyframes that represent the input video. This technique can determine keyframes from a video by clustering frames. Each cluster has a representative frame, and some clusters are not considered and left unprocessed on temporal grounds [20]. Guan et al. suggested a keypoint-based framework for selecting keyframes using local features. The resultant frames represent video without any redundancy [21]. Asade et al. suggested an algorithm to extract static video summaries. Their technique is based on fuzzy c-means clustering. The frame with the highest membership grade for any cluster is selected as a keyframe. Their method gives a lower error rate with a higher accuracy level [22].

Zhang et al. used the similarity distance of the adjacent frames to adjust the threshold input adaptive algorithm. They then used the Iterative Self-Organizing Data Analysis Technique (ISODATA) to cluster frames into classes automatically. Their algorithm focuses on different motion types reliably and efficiently. Their results were tested using metrics that analyzed for the reconstructed motion and the mean absolute error value [23]. Dong et al. suggested an algorithm for keyframe selection and recognition method for robust markerless real-time camera tracking. Their technique used one offline and one online module—offline uses a number of images and online uses a video to detect a pose. Their technique reduces redundancy and, at the same time, produces a best possible set of frames [24]. Kim et al. proposed a technique that generates panoramic images from web-based geographic information systems. Their algorithm performs data fusion, crowd sourcing, and recent advances in media processing. Their work shows that a great deal of time can be saved if "geospatial metadata" is used without any compromise on image quality [25].

Mei et al. generated audio streams, compressed images, and metadata for motion information and temporal structure. Their technique works at a very low compression rate. The proposed Near-Lossless Semantic Video Summarization (NLSS) method is effectively used for visualization, indexing, browsing, duplicate detection, concept detection, etc. The NLSS is tested on TREC Video Retrieval Evaluation (TRECVID) and other video collections, showing that it significantly reduces storage consumption while giving high-level semantic fidelity [26]. Shroung et al. used the image difference and classification theory to identify keyframes from video captured using ordinary mobile or laptop cameras, yielding a highly accurate video summary. These video frames are used for dynamic sign recognition [27]. Vázquez-Martín et al. utilized consecutive frames and their features. They built a graph using these features and used clustering to partition the graph [28]. Khurana et al. used the edge detection and the difference of this value between the consecutive frames. The frames matching a threshold are treated as keyframes [29]. Thakre et al. proposed a technique for keyframe selection of compressed video shots using the adaptive threshold method working on 200 plus video clips [30].

Wang et al. elaborated the important issues in information theory and discussed the use of these concepts in visualization in relating data communication to data visualization [31]. Entropy has been used for image segmentation by [32–35], covering various types of available entropy algorithms. Sabuncu discussed the use of different entropic measures that can be used for image registration [36]. Ratsamee et al. proposed finding a keyframe that is based on image quality measurements such as color, sharpness, noise, etc. However, a biosensor is required to determine human excitement [37]. Angadi et al. proposed a technique that uses a fuzzy c-means clustering algorithm. The technique merges keyframes in a timewise order [38]. Yuan et al. used a Deep Side Semantic Embedding (DSSE) model to select keyframes. They correlated two uni-modal autoencoders, yielding side information and video frames. They tested their work on the Title-based Video Summarization (TVSum50) dataset [39]. Chen et al. employed the visual and textual features of videos. Their technique uses their previously reviewed frames and posted comments [40]. Panda et al. used video-level annotation for summarizing web videos. They used Deep Convolutional Neural Network (3D CNN) architecture for video-level annotation [41]. Mahasseni et al. used a deep summarizer network that used a summarizer autoencoder named a Long Short-term Memory Network (LSTM) [42]. Jeoung et al. proposed a technique for a static summary of consumer videos. They completed the process in two steps: first they skimmed the video and then performed content-aware clustering with keyframe selection [43]. Yoon et al. proposed an approach based on learning principal person appearance [44].

De Avila et al. proposed Video SUMMarization (VSUMM) for producing static video summaries. The method is based on color feature extraction from video frames and a k-means clustering algorithm. The work was compared with manually created static summaries, demonstrating the high accuracy of the proposed VSUMM technique. The technique improves on visual features, their fusion, and the estimation of the number of clusters [45]. Kanehira et al. proposed Fisher's discriminant criteria for inner-summary, inner-group, and between-group variances defined on the feature representation of summary [46]. Manis et al. have proposed the Bubble Entropy to rank the elements inside the vectors for doing reallocation to sort these elements [47]. Athitsos et al. have designed the dataset ASL Lexicon Video Dataset to develop a computer vision system that helps in recognizing the meaning of an ASL sign. The dataset can be a benchmark for a variety of computer vision and machine learning methods [48]. PUN proposed a technique for threshold selection method to segment images using the entropy of the grey level histogram dividing them into two-level images [49]. Sluder and David have proposed to use averaging to reduce noise in an image. The magnitude of noise drops by the square root of the number of images averaged [50]. Panagiotakis et al. suggested using three iso-content principles (Iso-Content Error, Iso-Content Distance, and Iso-Content Distortion) so that the selected keyframes are generated according to the algorithm used. The technique used both Supervised and Unsupervised approaches. The proposed technique requires an improvement in the temporal order of frames from different shots [51]. Song et al. presented Title-based Video Summarization (TVSum), which is an unsupervised video summarization framework that uses video labeling to summarize the video. The co-archetypal analysis is done for canonical patterns between two sets of data. However, they need to improve the image collection procedure and to make use of metadata to produce the video summary [52]. Mei et al. proposed video summarization based on a constrained Minimum Sparse Reconstruction (MSR) model by recreating a video using keyframes generated with minimum possible frames. A Percentage of Reconstruction (POR) criterion is used to determine the length of the summary. Their technique summarizes both structured videos and the consumer videos [53]. Ajmal et al. used the Histogram of Oriented Gradient (HOG) using a Support Vector Machine (SVM) classifier. The Kalman filter in the algorithm determines the track of each person [54].

3. The Proposed Median of Entropy of Mean Frames (MME) Technique for Keyframe Selection

The proposed technique uses the concept of the mean and then applies the median of the entropy to the resultant images for video summarization. The resultant keyframes thus generated will be used for continuous gesture recognition. The technique uses the mean of k images. It then takes a group of $k/2$ mean frames at a time and determines their entropy. The median of the entropy measure is calculated to select the keyframes. The value of k is chosen such that $k/2$ is odd, for easy selection of keyframes.

3.1. Mean

The mean is a very important measure in digital image processing. It is used in spatial filtering and is helpful in noise reduction. The mean of k frames is defined as

$$\hat{f}_l(i,j) = \frac{\sum_{m=1}^{k}\sum_{i=1}^{n}\sum_{j=1}^{n} f_m(i,j)}{k}. \tag{1}$$

Here, $\hat{f}_l(i,j)$ shows the lth mean of k images. $\sum_{m=1}^{k}\sum_{i=1}^{n}\sum_{j=1}^{n} f_m(i,j)$ is the sum of k frames. $\sum_{i=1}^{n}\sum_{j=1}^{n} f_m(i,j)$ shows the mth frame.

3.2. Entropy

Entropy is the measure of randomness (or uncertainty) in an image. It is a measure of the information transmitted [5]. The concept was given by Claude Shannon and is called Shannon's entropy [35]. Maximum entropy, Renvi entropy, Tsallis entropy, spatial entropy, minimum entropy, conditional entropy, cross-entropy, relative entropy, and fuzzy entropy are used for image segmentation, image registration, image compression, image reconstruction, and edge detection in gray level images [33]. Bubble entropy investigates the rank of the members of the collection of data and determines a method of sorting these elements. Bubble entropy is considered a good option in biomedical signal analysis and interpretation [47].

Entropy is a measure of the spread of states in which a system can adjust. A system with low entropy will have a small number of such states, while a high entropy system will be spread over a large number of states. Suppose X is a random variable consisting of following $X_1, X_2..., X_l$. The variable X has a probability distribution $p(x) = (p_1, p_2, p_3..p_l)$, which is used for the calculation of the Shannon's entropy. The entropy of an l-state system is given as

$$H = -\sum_{k=0}^{l-1} p_k \log_b p_k \tag{2}$$

where p_i is the probability of occurrence of the event i and $\sum_{i=0}^{l-1} p_i = 1$. b is the base of the algorithm and is usually 2. If $P(x_i) = 0$ for some i, then the multiplier $0log_b0$ is considered as zero, which is consistent with the limit [36]. The term $\log(\frac{1}{p_i})$ shows the uncertainty associated with the corresponding outcome or can also be viewed as the amount of information gained by observing that outcome. Entropy represents the statistical average of uncertainty or information. The number of pixels in the image is n, while n_k represents a total number of pixels at level k.

$$P_k = \frac{n_k}{n} \tag{3}$$

where l is the total number of gray levels, and P_k is the probability associated with each gray level k. The value of entropy is highest when samples are equally likely and when $H(p1\ldots pl) \leq \log(l)$ [49].

The video summarization process involves the following stages:

- **Input Video**. This is the video that is to be converted into keyframes. It can be in any standard format.
- **Frame Extraction**. Every video is basically a sequence of a finite number of still images called frames. These frames occupy a large amount of memory. The frame rate is about 20–30 frames per second (FPS). Movies are shown at a rate of almost 24 fps. In some countries, it is 25 fps. In North America and Japan, the movies are shown at 29.97 fps. In other image processing applications, it is usually at 30 frames per second. Other common frame rates are usually multiples of these [19]. It has been found that, usually, 1–2 frames per second creates the illusion of movement. The rest of the frames show almost the same scene repeatedly [30].
- **Feature Extraction**. This process can be based on features such as colors, edges, or motion features. Some algorithms use other low-level features such as color histograms, frame correlations, and edge histograms [19].

Figure 3 elaborates the mechanism used in the proposed solution. It starts by capturing input video. The video is then converted to frames. Frames are then preprocessed and resized to an appropriate dimension. The proposed algorithm then takes the mean of k frames at a time, thus reducing images from n to n/k. After this, a sliding window of size $k/2$ is applied to the resultant frames, and in each window the frame with the median value of entropy is selected.

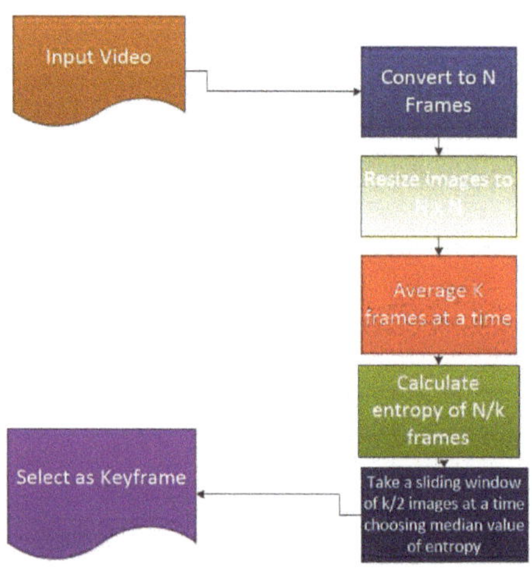

Figure 3. Flow of the procedure to extract keyframes.

3.3. Algorithm to Find Keyframes

The technique uses the following **pseudocode**:

Input: The video.
Output: kf_1,kf_2, $kf_3 \ldots kf_{tkfr}$, where tkfr represents total keyframes
procedure:
convert the video to frames $f_1, f_2, f_3..f_n$;
resize each frame to an image size of $n \times n$ (in this proposed technique, we chose $n = 100$);
initialize l to 1;
$\forall f_i \ldots 1 \leq i \leq n$;
$\mu f_l = \frac{\sum_{i=1}^{k} f_i}{k}$;
increment l by 1;

reduce the frame count to n/k;
consider 1st sliding window of size $k/2$;
calculate entropy of frame $l, l+1, l+2\ldots$;
compare the frames in the sliding windows;
choose a frame with the median value of entropy;
slide window to the next $k/2$ consecutive frames.

Algorithm 1: Keyframe Extraction through the proposed MME method.

Input: The Video converted to frames f_1,f_2, $f_3\ldots\forall f_i, where 1 \leq i \leq n$ and $k=5$ for the examples used in the proposed research.
Output:kf_j, where $1 \leq j \leq tkfr$.
$i \leftarrow 1$;
$l \leftarrow 1$;
while i<=n **do**
  $j \leftarrow 1$;
  while j<=k **do**
    $m[l] = m[l] + f[i+j-1]$;
    $j \leftarrow j+1$
  end
  $m[l] = \frac{1}{k} * m[l]$;
  $i = i+k$;
  $l = l+1$;
end
$i \leftarrow 1$;
while i<=l **do**
  calculate the entropy of each m[i];
  $i \leftarrow i+1$;
end
$winsize \leftarrow k/2$;
$i \leftarrow 1$;
$j \leftarrow 1$;
while i<=l **do**
  $kf[j] = median(entropy(m[i]), entropy(m[i+1]), \ldots entropy(m[i+winsize]))$;
  $i = i + winsize$;
  $j = j+1$;
end
$tkfr = j$;

The proposed Algorithm 1 can be used for any type of video, but it has also been tested rigorously for continuous gestures. We tested this algorithm on several videos. The complexity of the algorithm is $O(n^2)$. As a test case for the proposed technique, we took an example of a video of a gesture for the word **dress** in Pakistan Sign Language, which is 3 s long. It consists of almost 90 frames. In the first loop, five frames are averaged at a time, yielding 17 frames. We continued until all frames had been processed. Using computed entropy, we designed a sliding window of size 3 frames. Later on, the median of the entropies of the frames in the window was calculated. Using a 3 s video, we obtained six keyframes. The compression ratio (CR) is determined by

$$CR = keyframes/totalframes. \tag{4}$$

A low CR represents an efficient technique. Fidelity is another measure to determine the efficiency of the keyframe selection algorithm. It is the maximum of the minimum of the distance between keyframes and the individual frames.

$$d_j = min\{dis(kf_j, f_i)\}. \quad (5)$$

$$fidelity = max\{d_j\}. \quad (6)$$

Fidelity basically determines how effectively an algorithm maintains the global content of the original video [20].

4. Results and Analysis

The algorithm was tested on a number of videos, and a few examples are presented here. The technique was applied to the ASL LexiconVideo Dataset, containing thousands of distinct sign classes of American Sign Language [48]. Figure 4 shows extracted frames from a video of a gesture of the word **bird**, which is 3 s long.

Figure 4. Frames in the video of a gesture for the word **bird** from American Sign Language.

Figure 5 shows the 17 mean frames calculated by taking the mean of k frames using the video of the gesture for the word **bird** for $k = 5$.

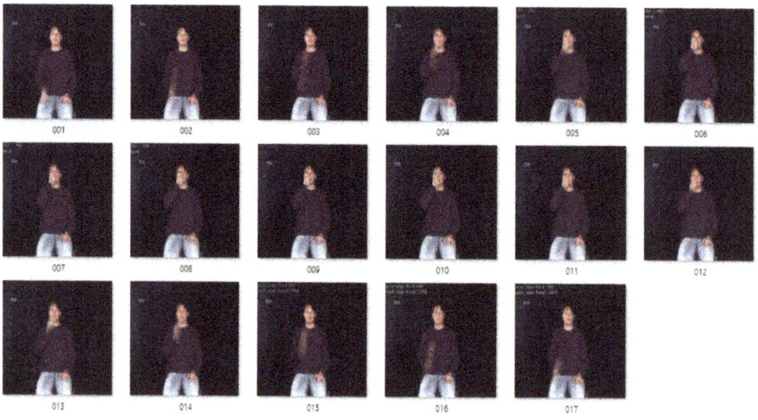

Figure 5. Average frames for the video of the gesture for the word **bird**.

Figure 6 shows the keyframes generated with the proposed MME technique using the median of the entropy of the mean frames using a sliding window of size $k/2$.

Figure 6. Keyframes generated by using the MME methodfor $k = 5$.

Figure 7 shows the median of entropy from the mean frames using a sliding window of $k/2$ while $k = 3$. The video of the gesture for the word **bird** changes frames at a faster pace. Therefore, for the faster videos, we decreased the value of the k; for the slower videos, we increased the value of k accordingly.

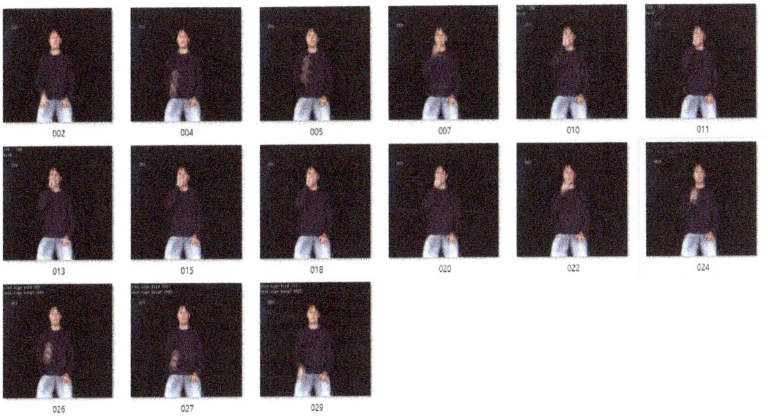

Figure 7. Keyframes generated by using the MME method for $k = 3$.

In another scenario, the video of the gesture for the word **dress** was converted to frames. It was also 3 s long. It was converted to 90 frames, as shown below in Figure 8.

Figure 8. Frames in the video of a gesture for the word **dress** in Pakistan Sign Language.

The mean for these frames was calculated taking 5 frames at a time, so we obtained 17 frames after the process was applied. Figure 9 represents the frames calculated using the mean of 5 frames at a time.

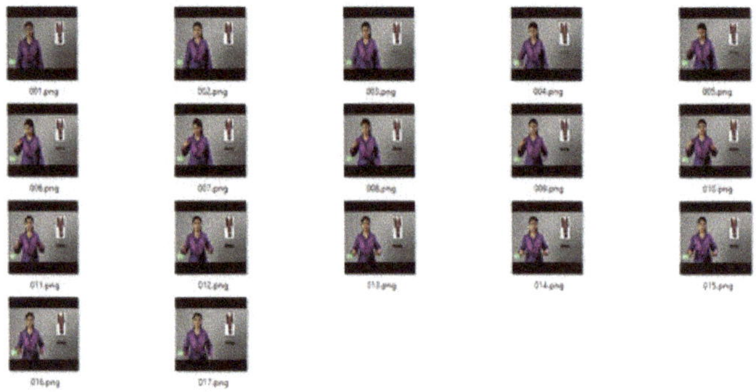

Figure 9. The mean of input frames of a video of the gestures representing the word **dress** in Pakistan Sign Language.

The entropy was calculated for all frames after taking the mean. A sliding window was applied to three frames at a time. The entropy value of all images was calculated, and the frame representing the median of these frames was selected as the keyframe. Once a keyframe was selected, the sliding window moved to the next three frames. We obtained six frames as the keyframes. Figure 10 shows selected keyframes.

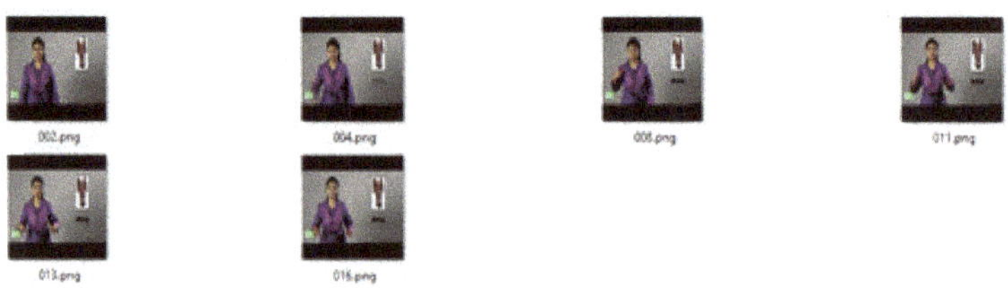

Figure 10. Keyframes generated by using the proposed MME technique.

In another example, a video of the gesture for the word **letter** in Pakistan Sign Language was chosen to test the proposed algorithm. For a video 2 s long, we had approximately 70 frames. Figure 11 shows the frames extracted from the video of the gesture for the word **letter**.

Figure 11. Frames extracted from the video of the gesture for the word **letter** in Pakistan Sign Language.

We obtained 13 frames after applying a mean 5 frames at a time. Figure 12 shows the resultant frames after taking the mean of the input frames.

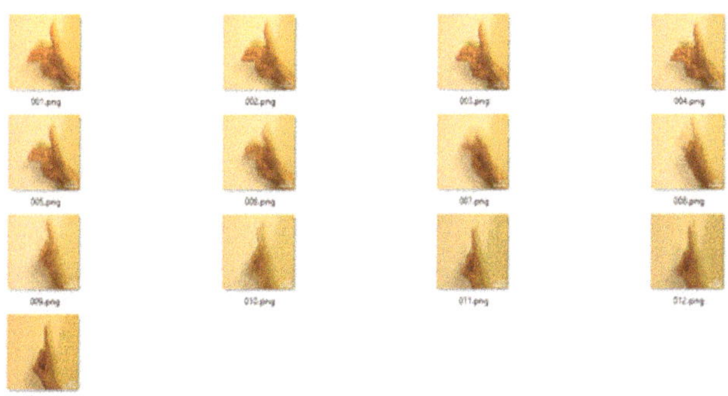

Figure 12. Average frames for the video of the gesture for the word **letter**.

In the last step, using the sliding window of size 3, we obtained five frames as keyframes. Figure 13 shows the keyframes after applying a median on the value of entropy.

Figure 13. Keyframes for the word letter using the proposed MME technique.

Table 1 shows the result of the proposed technique in various videos of Pakistan Sign Language gestures. Its first column shows the input video, and the second column shows the total number of frames from the input video. The table also provides the video duration in seconds, the number of frames after applying the mean, the number of frames after applying the median value of entropy, and the percentage of compression ratio.

Table 1. Videos converted to frames using MME.

Query Video	Frames	Video Duration	Frames after Taking Mean	Keyframes Extracted	Compression Ratio (%)
dress	90	3	17	6	6.67
letter	69	2	13	5	7.24
Apple	90	3	17	6	6.67
Banana	150	5	29	10	6.66
Raisin	130	4.5	25	9	6.6
Lychee	150	5	29	10	6.66
Shoe	150	5	29	10	6.66
mango	110	3.5	21	7	6.36

Table 2 shows the results of the proposed technique and the technique based on the simple mean and the threshold of entropy.

Table 2. Comparison of the Proposed MME with Simple Mean and Simple Entropy.

Query Video	Frames	Keyframes Extracted by the Proposed MME	Keyframes Using Mean	Keyframes Using Threshold of Entropy
dress	90	6	7	7
Khatt (letter)	69	5	6	2
Apple	90	6	7	3
Banana	150	10	12	6
Raisin	130	9	10	1
Lychee	150	10	12	4
Shoe	150	10	12	4
mango	110	7	9	5

Figure 14 shows the graph of the initial number of frames, the number of frames after the mean, and the number of frames after the sliding window operation. Blue bars show the total frames, and light blue bars show the number of frames after taking the mean. Yellow bars represent the resultant number of frames from the proposed MME technique. Figure 15 shows the keyframeextracted using the different techniques. The graph confirms that the proposed technique has an advantage over the techniques using simple mean or simple entropy threshold values. The simple mean can generate too many frames. Secondly, increasing the number of frames in calculating the mean beyond a certain limit is an expensive process, as it increases the required computational time. It grows with the number of images as well as the image size. The entropy threshold technique has its own weaknesses, as the selection of the threshold is a very challenging task. It may fail to deliver good results for certain videos. For the video of the dynamic gesture for the word **raisins**, a 4.5 s long video, the technique generated only one keyframe.

The results were also compared with other existing techniques. The proposed technique achieves accuracy comparable to those provided by [3,10,30,38,54]. It can be tested for qualitative as well as quantitative features, but an actual test is only valid if the video summary is used in the applications for which it was created. Table 3 shows the compression ratios of the proposed technique along with these other techniques. The proposed method performs fairly well in terms of this metric.

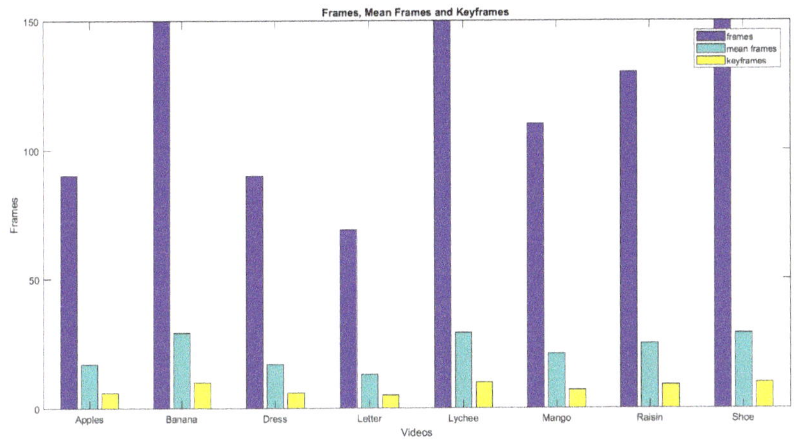

Figure 14. Frames, mean frames, and keyframes using the proposed MME.

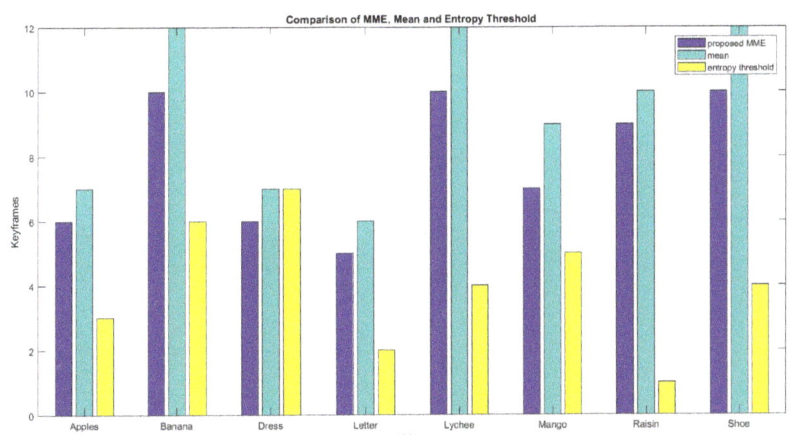

Figure 15. The amount of keyframes using the proposed MME, simple mean, and simple threshold of entropy methods.

Table 3. Comparison of some existing techniques.

Technique Name	Compression Ratio (%)
Analysis of Histograms of Video Frames using Statistical Methods [3]	7.08
Video Summarization Using Motion Activity Descriptors [10]	4.25
Keyframe Extraction of Compressed Video Shots using Adaptive Threshold Method [30]	4.5
Entropy-Based Fuzzy C-Means Clustering and KeyFrame Extraction [38]	8.4
Video Summarization for Sign Languages using MME	6.7
Human Motion Trajectory Analysis Based Video Summarization [54]	6.74

The proposed technique uses an average of k frames with a median of entropy using a sliding window of size $k/2$. The technique incorporates the advantage of taking the mean of consecutive frames. The noise is reduced at the rate of the square root of the number of frames that are used in taking the mean. Therefore, when we average n frames at a time, the noise in the images is reduced by $\sqrt{n}$ [50]. This implies that averaging 5 frames reduces noise by a factor of 2. However, it is a time-consuming process to take an average of frames depending on the capability of the device, as 1/30 of a second is required to average each frame. The technique based on simple averaging loses sharp transitions, and the selected keyframes might not produce an appropriate video summary. With simple entropy, selection of the threshold value is a difficult task. For faster videos such as those provided in the ASL LexiconVideo Dataset or videos of rapidly moving objects, we can select lower values of k for better results as shown in Figure 7.

5. Conclusions

Keyframe selection is an active area of research and plays a pivotal role in video summarization. A keyframe is one of the most efficient methods of obtaining a summary of the video content. The temporal order of the frames is maintained by the proposed MME technique. The frame average count k and sliding windows size $k/2$ can be changed depending on the nature of the video. For dynamic gestures, the value of k in the range 5–15 is preferential. Selecting a reasonable size of k rules out the possibility of missing frames in the proposed MME technique. Another distinct quality of the proposed technique is the distance of keyframes. We obtained keyframes that were at most $k \times k/2$ frames apart. A few redundant frames may have been added, but thereis a much lower chance of losing important information. The system is designed to provide input to video recognition in the form of images that tell the story of the input video by different signers. Adjusting the value of k accordingly or adding a mechanism to learn the value of k makes the system workable, even for very fast or very slow videos. The effectiveness of the MME method is, however, compromised for very rapid changes in a scene. In the future, methods with improved criteria in terms of the mean rate and the sliding window size will be designed. Integrating appropriate filters to remove noise from the final selected keyframes may improve the proposed technique.

Author Contributions: S.S. worked under the supervision of A.R.K., researched the reagents/materials/analysis tools, designed the methodology, and performed the experiments. A.R.K. played his part in the mathematical modeling of the problem and worked on the methodology. A.R.K. also helped in deriving the algorithm used in this work.

Funding: This research received no external funding.

Acknowledgments: The authors are highly indebted to the HonorableObaid Bin Zakria, HI(M), for his policies that continuously motivated our research. We are also grateful to Aasia Khanam, Forman Christian College, from the inception of this concept to the end of the research; her guidance has been invaluable. Adnan Khan, National College of Business Administration and Economics, Lahore, Pakistan, was a great source of support during the research. We are also thankful to Haroon as well as Khalid, Lahore Garrison University, for their guidance in completing the task.

Conflicts of Interest: The authors declare no conflict of interest.

References

1. Shazia, S.; Syed, A.R.K. Repository of Static and Dynamic Signs. *Int. J. Adv. Comput. Sci. Appl.* **2017**, *8*. [CrossRef]
2. Saqib, S.; Kazmi, S.A.R. Recognition of static gestures using correlation and cross-correlation. *Int. J. Adv. Appl. Sci.* **2018**, *5*, 11–18. [CrossRef]
3. Sheena, C.V.; Narayanan, N.K. Key-frame extraction by analysis of histograms of video frames using statistical methods. *Proc. Comput. Sci.* **2015**, *70*, 36–40.
4. Elkhattabi, Z.; Youness, T.; Abdelhamid, B. Video summarization: Techniques and applications. *World Acad. Sci. Eng. Technol.* **2015**, *9*, 928–933.
5. Tsai, D.Y.; Yongbum, L.; Eri, M. Information entropy measure for evaluation of image quality. *J. Digit. Imaging* **2008**, *21*, 338–347. [CrossRef] [PubMed]

6. Brigitte, F.; Patrick, B.; Patrick, G.; Fabien, S. A geometrical key-frame selection method exploiting dominant motion estimation in video. In *CIVR 2004: Image and Video Retrieval*; Springer: Berlin/Heidelberg, Germany, 2004; pp. 419–427.
7. Vasconcelos, N.; Andrew, L. Bayesian modeling of video editing and structure: Semantic features for video summarization and browsing. In Proceedings of the International Conference on Image Processing, Chicago, IL, USA, 7 October 1998.
8. Mikolajczyk, K.; Cordelia, S. A performance evaluation of local descriptors. *IEEE Trans. Pattern Anal. Mach. Intell.* **2005**, *27*, 1615–1630. [CrossRef] [PubMed]
9. Sebastian, T.; Jiby, J.P. A survey on video summarization techniques. *Int. J. Comput. Appl.* **2015**, *132*, 30–32. [CrossRef]
10. Supriya, K.; Rohan, M.; Aditya, M.; Abhishek, N. Key frame extraction for video summarization using motion activity descriptors. *IJRET* **2014**, *62*, 291–294.
11. Mentzelopoulos, M.; Alexandra, P. Key-frame extraction algorithm using entropy difference. In Proceedings of the 6th ACM SIGMM International Workshop on Multimedia information Retrieval, New York, NY, USA, 15–16 October 2004; pp. 39–45.
12. Cahuina, E.J.Y.C.; Guillermo, C.C. A new method for static video summarization using local descriptors and video temporal segmentation. In Proceeding of 2013 XXVI Conference on Graphics, Patterns and Images, Arequipa, Peru, 5–8 August 2013.
13. Yunyu, S.; Haisheng, Y.; Ming, G.; Xiang, L.; Xia, Y.X. A fast and robust key frame extraction method for video copyright protection. *J. Electric. Comput. Eng.* **2017**, *2017*. [CrossRef]
14. Zhao, Z.; Ahmed, M.E. Information Theoretic Key Frame Selection for Action Recognition. *Proc. BMVC.* **2008**, *2008*, 1–10.
15. Satoshi, H.; Makoto, N.; Shogo, M.; Hisakazu, K. Video key frame selection by clustering wavelet coefficients. In Proceedings of the 12th European Signal Processing Conference, Vienna, Austria, 6–10 September 2004; pp. 2303–2306.
16. Mahmoud, K.; Nagia, G.; Mohamed, I. VGRAPH: An effective approach for generating static video summaries. In Proceedings of the IEEE International Conference on Computer Vision Workshops, Sydney, Australia, 1–8 December 2013; pp. 811–817.
17. Ciocca, G.; Raimondo, S. Dynamic key-frame extraction for video summarization. *Int. Imaging VI* **2005**, *5670*, 137–143.
18. Ejaz, N.; Tayyab, B.T.; Sung, W.B. Adaptive key frame extraction for video summarization using an aggregation mechanism. *J. Visual Commun. Image Represent.* **2012**, *23*, 1031–1040. [CrossRef]
19. Rajendra, S.P.; Keshaveni, N. A survey of automatic video summarization techniques. *Int. J. Electron. Elect. Comput. Syst.* **2014**, *3*, 1–6.
20. Girgensohn, A.; John, B. Time-constrained keyframe selection technique. Multime. *Tools Appl.* **2000**, *11*, 347–358. [CrossRef]
21. Genliang, G.; Zhiyong, W.; Shiyang, L.; Jeremiah, D.D.; David, D.F. Keypoint-based keyframe selection. *IEEE Trans. Circuits Syst. Video Technol.* **2013**, *23*, 729–734.
22. Asadi, E.; Nasrolla, M.C. Video summarization using fuzzy c–means clustering. In Proceedings of the 20th Iranian Conference on Electrical Engineering (ICEE2012), Tehran, Iran, 15–17 May 2012; pp. 690–694.
23. Zhang, Q.; Yu, S.P.; Zhou, D.S.; Wei, X.P. An efficient method of key-frame extraction based on a cluster algorithm. *J. Human Kinet.* **2013**, *39*, 5–14. [CrossRef] [PubMed]
24. Dong, Z.; Zhang, G.F.; Jia, J.Y.; Bao, H.J. Keyframe-based real-time camera tracking. In Proceeding of 12th International Conference on Computer Vision, Kyoto, Japan, 29 September–2 October 2009; pp. 1538–1545.
25. Kim, S.H.; Lu, Y.; Shi, J.Y.; Alfarrarjeh, A.; Shahabi, S.; Wang, G.F.; Zimmermann, R. Key frame selection algorithms for automatic generation of panoramic images from crowdsourced geo-tagged videos. In *Web and Wireless Geographical Information Systems*; Springer: Berlin/Heidelberg, Germany, 2014; pp. 67–84.
26. Mei, T.; Tang, L.X.; Tang, J.H.; Hua, X.S. Near-lossless semantic video summarization and its applications to video analysis. *ACM Trans. Multime. Compu. Commun. Appl. (TOMM)* **2013**, *9*. [CrossRef]
27. Shu, R.L. Key Frame Detection Algorithm based on Dynamic Sign Language Video for the Non Specific Population. *Int. J. Signal Proc. Image Proc. Pattern Recognit.* **2015**, *8*, 135–148.
28. Ricardo, V.M.; Antonio, B. Spatio-temporal feature-based keyframe detection from video shots using spectral clustering. *Pattern Recogni. Lett.* **2013**, *34*, 770–779.

29. Khurana, K.; Chandak, M.B. Key frame extraction methodology for video annotation. *Int. J. Comput. Eng. Technol.* **2013**, *4*, 221–228.
30. Thakre, K.S.; Rajurkar, A.M.; Manthalkar, R.R. Video Partitioning and Secured Keyframe Extraction of MPEG Video. *Proc. Comput. Sci.* **2016**, *78*, 790–798. [CrossRef]
31. Wang, C.; Shen, H.W. Information theory in scientific visualization. *Entropy* **2011**, *13*, 254–273. [CrossRef]
32. Prasad, M.S.; Krishna, V.R.; Reddy, L.S. Investigations on Entropy Based Threshold Methods. *Asian J. Comput. Sci. Inf. Technol.* **2013**, *1*.
33. Chamoli, N. Kukreja, S.; Semwal, M. Survey and comparative analysis on entropy usage for several applications in computer vision. *Int. J. Comput. Appl.* **2014**, *97*, 1–5.
34. Qi, C. Maximum entropy for image segmentation based on an adaptive particle swarm optimization. *Appl. Math. Inf. Sci.* **2014**, *8*, 3129. [CrossRef]
35. Naidu, M.S.; Kumar, P.R.; Chiranjeevi, K. Shannon and fuzzy entropy based evolutionary image thresholding for image segmentation. *Alexandria Eng. J.* **2017**. [CrossRef]
36. Sabuncu, M.R. Entropy-Based Image Registration. Ph.D. Thesis, Princeton University, Princeton, NJ, USA, November 2006.
37. Ratsamee, P. Mae, Y.; Jinda, A.A.; Machajdik, J.; Ohara, K.; Kojima, M.; Sablatnig, R.; Arai, T. Lifelogging keyframe selection using image quality measurements and physiological excitement features. In Proceedings of the International Conference on Intelligent Robots and Systems, Tokyo, Japan, 3–7 Novenber 2013; pp. 5215–5220.
38. Angadi, S.; Naik, V. Entropy based fuzzy C means clustering and key frame extraction for sports video summarization. In Proceedings of the 5th International Conference on Signal and Image Processing, Chennai, India, 14–15 July 2018.
39. Yuan, Y.; Mei, T.; Cui, P.; Zhu, W. Video Summarization by Learning Deep Side Semantic Embedding. *IEEE Trans. Circuits Syst. Video Technol.* **2017**, 1. [CrossRef]
40. Chen, X.; Zhang, Y.; Ai, Q.; Xu, H.; Yan, J.; Qin, Z. Personalized key frame recommendation. In Proceedings of the 40th International ACM SIGIR Conference on Research and Development in Information Retrieval, Tokyo, Japan, 7–11 August 2017.
41. Panda, R.; Das, A.; Wu, Z.; Ernst, J.; Roy, C.A.K. Weakly supervised summarization of web videos. In Proceedings of the International Conference on Computer Vision, Venice, Italy, 22–29 October 2017.
42. Mahasseni, B.; Lam, M.; Todorovic, S. Unsupervised video summarization with adversarial lstm networks. In Proceedings of the Computer Vision and Pattern Recognition, Honolulu, HI, USA, 21–26 July 2017.
43. Jeong, D.J.; Yoo, H.J.; Cho, N.I. A static video summarization method based on the sparse coding of features and representativeness of frames. *EURASIP J. Image Video Proc.* **2017**, *1*. [CrossRef]
44. Yoon, S.; Khan, F.; Bremond, F. Efficient Video Summarization Using Principal Person Appearance for Video-Based Person Re-Identification. In Proceedings of the The British Machine Vision Conference, London, UK, 4 September 2017.
45. De Avila, S.E.; Lopes, A.P.; da Luz, J.A.; de Albuquerque, A.A. VSUMM: A mechanism designed to produce static video summaries and a novel evaluation method. *Pattern Recognit. Lett.* **2011**, *32*, 56–68. [CrossRef]
46. Kanehira, A.; Van Gool, L.; Ushiku, Y.; Harada, T. Viewpoint-aware Video Summarization. In Proceedings of the Computer Vision and Pattern Recognition, Honolulu, HI, USA, 21–26 July 2017; pp. 7435–7444.
47. Manis, G.; Aktaruzzaman, M.D.; Sassi, R. Bubble entropy: an entropy almost free of parameters. *IEEE Trans. Biomed. Eng.* **2017**, *64*, 2711–2718. [PubMed]
48. Athitsos, V.; Neidle, C.; Sclaroff, S.; Nash, J.; Stefan, A.; Yuan, Q.; Thangali, A. The american sign language lexicon video dataset. In Proceedings of the Computer Society Conference on Computer Vision and Pattern Recognition, Anchorage, AK, USA, 23–28 June 2008.
49. Pun, T. A new method for grey-level picture thresholding using the entropy of the histogram. *Signal process.* **1980**, *2*, 223–237. [CrossRef]
50. Sluder, G.; Wolf, D.E. *Digital Microscopy*; Academic Press: London, UK, 2013.
51. Panagiotakis, C.; Doulamis, A.; Tziritas, G. Equivalent key frames selection based on iso-content principles. *IEEE Trans. Circuits Syst. Video Technol.* **2009**, *19*, 447–451. [CrossRef]
52. Song, Y.; Vallmitjana, J.; Stent, A.; Jaimes, A. Tvsum: Summarizing web videos using titles. In Proceedings of the Computer Vision and Pattern Recognition, Boston, MA, USA, 7–25 June 2015.

53. Mei, S.; Guan, G.; Wang, Z.; Wan, S.; He, M.; Feng, D.D. Video summarization via minimum sparse reconstruction. *Pattern Recognit.* **2015**, *48*, 522–533. [CrossRef]
54. Ajmal, M.; Naseer, M.; Ahmad, F.; Saleem, A. Human Motion Trajectory Analysis Based Video Summarization. In Proceedings of the International Conference on Machine Learning and Applications, Cancun, Mexico, 18–21 December 2017.

entropy

MDPI

Article

Encryption Algorithm of Multiple-Image Using Mixed Image Elements and Two Dimensional Chaotic Economic Map

A. A. Karawia [1,2]

[1] Department of Mathematics, Faculty of Science, Mansoura University, Mansoura 35516, Egypt; abibka@mans.edu.eg or kraoieh@qu.edu.sa; Tel.: +966-55-305-9668

[2] Computer Science Unit, Deanship of Educational Services, Qassim University, P.O. Box 6595, Buraidah 51452, Saudi Arabia

Received: 15 September 2018; Accepted: 16 October 2018; Published: 18 October 2018

Abstract: To enhance the encryption proficiency and encourage the protected transmission of multiple images, the current work introduces an encryption algorithm for multiple images using the combination of mixed image elements (MIES) and a two-dimensional economic map. Firstly, the original images are grouped into one big image that is split into many pure image elements (PIES); secondly, the logistic map is used to shuffle the PIES; thirdly, it is confused with the sequence produced by the two-dimensional economic map to get MIES; finally, the MIES are gathered into a big encrypted image that is split into many images of the same size as the original images. The proposed algorithm includes a huge number key size space, and this makes the algorithm secure against hackers. Even more, the encryption results obtained by the proposed algorithm outperform existing algorithms in the literature. A comparison between the proposed algorithm and similar algorithms is made. The analysis of the experimental results and the proposed algorithm shows that the proposed algorithm is efficient and secure.

Keywords: image encryption; multiple-image encryption; two-dimensional chaotic economic map; security analysis

MSC: 68U10; 68P25; 94A60

---

1. Introduction

A huge number of images are produced in many fields, such as weather forecasting, military, engineering, medicine, science and personal affairs. Therefore, with the fast improvement of computer devices and the Internet, media security turns into a challenge, both for industry and academic research. Image transmission security is our target. Many authors have proposed many single-image encryption algorithms to solve this problem [1–8]. Single-image encryption algorithms involve those using a chaotic economic map [1,2], using a chaotic system [3], via one-time pads-a chaotic approach [4], via pixel shuffling and random key stream [5], using chaotic maps and DNA encoding [6] and using the total chaotic shuffling scheme [7]. In [8], the authors proposed two secret sharing approaches for 3D models using the Blakely and Thien and Lin schemes. Those approaches reduce share sizes and remove redundancies and patterns, which may ease image encryption. The authors in [9] concluded that the dynamic rounds chaotic block cipher can guarantee the security of information transmission and realize a lightweight cryptographic algorithm. A single-image can encrypt multiple images repeatedly, but the efficiency of that encryption is always unfavorable. Researchers have increased their attention towards multiple-image encryption because a high efficiency of secret information transmission is required for modern multimedia security technology. Many multiple-image algorithms have been presented. The authors of [10] presented a multiple-image algorithm via mixed image elements and chaos. A multiple-image algorithm using the pixel exchange operation and vector decomposition was proposed in [11]. In [12], the authors presented an algorithm using mixed permutation and image elements. The authors presented multiple-image encryption via computational ghost imaging in [13]. In [14], the authors proposed an algorithm using an optical asymmetric key cryptosystem. A multiple-image encryption algorithm based on spectral cropping and spatial multiplexing was presented in [15]. The authors of [16] proposed a multiple-image encryption algorithm based on the lifting wavelet transform and the XOR operation based on compressive ghost imaging scheme. Even with this large number of proposed algorithms, some practical problems still exist. For instance, some multiple-image algorithms have faced the problem that the original images cannot be recovered completely [17–19]. Those algorithms were used to encrypt multiple images, but the corresponding original images were not recovered completely. This leads to lossy algorithms, which are not appropriate for those applications needing images with high visual quality. Another problem is that the complex computations of some algorithms affect the encryption efficiency [20,21]. Therefore, good techniques are required for solving these problems [22]. In the current paper, a new efficient multiple-image encryption algorithm using mixed image elements (MIES) and a two-dimensional chaotic economic map is proposed. The advantages of this algorithm are that it is able to recover plain images completely and simplifies the computations. Experimental results demonstrate its practicality and high proficiency.

The rest of the paper is organized as follows. The pure image elements (PIES) and the MIES are defined in Section 2. In Section 3, a brief introduction to the two-dimensional chaotic economic map is presented. The secret key generation is presented in Section 4. In Section 5, a new encryption algorithm of multiple images is designed. Experimental results and analyses are introduced in Section 6. Section 7 presents a comparison between the proposed algorithm and the identical algorithms. Conclusions are given in Section 8.

2. PIES and MIES

Matrix theory can be used to divide a big matrix into many small matrices and vice versa. Furthermore, in the image processing field, it is simple to divide an image into many small images and vice versa. For instance, Figure 1 can be divided into 16 small images with an equal size, as displayed in Figure 2. Therefore, the original image can be retrieved from these 16 images.

Figure 1. Lena image with a 512 × 512 size.

Assume that $O1_{m\times n}, O2_{m\times n}, \cdots, Ok_{m\times n}$ are k original images. $O1_{m\times n}$ can be divided into a small images set, $\{B1_i\}$. Each element $B1_i \in \{B1_i\}$ is referred to as the pure image element. On the other hand, k sets of PIES $\{B1_i\}$, $\{B2_i\}$, $\cdots$, $\{Bk_i\}$ can be created, which correspond to

$O1_{m\times n}, O2_{m\times n}, \cdots, Ok_{m\times n}$, respectively. A large set $C = \{B1_i\} \cup \{B2_i\} \cup \cdots \cup \{Bk_i\}$ can be obtained by mixing all PIES together. Each element $C_i \in C$ is referred to as the mixed image element.

The current paper presents a new encryption algorithm of multiple images using MIES and the two-dimensional chaotic economic map. The secret key is very important to restore the original images from the MIES.

Figure 2. Pure image elements (PIES) of the Lena image with a 512 × 512 size.

3. The Two-Dimensional Chaotic Economic Map

The study of the following two-dimensional chaotic economic system (dynamical system) was introduced in [23]:

$$\left.\begin{aligned} \alpha_{n+1} &= \alpha_n + k\left[a - c - \frac{b\alpha_n}{\gamma_n} - b\log(\gamma_n)\right], \\ \beta_{n+1} &= \beta_n + k\left[a - c - \frac{b\beta_n}{\gamma_n} - b\log(\gamma_n)\right], \end{aligned}\right\} \tag{1}$$

where:

$$\gamma_n = \alpha_n + \beta_n, \quad n = 0, 1, 2, \ldots$$

There are six parameters in the chaotic economic map (1). These parameters have economic significance; the parameter $a > 0$ is used to capture the economic market size, while the market price slope is referred to by the parameter $b > 0$. To obtain a chaotic region, a must be greater than b and c. A fixed marginal cost parameter is denoted by $c \geq 0$, and the speed of adjustment parameter $k > 0$. The chaotic behavior of the chaotic economic map (1) at $a = 3, b = 1, c = 1, \alpha_0 = 0.19, \beta_0 = 0.15$ and $k \in [0, 6.0001]$ is shown in Figure 3. In the current paper, the parameters $a = 3, b = 1, c = 1$ and $k = 5.9$ of the map (1) have been chosen in the chaotic region having positive Lyapunov exponents, as displayed in Figure 4.

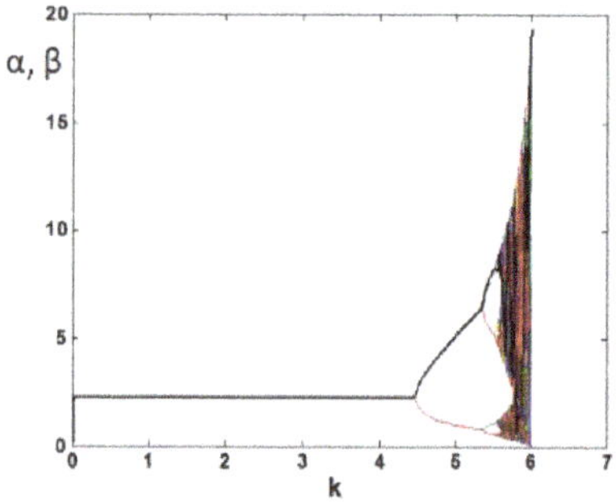

Figure 3. The chaotic behavior of the map (1) at $a = 3$, $b = 1$, $c = 1$, $\alpha_0 = 0.19$, $\beta_0 = 0.15$ and $k \in [0, 6.0001]$.

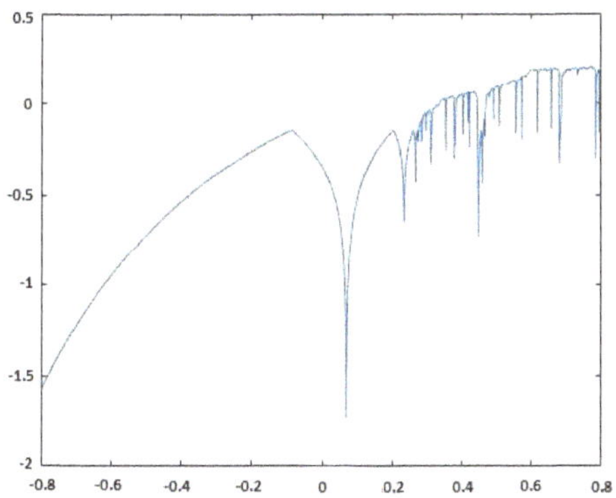

Figure 4. Lyapunov exponent for the chaotic economic map (1) at $a = 3,\ b = 1,\ c = 1,\ \alpha_0 = 0.19$, $\beta_0 = 0.15$ and $k \in [0, 6.0001]$.

4. The Secret Key Generation

Let $\mathbf{B} = (b_{ij}), i = 1, 2, ..., M, j = 1, 2, ..., N$, be the big image created by combining the k original images of size $m\times$, where b_{ij} refers to the pixel value at the position (i, j) and (M, N) is the size of the big image $\mathbf{B}$. The key mixing proportion factor can be used to calculate $K_z, z = 1, 2, 3, \cdots, 10$, as follows:

$$K_z = \frac{1}{256} mod \left(\sum_{i=\frac{(z-1)M}{8}+1}^{\frac{zM}{8}} \sum_{j=1}^{N} b_{ij}, 256 \right) \tag{2}$$

Then, update the initial condition Θ_0 using the following formula:

$$\Theta_0 \leftarrow \frac{(\Theta_0 + K)}{2}, \tag{3}$$

where $\Theta_0 = x_{10}, x_{20}, x_{30}, x_{40}, r_{10}, r_{20}, r_{30}, r_{40}, q_{10}, q_{20}$ and $K = K_j, j = 1, 2, \cdots, 10$, receptively.

After that, take four initial values, $x_{10}, x_{20}, x_{30}, x_{40}$, four parameters for the logistic map, $r_{10}, r_{20}, r_{30}, r_{40}$, two initial values for the system, q_{10}, q_{20}, and four system parameters, a, b, c, k.

5. The Proposed Multiple-Image Algorithm

To encrypt multiple images jointly, the current work presents a new encryption algorithm of multiple images using MIES and the two-dimensional chaotic economic map. The flowchart of the new encryption algorithm is shown in Figure 5.

The proposed algorithm is processed as follows:

In the multiple-image decryption, the same chaotic economic sequences are generated on the multiple-image encryption that will be used to recover the original images and using the inverse steps of Algorithm 1.

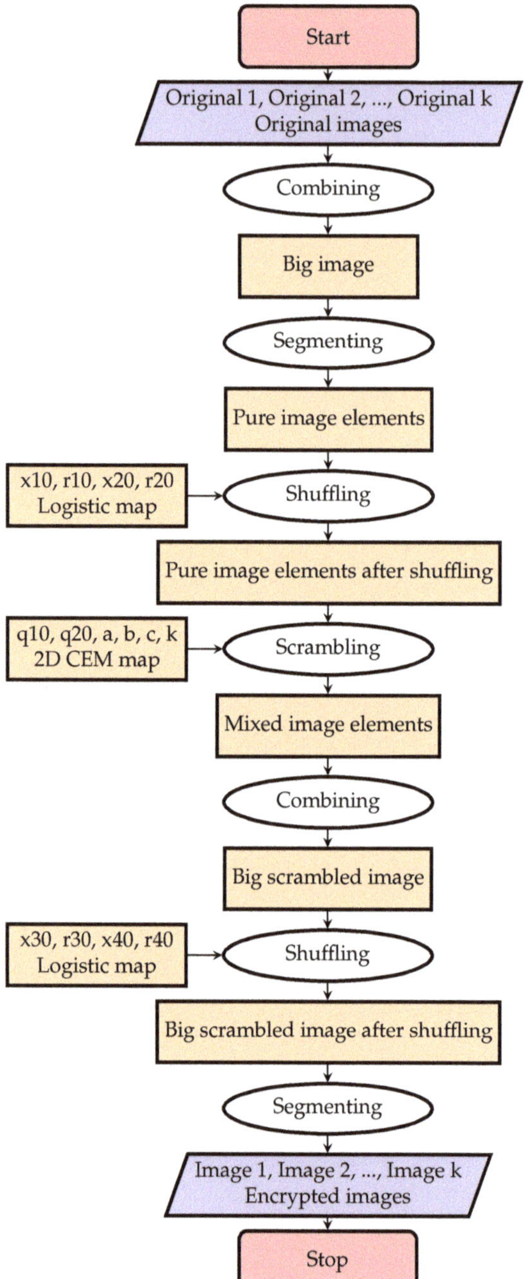

Figure 5. Flowchart of multiple-image encryption.

Algorithm 1 Multiple-image encryption

Input: k original images, $\mathbf{O1}, \mathbf{O2}, \cdots, \mathbf{Ok}$, $x_{i0}, r_{i0}, i = 1,2,3,4$ for logistic shuffling and $a, b, c, k, \alpha_0, \beta_0$ for the two-dimensional chaotic economic map (1).
Output: Encrypted images $\mathbf{Image\ 1}, \mathbf{Image\ 2}, \cdots, \mathbf{Image\ k}$.
Step 1: Create a big image by combining the k original images.
Step 2: Divide the big image into PIES of $m_1 \times n_1$ size such that $\mathbf{mod}(m, m_1) = 0$, $\mathbf{mod}(n, n_1) = 0$ and the original images with size $m \times n$.
Step 3: Shuffle the pixels of PIES using the logistic map: $x_n = rx_{n-1}(1 - x_{n-1}), n = 1,2,3,\cdots$, and use the parameters (x_{10}, r_{10}) and (x_{20}, r_{20}) for shuffling the rows and the columns, respectively.
Step 4: Generate the chaotic economic sequences using:

$$\alpha_{n+1} = \alpha_n + k\left[a - c - \frac{b\alpha_n}{\gamma_n} - b\log(\gamma_n)\right],$$

$$\beta_{n+1} = \beta_n + k\left[a - c - \frac{b\beta_n}{\gamma_n} - b\log(\gamma_n)\right],$$

where $n = 0,1,2,\cdots, a = 3, b = 1, c = 1, \alpha_0 = 0.001, \beta_0 = 0.002$ and $k = 5.9$.
Step 5: Do the following preprocessing for the generated values in **Step 4**: $\alpha_i = floor(mod(\alpha_i \times 10^{14}, 256))$ and $\beta_i = floor(mod(\beta_i \times 10^{14}, 256))$,
Step 6: Convert α_i and β_i into binary vectors, say A and B, respectively.
Step 7: Perform a bit-wise XOR between A and B, say C = bitxor(A,B).
Step 8: Convert the pixels of shuffled PIES into a binary vector, say D.
Step 9: Perform a bit-wise XOR between C and D, say E = bitxor(C,D).
Step 10: Combine these mixed scrambled PIES into a big scrambled image.
Step 11: Shuffle the pixels of the big scrambled image using the logistic map, and use the parameters (x_{30}, r_{30}) and (x_{40}, r_{40}) for shuffling the rows and the columns, respectively.
Step 12: Divide it into images of equal size $m \times n$. These images are viewed as encrypted images, say $\mathbf{Image\ 1}, \mathbf{Image\ 2}, \cdots, \mathbf{Image\ k}$.
Step 13: End.

6. Experimental Results and Analyses

To show the efficiency and robustness of the proposed algorithm, nine ($k = 9$) original gray images of a 512×512 size are shown in Figure 6. Let $x_{10} = 0.1, x_{20} = 0.2$ be the initial values and $r_{10} = 3.9985, r_{20} = 3.9988$ be the parameters of the logistic map for shuffling the PIES. Furthermore, let $x_{30} = 0.3, r_{30} = 3.9984$ and $x_{40} = 0.4, r_{40} = 3.9986$ be the initial values and the parameters of the logistic map for shuffling the big scrambled image. Let $\alpha_0 = 0.19, \beta_0 = 0.15, a = 3, b = 1, c = 1$ and $k = 5.9$ be the initial values and the control parameters of the chaotic economic map (1). All nine original gray images are combined into one big image, which is displayed in Figure 7. Figures 8–13 show the big scrambled images that correspond to the MIES of equal sizes 4×4, 8×8, 16×16, 32×32, 64×64 and 128×128, respectively. The corresponding encrypted images of MIES with size 64×64 are shown in Figure 14. Furthermore, the corresponding decrypted images are displayed in Figure 15. Experiments are performed with MATLAB R2016a software to execute the proposed algorithm on a laptop with the following characteristics: 2.40 GHz Intel Core i7-4700MQ CPU and 12.0 GB RAM memory.

The performance of the presented multiple-image encryption algorithm is investigated in detail as follows.

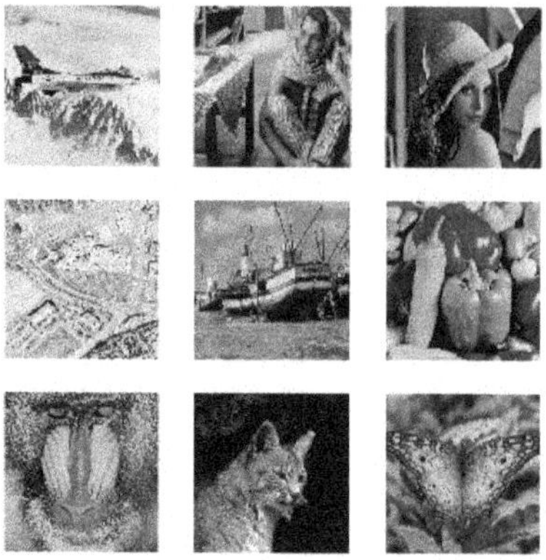

Figure 6. Original images.

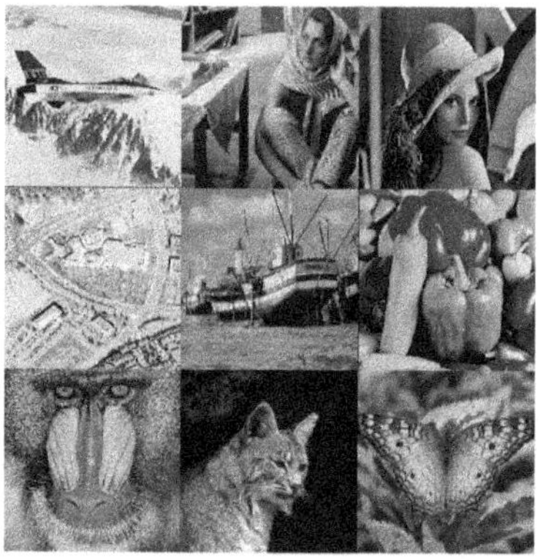

Figure 7. Big image.

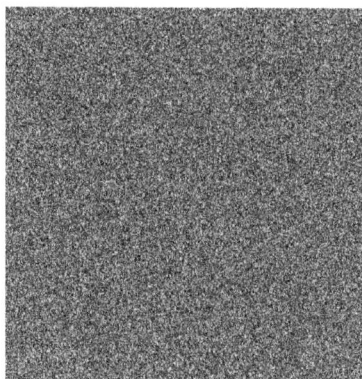

Figure 8. Mixed image elements (MIES) with equal size 4 × 4.

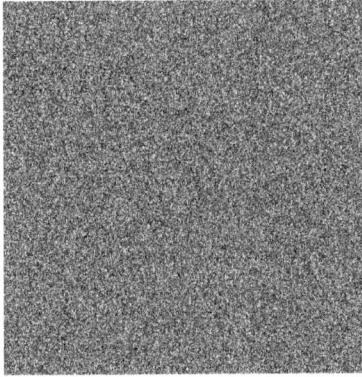

Figure 9. MIES with equal size 8 × 8.

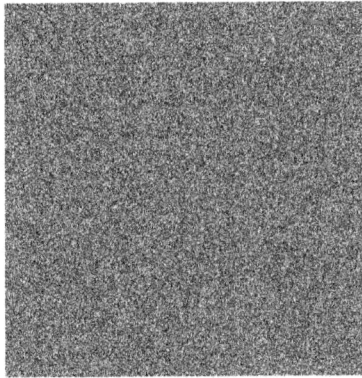

Figure 10. MIES with equal size 16 × 16.

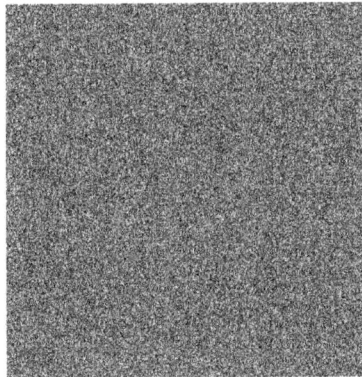

Figure 11. MIES with equal size 32 × 32.

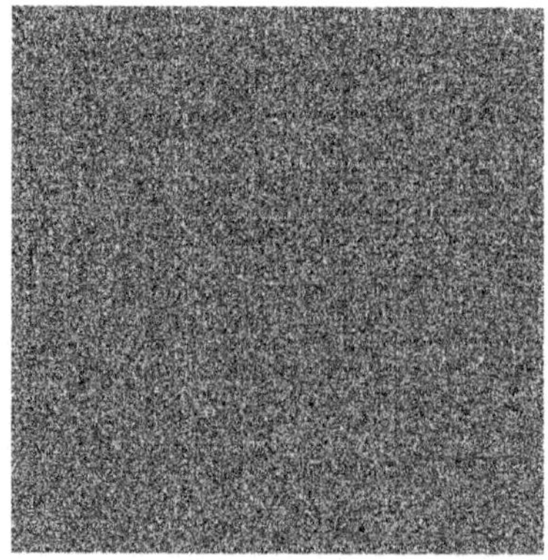

Figure 12. MIES with equal size 64 × 64.

Figure 13. MIES with equal size 128 × 128.

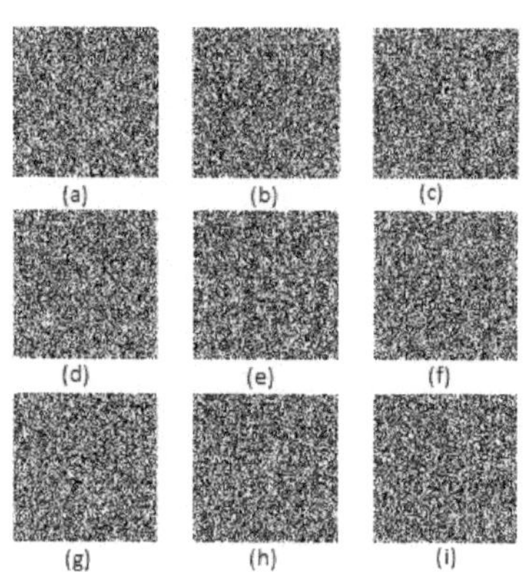

Figure 14. Encrypted images. (**a**) encrypted image of airplane. (**b**) encrypted image of barbara. (**c**) encrypted image of lena. (**d**) encrypted image of aerial. (**e**) encrypted image of boat. (**f**) encrypted image of peppers. (**g**) encrypted image of baboon. (**h**) encrypted image of cat. (**i**) encrypted image of butterfly.

Figure 15. Decrypted images.

6.1. *Analysis of the Key Space*

A large key space is required to make the brute-force attack infeasible [10]. In the proposed algorithm, the key space was selected as follows. In the logistic map, x_{10}, r_{10}, x_{20}, r_{20}, x_{30}, r_{30}, x_{40}, r_{40} were selected to shuffle rows and columns. $\alpha_0, \beta_0, a, b, c$ and k were selected for the chaotic economic map (1). Then, the key space size was $10^{15\times 14} = 10^{210}$ if the computer precision were 10^{-15}. Table 1 shows that the key spaces in [10,20,22] were less than the presented key space. Therefore, it was large enough to make the brute-force attack infeasible.

Table 1. Comparison of the current key space with other key spaces in the literature.

Algorithm	Proposed Algorithm	Ref. [10]	Ref. [20]	Ref. [22]
Key Space	10^{210}	10^{60}	$2^{451} = 5.8147 \times 10^{135}$	$1.964 \times 2^{428} = 1.3614 \times 10^{129}$

6.2. *Analysis of the Key Sensitivity*

An excellent multiple-image encryption algorithm should be very sensitive to modifying any key of the encryption and the decryption processes. Making a small modification to the key of the encryption, the output encrypted image (the second one) should be absolutely unlike the first encrypted image. Furthermore, if the encryption and decryption keys have a small difference, then the encrypted image cannot be restored correctly [23]. The restored images of the encrypted images in Figure 14 with a small change of the secret key, say $\alpha_0 = 0.190000000000001$ instead of $\alpha_0 = 0.19$, and the other parameters unchanged, are shown in Figure 16. The result shows that a small modification of the key can lead to completely different encrypted images, and the restoration of original images becomes very complicated. As the sensitivity of x_{10}, r_{10}, x_{20}, r_{20}, x_{30}, r_{30}, x_{40}, r_{40}, β_0, a, b, c and k was the same as α_0, their examples are omitted here.

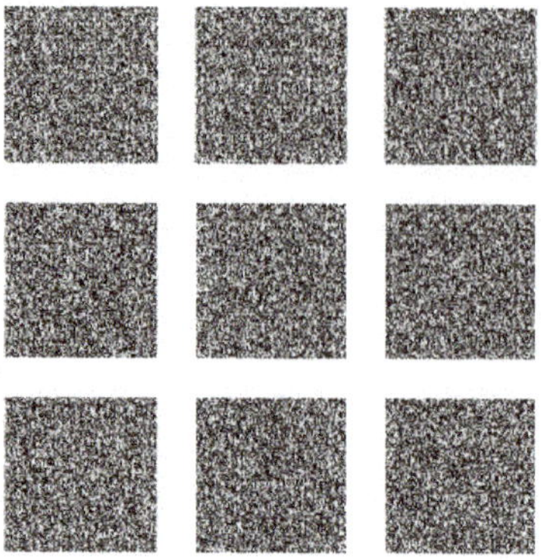

Figure 16. Decrypted images with the correct secret key, except $\alpha_0 = 0.190000000000001$, instead of $\alpha_0 = 0.19$.

6.3. *Analysis of the Histogram*

The original images' histograms are shown in Figure 17, while the corresponding encrypted images histograms are shown in Figure 18. Figures 16 and 18 display that the original images had different histograms, while the corresponding encrypted images histograms had a uniform distribution approximately. Therefore, the encryption process damaged the original images' features.

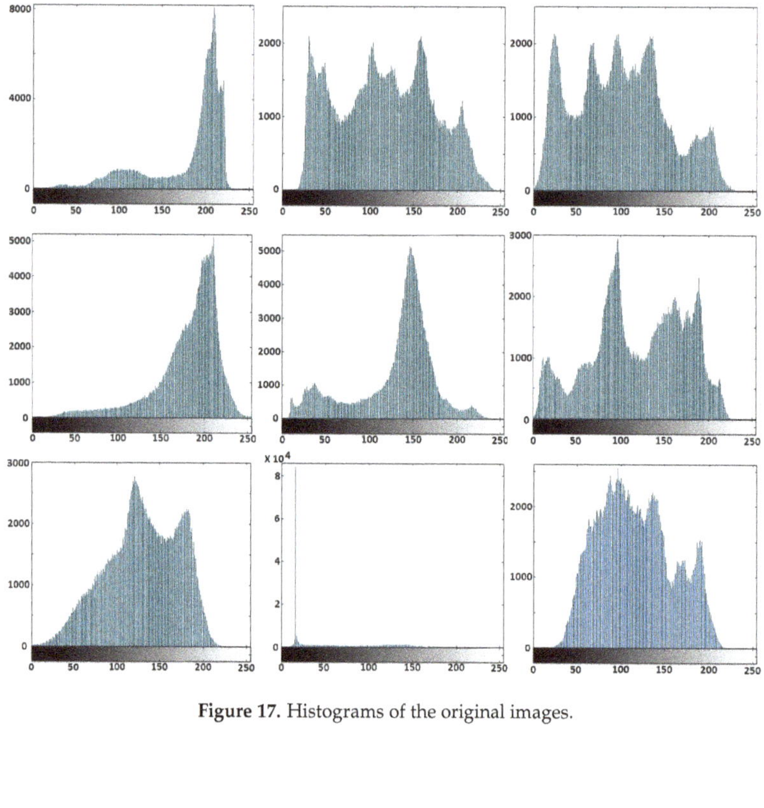

Figure 17. Histograms of the original images.

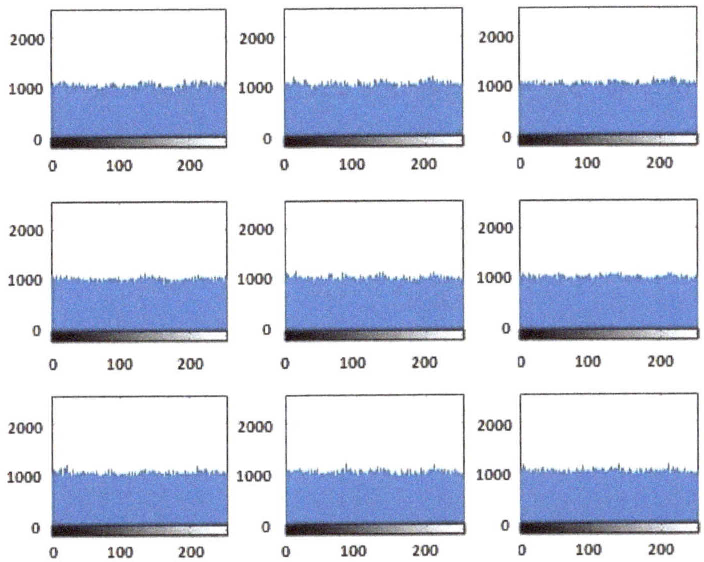

Figure 18. Histograms of the encrypted images.

6.4. Analysis of Histogram Variance

The histogram variance of a gray image is defined by:

$$Var(V) = \frac{1}{256}\sum_{i=0}^{255}[v_i - E(V)]^2, \tag{4}$$

where $E(V) = \frac{1}{256}\sum_{i=0}^{255} v_i$ and V is the pixel number vector of 256 gray levels.

This can clarify the impact of the encrypted image to some degree. In a perfect random image, all the gray levels have equal probabilities. Therefore, the histogram variance equals zero. Therefore, the histogram variance of the encrypted image via an effective encryption algorithm should tend to zero. Table 2 shows the values of the histogram variances of the encrypted images of the original images in Figure 19 via Tang's algorithm [20], Zhang's algorithm [10] and the proposed algorithm, respectively.

Table 2. Comparison of histogram variances between three algorithms.

Algorithm	Tang's Algorithm [20]	Zhang's algorithm [10]	Proposed Algorithm
Figure 19a	1261.8	1155.5	1055.5
Figure 19b	1192.3	989.6	984.8
Figure 19c	1213.1	1111.6	1079.7
Figure 19d	8710.3	929.6	916.9

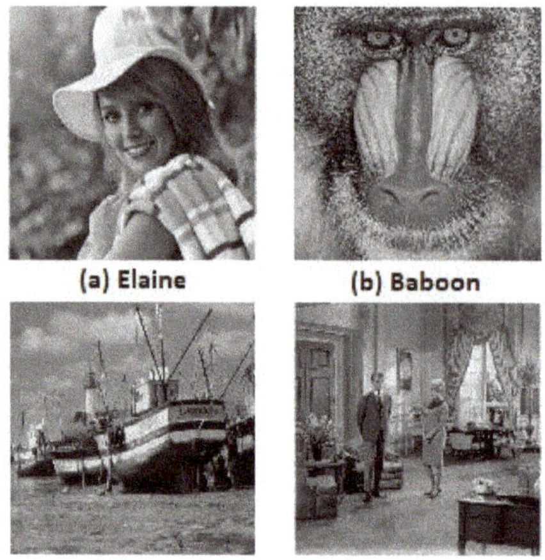

Figure 19. Input images. (**a**) Elaine; (**b**) Baboon; (**c**) Boat; (**d**) Couple.

6.5. Analysis of Information Entropy

In a digital image, the information entropy can be an indicator of the pixel values' distribution. For a perfect random image, $P(v_i) = \frac{1}{256}, i = 0, 1, 2, \cdots, 255$, where v_i is the i-th gray level of the image and $P(v_i)$ is the probability of v_i. Furthermore, it has information entropy = 8. Now, the information entropy is computed by [24]:

$$H(V) = -\sum_{i=0}^{255} P(v_i) log_2 P(v_i) \tag{5}$$

Table 3 lists the values of information entropy for the encrypted images in Figure 14. The information entropy of the encrypted images of the proposed algorithm is better than the information entropy of the encrypted images of the multiple-image encryption algorithm in [10]. Therefore, the efficiency and security of the proposed algorithm is clear.

Table 3. Information entropy for the encrypted images in Figure 14.

Images	(a)	(b)	(c)
Entropy	7.9984	7.9987	7.9986
Images	(d)	(e)	(f)
Entropy	7.9982	7.9986	7.9983
Images	(g)	(h)	(i)
Entropy	7.9986	7.9989	7.9986

6.6. Analysis of the Correlation Coefficients

In the image encryption, the correlation coefficient was used to measure the correlation between two neighboring pixels, horizontally, vertically and diagonally neighboring. It is evaluated by [25]:

$$R_{V_1 V_2} = \frac{COV(V_1, V_2)}{\sqrt{D(V_1)}\sqrt{D(V_2)}} \tag{6}$$

where:

$$COV(V1, V2) = \frac{1}{N}\sum_{i=1}^{N}(v1_i - E(V1))(v2_i - E(V2)),$$

$$D(V) = \frac{1}{N}\sum_{i=1}^{N}(v_i - E(V)),$$

and

$$E(V) = \frac{1}{N}\sum_{i=1}^{N} v_i.$$

Three thousand pairs of pixels were selected randomly in all three directions from the two images (original and encrypted); see Figures 19a and 21a, respectively. Then, the correlation coefficients of the two neighboring pixels were computed using Equation (4). The neighboring pixel correlation of Figures 19a and 20a are plotted in Figures 21 and 22. Their correlation coefficients are illustrated in Tables 4 and 5. The original images' correlation coefficients were approximately equal to one, while the corresponding ones of encrypted images were approximately equal to zero. The results conclude that the proposed algorithm can conserve the image information.

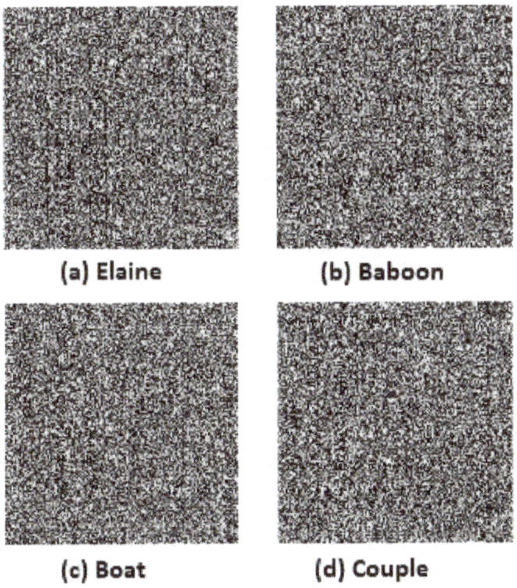

Figure 20. Encrypted images of the proposed algorithm. (**a**) Elaine; (**b**) Baboon; (**c**) Boat; (**d**) Couple.

Table 4. The original images' correlation.

Directions	Horizontal	Vertical	Diagonal
Figure 19a	0.9757	0.9729	0.9685
Figure 19b	0.9228	0.8597	0.8476
Figure 19c	0.9383	0.9715	0.9224
Figure 19d	0.9439	0.8687	0.8334

Table 5. The encrypted images' correlations.

Directions	Horizontal	Vertical	Diagonal
Figure 20a	−0.0035	0.0014	0.0007
Figure 20b	0.0036	−0.0005	0.0010
Figure 20c	0.0015	0.0013	−0.0017
Figure 20d	−0.0008	0.0008	0.0031

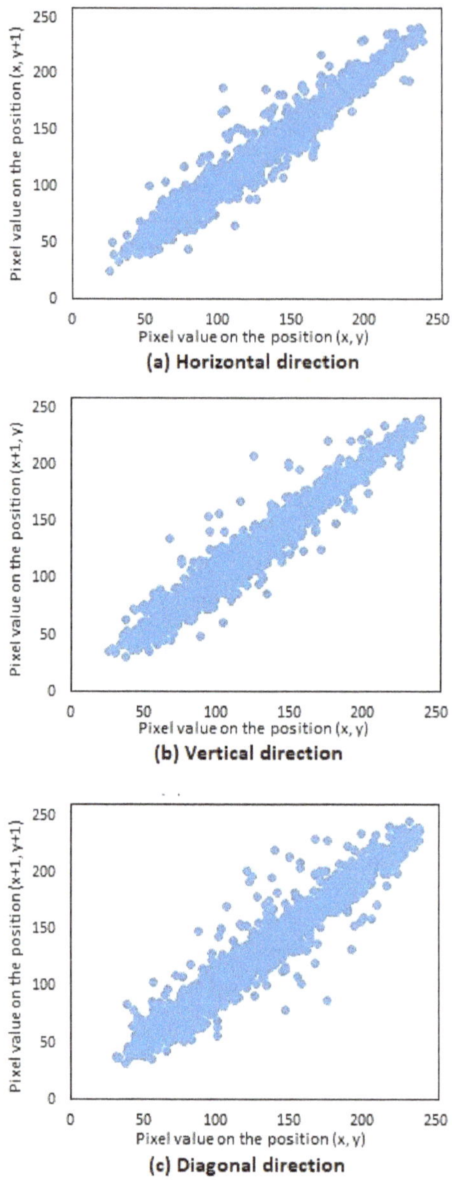

Figure 21. Neighboring pixel correlation of Figure 19a (original image). (**a**) Horizontal direction; (**b**) Vertical direction; (**c**) Diagonal direction.

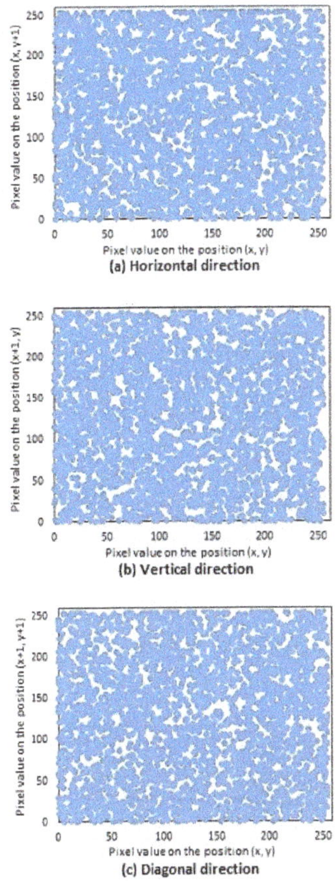

Figure 22. Neighboring pixel correlation of Figure 20a (encrypted image). (**a**) Horizontal direction; (**b**) Vertical direction; (**c**) Diagonal direction.

6.7. Analysis of Differential Attack

In the differential attack, the encryption algorithm was used to encrypt the original image before and after modification, then the two encrypted images were compared to discover the link between them [26]. Therefore, a good image encryption algorithm should be the desired property to spread the effect of a minor change in the original image of as much an encrypted image as possible. Number of pixels change rate (NPCR) and unified averaged changed intensity (UACI) are famous measurements, which were used to measure the resistance of the image encryption algorithm for differential attacks. The NPCR and UACI are defined as follows,

$$NPCR = \frac{\sum_{i,j} d(i,j)}{M \times N} \times 100\%, \tag{7}$$

$$UACI = \frac{1}{M \times N}\left[\sum_{i,j} \frac{|I_1(i,j) - I_2(i,j)|}{255}\right] \times 100\%. \tag{8}$$

where:

$$d(i,j) = \begin{cases} 0 & \text{if } I_1(i,j) = I_2(i,j), \\ 1 & \text{if } I_1(i,j) \neq I_2(i,j) \end{cases} \tag{9}$$

M and N are the width and height of the original and the encrypted images; I_1 and I_2 are the encrypted images before and after one pixel changed from the original image. For example, a pixel position $(71, 42)$ was selected randomly, and it has the value 159 in Figure 19a. The pixel value was modified to 244 to examine the ability to combat the differential attacks. Table 6 lists the results of Figure 19a–d. The results show that a small modification in the plain image will result in a big modification in the cipher image. Therefore, the proposed algorithm can face differential attacks.

Table 6. The values of number of pixels change rate (NPCR) and unified averaged changed intensity (UACI) for Figure 19.

Image	NPCR	UACI
Figure 19a	99.62%	33.44%
Figure 19b	99.61%	33.85%
Figure 19c	99.62%	33.42%
Figure 19d	99.60%	33.18%

6.8. Chosen/Known Plaintext Attack Analysis

Attackers have used two famous attacks called chosen-plaintext attack and known-plaintext attack for attacking any cryptosystem. The secret keys are not only dependent on the given initial values and system parameters, but also on the plain images. Therefore, when the plain images are changed, the secret keys will be changed in the encryption process. Therefore, attackers cannot take important information by encrypting some predesigned special images. Therefore, the proposed algorithm robustly resisted both attacks.

6.9. Noise Attack Analysis

The encrypted images in Figure 20 are distorted by adding Gaussian noise with mean = 0 and variance = 0.001 and salt and pepper noise with density = 0.05. The corresponding decrypted images are displayed in Figure 23. Moreover, Table 7 shows the mean squared error (MSE) and the peak signal-to-noise ratio (PSNR) between input images and decrypted images based on the proposed algorithm. Based on Table 7, we can conclude that the proposed algorithm had the highest resisting ability to salt and pepper noise since the PSNR was more than 65 (dB).

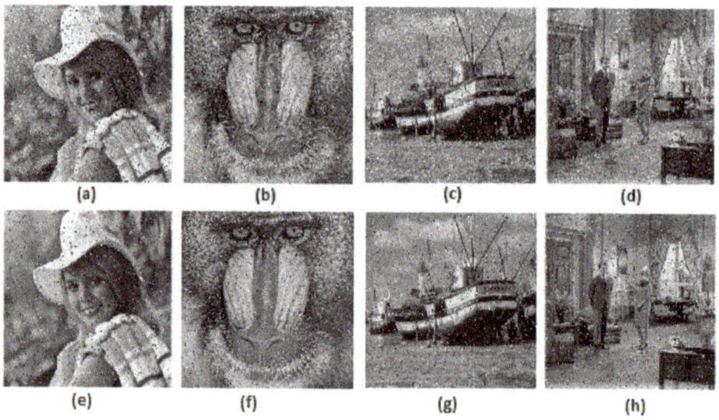

Figure 23. Results of noise attack analysis: (**a**–**d**) the decrypted images after adding Gaussian noise with mean = 0 and variance = 0.001; (**e**–**h**) the decrypted images after added salt and pepper noise with density = 0.05.

6.10. Analysis of Occlusion Attack

The current section is assigned to the analyses of occluded data decryption. Data that are occluded are hidden or ignored data inside the process. Firstly, $128 \times 128, 512 \times 512, 512 \times 1024$ and 512×1536

sized data occlusions of the horizontally concatenated encrypted image were performed. Secondly, the decrypted image of each one was analyzed. Figure 24 shows the results of the occlusion attack. Based on Figure 24, the decrypted images of $128 \times 128, 512 \times 512, 512 \times 1024$ sized occluded encrypted images were disfigured, but discernible by the human eye, while decrypted images of 512×1536 sized occluded encrypted images were not restored. Hence, the proposed algorithm could resist up to a 50% (512×1024) occlusion attack.

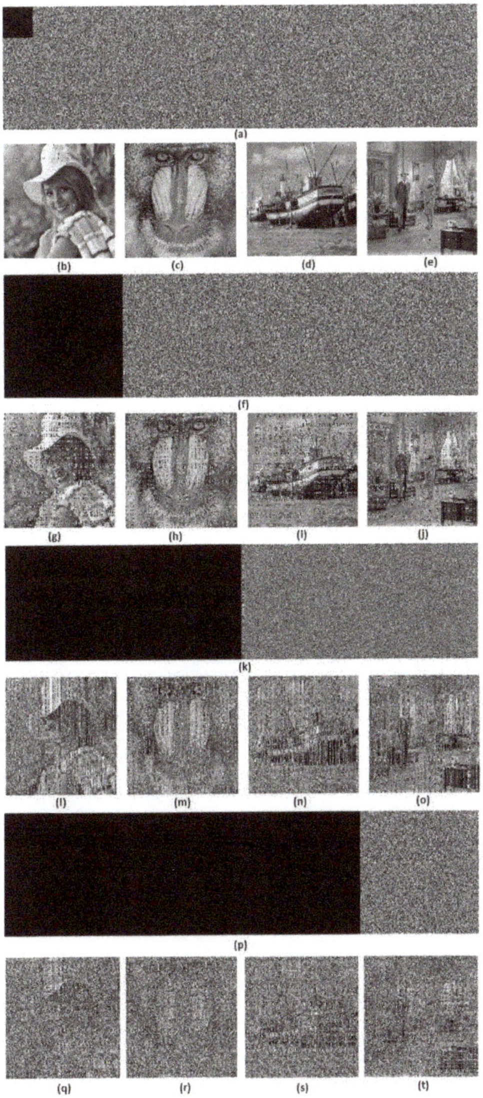

Figure 24. Results of occlusion attack analysis: (**a**,**f**,**k**,**p**) horizontally concatenated encrypted image with a $128 \times 128, 512 \times 512, 512 \times 1024$ and 512×1536 size of occlusion, respectively; (**b**–**e**), (**g**–**j**), (**l**–**o**) and (**q**–**t**) decrypted "Elaine", "Baboon", "Boat" and "Couple" images, respectively, when there is a $128 \times 128, 512 \times 512, 512 \times 1024$ and 512×1536 size of occlusion in the horizontally concatenated encrypted image.

Table 7. Measurements of the noise attacks of the proposed algorithm.

Image	Noise	MSE	PSNR
Figure 23a		0.0603	60.3255
Figure 23b	Gaussian	0.0602	60.3346
Figure 23c	variance = 0.001	0.0474	61.3691
Figure 23d		0.0560	60.6455
Figure 23a		0.0184	65.4921
Figure 23b	salt & pepper	0.0162	66.0291
Figure 23c	density = 0.05	0.0172	65.7719
Figure 23d		0.0155	66.2276

7. Comparison with Other Algorithms

A comparison between Tang's algorithm [20] and Zhang's algorithm was performed in [10]. The result of the comparison concluded that Zhang's algorithm was faster than Tang's algorithm. Therefore, a comparison between Zhang's algorithm and the proposed algorithm is presented. The same four original gray images are chosen as input images and are displayed in Figure 19. Furthermore, the size of MIES = 64×64 is selected. The encrypted images of the proposed algorithm and Zhang's algorithm are shown in Figures 20 and 25, respectively. The computational times of both algorithms are listed in Table 8. Although the time of Zhang's algorithm is less than the proposed algorithm, the encrypted images' histograms of the proposed algorithm are uniformly distributed, and the encrypted images histograms of Zhang's algorithm are not uniformly distributed (see Figure 13 in [10]). Therefore, the experimental results conclude that the proposed algorithm is efficient. The security of Zhang's algorithm is a little weaker than the proposed algorithm since the key space size of the proposed algorithm is larger than Zhang's algorithm and two additional shuffling operations are added to the proposed algorithm, one for PIES and one for the big scrambled image.

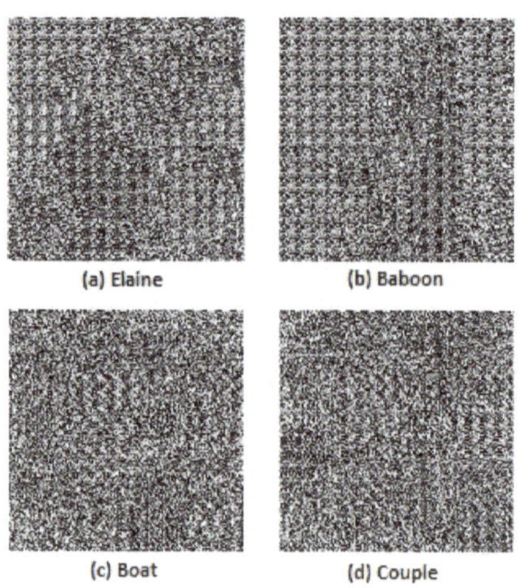

Figure 25. Encrypted images of Zhang's algorithm.

Table 8. Computational time (seconds).

Algorithm	Time
Zhang's algorithm [10]	2.169
Proposed algorithm	2.386

8. Conclusions

The current paper has proposed a new multiple-image encryption algorithm using combination of MIES and a two-dimensional chaotic economic map. The key space size of the proposed algorithm is 10^{210}. Therefore, it gives priority to the proposed algorithm to resist against brute-force attack. The experimental results have demonstrated that the proposed algorithm produced encrypted images that have histograms with uniform distributions. In addition, the proposed algorithm has demonstrated that the encrypted images have information entropies close to eight. It robustly resists chosen/known plaintext attacks, has the highest resisting ability to salt and pepper noise and can resist up to a 50% (512×1024) occlusion attack. Comparison experiments with Zhang's algorithm were performed. Furthermore, the analyses of the algorithm conclude that the proposed algorithm is secure and efficient. It can be applied in several fields like weather forecasting, military, engineering, medicine, science and personal affairs. In this paper, the proposed idea was simulated on grayscale images, which had the same size. In the future, the proposed idea will applied on grayscale images with different sizes.

Funding: This research received no external funding.

Acknowledgments: I deeply thank Shehzad Ahmed for his contribution in editing and proof reading the paper.

Conflicts of Interest: The author declares no conflict of interest.

References

1. Askar, S.S.; Karawia, A.A.; Alshamrani, A. Image encryption algorithm based on chaotic economic model. *Math. Probl. Eng.* **2015**, *2015*, 341729. [CrossRef]
2. Askar, S.S.; Karawia, A.A.; Alammar, F.S. Cryptographic algorithm based on pixel shuffling and dynamical chaotic economic map. *IET Image Process.* **2018**, *12*, 158–167. [CrossRef]
3. Cao, Y.; Fu, C. An image encryption scheme based on high dimension chaos system. In Proceedings of the 2008 International Conference on Intelligent Computation Technology and Automation, Changsha, China, 20–22 October 2008; pp. 104–108.
4. Jeyamala, J.; GrpiGranesh, S.; Raman, S. An image encryption scheme based on one time pads-a chaotic approach. In Proceedings of the 2010 Second International conference on Computing, Communication and Networking Technologies, Karur, India, 29–31 July 2010; pp. 1–6.
5. Sivakumar, T.; Venkatesan, R. Image encryption based on pixel shuffling and random key stream. *Int. J. Comput. Inf. Technol.* **2014**, *3*, 1468–1476.
6. Zhang, J.; Fang, D.; Ren, H. Image encryption algorithm based on DNA encoding and chaotic maps. *Math. Probl. Eng.* **2014**, *2014*, 917147. [CrossRef]
7. Vaferi, E.; Sabbaghi-Nadooshan, R. A new encryption algorithm for color images based on total chaotic shuffling scheme. *Opt.-Int. J. Light Electron Opt.* **2015**, *126*, 2474–2480. [CrossRef]
8. Elsheh, E.; Hamza, A. Secret sharing approaches for 3D object encryption. *Expert Syst. Appl.* **2011**, *38*, 13906–13911. [CrossRef]
9. Wang, J.; Ding, Q. Dynamic rounds chaotic block cipher based on keyword abstract extraction. *Entropy* **2018**, *20*, 693. [CrossRef]
10. Zhang, X.; Wang, X. Multiple-image encryption algorithm based on mixed image element and chaos. *Comput. Electr. Eng.* **2017**, *62*, 401–413. [CrossRef]
11. Xiong, Y.; Quan, C.; Tay, C.J. Multiple image encryption scheme based on pixel exchange operation and vector decomposition. *Opt. Lasers Eng.* **2018**, *101*, 113–121. [CrossRef]
12. Zhang, X.; Wang, X. Multiple-image encryption algorithm based on mixed image element and permutation. *Opt. Lasers Eng.* **2017**, *92*, 6–16. [CrossRef]
13. Wu, J.; Xie, Z.; Liu, Z.; Liu, W.; Zhang, Y.; Liu, S. Multiple-image encryption based on computational ghost imaging. *Opt. Commun.* **2016**, *359*, 38–43. [CrossRef]
14. Liu, W.; Xie, Z.; Liu, Z.; Zhang, Y.; Liu, S. Multiple-image encryption based on optical asymmetric key cryptosystem. *Opt. Commun.* **2015**, *335*, 205–211. [CrossRef]

15. Deng, P.; Diao, M.; Shan, M.; Zhong, Z.; Zhang, Y. Multiple-image encryption using spectral cropping and spatial multiplexing. *Opt. Commun.* **2016**, *359*, 234–239. [CrossRef]
16. Li, X.; Meng, X.; Yang, X.; Wang, Y.; Yin, Y.; Peng, X.; He, W.; Dong, G.; Chen, H. Multiple-image encryption via lifting wavelet transform and XOR operation based on compressive ghost imaging scheme. *Opt. Lasers Eng.* **2018**, *102*, 106–111. [CrossRef]
17. Lin, Q.; Yin, F.; Mei, T.; Liang, H. A blind source separation-based method for multiple images encryption. *Image Vis. Comput.* **2008**, *26*, 788–798. [CrossRef]
18. Wang, Q.; Guo, Q.; Zhou, J. Double image encryption based on linear blend operation and random phase encoding in fractional Fourier domain. *Opt. Commun.* **2012**, *285*, 4317–4323. [CrossRef]
19. Li, C.; Li, H.; Li, F.; Wei, D.; Yang, X.; Zhang, J. Multiple-image encryption by using robust chaotic map in wavelet transform domain. *Opt.-Int. J. Light Electron Opt.* **2018**, *171*, 277–286. [CrossRef]
20. Tang, Z.; Song, J.; Zhang, X.; Sun, R. Multiple-image encryption with bit-plane decomposition and chaotic maps. *Opt. Lasers Eng.* **2016**, *80*, 1–11. [CrossRef]
21. Parvin, Z.; Seyedarabi, H.; Shamsi, M. Breaking an image encryption algorithm based on the new substitution stage with chaotic functions. *Multimed. Tools Appl.* **2016**, *75*, 10631–10648. [CrossRef]
22. Patro, K.; Acharya, B. Secure multi–level permutation operation based multiple colour image encryption. *J. Inf. Secur. Appl.* **2018**, *40*, 111–133. [CrossRef]
23. Askar, S.S. Complex dynamic properties of Cournot duopoly games with convex and log-concave demand function. *Oper. Res. Lett.* **2014**, *42*, 85–90. [CrossRef]
24. Wang, W.; Tan, H.; Pang, Y.; Li, Z.; Ran, P.; Wu, J. A Novel encryption algorithm based on DWT and multichaos mapping. *J. Sens.* **2016**, *2016*, 2646205. [CrossRef]
25. Liu, H.; Wang, X. Color image encryption based on one-time keys and robust chaotic maps. *Comput. Math. Appl.* **2010**, *59*, 3320–3327. [CrossRef]
26. Belazi, A.; Abd El-Latif, A.; Belghith, S. A novel image encryption scheme based on substitution-permutation network and chaos. *Signal Process.* **2016**, *128*, 155–170. [CrossRef]

entropy

MDPI

Article

Improved Cryptanalysis and Enhancements of an Image Encryption Scheme Using Combined 1D Chaotic Maps

Congxu Zhu [1,2,3,*], Guojun Wang [2] and Kehui Sun [4]

1 School of Information Science and Engineering, Central South University, Changsha 410083, China
2 School of Computer Science and Technology, Guangzhou University, Guangzhou 510006, China; csgjwang@163.com
3 School of Computer and Science, Liaocheng University, Liaocheng 252059, China
4 School of Physics and Electronics, Central South University, Changsha 410083, China; kehui@csu.edu.cn
* Correspondence: zhucx@csu.edu.cn; Tel.: +86-0731-8882-7601

Received: 4 October 2018; Accepted: 31 October 2018; Published: 3 November 2018

Abstract: This paper presents an improved cryptanalysis of a chaos-based image encryption scheme, which integrated permutation, diffusion, and linear transformation process. It was found that the equivalent key streams and all the unknown parameters of the cryptosystem can be recovered by our chosen-plaintext attack algorithm. Both a theoretical analysis and an experimental validation are given in detail. Based on the analysis of the defects in the original cryptosystem, an improved color image encryption scheme was further developed. By using an image content–related approach in generating diffusion arrays and the process of interweaving diffusion and confusion, the security of the cryptosystem was enhanced. The experimental results and security analysis demonstrate the security superiority of the improved cryptosystem.

Keywords: image encryption; chaotic cryptography; cryptanalysis; chosen-plaintext attack; image information entropy

1. Introduction

The transmission of a digital image from the public network is becoming more and more frequent nowadays. Consequently, it is urgent to guarantee the security and privacy of image transmission, especially for military images and some sensitive content images. As an essential technical means, image encryption approaches are particularly important in image communications. However, traditional cryptography cannot quickly encrypt images with large amounts of data. As traditional cryptography relies on the complexity of computation, it is not easy to generate a large number of keys quickly. In this application background, chaotic encryption is a good complement to traditional cryptography, especially in image encryption. As chaotic signals have some excellent characteristics required by cryptography, chaotic systems have become a fine tool for information encryption [1], especially for image encryption applications. Due to this, chaotic systems have been widely used in designing image encryption algorithms. Entropy is an important measure of the chaotic characteristics of dynamical systems. Entropy, chaos and information theory are closely related [2–4].

Among many chaos-based algorithms for encrypting an image, the permutation and diffusion (PD) structure encryption algorithm, proposed by Fridrich [5], has become a typical model. This structure consists of a permutation (i.e., pixel position scrambling) procedure and a diffusion (i.e., pixel value alteration) procedure. Based on a typical model, researchers have tried many different ways of improving innovation. Some studies have proposed different image permutation strategies [6–12]. Some researchers have proposed novel image diffusion techniques [10,13–17]. Many researchers have attempted to improve the performance of image encryption systems through other improvements. References [18–24] improve the performance of secret key streams through a new chaotic system model. References [25–29] improve the anti-attack performance of a cryptographic algorithm by introducing a plaintext-related mechanism in generating the key streams. References [30–36] introduce the DNA coding principle in bioinformatics to enhance the security of the algorithms. References [37–41] focus on improving the speed of image encryption algorithms through comprehensive means. In References [42–44], the S-boxes are applied to the design of efficient image encryption algorithm and combined with transformation technology, the performance of the image encryption algorithm is improved. In References [45–47], wavelet analysis technology is introduced into the field of image encryption and the ideas are novel. In References [48,49], fractal analysis technology is investigated, which is related to chaos and has a good application potential in image encryption.

Another research direction closely related to encryption is cryptanalysis. The goal of cryptanalysis is to find a way to decipher secret keys or plaintext without knowing the secret keys of encryption systems [50–53]. Cryptanalysis can also find out flaws in encryption algorithms and can help cryptographic system designers to improve the security performance of cryptographic algorithms, which can avoid losses caused by potential vulnerabilities and make valuable contributions to encryption. Hence, cryptanalysis can also promote cryptography. Recent cryptanalysis research shows that some chaos-based image encryption algorithms have some security flaws and the cryptosystems can be broken by using various attack methods. For example, we launched a chosen-plaintext attack [54] on the scheme in Reference [55]. Wu [53] broke the encryption scheme in Reference [56] by a chosen-plaintext attack.

Except for the security performance, efficiency is another important issue of an image encryption scheme for practical applications. For this reason, image ciphers with a higher speed are consequently more desirable than those with a low speed. It is well known that low-dimensional discrete chaotic systems need less time than high-dimensional continuous-time chaotic systems to generate chaotic sequences of the same length. Therefore, using a low dimensional discrete chaotic system as a key generator of a cryptosystem has a higher speed. Furthermore, the complexity of discrete chaotic systems is much larger than those of the continuous-time chaotic systems [57–59] and the cipher image encrypted with a chaotic sequence of much larger complexity has a higher security. Therefore, using a 1D discrete chaotic system to encrypt color images, not only has the advantage of fast speed but also has the advantage of higher security. In Reference [60], Pak proposed a color image encryption scheme (denoted as Pak's cryptosystem hereinafter) by using combined 1D chaotic maps. Pak's system has the merits of a simple structure, high speed and a relatively high safety. It is a pity that Pak's algorithm cannot resist the chosen-plaintext attack. To the best of our knowledge, so far, only Wang [61] and Chen [62] have done cryptanalysis on Pak's scheme. Unfortunately, neither of the two previous analyses can crack all of the unknown parameters of Pak's encryption scheme due to the difficulty of the comprehensive cryptanalysis. Therefore, the previous cryptanalysis is incomplete. Moreover, Wang's cryptanalysis scheme has obvious problems and a very low efficiency, while Chen's cryptanalysis scheme did not give the specific process of deciphering the permutation secret keys. In order to overcome the shortcomings of the above cryptanalysis work, this paper presents a more comprehensive and efficient cryptanalysis on Pak's cryptosystem. With our improved cryptanalysis, both equivalent secret keys and all of the unknown parameters of Pak's cryptosystem can be completely deciphered.

Despite its security flaws, Pak's encryption scheme still has many advantages to carry forward. Its design idea is clear and novel, and its efficiency is relatively high. Therefore, it is worth preserving these advantages and improving their defects. For this reason, this paper further proposes an improved enhanced color image encryption scheme, which includes both an image content–related approach in generating diffusion arrays and the process of interweaving diffusion and confusion.

The rest of this paper is organized as follows. Section 2 describes briefly Pak's algorithm and the related cryptanalysis. The improved cryptanalysis and attacks on Pak's algorithm are presented in Section 3. An enhanced encryption scheme is proposed in Section 4. Some experimental results and analysis for the enhanced scheme are given in Section 5. Finally, some concluding remarks are given in Section 6.

2. Description of Pak's Scheme and the Related Cryptanalysis

Pak's algorithm includes three processing stages. (1) Confusion: Pixel level permutation; (2) Diffusion: Pixel values encryption; (3) Linear transformation. Before the encrypting process, the 3D RGB color image with M row N columns is converted into a 1D pixel array $\mathbf{P} = [p(1), p(2), \ldots, p(L)]$ according to the R, G, and B components successively, where $L = M \times N \times 3$. Each value of $p(i)$ is an integer in the range [0, 255]. The flow of Pak's encryption scheme can be visualized in Figure 1. Where, $\mathbf{P}$ is the plain image pixel array, and $\mathbf{C}'$ is the final cipher image pixel array. SSS represents the combined Sine-Sine chaotic System. $\mathbf{X}'$ is the permutation position array and $\mathbf{D}'$ is the diffusion array. Both $\mathbf{X}'$ and $\mathbf{D}'$ are generated by chaotic sequences.

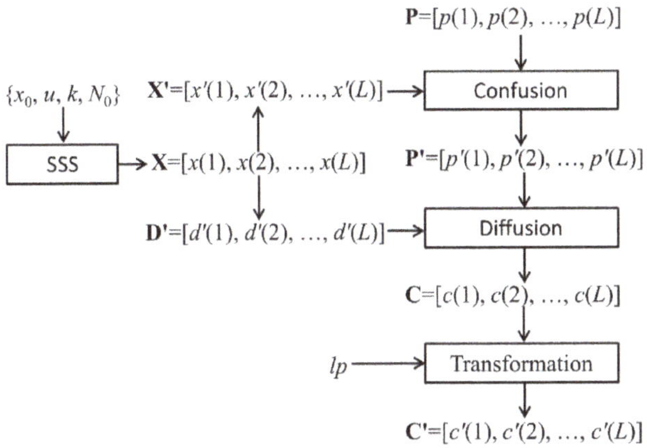

Figure 1. A flow chart of Pak's encryption scheme.

2.1. The New Chaotic System

The chaotic system adopted in Pak's encryption scheme is a newly discovered chaotic map by using the chaotic sine map, which is expressed as

$$x(n+1) = u \times \sin[\pi \times x(n)] \times 2^k - \lfloor u \times \sin[\pi \times x(n)] \times 2^k \rfloor \tag{1}$$

where, u is the control parameter of the system and $\{x(n), n = 0, 1, 2, \ldots\}$ is the output chaotic sequence with the initial value $x(0) = x_0$. $\lfloor x \rfloor$ is the largest integer that is smaller than or equal to x. System (1) is called a Sine-Sine system (SSS) [60], which is chaotic when $u \in (0, 10]$ and $k \in [8, 20]$. Parameters k, u and x_0 were used as secret keys.

2.2. The Confusion Process

In the confusion process, a permutation operation is performed on the pixel level with a position transformation. The operational process consists of the following steps:

Step 1: By using specified parameter values x_0, u and k, iterate the new chaotic system $(N_0 + L)$ times and select the rear L elements to make a sub chaotic sequence $\mathbf{X} = [x(1), x(2), \ldots, x(L)]$. Where N_0 is an integer used as a security key.

Step 2: Sequence $\mathbf{X}$ is sorted in ascending order. Then, one can obtain a sorted chaotic sequence $\mathbf{SX} = [sx(1), sx(2), \ldots, sx(L)]$ and a permutation position array $\mathbf{X}' = [x'(1), x'(2), \ldots, x'(L)]$, where $x'(i)$ are integers ranging from 1 to L. If $x(i) = sx(j)$, then $x'(i) = j$.

Step 3: Get the permuted image pixel sequence $\mathbf{P}' = [p'(1), p'(2), \ldots, p'(L)]$ by using the permutation position array $\mathbf{X}'$ and the plain image pixel sequence $\mathbf{P}$. The transformation relation is

$$p'(i) = p\,(x'(i)). \tag{2}$$

2.3. Diffusion Process

In the diffusion process, pixel value encryption is performed based on a diffusion array $\mathbf{D}'$. The operational process consists of the following two steps: The operational process consists of the following steps:

Step 1: Generate the diffusion array $\mathbf{D}' = [d'(1), d'(2), \ldots, d'(L)]$ from the chaotic sequence $\mathbf{X}$ as:

$$d'(i) = \mathrm{mod}(\lfloor x(i) \times 10^k \rfloor, 256). \tag{3}$$

Step 2: Get the temporary ciphered image pixel array $\mathbf{C} = [c(1), c(2), \ldots, c(L)]$ from the diffusion vector $\mathbf{D}'$ and the permuted image array $\mathbf{P}'$ according to the following diffusion equation:

$$\begin{cases} c(i) = \mathrm{mod}(p'(i) + d'(i), 256) \oplus seed, \text{if } i = 1, \\ c(i) = \mathrm{mod}(p'(i) + d'(i), 256) \oplus c(i-1), \text{if } i > 1, \end{cases} \tag{4}$$

where $\oplus$ denotes the binary *XOR* operator. $c(i - 1)$ is the previous cipher pixel, and *seed* is a preset constant.

2.4. Linear Transformation

Get the final cipher image pixel array $\mathbf{C}' = [c'(1), c'(2), \ldots, c'(L)]$ from the temporary cipher image pixel array $\mathbf{C}$ and a security number lp as

$$\begin{cases} c'(i - lp) = c(i), \text{if } i > lp, \\ c'(i - lp + L) = c(i), \text{if } i \leq lp, \end{cases} \tag{5}$$

where lp is used as a security key. In order to see the result of the linear transformation at a glance, we used a graph to express the linear transformation process, which is shown in Figure 2. There are two key points in this linear transformation process, which deserve our special attention. One, the first pixel in the array $\mathbf{C}$ was moved to the $(L - lp + 1)$ position in the array $\mathbf{C}'$, that is $c'(L - lp + 1) = c(1)$. Second, the original two adjacent pixels $c(lp)$ and $c(lp + 1)$ were moved to the end and start of the array $\mathbf{C}'$, that is, $c'(L) = c(lp)$, $c'(1) = c(lp + 1)$. If $lp = 0$ or $lp = L$, then $c'(i) = c(i)$. Hence, a reasonable range of lp is $0 < lp < L$.

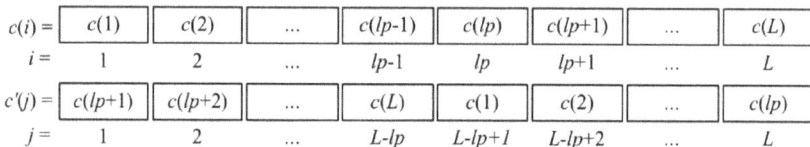

$c(i) =$	$c(1)$	$c(2)$	…	$c(lp\text{-}1)$	$c(lp)$	$c(lp+1)$	…	$c(L)$
$i =$	1	2	…	lp-1	lp	lp+1	…	L
$c'(j) =$	$c(lp+1)$	$c(lp+2)$	…	$c(L)$	$c(1)$	$c(2)$	…	$c(lp)$
$j =$	1	2	…	L-lp	L-lp+1	L-lp+2	…	L

Figure 2. The linear transformation operation.

The final cipher image was obtained by converting the 1D pixel vector $\mathbf{C}'$ into a 2D color image consisting of R, G and B components with the size of $M \times N$. The secret keys used in Pak's algorithm consists of five parameters $\{x_0, u, k, N_0, lp\}$.

The decryption process is the inverse operation of the encryption process and it was omitted here.

According to Kerchoff's principle, when analyzing an encryption algorithm, an assumption is made that the cryptanalyst knows exactly the design and working of the cryptosystem. Namely, the only thing the attacker does not know is the secret key. The definition of a chosen-plaintext attack can be described as follows: Attackers have the chance to use the encryption machine temporarily, hence they can select a special plaintext to encrypt and get its corresponding ciphertext without knowing the secret keys.

In Pak's algorithm, the permutation position array $\mathbf{X}'$ and the diffusion array $\mathbf{D}'$ are determined by parameters $\{x_0, u, k, N_0\}$ and have nothing to do with the plain image. Namely, $\mathbf{X}'$ and $\mathbf{D}'$ are static and do not change with different images to be encrypted. The secret key, *lp*, and the unknown parameter *seed* also have nothing to do with the plain image. Therefore, attackers can choose some special plaintext images to encrypt by using Pak's encryption machine when they temporarily obtain the opportunity to use Pak's encryption machine and obtain the corresponding ciphertext image to use these known plaintext-ciphertext image pairs to crack the equivalent key sequences $\mathbf{X}'$, $\mathbf{D}'$, parameters *lp* and *seed*. By using these equivalent key sequences $\mathbf{X}'$ and $\mathbf{D}'$, parameters *lp* and *seed*, any image encrypted by Pak's encryption machine can be decrypted without knowing the original keys of Pak's encryption machine. This is the basic principle of the chosen-plaintext attack model. According to this attack model, it is obvious that Pak's scheme cannot resist a chosen-plaintext attack.

2.5. The Related Cryptanalysis Work

In Wang's cryptanalysis scheme, the authors constructed an equivalent cryptosystem for Pak's cryptosystem. In the equivalent cryptosystem, they constructed the new permutation position array $\mathbf{X}''$ and diffusion array $\mathbf{D}''$ of the equivalent encryption scheme by transforming the original permutation position array $\mathbf{X}'$ and diffusion array $\mathbf{D}'$ with the secret parameter *lp* respectively. The relationships of the key streams between the equivalent cryptosystem and Pak's cryptosystem are as follows

$$\begin{cases} x''(i-lp) = x'(i), \text{ if } i \in (lp, L], \\ x''(i-lp+L) = x'(i), \text{ if } i \in [1, lp]. \end{cases} \tag{6}$$

$$\begin{cases} d''(i-lp) = d'(i), \text{ if } i \in (lp, L], \\ d''(i-lp+L) = d'(i), \text{ if } i \in [1, lp]. \end{cases} \tag{7}$$

Wang's equivalent encryption scheme contains only two processes: permutation and diffusion, which can be described by Equations (8) and (9) respectively.

$$p''(i) = p(x''(i)) \tag{8}$$

$$c''(i) = \text{mod}(p''(i) + d''(i), 256) \oplus c''(i-1) \tag{9}$$

where $\mathbf{P}'' = [p''(1), p''(2), \ldots, p''(L)]$ is the permuted image pixel sequence of Wang's equivalent cryptosystem, which has the following relations with $\mathbf{P}'$ in Pak's system

$$\begin{cases} p''(i-lp) = p'(i), \text{ if } i \in (lp, L], \\ p''(i-lp+L) = p'(i), \text{ if } i \in [1, lp]. \end{cases} \tag{10}$$

$\mathbf{C}'' = [c''(1), c''(2), \ldots, c''(L)]$ is the final cipher image pixel array of Wang's equivalent cryptosystem. The authors claim that $c''(i) = c'(i)$ will hold if the Equations (6)–(10) hold.

The operation process of Wang's chosen-plaintext attack scheme is divided into the following three stages.

(1) Extract the diffusion array $\mathbf{D}''$. Select a special plain-image $\mathbf{P}$ consisting of all 0 elements such that $p''(i) = 0$ and obtain the corresponding cipher-image $\mathbf{C}''$. According to Equation (9), the diffusion array $\mathbf{D}''$ is extracted as

$$d''(i) = c''(i) \oplus c''(i-1). \tag{11}$$

(2) Extract the permutation position array $\mathbf{X}''$. Select L special plain images with the 1D pixel arrays respectively denoted as $\mathbf{P}_1, \mathbf{P}_2, \ldots, \mathbf{P}_L$ and the jth element in the pixel array $\mathbf{P}_j$ is 1; all other elements are 0. Get the corresponding encrypted image arrays $\mathbf{C}_1, \mathbf{C}_2, \ldots, \mathbf{C}_L$. By using one plain

image $\mathbf{P}_j$ and the corresponding $\mathbf{C}_j$, only one element $x''(i)$ in $\mathbf{X}''$ can be obtained. All elements of $\{x''(1), x''(1), \ldots, x''(L)\}$ can be obtained when L pairs of ($\mathbf{P}_j$, $\mathbf{C}_j$) are used.

(3) Recover the original plain image. By using the new permutation position array $\mathbf{X}''$ and the new diffusion array $\mathbf{D}''$, recover the original plain image $\mathbf{P}$ from the target cipher image $\mathbf{C}$.

We find that Wang's cryptanalysis algorithm has the following issues:

(1) The authors assume that the attacker knows the parameter *seed* and used it as a known parameter in the equivalent encryption system. In fact, the *seed* parameter is a constant set in Pak's cryptosystem. Although the attacker can use Pak's encryption machine temporarily, the *seed* parameter is unknown to the attacker.

(2) Although the authors claim that the cipher image $\mathbf{C}''$ obtained from their equivalent cryptosystem is the same as the cipher image $\mathbf{C}'$ obtained by the original Pak's cryptosystem, no strict proof is given. In fact, the cipher image pixel array $\mathbf{C}''$ is not equivalent to $\mathbf{C}'$ due to the unknown parameter lp, which is not broken out by the authors. The proof procedure is as follows.

When encrypting the first pixel by Wang's equivalent cryptosystem, Equation (9) is degenerated into the form as $c''(1) = \mathrm{mod}(p''(1) + d''(1), 256) \oplus c''(0)$, where $c''(0)$ is not a pixel value of the array $\mathbf{C}''$ and $c''(0)$ may be the parameter *seed*. From Equations (7), (9) and (10), we can get $p''(1) = p'(lp + 1)$ and $d''(1) = d'(lp + 1)$. Then one can obtain $c''(1)$ as

$$c''(1) = \mathrm{mod}(p'(lp + 1) + d'(lp + 1), 256) \oplus seed. \tag{12}$$

while $c'(1)$ obtained by using Pak's algorithm is as

$$c'(1) = \mathrm{mod}(p'(lp + 1) + d'(lp + 1), 256) \oplus c'(L). \tag{13}$$

By comparing Equations (12) with (13), $c''(1) \neq c'(1)$.

When $i = L - lp + 1$, Equation (9) is degenerated into the form $c''(L - lp + 1) = \mathrm{mod}(p''(L - lp + 1) + d''(L - lp + 1), 256) \oplus c''(L - lp)$. From Equations (7), (9) and (10), we can get $p''(L - lp + 1) = p'(1)$ and $d''(L - lp + 1) = d'(1)$. As a result, $c''(L - lp + 1)$ is as

$$c''(L - lp + 1) = \mathrm{mod}(p'(1) + d'(1), 256) \oplus c''(L - lp). \tag{14}$$

while using Pak's algorithm, $c'(L - lp + 1)$ is as

$$c'(L - lp + 1) = \mathrm{mod}(p'(1) + d'(1), 256) \oplus seed. \tag{15}$$

Comparing Equations (14) with (15), $c''(L - lp + 1) \neq c'(L - lp + 1)$.

Based on $c''(1) \neq c'(1)$, one can deduce that $c''(i) \neq c'(i)$, $i = 2, 3, \ldots, L$.

In fact, there are some defects in Wang's cryptanalysis algorithm because the authors completely ignore the role of the parameter lp and do not break out lp. However, when the parameter lp is not known, one cannot know where the *seed* should be used to calculate $c''(i)$.

(3) The most serious problem in Wang's cryptanalysis scheme is that the number of chosen plain images is too high to reach $M \times N \times 3$ in extracting the permutation position array $\mathbf{X}''$. The use of one chosen plain image at a time can only break one element value of $\mathbf{X}''$, which is very inefficient, so Wang's cryptanalysis scheme is unrealistic.

In Chen's cryptanalysis scheme, unfortunately, the parameter *seed* is also not deciphered and used as a known parameter. Thus, reducing the difficulty of the cryptanalysis. In addition, Chen did not give the specific process of deciphering the permutation position array $\mathbf{X}'$.

3. The Improved Cryptanalysis Scheme

In order to provide a more comprehensive and efficient cryptanalysis method on Pak's encryption algorithm, we propose an improved chosen-plaintext attack algorithm to Pak's scheme. Suppose the target color cipher image to be decrypted has the size of $L = M \times N \times 3$. Firstly, we cracked the secret parameter lp and the diffusion array $[d'(2), d'(3), \ldots, d'(L)]$ except for $d'(1)$ by using two selected plain images and their corresponding cipher images. Secondly, we cracked the unknown parameter *seed* and $d'(1)$ by using one or more than one selected plain images. Thirdly, we cracked the permutation position array $\mathbf{X}'$ by using $\lceil (M \times N \times 3)/255 \rceil$ selected plain images and their corresponding cipher images, where $\lceil x \rceil$ is the smallest integer that is greater than or equal to x. Wang's cryptanalysis algorithm needs $M \times N \times 3$ selected plain images to decipher the permutation position array $\mathbf{X}'$, while our cryptanalysis algorithm only needs $\lceil (M \times N \times 3)/255 \rceil$ selected plain images to decipher the permutation position array $\mathbf{X}'$. Hence, the efficiency of our improved chosen-plaintext attack algorithm is about 255 times that of Wang's algorithm.

3.1. Recover the Secret Key lp and the Diffusion Array

According to Equations (4) and (5), $d'(i)$ can be calculated as

$$\begin{cases} d'(1) = c'(L - lp + 1) \oplus seed \dot{-} p'(1), \text{if } i = 1, \\ d'(i) = c'(L - lp + i) \oplus c'(L - lp + i - 1) \dot{-} p'(i), \text{if } 1 < i \leq lp, \\ d'(lp + 1) = c'(1) \oplus c'(L) \dot{-} p'(lp + 1), \text{if } i = lp + 1, \\ d'(i) = c'(i - lp) \oplus c'(i - lp - 1) \dot{-} p'(i), \text{if } lp + 1 < i \leq L, \end{cases} \quad (16)$$

where $x \dot{-} y = \mathrm{mod}(x - y + 256, 256)$. Obviously, if the *seed* in Equation (16) is replaced by $c'(L - lp)$, then the relationship between $d'(i)$ and $c'(j)$ can be expressed in Figure 3.

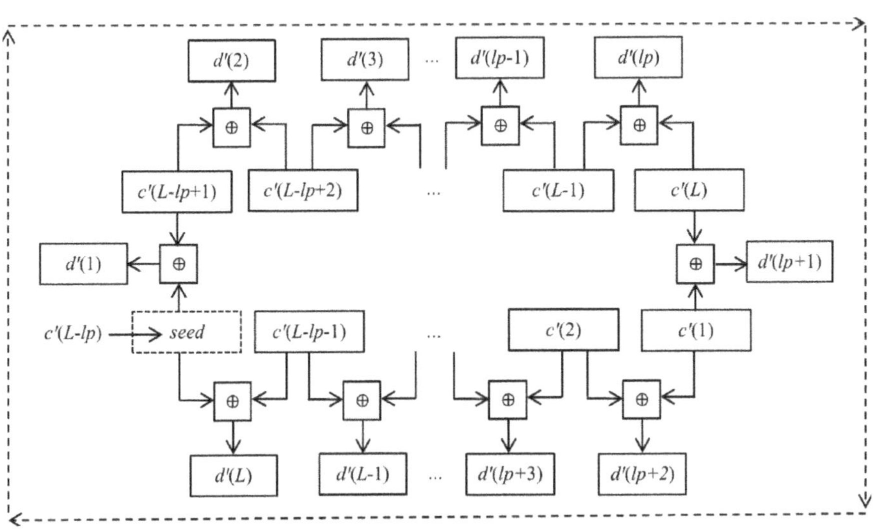

Figure 3. The diagram of the relationship between $\mathbf{D}'$ and $\mathbf{C}'$.

From Figure 3, one can see that each key $d'(i)$ is related to a pair of adjacent pixel values $\{c'(j), c'(j+1)\}$ or $\{c'(L), c'(1)\}$. To avoid the influence of the unknown parameter lp, we can select a specific plain image where all pixels $p'(i)$ have the same value q, then we can calculate a series of values by neighbors $\{c'(L), c'(1)\}, \{c'(1), c'(2)\}, \{c'(2), c'(3)\}, \ldots, \{c'(L-1), c'(L)\}$ and store these values in a temporary array $\mathbf{D} = [d(1), d(2), \ldots, d(L)]$, where $d(i)$ is as

$$\begin{cases} d(1) = c'(1) \oplus c'(L) \dot{-} q, \\ d(i) = c'(i) \oplus c'(i-1) \dot{-} q, \ i = 2, 3, \ldots, L. \end{cases} \quad (17)$$

Equation (17) brings us great convenience for computing $d(i)$ because it does not contain the unknown parameters lp and *seed*. Obviously, the equivalent relationship of the elements between $\mathbf{D}' = [d'(1), d'(2), \ldots, d'(L)]$ and $\mathbf{D} = [d(1), d(2), \ldots, d(L)]$ is as follows

$$\begin{cases} d'(i) = d(i - lp), \text{if } i > lp, \\ d'(i) = d(L + i - lp), \text{if } i \leq lp. \end{cases} \tag{18}$$

Namely, $d'(lp + 1) = d(1)$, $d'(lp + 2) = d(2), \ldots, d'(L - 1) = d(L - lp - 1)$, $d'(L) = d(L - lp)$; $d'(1) = d(L - lp + 1)$, $d'(2) = d(L - lp + 2), \ldots, d'(lp - 1) = d(L - 1)$, $d'(lp) = d(L)$.

It is worth noting that, except for $d'(1)$ or $d(L - lp + 1)$, the rest of the values $d(i)$ ($i \neq L - lp + 1$) obtained by Equation (17) are all right values. Namely, when calculating $d(L - lp + 1)$, if we do not use the parameter *seed* and use the $c'(L - lp)$ value instead of *seed*, then the result of $d(L - lp + 1)$ may be wrong. Considering the values of $d(i)$ or $d'(i)$ are determined by parameters $\{x_0, u, k, N_0\}$ and have nothing to do with the content of the image, if we choose two different plain images and get the corresponding cipher images, by using the two pairs of plaintext-ciphertext to calculate $d_1(i)$ and $d_2(i)$, then one can find the only position of ii that the value of $d_1(ii)$ and $d_2(ii)$ will not be identical but values of $d_1(i)$ and $d_2(i)$ at other locations i ($i \neq ii$) are definitely the same. Once the location ii is sought out, the value of lp can be determined, which is $lp = L + 1 - ii$.

Based on the above idea, we get the algorithm for deciphering the secret key parameter lp and the diffusion array $d'(i)$, which is described as follows:

Step 1: Let $q = 0$, and select a special plain image $\mathbf{PA} = [pa(1), pa(2), \ldots, pa(L)]$ that all pixels $pa(i)$ have the same value q and obtain the corresponding cipher image $\mathbf{CA}' = [ca'(1), ca'(2), \ldots, ca'(L)]$ by using Pak's encryption machinery. As $\mathbf{PA}' = [q, q, \ldots, q]$, then we can get a array $\mathbf{DA} = [da(1), da(2), \ldots, da(L)]$ by using Equation (17).

Step 2: Let $q = q + 1$, and select a special plain image $\mathbf{PB} = [pb(1), pb(2), \ldots, pb(L)]$ that all pixels $pb(i)$ have the same value q. Obtain the corresponding cipher-image $\mathbf{CB}' = [cb'(1), cb'(2), \ldots, cb'(L)]$ by using Pak's encryption machine. Because $\mathbf{PB}' = [q, q, \ldots, q]$, then we can get another array $\mathbf{DB} = [db(1), db(2), \ldots, db(L)]$ by using Equation (17).

Step 3: Compare $da(i)$ and $db(i)$ one by one for $i = 1, 2, \ldots, L$. If it exists at position $I = ii$ and meets the relationship $da(ii) \neq db(ii)$, then $L - lp + 1 = ii$, so lp is determined as $lp = L + 1 - ii$, and go to Step 4. Otherwise, repeat Step 2 to Step 3 until lp is determined.

Step 4: After the value of lp is ascertained, we can recover the diffusion array $\mathbf{D}'$ of Pak's cryptosystem by using Equation (18). Where only the value of $d'(1)$ is incorrect.

3.2. Recover $d'(1)$ and the Unknown Parameter Seed

According to the first formula in Equation (16), ($d'(1)$, *seed*) meets the following relationship

$$c'(L - lp + 1) = \text{mod}(d'(1) + p'(1), 256) \oplus seed. \tag{19}$$

Using the special chosen plain image $\mathbf{PA} = [0, 0, \ldots, 0]$ and $\mathbf{PB} = [1, 1, \ldots, 1]$, we have got a pair of ciphertext data ($ca'(L - lp + 1)$, $cb'(L - lp + 1)$) in the previous section. Therefore, $d'(1)$ and *seed* needs to satisfy the following equation:

$$\begin{cases} ca'(L - lp + 1) = \text{mod}(d'(1) + 0, 256) \oplus seed, \\ cb'(L - lp + 1) = \text{mod}(d'(1) + 1, 256) \oplus seed. \end{cases} \tag{20}$$

Consider such a fact that $seed \in \{0, 1, 2, \ldots, 255\}$ and $d'(1) \in \{0, 1, 2, \ldots, 255\}$, so the solution of Equation (20) can be easily obtained by the computer exhaustive algorithm. However, the solution

$[d'(1), seed]$ of Equation (20) is not unique because the equations in Equation (20) are not two linear equations. Suppose an equation for d' and *seed* has the following form:

$$\text{mod}(d' + q, 256) \oplus seed = c'. \tag{21}$$

Regarding the solutions of Equation (21), We have the following Proposition:

Proposition 1. *For any values of $q \in Z_{256}$ and $c' \in Z_{256}$, if $[d', seed]$ is a solution of Equation (21), then $[\text{mod}(d' + 128, 256), \text{mod}(seed + 128, 256)]$ is also a solution of Equation (21). Where $d' \in Z_{256}$ and $seed \in Z_{256}$.*

Proof. Suppose the binary value of $\text{mod}(d' + q, 256)$ is $(d_8d_7d_6d_5d_4d_3d_2d_1)_2$ and the binary value of *seed* is $(s_8s_7s_6s_5s_4s_3s_2s_1)_2$.

If $d_8 = 0$, then $\text{mod}(\text{mod}(d' + 128, 256) + q, 256) = \text{mod}(d' + q + 128, 256) = \text{mod}(\text{mod}(d' + q, 256) + 128, 256) = \text{mod}((0d_7d_6d_5d_4d_3d_2d_1)_2 + (10000000)_2, 256) = (1d_7d_6d_5d_4d_3d_2d_1)_2 = (\overline{d}_8d_7d_6d_5d_4d_3d_2d_1)_2$. Where, $\overline{x}$ represents the binary inverse value of x.

If $d_8 = 1$, then $\text{mod}(\text{mod}(d' + 128, 256) + q, 256) = \text{mod}(d' + q + 128, 256) = \text{mod}(\text{mod}(d' + q, 256) + 128, 256) = \text{mod}((1d_7d_6d_5d_4d_3d_2d_1)_2 + (10000000)_2, 256) = (0d_7d_6d_5d_4d_3d_2d_1)_2 = (\overline{d}_8d_7d_6d_5d_4d_3d_2d_1)_2$.

If $s_8 = 0$, then $\text{mod}(seed + 128, 256) = \text{mod}((0s_7s_6s_5s_4s_3s_2s_1)_2 + (10000000)_2, 256) = (1s_7s_6s_5s_4s_3s_2s_1)_2 = (\overline{s}_8s_7s_6s_5s_4s_3s_2s_1)_2$.

If $s_8 = 1$, then $\text{mod}(seed + 128, 256) = \text{mod}((1s_7s_6s_5s_4s_3s_2s_1)_2 + (10000000)_2, 256) = (0s_7s_6s_5s_4s_3s_2s_1)_2 = (\overline{s}_8s_7s_6s_5s_4s_3s_2s_1)_2$. □

Considering $\overline{d}_8 \oplus \overline{s}_8 = d_8 \oplus s_8$, we can obtain that $\text{mod}(\text{mod}(d' + 128, 256) + q, 256) \oplus \text{mod}(seed + 128, 256) = (\overline{d}_8d_7d_6d_5d_4d_3d_2d_1)_2 \oplus (\overline{s}_8s_7s_6s_5s_4s_3s_2s_1)_2 = (d_8d_7d_6d_5d_4d_3d_2d_1)_2 \oplus (s_8s_7s_6s_5s_4s_3s_2s_1)_2 = c'$. This means that $[\text{mod}(d' + 128, 256), \text{mod}(seed + 128, 256)]$ is also a solution of Equation (21).

Suppose Equation (20) has m groups of solutions ($m \geq 2$) as $[d_1'(1), seed_1]$, $[d_2'(1), seed_2]$, …, $[d_m'(1), seed_m]$. If $m = 2$, then the two groups of solutions are all the required results and the task of recovering $(d'(1), seed)$ has been completed. If $m > 2$, then we must select some other plain image $\mathbf{P} = [q, q, \ldots, q]$ and obtain the corresponding cipher image $\mathbf{C}' = [c'(1), c'(2), \ldots, c'(L)]$, where $q > 1$. In addition, we can obtain another equation as: $\text{mod}(d'(1) + q, 256) \oplus seed = c'(L - lp + 1)$. Under the constraint of the other equation, we can remove those superfluous solutions that do not satisfy all equations until the remaining solutions are only 2 groups. In this way, the unknown parameter *seed* and the secret key $d'(1)$ of the original encryption system can be deciphered. The concrete algorithm for recovering $d'(1)$ and *seed* is described as follows:

Step 1: Let m groups of solutions of Equation (20) be saved in the array $\mathbf{R} = [r(1), r(2), \ldots, r(m)]$ and $\mathbf{S} = [s(1), s(2), \ldots, s(m)]$ sequentially, Where $r(i) = d_i'(1)$, $s(i) = seed_i$, $i = 1, 2, \ldots, m$. Let $q = 1$.

Step 2: Check the value of m. If $m \leq 2$, then go to Step 9. If $m > 2$, then go to Step 3.

Step 3: $q = q + 1$.

Step 4: For $i = 1, 2, \ldots, m$, each groups of solutions $[r(i), s(i)]$ is assumed to be used to encrypt the plaintext pixel value q and calculate the corresponding ciphertext values as $cc(i) = \text{mod}(q + r(i), 256) \oplus s(i)$.

Step 5: For $i = 1, 2, \ldots, m$, Check whether the value of each element in the array $[cc(1), cc(2), \ldots, cc(m)]$ is exactly the same. If $cc(i)$ is exactly the same, then repeat Step 3 to Step 5. If $cc(i)$ is not exactly the same, then go to Step 6.

Step 6: Select a special plain image array $\mathbf{P} = [q, q, \ldots, q]$ and obtain the corresponding cipher image pixels array $\mathbf{C}' = [c'(1), c'(2), \ldots, c'(L)]$ by using Pak's encryption machine.

Step 7: For each solution group $[r(i), s(i)]$, calculate the values of $\text{mod}(r(i) + q, 256) \oplus s(i)$, $i = 1, 2, \ldots, m$. If $\text{mod}(r(i) + q, 256) \oplus s(i) \neq c'(L - lp + 1)$, then delete the i-th solution group $[r(i), s(i)]$ from $\mathbf{S}$ and $\mathbf{R}$ respectively.

Step 8: Modify the value of m, that is, $m = \text{size}(\mathbf{R})$, and return to Step 2.

Step 9: Output the final values of $[d'(1), seed]$, that is $[d'(1), seed] = [r(1), s(1)]$ or $[d'(1), seed] = [r(2), s(2)]$.

3.3. Recover the Permutation Position Array $\mathbf{X}'$

After the RGB image matrix is converted into a 1D gray image pixel sequence $\mathbf{P} = [p(1), p(2), \ldots, p(L)]$, array $\mathbf{P}$ has L pixels and $L = M \times N \times 3$. Each value of $p(i)$ is an integer in the range of [0, 255]. If $L \leq 255$, then only one chosen-plain image $\mathbf{P} = [1, 2, \ldots, L]$ is necessary to recover the permutation position array $\mathbf{X}'$, so that each pixel in the chosen plain image has different values in $\{1, 2, \ldots, L\}$. If $L > 255$, then n chosen plain images are required to recover the permutation position array $\mathbf{X}'$, where $n = \lceil L/255 \rceil > 1$. In this case, we select a series of special color plain images $(\mathbf{P}_1, \mathbf{P}_2, \ldots, \mathbf{P}_n)$ and $\mathbf{P}_j = [p_j(1), p_j(2), \ldots, p_j(L)]$. We divide $\mathbf{P}_j$ into n groups and each group contains 255 pixels except for the last one and the last group contains q pixels ($q \leq 255$). For the j-th chosen-plain image pixel array $\mathbf{P}_j$, we assign each element of the j-th group a distinct value between 1 to 255 and the others are assigned the value of 0. The patterns of elements in each chosen plain image pixel array $\mathbf{P}_j$ are shown in Figure 4.

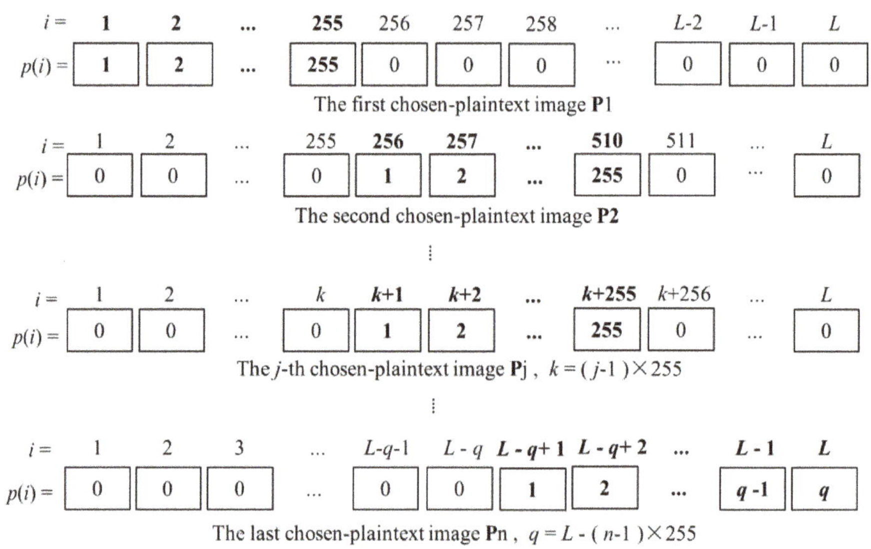

Figure 4. The patterns of elements in each chosen plain image pixel array $\mathbf{P}_j$.

We then obtain the corresponding series of cipher images $(\mathbf{C}'_1, \mathbf{C}'_2, \ldots, \mathbf{C}'_n)$ by using Pak's encryption machine. Where, $\mathbf{C}'_j = [c'_j(1), c'_j(2), \ldots, c'_j(L)]$. Then, we can decrypt $\mathbf{C}'_j$ to obtain $\mathbf{P}'_j = [p'_j(1), p'_j(2), \ldots, p'_j(L)]$, where $p'_j(i)$ can be obtained by using Equation (16).

Finally, because of the relationship $p_j'(i) = p_j(x'(i))$ ($i \in [1, L]$), $\mathbf{X}'$ can be determined by comparing $\mathbf{P}'_j$ and $\mathbf{P}_j$. Namely, if $p'_j(i) = p_j(k)$, then $x'(i) = k$.

3.4. Recover the Original Plain Image

In Section 3.1 to 3.3, we obtained the secret keys $\{lp, \mathbf{X}', \mathbf{D}'\}$ and the unknown parameter *seed*, which are unrelated to the plain image or ciphertext image. Therefore, we can decrypt any other ciphertext image **CI** by using the parameter set $\{seed, lp, \mathbf{X}', \mathbf{D}'\}$. The decryption process to recover the plain image **PI** from the target ciphertext image **CI** is exactly the same as the decryption process of Pak's scheme, which can be described as follows:

Step 1: Convert the color ciphertext image **CI** with a size of $M \times N \times 3$ into a 1D pixel array $\mathbf{C}' = [c'(1), c'(2), \ldots, c'(L)]$, where $L = M \times N \times 3$.

Step 2: Obtain the intermediary cipher image pixel array $\mathbf{C} = [c(1), c(2), \ldots, c(L)]$ from the final cipher pixel array $\mathbf{C}' = [c'(1), c'(2), \ldots, c'(L)]$ by performing the inverse transformation of Equation (5).

Step 3: Recover the permuted image pixel array $\mathbf{P}' = [p'(1), p'(2), \ldots, p'(L)]$ by performing the inverse diffusion process of Equation (4).

Step 4: Do inverse permutation on $\mathbf{P}'$ to obtain $\mathbf{P}$ by using the inverse permutation process of Equation (2).

Step 5: Convert the 1D array $\mathbf{P}$ into a 3D matrix with a size of $M \times N \times 3$ and the original color plain image **PI** is recovered.

3.5. Examples of the Improved Cryptanalysis Scheme

Suppose the right values of original secret keys in Pak's cryptosystem are as follows: $x_0 = 0.456$, $u = 5.4321$, $k = 14$, $N_0 = 1000$, $lp = 5$, and $seed = 250$.

Example 1. *In this example, the plain image* **P** *is the color peppers with a size of* $256 \times 256 \times 3$. *The plain image and its cipher image encrypted by using Pak's encryption machine are shown in Figure 5a,b respectively. The deciphered image by using our chosen-plaintext attack is shown in Figure 5c, which is exactly the same as the original plain image in Figure 5a. Through the image peppers as an example, our attack attains demonstration.*

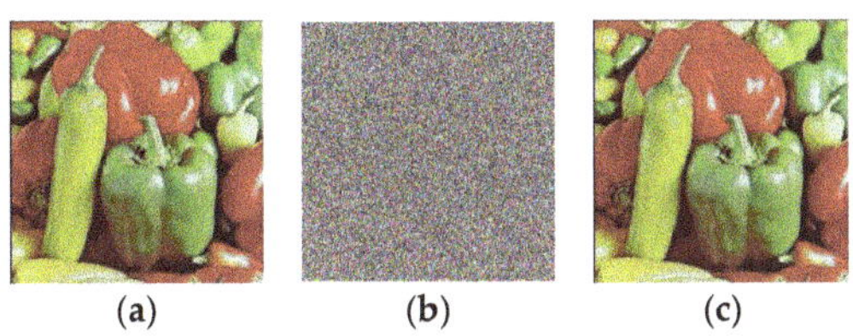

Figure 5. The experimental results of the chosen-plaintext attacks. (**a**) The plain image; (**b**) the cipher image; (**c**) the cracked image.

Example 2. *The secret key parameters are the same as those of Example 1. In order to verify the correctness of our chosen-plaintext attack scheme more intuitively, this example shows a simple and specific numerical experiment. In this example, the plain image* **PI** *is the color image with size of* $2 \times 2 \times 3$ *(*$L = M \times N \times 3 = 12$*), and its components are as*

$$\mathbf{P}_R = \begin{bmatrix} 11 & 13 \\ 12 & 14 \end{bmatrix}, \mathbf{P}_G = \begin{bmatrix} 21 & 23 \\ 22 & 24 \end{bmatrix}, \mathbf{P}_B = \begin{bmatrix} 31 & 33 \\ 32 & 34 \end{bmatrix}. \tag{22}$$

Its corresponding 1D pixel array **P** is:

$$\mathbf{P} = [11, 12, 13, 14, 21, 22, 23, 24, 31, 32, 33, 34]. \tag{23}$$

As the result, the 1D pixel array $\mathbf{C}'$ encrypted by Pak's encryption machine is:

$$C' = [246, 16, 1, 6, 37, 3, 137, 197, 162, 215, 51, 22]. \tag{24}$$

By choosing two special plain-image array **PA** = [0, 0, 0, 0, 0, 0, 0, 0, 0, 0, 0, 0] and **PB** = [1, 1, 1, 1, 1, 1, 1, 1, 1, 1, 1, 1], we obtain the corresponding cipher image arrays as $\mathbf{CA}'$ = [173, 117, 137, 108, 97, 110, 17, 229, 170, 195, 20, 18] and $\mathbf{CB}'$ = [255, 38, 219, 61, 51, 35, 163, 218, 138, 224, 56, 63]. According to Equation (16), we obtain **DA** and **DB** as: **DA** = [191, 216, 252, 229, 13, 15, 127, **244**, 79, 105, 215, 6], **DB** = [191, 216, 252, 229, 13, 15, 127, **120**, 79, 105, 215, 6]. By comparing **DA** and **DB**, we find that $da(8) \neq db(8)$, then $ii = 8$, and $lp = L + 1 - ii = 12 + 1 - 8 = 5$. Then, Equation (19) is changed into the following form

$$\begin{cases} ca'(8) = 229 = \mathrm{mod}(d'(1) + 0, 256) \oplus seed \\ cb'(8) = 218 = \mathrm{mod}(d'(1) + 1, 256) \oplus seed \end{cases},$$

which has four groups of solution: $[d'(1), seed]$ = {[31, 250], [95, 186], [159, 122], [223, 58]}. For $q = 2$, 3, ... , check the values of "mod($d'(1) + q$, 256)⊕*seed*" with the four groups of solution. When $q = 33$, we find that "mod($d'(1) + q$, 256)⊕*seed*" has different values (186, 58, 186, 58) corresponding to the four groups of solution. We then select a special color plain image **P** = [33, 33, ... , 33] and obtain the corresponding cipher image **C**′ = [127, 134, 155, 157, 179, 131, 35, **186**, 202, 64, 184, 159] by using Pak's encryption machine, in which $c'(8) = 186$. Then we can determine that (31, 250) and (159, 122) are two right groups of secret keys to $[d'(1), seed]$. If we adopt $[d'(1), seed]$ = [159, 122] as the secret keys, then we can obtain **D**′ from **DA** or **DB** by using Equation (18), that is, **D**′ = [**159**, 79, 105, 215, 6, 191, 216, 252, 229, 13, 15, 127].

To recover the permutation position array **X**′, we select a special color plain image **P** = [1, 2, 3, 4, 5, 6, 7, 8, 9, 10, 11, 12] and obtain the corresponding cipher image **C**′ = [240, 44, 45, 220, 207, 217, 89, **211, 132, 239, 53, 58**] by using Pak's encryption machine. Then we can obtain its intermediary ciphertext array **C** according to Equation (5) as **C** = [**211, 132, 239, 53, 58**, 240, 44, 45, 220, 207, 217, 89]. Then we can obtain the permutated pixel array **P**′ from **D**′ by using Equation (4), that is, **P**′ = [10, 8, 2, 3, 9, 11, 4, 5, 12, 6, 7, 1]. By comparing **P** and **P**′, the permutation array **X**′ is recovered as **X**′ = [10, 8, 2, 3, 9, 11, 4, 5, 12, 6, 7, 1].

For the target ciphertext array C′ of Equation (24), we obtain its intermediary cipher pixel array **C** according to Equation (5) as **C** = [197, 162, 215, 51, **22**, 246, 16, 1, 6, 37, 3, 137]. Then we obtain the permutated pixel array **P**′ from **D**′ by using Equation (4), that is, **P**′ = [32, 24, 12, 13, 31, 33, 14, 21, 34, 22, 23, 11]. Finally, according to **X**′, **P** is recovered as **P** = [11, 12, 13, 14, 21, 22, 23, 24, 31, 32, 33, 34], which coincides with the original plain image array of Equation (23).

Through the two examples, our attack attains demonstration. Therefore, Pak's encryption scheme cannot resist the chosen-plaintext attacks and the security of the algorithm is not high enough.

4. The Improved Cryptosystem

In Pak's encryption scheme, the diffusion array **D**′ and the permutation position array **X**′ are used separately in the diffusion and permutation stage. Accordingly, the diffusion array **D**′ and the permutation position array **X**′ are easily deciphered separately by the attackers. This is a weakness of Pak's encryption scheme. In Wang's improved encryption scheme, a parameter E determined by the plaintext image is introduced. In order to obtain the value of the E parameter, it is necessary to calculate the average value of all the pixels of the image, which obviously increases the time overhead of the algorithm. In addition, the linear transformation operation of Wang's algorithm is changed to the binary shift operation to each pixel, which makes encryption speed very slow.

Our improved algorithm retains the advantages of the speed of the original algorithm and overcomes its shortcomings. It includes two rounds of synchronous operations of diffusion and confusion. Two diffusion arrays **D**′ and **D** are generated by using the chaotic sequence **X** and the previously encrypted pixel value. **D**′ and **D** are used to encrypt the image pixels respectively in the two rounds of synchronous operation.

4.1. Encryption Process

Step 1: Input the secret parameters $\{x_0, u, k, N_0, C_0\}$ and the color image **PI** with the size of $M \times N \times 3$, and **PI** is reshaped to a one-dimensional grayscale image array **P** = $[p(1), p(2), \ldots, p(L)]$, where $L = M \times N \times 3$.

Step 2: By using the parameters of $\{x_0, u, k, N_0\}$, iterate the new chaotic Sine-Sine system $(L + N_0)$ times and abandon the front N_0 elements to make the chaotic sequence **X** = $[x(1), x(2), \ldots, x(L)]$.

Step 3: Get the permutation position matrix **X**′ = $[x'(1), x'(2), \ldots, x'(L)]$ by sorting the chaotic sequence **X** in ascending order. Where, $x'(i)$ are integers ranging from 1 to L, $i = 1, 2, \ldots, L$.

Step 4: Perform the permutation and diffusion operations on array **P** simultaneously and obtain the temporary cipher image pixel array $\mathbf{C}' = [c'(1), c'(2), \ldots, c'(L)]$ as

$$\begin{cases} d'(1) = \mathrm{mod}(\mathrm{floor}(\frac{x(1)+C_0/256}{2} \times 10^{10}), 256) \\ c'(1) = \mathrm{mod}(p(x'(1)) + d'(1) + C_0, 256) \end{cases} \tag{25}$$

$$\begin{cases} d'(i) = \mathrm{mod}(\mathrm{floor}(\frac{x(i)+c'(i-1)/256}{2} \times 10^{10}), 256), \\ c'(i) = \mathrm{mod}(p(x'(i)) + d'(i) + c'(i-1), 256), i > 1. \end{cases} \tag{26}$$

where, $\mathbf{D}' = [d'(1), d'(2), \ldots, d'(L)]$ is the first diffusion array.

Step 5: Obtain the final cipher image pixel array $\mathbf{C} = [c(1), c(2), \ldots, c(L)]$ from the second diffusion array **D**, permutation position matrix $\mathbf{X}'$ and the temporary cipher image pixel array $\mathbf{C}'$ as

$$\begin{cases} d(1) = \mathrm{mod}(\mathrm{floor}(\frac{x(1)+c'(L)/256}{2} \times 10^{10}), 256) \\ c(1) = \mathrm{mod}(c'(1) + d(1) + x'(1) + c'(L), 256) \end{cases} \tag{27}$$

$$\begin{cases} d(i) = \mathrm{mod}(\mathrm{floor}(\frac{x(i)+c(i-1)/256}{2} \times 10^{10}), 256), \\ c(i) = \mathrm{mod}(c'(i) + d(i) + x'(i) + c(i-1), 256), i > 1. \end{cases} \tag{28}$$

where, $\mathbf{D} = [d(1), d(2), \ldots, d(L)]$ is the second diffusion array.

Step 6: Transform the 1D vector **C** into a 3D matrix with a size of $M \times N \times 3$, then the ciphered color image **CI** is obtained.

4.2. Decryption Process

To decrypt the cipher image **CI** with the secret keys $\{x_0, u, k, N_0, C_0\}$, the following decryption operations can be executed.

Step 1: Transform the 3D matrix **CI** into a gray scale image pixel sequence **C**.

Step 2: Similar to Step 2 of the encryption process, generate the chaotic sequence $\mathbf{X} = [x(1), x(2), \ldots, x(L)]$.

Step 3: Similar to Step 3 of the encryption process, get the permutation position matrix $\mathbf{X}' = [x'(1), x'(2), \ldots, x'(L)]$ by sorting **X**.

Step 4: Obtain the temporary cipher image pixel array $\mathbf{C}' = [c'(1), c'(2), \ldots, c'(L)]$ as

$$\begin{cases} d(i) = \mathrm{mod}(\mathrm{floor}(\frac{x(i)+c(i-1)/256}{2} \times 10^{10}), 256), \\ c'(i) = \mathrm{mod}(c(i) - d(i) - x'(i) - c(i-1), 256), i > 1. \end{cases} \tag{29}$$

$$\begin{cases} d(1) = \mathrm{mod}(\mathrm{floor}(\frac{x(1)+c'(L)/256}{2} \times 10^{10}), 256) \\ c'(1) = \mathrm{mod}(c(1) - d(1) - x'(1) - c'(L), 256) \end{cases}. \tag{30}$$

Step 5: Obtain the recovered plain image pixel array $\mathbf{P} = [p(1), p(2), \ldots, p(L)]$ as

$$\begin{cases} d'(1) = \mathrm{mod}(\mathrm{floor}(\frac{x(1)+C_0/256}{2} \times 10^{10}), 256) \\ p(x'(1)) = \mathrm{mod}(c'(1) - d'(1) - C_0, 256) \end{cases} \tag{31}$$

$$\begin{cases} d'(i) = \mathrm{mod}(\mathrm{floor}(\frac{x(i)+c'(i-1)/256}{2} \times 10^{10}), 256), \\ p(x'(i)) = \mathrm{mod}(c'(i) - d'(i) - c'(i-1), 256), i > 1. \end{cases} \tag{32}$$

Step 6: Transform **P** into a 3D matrix, and the decrypted color image **PI** is obtained.

5. Tests and Analysis for the Improved Cryptosystem

To examine the performance of the improved cryptosystem, we carried out a simulation experiment. The secret keys were set as (x_0 = 0.4563, u = 5.4321, k = 14, N_0 = 1000, C_0 = 98). The encryption and decryption algorithms were run on the platform Matlab R2016b in a computer with 3.3 GHz CPU, 4 GB memory and a 64 bit Microsoft Windows 7 operating system. The plain image used in the experiments was the color image lena. Figure 6 shows the original plain image and its cipher image encrypted by the improved scheme. The results reveal that the improved scheme has reliable encryption and decryption effect.

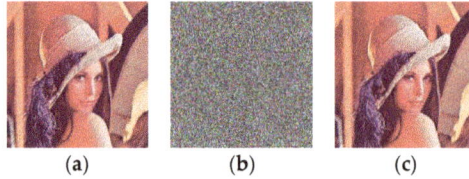

(a) (b) (c)

Figure 6. The encryption and decryption effect of the improved scheme. (**a**) The plain image; (**b**) the cipher image; (**c**) the decrypted image.

5.1. Resistance to Chosen-Plaintext Attacks

In our improved scheme, the diffusion matrices $\mathbf{D}'$ and $\mathbf{D}$ are related to the temporary and final ciphertext image, which is evident from Equations (25)–(28). It means that images with different contents are encrypted with different diffusion matrices. Furthermore, by using two rounds of diffusion processes, the change of the pixel value at any position in the image will affect all cipher pixel values. Even if the opponent cracked the key streams $\mathbf{D}'$ and $\mathbf{D}$ with some specially selected plain images, the key streams $\mathbf{D}'$ and $\mathbf{D}$ cannot be used to decrypt the target cipher image because the key streams of the target cipher image are different from the cracked key streams. Moreover, it is difficult to decipher the key streams $\mathbf{D}'$ and $\mathbf{D}$ directly by using chosen-plaintext attacks. Therefore, the improved scheme can well resist the chosen-plaintext attacks.

5.2. Key Space Analyses

In order to resist a brute-force attack, a cryptographic system must have enough large key space. In our improved cryptosystem, the secret keys include: x_0, u, k, N_0, C_0, so its key space is 2^{128}, which is the same as those in Reference [60]. Under the current computing power, the key space is large enough to resist a brute-force attack. The size of the key space depends not only on the number of keys but also on the number of possible values for each key. The problem of numerical chaotic systems is that the finite precision of the machines (e.g., computers) leads to performance degradation [63–66], such as the key space is reduced, some weak keys appear, and the randomness of the sequence is reduced. In order to identify and avoid weak keys, we need to calculate the Lyaponuv exponents of chaotic systems or plot the phase space trajectories of the system.

5.3. Statistical Analysis

5.3.1. Histogram Analysis

An image histogram displays the distribution of the values of its pixels and provides some statistical information about the image. The histograms of each component of the color lena image and its cipher image are shown in Figure 7. The experimental results in Figure 7 show objectively the statistical distribution of plaintext and ciphertext pixels. The histogram of the cipher image shows that the pixel distribution in the cipher image is very uniform, which means that our improved algorithm has excellent performance in resisting statistical attacks.

The variance of a histogram can quantitatively describe the distribution of pixel values, which is calculated by [54]

$$\mathrm{var}(\mathbf{Z}) = \frac{1}{n^2}\sum_{i=1}^{n}\sum_{j=1}^{n}\frac{1}{2}(z_i - z_j)^2. \tag{33}$$

where $\mathbf{Z}$ is a vector and $\mathbf{Z} = \{z_1, z_2, \ldots, z_{256}\}$, z_i and z_j are the numbers of pixels with gray values equal to i and j respectively. The lower value of variance indicates the higher uniformity of ciphered images.

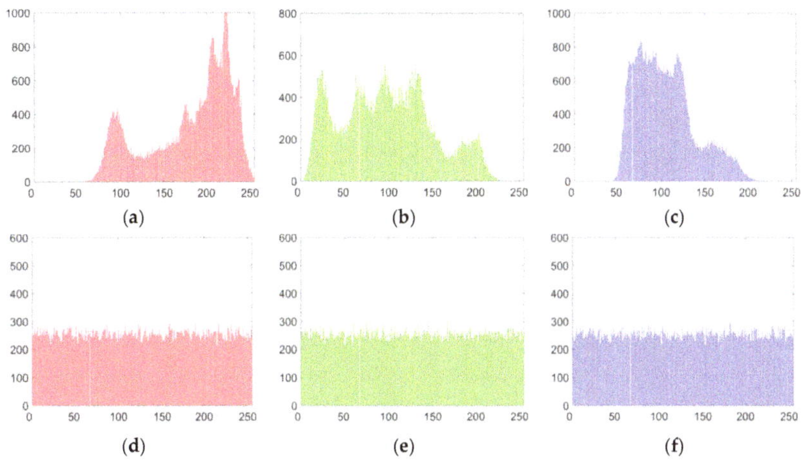

Figure 7. Encryption results for lena. (**a**) The histograms of R component of Figure 6a; (**b**) the histograms of G component of Figure 6a; (**c**) the histograms of B component of Figure 6a; (**d**) the histograms of R component of Figure 6b; (**e**) the histograms of G component of Figure 6b; (**f**) the histograms of B component of Figure 6b.

In the experimental tests, the variances of the histograms of the lena plain image (size of 256 × 256 × 3) and its cipher image were calculated by using Equation (33). The results obtained using two different algorithms are listed in Table 1. From Table 1, one can see that the average variance of the cipher image lena obtained with the proposed improved algorithm is 241.4141, which is much less than that of Wang's algorithm [61]. Thus, our improved algorithm has better performance in resisting statistical attacks.

Table 1. Variances of the histograms of the Lena image.

Channel	Plain Image	Cipher Image [61]	Cipher Image
R	63,888.1328	527.3242	244.6797
G	28,546.0078	504.7522	239.7656
B	86,487.8906	501.6874	239.7969
Average	57,516.9492	511.2546	241.4141

5.3.2. Correlation of Two Adjacent Pixels

Adjacent pixels in images usually have a strong correlation. A good encryption algorithm should break the correlation of adjacent pixels in an image. In order to directly describe the correlation of adjacent pixels in an image, based on 5000 randomly selected pairs of pixels (in horizontal, vertical and diagonal directions), the correlation distribution graphs of the lena plain image and its corresponding cipher image are drawn in Figures 8 and 9. It can be seen that the adjacent pixels in three directions in the plain image have a strong correlation, while those in the cipher image have almost no correlation and it is a random pattern. The results mean that our improved scheme has greatly eliminated the correlation of adjacent pixels.

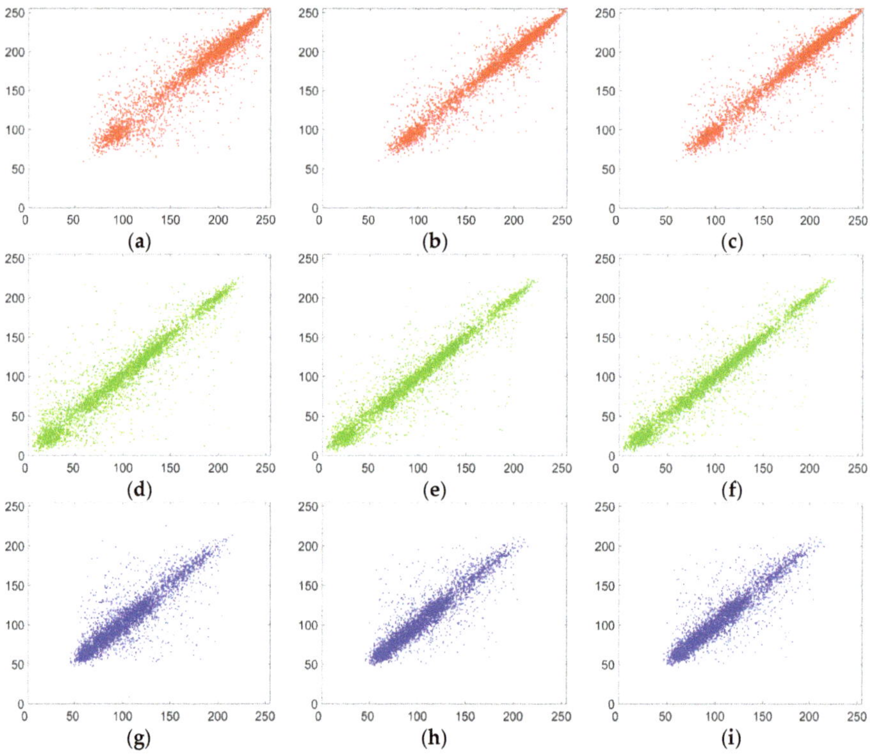

Figure 8. Correlation analysis of the plain image. (**a**) Horizontal correlation in R channel; (**b**) vertical correlation in R channel; (**c**) diagonal correlation in R channel; (**d**) horizontal correlation in G channel; (**e**) vertical correlation in G channel; (**f**) diagonal correlation in G channel; (**g**) horizontal correlation in B channel; (**h**) vertical correlation in B channel; (**i**) diagonal correlation in B channel.

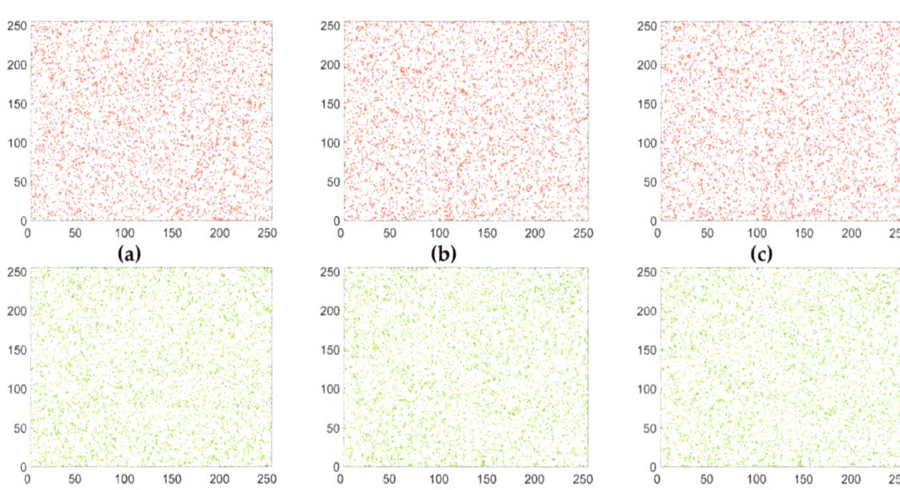

Figure 9. *Cont.*

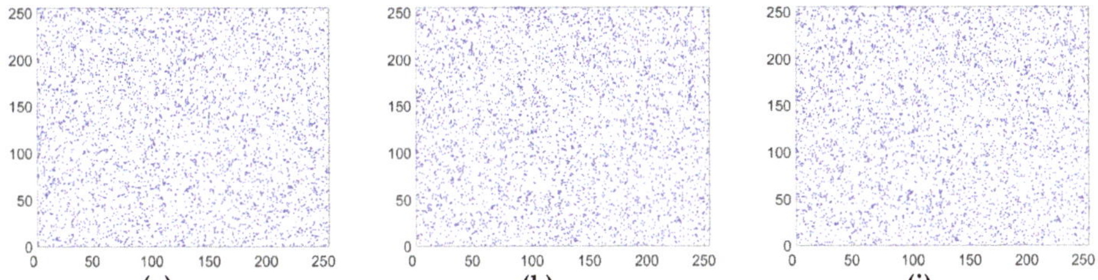

Figure 9. Correlation analysis of the corresponding cipher image. (**a**) Horizontal correlation in R channel; (**b**) vertical correlation in R channel; (**c**) diagonal correlation in R channel; (**d**) horizontal correlation in G channel; (**e**) vertical correlation in G channel; (**f**) diagonal correlation in G channel; (**g**) horizontal correlation in B channel; (**h**) vertical correlation in B channel; (**i**) diagonal correlation in B channel.

In order to quantitatively depict the correlation of adjacent pixels of an image, we introduce correlation coefficient index r_{XY}, which is calculated as follows:

$$r_{XY} = \text{cov}(X,Y)/\sqrt{D(X)}\sqrt{D(Y)} \tag{34}$$

$$E(X) = \frac{1}{N}\sum_{i=1}^{N} x_i \tag{35}$$

$$D(X) = \frac{1}{N}\sum_{i=1}^{N}(x_i - E(X))^2 \tag{36}$$

$$\text{cov}(X,Y) = \frac{1}{N}\sum_{i=1}^{N}(x_i - E(X))(y_i - E(Y)). \tag{37}$$

where X and Y are gray-scale values of two adjacent pixels in the images. For the color lena image, the correlation coefficients of adjacent pixels in R component of plaintext image and R component of ciphertext image were calculated respectively. The results are listed in Table 2. From Table 2, we can see that the correlation coefficients of adjacent pixels in R component of plaintext image are close to 1 while those of the cipher image are close to 0. The experimental results also show that our improved algorithm has smaller absolute values of correlation coefficient than Wang's algorithm in the vertical and diagonal directions and Pak's algorithm in all three directions.

Table 2. Correlation coefficients of the plain image and cipher images of lena in the R channel.

Directions	**Plain Image**	**Cipher Image**		
		R	**Reference [61]**	**Ours**
H	0.9567	−0.0026	0.00037	0.00063
V	0.9239	−0.0038	−0.00540	−0.00052
D	0.8888	0.0017	0.00166	−0.00012

5.3.3. Sensitivity Analysis

In order to resist differential attacks, the algorithm must be sensitive to the secret keys and plain images. To measure the sensitivity of an algorithm to tiny changes in key or plain image, we cite two metrics. One is the number of pixel changing rate (NPCR), another is the unified averaged changed intensity (UACI). The definitions of NPCR and UACI are

$$\mathrm{NPCR} = \frac{1}{m \times n}\sum_{i=1}^{m}\sum_{j=1}^{n}\delta(i,j) \times 100\%, \tag{38}$$

$$\mathrm{UACI} = \frac{1}{m \times n}\left(\sum_{i=1}^{m}\sum_{j=1}^{n}\frac{|c_1(i,j) - c_2(i,j)|}{255}\right) \times 100\%. \tag{39}$$

where m, n represent the pixel rows and columns of an image, respectively. Here, $\mathbf{C}_1 = [\,c_1(i,j)]$ and $\mathbf{C}_2 = [c_2(i,j)]$ express two encrypted images corresponding to two security keys or two plain images, and $\delta(i,j)$ is computed by

$$\delta(i,j) = \begin{cases} 1, \ if\ c_1(i,j) \neq c_2(i,j), \\ 0, \ if\ c_1(i,j) = c_2(i,j). \end{cases} \tag{40}$$

The desired value of NPCR is 1 and the desired value of UACI is 0.3346 [54].

To measure the sensitivity of our improved algorithm for the plain image, the color lena image (size 256 × 256 × 3) is chosen as the plain image one, and the plain image two is obtained by changing only one pixel of the plain image one. Then, two encrypted images are obtained by executing the improved encryption algorithm with the same secret keys, respectively. NPCR and UACI values are computed with two cipher images, and the results are listed in Table 3. The results indicate that our improved encryption algorithm is very sensitive to the plain image.

Table 3. Values of NPCR and UACI of Lena cipher images.

Channel	NPCR [61]	NPCR	UACI [61]	UACI
R	0.996413	1	0.334801	0.3341
G	0.996328	1	0.334791	0.3363
B	0.996250	0.9974	0.334558	0.3346

To measure the sensitivity of the improved algorithm to the secret keys, two different keys with a tiny difference are used to encrypt the same plain image lena and the two cipher images, $\mathbf{C}_1$ and $\mathbf{C}_2$, are obtained. The tiny change (10^{-14}) is introduced to one of the secret keys (x_0, u) while keeping all the others unchanged. Similarly, k is changed to $k + 1$, N_0 is changed to $N_0 + 1$, C_0 is changed to $C_0 + 1$, while keeping all the others unchanged. The NPCR and UACI of the cipher images $\mathbf{C}_1$ and $\mathbf{C}_2$ are given in Tables 4 and 5. The experimental results indicate that our improved algorithm is very sensitive to any slight change in each secret key.

Table 4. NPCR of the improved algorithm with a slight change in the secret keys.

Channel	$x_0 + 10^{-14}$	$u + 10^{-14}$	$k + 1$	$N_0 + 1$	$C_0 + 1$
R	0.9961	0.9960	0.9958	0.9959	0.9961
G	0.9959	0.9964	0.9962	0.9962	0.9961
B	0.9963	0.9961	0.9960	0.9964	0.9961

Table 5. UACI of the improved algorithm with a slight change in the secret keys.

Channels	$x_0 + 10^{-14}$	$u + 10^{-14}$	$k + 1$	$N_0 + 1$	$C_0 + 1$
R	0.3334	0.3356	0.3353	0.3370	0.3344
G	0.3350	0.3355	0.9962	0.3348	0.3348
B	0.3343	0.3355	0.3352	0.3340	0.3337

5.3.4. Information Entropy Analysis

Image information entropy is an important way to measure the randomness of the pixel distribution. Let I be an image and its information entropy can be calculated as:

$$H(I) = -\sum_{i=0}^{2^n-1} P(I_i)\log_2[P(I_i)], \quad (41)$$

where $P(I_i)$ denotes the occurrence probability of gray level i, $I_i = i$, and $i = 0, 1, 2, \ldots, 2^n$. Here, 2^n is the number of grayscale levels of an image. If $P(I_i) = 1/2^n$, then the image is completely random. For an image with 256 gray-scales, $n = 8$ and the image has 2^8 grayscale levels, so the ideal value of information entropy is 8. For an encrypted image, the closer the entropy is to 8, the closer the image is to a randomly distributed image. We experimentally tested the information entropy of the color lena ciphertext images obtained by three kinds of encryption algorithms. The results of the information entropy corresponding to the R, G and B channels are listed in Table 6. From Table 6, one can see that all the entropy values are significantly closer to 8, so the randomness is satisfactory. Among these three algorithms, our improved algorithm has the largest average entropy value. Hence, our improved encryption scheme is more capable of resisting information entropy-based attacks.

Table 6. Entropies of the encrypted lena image by three encryption schemes.

Channels	Reference [60]	Reference [61]	Ours
R	7.9971	7.9970	7.9973
G	7.9972	7.9965	7.9973
B	7.9974	7.9973	7.9974
Average	7.9972	7.9969	7.9973

5.3.5. Cropping and Noise Attack

To test the performance of our improved scheme in resisting data loss and noise attacks. The encrypted lena image (Figure 10a) was attacked by a data cut with a size of 64 × 64 (Figure 10b) and a 3% "salt & pepper" noise attack (Figure 10c), respectively. Then, these cipher images were decrypted respectively and the results of the decryption are given in Figure 10d–f. The results indicate that our improved scheme can resist cutting and noise pollution attacks.

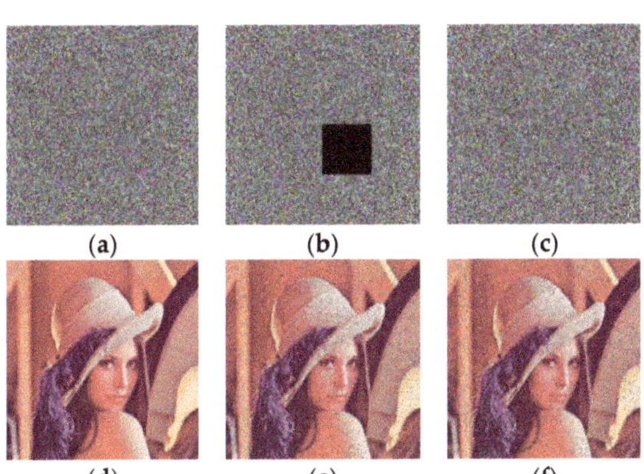

Figure 10. Data loss and noise attack. (**a**) The original cipher image; (**b**) the cipher images with data loss; (**c**) the cipher image added with 3% "salt & pepper" noise; (**d**) the decrypted image of (**a**); (**e**) the decrypted image of (**b**); (**f**) the decrypted image of (**c**).

5.4. Analysis of Speed

A practical encryption algorithm should be efficient in terms of encryption speed. To test the encryption speed of the improved scheme, three RGB color images with different size have been used for the encryption. The simulation experiments were run on a desktop PC with Intel(R) Core i5-4590 3.30 GHz CPU, 4 GB RAM and 500 GB hard disk. The operating system was 64 bits Microsoft Windows 7 and the computational platform was Matlab R2016b. The average encryption/decryption time taken by Pak's algorithm, Wang's algorithm and our improved algorithm for processing the images with different size are shown in Table 7. The results show that our algorithm has the fastest speed. This is because our encryption algorithm has abandoned binary XOR operations.

Table 7. The time cost tests.

Image size	Reference [60]	Reference [61]	Ours
256×256	0.5693 s	8.2328 s	0.3873 s
512×512	2.2340 s	32.7673 s	1.5145 s
1024×1024	8.9055 s	131.6625 s	6.0163 s

6. Conclusions

In this paper, an improved cryptanalysis on a color image cryptosystem is presented. It has been shown that the equivalent secret key and all the unknown parameters of the cryptosystem can be recovered by our chosen-plaintext attack algorithm. Furthermore, based on the analysis of defects in the original cryptosystem, an improved color image encryption scheme is proposed. The contributions of this paper include two aspects: First, a more complete and efficient method to comprehensively crack Pak's encryption scheme is proposed, which further enriches the research of cryptanalysis. The validity and correctness of the cryptanalysis algorithm were verified by theoretical analysis and experimental results. Second, a new color image encryption algorithm with a higher security and a higher encryption efficiency is proposed. In the new encryption scheme, the generation of diffusion arrays depends on the content of the image itself and the permutation position array. In the process of diffusion, two effects of ciphertext feedback and pixel scrambling are also implemented simultaneously. Using these methods, the security of the cryptosystem is enhanced. Experimental results and security analysis demonstrate that the improved cryptosystem can achieve a satisfactory security level after two rounds of diffusion encryption.

Looking to the future in image encryption field, some new research directions are worth considering, such as efficient image encryption technology in the resource-constrained mobile social network [67] or sensor network communication environment [68]. Another interesting form of encryption is searchable encryption [69], which is a very promising direction in the field of cloud computing.

Author Contributions: Conceptualization, C.Z. and G.W.; methodology, C.Z.; software, C.Z.; validation, C.Z., G.W. and K.S.; formal analysis, C.Z.; investigation, C.Z.; resources, C.Z.; data curation, K.S.; writing—original draft preparation, C.Z.; writing—review and editing, G.W.; visualization, K.S.; supervision, G.W.; project administration, K.S.; funding acquisition, G.W.

Funding: This research was funded by [the Open Project of Guangxi Colleges and Universities Key Laboratory of Complex System Optimization and Big Data Processing] grant number [No. 2016CSOBDP0103]; [the National Natural Science Foundation of China] grant number [Nos. 61472451 and 61632009].

Acknowledgments: The authors are thankful to the reviewers for their comments and suggestions to improve the quality of the manuscript.

Conflicts of Interest: The authors declare no conflict of interest.

References

1. Alvarez, G.; Li, S. Some basic cryptographic requirements for chaos-based cryptosystems. *Int. J. Bifurc. Chaos* **2006**, *16*, 2129–2151. [CrossRef]
2. Zanette, D.H. Generalized kolmogorov entropy in the dynamics of the multifractal generation. *Phys. A Stat. Mech. Appl.* **1996**, *223*, 87–98. [CrossRef]
3. Crutchfield, J.P.; Packard, N.H. Symbolic dynamics of noisy chaos. *Phys. D* **1983**, *7*, 201–223. [CrossRef]
4. Crutchfield, J.P.; Feldman, D.P. Regularities unseen, randomness observed: Levels of entropy convergence. *Chaos* **2003**, *13*, 25–54. [CrossRef] [PubMed]
5. Fridrich, J. Symmetric ciphers based on two-dimensional chaotic maps. *Int. J. Bifurc. Chaos* **1998**, *8*, 1259–1284. [CrossRef]
6. Zhang, Y.; Xiao, D. An image encryption scheme based on rotation matrix bit-level permutation and block diffusion. *Commun. Nonlinear Sci. Numer. Simul.* **2014**, *19*, 74–82. [CrossRef]
7. Zhang, Y.; Xiao, D. Double optical image encryption using discrete chirikov standard map and chaos-based fractional random transform. *Opt. Lasers Eng.* **2013**, *51*, 472–480. [CrossRef]
8. Gan, Z.H.; Chai, X.L.; Han, D.J.; Chen, Y.R. A chaotic image encryption algorithm based on 3-D bit-plane permutation. *Neural Comput. Appl.* **2018**, 1–20. [CrossRef]
9. Hu, G.; Xiao, D.; Zhang, Y.; Xiang, T. An efficient chaotic image cipher with dynamic lookup table driven bit-level permutation strategy. *Nonlinear Dyn.* **2016**, *87*, 1359–1375. [CrossRef]
10. Ye, G.; Zhao, H.; Chai, H. Chaotic image encryption algorithm using wave-line permutation and block diffusion. *Nonlinear Dyn.* **2016**, *83*, 2067–2077. [CrossRef]
11. Abd-El-Hafiz, S.K.; AbdElHaleem, S.H.; Radwan, A.G. Novel permutation measures for image encryption algorithms. *Opt. Lasers Eng.* **2016**, *85*, 72–83. [CrossRef]
12. Li, Y.; Wang, C.; Chen, H. A hyper-chaos-based image encryption algorithm using pixel-level permutation and bit-level permutation. *Opt. Lasers Eng.* **2017**, *90*, 238–246. [CrossRef]
13. Zhang, Y.; Xiao, D.; Shu, Y.; Li, J. A novel image encryption scheme based on a linear hyperbolic chaotic system of partial differential equations. *Signal Process. Image Commun.* **2013**, *28*, 292–300. [CrossRef]
14. Wang, X.; Liu, C.; Zhang, H. An effective and fast image encryption algorithm based on chaos and interweaving of ranks. *Nonlinear Dyn.* **2016**, *84*, 1595–1607. [CrossRef]
15. Xu, L.; Gou, X.; Li, Z.; Li, J. A novel chaotic image encryption algorithm using block scrambling and dynamic index based diffusion. *Opt. Lasers Eng.* **2017**, *91*, 41–52. [CrossRef]
16. Hua, Z.; Yi, S.; Zhou, Y. Medical image encryption using high-speed scrambling and pixel adaptive diffusion. *Signal Process.* **2018**, *144*, 134–144. [CrossRef]
17. Huang, H.; He, X.; Xiang, Y.; Wen, W.; Zhang, Y. A compression-diffusion-permutation strategy for securing image. *Signal Process.* **2018**, *150*, 183–190. [CrossRef]
18. Cao, C.; Sun, K.; Liu, W. A novel bit-level image encryption algorithm based on 2D-LICM hyperchaotic map. *Signal Process.* **2018**, *143*, 122–133. [CrossRef]
19. Chai, X. An image encryption algorithm based on bit level brownian motion and new chaotic systems. *Multimed. Tools Appl.* **2017**, *76*, 1159–1175. [CrossRef]
20. Hua, Z.; Jin, F.; Xu, B.; Huang, H. 2D Logistic-Sine-coupling map for image encryption. *Signal Process.* **2018**, *149*, 148–161. [CrossRef]
21. Hua, Z.; Zhou, Y. Image encryption using 2D Logistic-adjusted-Sine map. *Inf. Sci.* **2016**, *339*, 237–253. [CrossRef]
22. Kaur, M.; Kumar, V. Efficient image encryption method based on improved lorenz chaotic system. *Electron. Lett.* **2018**, *54*, 562–564. [CrossRef]
23. Liu, J.; Yang, D.; Zhou, H.; Chen, S. A digital image encryption algorithm based on bit-planes and an improved logistic map. *Multimed. Tools Appl.* **2018**, *77*, 10217–10233. [CrossRef]
24. Zhu, C. A novel image encryption scheme based on improved hyperchaotic sequences. *Opt. Commun.* **2012**, *285*, 29–37. [CrossRef]
25. Zhang, Y.; Tang, Y. A plaintext-related image encryption algorithm based on chaos. *Multimed. Tools Appl.* **2018**, *77*, 6647–6669. [CrossRef]
26. Ye, G.; Huang, X. A secure image encryption algorithm based on chaotic maps and SHA-3. *Secur. Commun. Netw.* **2016**, *9*, 2015–2023. [CrossRef]

27. Wu, X.; Kan, H.; Kurths, J. A new color image encryption scheme based on DNA sequences and multiple improved 1D chaotic maps. *Appl. Soft Comput.* **2015**, *37*, 24–39. [CrossRef]
28. Wang, X.; Zhu, X.; Wu, X.; Zhang, Y. Image encryption algorithm based on multiple mixed hash functions and cyclic shift. *Opt. Lasers Eng.* **2017**, *107*, 370–379. [CrossRef]
29. Chai, X.; Gan, Z.; Zhang, M. A fast chaos-based image encryption scheme with a novel plain image-related swapping block permutation and block diffusion. *Multimed. Tools Appl.* **2016**, *76*, 15561–15585. [CrossRef]
30. Chai, X.; Chen, Y.; Broyde, L. A novel chaos-based image encryption algorithm using DNA sequence operations. *Opt. Lasers Eng.* **2017**, *88*, 197–213. [CrossRef]
31. Guesmi, R.; Farah, M.A.B.; Kachouri, A.; Samet, M. A novel chaos-based image encryption using DNA sequence operation and secure hash algorithm SHA-2. *Nonlinear Dyn.* **2016**, *83*, 1123–1136. [CrossRef]
32. Hu, T.; Liu, Y.; Gong, L.-H.; Guo, S.-F.; Yuan, H.-M. Chaotic image cryptosystem using DNA deletion and DNA insertion. *Signal Process.* **2017**, *134*, 234–243. [CrossRef]
33. Wang, X.; Liu, C. A novel and effective image encryption algorithm based on chaos and DNA encoding. *Multimed. Tools Appl.* **2016**, *76*, 6229–6245. [CrossRef]
34. Wang, X.-Y.; Li, P.; Zhang, Y.-Q.; Liu, L.-Y.; Zhang, H.; Wang, X. A novel color image encryption scheme using DNA permutation based on the lorenz system. *Multimed. Tools Appl.* **2017**, *77*, 6243–6265. [CrossRef]
35. Wang, X.-Y.; Zhang, Y.-Q.; Bao, X.-M. A novel chaotic image encryption scheme using DNA sequence operations. *Opt. Lasers Eng.* **2015**, *73*, 53–61. [CrossRef]
36. Zhang, L.-M.; Sun, K.-H.; Liu, W.-H.; He, S.-B. A novel color image encryption scheme using fractional-order hyperchaotic system and DNA sequence operations. *Chin. Phys. B* **2017**, *26*, 100504. [CrossRef]
37. Zhang, D.; Liao, X.; Yang, B.; Zhang, Y. A fast and efficient approach to color-image encryption based on compressive sensing and fractional fourier transform. *Multimed. Tools Appl.* **2018**, *77*, 2191–2208. [CrossRef]
38. Wang, X.; Wang, Q.; Zhang, Y. A fast image algorithm based on rows and columns switch. *Nonlinear Dyn.* **2015**, *79*, 1141–1149. [CrossRef]
39. Tong, X.-J.; Zhang, M.; Wang, Z.; Liu, Y.; Xu, H.; Ma, J. A fast encryption algorithm of color image based on four-dimensional chaotic system. *J. Vis. Commun. Image Represent.* **2015**, *33*, 219–234. [CrossRef]
40. Liu, H.; Kadir, A.; Sun, X. Chaos-based fast colour image encryption scheme with true random number keys from environmental noise. *IET Image Process.* **2017**, *11*, 324–332. [CrossRef]
41. Liu, W.; Sun, K.; Zhu, C. A fast image encryption algorithm based on chaotic map. *Opt. Lasers Eng.* **2016**, *84*, 26–36. [CrossRef]
42. Bibi, N.; Farwa, S.; Muhammad, N.; Jahngir, A.; Usman, M. A novel encryption scheme for high-contrast image data in the fresnelet domain. *PLoS ONE* **2018**, *13*, e0194343.
43. Farwa, S.; Muhammad, N.; Shah, T.; Ahmad, S. A novel image encryption based on algebraic s-box and arnold transform. *3D Res.* **2017**, *8*, 26. [CrossRef]
44. Farwa, S.; Shah, T.; Muhammad, N.; Bibi, N.; Jahangir, A.; Arshad, S. An image encryption technique based on chaotic s-box and arnold transform. *Int. J. Adv. Comput. Sci. Appl.* **2017**, *8*, 360–364. [CrossRef]
45. Martin, K.; Lukac, R.; Plataniotis, K.N. Efficient encryption of wavelet-based coded color images. *Pattern Recognit.* **2005**, *38*, 1111–1115. [CrossRef]
46. Shahed, M.A. Wavelet based fast technique for images encryption. *Basrah J. Sci.* **2007**, *25*, 126–141.
47. Gao, H.J.; Zhang, Y.S.; Liang, S.Y.; Li, D.Q. A new chaotic algorithm for image encryption. *Chaos Solitons Fractals* **2006**, *29*, 393–399. [CrossRef]
48. Guariglia, E. Entropy and fractal antennas. *Entropy* **2016**, *18*, 84. [CrossRef]
49. Guariglia, E. Harmonic sierpinski gasket and applications. *Entropy* **2018**, *20*, 714. [CrossRef]
50. Li, C.; Lin, D.; Lu, J. Cryptanalyzing an image-scrambling encryption algorithm of pixel bits. *IEEE Multimed.* **2017**, *24*, 64–71. [CrossRef]
51. Li, C.; Liu, Y.; Xie, T.; Chen, M.Z.Q. Breaking a novel image encryption scheme based on improved hyperchaotic sequences. *Nonlinear Dyn.* **2013**, *73*, 2083–2089. [CrossRef]
52. Wang, X.; Luan, D.; Bao, X. Cryptanalysis of an image encryption algorithm using chebyshev generator. *Digit. Signal Prog.* **2014**, *25*, 244–247. [CrossRef]
53. Wu, J.; Liao, X.; Yang, B. Cryptanalysis and enhancements of image encryption based on three-dimensional bit matrix permutation. *Signal Process.* **2018**, *142*, 292–300. [CrossRef]
54. Zhu, C.; Sun, K. Cryptanalyzing and improving a novel color image encryption algorithm using rt-enhanced chaotic tent maps. *IEEE Access* **2018**, *6*, 18759–18770. [CrossRef]

55. Wu, X.; Zhu, B.; Hu, Y.; Ran, Y. A novel colour image encryption scheme using rectangular transform-enhanced chaotic tent maps. *IEEE Access* **2017**, *5*, 6429–6436. [CrossRef]
56. Zhang, W.; Yu, H.; Zhao, Y.L.; Zhu, Z.L. Image encryption based on three-dimensional bit matrix permutation. *Signal Process.* **2016**, *118*, 36–50. [CrossRef]
57. Sun, K.H.; He, S.B.; Yin, L.Z.; Duo, L.K. Application of fuzzyen algorithm to the analysis of complexity of chaotic sequence. *Acta Phys. Sin.* **2012**, 130507.
58. Sun, K.H.; He, S.B.; He, Y.; Yin, L.Z. Complexity analysis of chaotic pseudo-random sequences based on spectral entropy algorithm. *Acta Phys. Sin.* **2013**, *62*, 010501.
59. He, S.B.; Sun, K.H.; Zhu, C.X. Complexity analyses of multi-wing chaotic systems. *Chin. Phys. B* **2013**, 220–225. [CrossRef]
60. Pak, C.; Huang, L. A new color image encryption using combination of the 1d chaotic map. *Signal Process.* **2017**, *138*, 129–137. [CrossRef]
61. Wang, H.; Xiao, D.; Chen, X.; Huang, H. Cryptanalysis and enhancements of image encryption using combination of the 1d chaotic map. *Signal Process.* **2018**, *144*, 444–452. [CrossRef]
62. Chen, J.; Han, F.; Qian, W.; Yao, Y.-D.; Zhu, Z.L. Cryptanalysis and improvement in an image encryption scheme using combination of the 1d chaotic map. *Nonlinear Dyn.* **2018**, *93*, 2399–2413. [CrossRef]
63. Li, S.; Chen, G.; Mou, X. On the dynamical degradation of digital piecewise linear chaotic maps. *Int. J. Bifurc. Chaos* **2015**, *15*, 3119–3151. [CrossRef]
64. Li, S.; Chen, G.; Wong, K.-W.; Mou, X.; Cai, Y. Baptista-type chaotic cryptosystems: Problems and countermeasures. *Phys. Lett. A* **2004**, *332*, 368–375. [CrossRef]
65. Curiac, D.I.; Volosencu, C. Chaotic trajectory design for monitoring an arbitrary number of specified locations using points of interest. *Math. Probl. Eng.* **2012**, *2012*, 940276. [CrossRef]
66. Curiac, D.I.; Iercan, D.; Dragan, F.; Banias, O. Chaos-based cryptography: End of the road? In In Proceedings of the International Conference on Emerging Security Information, System and Technologies, Valencia, Spain, 14–20 October 2007; pp. 71–76.
67. Zhang, S.; Wang, G.; Liu, Q.; Abawajy, J.H. A trajectory privacy-preserving scheme based on query exchange in mobile social networks. *Soft Comput.* **2018**, *22*, 6121–6133. [CrossRef]
68. Bhuiyan, M.Z.A.; Wang, G.; Wu, J.; Cao, J.; Liu, X.; Wang, T. Dependable structural health monitoring using wireless sensor networks. *IEEE Trans. Dependable Secur.* **2017**, *14*, 363–376. [CrossRef]
69. Zhang, Q.; Liu, Q.; Wang, G. PRMS: A personalized mobile search over encrypted outsourced data. *IEEE Access* **2018**, *6*, 31541–31552. [CrossRef]

entropy

MDPI

Article

Blind Image Quality Assessment of Natural Scenes Based on Entropy Differences in the DCT Domain

Xiaohan Yang [1], Fan Li [1,*], Wei Zhang [2] and Lijun He [1]

[1] School of Electronic and Information Engineering, Xi'an Jiaotong University, Xi'an 710049, China; yangxiaohan@stu.xjtu.edu.cn (X.Y.); jzb2016125@mail.xjtu.edu.cn (L.H.)

[2] The State Key Laboratory of Integrated Services Networks, Xidian University, Xi'an 710071, China; wzhang@xidian.edu.cn

* Correspondence: lifan@mail.xjtu.edu.cn

Received: 29 October 2018; Accepted: 16 November 2018; Published: 17 November 2018

Abstract: Blind/no-reference image quality assessment is performed to accurately evaluate the perceptual quality of a distorted image without prior information from a reference image. In this paper, an effective blind image quality assessment approach based on entropy differences in the discrete cosine transform domain for natural images is proposed. Information entropy is an effective measure of the amount of information in an image. We find the discrete cosine transform coefficient distribution of distorted natural images shows a pulse-shape phenomenon, which directly affects the differences of entropy. Then, a Weibull model is used to fit the distributions of natural and distorted images. This is because the Weibull model sufficiently approximates the pulse-shape phenomenon as well as the sharp-peak and heavy-tail phenomena of natural scene statistics rules. Four features that are related to entropy differences and human visual system are extracted from the Weibull model for three scaling images. Image quality is assessed by the support vector regression method based on the extracted features. This blind Weibull statistics algorithm is thoroughly evaluated using three widely used databases: LIVE, TID2008, and CSIQ. The experimental results show that the performance of the proposed blind Weibull statistics method is highly consistent with that of human visual perception and greater than that of the state-of-the-art blind and full-reference image quality assessment methods in most cases.

Keywords: blind image quality assessment (BIQA); information entropy, natural scene statistics (NSS); Weibull statistics; discrete cosine transform (DCT)

---

1. Introduction

The human visual system (HVS) is important for perceiving the world. As an important medium of information transmission and communication, images play an increasingly vital role in human life. Since distortions can be introduced during image acquisition, compression, transmission and storage, the image quality assessment (IQA) method is widely studied for evaluating the influence of various distortions on perceived image quality [1,2].

In principal, subjective assessment is the most reliable way to evaluate the visual quality of images. However, this method is time-consuming, expensive, and impossible to implement in real-world systems. Therefore, objective assessment of image quality has gained growing attention in recent years. Depending on to what extent a reference image is used for quality assessment, existing objective IQA methods can be classified into three categories: full-reference (FR), reduced-reference (RR) and no-reference/blind (NR/B) methods. Accessing all or part of the reference image information is unrealistic in many circumstances [3–8], hence it has become increasingly important to develop effective blind IQA (BIQA) methods.

Many NR IQA metrics focus on assessing a specific type of visual artifact, such as blockiness artifacts [9], blur distortion [10], ringing distortion [11] and contrast distortion [12]. The main limitation is that the distortion type must be known in advance. However, generic NR-IQA metrics have recently become a research hotspot because of their general applicability.

According to the dependency on human opinion scores, the generic NR approaches can be roughly divided into two categories [13]: distance-based methods and learning-based methods. Distance-based methods express the image distortion as a simple distance between the model statistics of the pristine image and those of the distorted image [14–16]. For example, Saha et al. [16] proposed a completely training-free model based on the scale invariance of natural images.

Learning-based methods have attracted increasing attention with the development of artificial intelligence. The basic strategy is to learn a regression model that maps the image features directly to a quality score. Various regression methods, including support vector regression (SVR) [17], neural network [18–20], random forest regression [21] and deep learning framework [22,23], are widely used for model learning. More importantly, after pre-processing of image [24], image features, which are extracted for model learning, are directly related to the accuracy of the IQA. The codebook-based method [25] aims at extracting Gabor filter-based local features, which describe changes of texture information. Moreover, NSS-based methods are also widely used to extract features [26–33]. In [26,27], these methods use the difference of NSS histogram of natural and distorted images to extract image features. In [28–33], they aim to establish NSS model to extract features. The Laplace model [28], the Generalized Gaussian distribution (GGD) model [29,32,33], the generalized gamma model [30] and Gaussian scale mixture model [31] are widely used as NSS model to extract features in different domains. In addition, Ghadiyaram et al. combined histogram features and NSS model features to achieve good-quality predictions on authentically distorted images [34].

NSS model-based methods have achieved promising results. The NSS model assumes that natural images share certain statistical regularities and various distortions may change these statistics. Therefore, the NSS model is capable of fitting statistics of natural and distorted images. The GGD model is a typical NSS model that is widely studied and applied. The GGD model in the DCT domain is able to follow the heavy-tail and sharp-peak characteristics of natural images. By using the GGD features, the distorted image quality can be estimated. However, the GGD model has some shortcomings in fitting the statistics of distorted images because the distribution of DCT coefficients shows a pulse-shape phenomenon for distorted images, which is described as the rapid increase of discontinuity. The discontinuity is derived from the differences between the high- and low-frequency coefficients of distorted images. Thus, the pulse-shape phenomenon cannot be fitted by the GGD model, which leads to inaccurate quality assessments.

In this paper, an effective blind IQA approach of natural scenes related to entropy differences is developed in the DCT domain. The differences of entropy can be described by probability distribution in distorted images. We find the DCT coefficients' distribution of distorted images shows a pulse-shape phenomenon in addition to the heavy-tail and sharp-peak phenomena. Since the pulse-shape phenomenon is often neglected in NSS model, image structure cannot be fully presented by image entropy. Therefore, the performance of the IQA methods based on such NSS model can be affected to some extent. To this end, the Weibull model is proposed in this paper to overcome the under-fit caused by the pulse shape phenomenon of traditional GGD model. Furthermore, we prove that the Weibull model correlates well with the human visual perception. Based on the Weibull model, corresponding features are extracted in different scales and the prediction model is derived using the SVR method. Experimental results show that the Weibull statistics (BWS) method consistently outperforms the state-of-the-art NR and FR IQA methods over different image databases. Moreover, the BWS method is a generic image quality algorithm, which is applicable to multiple distortion types.

The novelty of our work lies in that we find the pulse-shape phenomenon when using existing GGD model to characterize image distortions. Then, we propose a useful Weibull model to overcome under-fit the pulse-shape phenomenon and extract features related to visual perception from the

Weibull model to evaluated image quality. Furthermore, the proposed method has the advantage of high prediction accuracy and high generalization ability.

The rest of the paper is organized as follows. Section 2 presents the NSS model based on entropy differences. Section 3 presents the proposed BWS algorithm in details. Section 4 evaluates the performance of the BWS algorithm from various aspects. Section 5 concludes the paper.

2. NSS Model Based on Entropy Differences

Information entropy indicates the amount of information contained within an image and the changes of entropy are highly sensitive to the degrees and types of image distortions. Our method utilizes distribution difference of DCT coefficients that directly affects entropy changes to assess image quality. As reported in the literature [35,36], natural images exhibit specific statistical regularities in spatial and frequency domains and are highly structured, and the statistical distribution remain approximately the same under scale and content changes. The characteristics of natural images are essential for IQA.

2.1. Under-Fitting Effect of GGD

The GGD model, as a typical NSS model, is widely used to fit distribution of AC coefficients in the DCT domain for natural images [29], and it can well simulate non-Gaussian behaviors, including sharp-peak and heavy-tail phenomena of the distribution of AC coefficients of natural images [35–37].

Figure 1 shows a natural image (i.e., "stream") included in the LIVE database [38], and two of its JPEG-compressed versions. The subjective qualities are scored by the Difference Mean Opinion Score (DMOS), which returned values of 0, 63.649 and 29.739 for the three images. A larger DMOS indicates lower visual quality. Figure 2 shows the distribution of AC coefficients for the corresponding images in Figure 1 and the GGD fitting curve. As shown in Figure 2a, the GGD model is capable of fitting the natural image, especially the sharp-peak and heavy-tail phenomena. For the distorted images, however, distinct deviations occur around the peak, as shown in Figure 2b,c. The underlying reason for the misfits is that the structure information of the distorted images has been changed including smoothness, texture, edge information. The number of the AC coefficients in the value of zero is increased rapidly, which triggers the pulse-shape. The pulse-shape phenomenon enhances along the increase of the distortion level. Thus, the GGD model fails at simulating the pulse-shape phenomenon, and it under-fits the distorted image distribution.

Figure 1. One natural image ("stream") and two of its JPEG-compressed versions in the LIVE database: (**a**) image with DMOS = 0; (**b**) image with DMOS = 63.649; and (**c**) image with DMOS = 29.739.

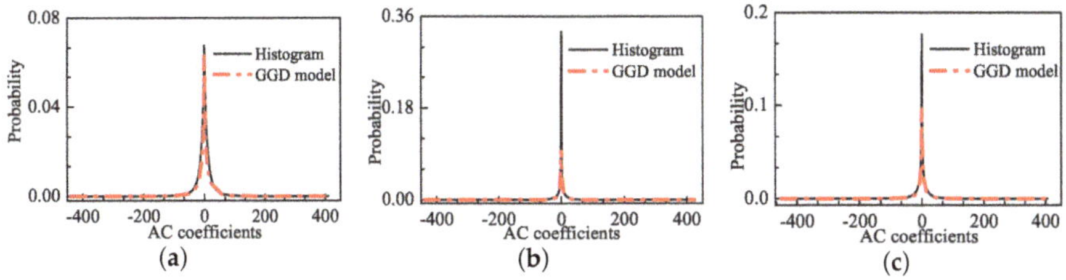

Figure 2. The distribution of AC coefficients for corresponding images and the GGD fitting curves: (**a**) image with DMOS = 0; (**b**) image with DMOS = 63.649; and (**c**) image with DMOS = 29.739.

Weibull Distribution Model

To well fit the distribution of AC coefficients for a distorted image, the Weibull distribution model is employed, which is given by [39]:

$$f_X(x) = \frac{a}{m}(\frac{x}{m})^{a-1}\exp[-(\frac{x}{m})^a] \qquad x > 0 \tag{1}$$

where a and m are the shape parameter and the scale parameter. The family of Weibull distributions includes the Exponential distribution ($a = 1$) and Rayleigh distribution ($a = 2$). When $a < 1$, $f_X(x)$ approaches infinity as x approaches zero. The characteristic can be used to describe the pulse-shape phenomenon for the distribution of AC coefficients. Figure 3 shows the change of Weibull distribution in different parameter settings. When m is fixed and a is less than 1, a larger shape parameter a corresponds to a slower change of $f_X(x)$. When a is fixed, a larger scale parameter m corresponds to a faster change of $f_X(x)$.

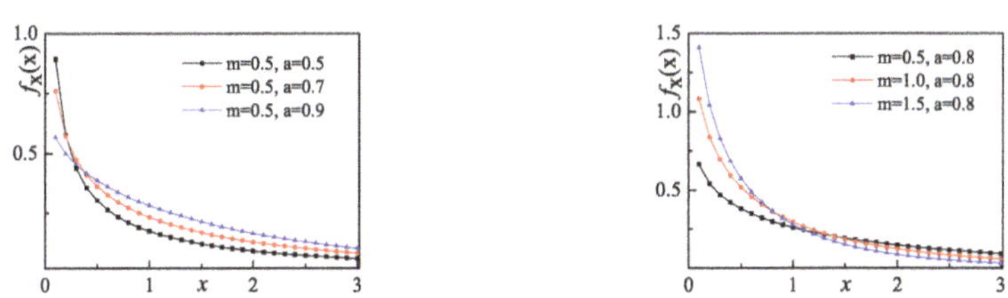

Figure 3. The change of Weibull distribution in different parameter settings.

Figure 4 shows the distribution of the absolute values of AC coefficients for the corresponding images in Figure 1 and the fitting curves of the Weibull model. The Weibull model well fits the pulse-shape in addition to the sharp-peak and heavy-tail phenomena. The faster the Weibull distribution changes, the larger the distortion is. We use the Mean Square Error (MSE) to express the fitting error:

$$MSE = \frac{1}{n}\sum_{n=1}^{n}(P_i - Q_i)^2 \tag{2}$$

where P_i is the value of the histogram for the absolute values of AC coefficients, and Q_i is the statistical distribution density of the fitting functions. The MSE values for Weibull model are 6.07×10^{-6} and 3.38×10^{-7} in Figure 4b,c, respectively, and these values are much lower than the MSE values of the GGD model, which are 1.1×10^{-4} and 3.36×10^{-5} in Figure 2b,c, respectively.

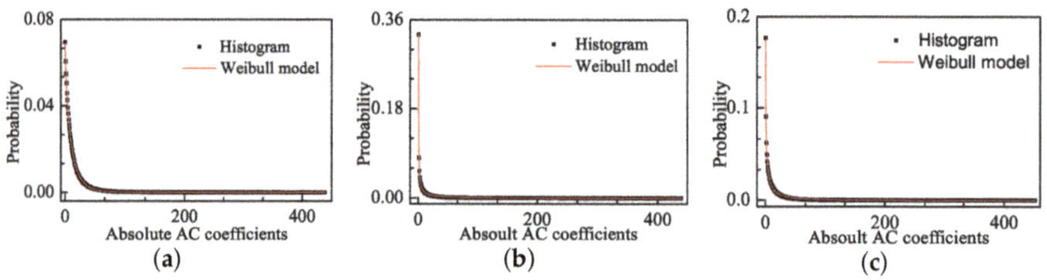

Figure 4. The distribution of absolute AC coefficients corresponding images and the Weibull fitting curves: (**a**) image with DMOS = 0; (**b**) image with DMOS = 63.649; and (**c**) image with DMOS = 29.739.

We also develop a Weibull model that fits five representative distortion types in the LIVE database: JP2K compression (JP2K), JPEG compression (JPEG), white noise (WN), Gaussian blur (GB), and fast-fading (FF) channel distortions [38]. Figure 5 presents the MSE comparison of two models for each distortion type images. Table 1 lists the average MSE of each distortion type. The fitting error of the Weibull model is obviously smaller than that of the GGD model. Therefore, the Weibull model is employed to evaluating the image quality in this paper.

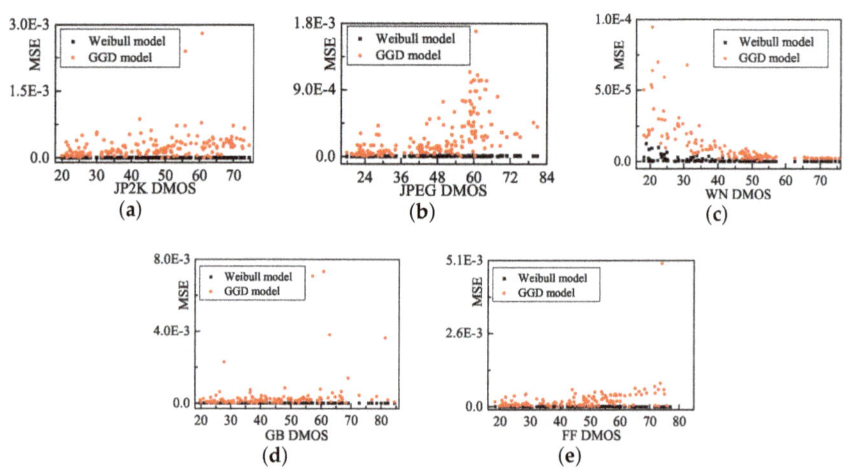

Figure 5. Fitting error of GGD model and Weibull model in LIVE database.

Table 1. The average MSE of each distortion type.

Types	Weibull	GGD
JP2K	4.39×10^{-6}	2.51×10^{-4}
JPEG	4.79×10^{-6}	2.73×10^{-4}
WN	1.06×10^{-6}	1.19×10^{-5}
GB	2.29×10^{-6}	1.30×10^{-3}
FF	2.88×10^{-6}	5.73×10^{-4}

3. Proposed BWS Method

In this section, we describe the proposed BWS method in detail. The framework is illustrated in Figure 6. First, block DCT processing is applied to images of different scales. The goal is not only to conform to the decomposition process of local information in the HVS [40] but also to reflect the image structure correlation [4]. In the DCT domain, the magnitudes of the block AC coefficients and those in the orientation and frequency sub-regions are extracted to describe HVS characteristics. Then, a Weibull model is employed to fit these values. From the Weibull model, the following four perceptive features are extracted: the shape-scale parameter feature, the coefficient of variation feature, the frequency sub-band feature, and the directional sub-band feature. Finally, by using the SVR learning method, the relationship between the image features and subjective quality scores in high-dimensional space is obtained.

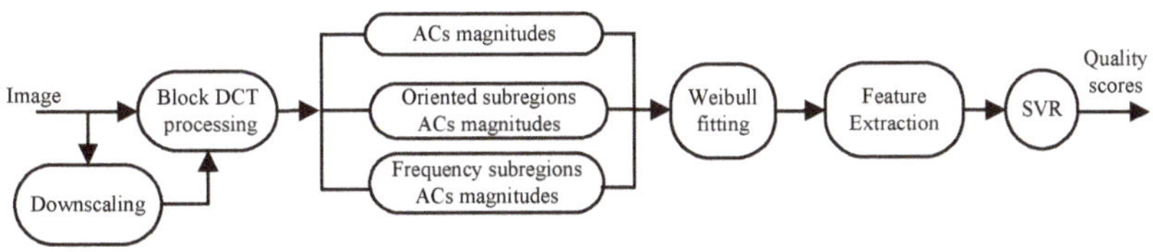

Figure 6. Framework of the BWS algorithm.

One advantage of this approach is that we consider only the change of luminance information in the BWS algorithm because neuroscience research shows that the HVS is highly sensitive to changes in image luminance information [37]. In addition, DCT processing is useful in IQA. The presentation of features can be enhanced by treating different frequency components with different distortion levels. The computational convenience is another advantage [41].

3.1. Relationship Between HVS and Perceptual Features

IQA modeling must be able to satisfy human perceptual requirements, which are closely related to HVS. Designing an HVS-based model for directly predicting image quality is infeasible because of the complexity. Therefore, in this paper, the features related to the HVS are extracted from the Weibull model and used to predict perceptual image quality scores by a learning method.

The visual perception system has been shown to be highly hierarchical [42]. Visual properties are processed in areas V1 and V2 of the primate neocortex, which occupies a large region of the visual cortex. V1 is the first visual perception cortical area, and the neurons in V1 can achieve succinct descriptions of images in terms of the local structural information pertaining to the spatial position, orientation, and frequency [43]. Area V2 is also a major visual processing area in the visual cortex [44], and the neurons in V2 have the property of the scale invariance [45]. Therefore, our proposed model is related to the properties of the HVS.

3.2. Shape-Scale Parameter Feature

Before extracting features, we divide the image into 5×5 blocks with a two-pixel overlap between adjacent blocks to remove the redundancy of image blocks and better reflect the correlation information among blocks. For each block, DCT is performed to extract the absolute AC coefficients. Then, the Weibull model is used to fit the magnitudes of the AC coefficients of each block. Theories in fragmentation posit that the scale parameter m and shape parameter a in the Weibull distribution are strongly correlated with brain responses. The experiments on brain responses showed that a and m explain up to 71 % of variance of the early electroencephalogram signal [39]. These parameters can also be estimated from the outputs of X-cells and Y-cells [46]. In addition, the two parameters can accurately describe the image structure correlation because a difference in the image distribution of quality degradation, which depends on the image structural information, results in a different shape of the Weibull distribution, thereby resulting in different values of a and m. In other words, the response of the brain to external image signals is highly correlated with the parameters a and m of the Weibull distribution. Thus, we defined the shape-scale parameter feature $\zeta = (1/m)^a$. The parameters that directly determine the form of the Weibull distribution are a and $(1/m)^a$, as shown in Equation (3). The Equation (3) is the deformation of Weibull distribution. Because the human subjects are viewing natural images that are correlated with a and m, considering only the influence of a while ignoring the effect of m on the Weibull distribution does not produce accurate results. Therefore, we chose $(1/m)^a$ as feature ζ for assessing image quality. The advantage of this feature is that it provides an intuitive expression of the Weibull equation as well as a monotonic function of distortion, which can be used to represent the levels of distortion in images.

$$f_X(x) = a(\frac{1}{m}^a)x^{a-1}\exp[-(\frac{1}{m})^a x^a] \qquad x > 0 \tag{3}$$

The efficiency of the features is verified in the LIVE database. We extracted the average value of the highest 10% and 100% (all block coefficients) of the shape-scale parameter features (ζ) in all blocks of the image. These two percentages correspond to image distortion of the local worst regions and the global regions, respectively. It may be inappropriate to only focus on distortion of local worst regions or the overall regions [47–49]. Thus, it is necessary to combine the local distortion with the global distortion.

Table 2 shows the Spearman Rank Order Correlation Coefficient (SROCC) values between the DMOS scores and the average values of the highest 10% of ζ as well as the DMOS values and the average values of the 100% of ζ in LIVE database. The SROCC value is larger than 0.7, which indicates a significant correlation with the subjective scores [50]. Therefore, these features can be effectively used as perceptual features for IQA.

Table 2. SROCC correlation of DMOS vs. ζ.

LIVE Subset	10%ζ	100%ζ
JP2K	0.824	0.772
JPEG	0.812	0.819
WN	0.981	0.988
GB	0.926	0.882
FF	0.760	0.786

3.3. Coefficient of Variation Feature

Natural images are known to be highly correlated [36,37]. The correlation can be affected by distortions in different forms. In the DCT domain, distortions change the distribution of the AC coefficients. For JP2K, JPEG, GB, and FF [38], the distortion increases the differences among low-, middle- and high-frequency information. Then, the standard deviation becomes larger under the unit mean than that in the natural image. Thus, the large variation of the standard deviation represents a large distortion. In contrast, for WN distortions, the increased random noise causes high-frequency information to increase rapidly, thereby reducing the differences among different frequency information. Thus, a small variation of the standard deviation corresponds to a large distortion.

Therefore, we define the coefficient of variation feature ξ, which describes the variation of the standard deviation under the unit mean as follows:

$$\xi = \frac{\sigma_X}{\mu_X} = \sqrt{\frac{\Gamma(1+\frac{2}{a})}{\Gamma^2(1+\frac{1}{a})} - 1} \tag{4}$$

where the mean μ_X and variance σ_X^2 of the Weibull model can be obtained as follows:

$$\mu_X = \int_0^\infty x f_X(x)\,dx = m\Gamma(1+\frac{1}{a}) \tag{5}$$

$$\sigma_X^2 = \int_0^\infty x^2 f_X(x)\,dx - \mu^2 = m^2\Gamma(1+\frac{2}{a}) - m^2\Gamma^2(1+\frac{1}{a}) \quad (6)$$

where Γ denotes the gamma function. This parameter is defined as follows:

$$\Gamma(z) = \int_0^\infty t^{z-1}e^{-t}\,dt \quad (7)$$

We calculated the average value of the highest 10% of ξ and the average value of 100% of ξ in all blocks across the image. Table 3 shows the SROCC values, which are verified in the LIVE database. The correlation is also significant in most distortion types. Thus, the features can be used to assess image quality.

Table 3. SROCC correlation of DMOS vs. ξ.

LIVE Subset	10%ξ	100%ξ
JP2K	0.922	0.899
JPEG	0.799	0.047
WN	0.961	0.937
GB	0.933	0.862
FF	0.832	0.843

3.4. Frequency Sub-Band Feature

Natural images are highly structured in the frequency domain. Image distortions often modify the local spectral properties of an image so that these properties are dissimilar to those of in natural images [29]. For JP2K, JPEG, GB, and FF distortions, distortions trigger a rapid increase of differences among the coefficients of variation of the frequency sub-bands coefficients. A large difference represents a large distortion. However, with the WN distortion type, the opposite change trend is observed.

To measure this difference, we defined the frequency sub-band feature f. According to the method in [29], we divided each 5 × 5 image block into three different frequency sub-bands, as shown in Table 4. Then, the Weibull fit was obtained for each of the sub-regions, and the coefficient of variation ξ_f was calculated using Equation (4) in the three sub-bands. Finally, the variance of ξ_f was calculated as the frequency sub-band feature f.

Table 4. DCT coefficients of three bands.

DC	C_{12}	C_{13}	C_{14}	C_{15}
C_{21}	C_{22}	C_{23}	C_{24}	C_{25}
C_{31}	C_{32}	C_{33}	C_{34}	C_{35}
C_{41}	C_{42}	C_{43}	C_{44}	C_{45}
C_{51}	C_{52}	C_{53}	C_{54}	C_{55}

The feature was pooled by calculating the average value of the highest 10% of f, and the average value of 100% of f in all blocks across the image in the LIVE database. In Table 5, we report how well the features are correlated with the subjective scores. The SROCC is clearly related to the subjective scores, which means that these features can be used to describe subjective perception.

Table 5. SROCC correlation of DMOS vs. f.

LIVE Subset	10%f	100%f
JP2K	0.804	0.887
JPEG	0.845	0.862
WN	0.935	0.916
GB	0.713	0.821
FF	0.801	0.869

3.5. Directional Sub-Band Feature

The HVS also has different sensitivities to sub-bands in different directions [40]. Image distortion often changes the correlation information of sub-bands in different directions, which makes the HVS highly sensitive to this change. For the JP2K, JPEG, GB, and FF distortion types, distortions modify the inconsistencies among coefficients of variation in sub-bands in different directions. A large inconsistency reflects a large distortion. Note that this effect has a reverse relationship to the WN distortion.

Therefore, this inconsistency can be described by the orientation sub-band feature S_o. We divided sub-bands into three different orientations for each block, as shown in Table 6. This decomposition approach is similar to the approach in [29]. Then, the Weibull model was fitted to the absolute AC coefficients within each shaded region in the block. The coefficient of variation ξ_o was also calculated, as shown in Equation (4), in three directional sub-bands. Finally, the directional sub-band feature S_o can be obtained from the variance of ξ_o.

Table 6. DCT coefficient collected along three orientations.

DC	C_{12}	C_{13}	C_{14}	C_{15}
C_{21}	C_{22}	C_{23}	C_{24}	C_{25}
C_{31}	C_{32}	C_{33}	C_{34}	C_{35}
C_{41}	C_{42}	C_{43}	C_{44}	C_{45}
C_{51}	C_{52}	C_{53}	C_{54}	C_{55}

The average values of the highest 10% and 100% of S_o for all blocks across images were collected. We report the SROCC values between DMOS scores and features in Table 7 and demonstrate an obvious correlation with human perception.

Table 7. SROCC correlations of DMOS vs. S_o.

LIVE Subset	$10\% S_o$	$100\% S_o$
JP2K	0.887	0.903
JPEG	0.813	0.748
WN	0.954	0.951
GB	0.928	0.923
FF	0.865	0.866

3.6. Multi-Scale Feature Extraction

Previous research has demonstrated that the incorporation of multi-scale information can enhance the prediction accuracy [5]. The statistical properties of a natural image are the same at different scales, whereas distortions affect the image structure across different scales. The perception of image details depends on the image resolution, the distance from the image plane to the observer and the acuity of the observer's system [40]. A multi-scale evaluation accounts for these variable factors. Therefore, we extracted 24 perceptual features across three scales. In addition to the original-scale image, the second-scale image was constructed by low-pass filtering and down-sampling the original image by a factor of two. Then, the third-scale image was obtained in the same way from the second-scale image. As listed in Table 8, each scale includes eight features. The extraction process is as described in Sections 3.2–3.5.

Table 8. Features used for BWS algorithm.

Scale	Feature Set
The first scale	$\{10\%\zeta, 100\%\zeta, 10\%\xi, 100\%\xi, 10\%f, 100\%f, 10\%o, 100\%S_o\}$
The second scale	$\{10\%\zeta, 100\%\zeta, 10\%\xi, 100\%\xi, 10\%f, 100\%f, 10\%o, 100\%S_o\}$
The third scale	$\{10\%\zeta, 100\%\zeta, 10\%\xi, 100\%\xi, 10\%f, 100\%f, 10\%o, 100\%S_o\}$

3.7. Prediction Model

After extracting features in three scales, we learned the relationship between image features and subjective scores. In the literature, SVR is widely adopted as the mapping function for learning this relationship [51,52]. Considering a set of training data $\{(x_1, y_1), \ldots, (x_l, y_l)\}$, where $x_i \in R^n$ is the extracted image feature and y_i is the corresponding DMOS, a regression function can be learned to map the feature to the quality score, i.e., $y_i = SVR(x_i)$. We used the LIBSVM package [53] to implement the SVR with a Radial Basis Function (RBF) kernel in our metric. Once the regression model was learned, we could use it to estimate the perceptual quality of any input image.

4. Experiments and Results

4.1. Experimental Setup

The performance of the blind IQA algorithms was validated using subjective image quality databases, where each image is associated with a human score (e.g., a (Difference) Mean Opinion Scores (DMOS/MOS)). The performance describes how well the objective metric is correlated with human ratings. Several subjective image quality evaluation databases have been established. We employed three widely used databases, namely, the LIVE database [38], the TID2008 database [54] and the CSIQ database [55], in our research. These three databases are summarized as follows:

(1) The LIVE database includes 29 reference images and 779 distorted images corrupted by five types of distortions: JP2K, JPEG, WN, GB and FF. Subjective quality scores are provided in the form of DMOS ranging from 0 to 100. Each of the distorted images is associated with a DMOS, representing the subjective quality of this image.

(2) The TID2008 database covers 17 distortion types, each of which consists of 100 distorted versions from 25 reference images. Subjective quality scores are provided for each image in the form of MOS, ranging from 0 to 9. Note that there is one artificial image and its distortion images in the TID2008 database. We discarded these images when evaluating the performance of the BWS method because the proposed method was designed to evaluate the quality of natural scenes. In addition, we mainly considered the subsets with JP2K, JPEG, WN and GB distortion types that appear in LIVE database. These four distortion types are also the most commonly encountered distortions in practical applications.

(3) The CSIQ database consists of 30 reference images and 866 distorted images corrupted by six types of distortions: JPEG, JP2K, WN, GB, pink Gaussian noise and global contrast decrements. Each distorted image has five different distortion levels. Subjective quality scores are provided in the form of DMOS ranging from 0 to 1. Similarly, we mainly considered the same four distortion types as included the LIVE database.

To evaluate the performance of the BWS algorithm, two correlation coefficients, the Pearson Linear Correlation Coefficient (PLCC) and the SROCC, are used as the criteria. The PLCC measures the prediction accuracy, whereas the SROCC represents the prediction monotonicity. Before calculating the PLCC, the algorithm scores are mapped using a logistic non-linearity as described in [56]. Both the SROCC and PLCC lie in the range $[-1,1]$. SROCC and PLCC values that are closer to "1" or "-1"correspond to better predictions of this algorithm.

4.2. Performance on Individual Databases

First, we evaluated the overall performance of the BWS method and other competing IQA methods, namely, BLIINDS-II [29], DIIVINE [31], BRISQUE [32], BIQI [57], PSNR and SSIM [4], on each database. The first four methods are NR IQA algorithms, and the latter two are FR IQA algorithms.

Because the BWS approach is based on SVR, we randomly divided each image database into two sets: a training set and a testing set. The training set is used to train the prediction model and the testing set is used to test the prediction results. In our experimental setup, 80% of the distorted images for each database are used as the training set and the remaining 20% of the images are used as the testing set. Content does not overlap between these two sets. We repeat the training–testing procedure 1000 times, and the median value of the obtained SROCCs and PLCCs is reported as the final performance of the proposed metric. Meanwhile, we adapted the same experimental setup for comparison algorithms. Although the FR IQA approaches PSNR and SSIM do not require training on a database, for a fair comparison, we also conducted the experiments on the randomly partitioned testing set and recorded the median value of the PLCC and SROCC.

Table 9 shows the performances of different methods on the LIVE, TID2008 and CSIQ databases. For each database, the top two IQA methods are highlighted in bold. For the LIVE database, the overall performance of the proposed BWS method is better than those of the other IQA methods. For the TID2008 database and the CSIQ database, the BWS algorithm outperforms the other NR and FR IQA methods. It is concluded that the BWS algorithm outperforms the other competitors overall on different databases. Although other methods may work well on some databases, they fail to deliver good results on other databases. For example, DIIVINE obtains a good result on the LIVE database but performs poorly on the CSIQ and TID2008 databases.

Table 9. Overall performance on three databases.

Algorithms	LIVE		TID2008		CSIQ	
	SROCC	PLCC	SROCC	PLCC	SROCC	PLCC
PSNR	0.867	0.859	0.877	0.863	0.905	0.904
SSIM	0.913	0.907	0.780	0.755	0.834	0.835
BIQI	0.819	0.821	0.803	0.852	0.905	0.892
DIIVINE	0.912	0.917	0.898	0.893	0.878	0.896
BLIINDS-II	0.931	0.930	0.889	**0.916**	**0.911**	0.926
BRISQUE	**0.940**	**0.942**	**0.906**	0.914	0.902	**0.927**
BWS	**0.934**	**0.943**	**0.921**	**0.942**	**0.931**	**0.934**

Moreover, we present weighted-average SROCC and PLCC results of competing IQA methods on all three databases in Table 10. The weight that is assigned to each database depends on the number of distorted images that the database contains [58,59]. BWS still performs best among the IQA methods. Hence, we conclude that the objective scores that are predicted by BWS correlate much more consistently with subjective evaluation than those that are predicted by other IQA metrics.

Table 10. Performance of weighted average over three databases.

Algorithms	Weighted Average	
	SROCC	PLCC
PSNR	0.882	0.875
SSIM	0.914	0.916
BIQI	0.845	0.852
DIIVINE	0.897	0.905
BLIINDS-II	0.915	0.926
BRISQUE	0.920	0.931
BWS	**0.930**	**0.940**

To determine whether the superiority of the BWS method over its counterparts is statistical significance, we conducted statistical analysis to validate their differences in performance. The hypothesis testing, which was based on the t-test [60], was presented, which measures the equivalence of the mean values of two independent samples. Experiments are conducted by randomly splitting the database into a training set and a testing set and the SROCC values are reported for 1000 training-testing trials. Thus, we applied the t-test between the SROCCs that were generated by each of the two algorithms and tabulated the results in Tables 11–13. Each table shows the results of the t-test on each database. A value of "1" in the tables indicates that the row algorithm is statistically superior to the column algorithm, whereas a value of "−1" indicates that the row algorithm is statistically inferior to the column algorithm. A value of "0" indicates that the row and column algorithms are statistically equivalent. From the experimental results in Tables 10–12, the BWS method was found to be statistically superior to FR approaches PSNR and SSIM and NR IQA approach BLIINDS-II.

Table 11. Statistical significance test on LIVE.

LIVE	BWS	BLIINDS-II	SSIM	PSNR
BWS	0	1	1	1
BLIINDS-II	−1	0	1	1
SSIM	−1	−1	0	1
PSNR	−1	−1	−1	0

Table 12. Statistical significance test on TID2008.

TID2008	BWS	BLIINDS-II	SSIM	PSNR
BWS	0	1	1	1
BLIINDS-II	−1	0	1	1
SSIM	−1	−1	0	1
PSNR	−1	−1	−1	0

Table 13. Statistical significance test on CSIQ.

CSIQ	BWS	BLIINDS-II	SSIM	PSNR
BWS	0	1	1	1
BLIINDS-II	−1	0	1	1
SSIM	−1	−1	0	1
PSNR	−1	−1	−1	0

4.3. Performance on Individual Distortion Type

In this section, we tested the performances of the proposed BWS method and other competing IQA methods on individual distortion type over the LIVE, TID2008 and CSIQ databases. For NR IQA, we trained on 80% of the distorted images with various distortion types randomly and tested on the remaining 20% of the distorted images with a specific distortion type. The SROCC and PLCC comparisons on each database are illustrated in Tables 14 and 15.

Table 14. The SROCC comparison on individual distortion types.

Databases	Types	PSNR	SSIM	BIQI	DIIVINE	BLIINDS-II	BRISQUE	BWS
LIVE	JP2K	0.865	0.939	0.856	**0.932**	0.928	0.914	**0.929**
	JPEG	0.883	0.947	0.786	**0.948**	0.942	**0.965**	0.895
	WN	0.941	0.964	0.932	**0.982**	0.969	**0.979**	0.976
	GB	0.752	0.905	0.911	0.921	0.923	**0.951**	**0.932**
	FF	0.874	0.939	0.763	0.871	**0.889**	0.877	**0.893**
TID2008	JP2K	0.854	0.900	0.857	0.826	**0.922**	0.895	**0.933**
	JPEG	0.886	0.931	0.859	0.913	**0.918**	0.910	**0.914**
	WN	0.923	0.836	0.798	**0.896**	0.805	0.862	**0.868**
	GB	0.944	0.954	**0.901**	**0.901**	0.868	0.890	0.892
CSIQ	JP2K	0.942	0.929	0.901	0.904	0.900	**0.951**	**0.912**
	JPEG	0.893	0.934	0.906	0.879	0.920	**0.925**	**0.922**
	WN	0.938	0.936	**0.921**	0.897	0.913	0.878	**0.937**
	GB	0.940	0.906	0.874	0.866	**0.941**	0.902	**0.954**

Table 15. The PLCC comparison on individual distortion types.

Databases	Types	PSNR	SSIM	BIQI	DIIVINE	BLIINDS-II	BRISQUE	BWS
LIVE	JP2K	0.876	0.941	0.809	0.922	**0.935**	0.923	**0.945**
	JPEG	0.903	0.946	0.901	0.921	**0.968**	**0.973**	0.923
	WN	0.917	0.982	0.939	**0.988**	0.980	**0.985**	0.984
	GB	0.780	0.900	0.829	0.923	0.938	**0.951**	**0.952**
	FF	0.880	0.951	0.733	0.868	0.896	**0.903**	**0.903**
TID2008	JP2K	0.906	0.906	0.891	0.810	**0.946**	0.905	**0.957**
	JPEG	0.896	0.961	0.883	0.906	**0.952**	0.923	**0.961**
	WN	0.953	0.852	0.823	**0.908**	0.840	0.862	**0.893**
	GB	0.950	0.955	**0.929**	0.898	0.906	0.896	**0.909**
CSIQ	JP2K	0.950	0.943	0.897	0.918	0.930	**0.957**	**0.938**
	JPEG	0.905	0.958	0.884	0.896	0.931	**0.956**	**0.940**
	WN	0.952	0.940	**0.929**	0.921	0.917	0.906	**0.942**
	GB	0.959	0.913	0.875	0.887	**0.926**	0.920	**0.944**

We observed the advantages of the individual distortion type of the BWS algorithm on each database. For the LIVE database, we clearly found that the proposed metric outperforms other NR metrics in FF. In particular, compared with the BLIINDS-II method, implemented in the same DCT domain, BWS performs better on the JP2K, WN, GB and FF distortion types. For FR metrics, although these methods require the complete information of the reference images, our algorithm still outperforms PSNR on all distortion types and outperforms SSIM on WN and GB distortions. For the TID2008 database, the performance of our method is superior on JP2K distortions relative to other NR metrics. Similarly, the performances on JP2K, WN, and GB distortion are better than those of the BLIINDS-II method. Compared with the FR metric, the performance of the BWS algorithm on the JP2K distortion type is better than those of PSNR and SSIM. For the CSIQ database, our method outperforms the remaining NR metrics on WN and GB distortions. Compared with the BLIINDS-II algorithm, the BWS method consistently performs better. Compared with the FR metric, the performances of BWS on the JPEG and GB types are better than those of PSNR, and on GB and WN the distortion is better than with SSIM. Therefore, we found that the performance of the BWS algorithm is superior on some specific distortion types in each database.

To facilitate a comparison of the effects between BWS method and the other IQA methods, 13 groups of distorted images were considered in the three databases. The best two results of the SROCC and PLCC are highlighted in boldface for the NR IQA methods. We calculated the number of times that each method was ranked in the top two in terms of the SROCC values and PLCC values for each distortion type. For the 13 groups of distorted images in the three databases, the BWS algorithm was ranked in the top two the most times, with 10 times for the SROCC and 11 times for the PLCC. We also report the weighted means and standard deviations (STDs) of competing IQA methods across all distortion groups in Table 16. The BWS has higher average and lower STD across all distortion groups. Hence, the BWS method achieves a consistently better performance on most commonly encountered distortion types.

Table 16. Performance of weighted average and STD across all distortion groups.

Algorithms	Weighted Average		Weighted STD	
	SROCC	PLCC	SROCC	PLCC
PSNR	0.894	0.908	0.188	0.169
SSIM	**0.927**	**0.937**	**0.113**	**0.118**
BIQI	0.867	0.871	0.205	0.207
DIIVINE	0.906	0.908	0.137	0.139
BLIINDS-II	0.915	0.930	0.141	0.122
BRISQUE	0.919	0.932	0.128	0.122
BWS	**0.922**	**0.938**	**0.099**	**0.087**

To visually show the correlation of the BWS method between the predicted quality scores and the subjective scores, we present scatter plots (for each distortion type and for the entire LIVE, TID2008 and CSIQ databases) of the predicted scores and the subjective scores in Figures 7–9. These figures show that a strong linear relationship occurs between the predicted scores of BWS and the subjective human ratings, which indicates a high prediction accuracy of the BWS algorithm.

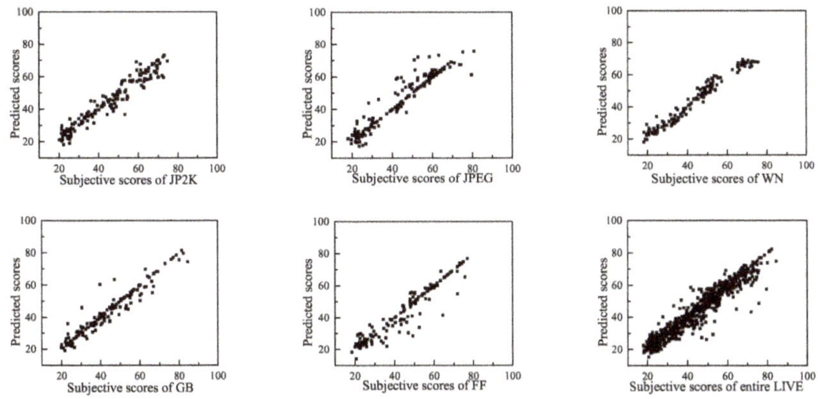

Figure 7. Predicted scores versus subjective scores on LIVE database.

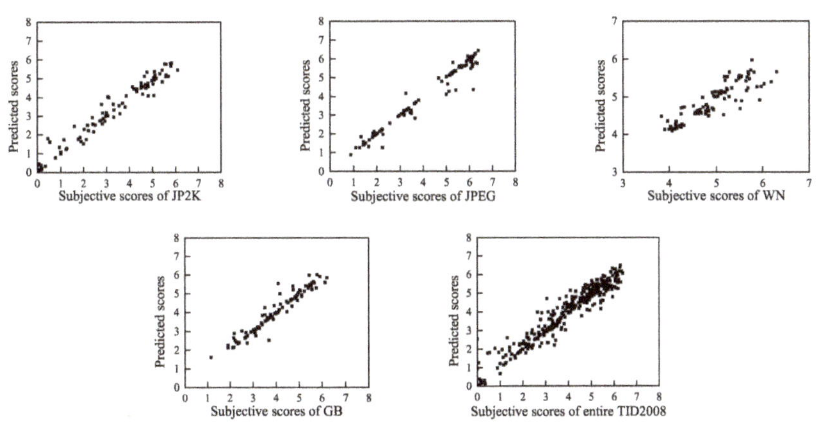

Figure 8. Predicted scores versus subjective scores on TID2008 database.

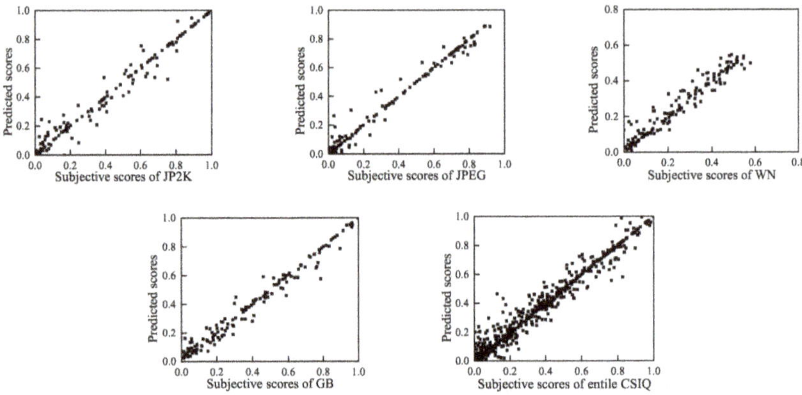

Figure 9. Predicted scores versus subjective scores on CSIQ database.

4.4. Cross-Database Validation

In our previous experiments, the training samples and test samples were selected from the same database. It is expected that an IQA model that has learned on one image quality database should be able to accurately assess the quality of images in other databases. Therefore, to demonstrate the generality and robustness of the proposed BWS method, the following experiments were conducted. We trained all distorted images from one database to obtain a prediction model and used this model to test the scores of distorted images from other databases. With the three databases, six combinations of training and testing database pairs were created. We compared our proposed BWS algorithm and BLIINDS-II method on the same DCT domain. The SROCC results of the cross-database validation are tabulated in Table 17. The results indicate that the BWS algorithm performs better than the BLIINDS-II method in most cases. Therefore, the cross-database validation has demonstrated the generality and robustness of the proposed BWS method in DCT domain.

Table 17. SROCC comparison on cross-database validation.

Train	Test	BWS	BLIINDS-II
LIVE	TID2008	**0.889**	0.844
LIVE	CISQ	0.839	0.868
TID2008	LIVE	**0.867**	0.742
TID2008	CISQ	0.828	0.853
CISQ	LIVE	**0.884**	0.833
CISQ	TID2008	**0.869**	0.832

4.5. Discussion

4.5.1. Model Selection

In Section 2, we analyze Weibull model as NSS model instead of a typical GGD model can well simulate the statistical regularities of distorted natural images in DCT domain. However, the changes of Weibull distribution model are similar to the simpler exponential distribution model. It is necessary to judge whether the exponential model is superior to Weibull model for IQA. Figure 10 shows the distribution of AC coefficients for the corresponding images in Figure 1 and the exponential fitting curve. We found that, if we used the exponential distribution model to fit statistics of natural and distorted images in DCT domain, it unfortunately failed at fitting these phenomena. Moreover, we used MSE to present fitting error comparison of three models for each distortion type images, as shown in Table 18. The fitting errors of Weibull model is minimum. Therefore, it is not appropriate to use exponential model instead of Weibull model.

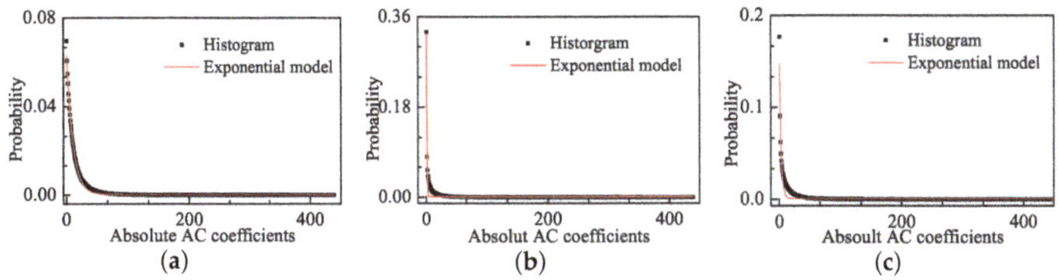

Figure 10. The distribution of absolute AC coefficients corresponding images and the Exponential fitting curves: (**a**) image with DMOS = 0; (**b**) image with DMOS = 63.649; and (**c**) image with DMOS = 29.739.

Table 18. The average MSE of each distortion type.

Types	Weibull	GGD	Exponential
JP2K	$\mathbf{4.39 \times 10^{-6}}$	2.51×10^{-4}	1.32×10^{-4}
JPEG	$\mathbf{4.79 \times 10^{-6}}$	2.73×10^{-4}	1.09×10^{-4}
WN	$\mathbf{1.06 \times 10^{-6}}$	1.19×10^{-5}	1.21×10^{-4}
GB	$\mathbf{2.29 \times 10^{-6}}$	1.30×10^{-3}	1.41×10^{-4}
FF	$\mathbf{2.88 \times 10^{-6}}$	5.73×10^{-4}	1.31×10^{-4}

4.5.2. The Block Size Selection

In the BWS method, we selected 5×5 block for DCT. On the one hand, if a smaller block size is selected, the correlation between blocks is very large, thus it is difficult to distinguish the difference of extracted features. It affects the prediction accuracy. Similarly, if a bigger block size is selected, the extracted features lack the correlation information between blocks, which leads to inaccurate evaluation performance. Meanwhile, we used experiments to prove 5×5 block size is better in our method, as shown in Table 19. On the other hand, the 5×5 block is very common in image quality assessment [29,61,62]. Therefore, selecting 5×5 block is reasonable and can improve predicted performance.

Table 19. SROCC correlation of DMOS vs. 100%ζ.

LIVE Subset	3 × 3	5 × 5	8 × 8
JP2K	0.719	**0.772**	0.702
JPEG	0.801	**0.819**	0.768
WN	0.969	**0.988**	0.976
GB	0.865	**0.882**	0.875
FF	0.754	**0.768**	0.732

4.5.3. The Pooling Strategy and Multi-Scales Selection

Pooling strategy has been studied recently as an important factor to the accuracy of objective quality metrics. In our method, we calculated the average value of the highest 10% of features and that of 100% of features. The highest 10% and the 100% features of all blocks in image describe image distortion of the worst 10% regions and the overall image regions. The averaging of features in all the local regions is one of the widely used methods in image quality metrics. It describes the global distortion of the image. However, when only a small region in an image is corrupted with extremely annoying artifacts, human subjects tend to pay more attention to the low-quality region. Thus, we also considered the image distortion of the worst 10% regions because the human eye is more annoyed by the local distortion of the image and the subjects are likely to focus their quality opinions on the worst 10% of the whole image [63].

We extracted three feature sets: the average values of the highest 10% features, the average values of the 100% features and the combination of the first two feature sets in three scales from the LIVE database. The feature extraction method and the experimental setup were the same as those reported in Sections 3 and 4. The median SROCC value of 1000 training-testing trials was utilized to evaluate the performance. The results of this experiment are shown in Table 20, and they indicate that the performance of the combination feature set is better than that when only one factor is considered. This finding demonstrates that the joint pooling strategy can improve the prediction performance of image quality measures and human intuition of images is the synthesis of local and global perception.

Table 20. SROCC performance with pooling strategy.

Types	Highest 10% Set	100% Set	Combination Set
JP2K	0.900	0.915	**0.929**
JPEG	0.850	0.872	**0.895**
WN	0.975	0.974	**0.976**
GB	0.906	0.921	**0.932**
FF	0.851	0.878	**0.893**
ALL	0.901	0.913	**0.934**

A multi-scale segmentation method has been developed and implemented for IQA. We proposed multi-scale features for predicting image quality because the perceptibility of image details depends on the viewing conditions. The subjective evaluation of a given image varies with these factors. Therefore, a multi-scale method is an effective method of incorporating image details at different resolutions.

We conducted an experiment to determine the impact of scale on IQA. In our method, we selected three scale images for extracting features because no significant gain in performance is obtained beyond the third scale of feature extraction. The methods of scale image segmentation, feature extraction and experiment setup were the same as the previous operation. We report the correlations of different scales between the predicted scores and the subjective scores on the LIVE database in Table 21. The experimental results show that the performance of BWS method in three scales outperform the other cases. It proves that the approach of multi-scale can improve the performance of our metric.

Table 21. SROCC performance with multi-scale selection.

Types	One Scale	Two Scales	Three Scales
JP2K	0.905	0.926	**0.929**
JPEG	0.857	0.877	**0.895**
WN	0.973	**0.979**	0.976
GB	0.923	0.928	**0.932**
FF	0.837	0.864	**0.893**
ALL	0.904	0.921	**0.934**

4.6. Computational Complexity

In many practical applications, the prediction accuracy and the algorithm complexity need to be considered comprehensively. Therefore, we evaluated the computational complexity of all competing methods in Table 22. One can see that, although the proposed BWS algorithm does not have the lowest complexity, the assessment accuracy is better among all the competing models, as shown in Tables 10 and 16. For example, the BIQI is the lowest with complexity (N), where N is the total number of image pixels. However, its performance is the worst. The DIIVINE is the most complex, but the performance is inferior to our algorithm. Therefore, the proposed BWS algorithm is a reasonable trade-off between assessment accuracy and computational complexity.

Table 22. Computational complexity. N is the total number of pixels in a test image.

Algorithms	Computational Complexity	Notes
BIQI	$O(N)$	
DIIVINE	$O(Nlog(N) + m^2 + N + 392b)$	m:neighborhood size in DNT; b:2D histogram bin number
BRISQUE	$O(d^2N)$	d:block size
BLIINDS-II	$O((N/d^2)log(N/d^2))$	d:block size
BWS	$O((N/d^2)log(N/d^2))$	d:block size

5. Conclusion

Since there exists a strong relationship between the changes of entropy and distribution of images, existing NR IQA models typically use the GGD model to fit the distribution of a natural image in the DCT domain and extract features from this model to learn a quality prediction model. However, the difference between the distribution of the distorted image and that of its natural image, which includes the pulse-shape phenomenon, is neglected. In this paper, we propose the BWS method, which is a new NR IQA method based on the Weibull model in the DCT domain. The most important pulse-shape phenomenon is first considered in the distorted image distribution. The proposed Weibull model not only overcomes the disadvantages of the GGD model but also reflects HVS perception. Our research findings suggest that the Weibull model plays an important role in quality assessment tasks. Extensive experimental results on three public databases have demonstrated that the proposed BWS method is highly correlated with human visual perception and competitive with the state-of-the-art NR and FR IQA methods in terms of prediction accuracy and database independence.

Author Contributions: X.Y. designed the methodology and conducted the experiments. F.L. drafted the manuscript. L.H. performed the statistical analysis. W.Z. modified the manuscript. All authors read and approved the final manuscript.

Funding: This work is supported by the National Science Foundation of China Project: 61671365, the Joint Foundation of Ministry of Education of China: 6141A020223, and the Natural Science Basic Research Plan in Shaanxi Province of China: 2017JM6018.

Conflicts of Interest: The authors declare no conflicts of interest.

Abbreviations

The following abbreviations are used in this manuscript:

FR	Full-Reference
RR	Reduced-Reference
NR	No-Reference
IQA	Image Quality Assessment
NSS	Natural Scene Statistics
BWS	Blind Weibull Statistics
DCT	Discrete Cosine Transform
AC	Alternating Current
HVS	Human Visual System
SVR	Support Vector Regression
GGD	Generalized Gaussian Distribution
DMOS	Difference Mean Opinion Score
MSE	Mean Square Error
JP2K	JP2K compression
JPEG	JPEG compression
WN	White Noise
GB	Gaussian Blur

FF	Fast Fading
MOS	Mean Opinion Score
SROCC	Spearman Rank Order Correlation
PLCC	Pearson Linear Correlation Coefficient
STDs	Standard Deviations
PSNR	Peak Signal to Noise Ratio

References

1. Shahid, M.; Rossholm, A.; Zepernick, A. No-reference image and video quality assessment: A classification and review of recent approaches. *J. Image Video Proc.* **2014**, *40*, 1–32. [CrossRef]
2. Li, F.; Fu, S.; Liu, Z.-Y.; Qian, X.-M. A cost-constrained video quality satisfaction study on mobile devices. *IEEE Trans. Multimed.* **2018**, *20*, 1154–1168. [CrossRef]
3. Gu, K.; Wang, S.-Q.; Yang, H.; Lin, W.-S.; Zhai,G.-T.; Yang, X.-K.; Zhang, W.-J. Saliency-guided quality assessment of screen content images. *IEEE Trans. Multimed.* **2016**, *18*, 1–13. [CrossRef]
4. Wang, Z.; Bovik, A.C.; Sheikh, H.R.; Simoncelli, E.P. Image quality assessment: From error visibility to structural similarity. *IEEE Trans. Image Proc.* **2016**, *13*, 600–612. [CrossRef]
5. Yang, G.-Y.; Li, D.-S.; Lu, F.; Yang, W. RVSIM: A feature similarity method for full-reference image quality assessment. *J. Image Video Proc.* **2018**, *6*, 1–15. [CrossRef]
6. Wu, J.-J.; Lin, W.-S.; Shi, G.-M.; Liu, A. Reduced-reference image quality assessment with visual information fidelity. *IEEE Trans. Multimed.* **2013**, *15*, 1700–1705. [CrossRef]
7. Soundararajan, R.; Bovik, A.C. RRED indices: Reduced reference entropic differencing for image quality assessment. *IEEE Trans. Image Proc.* **2012**, *21*, 517–526. [CrossRef] [PubMed]
8. Wang, S.-Q.; Gu, K.; Zhang, X.; Lin, W.-S.; Zhang, L.; Ma, S.-W.; Gao, W. Subjective and objective quality assessment of compressed screen content images. *IEEE J. Emerg. Sel. Top. Circuits Syst.* **2016**, *6*, 532–543. [CrossRef]
9. Li, L.-D.; Zhu, H.-C.; Yang, G.-B.; Qian, G.-S. Referenceless measure of blocking artifacts by Tchebichef kernel analysis. *IEEE Signal Proc. Lett.* **2014**, *21*, 122–125. [CrossRef]
10. Li, L.-D.;Lin, W.-S.; Wang, X.-O.; Yang, G.-B.; Bahrami, K.; Kot, A.C. No-reference image blur assessment based on discrete orthogonal moments. *IEEE Trans. Cybern.* **2016**, *46*, 39–50. [CrossRef] [PubMed]
11. Zhu, T.; Karam, L. A no-reference objective image quality metric based on perceptually weighted local noise. *J. Image Video Proc.* **2014**, *5*, 1–8. [CrossRef]
12. Fang, Y.M.; Ma, K.-D.; Wang, Z.; Lin, W.-S.; Fang, Z.-J.; Zhai, G.-T. No-reference quality assessment of contrast-distorted images based on natural scene statistics. *IEEE Signal Proc. Lett.* **2015**, *22*, 838–842. [CrossRef]
13. Jiang, Q.-P.; Shao, F.; Lin, W.-S.; Gu, K.; Jiang, G.-Y.; Sun, H.-F. Optimizing multi-stage discriminative dictionaries for blind image quality assessment. *IEEE Trans. Multimed.* **2017**, *99*, 1–14. [CrossRef]
14. Ye, P.; Kumar, J.; Doermann, D. Beyond human opinion scores blind image quality assessment based on synthetic scores. In Proceedings of the 27th IEEE Conference on Computer Vision and Pattern Recognition, Columbus, OH, USA, 24–27 June 2014.
15. Mittal, A.; Soundararajan, R.; Bovik, A.C. Making a completely blind image quality analyzer. *IEEE Signal Proc. Lett.* **2013**, *20*, 209–212. [CrossRef]
16. Saha, A.; Ming, Q.; Wu, J. Utilizing image scales towards totally training free blind image quality assessment. *IEEE Trans. Image Proc.* **2015**, *24*, 1879–1892. [CrossRef] [PubMed]
17. Nizami, I.-F.; Majid, M.; Khurshid, K. Efficient feature selection for Blind Image Quality Assessment based on natural scene statistics. In Proceedings of the 14th International Bhurban Conference on Applied Sciences and Technology, Islamabad, Pakistan, 10–14 January 2017.
18. Bosse, S.; Maniry, D.; Wiegand, T.; Samek, W. A deep neural network for image quality assessment. In Proceedings of the International Conference on Image Processing, Phoenix, AZ, USA, 25–28 September 2016.
19. Zuo, L.-X.; Wang, H.-L.; Fu, J. Screen content image quality assessment via convolutional neural network. In Proceedings of the International Conference on Image Processing, Phoenix, AZ, USA, 25–28 September 2016.

20. Fu, J.; Wang, H.-L.; Zuo, L.-X. Blind image quality assessment for multiply distorted images via convolutional neural networks. In Proceedings of the International Conference on Acoustics, Speech and Signal Processing, Shanghai, China, 20–25 March 2016.
21. Zhang, L.; Gu, Z.-G.; Liu, X.-X.; Li, H.-Y.; Lu, J.-W. Training quality-aware filters for no-reference image quality assessment. *IEEE Trans. Multimed.* **2014**, *21*, 67–75. [CrossRef]
22. Hou, W.-L.; Gao, X.-B.; Tao, D.-C.; Li, X.-L. Blind image quality assessment via deep learning. *IEEE Trans. Neural Netw. Learn. Syst.* **2015**, *26*, 1275–1286. [PubMed]
23. Kang, L.; Ye, P.; Li, Y.; Doermann, D. Convolutional neural networks for no-reference image quality assessment. In Proceedings of the 27th IEEE Conference on Computer Vision and Pattern Recognition, Columbus, OH, USA, 24–27 June 2014.
24. Panagiotakis, C.; Papadakis, H.; Grinias, E.; Komodakis, N.; Fragopoulou, P.; Tziritas, G. Interactive Image Segmentation Based on Synthetic Graph Coordinates, *Pattern Recognit.* **2013**, *46*, 2940–2952. [CrossRef]
25. Ye, P.; Doermann, D. No-reference image quality assessment using visual codebooks. *IEEE Trans. Image Proc.* **2012**, *21*, 3129–3138.
26. Li, Q.-H.; Lin, W.-S.; Xu, J.-T.; Fang, Y.-M. Blind image quality assessment using statistical structural and luminance features. *IEEE Trans. Multimed.* **2016**, *18*, 2457–2469. [CrossRef]
27. Wu, Q.-B.; Li, H.-L.; Meng, F.-M.; Ngan, K.N.; Luo, B.; Huang, C.; Zeng, B. Blind image quality assessment based on multichannel feature fusion and label transfer. *IEEE Trans. Circuits Syst. Video Technol.* **2016**, *26*, 425–440. [CrossRef]
28. Brandao, T.; Queluz, M.P. No-reference image quality assessment based on DCT domain statistics. *Signal Proc.* **2008**, *88*, 822–833. [CrossRef]
29. Saad, M.A.; Bovik, A.C.; Charrier, C. Blind image quality assessment: A natural scene statistics approach in the DCT domain. *IEEE Trans. Image Proc.* **2012**, *21*, 3339–3352. [CrossRef] [PubMed]
30. Chang, J.-H.; Shin, J.-W.; Kim, N.S.; Mitra, S. Image probability distribution based on generalized gamma function. *IEEE Signal Proc. Lett.* **2005**, *12*, 325–328. [CrossRef]
31. Moorthy, A.K.; Bovik, A.C. Blind image quality assessment: From natural scene statistics to perceptual quality. *IEEE Trans. Image Proc.* **2011**, *20*, 3350–3364. [CrossRef] [PubMed]
32. Mittal, A.; Moorthy, A.K.; Bovik, A.C. No-reference image quality assessment in the spatial domain. *IEEE Trans. Image Proc.* **2012**, *21*, 4695–4708. [CrossRef] [PubMed]
33. Zhang, L.; Zhang, L.; Bovik, A.C. A feature-enriched completely blind image quality evaluator. *IEEE Trans. Image Proc.* **2015**, *24*, 2579–2591. [CrossRef] [PubMed]
34. Ghadiyaram, D.; Bovik, A.C. Perceptual quality prediction on authentically distorted images using a bag of features approach. *J. Vision* **2016**, *17*, 1–29. [CrossRef] [PubMed]
35. Ruderman, D.L.; The statistics of natural images. *Netw. Comput. Neural Syst.* **1994**, *5*, 517–548. [CrossRef]
36. Srivastava, A.; Lee, A.B.; Simoncelli, E.P.; Zhu, S.-C. On advances in statistical modeling of natural images. *J. Math. Imag. Vision* **2003**, *18*, 17–22. [CrossRef]
37. Geisler, W.S. Visual perception and the statistical properties of natural scenes. *Annu. Rev. Psychol.* **2008**, *59*, 167–192. [CrossRef] [PubMed]
38. Sheikh, H.R.; Wang, Z.; Cormack, L.; Bovik, A.C. LIVE Image Quality Assessment Database. Available onlie: http://live.ece.utexas.edu/research/quality/subjective.htm (accessed on 16 November).
39. Scholte, H.S.; Ghebreab, S.; Waldorp, L.; Smeulders, A.; Lamme, V. Brain responses strongly correlate with Weibull image statistics when processing natural images. *J. Vision* **2009**, *9*, 1–15. [CrossRef] [PubMed]
40. Willmore, D.B.; Prenger, R.J.; Gallant, J.L. Neural representation of natural images in visual area V2. *J. Neurosci.* **2010**, *30*, 2102–2114. [CrossRef] [PubMed]
41. Karklin, Y.; Lewicki, M.S. Emergence of complex cell properties by learning to generalize in natural scenes. *Nature* **2009**, *457*, 83–86. [CrossRef] [PubMed]
42. Blake, R.; Sekuler, R. *Perception*, 5th ed.; McGraw Hill: New York, NY, USA, 2006.
43. Cho, N.I.; S. U. Lee, S.U. Fast algorithm and implementation of 2-D discrete cosine transform. *IEEE Trans. Circuits Syst.* **1991**, *38*, 297–305.
44. Felleman, D.J.; Van D.C. Distributed hierarchical processing in the primate cerebral cortex. *Cerebral Cortex* **1991**, *1*, 1–47. [CrossRef] [PubMed]

45. Kruger, N.; Janssen, P.; Kalkan, S.; Lappe, M.; Leonardis, A. Deep hierarchies in the primate visual cortex: What can we learn for computer vision? *IEEE Trans. Pattern Anal. Mach. Intell.* **2013**, *35*, 1847–1871. [CrossRef] [PubMed]
46. Xue W.-F.; Mou, X.-Q. Reduced reference image quality assessment based on Weibull statistics. In Proceedings of the Second International Workshop on Quality of Multimedia Experience, Trondheim, Norway, 21–23 June 2010.
47. Moorthy, A.K.; Bovik, A.C. Visual importance pooling for image quality assessment. *IEEE J. Sel. Top. Signal Proc.* **2009**, *3*, 193–201. [CrossRef]
48. Liu, L.; Liu, B.; Huang, H.; Bovik, A.C. No-reference image quality assessment based on spatial and spectral entropies. *Signal Proc. Image Commun.* **2014**, *29*, 856–863. [CrossRef]
49. Zhang, Y.; Lin, W.-S.; Li, Q.-H.; Cheng, W.; Zhang, X. Multiple-level feature based measure for retargeted image quality. *IEEE Trans. Image Proc.* **2018**, *27*, 451–463. [CrossRef] [PubMed]
50. Parscale, S.L.; Dumont, J.F.; Plessner, V.R. The effect of quality management theory on assessing student learning outcomes. *Sam Adv. Mange. J.* **2015**, *80*, 19–30.
51. Smola, A.; Scholkopf, A.B. A tutorial on support vector regression. Stat. Comput. **2004**, *14*, 199–222. [CrossRef]
52. Schölkopf .B.; Smola, A. *Learning With Kernels–Support Vector Machines, Regularization, Optimization and Beyond*; MIT Press:Cambridge, MA, USA, 2001.
53. Chang, C.-C.; Lin, C.-J. LIBSVM: A Library for Support Vector Machines. Available online: http://www.csie.ntu.edu.tw/~cjlin/libsvm/ (accessed on 16 November 2018).
54. Ponomarenko, N.; Lukin, V.; Zelensky, A.; Egiazarian, K.; Carli, M.; Battisti, F. TID 2008 a database for evaluation of full-reference visual quality assessment metrics. *Adv. Mod. Radioelectron* **2009**, *10*, 30–45.
55. Larson, E.C.; Chandler, D. Categorical image quality (CSIQ) database. Available online: http://vision.okstate.edu/csiq (accessed on 16 November).
56. Sheikh, H.R.; Sabir, M.F.; Bovik, A.C. A statistical evaluation of recent full reference image quality assessment algorithms. *IEEE Trans. Image Proc.* **2006**, *15*, 3440–3451. [CrossRef]
57. Moorthy, A.K.; Bovik, A.C. A two-step framework for constructing blind image quality indices. *IEEE Signal Proc. Lett.* **2010**, *17*, 2010, 17, 513–516. [CrossRef]
58. Zhang, L.; Mou, X.-Q.; Zhang, D. FSIM: A feature similarity index for image quality assessment. *IEEE Trans. Image Proc.* **2011**, *20*, 2378–2386. [CrossRef] [PubMed]
59. Zhang, L.; Shen, Y.; Li, H. VSI: A visual saliency-induced index for perceptual image quality assessment. *IEEE Trans. Image Proc.* **2014**, *23*, 4270–4281. [CrossRef] [PubMed]
60. Sheskin, D. *Handbook of Parametric and Nonparametric Statistical Procedures*; Chapman & Hall: London, UK, 2004.
61. Liu,Z.; and Laganière,R. Phase congruence measurement for image similarity assessment. *Pattern Recognit. Lett.* **2012**, *28*, 166–172. [CrossRef]
62. Zhang, L.; Zhang, L.; Mou, X.; Zhang, D. FSIM: A feature similarity index for image quality assessment. *IEEE Trans. Image Proc.* **2011**, *20*, 2378–2386. [CrossRef] [PubMed]
63. Petrovic, V.; Dimitrijevic, V. Focused pooling for objective quality estimation information fusion. In Proceedings of the International Conference on Image Processing, Melbourne, VIC, Australia, 15–18 September 2013.

entropy

MDPI

Article

Ultrasound Entropy Imaging of Nonalcoholic Fatty Liver Disease: Association with Metabolic Syndrome

Ying-Hsiu Lin [1], Yin-Yin Liao [2], Chih-Kuang Yeh [3], Kuen-Cheh Yang [4,5,*] and Po-Hsiang Tsui [1,6,7,*]

[1] Department of Medical Imaging and Radiological Sciences, College of Medicine, Chang Gung University, Taoyuan 33302, Taiwan; suntfg0710@gmail.com
[2] Department of Biomedical Engineering, Hungkuang University, Taichung 43302, Taiwan; g9612536@sunrise.hk.edu.tw
[3] Department of Biomedical Engineering and Environmental Sciences, National Tsing Hua University, Hsinchu 30013, Taiwan; ckyeh@mx.nthu.edu.tw
[4] Department of Family Medicine, National Taiwan University Hospital, Beihu Branch, Taipei 10800, Taiwan
[5] Health Science & Wellness Center, National Taiwan University, Taipei 10617, Taiwan
[6] Department of Medical Imaging and Intervention, Chang Gung Memorial Hospital at Linkou, Taoyuan 33305, Taiwan
[7] Medical Imaging Research Center, Institute for Radiological Research, Chang Gung University and Chang Gung Memorial Hospital at Linkou, Taoyuan 33302, Taiwan
* Correspondence: quintino.yang@gmail.com (K.-C.Y.); tsuiph@mail.cgu.edu.tw (P.-H.T.); Tel.: +886-2-2371-7101 (K.-C.Y.); +886-3-211-8800 (ext. 3795) (P.-H.T.)

Received: 17 October 2018; Accepted: 20 November 2018; Published: 22 November 2018

Abstract: Nonalcoholic fatty liver disease (NAFLD) is the leading cause of advanced liver diseases. Fat accumulation in the liver changes the hepatic microstructure and the corresponding statistics of ultrasound backscattered signals. Acoustic structure quantification (ASQ) is a typical model-based method for analyzing backscattered statistics. Shannon entropy, initially proposed in information theory, has been demonstrated as a more flexible solution for imaging and describing backscattered statistics without considering data distribution. NAFLD is a hepatic manifestation of metabolic syndrome (MetS). Therefore, we investigated the association between ultrasound entropy imaging of NAFLD and MetS for comparison with that obtained from ASQ. A total of 394 participants were recruited to undergo physical examinations and blood tests to diagnose MetS. Then, abdominal ultrasound screening of the liver was performed to calculate the ultrasonographic fatty liver indicator (US-FLI) as a measure of NAFLD severity. The ASQ analysis and ultrasound entropy parametric imaging were further constructed using the raw image data to calculate the focal disturbance (FD) ratio and entropy value, respectively. Tertiles were used to split the data of the FD ratio and entropy into three groups for statistical analysis. The correlation coefficient r, probability value p, and odds ratio (OR) were calculated. With an increase in the US-FLI, the entropy value increased ($r = 0.713$; $p < 0.0001$) and the FD ratio decreased ($r = -0.630$; $p < 0.0001$). In addition, the entropy value and FD ratio correlated with metabolic indices ($p < 0.0001$). After adjustment for confounding factors, entropy imaging (OR = 7.91, 95% confidence interval (CI): 0.96–65.18 for the second tertile; OR = 20.47, 95% CI: 2.48–168.67 for the third tertile; $p = 0.0021$) still provided a more significant link to the risk of MetS than did the FD ratio obtained from ASQ (OR = 0.55, 95% CI: 0.27–1.14 for the second tertile; OR = 0.42, 95% CI: 0.15–1.17 for the third tertile; $p = 0.13$). Thus, ultrasound entropy imaging can provide information on hepatic steatosis. In particular, ultrasound entropy imaging can describe the risk of MetS for individuals with NAFLD and is superior to the conventional ASQ technique.

Keywords: ultrasound; hepatic steatosis; Shannon entropy; fatty liver; metabolic syndrome

1. Introduction

Nonalcoholic fatty liver disease (NAFLD) is characterized by excess and abnormal intracellular accumulation of triglycerides in hepatocytes. Histologically, NAFLD refers to macrovesicular steatosis and is the leading cause of nonalcoholic steatohepatitis, fibrosis, cirrhosis, and hepatocellular carcinoma [1,2]. Therefore, NAFLD may be considered a critical health problem, and its early detection, follow-up, and management can help arrest the progression of advanced liver diseases [3,4].

Currently, liver biopsy is the gold standard for diagnosing NAFLD [5]. However, liver biopsy is an invasive procedure and can lead to serious complications (e.g., bleeding), and its diagnosis may be inconsistent between pathologists [6,7]. Moreover, sampling errors limit the use of liver biopsy in clinical practice. Additionally, most patients with NAFLD have no significant clinical symptoms, and performing liver biopsies on such patients is ethically controversial. To resolve this dilemma, noninvasive imaging modalities such as ultrasound, computed tomography, magnetic resonance imaging, and magnetic resonance spectroscopy (MRS) are commonly used for the assessment of hepatic steatosis [8]. Ultrasound imaging provides several advantages, including ease of routine examination, cost-effectiveness, portability, and nonionizing imaging principles, and thus it is currently the first-line modality for assessing hepatic steatosis and evaluating NAFLD.

Ultrasound performs well in detecting moderate to severe hepatic steatosis [9,10]. However, its diagnostic accuracy for detecting mild hepatic steatosis is limited. Furthermore, qualitative descriptions, operator experience, and interobserver and intraobserver variability degrade the sonographic assessment of fatty liver [11,12]. Quantitative analysis of ultrasound images may provide additional clues to improve the diagnosis of mild NAFLD. Essentially, liver parenchyma can be modeled as a scattering medium consisting of numerous acoustic scatterers [13,14] that interact with the incident wave to form ultrasound backscattered signals. Different scatterer properties result in different waveforms of backscattered signals, and thus the corresponding statistical properties may depend on information associated with changes in liver microstructures [13].

Considering the randomness of ultrasound backscattering, statistical distributions are widely used to model backscattered statistics for tissue characterization [15]. Nakagami [16,17] and homodyned-K distributions [18] have been applied to model ultrasound backscattered statistics for the assessment of hepatic steatosis. However, acoustic structure quantification (ASQ) based on Chi-squared testing of backscattered envelopes is the only technique that has been commercialized in ultrasound scanners (Toshiba machine) by using the concept of statistical distribution. Initially, ASQ was developed to quantify the difference between backscattered statistics and Rayleigh distribution [19]. ASQ has been validated as having high performance in evaluating NAFLD because fat accumulation in the liver tends to make the statistics of backscattered data follow the Rayleigh distribution [20–24].

When using ASQ or model-based methods to characterize tissue, the data used to estimate the parameters must conform to the used statistical distribution [25,26]. This requirement may not always be satisfied, because adjusting the settings in an ultrasound system or using nonlinear signal-processing approaches (e.g., logarithmic compression) may alter the statistical distribution of raw data. This limitation has motivated researchers to consider non-model-based statistical approaches. Among all possible approaches, Shannon entropy—an estimate of signal uncertainty and complexity proposed in information theory [27]—has the highest potential and flexibility for analyzing ultrasound backscattering. Hughes first proposed using information (Shannon) entropy to analyze ultrasound signals, indicating that entropy can be used to quantitatively depict changes in the microstructures of scattering media [28,29]. In particular, one report demonstrated that information entropy can describe ultrasound backscattered statistics without considering the statistical properties of ultrasound data [30]. Recent studies have further indicated that entropy parametric imaging enables visualization and characterization of hepatic steatosis, thereby making it possible to implement non-model-based structure quantification of NAFLD [31–33].

While non-model-based entropy imaging plays an increasingly key role in physically describing changes in the microstructures of fatty liver, its meanings require further biological explanation.

The establishment and validation of ultrasound entropy imaging to characterize hepatic steatosis are based on the association of entropy value with hepatic histological changes [33]. However, NAFLD is not only a change in liver microstructures caused by fat accumulation but also strongly related to obesity, hypertension, type 2 diabetes mellitus, and dyslipidemia, all of which are metabolic abnormalities and can be considered hepatic manifestations of metabolic syndrome (MetS) [34,35]. MetS is typically caused by insulin resistance, and although glucose clamp is the gold standard for quantifying insulin resistance, it is a complex procedure that is unsuitable for routine use. For this reason, Matthews et al. developed the homeostatic model assessment for insulin resistance (HOMA-IR) index, which is calculated using fasting insulin and blood glucose for a general evaluation of MetS [36]. The HOMA-IR index correlates with the conventional ultrasound B-scan image features of hepatic steatosis [37,38], implying that ultrasound imaging can depict metabolic information. Therefore, we explored the relationship between MetS and quantitative ultrasound analysis of NAFLD by using entropy imaging.

This study had two objectives: (i) investigating the association of ultrasound entropy imaging of NAFLD with MetS to endow entropy images with new biological insights, and (ii) comparing the performance of entropy imaging in predicting the risks of suffering from MetS with that of conventional ASQ to determine whether non-model-based approaches are at all superior for evaluating MetS. The results showed that ultrasound entropy imaging performed well in describing the metabolic behavior of patients with NAFLD. Moreover, ultrasound entropy imaging was superior to ASQ in risk evaluation for MetS.

2. Materials and Methods

2.1. Subjects

This study was conducted following approval by the Institutional Review Board of National Taiwan University Hospital. All participants were asked to complete standardized questionnaires and provided informed consent. Participants with the following conditions were excluded: excessive alcohol intake (>20 g/day for women and >30 g/day for men) and chronic liver disease (chronic hepatitis, autoimmune, drug-induced, vascular, or inherited hemochromatosis or Wilson disease). A total of 394 patients were recruited.

2.2. Anthropometric Indices and Biochemical Analyses

Routine physical examinations and blood tests were conducted for each participant. Body mass index (BMI) was calculated as weight divided by height squared. Waist circumference (WC) was measured at the middle between the costal margin and iliac crest. Systolic blood pressure (SBP) and diastolic blood pressure (DBP) were recorded. Fasting plasma glucose (FPG), total cholesterol (TCH), triglycerides (TG), high-density lipoprotein (HDL-C), low-density lipoprotein (LDL-C), aspartate aminotransferase (AST), alanine aminotransferase (ALT), and insulin were measured after 8 hours of overnight fasting. Using FPG and insulin, the HOMA-IR index was calculated to examine insulin resistance [36].

2.3. Diagnosis of MetS

Data obtained from anthropometric and blood examinations were further used to identify MetS. According to the modified National Cholesterol Education Program Adult Treatment Panel III Criteria (NCEP-ATP III), MetS (for the Taiwanese population) is diagnosed when at least three of the following criteria are satisfied [39]: (i) WC $\geq$ 90 cm in men and $\geq$ 80 cm in women; (ii) SBP $\geq$ 130 mmHg or DBP $\geq$ 85 mmHg or use medication for hypertension; (iii) hyperglycemia (FPG $\geq$ 100 mg/dL) or the use of medication for diabetes; (iv) hypertriglyceridemia (TG $\geq$ 150 mg/dL) or use of medication for hyperlipidemia; and (v) low HDL-C ($\leq$40 mg/dL in men and $\leq$50 mg/dL in women).

2.4. Ultrasound Examinations for NAFLD Evaluation

After blood withdrawal, standard abdominal ultrasound screening of the liver was performed immediately by three physicians, each with more than 20 years' experience. A clinical ultrasound scanner (Model 3000; Terason, Burlington, MA, USA) equipped with a convex transducer (Model 5C2A; Terason) of 3 MHz was used; the transducer had 128 elements and the pulse length of the incident wave was approximately 2.3 mm. For each participant, the ultrasonographic fatty liver indicator (US-FLI) was used as a semiquantitative measure of severity of NAFLD [40]. Specifically, the US-FLI was calculated using the following criteria: (i) presence of liver/kidney contrast graded as mild/moderate (score 2) or severe (score 3); (ii) presence (score 1 each) or absence (score 0 each) of posterior attenuation of ultrasound beam, vessel blurring, difficult visualization of the gallbladder wall, difficult visualization of the diaphragm, and areas of focal sparing. NAFLD was diagnosed if the score $\geq$2 [40].

2.5. Quantitative Analysis using ASQ and Entropy Imaging

Except for the standard abdominal scans, all physicians followed the same protocols and system settings for data acquisition and quantitative analysis. For each patient, the same scanner was used to scan the liver through the subcostal scanning approach. It has been shown that a signal-to-noise ratio (SNR) > 11 dB allows reliable descriptions of ultrasound backscattered statistics [41]. For this consideration, the system gain index was set at 6, corresponding to a SNR of approximately 30 dB, which was obtained from the calibrations in the previous study [42]. Such a high SNR implies that no significant noise components exist in the backscattered signals, ensuring the quality of parameter estimation in the ASQ analysis. The imaging depth was 16 cm and the focal zone corresponded to the central part of the liver to reduce the effect of beam diffraction. Raw image data consisting of 128 scan lines of backscattered radio frequency signals at a sampling rate of 30 MHz were obtained using the software kit provided by Terason. The envelope image of each raw image raw datum was constructed by taking the absolute value of the Hilbert transform of each scan line. The grayscale B-mode image was formed based on the logarithm-compressed envelope data at a dynamic range of 40 dB.

In ASQ, the Chi-squared test is used to evaluate the difference between the sample and the population data. The following equation is used to define parameter C_m^2 [19]:

$$C_m^2 = \frac{\sigma_m^2}{\sigma_R^2(\mu_m)} = \left[\frac{\pi}{4-\pi}\right]\frac{\sigma_m^2}{\mu_m^2} \tag{1}$$

where μ_m and σ_m^2 are the average and variance of the measured backscattered envelopes, respectively. The value of $\sigma_R^2(\mu_m)$ indicates the variance of the Rayleigh-distributed data estimated using μ_m. In this study, the sliding window technique was used to obtain a C_m^2 parametric map. In brief, a window was created to move across the entire envelope image in steps representing the number of pixels corresponding to the window overlap ratio (WOR); during this process, local parameters were successively estimated using local envelope data within the window so that a parametric map could eventually be constructed. The window side length (WSL) was three times the pulse length, which is an appropriate size for stably estimating ultrasound statistical parameters [17]. The WOR was 50% to provide a tradeoff between the parametric image resolution and computational time [43]. A region of interest (ROI) manually outlined on the B-mode image of the liver parenchyma was used for analysis of the C_m^2 parametric map. Some basic criteria suggested previously were used for determining the ROI [17]: (i) visible blood vessels were excluded in the ROI to reduce the bias of characterizing liver parenchyma. The size of the ROI was set 3 $\times$ 3 cm^2; (ii) the ROI was located at the focal zone, reducing the effects of attenuation and diffraction on the backscattered signals.

Referring to a previous study [44], the histogram of C_m^2 in the ROI revealed a narrow distribution when the tissue was homogeneous. A relatively broad distribution represented either diffusely inhomogeneous (consisting of microstructures) or focally inhomogeneous (consisting of macrostructures such as vessels) tissue. To eliminate macrostructural information, sliding window processing of the envelope image was performed again (generating a second C_m^2 map denoted as rC_m^2 map), where local data in the window were excluded if the amplitude > ($\mu + \alpha\sigma$) (μ: mean value of envelope data in the ROI, σ: standard deviation of envelope data in the ROI, and α: a removal coefficient). For each pixel location in the ROI, if the ratio of C_m^2/rC_m^2 was lower than the threshold k, C_m^2 was considered to exhibit no significant changes after rejecting the outliers of the envelope signals. In this condition, C_m^2 was assigned to the pixel location. If C_m^2/rC_m^2 was greater than k, rC_m^2 was used. Finally, the values of C_m^2 and rC_m^2 in the ROI were separated to construct two histograms to represent microstructure (diffuse inhomogeneity or homogeneity) and macrostructure (focal inhomogeneity) curves. The focal disturbance (FD) ratio was defined as the ratio of the area under the curve for C_m^2 and rC_m^2 histograms, expressed as

$$FD - ratio = \frac{AUC(rC_m^2\ histogram)}{AUC(C_m^2\ histogram)} \quad (2)$$

When the resolution cell of the transducer contains a large number of randomly distributed scatterers, the statistical distribution of ultrasound backscattered envelopes exhibits the Rayleigh distribution [13–15]. This condition represents that no macrostructures exist in the tissue to generate information of rC_m^2, and thus the FD ratio is theoretically equal to zero. On the contrary, the FD ratio increases with the degree of deviation from Rayleigh statistics [44]. Please note that the removal coefficient $\alpha = 7$ [19] and the threshold $k = 1.2$ [44] were suggested previously but could be empirically determined [19,45]. Because the initial equipment (Toshiba system with software package) was unavailable in this study, we fine-tuned the parameters for the used Terason system. The values of α and k were set at 3 and 1.1, respectively. The algorithmic scheme of estimating the FD ratio is illustrated in Figure 1.

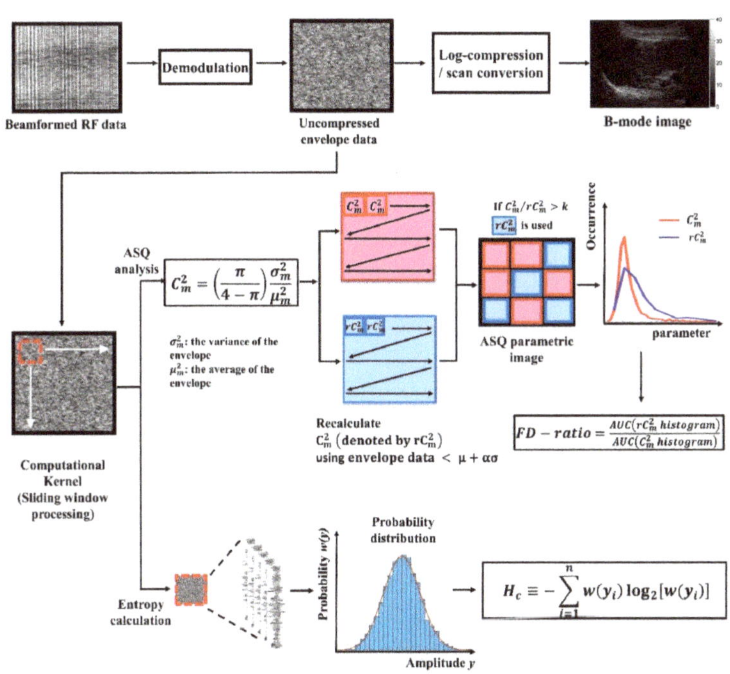

Figure 1. Computational flowchart for ultrasound acoustic structure quantification (ASQ) and entropy estimations.

The algorithm for ultrasound entropy imaging is also based on the sliding window technique to process the envelope image and is illustrated in Figure 1. Because the acquired ultrasound backscattered signals digitalized by the imaging system belong to discrete signals, the Shannon entropy of a discrete random variable Y with possible values $\{y_1, y_2, \ldots, y_n\}$ (i.e., the envelope data points included within the sliding window) was calculated using the following discrete form:

$$H_c \equiv -\sum_{i=1}^{n} w(y_i)\log_2[w(y_i)] \quad (3)$$

where $w(\cdot)$ represents the function of probability distribution. In this study, the statistical histogram of the data (bins = 200) was used as an alternative $w(\cdot)$ for estimation [31,32]. To compare the results of entropy with those of ASQ, ultrasound entropy imaging was constructed using the same WSL (6.9 mm) and WOR (50%). The ROI used in the ASQ analysis was directly applied to the entropy parametric image to calculate the average entropy value.

2.6. Statistical Analysis

The Kolmogorov-Smirnov, Anderson-Darling, Cramer-Von Mises, and Shapiro-Wilk tests of the data (the US-FLI, FD ratio, and entropy) were used for normality testing. Tertiles were used to split the data of the FD ratio and entropy into three groups. For each group, the categorical data were presented as percentages and the continuous variables were expressed as mean ± standard deviation. Initially, the interrelationships between the US-FLI, FD ratio, and entropy value were plotted to calculate the Pearson correlation coefficient r and probability value p. Then, the categorical data were analyzed using the Chi-squared test. The continuous variables in each group were compared using analyses of variance. The Cochran-Armitage trend test was conducted to test for trends in the anthropometric and metabolic factors by using tertiles of the FD ratio and entropy value. The associations of the FD ratio and entropy value with MetS were assessed using a multiple logistic regression model adjusted for age, sex, alcohol consumption, smoking, betel nut chewing, hours of exercise per week, menopause status (women only), BMI, and HOMA-IR. To further compare the abilities of the FD ratio and entropy in predicting the risk of suffering from MetS, the odds ratio (OR) and 95% confidence interval (CI) were calculated. The significant difference was set at $p < 0.05$. All statistical analyses were conducted using SAS version 9.3 (SAS Inc., Cary, NC, USA).

3. Results

3.1. Baseline Characteristics of the Participants

The baseline characteristics of the participants are shown in Table 1. A total of 394 participants was recruited, comprising 151 (38.3%) men and 243 (61.7%) women (mean age: 40.5 ± 11.3 years). According to information obtained from questionnaires, anthropometric examinations, blood tests, and ultrasound evaluations of NAFLD, the overall prevalence of MetS was 19.3% and the US-FLI, FD ratio, and entropy value of the participants were 2.22 ± 2.25, 0.96 ± 0.44, and 3.99 ± 0.06, respectively. To observe how the statistical properties of backscattered signal varied with the severity of NAFLD, dot and box plots of the FD ratio and entropy value corresponding to each US-FLI were plotted (Figure 2). Based on observations of the data, exponential increasing and decreasing functions were used for fitting dot plots of the entropy and FD ratio, respectively. With an increase in the US-FLI, the FD ratio decreased ($r = -0.630$; $p < 0.0001$) and the entropy value monotonically increased ($r = 0.713$; $p < 0.0001$). Box plots further identified outliers for the entropy and FD ratio. Some outliers were found to exist in the data distributions of the entropy and FD ratio. This is acceptable and reasonable, especially for a large amount of biodata (total n = 394). The US-FLI underestimating the extent of NAFLD [16] is another possible reason for the outliers of entropy and FD ratio. On the other hand, the normality tests based on four kinds of methods (as described in Section 2.6) indicated that the data of the US-FLI, FD ratio, and entropy did not follow the normal distribution ($p < 0.0001$). However, the data distribution does not affect the subsequent analysis (using the OR to evaluate the risk of MetS) because the OR estimations were based on the tertiles of entropy and FD ratio.

Table 1. Patient characteristics (n = 394).

Variables	Value *
Questionnaires	
Gender F/M	243/151
Age (yrs)	40.5 ± 11.3 (20–72)
Menopause	25 (6.4)
Smoking	
Never	336 (85.3)
Current	42 (10.7)
Previous	16 (4.1)
Alcohol	
Never	322 (81.7)
Current	64 (16.2)
Previous	8 (2)
Betel Nuts	
Never	375 (95.2)
Current	19 (4.8)
Exercise time (mins/per week)	99.6 ± 189.4 (0–1500)
Anthropometric variable	
BMI (kg/m^2)	24.1 ± 4.6 (14.8–43.7)
Waist (cm)	81.9 ± 11.3 (55–123)
SBP (mmHg)	122.5 ± 16.3 (86–180)
DBP (mmHg)	77.9 ± 11.9 (50–133)
Biochemistry parameters	
FPG (mg/dL)	87.7 ± 17.6 (58–272)
TCH (mg/dL)	192.9 ± 35.5 (101–320)
TG (mg/dL)	112.4 ± 90.3 (25–888)
HDL-C (mg/dL)	57.3 ± 15.8 (25–120)
LDL-C (mg/dL)	120.8 ± 32.5 (47–238)
AST (U/L)	22.9 ± 8.9 (11–68)
ALT (U/L)	26.5 ± 21.4 (2–151)
Insulin (μU/mL)	9.1 ± 8.2 (2–84.4)
HOMA-IR	1.17 ± 1.03 (0.26–10.2)
MetS (%)	76 (19.3)
Ultrasound parameters	
US-FLI Score	2.22 ± 2.25 (0–8)
ASQ FD-ratio	0.96 ± 0.44 (0.21–2.89)
Entropy	3.99 ± 0.06 (3.80–4.07)

* Categorical data are expressed as numbers (percentage); continuous variables are expressed as mean ± SD (range). BMI: body mass index; SBP: systolic blood pressure; DBP: diastolic blood pressure; FPG: fasting plasma glucose; TCH: total cholesterol; TG: triglycerides; HDL-C: high-density lipoprotein cholesterol; LDL-C: low-density lipoprotein cholesterol; AST: aspartate aminotransferase; ALT: alanine aminotransferase; HOMA-IR: homeostasis model assessment for insulin resistance; MetS: metabolic syndrome.

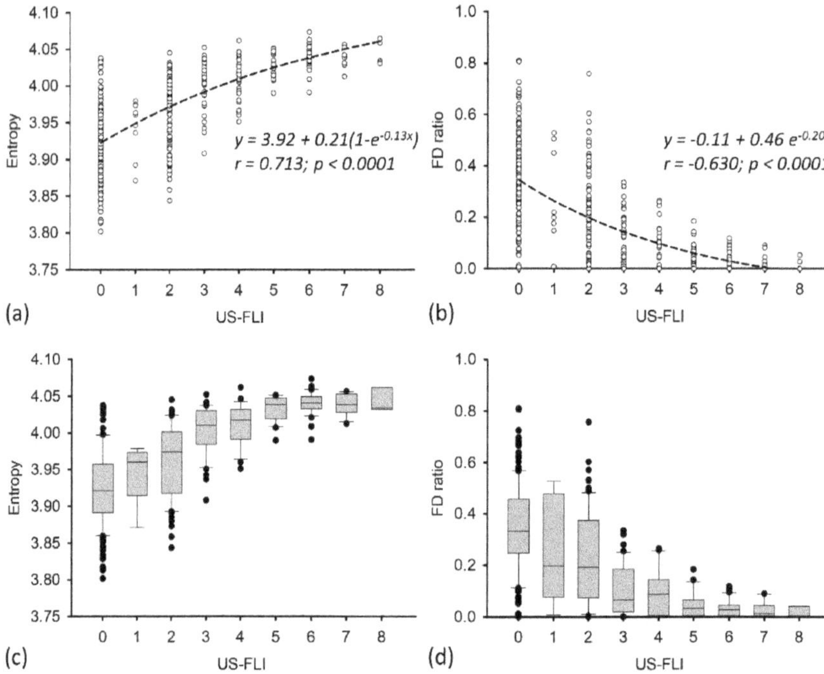

Figure 2. (**a**) and (**b**) Dot plots of entropy value (**left**)/FD (focal disturbance) ratio (**right**) corresponding to each US-FLI. (**c**) and (**d**) Box plots of entropy value (**left**)/FD ratio (**right**) corresponding to each US-FLI.

3.2. Characteristics of Participants in Different Tertiles

Table 2 shows the characteristics of participants in different tertiles of the FD ratio and entropy. No significant difference in age was found ($p = 0.2602$). The percentage in men decreased with an increase in the FD ratio ($p < 0.0001$). Compared with the patients in higher tertiles, those in lower tertiles (lower FD ratios) exhibited lower HDL-C ($p < 0.0001$) and higher WC, BMI, body fat, SBP, DBP, FPG, TCH, TG, LDL-C, MetS, insulin, HOMA-IR, and abnormal liver function ($p = 0.0353$ for TCH; $p = 0.0022$ for LDL-C; $p < 0.0001$ for the others). Similar results were found in the tertiles of entropy. With an increase in entropy (from lower to higher tertiles), WC, BMI, body fat, SBP, DBP, FPG, TCH, TG, LDL-C, MetS, insulin, HOMA-IR, and abnormal liver function increased (all $p < 0.0001$), and HDL-C decreased ($p < 0.0001$). These results revealed that the ultrasound entropy value and FD ratio correlate with MetS.

Table 2. Characteristics of participants in different tertiles of ultrasound quantitative parameters.

	Entropy				ASQ FD-ratio			
Variables	**1_{st} tertile**	**2_{nd} tertile**	**3_{rd} tertile**	***p*-value**	**1_{st} tertile**	**2_{nd} tertile**	**3_{rd} tertile**	***p*-value**
No. of participants	131	131	132		131	131	132	
Gender F/M	113/18	76/55	54/78	<0.0001	65/66	77/54	101/31	<0.0001
Age (yrs)	38.33 ± 9.87	41.6 ± 11.86	41.7 ± 11.81	0.009	41.40 ± 11.79	40.79 ± 11.02	39.40 ± 11.07	0.2602
Waist (cm)	73.31 ± 7.28	81.96 ± 9.9	90.43 ± 9.18	<0.0001	88.24 ± 10.98	82.61 ± 9.99	74.79 ± 8.32	<0.0001
BMI * (kg/m^2)	20.83 ± 2.35	24.2 ± 4.05	27.29 ± 4.41	<0.0001	26.35 ± 4.47	24.61 ± 4.70	21.40 ± 2.78	<0.0001
SBP (mmHg)	114.88 ± 13.1	121.98 ± 16.97	130.58 ± 14.81	<0.0001	128.76 ± 16.74	122.99 ± 15.48	115.76 ± 14.06	<0.0001
DBP (mmHg)	73.58 ± 9.93	76.81 ± 12.44	83.36 ± 11.27	<0.0001	81.56 ± 11.95	77.82 ± 12.00	74.42 ± 10.86	<0.0001
FPG (mg/dL)	81.64 ± 8.36	86.95 ± 11.45	94.47 ± 25.39	<0.0001	92.85 ± 24.77	87.59 ± 13.60	82.70 ± 9.13	<0.0001
TCH (mg/dL)	181.29 ± 31.12	194.95 ± 35.77	202.33 ± 36.34	<0.0001	198.20 ± 36.30	195.49 ± 36.13	185.01 ± 32.87	0.0353
TG (mg/dL)	65.44 ± 28.39	106.46 ± 67.7	164.86 ± 118.71	<0.0001	151.11 ± 121.47	111.71 ± 72.79	74.63 ± 40.21	<0.0001
HDL-C (mg/dL)	65.66 ± 13.78	57.03 ± 14.59	49.17 ± 14.63	<0.0001	50.66 ± 12.33	56.26 ± 17.57	64.82 ± 13.83	<0.0001
LDL-C (mg/dL)	106.73 ± 27.43	123.6 ± 31.9	131.98 ± 32.79	<0.0001	127.57 ± 32.65	124.27 ± 31.24	110.64 ± 31.23	0.0022
AST (U/L)	19.60 ± 5.92	22.59 ± 8.48	26.37 ± 10.42	<0.0001	26.68 ± 10.72	22.49 ± 8.50	19.45 ± 5.06	<0.0001
ALT (U/L)	15.84 ± 7.53	25.01 ± 18.65	38.67 ± 26.67	<0.0001	39.51 ± 28.02	24.93 ± 16.31	15.26 ± 5.99	<0.0001
Insulin (μU/mL)	5.52 ± 3.4	9.06 ± 9.29	12.05 ± 8.7	<0.0001	10.71 ± 7.54	10.44 ± 10.63	5.7 ± 3.49	<0.0001
HOMA-IR	0.71 ± 0.44	1.16 ± 1.14	1.56 ± 1.10	<0.0001	1.39 ± 0.98	1.33 ± 1.31	0.73 ± 0.45	<0.0001
MetS (%)	1 (0.8%)	20 (15.3%)	55 (41.7%)	<0.0001	44 (33.6%)	24 (18.3%)	8 (6.1%)	<0.0001
US-FLI Score	0.53 ± 0.9	1.71 ± 1.55	4.42 ± 2.0	<0.0001	3.82 ± 2.40	2.18 ± 1.82	0.67 ± 1.09	<0.0001

* BMI: body mass index; SBP: systolic blood pressure; DBP: diastolic blood pressure; FPG: fasting plasma glucose; TCH: total cholesterol; TG: triglycerides; HDL-C: high-density lipoprotein cholesterol; LDL-C: low-density lipoprotein cholesterol; AST: aspartate aminotransferase; ALT: alanine aminotransferase; HOMA-IR: homeostasis model assessment for insulin resistance; MetS: metabolic syndrome.

3.3. *The Risks of Metabolic Syndrome in Different Tertiles for the FD Ratio and the Entropy Value*

The risk of metabolic syndrome in each tertile are compared in Table 3. For the FD ratio, the second tertile (lower FD ratios) exhibited a higher risk of MetS (OR = 0.48; 95% CI: 0.26–0.89) than did the third tertile (OR = 0.04; 95% CI: 0.01–0.14) after use of model 1 adjusted for age, sex, smoking, alcohol consumption, betel nut chewing, hours of exercise per week, and menopause status ($p < 0.0001$). Following further adjustment for BMI (model 2), the ORs in the second and third tertiles were 0.59 (95% CI: 0.30–1.18) and 0.41 (95% CI: 0.16–1.05), respectively ($p = 0.1144$). After use of HOMA-IR to further adjust the OR (model 3), the ORs in the second and third tertile were 0.55 (95% CI: 0.27–1.14) and 0.42 (95% CI: 0.15–1.17), respectively ($p = 0.13$). Notably, entropy improved the performance of predicting the risk of MetS. Through use of model 1, the OR of entropy in the third tertile (85.57; 95% CI: 11.25–650.56) was larger than that in the second tertile (51.29; 95% CI: 2.76–164.43) ($p < 0.0001$). After adjustment using model 2, the OR of entropy in the third tertile (26.84; 95% CI: 3.34–215.4) was higher than that in the second tertile (10.27; 95% CI: 1.29–82.14) ($p = 0.0007$), as in model 3 (OR = 7.91, 95% CI: 0.96–65.18 for the second tertile; OR = 20.47, 95% CI: 2.48–168.67 for the third tertile; $p = 0.0021$). The results indicated that non-model-based entropy provides a stronger link to biologically metabolic information than does conventional ASQ.

Table 3. ORs in each tertile of entropy and the FD (focal disturbance) ratio for evaluating the risk of MetS.

	Entropy				ASQ FD-ratio			
	1_{st} tertile (n = 131)	2_{nd} tertile (n = 131)	3_{rd} tertile (n = 132)	*p*-value	1_{st} tertile (n = 131)	2_{nd} tertile (n = 131)	3_{rd} tertile (n = 132)	*p*-value
Model 1 *	ref	51.29 (2.76–164.43)	85.57 (11.25–650.56)	<0.0001	ref	0.48 (0.26–0.89)	0.04 (0.01–0.14)	<0.0001
Model 2	ref	10.27 (1.29–82.14)	26.84 (3.34–215.4)	0.0007	ref	0.59 (0.30–1.18)	0.41 (0.16–1.05)	0.1144
Model 3	ref	7.91 (0.96–65.18)	20.47 (2.48–168.67)	0.0021	ref	0.55 (0.27–1.14)	0.42 (0.15–1.17)	0.13

* Model 1: adjusted for age, gender, smoking, alcohol consumption, betel nut chewing, hours of exercise per week, and menopause status (women only). Model 2: same as model 1 plus further adjustment for BMI. Model 3: model 2 plus further adjustment for HOMA-IR.

4. Discussion

4.1. *Significance of This Study*

With the development and commercialization of ultrasound statistical models and parametric imaging, physicians gradually have a new choice for diagnosing NAFLD. As stated in the Introduction, more than one statistical distribution can be used to assess hepatic steatosis, and the ASQ technique has the clinical benefit of using the model-based method to analyze the statistical properties of backscattered signals from fatty liver. The best statistical distribution for modeling the backscattered statistics of NAFLD is yet to be determined. However, ultrasound entropy imaging based on information theory is more adaptive to various signal characteristics because the calculation of entropy does not need to consider the statistical properties of the signal itself. Therefore, when viewing entropy imaging as a new approach for NAFLD diagnosis, it is necessary to not only perform pathological validations but also to explore the metabolic meanings of entropy. Studies have confirmed that the value of ultrasonic entropy is closely related to the pathological changes of hepatic steatosis [32,33]. However, the present study expands our understanding and domain knowledge of ultrasound entropy imaging; we demonstrated that ultrasound entropy imaging can describe the risk of MetS for those with NAFLD and is superior to the conventional ASQ technique.

4.2. Effects of NAFLD on FD Ratio and Entropy

The US-FLI was used as a semiquantitative measure of NAFLD in this study. Our results showed that both the FD ratio obtained from ASQ analysis and the entropy value of ultrasound entropy imaging correlated with the US-FLI, indicating that these two parameters vary with the progress of NAFLD because macrovesicular steatosis is the major pathological change of NAFLD. Macrovesicular steatosis refers to the presence of a single large fat droplet in a hepatocyte that pushes the nucleus to the periphery. In this scenario, the number of acoustic scatterers (fat droplets) increases equivalently in the scattering medium (liver parenchyma), and the enhancement of constructive wave interference results in changes in the waveforms of the backscattered signals, making the corresponding backscattered statistics vary from pre-Rayleigh (backscattered statistics for healthy livers in practice) to Rayleigh distribution (hepatic steatosis) [16,17,21]. This explains why the FD ratio ASQ parameter monotonically decreases with an increase in the degree of hepatic steatosis. Concurrently, the effect of constructive wave interference leads to increases in signal uncertainty and complexity, making the entropy value [31,32].

4.3. Insulin Resistance: Bidirectional Link between MetS and NAFLD

In general, the increased prevalence of MetS is primarily a result of overnutrition and a sedentary lifestyle. MetS is a key risk factor for cardiovascular disease incidence and mortality, as well as for all-cause mortality [46]. The central etiological cause of MetS is commonly considered to be insulin resistance, which is defined as the failure of insulin to stimulate glucose transport to its target cells [47]. Insulin is a pleiotropic hormone that regulates several cell functions, including stimulation of glucose transport, cell growth, energy balance, and regulation of gene expression [48]. The functions of insulin are associated with two signal pathways: the phosphatidylinositol 3-kinase-protein kinase B pathway and mitogen-activated protein kinase pathway [49].

Once these signal pathways have been altered, insulin resistance is initialized. Free fatty acids (FFAs) play a key role in the development of insulin resistance [49]. As insulin resistance develops, a large quantity of plasma FFAs are released by white adipose tissues into the liver, leading to hepatic fat accumulation [50]. At the same time, overnutrition and a sedentary lifestyle closely correlate with the occurrence of NAFLD. In those who suffer from NAFLD, hepatic fat accumulation can result in hepatic insulin resistance to strengthen the behavior of MetS [49]. Therefore, insulin resistance could be considered the bidirectional link between MetS and NAFLD [49,51].

4.4. Superiority of Entropy in the Assessment of NAFLD and MetS

Several studies have clearly indicated that NAFLD is not only a cause of liver disease but also a key risk indicator of cardiovascular disease [52–54]. Patients with both NAFLD and MetS have an increased risk of cardiovascular disease [55]. For these reasons, a quantitative ultrasound parameter used for evaluating NAFLD should satisfy two requirements: (i) changes in liver microstructures during fatty infiltration can be described and explained from a histological viewpoint, and (ii) significant metabolic information can be reflected to satisfy a variety of clinical applications. In this study, both the ASQ and entropy imaging were shown to able to characterize NAFLD and MetS. Compared with the ASQ, however, entropy imaging better fulfills the above two requirements, as supported by the current results. First, the entropy value of ultrasound entropy imaging is more relevant than the FD ratio of the ASQ to the US-FLI (Figure 2), representing that the entropy image characterizes NAFLD more effectively. Second, the entropy value better predicted the risk of MetS than the FD ratio did (Table 3), demonstrating that entropy imaging links metabolic information more strongly.

Possible mechanisms for why ultrasound entropy imaging provides improved performances in evaluating NAFLD and MetS are discussed below. As mentioned in Section 4.3, NAFLD and MetS interact with each other. Consequently, as long as ultrasound parameters can robustly and precisely describe changes in the backscattered statistics during the process of fatty infiltration in the liver, the opportunities to show more metabolic information increase. From this viewpoint, ultrasound entropy imaging is superior to the ASQ technique. As reviewed in the Introduction, the ASQ technique based on the analysis of ultrasound backscattered statistics is gaining attention for the diagnosis of NAFLD. Some animal studies have revealed that ASQ has a high ability to detect hepatic steatosis [20,24,45]; however, its value in quantifying the degree of hepatic steatosis in human liver remains in dispute because of inconsistent findings. For example, Son et al. demonstrated that the FD ratio correlated with the hepatic fat fraction (HFF) measured using MRS ($r = -0.87$; $p < 0.001$) [21], whereas Karlas et al. found that the FD ratio did not significantly correlate with the HFF ($r = -0.43$; $p = 0.004$) [22]. Failure to use the same procedures and settings for ASQ measurements may be one cause for inconsistent findings. The criteria used for rejecting envelope signals and comparing C_m^2 and rC_m^2 in the ASQ algorithm may also result in uncertainty in ASQ analysis [33] because these criteria are empirically determined in practical applications [19,45]. By contrast, ultrasound entropy imaging does not require additional signal rejection criteria, and thus it is less influenced by the effects of computational settings and parameter tuning. In addition, the advantages of information entropy lie in entropy estimation no longer being limited by the statistical properties of signals [30,32], implying that entropy is a data-adaptive parameter for ultrasound tissue characterization. A relatively simple but more adaptive computational scheme enables ultrasound entropy imaging to robustly and stably perform NAFLD evaluations, as supported by histopathological validations of both the animal model [32] and clinical trials [31,33]. These reasons explain why ultrasound entropy imaging correlates with MetS more significantly than does ASQ. In other words, when entropy works for characterizing NAFLD, it simultaneously provides significantly metabolic meanings that benefit evaluations in various aspects.

4.5. Comparison with Related Studies

A novel parameter named the controlled attenuation parameter (CAP) has been developed based on the properties of ultrasonic signals acquired by transient elastography (Fibroscan®). The CAP was demonstrated to correlate with fat accumulation in the liver [56,57] and facilitate the diagnosis of hepatic steatosis [58,59]. Furthermore, one study found that the CAP correlated with several MetS components [60]. However, the question of whether the CAP can perform well in NAFLD diagnosis remains unanswered because unfavorable diagnoses have been reported [61–63]. This is likely because the meaning of the CAP corresponds to the viscoelastic properties of the liver but does not provide information on changes in the microstructure, which is crucial in the clinical evaluation of hepatic steatosis. In the future, combining entropy imaging with the CAP may be a feasible strategy for a more complete evaluation of NAFLD and MetS than either one alone.

4.6. Limitations of This Study

This study had two limitations. First, the effect of body habitus on the association of entropy and ASQ analysis with MetS was not investigated. For instance, obesity may restrain quantitative measurements of ultrasonography. Second, the original equipment (Toshiba systems and software packages) was unavailable for ASQ analysis. Therefore, we implemented ASQ analysis by using backscattered envelope data acquired from the Terason system. We cannot deny that some bias of estimation accuracy of the FD ratio may have existed among system platforms. However, prior to this study, we fine-tuned the parameters of the algorithm to mitigate this concern.

5. Conclusions

In this study, we performed ultrasound entropy imaging of NAFLD and studied its association with MetS through comparisons with biochemical examinations and HOMA-IR (the measure of insulin resistance). In addition, the dependency of entropy imaging on MetS was compared with that of the conventional ASQ technique to explore possible strengths. The clinical results showed that the entropy value was more closely correlated with the US-FLI than the FD ratio of ASQ analysis, indicating that ultrasound entropy imaging improved the performance of NAFLD evaluation. Moreover, the entropy value and FD ratio correlated with metabolic indices and HOMA-IR, thereby confirming metabolic meanings of quantitative ultrasound. Notably, after adjustment for confounding factors, ultrasound entropy still provided a stronger link to the risk of MetS than did the FD ratio according to statistical OR analysis. Compared with model-based ASQ, ultrasound entropy imaging not only characterizes liver microstructures but also reflects metabolic information of NAFLD more significantly. Thus, ultrasound entropy imaging—a data-adaptive approach for tissue characterization—may play a key role in the evaluation of NAFLD and MetS.

Author Contributions: Experimental design: K.-C.Y. and C.-K.Y.; methodology: P.-H.T.; software: Y.-H.L.; data collection: Y.-Y.L.; data analysis: K.-C.Y.; manuscript preparation: Y.-H.L. and P.-H.T. All the authors have read and approved the final manuscript.

Funding: This work was supported by the Ministry of Science and Technology in Taiwan (Grant Nos. MOST 106-2221-E-182-023-MY3, 106-2314-B-002-237-MY3, and 107-2218-E-182-004) and Chang Gung Memorial Hospital, Linkou, Taiwan (Grant No. CMRPD1H0381).

Conflicts of Interest: The authors declare no conflict of interest.

References

1. Yu, A.S.; Keeffe, E.B. Nonalcoholic fatty liver disease. *Rev. Gastroenterol. Disord.* **2002**, *2*, 11–19. [CrossRef] [PubMed]
2. Loomba, R.; Abraham, M.; Unalp, A.; Wilson, L.; Lavine, J.; Doo, E.; Bass, N.M. Association between diabetes, family history of diabetes, and risk of nonalcoholic steatohepatitis and fibrosis. *Hepatology* **2012**, *56*, 943–951. [CrossRef] [PubMed]
3. Rezvani, M.; Shaaban, A.M. Patterns of fatty liver disease. *Curr. Radiol. Rep.* **2016**, *4*, 26. [CrossRef]
4. Beeman, S.C.; Garbow, J.R. Fatty Liver Disease. In *Imaging and Metabolism*; Springer International Publishing AG: Cham, Switzerland, 2018; pp. 223–241.
5. Bravo, A.A.; Sheth, S.G.; Chopra, S. Liver biopsy. *N. Engl. J. Med.* **2001**, *344*, 495–500. [CrossRef] [PubMed]
6. Sumida, Y.; Nakajima, A.; Itoh, Y. Limitations of liver biopsy and non-invasive diagnostic tests for the diagnosis of nonalcoholic fatty liver disease/nonalcoholic steatohepatitis. *World J. Gastroenterol.* **2014**, *20*, 475–485. [CrossRef] [PubMed]
7. Nalbantoglu, I.L.; Brunt, E.M. Role of liver biopsy in nonalcoholic fatty liver disease. *World J. Gastroenterol.* **2014**, *20*, 9026–9037. [CrossRef] [PubMed]
8. Ma, X.Z.; Holalkere, N.S.; Kambadakone, R.A.; Mari, M.K.; Hahn, P.F.; Sahani, D.V. Imaging-based quantification of hepatic fat: Methods and clinical applications. *Radiographics* **2009**, *29*, 1253–1280. [CrossRef] [PubMed]
9. Saadeh, S.; Younossi, Z.M.; Remer, E.M.; Gramlich, T.; Ong, J.P.; Hurley, M. The utility of radiological imaging in nonalcoholic fatty liver disease. *Gastroenterology* **2002**, *123*, 745–750. [CrossRef] [PubMed]
10. Lee, S.S.; Park, S.H.; Kim, H.J.; Kim, S.Y.; Kim, M.Y.; Kim, D.Y. Non-inva¬sive assessment of hepatic steatosis: Prospective comparison of the accuracy of imaging examinations. *J. Hepatol.* **2010**, *52*, 579–585. [CrossRef] [PubMed]
11. Strauss, S.; Gavish, E.; Gottlieb, P.; Katsnelson, L. Interobserver and intraobserver variability in the sonographic assessment of fatty liver. *AJR* **2007**, *189*, 320–323. [CrossRef] [PubMed]
12. Cengiz, M.; Sentürk, S.; Cetin, B.; Bayrak, A.H.; Bilek, S.U. Sonographic assessment of fatty liver: Intraobserver and interobserver variability. *Int. J. Clin. Exp. Med.* **2014**, *7*, 5453–5460. [PubMed]
13. Mamou, J.; Oelze, M.L. *Quantitative Ultrasound in Soft Tissues*; Springer: Dordrecht, The Netherlands; New York, NY, USA, 2013.
14. Tsui, P.H.; Zhou, Z.; Lin, Y.H.; Hung, C.M.; Chung, S.J.; Wan, Y.L. Effect of ultrasound frequency on the Nakagami statistics of human liver tissues. *PLoS ONE* **2017**, *12*, e0181789. [CrossRef] [PubMed]
15. Destrempes, F.; Cloutier, G. A critical review and uniformized representation of statistical distributions modeling the ultrasound echo envelope. *Ultrasound Med. Biol.* **2010**, *36*, 1037–1051. [CrossRef] [PubMed]

16. Liao, Y.Y.; Yang, K.C.; Lee, M.J.; Huang, K.C.; Chen, J.D.; Yeh, C.K. Multifeature analysis of an ultrasound quantitative diagnostic index for classifying nonalcoholic fatty liver disease. *Sci. Rep.* **2016**, *6*, 35083. [CrossRef] [PubMed]
17. Wan, Y.L.; Tai, D.I.; Ma, H.Y.; Chiang, B.H.; Chen, C.K.; Tsui, P.H. Effects of fatty infiltration in human livers on the backscattered statistics of ultrasound imaging. *Proc. Inst. Mech. Eng. H* **2015**, *229*, 419–428. [CrossRef] [PubMed]
18. Fang, J.; Zhou, Z.; Chang, N.F.; Wan, Y.L.; Tsui, P.H. Ultrasound parametric imaging of hepatic steatosis using the homodyned-K distribution: An animal study. *Ultrasonics* **2018**, *87*, 91–102. [CrossRef] [PubMed]
19. Toyoda, H.; Kumada, T.; Kamiyama, N.; Shiraki, K.; Takase, K.; Yamaguchi, T.; Hachiya, H. B-mode ultrasound with algorithm based on statistical analysis of signals: Evaluation of liver fibrosis in patients with chronic hepatitis C. *Am. J. Roentgenol.* **2009**, *193*, 1037–1043. [CrossRef] [PubMed]
20. Kuroda, H.; Kakisaka, K.; Kamiyama, N.; Oikawa, T.; Onodera, M.; Sawara, K.; Oikawa, K.; Endo, R.; Takikawa, Y.; Suzuki, K.; et al. Non-invasive determination of hepatic steatosis by acoustic structure quantification from ultrasound echo amplitude. *World J. Gastroenterol.* **2012**, *18*, 3889–3895. [CrossRef] [PubMed]
21. Son, J.Y.; Lee, J.Y.; Yi, N.J.; Lee, K.W.; Suh, K.S.; Kim, K.G.; Lee, J.M.; Han, J.K.; Choi, B.I. Hepatic steatosis: Assessment with acoustic structure quantification of US imaging. *Radiology* **2016**, *278*, 257–264. [CrossRef] [PubMed]
22. Karlas, T.; Berger, J.; Garnov, N.; Lindner, F.; Busse, H.; Linder, N.; Schaudinn, A.; Relke, B.; Chakaroun, R.; Tröltzsch, M.; et al. Estimating steatosis and fibrosis: Comparison of acoustic structure quantification with established techniques. *World J. Gastroenterol.* **2015**, *21*, 4894–4902. [CrossRef] [PubMed]
23. Keller, J.; Kaltenbach, T.E.; Haenle, M.M.; Oeztuerk, S.; Graeter, T.; Mason, R.A. Comparison of acoustic structure quantification (ASQ), shearwave elastography and histology in patients with diffuse hepatopathies. *BMC Med. Imaging* **2015**, *15*, 58. [CrossRef] [PubMed]
24. Lee, D.H.; Lee, J.Y.; Lee, K.B.; Han, J.K. Evaluation of hepatic steatosis by using acoustic structure quantification US in a rat model: comparison with pathologic examination and MR spectroscopy. *Radiology* **2017**, *285*, 445–453. [CrossRef] [PubMed]
25. Shankar, P.M. A general statistical model for ultrasonic backscattering from tissues. *IEEE Trans. Ultrason. Ferroelectr. Freq. Control.* **2000**, *47*, 727–736. [CrossRef] [PubMed]
26. Smolikova, R.; Wachowiak, M.P.; Zurada, J.M. An information-theoretic approach to estimating ultrasound backscatter characteristics. *Comput. Biol. Med.* **2004**, *34*, 355–370. [CrossRef]
27. Shannon, C.E. A Mathematical Theory of Communication. *Bell Syst. Tech. J.* **1948**, *27*, 379–423. [CrossRef]
28. Hughes, M.S. Analysis of digitized waveforms using Shannon entropy. *J. Acoust. Soc. Am.* **1993**, *93*, 892–906. [CrossRef]
29. Hughes, M.S.; McCarthy, J.E.; Marsh, J.N.; Arbeit, J.M.; Neumann, R.G.; Fuhrhop, R.W.; Wallace, K.D.; Znidersic, D.R.; Maurizi, B.N.; Baldwin, S.L.; et al. Properties of an entropy-based signal receiver with an application to ultrasonic molecular imaging. *J. Acoust. Soc. Am.* **2007**, *121*, 3542–3557. [CrossRef] [PubMed]
30. Tsui, P.H. Ultrasound detection of scatterer concentration by weighted entropy. *Entropy* **2015**, *17*, 6598–6616. [CrossRef]
31. Tsui, P.H.; Wan, Y.L. Effects of fatty infiltration of the liver on the Shannon entropy of ultrasound backscattered signals. *Entropy* **2016**, *18*, 341. [CrossRef]
32. Fang, J.; Chang, N.F.; Tsui, P.H. Performance evaluations on using entropy of ultrasound log-compressed envelope images for hepatic steatosis assessment: An in vivo animal study. *Entropy* **2018**, *20*, 120. [CrossRef]
33. Zhou, Z.; Tai, D.I.; Wan, Y.L.; Tseng, J.H.; Lin, Y.R.; Wu, S.; Yang, K.C.; Liao, Y.Y.; Yeh, C.K.; Tsui, P.H. Hepatic steatosis assessment with ultrasound small-window entropy imaging. *Ultrasound Med. Biol.* **2018**, *44*, 1327–1340. [CrossRef] [PubMed]
34. Leite, N.C.; Salles, G.F.; Araujo, A.L.; Cristiane, V.N.; Cardoso, C.R. Prevalence and associated factors of nonalcoholic fatty liver disease in patients with type-2 diabetes mellitus. *Liver Int* **2009**, *29*, 113–119. [CrossRef] [PubMed]
35. Fabbrini, E.; Sullivan, S.; Klein, S. Obesity and nonalcoholic fatty liver disease: biochemical, metabolic, and clinical implications. *Hepatology* **2010**, *51*, 679–689. [CrossRef] [PubMed]

36. Matthews, D.R.; Hosker, J.P.; Rudenski, A.S.; Naylor, B.A.; Treacher, D.F.; Turner, R.C. Homeostasis model assessment: Insulin resistance and beta-cell function from fasting plasma glucose and insulin concentrations in man. *Diabetologia* **1985**, *28*, 412–419. [CrossRef] [PubMed]
37. Cruz, M.A.F.; Cruz, J.F.; Macena, L.B.; Santana, D.S.; Oliveira, C.C.; Lima, S.O.; Franca, A.V. Association of the nonalcoholic hepatic steatosis and its degrees with the values of liver enzymes and homeostasis model assessment-insulin resistance index. *Gastroenterol. Res.* **2015**, *8*, 260–264. [CrossRef] [PubMed]
38. Isaksen, V.T.; Larsen, M.A.; Goll, R.; Florholmen, J.R.; Paulssen, E.J. Hepatic steatosis, detected by hepatorenal index in ultrasonography, as a predictor of insulin resistance in obese subjects. *BMC Obes.* **2016**, *3*, 39. [CrossRef] [PubMed]
39. Yang, K.C.; Hung, H.F.; Lu, C.W.; Chang, H.H.; Lee, L.T.; Huang, K.C. Association of non-alcoholic fatty liver disease with metabolic syndrome independently of central obesity and insulin resistance. *Sci. Rep.* **2016**, *6*, 27034. [CrossRef] [PubMed]
40. Ballestri, S.; Lonardo, A.; Romagnoli, D.; Carulli, L.; Losi, L.; Day, C.P.; Loria, P. Ultrasonographic fatty liver indicator, a novel score which rules out NASH and is correlated with metabolic parameters in NAFLD. *Liver Int.* **2012**, *32*, 1242–1252. [CrossRef] [PubMed]
41. Tsui, P.H.; Wang, S.H.; Huang, C.C.; Chiu, C.Y. Quantitative analysis of noise influence on the detection of scatterer concentration by Nakagami parameter. *J. Med. Biol. Eng.* **2005**, *25*, 45–51.
42. Zhou, Z.; Wu, W.; Wu, S.; Jia, K.; Tsui, P.H. Empirical mode decomposition of ultrasound imaging for gain-independent measurement on tissue echogenicity: A feasibility study. *Appl. Sci. Basel* **2017**, *7*, 324. [CrossRef]
43. Tsui, P.H.; Chen, C.K.; Kuo, W.H.; Chang, K.J.; Fang, J.; Ma, H.Y.; Chou, D. Small-window parametric imaging based on information entropy for ultrasound tissue characterization. *Sci. Rep.* **2017**, *7*, 41004. [CrossRef] [PubMed]
44. Yakoshi, Y.; Kudo, D.; Toyoki, Y.; Isido, K.; Kimura, N.; Wakiya, T.; Sakuraba, S.; Yoshizawa, T.; Sakamoto, Y.; Kijima, H.; et al. Non-invasive quantification of liver damage by a novel application for statistical analysis of ultrasound signals. *Hirosaki Med. J.* **2014**, *65*, 199–208.
45. Shen, C.C.; Yu, S.C.; Liu, C.Y. Using high-frequency ultrasound statistical scattering model to assess nonalcoholic fatty liver disease (NAFLD) in mice. *IEEE Ultrason. Symp. Proc.* **2016**, *1*, 379–382. [CrossRef]
46. Galassi, A.; Reynolds, K.H.J. Metabolic syndrome and risk of cardiovascular disease: A meta-analysis. *Am. J. Med.* **2006**, *119*, 812–819. [CrossRef] [PubMed]
47. Bugianesi, E.; McCullough, A.J.; Marchesini, G. Insulin resistance: A metabolic pathway to chronic liver disease. *Hepatology* **2005**, *42*, 987e1000. [CrossRef] [PubMed]
48. Carl, L.; Olefsky, J.M. Inflammation and insulin resistance. *FEBS Lett.* **2008**, *582*, 97e105. [CrossRef]
49. Asrih, M.; Jornayvaz, F.R. Metabolic syndrome and nonalcoholic fatty liver disease: Is insulin resistance the link? *Mol. Cell Endocrinol.* **2015**, *418*, 55–65. [CrossRef] [PubMed]
50. Donnelly, K.L.; Smith, C.I.; Schwarzenberg, S.J.; Jessurun, J.; Boldt, M.D.; Parks, E.J. Sources of fatty acids stored in liver and secreted via lipoproteins in when cells fail to respond normally to the hormone insulin in patients with nonalcoholic fatty liver disease. *J. Clin. Invest* **2005**, *115*, 1343–1351. [CrossRef] [PubMed]
51. Bugianesi, E.; Moscatiello, S.; Ciaravella, M.F.; Marchesini, G. Insulin resistance in nonalcoholic fatty liver disease. *Curr. Pharm. Des.* **2010**, *16*, 1941–1951. [CrossRef] [PubMed]
52. Luo, J.; Xu, L.; Li, J.; Zhao, S. Nonalcoholic fatty liver disease as a potential risk factor of cardiovascular disease. *Eur. J. Gastroenterol. Hepatol.* **2015**, *27*, 193–199. [CrossRef] [PubMed]
53. Pisto, P.; Santaniemi, M.; Bloigu, R.; Ukkola, O.; Kesäniemi, Y.A. Fatty liver predicts the risk for cardiovascular events in middle-aged population: A population-based cohort study. *BMJ Open* **2014**, *4*, e004973. [CrossRef] [PubMed]
54. Motamed, N.; Rabiee, B.; Poustchi, H.; Dehestani, B.; Hemasi, G.R.; Khonsari, M.R.; Maadi, M.; Saeedian, F.S.; Zamani, F. Non-alcoholic fatty liver disease (NAFLD) and 10-year risk of cardiovascular diseases. *Clin. Res. Hepatol. Gastroenterol.* **2017**, *41*, 31–38. [CrossRef] [PubMed]
55. Misra, V.L.; Khashab, M.; Chalasani, N. Non-alcoholic fatty liver disease and cardiovascular risk. *Curr. Gastroenterol. Rep.* **2009**, *11*, 50–55. [CrossRef] [PubMed]
56. Wang, Y.; Fan, Q.; Wang, T.; Wen, J.; Wang, H.; Zhang, T. Controlled attenuation parameter for assessment of hepatic steatosis grades: A diagnostic meta-analysis. *Int. J. Clin. Exp. Med.* **2015**, *8*, 17654–17663. [PubMed]

57. Sasso, M.; Beaugrand, M.; Ledinghen, V.; Douvin, C.; Marcellin, P.; Poupon, R.; Sandrin, L.; Miette, V. Controlled attenuation parameter (CAP): A novel VCTE guided ultrasonic attenuation measurement for the evaluation of hepatic steatosis: preliminary study and validation in a cohort of patients with chronic liver disease from various causes. *Ultrasound Med. Biol.* **2010**, *36*, 1825–1835. [CrossRef] [PubMed]
58. Ledinghen, V.; Vergniol, J.; Foucher, J.; Merrouche, W.; Bail, B. Non-invasive diagnosis of liver steatosis using controlled attenuation parameter (CAP) and transient elastography. *Liver Int.* **2012**, *32*, 911–918. [CrossRef] [PubMed]
59. Mikolasevic, I.; Orlic, L.; Franjic, N.; Hauser, G.; Stimac, D.; Milic, S. Transient elastography (FibroScanR) with controlled attenuation parameter in the assessment of liver steatosis and fibrosis in patients with nonalcoholic fatty liver disease: Where do we stand? *World J. Gastroenterol.* **2016**, *22*, 7236–7251. [CrossRef] [PubMed]
60. Mikolasevic, I.; Milic, S.; Orlic, L.; Stimac, D.; Franjic, N.; Targher, G. Factors associated with significant liver steatosis and fibrosis as assessed by transient elastography in patients with one or more components of the metabolic syndrome. *J. Diabetes Complic.* **2016**, *30*, 1347–1353. [CrossRef] [PubMed]
61. Imajo, K.; Kessoku, T.; Honda, Y.; Tomeno, W.; Ogawa, Y.; Mawatari, H.; Fujita, K.; Yoneda, M.; Taguri, M.; Hyogo, H.; et al. Magnetic resonance imaging more accurately classifies steatosis and fibrosis in patients with nonalcoholic fatty liver disease than transient elastography. *Gastroenterology* **2016**, *150*, 626–637. [CrossRef] [PubMed]
62. Myers, R.P.; Pollett, A.; Kirsch, R.; Pomier-Layrargues, G.; Beaton, M.; Levstik, M.; Duarte-Rojo, A.; Wong, D.; Crotty, P.; Elkashab, M. Controlled attenuation parameter (CAP): A noninvasive method for the detection of hepatic steatosis based on transient elastography. *Liver Int.* **2012**, *32*, 902–910. [CrossRef] [PubMed]
63. Kumar, M.; Rastogi, A.; Singh, T.; Behari, C.; Gupta, E.; Garg, H.; Kumar, R.; Bhatia, V.; Sarin, SK. Controlled attenuation parameter for non-invasive assessment of hepatic steatosis: Does etiology affect performance? *J. Gastroenterol. Hepatol.* **2013**, *28*, 1194–1201. [CrossRef] [PubMed]

entropy

MDPI

Article

A Novel Multi-Exposure Image Fusion Method Based on Adaptive Patch Structure

Yuanyuan Li [1,2], Yanjing Sun [1,*], Mingyao Zheng [2], Xinghua Huang [3], Guanqiu Qi [4,5], Hexu Hu [2] and Zhiqin Zhu [2]

1 School of Information and Electrical, China University of Mining and Technology, Xuzhou 221116, China; liyy@cqupt.edu.cn
2 College of Automation, Chongqing University of Posts and Telecommunications, Chongqing 400065, China; ZMYzhengmingyao@126.com (M.Z.); cqgmhxy@163.com (H.H.); zhuzq@cqupt.edu.cn (Z.Z.)
3 College of Automation, Chongqing University, Chongqing 400044, China; reilove2@gmail.com
4 Department of Mathematics and Computer Information Science, Mansfield University of Pennsylvania, Mansfield, PA 16933, USA
5 School of Computing, Informatics, and Decision Systems Engineering, Arizona State University, Tempe, AZ 85287, USA; guanqiuq@asu.edu
* Correspondence: yjsun@cumt.edu.cn

Received: 29 October 2018; Accepted: 3 December 2018; Published: 6 December 2018

Abstract: Multi-exposure image fusion methods are often applied to the fusion of low-dynamic images that are taken from the same scene at different exposure levels. The fused images not only contain more color and detailed information, but also demonstrate the same real visual effects as the observation by the human eye. This paper proposes a novel multi-exposure image fusion (MEF) method based on adaptive patch structure. The proposed algorithm combines image cartoon-texture decomposition, image patch structure decomposition, and the structural similarity index to improve the local contrast of the image. Moreover, the proposed method can capture more detailed information of source images and produce more vivid high-dynamic-range (HDR) images. Specifically, image texture entropy values are used to evaluate image local information for adaptive selection of image patch size. The intermediate fused image is obtained by the proposed structure patch decomposition algorithm. Finally, the intermediate fused image is optimized by using the structural similarity index to obtain the final fused HDR image. The results of comparative experiments show that the proposed method can obtain high-quality HDR images with better visual effects and more detailed information.

Keywords: multi-exposure image fusion; texture information entropy; adaptive selection; patch structure decomposition

1. Introduction

Due to the limited dynamic range of imaging devices, it is not possible to capture all the details in one scene by a single exposure with existing imaging devices [1,2]. This seriously affects image visualization and the demonstration of key information. Figure 1a shows an over-exposed image. When shooting requires a long exposure time, the imaging device can effectively capture the information from the dark part. However, due to the over-exposure, the details of the bright part get severely lost. On the contrary, when the exposure time is short, the information of the bright part is captured, but the information of the dark part is lost. As shown in Figure 1b, the under-exposure phenomenon is caused by the mismatch of the dynamic range between the human visual system and electronic imaging devices [1].

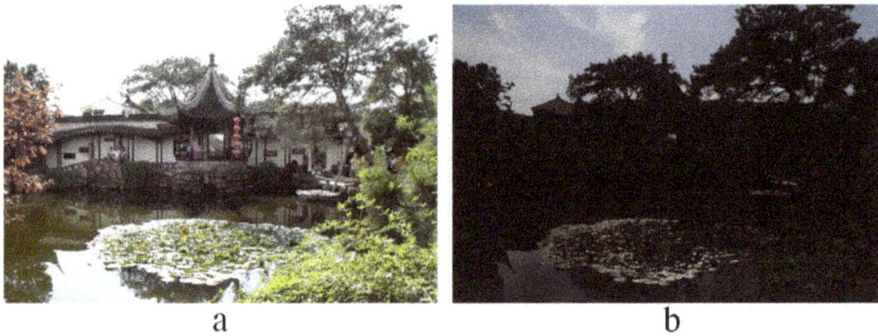

Figure 1. (**a**) Over-exposure image; (**b**) under-exposure image.

Multi-exposure image fusion (MEF) methods provide an effective way to solve the mismatch of dynamic range among existing imaging devices, display equipment, and human eyes' response to real scenes. It takes source image sequences of different exposure levels as the inputs. An informative and perceptive high-dynamic-range (HDR) image is generated by synthesizing the information of the luminance conforming to the human visual system [3]. The fused image contains more abundant scene luminance, color, and detailed information, which make the image correspond to the real scene observed by the human eye [4,5]. In addition, it also provides more information for subsequent image processing [6]. MEF algorithms are mainly categorized as transform domain- and spatial domain-based fusion algorithms.

Transform domain-based fusion methods have three main steps: First, source image sequences are decomposed into the transform domain. Then, fusion coefficients from the source images are selected according to the fusion rules. Finally, the fused image is obtained by inversely transforming the fusion coefficients [7–9]. Based on the Laplacian pyramid, Mertens proposed multiple resolutions to fuse an exposure sequence into an HDR image [10]. The weighted value determined by contrast, saturation, and well-exposedness took a weighted average to obtain the pyramid coefficients. The obtained pyramid coefficients were reconstructed to get the final image. Li performed two-scale decomposition on source images to obtain the image base layer and the detail layer first. Then, the spatial consistency was applied to fuse the obtained base and detail layers to get the fused image [11]. Bruce introduced nonlinearity to balance the visible details and smoothness of the fused result to capture the information present in the source images [12]. Kou applied the weighted-least-squares-based image smoothing algorithm to one MEF algorithm for detail extraction in an HDR scene [13]. The extracted details were then used in the multi-scale exposure fusion algorithm to achieve image fusion. Based on the hybrid exposure weight and the novel boosting Laplacian pyramid (BLP), an exposure fusion approach proposed by Shen considered gradient vectors among different exposure source images [14]. This method used the improved Laplacian pyramid to decompose input signals into base and detail layers. Then, a fused image with rich color and detailed information could be obtained. As a shortcoming of transform domain-based fusion algorithms, when the luminance range of the real scene is large, the useful details of the over-exposure and under-exposure regions are lost. This seriously affects the visual effects of the fused image.

Source images from different exposures can also be fused by spatial domain-based fusion methods [15–17]. Spatial domain fusion methods have two main types: image patch- and pixel-based fusion [5,18,19]. As a pixel-based fusion algorithm, Shen proposed a probabilistic model for MEF [20]. Subject to two quality measures, such as the local contrast and color consistency of source image sequence, the proposed model calculated an optimal set of probabilities by using a generalized random walk framework. The probability sets were then used as weights to achieve image fusion. Based on the probabilistic model [20,21], Shen proposed another MEF method by integrating the perceptual quality measure [22]. In this method, the probability of the human visual system was modeled by contrast and color information, as well as the optimal fusion weight obtained by using the hierarchical multivariate Gaussian conditional random field. This method can improve MEF performance and provide a better visual experience for audiences. Gu introduced a new method for MEF that obtained the gradient value of the pixel by maximizing the structure tensor [23]. The local gradient was used to represent pixel contrast first. Then, the fused image was obtained by the inverse transformation of the gradient field. Li established a multi-exposure image fusion model based on median filtering and recursive filtering [24]. It made a comprehensive evaluation of different regions of the multi-exposure image and fused those pixels from median filtering and recursive filtering over contrast, color, and brightness exposure. This method reduced the computational complexity of effective fusion. On the basis of the rank-one structure of low-dynamic-range (LDR) images, Tae-Hyun proposed a MEF image algorithm [6]. This algorithm formulated HDR generation as a rank minimization problem (RMP) that simultaneously estimated a set of geometric transformations to align LDR images and detected both moving objects and under-/over-exposed regions. Since pixel-based MEF algorithms involved averaging pixels to obtain fused pixels, this method reduced the sharpness and contrast, which affected the visual quality of fused image.

Owing to the image patches, Song proposed a fusion method to suppress gradient inversion by integrating the details of the local adaptive scene [25]. A variational method that combined color matching and gradient information was proposed by Bertalmio to achieve image fusion [26]. It used short- and long-exposure images to measure differences in edge information and local chromatic aberrations, respectively. Zhang used the contrast standard to measure the quality of details in the exposure and kept the details in intermediate images [27]. By combining intermediate images seamlessly, an HDR image with rich details could be generated. According to the principle of the structural similarity (SSIM) index [28], Ma proposed a structural patch decomposition-based MEF for image fusion [3]. This method can produce noise-free weighted mapping, more natural color information, and a high-quality fused image. Based on the decomposition of the image patch structure [3], Ma introduced a color structural similarity (SSIMc) index to achieve multi-exposure image fusion [29]. The source image spaces were explored in an iterative process by the gradient ascent algorithm to search for an image with optimized SSIMc. Finally, a high-quality fused image with a realistic structure and vivid color was obtained. Compared with pixel-based fusion methods, patch-based fusion methods do not average pixels, but can obtain a fused image with better sharpness and contrast. Since image patch size in patch-based fusion algorithms is fixed, it causes the fused image to lose fine-detail information of the structure and texture in the multi-exposure source image sequence.

In order to obtain high-quality HDR images, this paper proposes an MEF algorithm based on adaptive patch structure (APS-MEF), which retains more detailed information of the scene. First, input multi-exposure source images are subjected to cartoon-texture decomposition, and the adaptive decomposition of the image patch is realized by calculating the entropy of the image texture [30]. This is helpful to improve the robustness of image patch decomposition. Then, three components, the signal strength, signal structure, and mean intensity, are obtained by applying structural patch decomposition. The initial fused image is obtained by processing three components separately. The decomposition algorithm of the structural patch processes three color channels at the same time, which can capture more color information of the source scene and obtain a more vivid fused image. Finally, the SSIMc is applied to optimize the initial fused image to balance both local and overall brightness. The fused image is consistent with the human visual system. Compared with five MEF methods in 24 different scenes, the experiment results confirm that the proposed APS-MEF method can retain more detailed information and generate high-quality fused images. Based on the adaptive selection of the image patch size, the proposed fusion algorithm brings three main contributions to traditional MEF methods:

- It uses texture entropy to evaluate image information, which has strong adaptability and robustness.
- It implements the adaptive selection of image patch size by measuring texture entropy, which enables the fused image to retain more detailed information of the source images.

- It combines image cartoon-texture decomposition, image patch structure decomposition, and SSIM index optimization to adjust the local brightness and makes fused images sharper and more smooth.

The rest of this paper is organized as follows: Section 2 presents the APS-MEF algorithm in detail; Section 3 discusses comparative experiments; and Section 4 concludes this paper.

2. MEF Framework Based on the Adaptive Selection of Image Patch Size

Based on the adaptive selection of image patch size, a novel MEF framework as shown in Figure 2 is proposed. First, texture-cartoon decomposition is applied to obtain image texture components. Then, image texture entropy is calculated to achieve the adaptive selection of image patch size. Third, the structural patch decomposition approach is utilized to obtain the initial fused image. Finally, the color MEF structural similarity index is used to iteratively optimize the initial fused image to get the final fused image.

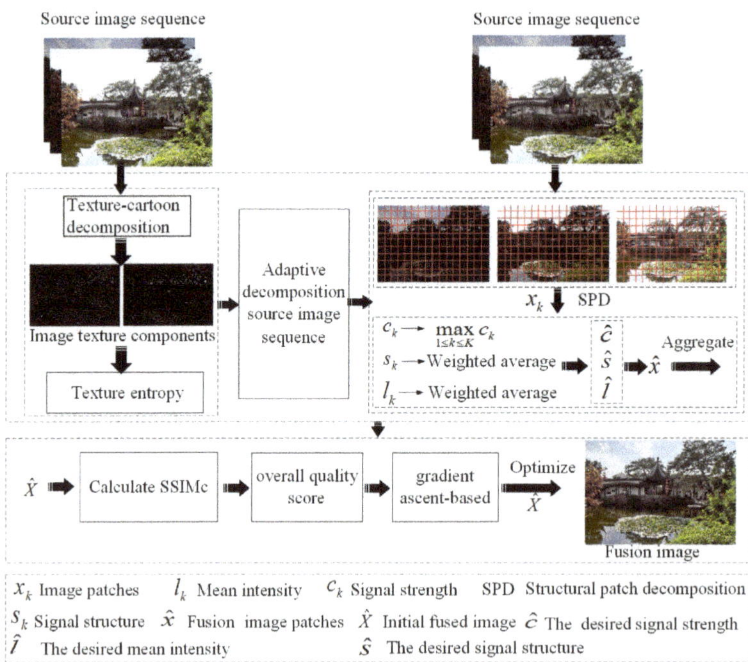

Figure 2. The proposed adaptive patch structure multi-exposure image fusion (APS-MEF) framework. SSIMc: color structural similarity.

2.1. Image Cartoon-Texture Decomposition

The texture and cartoon components of the image describe the detailed and structured information, respectively [31,32]. In the proposed fusion framework, texture and cartoon components are obtained by decomposing input images. In this work, image texture decomposition is achieved by implementing the regularization Vese-Osher (VO) model [33,34]. The VO model is shown as Equation (1):

$$\inf_{u,\vec{g}} \left\{ VO_P(u,\vec{g}) = |u|_{TV} + \lambda \|f - udiv(\vec{g})\|_{L^2}^2 + \mu \|\vec{g}\|_{L^P} \right\} \tag{1}$$

where $\|\vec{g}\|_{L^P} = \left[\int \left(\sqrt{g_1^2 + g_2^2}\right)^P dxdy\right]^{\frac{1}{P}}$ represents the L^P norm of $\vec{g}$, the value of p being between one and 10. The $\mathbf{g} = (g_1, g_2)$ is a vector to represent digital images in G space, and λ and μ are regularization parameters. In Equation (1, the first term u is the cartoon component of the image. The second term $f - u - div(\vec{g})$ together with the third term $\|\vec{g}\|$ ensures that $v = f - u \approx div(\vec{g})$, which represents the image minus the rest of the cartoon. v is the texture component of the image. When $\lambda \to \infty$ and $P \to \infty$, then in the limit, $f - u = div(\vec{g})$ almost everywhere for those $\vec{g}$. Therefore, in the limit, the middle part in Equation (1) will disappear, and the third part becomes $\|f - u\|$, which represents the texture component of the image. f represents the input image. The cartoon component of image u can be obtained by the Euler–Lagrange equation, shown in Equation (2): Once the cartoon component is obtained, the texture component can be simply calculated using $v = f - u$. Figure 3a represents the source image, and Figure 3b is the texture component obtained by the *VO* model. For more information, please refer to [34].

$$u = f - \partial_x g_1 - \partial_y g_2 + \frac{1}{2\lambda} div\left(\frac{\nabla u}{|\nabla u|}\right) \tag{2}$$

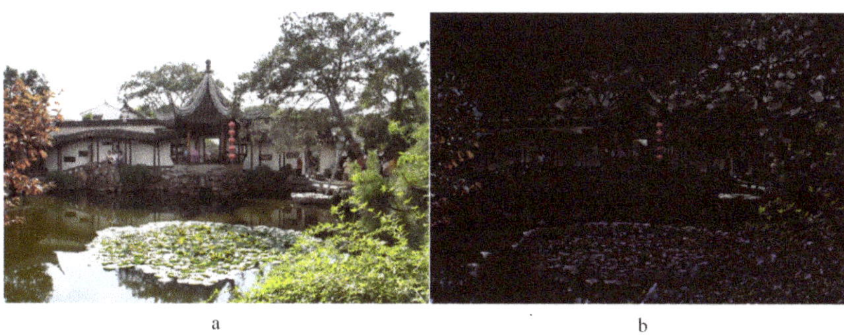

Figure 3. (**a**) The source image; (**b**) the texture component.

2.2. *Adaptive Selection of Image Patch Size*

The proposed adaptive selection of image patch size (APS) algorithm applies the statistical method to the grayscale difference to calculate the entropy value of image texture features. The details of the proposed APS algorithm are shown as the following steps:

Step 1: This converts image texture components into a grayscale image. The grayscale image is shown in Figure 4a. (x, y) denotes a point in the image. A point that has quite a small distance from (x, y) is denoted as $(x + \Delta x, y + \Delta y)$. Its grayscale difference value can be represented as Equation (3). Then, a grayscale differential image is obtained.

$$g_\Delta(x, y) = g(x, y) - g(x + \Delta x, y + \Delta y) \tag{3}$$

where g_Δ denotes the gray-value difference. Letting (x, y) move over the entire image, then a grayscale differential image is obtained. Figure 4b is a grayscale differential image obtained by the gray difference algorithm.

Step 2: Assuming that all possible values of the grayscale difference have m levels, it calculates the entropy value of the image texture features. A histogram of g_Δ is obtained by letting (x, y) move over the entire image and counting the number of times for each value of g_Δ. $p(i)$ is the probability value of each gray-level difference obtained from histogram statistics. The entropy value of image texture is obtained by Equation (4).

$$ent = -\sum_{i=0}^{m} p(i) \log_2[p(i)] \tag{4}$$

Step 3: After iterating the above processes for all input images, this algorithm obtains the entropies of all image texture features as $\{ent_1, ent_2, \ldots, ent_n\}$, where n represents the number of input images.

Step 4: Adaptive selection of the image patch size is obtained, according to the entropy value of the image texture feature. Based on the gray-level co-occurrence matrix, the obtained texture entropy value can reflect the coarseness of image texture to a large extent. When the entropy value becomes smaller, the texture is finer. On the contrary, when the entropy value becomes bigger, the texture is coarser [35]. The optimal image patch size is closely related to the coarseness of image texture. When image texture becomes coarser, a larger image patch size should be selected in image decomposition to ensure the good performance of the texture structure in decomposed components. Conversely, when image texture gets finer, a smaller image patch size should be selected to achieve a better texture synthesis effect. In this paper, the coarseness of image texture is characterized according to the entropy value of image texture, and the optimal image patch size is automatically selected. When the image texture entropy value is small, this indicates that the texture is fine, and a larger image patch size should be selected. When the image texture entropy value is larger, this indicates that the texture is rough, and a smaller image patch size should be selected [36]. For the 24 sets of multi-exposure source image sequences used in the experiment, a 16 size image patch size was selected for image fusion. For each set of texture images, the image patch size is adjusted from large to small, and the fusion result is compared to find a reasonable parameter range of the image patch size, as shown in Figure 5. The abscissa represents the entropy of different texture images, arranged from small to large, and the ordinate represents the optimal image patch parameters of the corresponding texture image. It can be seen from Figure 5 that as the image texture entropy value increases, the optimal image patch size parameter decreases, and the hyperbolic function can be used for fitting. The empirical formula for the optimal image patch size is shown as Equation (5).

$$wSize = pSize \times 0.1 \times \frac{(ENT/10)^{ENT} - (ENT/10)^{-ENT}}{(ENT/10)^{ENT} + (ENT/10)^{-ENT}} + pSize \times e^{-ENT} \tag{5}$$

where ENT is the mean of image texture entropy values, $ENT = \frac{1}{n}\sum_{i=1}^{n} ent_i$, $pSize$ is the preset image patch decomposition size, and $pSize = 21$. Thus,the corresponding optimal matching patch size is $wSize \times wSize$.

a b

Figure 4. (**a**) The texture image in gray space; (**b**) the differential image.

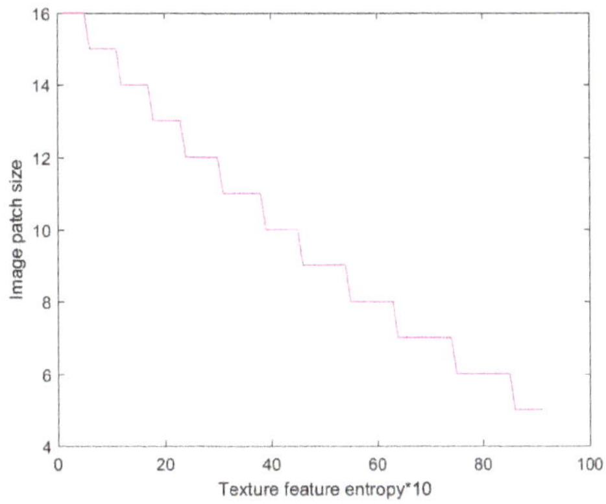

Figure 5. Relationship between image texture entropy and image patch size.

2.3. Structure Patch Decomposition and Structural Similarity Optimization for MEF

The image patches obtained by the APS algorithm are decomposed into three components to obtain an initial fused image by using the structure patch decomposition algorithm [3,37]. Then, the initial fused image is optimized by the MEF-SSIMc algorithm [29] to obtain the final fused image. Specifically, the algorithm details are shown as follows:

Step 1: Patches of the same spatial position are extracted from the image sequence processed by the APS algorithm using a dynamic stride D.

Step 2: Three conceptually-irrelevant components, signal strength c_i, signal structure s_i, and mean intensity l_i, are obtained by applying the structure patch decomposition approach to an image patch.

Step 3: Three components are processed separately. The signal strength component is processed first. As shown in Equation (6), the maximum signal strength of all source image patches in the same spatial location is selected as the signal strength of the fused image patch. The local contrast determines the texture and structure visibility of the local image patch. Generally, when local contrast becomes higher, the visibility of the local image patch is better. In this paper, local contrast is directly related to signal strength.

$$\hat{c} = \max_{1 \le i \le n} c_i = \max_{1 \le i \le n} \|\tilde{x}_i\| \tag{6}$$

where $\|\cdot\|$ represents the l_2 and $\tilde{x}_i$ is local contrast.

For signal structure s_i, in order to make the fused image patch represent the structures of all source image patches, a weighted average processing signal structure as shown in Equation (7) is applied.

$$\hat{s} = \frac{\sum_i^n S(\tilde{x}_i)\, s_i / \sum_i^n S(\tilde{x}_i)}{\left\| \sum_i^n S(\tilde{x}_i)\, s_i / \sum_i^n S(\tilde{x}_i) \right\|} \tag{7}$$

where $S(\cdot)$ is a weighting function defined as $S(\tilde{x}_i) = \|\tilde{x}_i\|^4$. Similar to Equation (7), a weighted average process as shown in Equation (8) is applied to mean intensity l_i.

$$\hat{l} = \frac{\sum_{i=1}^n L(\mu_i, l_i)\, l_i}{\sum_{i=1}^n L(\mu_i, l_i)} \tag{8}$$

where $L(\cdot)$ defines a weighting function. It quantifies the well-exposedness of the local image patch in the source image and performs the calculation using the Gaussian model.

$$L(\mu_i, l_i) = \exp\left(-\left(\frac{\mu_i - 0.5}{2\sigma_g^2}\right)^2 - \left(\frac{l_i - 0.5}{2\sigma_l^2}\right)^2\right) \tag{9}$$

where μ_i and l_i denote the global mean intensity of the source image and the local mean intensity of current patch, respectively, and σ_g and σ_l are the Gaussian standard deviation. In this paper, σ_g and σ_l are 0.2 and 0.5, respectively.

Step 4: When $\hat{c}$, $\hat{s}$, and $\hat{l}$ are calculated, a new uniquely fused image patch can be obtained by recombination.

$$\hat{x} = \hat{c} \cdot \hat{s} + \hat{l} \tag{10}$$

Step 5: In this fused framework, all source image sequences iterate Processes 1–4 above to obtain all fused patches. Then, the fused patches are aggregated to achieve the initial fused image.

Step 6: In this step, the MEF-SSIMc algorithm [29] defines the structural similarity index (SSIM) to evaluate the image patch quality.

$$S(\{x_i\}, y) = \frac{(2\mu_{\hat{x}}\mu_y + C_1)(2\sigma_{\hat{x}y} + C_2)}{\left(\mu_{\hat{x}}^2 + \mu_y^2 + C_1\right)\left(\sigma_{\hat{x}}^2 + \sigma_y^2 + C_2\right)} \tag{11}$$

where $\{x_i\} = \{x_i | 1 \le i \le n\}$ represents the group of image patches at the same location in the sequence of source images, $\mu_{\hat{x}}$ and μ_y represent the average intensity of fused image patch $\hat{x}$ and the referenced image patch y, respectively, $\sigma_{\hat{x}}^2$ and σ_y^2 represent the local deviation of $\hat{x}$ and y respectively, $\sigma_{\hat{x}y}$ is the local covariance between $\hat{x}$ and y, and C_1 and C_2 are small constants that satisfy $C_1 > 0, C_2 > 0$ to avoid the algorithm becoming unstable when the denominator approaches zero.

Step 7: As shown in Equation (12), the fused image overall quality score is obtained by averaging the SSIM index of fused image patches.

$$Q(\{x_i\}, Y) = \frac{1}{N}\sum_{j=1}^{N} S(\{R_j X_i\}, R_j Y) \tag{12}$$

where N represents the number of image patches, R_j denotes a binary matrix, the number of its columns is equal to the image dimension, the number of rows is equal to the patch size CN^2, C is the number of color channels, and N represents the patch size.

Step 8: The SSIM index is updated by using gradient iterations. As illustrated in Equation (13), the SSIM index Y_i obtained by the i-th iteration is improved by using the gradient ascent algorithm to achieve image optimization.

$$Y_{i+1} = Y_i + \lambda \nabla_Y Q(\{X_i\}, Y)|_{Y=Y_i} \tag{13}$$

where $\nabla_Y Q(\{X_i\}, Y)$ denotes the gradient of $Q(\{X_i\}, Y)$, and the details of calculating $\nabla_Y Q(\{X_i\}, Y)$ are described in [29]. λ denotes a step parameter controlling the speed of movement in the image. When $|Q_{i+1} - Q_i| < \varepsilon = 10^{-6}$ is satisfied, the iteration stops, and the final fused image is obtained.

3. Experiments and Analyses

3.1. Experiment Preparation

In this section, 24 sets of multi-exposure source image sequences that describe diverse scenes containing different shades regions with disparate colors were used in comparative experiments. All the source image sequences were collected by Ma [29] and can be downloaded form https://ece.uwaterloo.ca/~k29ma/. Ten different MEF methods, such as Bruce13 [12], Gu12 [23], Mertens07 [10],

Shen14 [14], Ma17 [3], SSIM-MEF [3,29], Proposed-8, Proposed-16, Proposed-24, and the proposed APS-MEF solution were applied to 24 sets of multi-exposure source image sequences for comparison. All of the fused images were either collected by Ma [29] or generated by open source codes. All the experiments were programmed in MATLAB 2016a (MathWorks, Natick, MA, USA) on an Intel® Core™ i7-7700k CPU @ 4.20-GHz desktop with 16.00 GB RAM.

Objective Evaluation Metrics

To evaluate quantitatively the quality of a fused image, a single evaluation metric cannot fully reflect the quality of the fused image. Therefore, several metrics were applied to make as comprehensive an evaluation as necessary. In these experiments, three objective evaluation indexes, $Q^{AB/F}$ [38,39], MI [40,41], and Q^{CB} [42,43], were selected to quantify the fused results of different MEF methods.

As a gradient-based quality index, $Q^{AB/F}$ [38,39] and MI [40,41] were used to measure the edge information and the similarity between the fused image and the source images, respectively. Q^{CB} [42,43], as a human perception-inspired fusion metric, was used to evaluate the human visualization performance of the fused image.

3.2. Experiment Results and Analyses

Experiment Results of Six MEF Methods

We conducted the following comparative experiments to prove that the proposed APS-MEF algorithm can achieve excellent fusion performance in human visual observation. Twenty four sets of multi-exposure source images were fused by ten MEF methods, which were Bruce13 [12], Gu12 [23], Mertens07 [10], Shen14 [14], Ma17 [3], SSIM-MEF [3,29], Proposed-8, Proposed-16, Proposed-24, and the proposed APS-MEF. The SSIM-MEF method is the result of the optimization of Ma17 [3] using SSIM-MEF [29]. Proposed-8, Proposed-16, and Proposed-24 represent the fusion results of the proposed algorithm using 8×8, 16×16,and 24×24 fixed image patch sizes, respectively. The selected patch sizes of the 24 sets of multi-exposure images in the proposed APS-MEF algorithm are shown in Table 1. In total, 240 fused images were obtained and divided into 24 groups according to the scene content. Five sets were selected from the total of 24 sets of fused images for demonstration in this paper.

Table 1. The selected patch sizes of the 24 sets of multi-exposure images.

Image Set	Patch Size	Image Set	Patch Size
Arno	13	Balloons	15
BelgiumHouse	8	Cave	10
Chinese Garden	14	Church	9
Farmhouse	17	House	10
Kluki	12	Lamp	13
Landscape	12	Laurenziana	12
Lighthouse	12	MadisonCapitol	9
Mask	10	Office	17
Ostrow	18	Room	15
Set	13	Studio	12
Tower	15	Venice	10
Window	15	Yello wHall	18

All fused images of "Chinese Garden" obtained by ten different methods are illustrated in Figure 6. Compared with Bruce13 [12], the fused image obtained by the proposed ASP-MEF contained more structure and texture details in the pool and corridor areas, and it achieved excellent performance in global contrast. Moreover, the fused image obtained by the proposed method had more vivid color and comfortable visual effects than the fused ones of Gu12 [23] and Shen14 [14]. The color of the fused image obtained by Gu12 [23] was distorted; for example, the sky is gray in Figure 6c. The fused image produced by Shen14 [14] had sharp intensity changes and unnatural colors that were either saturated or pale. According to the details of the fused images shown in Figure 6, the plants shown in the fused images by Bruce13 [12] and Ma17 [3] had unclear structure and texture details. Although images fused by Gu12 [23] and Shen14 [14] had good local details, the global visual effects were poor. The fused image obtained by the SSIM-MEF [3,29] method had high saturation. Compared with other MEF methods, the proposed APS-MEF method not only ensured the articulation of local details, but also achieved contrast and color saturation that conformed to the human visual observation. In addition, compared to the Proposed-8, Proposed-16, and Proposed-24 fusion methods, our proposed ASP-MEF performed better with respect to the human visual system.

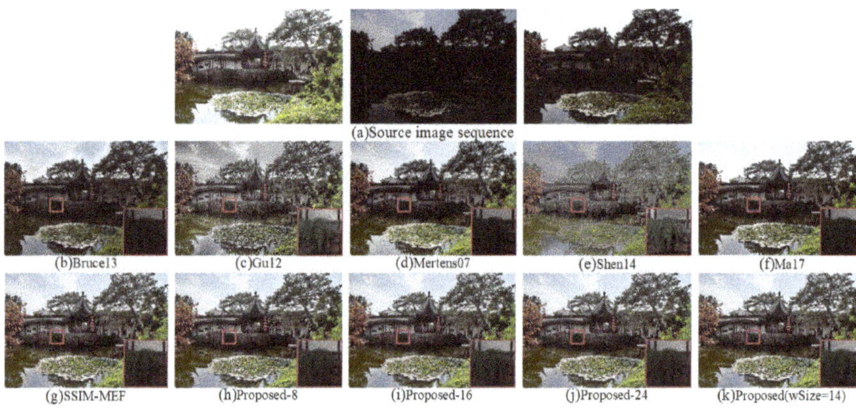

Figure 6. Visual comparison of the fused "Chinese Garden" between the proposed method and nine existing methods.

The fused "Yellow Hall" images by the ten methods are shown in Figure 7. The fused image obtained by Bruce13 [12] had poor overall brightness, and the details of dark regions shown in the source images could not be well represented. Although Figure 7c,d shows good performance in global contrast, both of them had distortions to varying degrees. The color saturation of stair areas shown in the fused result of Gu12 [23] was poor. Due to the high sharpness of the fused image by Shen14 [14], the edges of wall were unsmooth. The color appearances of the wall and stair areas in the fused images by Mertens07 [10], Ma17 [3], and APS-MEF were relatively natural and consistent with the source images. However, the edge details of portraits shown in the local enlarged areas of Figure 7d,f were blurred, and the brightness was dark. The color saturation of the wall relief of the Proposed-24 method fusion result was slightly worse than that of the proposed APS-MEF, SSIM-MEF [3,29], Proposed-8, and Proposed-16 methods. Compared to the proposed APS-MEF and SSIM-MEF [3,29] methods, the wall relief edge of the Proposed-8 and Proposed-16 fusion results was smoother. Although the proposed APS-MEF and SSIM-MEF [3,29] method performed excellent in the color and local detail of the image, the overall appearance of the proposed APS-MEF method fusion result was brighter. Therefore, compared to the other nine fused results, the image fused by the proposed APS-MEF method had clearer local texture details, as well as a brighter and warmer overall appearance.

The fused "Window" images are demonstrated in Figure 8. In the fused image obtained by Bruce13 [12], the light brightness was weak, and the local area details of the window were blurred. Compared to the fused images of Shen14 [14] and APS-MEF, there were obviously black shadows from the lights in Figure 8e and a black shadow on the wall. Besides, the structure and texture of the scene outside the window were not obvious. In the fusion result of Gu12 [23], the colors of the bed, wall, and portrait were obviously distorted. The chair shown in Figure 8d had a weak brightness and blurry edge. The local enlargement of the fused image obtained by the Ma17 [3] method had a high exposure, and the detail information was blurred. Compared to the Proposed-8 and Proposed-16 methods, the proposed method was moderately bright. The fused images of the proposed method, SSIM-MEF [3,29], and Proposed-24 were natural in color and brightness. The proposed method had much clearer edge and structure details of the scene outside the window. Therefore, compared with the other nine methods, the fused image obtained by the proposed method was more natural with respect to human vision and had better local details.

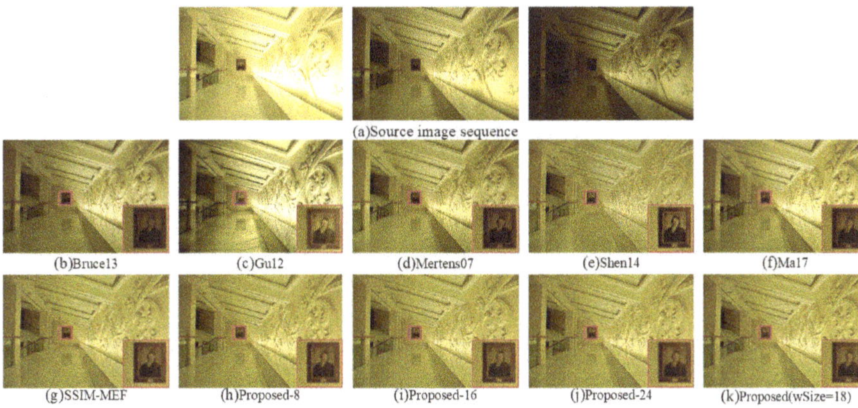

Figure 7. Visual comparison of the fused "Yellow Hall" between the proposed method and nine existing methods.

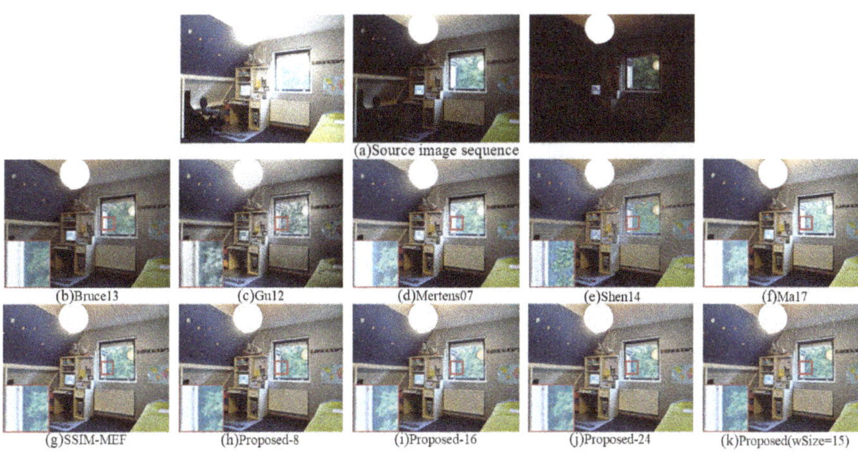

Figure 8. Visual comparison of the fused "Window" between the proposed method and nine existing methods.

Figure 9 shows ten fused images of "Tower" generated by the ten different methods. Towers in the fusion results of Bruce13 [12] and Mertens07 [10] were black and indistinct. The magnified local images clearly show that the interior details of the tower were missing. Although the details of the tower in Figure 9d are clear, the brightness of the clouds is overexposed, as well as the colors of the clouds and sky are obviously distorted. The edge of the cloud in the fused image of Shen14 [14] was too sharp. Moreover, the overall color was not soft, and the visual effect was poor. The fused image obtained by Ma17 [3] had a high exposure for the the clouds, which weakened the detailed texture information. The fused images obtained by the Proposed-16 and Proposed-24 methods were similar to human visual perception. SSIM-MEF [3,29], Proposed-8, and the proposed method reached the best overall visual effect. It was difficult to distinguish the difference between the Proposed-16, Proposed-24, SSIM-MEF [3,29], Proposed-8, and the proposed APS-MEF fusion results by human visual observation.

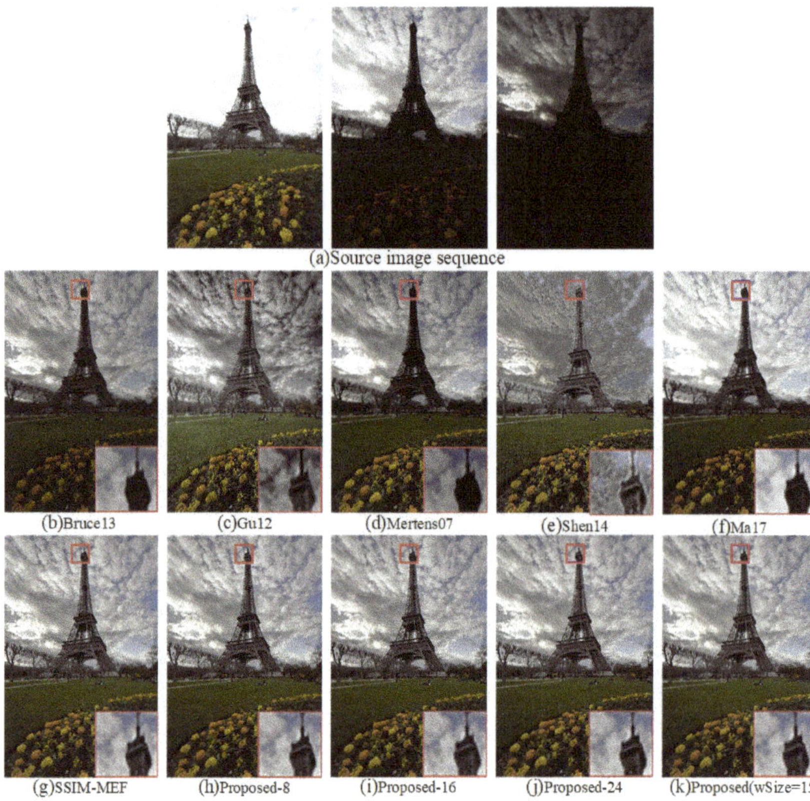

Figure 9. Visual comparison of the fused "Tower" between the proposed method and nine existing methods.

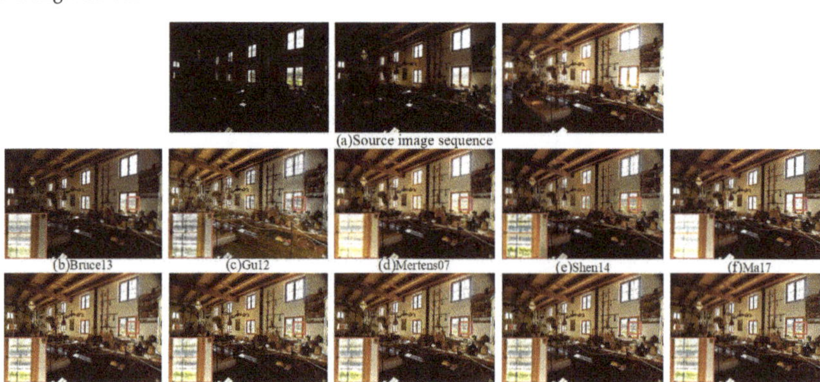

Figure 10. Visual comparison of the fused "Farmhouse" between the proposed method and nine existing methods.

The ten fused images of "Farmhouse" are shown in Figure 10. As shown in Figure 10b, the overall brightness of Bruce 13 [12] is weak, and the details of some dark areas cannot be well demonstrated. The overall color and brightness of Gu12 [23] was natural, but the color of the marked area outside the window was obviously distorted. The brightness of the bottom part shown in the fused images obtained by Mertens07 [10] and Shen14 [14] was dark. At the same time, the colors of the small ornaments were not natural. Compared with Ma17 [3], the fused image of the proposed method presented more details outside the window. Compared with the SSIM-MEF [3,29] and Proposed-24 methods, the proposed method had moderate brightness. The fused images obtained by the Proposed-8 and Proposed-16 methods performed similarly to the human visual system. The fused image obtained by the proposed method had the best overall performance in brightness, local detail processing, and visual effect among all ten image fusion methods.

3.3. Objective Evaluation Metrics

In this paper, four objective evaluation indicators were used to evaluate the fusion performance objectively from four aspects: edge preservation, similarity, human visual effects, and calculation time. The average values of the 24 sets for the comparative experiments obtained by ten different methods in three objective evaluation indexes are shown in Table 1. The objective evaluation results of the 24 sets of multi-exposure fusion images are shown in Figure 11. In Figure 11, the three bar charts represent $Q^{AB/F}$, MI and Q^{CB} objective values, respectively. As can be seen from Figure 11, in all of the comparison methods, our method ranks as the top two in most cases for the three objective indicators and all 24 groups of multi-exposure image fusion problems. Concretely, the $Q^{AB/F}$ index scores of the proposed fusion results were ranked as top three in 20 groups. There were 16 groups in which the MI scores of the proposed fusion results were ranked in the top three place. For the Q^{CB}-index score, the proposed method had the top three highest scores in 22 groups. From Figure 11, among the three objective evaluation indicators of all fusion results, the objective evaluation scores of the proposed method were in the top three for 80% of the results, and the remaining 20% of the results of the fusion objective scores were in the middle. This indicated that the proposed method could better preserve the details of the source scene and obtain better human visual effects.

From Table 2, it is indicated that the fused image obtained by Shen14 [14] had the lowest value for $Q^{AB/F}$ and Q^{CB}. Except Gu12 [23], the MI value of Shen14 [14] was also lower than the other eight fusion methods. This means that the fusion result of Shen14 [14] did not have good performance in image edge processing and human visual effects. In Figures 6–8, the fused images of Shen14 [14] have excessive sharpening, which leads to a poor visual effect and poor edge detail processing of the local magnified region. The objective and subjective evaluations of Shen14 [14] were almost the same. Compared with the other nine fusion methods, the MI value of Gu12 [23] was the lowest, which indicates that the similarity between the fused image and source images was objectively the worst. According to the previous subjective comparison, the fused images of Gu12 [23] shown in Figures 6, 8 and 9 had a color distortion issue. The objective evaluation values indicate the same result as the subjective comparison. The values of SSIM-MEF [3,29] in the chart were the result of the optimization of Ma17 [3] using SSIM-MEF. It can be observed that the $Q^{AB/F}$ and Q^{CB} evaluation scores of the optimized image were higher than the unoptimized image, which implies that the quality of the image of Ma17 [3] was good after SSIM-MEF [3,29] optimization. As can be seen in Figure 6, the color of the fusion image of SSIM-MEF was more natural than that of Ma17 [3], and the edge detail of the tree were clearer. The same conclusion can be drawn from both subjective and objective evaluation. Proposed-8, Proposed-16, and Proposed-24 differed with the proposed method only in the size selection of image patches, but the objective indices of fused images were completely different. Their values were lower than those of the proposed method and better than those of a few other comparative experiments. It can be inferred that the method of adaptive structure selection was superior to the method of fixed structure. The proposed ASP-MEF obtained the maximum values in all three indexes. This confirms that the proposed method had good performance in edge detail processing and visual effect, and the fused image of proposed method had high similarity to the source images. From Table 2, the calculation time of Shen14 [14] was the longest, Proposed-24 the second, and the time difference of other methods was not obvious. Therefore, except for the Shen14 [14] method, the difference in fusion complexity and efficiency of the remaining algorithms was not significant. However, the proposed method performed better than the other methods in the other three objective evaluation indicators. In addition, compared with the other nine methods, subjectively, the proposed method also achieved optimal performance in color softness, brightness, and local detail processing. In conclusion, the proposed method was the best in terms of both subjective comparison and objective evaluation.

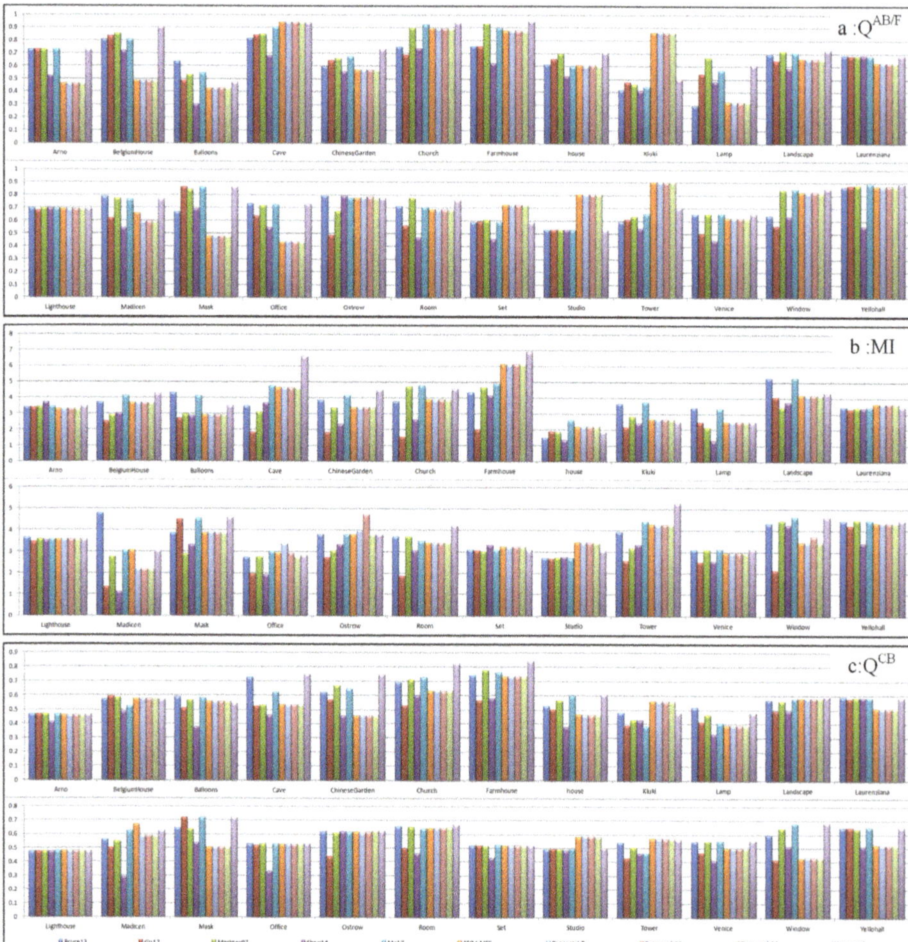

Figure 11. Objective evaluations of 24 source image sets in the MEF experimentation.

Table 2. Objective evaluation of the ten MEF methods.

	$Q^{AB/F}$	MI	Q^{CB}	Time
Bruce13	0.66684	3.67199	0.57956	17.30 s
Gu12	0.64301	2.61998	0.50975	13.60 s
Mertens07	0.71941	3.26387	0.57021	10.20 s
Shen14	0.57109	2.93935	0.46300	57.27 s
Ma17	0.71470	3.85767	0.57580	13.64 s
SSIM-MEF	0.72586	3.67061	0.57730	15.93 s
Proposed-8	0.65852	3.50575	0.53225	**9.50 s**
Proposed-16	0.65863	3.53180	0.53234	14.11 s
Proposed-24	0.65814	3.47528	0.53253	21.13 s
Proposed	**0.73623**	**3.91869**	**0.60737**	14.12 s

4. Conclusions

This paper proposes a novel MEF method named the adaptive patch structure-based MEF (APS-MEF). First, texture-cartoon decomposition is applied to obtain image texture components. Second, the image texture entropy is calculated to achieve the adaptive selection of image patch size. Then, the structural patch decomposition approach is utilized to obtain the initial fused image. Finally, the initial fused image is iteratively optimized by applying the color MEF structural similarity index to obtain the final fused image. The proposed algorithm evaluates the local information by texture entropy and adaptively selects image patch size, which allows the fused image to contain more detailed information. The visual quality of the fused image is improved by combining the decomposition of the image patch structure and the similarity index algorithm of the color image structure. The comparative experiments show that the proposed APS-MEF fusion method can preserve more detailed information and obtain better human visual effects.

The proposed APS-MEF uses SSIMc-MEF as the iterative optimization algorithm. However, the iterative optimization algorithm is not suitable for real-time applications. In the future, an efficient non-iterative optimization algorithm will be adopted to improve the efficiency of the fusion algorithm.

Author Contributions: Data curation, M.Z.; funding acquisition, Y.L. and Y.S.; investigation, Y.L.; methodology, Y.L. and M.Z.; project administration, Y.S.; resources, X.H.; software, Z.Z. and G.Q.; supervision, Y.S.; visualization, H.H.; writing, original draft, M.Z. and G.Q.; writing, review and editing, G.Q.

Funding: This research is funded by the National Natural Science Foundation of China under Grants 61803061, 61703347; the Science and Technology Research Program of Chongqing Municipal Education Commission (Grant No. KJQN201800603); the Chongqing Natural Science Foundation Grant cstc2016jcyjA0428; the Common Key Technology Innovation Special of Key Industries of the Chongqing Science and Technology Commission under Grant Nos. cstc2017zdcy-zdyf0252 and cstc2017zdcy-zdyfX0055; the Artificial Intelligence Technology Innovation Significant Theme Special Project of the Chongqing Science and Technology Commission under Grant Nos. cstc2017rgzn-zdyf0073 and cstc2017rgzn-zdyf0033; the China University of Mining and Technology Teaching and Research Project (2018ZD03, 2018YB10).

Conflicts of Interest: The authors declare no conflict of interest. The funders had no role in the design of the study; in the collection, analyses, or interpretation of data; in the writing of the manuscript, or in the decision to publish the results.

References

1. Reinhard, E.; Ward, G.; Pattanaik, S.; Debevec, P.E. *High Dynamic Range Imaging: Acquisition, Display, and Image-Based Lighting*; Princeton University Press: Princeton, NJ, USA, 2005; pp. 2039–2042.
2. Zhu, Z.; Qi, G.; Chai, Y.; Yin, H.; Sun, J. A Novel Visible-infrared Image Fusion Framework for Smart City. *Int. J. Simul. Process Model.* **2018**, *13*, 144–155. [CrossRef]
3. Ma, K.; Li, H.; Yong, H.; Wang, Z.; Meng, D.; Zhang, L. Robust Multi-Exposure Image Fusion: A Structural Patch Decomposition Approach. *IEEE Trans. Image Process.* **2017**, *26*, 2519–2532. [CrossRef] [PubMed]
4. Artusi, A.; Richter, T.; Ebrahimi, T.; Mantiuk, R.K. High Dynamic Range Imaging Technology [Lecture Notes]. *IEEE Signal Process. Mag.* **2017**, *34*, 165–172. [CrossRef]

5. Qi, G.; Zhu, Z.; Chen, Y.; Wang, J.; Zhang, Q.; Zeng, F. Morphology-based visible-infrared image fusion framework for smart city. *Int. J. Simul. Process Model.* **2018**, *13*, 523–536. [CrossRef]
6. Oh, T.H.; Lee, J.Y.; Tai, Y.W.; Kweon, I.S. Robust High Dynamic Range Imaging by Rank Minimization. *IEEE Trans. Pattern Anal. Mach. Intell.* **2015**, *37*, 1219–1232. [CrossRef] [PubMed]
7. Li, H.; Qiu, H.; Yu, Z.; Li, B. Multifocus image fusion via fixed window technique of multiscale images and non-local means filtering. *Signal Process.* **2017**, *138*, 71–85. [CrossRef]
8. Li, S.; Kang, X.; Fang, L.; Hu, J.; Yin, H. Pixel-level image fusion: A survey of the state of the art. *Inf. Fusion* **2017**, *33*, 100–112. [CrossRef]
9. Zhu, Z.; Chai, Y.; Yin, H.; Li, Y.; Liu, Z. A novel dictionary learning approach for multi-modality medical image fusion. *Neurocomputing* **2016**, *214*, 471–482. [CrossRef]
10. Mertens, T.; Kautz, J.; Van Reeth, F. Exposure Fusion: A Simple and Practical Alternative to High Dynamic Range Photography. *Comput. Graph. Forum* **2010**, *28*, 161–171. [CrossRef]
11. Li, S.; Kang, X.; Hu, J. Image fusion with guided filtering. *IEEE Trans. Image Process.* **2013**, *22*, 2864–2875. [PubMed]
12. Bruce, N.D.B. ExpoBlend: Information preserving exposure blending based on normalized log-domain entropy. *Comput. Graph.* **2014**, *39*, 12–23. [CrossRef]
13. Kou, F.; Wei, Z.; Chen, W.; Wu, X.; Wen, C.; Li, Z. Intelligent Detail Enhancement for Exposure Fusion. *IEEE Trans. Multimed.* **2017**, *20*, 484–485. [CrossRef]
14. Shen, J.; Zhao, Y.; Yan, S.; Li, X. Exposure Fusion Using Boosting Laplacian Pyramid. *IEEE Trans. Cybern.* **2014**, *44*, 1579–1590. [CrossRef] [PubMed]
15. Li, Y.; Sun, Y.; Huang, X.; Qi, G.; Zheng, M.; Zhu, Z. An Image Fusion Method Based on Sparse Representation and Sum Modified-Laplacian in NSCT Domain. *Entropy* **2018**, *20*, 522. [CrossRef]
16. Liu, Y.; Liu, S.; Wang, Z. A general framework for image fusion based on multi-scale transform and sparse representation. *Inf. Fusion* **2015**, *24*, 147–164. [CrossRef]
17. Huafeng, L.; Jinting, Z.; Dapeng, T. Asymmetric Projection and Dictionary Learning with Listwise and Identity Consistency Constraints for Person Re-Identification. *IEEE Access* **2018**, *6*, 37977–37990.
18. Zhu, Z.; Qi, G.; Chai, Y.; Li, P. A Geometric Dictionary Learning Based Approach for Fluorescence Spectroscopy Image Fusion. *Appl. Sci.* **2017**, *7*, 161. [CrossRef]
19. Wang, K.; Qi, G.; Zhu, Z.; Chai, Y. A Novel Geometric Dictionary Construction Approach for Sparse Representation Based Image Fusion. *Entropy* **2017**, *19*, 306. [CrossRef]
20. Shen, R.; Cheng, I.; Shi, J.; Basu, A. Generalized Random Walks for Fusion of Multi-Exposure Images. *IEEE Trans. Image Process.* **2011**, *20*, 3634–3646. [CrossRef]
21. Qi, G.; Zhu, Z.; Erqinhu, K.; Chen, Y.; Chai, Y.; Sun, J. Fault-diagnosis for reciprocating compressors using big data and machine learning. *Simul. Model. Pract. Theory* **2018**, *80*, 104–127. [CrossRef]
22. Shen, R.; Cheng, I.; Basu, A. QoE-based multi-exposure fusion in hierarchical multivariate Gaussian CRF. *IEEE Trans. Image Process.* **2013**, *22*, 2469–2478. [CrossRef] [PubMed]
23. Gu, B.; Li, W.; Wong, J.; Zhu, M.; Wang, M. Gradient field multi-exposure images fusion for high dynamic range image visualization. *J. Vis. Commun. Image Represent.* **2012**, *23*, 604–610. [CrossRef]
24. Li, S.; Kang, X. Fast multi-exposure image fusion with median filter and recursive filter. *IEEE Trans. Consum. Electron.* **2012**, *58*, 626–632. [CrossRef]
25. Song, M.; Tao, D.; Chen, C.; Bu, J.; Luo, J.; Zhang, C. Probabilistic Exposure Fusion. *IEEE Trans. Image Process.* **2012**, *21*, 341. [CrossRef] [PubMed]
26. Bertalmío, M.; Levine, S. Variational approach for the fusion of exposure bracketed pairs. *IEEE Trans. Image Process.* **2013**, *22*, 712–723. [CrossRef] [PubMed]
27. Zhang, W.; Hu, S.; Liu, K. Patch-Based Correlation for Deghosting in Exposure Fusion. *Inf. Sci.* **2017**, *415*, 19–27. [CrossRef]
28. Wang, Z.; Bovik, A.C.; Sheikh, H.R.; Simoncelli, E.P. Image quality assessment: From error visibility to structural similarity. *IEEE Trans. Image Process.* **2004**, *13*, 600–612. [CrossRef]
29. Ma, K.; Duanmu, Z.; Yeganeh, H.; Wang, Z. Multi-Exposure Image Fusion by Optimizing A Structural Similarity Index. *IEEE Trans. Comput. Imag.* **2017**, *4*, 60–72. [CrossRef]
30. Qi, G.; Wang, J.; Zhang, Q.; Zeng, F.; Zhu, Z. An Integrated Dictionary-Learning Entropy-Based Medical Image Fusion Framework. *Future Internet* **2017**, *9*, 61. [CrossRef]

31. Li, H.; Li, X.; Yu, Z.; Mao, C. Multifocus image fusion by combining with mixed-order structure tensors and multiscale neighborhood. *Inf. Sci.* **2016**, *349–350*, 25–49. [CrossRef]
32. Li, H.; He, X.; Tao, D.; Tang, Y.; Wang, R. Joint medical image fusion, denoising and enhancement via discriminative low-rank sparse dictionaries learning. *Pattern Recognit.* **2018**, *79*, 130–146. [CrossRef]
33. Zhu, Z.Q.; Yin, H.; Chai, Y.; Li, Y.; Qi, G. A Novel Multi-modality Image Fusion Method Based on Image Decomposition and Sparse Representation. *Inf. Sci.* **2018**, *432*, 516–529. [CrossRef]
34. Vese, L.A.; Osher, S.J. Image Denoising and Decomposition with Total Variation Minimization and Oscillatory Functions. *J. Math. Imaging Vis.* **2004**, *20*, 7–18. [CrossRef]
35. Chamorro-Martinez, J.; Martinez-Jimenez, P. A comparative study of texture coarseness measures. In Proceedings of the IEEE International Conference on Image Processing, Cairo, Egypt, 7–10 November 2010; pp. 1329–1332.
36. Zhang, W.; He, K.; Meng, C. Texture synthesis method by adaptive selecting size of patches. *Comput. Eng. Appl.* **2012**, *48*, 170–173.
37. Qi, G.; Zhang, Q.; Zeng, F.; Wang, J.; Zhu, Z. Multi-focus image fusion via morphological similarity-based dictionary construction and sparse representation. *CAAI Trans. Intell. Technol.* **2018**, *3*, 83–94. [CrossRef]
38. Petrović, V. Subjective tests for image fusion evaluation and objective metric validation. *Inf. Fusion* **2007**, *8*, 208–216. [CrossRef]
39. Zhu, Z.; Qi, G.; Chai, Y.; Chen, Y. A Novel Multi-Focus Image Fusion Method Based on Stochastic Coordinate Coding and Local Density Peaks Clustering. *Future Internet* **2016**, *8*, 53. [CrossRef]
40. Qu, G.; Zhang, D.; Yan, P. Information measure for performance of image fusion. *Electron. Lett.* **2002**, *38*, 313–315. [CrossRef]
41. Zhu, Z.; Sun, J.; Qi, G.; Chai, Y.; Chen, Y. Frequency Regulation of Power Systems with Self-Triggered Control under the Consideration of Communication Costs. *Appl. Sci.* **2017**, *7*, 688. [CrossRef]
42. Liu, Z.; Blasch, E.; Xue, Z.; Zhao, J.; Laganiere, R.; Wu, W. Objective Assessment of Multiresolution Image Fusion Algorithms for Context Enhancement in Night Vision: A Comparative Study. *IEEE Trans. Pattern Anal. Mach. Intell.* **2011**, *34*, 94–109. [CrossRef] [PubMed]
43. Chen, Y.; Blum, R.S. A new automated quality assessment algorithm for image fusion. *Image Vis. Comput.* **2009**, *27*, 1421–1432. [CrossRef]

entropy

Article

An Image Encryption Algorithm Based on Time-Delay and Random Insertion

Xiaoling Huang and Guodong Ye *

Faculty of Mathematics and Computer Science, Guangdong Ocean University, Zhanjiang 524088, China; xyxhuang@gdou.edu.cn
* Correspondence: guodongye@hotmail.com or yegd@gdou.edu.cn; Tel.: +86-759-2383064

Received: 19 November 2018; Accepted: 13 December 2018; Published: 15 December 2018

Abstract: An image encryption algorithm is presented in this paper based on a chaotic map. Different from traditional methods based on the permutation-diffusion structure, the keystream here depends on both secret keys and the pre-processed image. In particular, in the permutation stage, a middle parameter is designed to revise the outputs of the chaotic map, yielding a temporal delay phenomena. Then, diffusion operation is applied after a group of random numbers is inserted into the permuted image. Therefore, the gray distribution can be changed and is different from that of the plain-image. This insertion acts as a one-time pad. Moreover, the keystream for the diffusion operation is designed to be influenced by secret keys assigned in the permutation stage. As a result, the two stages are mixed together to strengthen entirety. Experimental tests also suggest that our algorithm, permutation–insertion–diffusion (PID), performs better when expecting secure communications for images.

Keywords: image encryption; time-delay; random insertion; information entropy; chaotic map

1. Introduction

With fast development of computer and network technologies, digital information (multimedia) modalities (such as images, video, and audio) have been widely adopted for daily communication. Among these, image analysis is a most direct and simple way to learn and understand the natural world. Images are increasingly transformed over networks every day, according to the Google analysis. Images and applications utilizing image processing are used in many fields, such as medicine, education, and aerospace, to name a few. However, illegal attackers may visit, read, or intercept our transmitted information.

Cryptology can be utilized to develop methods for secure transmission of images. However, images are different from text files, and have many unique characteristics, such as bulk data capacity, high redundancy, and strong inter-pixel correlation. As a result, traditional encryption algorithms such as DES, AES, and IDEA are not suitable for secure encoding of images. Development of algorithms for effective image encryption remains an important priority in the fields of computer science and communications. Recently, chaos-based image encryption schemes have received considerable attention; these methods allow for hiding image-related information accounting for the desirable properties [1,2] of extreme sensitivity to initial conditions, ergodicity, and pseudo-randomness of chaos systems (maps). As early as 1998, Fridrich [3] proposed an image encryption method that used a two-dimensional chaotic map. In what follows, many encryption algorithms have been designed, which fully or partially utilize the Fridrich structure (i.e., permutation–diffusion). For example, a bit-level image encryption algorithm [4] was proposed based on piecewise linear chaotic maps, in which a diffusion strategy was introduced followed by a permutation of bits for each value. Quantum chaotic map [5] with a diffusion-permutation architecture-based image encryption algorithm has been presented. Norouzi et al. designed diffusion-only image encryption schemes [6,7]. The test results show high sensitivity and high complexity. The behavior of quantum walks was proved [8] to be chaotic, and a permutation-based image encryption algorithm has been proposed. It calculates the sum of the plain-image and uses the resulting value to diffuse the image's pixels. Furthermore, to enhance the sensitivity of the encryption method, a quantum hash function is taken to act as a hash function for the privacy amplification process [9]. Exclusive OR (XOR) as a diffusion operation and shuffling as a permutation are then applied to the plain-image and yield a cipher-image with a new encryption structure. A unique and more distinctive encryption algorithm is proposed based on the complexity of a highly nonlinear S box in Flesnelet domain [10]. DNA-based image encryption methods [11–13] and other similar architectures [14–22] have also been presented as encryption techniques to ensure communication of images.

However, some schemes have been found to be insecure. For example, Li [23] evaluated a class of permutation-only encryption algorithms. Using a known(chosen)-plaintext attack, the plain-image could be recovered if the encryption algorithm [24] was used. Furthermore, it was shown how permutation-only image encryption schemes can be broken with little computation complexity [25,26]. Eslami and Bakhshandeh [27] designed a new image encryption to promote the plain-text sensitivity and to enhance the diffusion performance. However, the keystream used in that diffusion was not related to the plain-image. As a result, Akhavan et al [28] re-evaluated the security and broke it successfully using a chosen plain-text attack [27]. Other cryptanalysis methods [29–32] have been proposed as well.

To solve the above security problem and to enhance the connection between the plain-image and the keystream, a novel chaotic image encryption scheme, named permutation–insertion–diffusion (PID), is proposed in this paper. A middle parameter is designed to revise the outputs of the chaotic map, acting like a time-delay phenomena. To enhance the security of the Fridrich structure, especially the shortcoming of unchanged gray values before the diffusion operation [33,34], a group of random numbers are inserted in the pre-encrypted image to rewrite the gray distribution followed by the diffusion encryption. As a result, the proposed algorithm can be seen as a one-time pad. The rest of this paper is organized as follows. The proposed cryptosystem is described after an introduction of a chaotic map. Then, some experimental results are shown by using our method. After that, security analyses are evaluated to explain the better performance of our scheme. Finally, conclusions are drawn followed by a discussion.

2. The Proposed Cryptosystem

A two-dimensional (2D) chaotic map, called a 2D Sine Logistic modulation map (2D-SLMM), was studied in [34]. The map is defined by

$$\begin{cases} x_{i+1} = u(\sin(\pi y_i) + v)x_i(1 - x_i), \\ y_{i+1} = u(\sin(\pi x_{i+1}) + v)y_i(1 - y_i), i = 0, 1, 2 \cdots, \end{cases} \tag{1}$$

where $u \in [0, 1]$, $v \in [0, 3]$. To enhance the nonlinearity and the randomness, parameter v is set to modulate the output of the Logistic map. When we let v be close to 3, the output pairs (x_{i+1}, y_{i+1}) of 2D-LASM distribute in the whole data range of the 2D phase plane. Thus, v is set to be 3 [34] in 2D-SLMM to display good chaotic performance. Figure 1 shows the chaotic orbit for the 2D-SLMM output. A detailed description of the map is provided in [34].

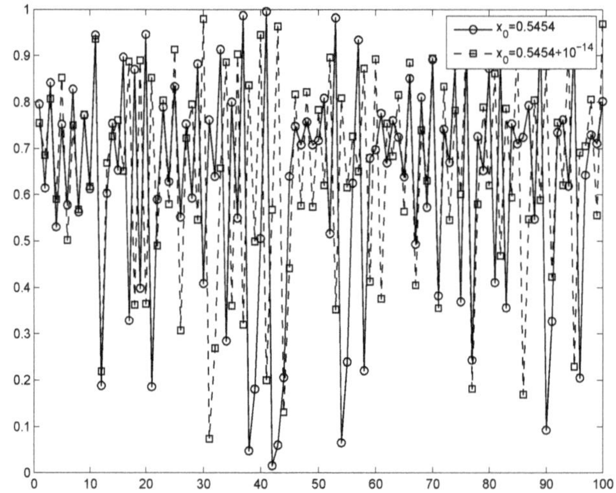

Figure 1. Chaotic dynamics in the 2D-SLMM map.

2.1. Image Cryptosystem

To deduce the strong correlation among adjacent pixels in the plain-image, pixel shuffling is considered as a first step. Let P be an $m \times n$ plain-image, and randomly set initial conditions $u = 0.9966$, $v = 3.000$, $x_0 = 0.4237$ and $y_0 = 0.1784$ in the permutation stage. Then, after a certain number of iterations, two sets $\{x_i\}$ and $\{y_i\}$ are obtained. To generate a keystream with a time-delay-like phenomenon, the following operations are performed:

$$\begin{cases} s = 1 + \sum P_{i,j}, \\ t = \lceil y_i \times 10^{14} + s \rceil \bmod 7 + 1, \\ \overline{x}_i = x_{i+t}, \\ h_i = \lceil 3\overline{x}_i \times 10^{14} \rceil \bmod n + 1, i = 1, 2, \cdots, \\ l_j = \lceil 5\overline{x}_j \times 10^{14} \rceil \bmod m + 1, j = 1, 2, \cdots, \end{cases} \quad (2)$$

where $\lceil a \rceil$ corresponds to the floor operation on a, and t is a time-delay factor. As a result, we obtain $H = \{h_1, h_2, \cdots, h_m\}$ and $L = \{l_1, l_2, \cdots, l_n\}$ for circular permutations of row and column. Assume that the permuted image is T after permutation encryption.

If permutation-only operation is applied to a plain-image, then the gray distribution of the permuted image is the same as that of the plain-image. Moreover, encryption schemes in this family were found to be insecure. To enhance the security level of the proposed algorithm, random numbers are inserted into the image T before the first row with a random row a and the first column with a random column b. A new image B is obtained, with the dimensions of $(m+1) \times (n+1)$. As a result, the gray distribution of B is different from that of image P. Fortunately, the insertion function acts as a one-time pad owing to the random numbers being generated anew each time. For example, vector $a = \{3, 8, 9, 20\}$ may become $a = \{11, 34, 5, 7\}$ randomly with a four-dimension. Thus, the obtained cipher-images are different, even if the encryption is performed on the same plain-image in different communications.

To determine the relationship between the different pixels in the cipher-image, diffusion is further used to encrypt the permuted image B. Again, random initial conditions are set as $u = 0.9966$, $v = 3.000$, $\hat{x}_0 = 0.6028$ and $\hat{y}_0 = 0.1883$ in the diffusion stage, and the chaotic map is iterated. Then, a chaotic matrix M with the same size as B is obtained after a certain number of iterations. To revise the gray distribution, the following operation is performed on the matrix rows:

$$\begin{cases} D = B + M \bmod 256, \\ r = \lceil (x_0 + y_0) \times 10^{14} \rceil \bmod 7 + 1, \\ C_i = C_{i-1} + rM_i + D_i \bmod 256, i = 1, 2, \cdots, m+1, \end{cases} \tag{3}$$

where C_i, M_i, and D_i represent the row vectors of C, M, and D, respectively. C_0 is a constant vector. Finally, a cipher-image C is obtained. It is noted that, before the diffusion operation, a rewriting operation for the permuted image B should be performed, which overcomes the shortcoming of the Fridrich structure and enhances the encryption security. Considering a similar function in the case of columns, the above function by row is applied again, this time on the columns of image C, and the following cipher-image E is obtained:

$$\begin{cases} F = C + M \bmod 256, \\ E_j = E_{j-1} + rM_j + F_j \bmod 256, j = 1, 2, \cdots, n+1. \end{cases} \tag{4}$$

2.2. Encryption Steps

As described above, the proposed encryption scheme can be summarized in the following steps, with the symmetric PID structure:

Step 1. Read the plain-image as P and obtain its size $m \times n$.
Step 2. Compute the sum s over the plain-image.
Step 3. Generate the two sets H and L by simulating a time-delay phenomena.
Step 4. Apply circular permutation to both rows and columns, and obtain T.
Step 5. Insert random numbers into the permuted image T and obtain B by simulating a one-time pad.
Step 6. Iterate the chaotic map again and obtain matrix M.
Step 7. Apply the diffusion operation to revise the gray distribution, on both row and column dimensions.
Step 8. Obtain the cipher-image E.

2.3. Decryption

Owing to the symmetric cryptosystem nature of our method, image decryption can be performed by applying the same steps in reverse, starting from the ciphered image and ending up with the plain image. Using correct keys, the diffusion operation is firstly applied, followed by the permutation operations in the reverse order.

3. Experimental Results

Three images were randomly chosen, tests were performed using our proposed method, and the results are reported in this section. The test was implemented in Matlab 2011b running on Windows 7 (Notebook with Intel(R) Core(TM) i3-2350, 2.30 GHz CPU). To increase the security, the PID process was applied twice to each image. Then, the former 100 iteration results for the chaotic map were deleted to avoid harmful effects. Figure 2 shows the plain-images, corresponding cipher-images, and their decrypted results. It is clear that these cipher-images contain no useful image-related information, compared with their corresponding plain-images.

Figure 2. Experimental tests: (a) plain-image of Tree; (b) cipher-image of Tree; (c) decrypted image of Tree; (d) plain-image of Lake; (e) cipher-image of Lake; (f) decrypted image of Lake; (g) plain-image of Building; (h) cipher-image of Building; (i) decrypted image of Building.

4. Security Analyses

4.1. Key Space Analysis

The key space corresponds to the space of all combinations of keys that can be used in a certain encryption scheme. Here, there are four keys, i.e., x_0, y_0, $\hat{x}_0$, $\hat{y}_0$, not including the parameters u, v. The key space becomes as large as $10^{56} \approx 2^{186}$ if the precision is set to 10^{-14}. As a result, it is difficult to conduct a successful brute-force.

4.2. Histogram Analysis

To reduce the chance of attack and to efficiently hide the information of a plain-image, the histogram of the corresponding cipher-image should be uniform and significantly distinct from that of the plain-image. Figure 3 shows the histograms for the images of Lena, Baboon, Boat, and Peppers before and after using our encryption scheme. It is clear that the histograms of the encrypted images are flat. Thus, successful attacks are impossible.

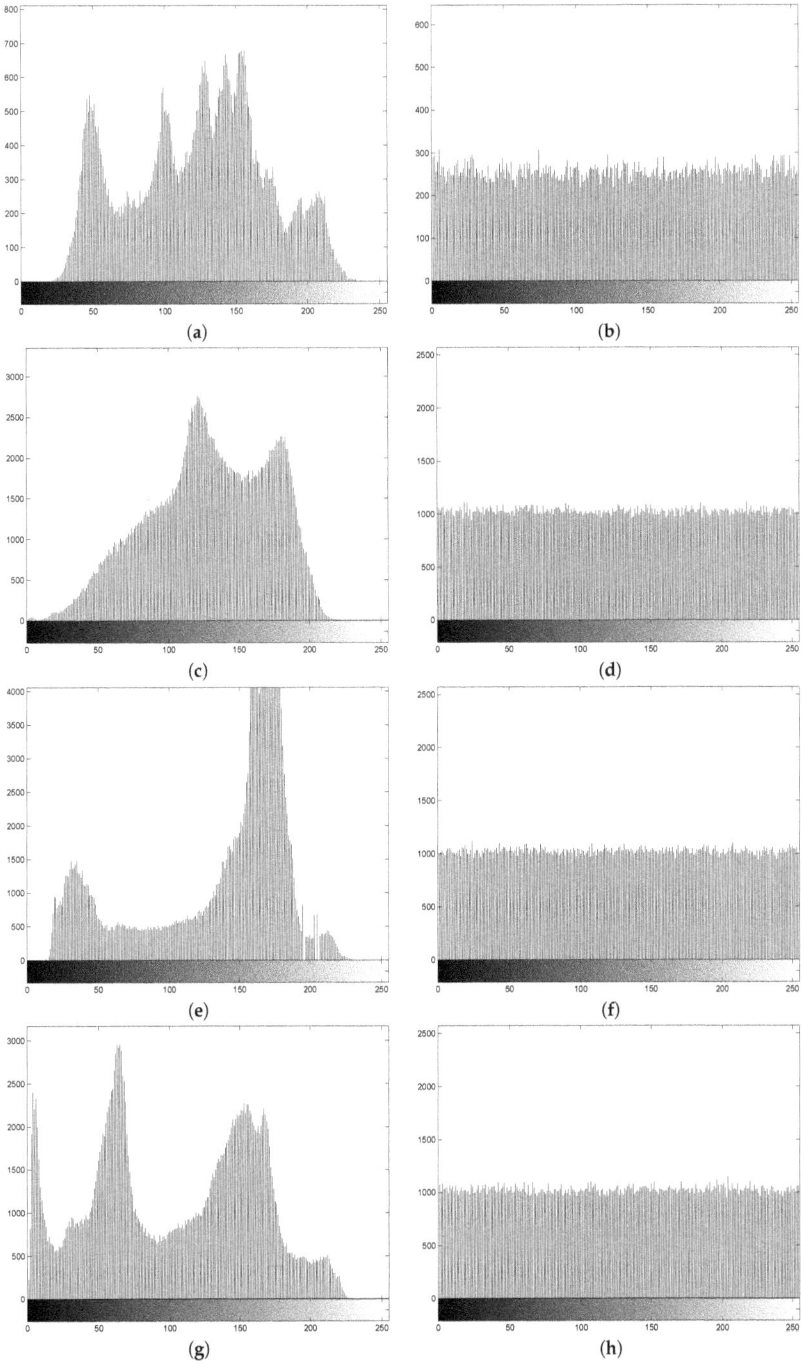

Figure 3. Histograms of: (**a**) the plain-image of Lena; (**b**) the cipher-image of Lena; (**c**) the plain-image of Baboon; (**d**) the cipher-image of Baboon; (**e**) the plain-image of Boat; (**f**) the cipher-image of Boat; (**g**) the plain-image of Peppers; (**h**) the cipher-image of Peppers.

4.3. *Information Entropy Analysis*

Information entropy [1] is an efficient measure of the randomness of an input image (message). This measure can be defined using the following equation:

$$E(m) = \sum_{i=0}^{L-1} p(m_i) log_2 \frac{1}{p(m_i)}, \quad (5)$$

where $L = 2^k$ is the total number of states of the tested message ($k = 8$ for a gray level image). Here, we tested four images, and the results are listed in Table 1 (using code "entropy" in Matlab). We conclude that the information entropy indicates that it is difficult to conduct a successful attack because the values of the information entropy for the cipher-images are close to a theoretical value of 8 [35,36]. The random numbers inserted into the image in each encryption, so the values of the information entropy would be changed very slightly each time. Figure 4 shows the results for encrypting Lena and Boat at different times.

Table 1. Information entropy tests.

Test Images	Plain-Image	Cipher-Image
Lena	7.4532	7.9970
Boat	7.1238	7.9993
Peppers	7.5715	7.9992
Baboon	7.3579	7.9993

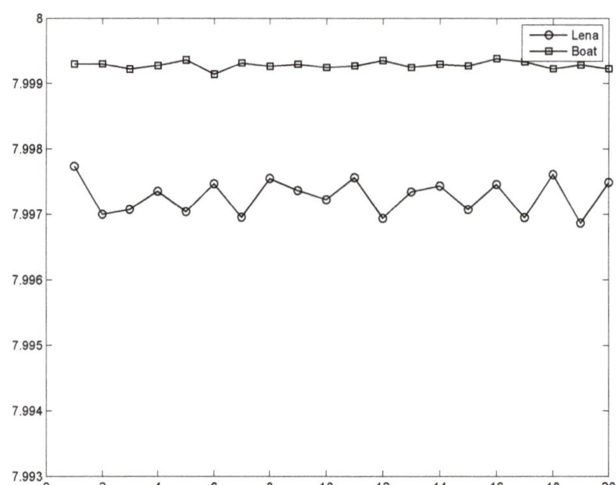

Figure 4. Information entropy at different times of encryption.

4.4. *Key Sensitivity Analysis*

A good image encryption algorithm should be very sensitive to all of the keys used. We tested our algorithm on the image of Lena, and the results are listed in Figure 5. Figure 5a–d shows the incorrect decryption from the cipher-image with a small change (i.e., 10^{-14}) added in keys x_0, y_0, $\hat{x}_0$, and $\hat{y}_0$, respectively. Therefore, the proposed image encryption algorithm possesses a high key sensitivity and can frustrate brute-force attackers.

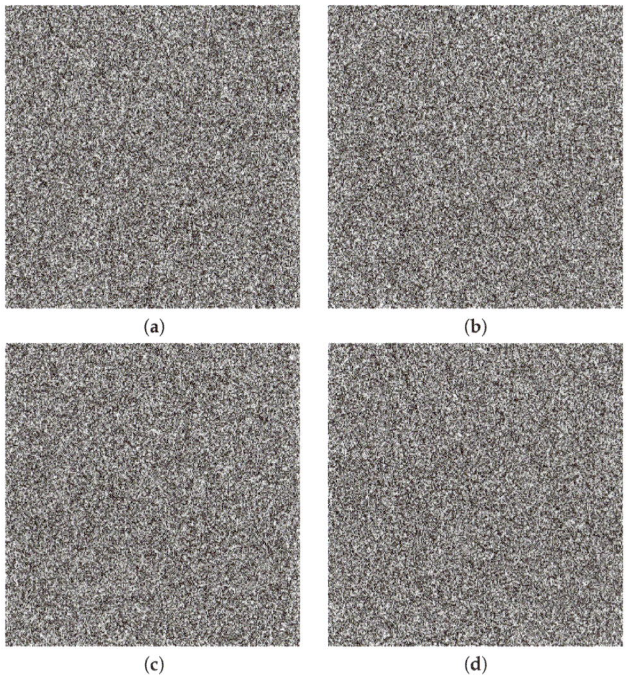

Figure 5. Key sensitivity tests for Lena: (**a**) decryption with $x_0 + 10^{-14}$; (**b**) decryption with $y_0 + 10^{-14}$; (**c**) decryption with $\hat{x}_0 + 10^{-14}$; (**d**) decryption with $\hat{y}_0 + 10^{-14}$.

4.5. *Differential Analysis*

To test the sensitivity of the proposed encryption method to a small change, even one bit, in the plain-image, we used two common measures [37,38], the number of pixel change rate (NPCR) and the unified average changing intensity (UACI). The measures are defined as follows:

$$NPCR = \frac{\sum_{i,j} D(i,j)}{m \times n} \times 100\%, \tag{6}$$

$$UACI = \frac{1}{m \times n} \sum_{i,j} \frac{|C'(i,j) - C(i,j)|}{255} \times 100\%, \tag{7}$$

$$D(i,j) = \begin{cases} 0, \text{ if } C'(i,j) = C(i,j), \\ 1, \text{ otherwise}, \end{cases} \tag{8}$$

where C' and C are two cipher-images corresponding to the same plain-images differing only in one bit. The results of this test are listed in Table 2 for a change in the value of the $(100, 89)$ position. The results in Table 2 show that our method has high sensitivity to changes in the plain-images because the values are nearly ideal [39].

Table 2. Sensitivity tests.

Test Images	UACI	NPCR
Lena	33.3537	99.6109
Boat	33.4899	99.5900
Peppers	33.5186	99.6044
Baboon	33.5280	99.6136

UACI: unified average changing intensity; NPCR: number of pixel change rate.

4.6. Run Test for Randomness

The run test mainly examines whether the probability of an event is random. In Matlab software, "runstest" performs a run test on a given sequence X. This is a test of the hypothesis that the values in X come in a random order. If the sequence is random, then the test result is 0, or the result is 1. By using our algorithm, the test results are listed in Table 3. Therefore, the outputs show good statistical randomness.

Table 3. Run test for randomness.

Images	Lena	Peppers	Boat	Baboon
Results	0	0	0	0
Randomness	Pass	Pass	Pass	Pass

4.7. Comparisons

To make a comparison, information entropy was taken to measure the randomness of different plain-images and their corresponding cipher-images. Here, a color image of Lena was selected for comparison with some methods [1,38,40]. The results are given in Table 4 for tests on cipher-images. Obviously, the information entropy values are close to the ideal value of 8 for our proposed scheme. Furthermore, computational complexity is an important metric for measuring the efficiency of the designed algorithm. Table 5 compares the proposed algorithm with some recent references, for different sizes. Considering key size, information entropy, and running speed, Table 6 displays the comparisons for some of other methods [41–45], where information entropy is tested for a cipher-image. Thus, our method can show good performance to satisfy a real-time communication.

Table 4. Comparisons of information entropy.

Channels	R	G	B	Average
Ref. [1]	7.9903	7.9890	7.9893	7.9895
Ref. [38]	7.9871	7.9881	7.9878	7.9877
Ref. [40]	7.9278	7.9744	7.9705	7.9576
Ref. [46]	7.9969	7.9974	7.9970	7.9971
Ref. [47]	7.9895	7.9897	7.9893	7.9895
Ref. [48]	7.9968	7.9970	7.9972	7.9970
Ours	7.9977	7.9973	7.9975	7.9975

Table 5. Comparisons of speed performance.

Sizes	Ref. [46]	Ref. [47]	Ref. [48]	Ours
256 × 256	0.1641 s	0.0552 s	0.0671 s	0.0312 s
512 × 512	0.6630 s	0.2031 s	0.2293 s	0.1373 s

Table 6. Comparisons by gray Boat image.

Sizes	Key Size	Information Entropy	Running Speed	Software
[41]	2^{298}	7.9993	21.684 s	Matlab
[44]	2^{128}	7.9993	5.960 s	Matlab
ours	2^{186}	7.9992	0.137 s	Matlab

5. Discussion and Conclusions

In this paper, an image encryption scheme was proposed that utilizes a chaotic map. This paper makes four significant and novel contributions: (1) The keystream used in the permutation stage is affected by the plain-image, (2) a time-delay phenomenon is simulated and constructed for choosing chaotic outputs, (3) a group of random numbers are inserted into the permuted image before diffusion, and (4) the keystream used in the diffusion stage is affected by the keys assigned in the permutation stage. According to the results of some tests and security analyses, the proposed image encryption scheme exhibits a good performance and is suitable for application in secure communications.

An image encryption algorithm based on time-delay and random insertion with a PID structure was investigated in this paper. With the help of chaotic map as key generator and its inherent properties, time-delay was simulated by outputs of chaotic map. Then, random numbers are inserted before diffusion operation to remedy the shortcoming of Fridrich structure. Compared with previous works, the proposed image encryption algorithm has the following features:

(1) High sensitivity to keys and the plain-image.
(2) Time-delay phenomenon is simulated according to outputs of the chaotic map.
(3) One-time pad is designed by inserting random numbers before diffusion.
(4) The keystream used in the diffusion stage is affected by keys assigned in the permutation stage.
(5) Faster speed to implement the encryption.

Author Contributions: X.H. proposed the main idea of random insertion; G.Y. performed the experiments and then wrote the paper.

Acknowledgments: This work was fully supported by the National Natural Science Foundations of China (No. 61602124, No. 61702116), the Science and Technology Planning Project of Guangdong Province of China (No. 2017A010101025), the Natural Science Foundations of Guangdong Province of China (No. 2016A030310333), the Program for Scientific Research Start-up Funds of Guangdong Ocean University of China (No. R17037), the Special Funding Program for Excellent Young Scholars of Guangdong Ocean University of China (No. HDYQ2017006), and the Supporting funding Projects of Guangdong Ocean University of China (No. P15238, No. P16084).

Conflicts of Interest: The authors declare no conflict of interest.

References

1. Wu, X.J.; Wang, K.S.; Wang, X.Y.; Kan, H.B. Lossless chaotic color image cryptosystem based on DNA encryption and entropy. *Nonlinear Dyn.* **2017**, *90*, 855–875. [CrossRef]
2. Huang, X.L.; Ye, G.D. An efficient self-adaptive model for chaotic image encryption algorithm. *Commun. Nonlinear Sci.* **2014**, *19*, 4094–4104. [CrossRef]
3. Fridrich, J. Symmetric ciphers based on two-dimensional chaotic maps. *Int. J. Bifurc. Chaos* **1998**, *8*, 1259–1284. [CrossRef]
4. Xu, L.; Li, Z.; Li, J.; Hua, W. A novel bit-level image encryption algorithm based on chaotic maps. *Opt. Laser Eng.* **2016**, *78*, 17–25. [CrossRef]
5. Seyedzadeh, S.M.; Norouzi, B.; Mosavi, M.R.; Mirzakuchaki, S. A novel color image encryption algorithm based on spatial permutation and quantum chaotic map. *Nonlinear Dyn.* **2015**, *81*, 511–529. [CrossRef]
6. Norouzi, B.; Seyedzadeh, S.M.; Mirzakuchaki, S.; Mosavi, M.R. A novel image encryption based on hash function with only two-round diffusion process. *Multimed. Syst.* **2014**, *20*, 45–64. [CrossRef]
7. Norouzi, B.; Mirzakuchaki, S.; Seyedzadeh, S.M.; Mosavi, M.R. A simple, sensitive and secure image encryption algorithm based on hyper-chaotic system with only one round diffusion process. *Multimed. Tools Appl.* **2014**, *71*, 1469–1497. [CrossRef]

8. Yang, Y.G.; Pan, Q.X.; Sun, S.J.; Xu, P. Novel image encryption based on quantum walks. *Sci. Rep.* **2015**, *5*, 7784. [CrossRef]
9. Yang, Y.G.; Xu, P.; Yang, R.; Zhou, Y.H.; Shi, W.M. Quantum Hash function and its application to privacy amplification in quantum key distribution, pseudo-random number generation and image encryption. *Sci. Rep.* **2016**, *6*, 19788. [CrossRef]
10. Bibi, N.; Farwa, S.; Muhammad, N.; Jahngir, A.; Usman, M. A novel encryption scheme for high-contrast image data in the Fresnelet domain. *PLoS ONE* **2018**, *13*, e0194343. [CrossRef]
11. Guesmi, R.; Farah, M.A.B.; Kachouri, A.; Samet, M. A novel chaos-based image encryption using DNA sequence operation and Secure Hash Algorithm SHA-2. *Nonlinear Dyn.* **2016**, *83*, 1123–1136. [CrossRef]
12. Chen, J.X.; Zhu, Z.L.; Zhang, L.B.; Zhang, Y.S.; Yang, B.Q. Exploiting self-adaptive permutation-diffusion and DNA random encoding for secure and efficient image encryption. *Inf. Sci.* **2016**, *345*, 257–270. [CrossRef]
13. Huang, X.L.; Ye, G.D. An image encryption algorithm based on hyper-chaos and DNA sequence. *Multimed. Tools Appl.* **2014**, *72*, 57–70. [CrossRef]
14. Shen, Q.; Liu, W.B. A novel digital image encryption algorithm based on orbit variation of phase diagram. *Int. J. Bifurc. Chaos* **2017**, *27*, 1750204. [CrossRef]
15. Zhang, Y.S.; Xiao, D. Double optical image encryption using discrete Chirikov standard map and chaos-based fractional random transform. *Opt. Laser Eng.* **2013**, *51*, 472–480. [CrossRef]
16. Ghebleh, M.; Kanso, A.; Noura, H. An image encryption scheme based on irregularly decimated chaotic maps. *Signal Process.-Image* **2014**, *29*, 618–627. [CrossRef]
17. Hua, Z.Y.; Zhou, B.H.; Zhou, Y.C. Sine-transform-based chaotic system with FPGA implementation. *IEEE Trans. Ind. Electron.* **2018**, *65*, 2557–2566. [CrossRef]
18. Hua, Z.Y.; Yi, S.; Zhou, Y.C. Medical image encryption using high-speed scrambling and pixel adaptive diffusion. *Signal Process.* **2018**, *144*, 134–144. [CrossRef]
19. Karawia, A.A. Encryption algorithm of multiple-image using mixed image elements and two dimensional chaotic economic map. *Entropy* **2018**, *20*, 801. [CrossRef]
20. Abdallah, E.E.; Hamza, A.B.; Bhattacharya, P. Video watermarking using wavelet transform and tensor algebra. *Signal Image Video Process.* **2010**, *4*, 233–245. [CrossRef]
21. Abdallah, E.E.; Hamza, A.B.; Bhattacharya, P. MPEG video watermarking using tensor singular value decomposition. In Proceedings of the 2007 International Conference Image Analysis and Recognition, Montreal, QC, Canada, 22–24 August 2007; pp. 772–783.
22. Li, S.L.; Ding, W.K.; Yin, B.S.; Zhang, T.F.; Ma, Y.D. A novel delay linear coupling logistics map model for color image encryption. *Entropy* **2018**, *20*, 463. [CrossRef]
23. Li, C.Q. Cracking a hierarchical chaotic image encryption algorithm based on permutation. *Signal Process* **2016**, *118*, 203–210. [CrossRef]
24. Yeo, J.C.; Guo, J.I. Efficient hierarchical chaotic image encryption algorithm and its VLSI realization. *IEE Proc.-Vis. Image Signal Process.* **2000**, *147*, 167–175.
25. Jolfaei, A.; Wu, X.W.; Muthukkumarasamy, V. On the security of permutation-only image encryption schemes. *IEEE Trans. Inf. Forensics Sec.* **2016**, *11*, 235–246. [CrossRef]
26. Li, C.Q.; Lin, D.D.; Lü, J.H. Cryptanalyzing an image-scrambling encryption algorithm of pixel bits. *IEEE Multimed.* **2017**, *24*, 64–71. [CrossRef]
27. Eslami, Z.; Bakhshandeh, A. An improvement over an image encryption method based on total shuffling. *Opt. Commun.* **2013**, 286, pp. 51–55. [CrossRef]
28. Akhavan, A.; Samsudin, A.; Akhshani, A. Cryptanalysis of "an improvement over an image encryption method based on total shuffling". *Opt. Commun.* **2015**, *350*, 77–82. [CrossRef]
29. Hermassi, H.; Belazi, A.; Rhouma R.; Belghith, S.M. Security analysis of an image encryption algorithm based on a DNA addition combining with chaotic maps. *Multimed. Tools Appl.* **2014**, *72*, 2211–2224. [CrossRef]
30. Liu, Y.S.; Tang, J.; Xie, T. Cryptanalyzing a RGB image encryption algorithm based on DNA encoding and chaos map. *Opt. Laser Technol.* **2014**, *60*, 111–115. [CrossRef]
31. Zhu, C.X.; Sun, K.H. Cryptanalyzing and improving a novel color image encryption algorithm using RT-enhanced chaotic tent maps. *IEEE Access* **2018**, *6*, 18759–18770. [CrossRef]
32. Xie, E.Y.; Li, C.Q.; Yu, S.M.; Lü, J.H. On the cryptanalysis of Fridrich's chaotic image encryption scheme. *Signal Process* **2017**, *132*, 150–154. [CrossRef]

33. Solak, E.; Çokal, C.; Yildiz, O.T.; Biyikoğu, T. Cryptanalysis of Fridrich's chaotic image encryption. *Int. J. Bifurc. Chaos* **2010**, *20*, 1405–1413. [CrossRef]
34. Hua, Z.Y.; Zhou, Y.C.; Pun, C.M.; Chen, C.L.P. 2D Sine Logistic modulation map for image encryption. *Inf. Sci.* **2015**, *297*, 80–94. [CrossRef]
35. Zhu, C.X. A novel image encryption scheme based on improved hyperchaotic sequences. *Opt. Commun.* **2012**, *285*, 29–37. [CrossRef]
36. Hua, Z.Y.; Zhou, B.H.; Zhou, Y.C. Sine chaotification model for enhancing chaos and its hardware implementation. *IEEE Trans. Ind. Electron.* **2019**, *66*, 1273–1284. [CrossRef]
37. Ye, G.D.; Huang, X.L. An efficient symmetric image encryption algorithm based on an intertwining logistic map. *Neurocomputing* **2017**, *251*, 45–53. [CrossRef]
38. Liu, H.J.; Wang, X.Y. Color image encryption using spatial bit level permutation and high-dimension chaotic system. *Opt. Commun.* **2011**, *284*, 3895–3903. [CrossRef]
39. Dăscălescu, A.C.; Boriga, R.E. A novel fast chaos-based algorithm for generating random permutations with high shift factor suitable for image scrambling. *Nonlinear Dyn.* **2013**, *74*, 307–318. [CrossRef]
40. Kadir, A.; Hamdulla, A.; Guo, W.Q. Color image encryption using skew tent map and hyper chaotic system of 6th-order CNN. *Optik* **2014**, *125*, 1671–1675. [CrossRef]
41. Stoyanov, B.; Kordov, K. Image encryption using chebyshev map and rotation equation. *Entropy* **2015**, *17*, 2117–2139. [CrossRef]
42. Stoyanov, B.; Kordov, K. Novel image encryption scheme based on chebyshev polynomial and duffing map. *Sci. World J.* **2014**, *2014*, 283639. [CrossRef] [PubMed]
43. Seyedzade, S.M.; Mirzakuchaki, S.; Atani, R.E. A novel image encryption algorithm based on hash function. In Proceedings of the 2010 Iranian Conference on Machine Vision and Image Processing, Isfahan, Iran, 27–28 October 2010.
44. Chai, X.L.; Gan, Z.H.; Yuan, K.; Lu, Y.; Chen, Y.R. An image encryption scheme based on three-dimensional Brownian motion and chaotic system. *Chin. Phys. B* **2017**, *26*, 020504. [CrossRef]
45. Ramadan, N.; Ahmed, H.H.; El-khamy, S.E.; El-Samie, F.E.A. Permutation-substitution image encryption scheme based on a modified chaotic map in transform domain. *J. Cent. South Univ.* **2017**, *24*, 2049–2057. [CrossRef]
46. Huang, X.L. Image encryption algorithm using chaotic Chebyshev generator. *Nonlinear Dyn.* **2012**, *67*, 2411–2417. [CrossRef]
47. Ye, G.D.; Huang, X.L. A novel block chaotic encryption scheme for remote sensing image. *Multimed. Tools Appl.* **2016**, *75*, 11433–11446. [CrossRef]
48. Fouda, J.S.A.E.; Effa J.Y.; Sabat S.L.; Ali, M. A fast chaotic block cipher for image encryption. *Commun. Nonlinear Sci. Numer. Simul.* **2014**, *9*, 578–588. [CrossRef]

entropy

Article

Uncertainty Assessment of Hyperspectral Image Classification: Deep Learning vs. Random Forest

Majid Shadman Roodposhti [1,*], Jagannath Aryal [1], Arko Lucieer [1] and Brett A. Bryan [2]

[1] Discipline of Geography and Spatial Sciences, School of Technology, Environments and Design, University of Tasmania, Hobart 7018, Australia; jagannath.aryal@utas.edu.au (J.A.); arko.lucieer@utas.edu.au (A.L.)

[2] Centre for Integrative Ecology, School of Life and Environmental Sciences, Deakin University, Burwood 3125, Australia; b.bryan@deakin.edu.au

* Correspondence: majid.shadman@utas.edu.au

Received: 16 December 2018; Accepted: 10 January 2019; Published: 16 January 2019

Abstract: Uncertainty assessment techniques have been extensively applied as an estimate of accuracy to compensate for weaknesses with traditional approaches. Traditional approaches to mapping accuracy assessment have been based on a confusion matrix, and hence are not only dependent on the availability of test data but also incapable of capturing the spatial variation in classification error. Here, we apply and compare two uncertainty assessment techniques that do not rely on test data availability and enable the spatial characterisation of classification accuracy before the validation phase, promoting the assessment of error propagation within the classified imagery products. We compared the performance of emerging deep neural network (DNN) with the popular random forest (RF) technique. Uncertainty assessment was implemented by calculating the Shannon entropy of class probabilities predicted by DNN and RF for every pixel. The classification uncertainties of DNN and RF were quantified for two different hyperspectral image datasets—Salinas and Indian Pines. We then compared the uncertainty against the classification accuracy of the techniques represented by a modified root mean square error (RMSE). The results indicate that considering modified RMSE values for various sample sizes of both datasets, the derived entropy based on the DNN algorithm is a better estimate of classification accuracy and hence provides a superior uncertainty estimate at the pixel level.

Keywords: uncertainty assessment; deep neural network; random forest; Shannon entropy

---

1. Introduction

Assessing and mapping the state of the Earth's surface is a key requirement for many global researches in the context of natural resources management [1], natural hazards modelling [2,3], urban planning [4,5] etc., where all these mapping products need to be validated [6,7]. With the initiation of more advanced digital satellite remote sensing techniques, accuracy assessment of emerging methods has received major interest [6]. The conventional way to report classification and/or prediction of map accuracy is through an error matrix estimated from a test dataset, which is independent of the training process [8]. Accuracy metrics such as Cohen's Kappa coefficient [9], overall accuracy (OA) [7] and class-specific measures such as user's and producer's accuracies are usually estimated based on an error matrix [10]. However, it is not clear how these accuracy metrics relate to per-pixel accuracy [11] as these types of accuracy metrics are incapable of understanding the spatial variation of classification accuracies despite its importance in modelling spatial phenomena [12,13].

Different approaches have been proposed to characterise the quality of classified maps at the local scale [8]. One method is to apply empirical models to link classification accuracy (dependent variable) to different independent (predictor) variables, such as land cover class [14,15]. As the dependent variable is dichotomous (i.e., classified correctly or not), logistic regression is the most frequently applied algorithm for this purpose. Another approach to characterizing map quality at the local scale involves spatial interpolation of classification accuracy of the test dataset [16]. The most recent approach is introduced by Khatami et al. [8], built on Stehman [17]. Here, a per-pixel accuracy prediction is implemented by applying different accuracy prediction methods based on four factors, including predictive domain (spatial or spectral), interpolation function (constant, linear, Gaussian, and logistic), incorporation of class information (interpolating each class separately versus grouping them together), and sample size. The fourth and most popular approach [8] is to use the probabilities of class memberships or prediction strength (i.e., tree votes in the random forest (RF) or probabilities in neural networks) as indicators of classification uncertainty. The idea is that for a certain pixel, the greater the probability of class membership for a given labelled class, the lower the uncertainty associated with that class and analytical functions can be used to quantify the uncertainty measures instead of using only the membership value of the most probable class. Examples of these functions include ignorance uncertainty [18], Shannon entropy [19,20], and α-quadratic entropy and maximum probability [21], where entropy summarizes the information from membership values of all classes.

Uncertainty assessment techniques can provide an uncertainty map as a spatial approximator of classification accuracy, which can be used to locate and segregate unreliable pixel-level class allocations from reliable ones. In addition, this approach is independent of test data availability. This uncertainty assessment may be implemented using two types of classification approaches: unsupervised schemes using no training dataset [22,23], and supervised schemes [19,24–26]. Although unsupervised approaches can be applied regardless of the training dataset availability (i.e., by applying unsupervised algorithms), their relevant uncertainty assessment results may be misleading due to incorrect classification of pixels. In terms of supervised methods, various algorithms have been applied to evaluate the uncertainty of correct/incorrect classified pixels including RF as one of the most popular algorithms. RF [27,28] has a rich and successful history in machine learning including applications in hyperspectral image classification [29–33] and uncertainty assessment [34–36]. It has been demonstrated to outperform most state-of-the-art learners when it comes to handling high-dimensional data [37], such as hyperspectral image datasets. Nonetheless, we assumed that considering high-dimensional hyperspectral data, newly emerging deep learning algorithms may be efficient for uncertainty assessment, but they have been rarely applied for this purpose. On the other hand, the deep learning algorithms have also been found to be more accurate than traditional algorithms, especially for image classification [38–40]. Further, with multiple layers of processing, they may extract more abstract, invariant features of data, which is considered beneficial for uncertainty assessment studies.

Uncertainty assessment techniques have been repeatedly applied to assess the quality of hyperspectral image classification [23,41,42]. While deep learning has attracted broad attention as a classification algorithm [43–46], it has not been applied to uncertainty assessment of hyperspectral image classification nor compared to other methods. Thus, here we aim to apply deep neural network (DNN) for uncertainty assessment of correct/incorrect classification for every pixel and then compare it with RF. Due to its high performance in uncertainty assessment studies, the RF algorithm provides an appropriate benchmark for comparing the performance of uncertainty assessment derived from deep learning. This paper aims to explore, quantify and compare the capability of DNN and RF algorithms for uncertainty assessment of hyperspectral imagery using two different hyperspectral datasets. To this end, by applying DNN in this study, we compare the uncertainty assessment of hyperspectral image classification using probability values derived from deep learning neurons and popularity votes of RF trees combined with uncertainty values using Shannon entropy.

2. Methods and Dataset

2.1. Method

This study followed two major steps (Figure 1). In step 1, the whole dataset was randomly divided into training (50%) and test data (50%). For each dataset, the hyper-parameters of the optimum DNN and RF algorithms (Table 1) were configured using a 5-fold cross-validation of the training data in the pre-processing stage. This was done only for hyper-parameters with a significant effect on the datasets and the remaining hyper-parameters were kept at the default values. Although test data were always a constant sub-set of the whole dataset, the training procedure was done using different portions of training sample (i.e., 10%, 20%, ... , 100%) to assess the effects of training sample size in uncertainty assessment. Thus, the training sample itself was sliced into 10 equal random portions, and then applied for training the tuned algorithms. The algorithms were then trained 10 times each, from 10% to 100%, every time by a 10% increase of training samples, i.e., x = {10%, 20%, ... , 100%}, where *x* is a set of applied training samples. Here, the test dataset was always the same. In addition, to achieve more consistent results and to account for sensitivity analysis, each algorithm was applied in five consecutive runs, where the sampling strategy was the same but the locations of initial sampling seeds (i.e., random training (50%) and test data (50%)) were modified by a different random function. As the hyper-parameters of the DNN and RF algorithms were optimised using a validation sample, they were not modified for the other sample sizes. Here, for both DNN and RF, the probability of belonging to each possible class was estimated for every pixel and used to compute the uncertainty of classification for the pixel using Shannon entropy [20], where entropy represents uncertainty in this research [8].

Table 1. The optimised hyper-parameters of DNN and RF using 5-fold cross-validation data for uncertainty assessment.

Algorithm	Hyper-Parameter	Description	Salinas	Indian Pines
DNN	hidden	Hidden layer sizes	(100, 100)	(200, 200)
DNN	epoch	How many times the dataset should be iterated (streamed)	300	300
DNN	activation	Activation function for non-linear transformation.	"Maxout"	"Maxout"
DNN	stopping metric	A metric that is used as a stopping criterion	"RMSE"	"RMSE"
DNN	l1	Only allows strong values to survives	0.0001	0.0001
DNN	l2	Prevents any single weight from getting too big	0.001	0.001
DNN	epsilon	Prevents getting stuck in local optima	$1 \times e^{-10}$	$1 \times e^{-10}$
RF	ntree	Number of trees to grow	100	100
RF	mtry	Number of variables available for splitting at each tree node	14	15

* For deep learning, this optimisation is done using "Grid Search" by h20.grid() function, and for random forest it has been done manually for the number of trees while tunerf() function is used to optimise mtry.

In step 2, for a better demonstration of classification performance considering the low and high uncertainty values, we mapped the uncertainty outputs along with the mode of correct/incorrect classified test pixels for all applied training samples (i.e., from 10% to 100%). Whenever an optimised algorithm is applied in the context of uncertainty assessment, the uncertainty value for a correctly classified pixel should be minimised (i.e., "0") while it should be maximised (i.e., "1") for misclassified pixels. Thus, we then calculated root mean square error (RMSE) of every prediction implemented by each algorithm [20] to quantify the degree of deviation from this optimum state. For this purpose, entropy values were normalised between 0 and 1. This whole process was implemented in R [47] using three major packages namely "H2O" [48], "randomforest" [49], and "entropy" [50].

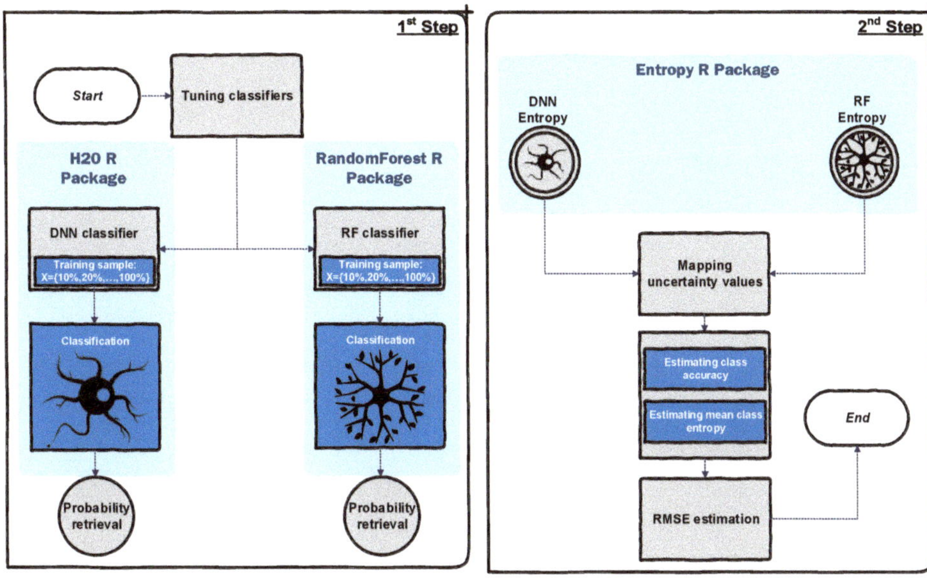

Figure 1. Flowchart of methodology implementation labelled with the main R packages utilized.

2.1.1. Supervised Uncertainty Assessment Approach

The most popular and accurate way of uncertainty assessment is based on a supervised scheme using a machine learning algorithm. Here, we implemented a model that can assess the uncertainty values of a classified hyperspectral image containing various class labels. We first collected ground truth data labelled with their class categories such as corn, grass, hay, oats, and soybean. During training, the algorithm was provided with a training example and produced a response in the form of a vector of probabilities, one for each class. Then, the best-case scenarios would be the highest probability score for one class and the lowest possible probability score for the other remaining classes. The least desirable case, on the other hand, would be equal probability scores for all the existing class labels (Figure 2). We then computed the uncertainty of probability scores for all potential class labels for a pixel by using entropy. An ideal algorithm, for uncertainty assessment, is not only capable of classifying input data with the highest possible accuracy but also capable of producing class labels with low uncertainty for correctly classified pixels and vice versa.

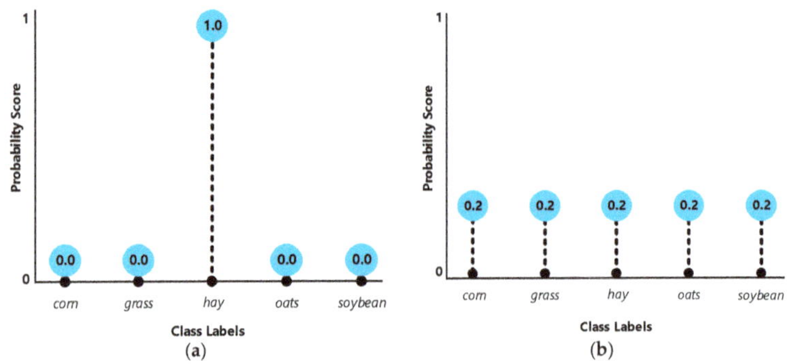

Figure 2. The best-case scenarios for every pixel representing low uncertainty (**a**) versus the worst-case scenario denoting high uncertainty (**b**). The other instances would be intermediate states of these two.

In this study, the uncertainty derived from deep learning neurons and popularity votes of RF trees was quantified using Shannon entropy [51]. Entropy summarizes the information from membership values of all classes using Equation (1):

$$e_x = -\sum_{i=1}^{h} P_i \log_2 P_i \tag{1}$$

where P_i is the probability of class membership for h class labels. Further, the selection of the logarithm base is unimportant, as it only affects the units of entropy [25].

2.1.2. Deep Neural Network (DNN)

The deep learning algorithm applied in this research is based on R studio deep neural network (DNN) from H2O package [48], which is a feed-forward artificial neural network, trained with stochastic gradient descent using backpropagation. Here, multiple layers of hidden units were applied between the inputs and the outputs of the model [52–54].

Each hidden unit, j, typically uses the logistic function β the closely related hyperbolic tangent is also often used and any function with a well-behaved derivative can be used) to map its outputsing y_j total input from x_j:

$$y_i = \beta(x_j) = \frac{1}{1+e^{-x_j}} \tag{2}$$

For multiclass classification, such as our problem of hyperspectral image classification, output unit j converts its total input, x_j, into a class probability, P_j, by using a normalised exponential function named "softmax":

$$P_j = \frac{\exp(X_j)}{\sum_h \exp(X_h)} \tag{3}$$

where h is an index over all classes. DNNs are discriminatively trained by backpropagating derivatives of a cost function that measure the discrepancy between the target outputs and the actual outputs produced for each training case [55]. When using the softmax output function, the natural cost function C is the cross-entropy between the target probabilities d and the softmax outputs, P:

$$C = -\sum_{i} d_j \ln P_j \tag{4}$$

where the target probabilities, typically taking values of one or zero, are the supervised information provided to train the DNN algorithm.

2.1.3. Random Forests as a Benchmark

To measure and quantify DNN performance for uncertainty assessment of hyperspectral classification, we implemented the RF algorithm applied to the same datasets [49]. The RF algorithm provides an appropriate benchmark for assessing the performance of the DNN scheme because of its high performance found in hyperspectral data classification [30–32,56,57]. RF is also computationally efficient and suitable for training datasets with many variables and can solve multiclass classification problems [58]. We compared the uncertainty assessment results of DNN and RF using two different datasets.

2.1.4. RMSE of Uncertainty Assessment

RMSE is the standard deviation of the residuals (prediction errors). Here, RMSE demonstrates standard deviation of prediction for correct and erroneous estimates of test dataset. In other words, it explains how concentrated the data are around the line of best fit considering entropy of correct and erroneous estimates:

$$\mathrm{RMSE} = \sqrt{\sum_{i=1}^{n}(e-o)^2/n} \tag{5}$$

where e represents the estimated entropy value from "0" (minimum entropy value) to "1" (maximum entropy value) after normalisation; o represents classification result for the observed values, which is "1" for erroneous predictions and "0" for correct answers. Here, RMSE is applied as a goodness of fit for uncertainty assessment results. Therefore, the best-case scenarios would be those classification cases where the algorithm is at both the maximum confidence and accuracy ($e = 0$ and $o = 0$) or minimum confidence and minimum accuracy ($e = 1$ and $o = 1$). The worst-case scenarios, however, occurs when the algorithm is at minimum confidence and maximum accuracy ($e = 1$ and $o = 0$) or vice versa ($e = 0$ and $o = 1$). Table 2 demonstrates the intuitions behind the proposed RMSE.

Table 2. The intuition behind the proposed RMSE.

Best-Case Scenarios	e	o	RMSE	Worst-Case Scenarios	e	o	RMSE
Positive	0	0	0	Positive	0	1	1
Negative	1	1	0	Negative	1	0	1

* All other instances fall within intermediate states.

2.2. Datasets

In this study, two widely used hyperspectral datasets including the Salinas [59–61] and Indian Pines [59,62,63] image datasets were used (Table 3) and divided into validation, train and test samples (Figure 3). Both datasets contain noisy bands due to dense water vapour, atmospheric effects, and sensor noise. These datasets are all available at http://www.ehu.eus/ccwintco/index.php?title%20=%20Hyperspectral_Remote_Sensing_Scenes.

Table 3. The major attributes of the hyperspectral datasets.

Dataset	Sensor	Total Bands	Excluded Bands	Number of Classes	Dimension	Resolution
Salinas	AVIRIS	224	20	16	512 × 217	20 metre
Indian Pines	AVIRIS	224	24	16	145 × 145	20 metre

The Salinas image consists of 224 bands and each band contains 512 × 217 pixels covering 16 classes comprising different sub-classes of vegetables (nine sub-classes), bare soils (three sub-classes) and vineyard (four sub-classes). It was recorded by AVIRIS sensor over the South of the city of Greenfield in the Salinas Valley, CA, USA on October 9, 1998. This dataset is characterised by a spatial resolution of 3.7 m, and the spectral information ranges from 0.4 to 2.5 μm. As shown in Figure 3, the ground truth is available for nearly two-thirds of the entire scene. We used 204 bands, after removing bands of the water absorption features.

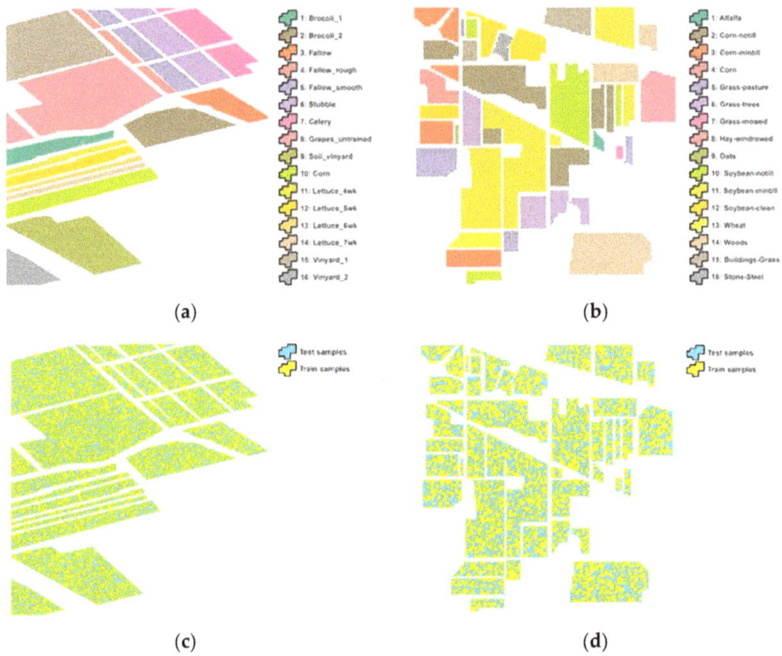

Figure 3. Ground truth data of two datasets including the Salinas (**a**) and the Indian Pines (**b**). The bottom images represent the location of the train and test data for the Salinas (**c**) and the Indian Pines (**d**).

The Indian Pines dataset is also an AVIRIS image collected over the Indian Pines test site location, Western Tippecanoe County, Indiana, USA on June 12, 1992. This dataset consists of 220 spectral bands in the same wavelength range as the Salinas dataset; however, four spectral bands are removed as they contain no data. This scene is a subset of a larger scene and it contains 145 $\times$ 145 pixels covering 16 ground truth classes (Figure 3). The ground-truthing campaign consists of approximately 10,000 samples which are distributed over the area of 2.5 km by 2.5 km. The ground truth data were collected by walking through the fields in the image. Plant species, as well as some more characteristics, were recorded along with photos of sites in the field. In the present research experiment, 20 spectral bands were removed because of the water absorption phenomena and noise.

3. Results

3.1. Salinas Simulation Experiments

The results of uncertainty assessment for the Salinas dataset using DNN and RF are presented in Figure 4. However, to avoid redundancy in the representation of the results, only half of the achieved uncertainty images are displayed (i.e., 10%, 30%, 50%, 70% and 90%). Regardless of the classification scheme and/or training sample size, classes 8 (8: Grapes_untrained) and 15 (15: Vinyard_1) belonged to the highest uncertainty level among all the available class labels. For both algorithms, this was followed by concentration of incorrect predictions within the high-uncertainty areas, which are identified as false values in the mode of correct/incorrect classified test data based on all training samples from 10% to 100%.

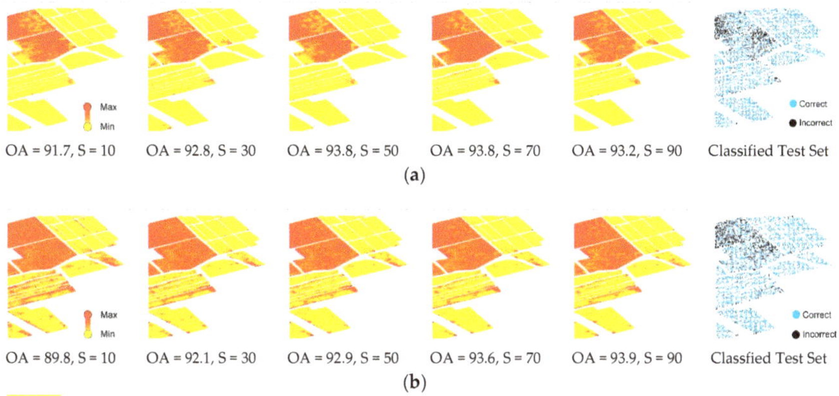

Figure 4. Results of uncertainty assessment for DNN (**a**) and RF (**b**) using different portions of training sample (S, in %) and mode of correct/incorrect classified test data for the Salinas dataset. The estimated overall accuracy (OA, in %) of the whole classification scheme is also demonstrated for each training sample.

RF and DNN were comparable in terms of achieved OA of classification for the majority of sample sizes, while the areas covered with the high uncertainty values were less obvious within DNN results. This was observable for all corresponding sample sizes. Further, to quantify the capabilities of DNN and RF for uncertainty assessment and potential application as an estimate of accuracy, we then calculated the root mean square error (RMSE) of every sample size (Equation (5)) applied for image classification for each algorithm. Following the mapping of uncertainty values, which represent uncertainty levels, we plotted the RMSE of the classification (y-axis) of the test data for various training sample sizes (x-axis). Here, lower RMSE values indicate better estimates of uncertainty (Table 2), and vice versa. For the Salinas dataset, RMSE values for the DNN algorithm were lower than RF values for all sample sizes while RMSE values derived from RF are more consistent (Figure 5).

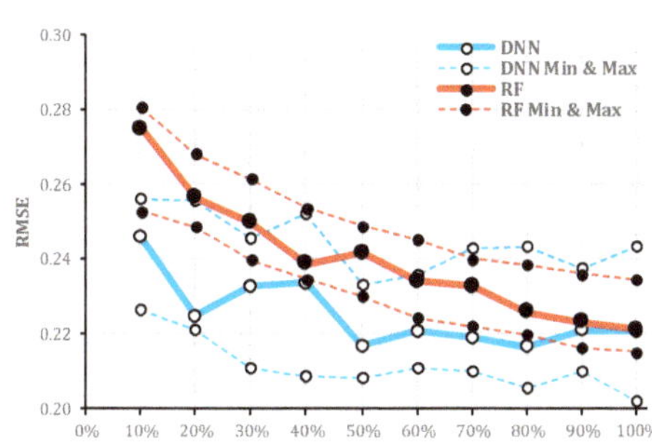

Figure 5. The estimated RMSE values of uncertainty assessment for test datasets (y-axis) where the algorithm is trained with different portions of the training sample (x-axis) of Salinas dataset. Dashed lines represent the minimum and maximum RMSE values for each sample size achieved in five consecutive simulation runs.

Further, to better understand the capability of uncertainty measures as an estimate of accuracy, we plotted the correspondence between mean class uncertainty (i.e., entropy) and class accuracy (Figure 6). Nonetheless, to avoid unnecessary repetition of results, only the 50% training sample was plotted, which confirmed the accuracy of classification within the majority of image classes will be reduced by an increase in the uncertainty of pixels belonging to these classes and vice versa. In accordance with Figure 4, it was also demonstrated classes 8 and 15 of Salinas dataset with the highest mean uncertainty values belong to the least accurate estimation.

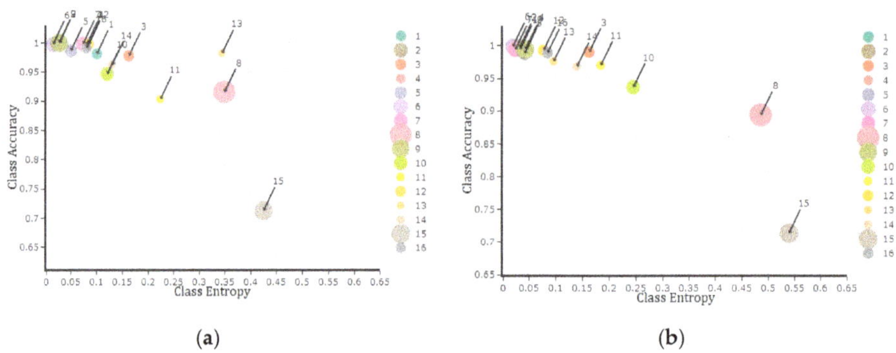

Figure 6. Class entropy/uncertainty (x-axis) versus class accuracy (y-axis) plots of Salinas dataset using DNN (**a**, left) and RF (**b**, right) algorithms observed by applying 50% of training data. The bubble sizes represent the frequency of land use class labels while bigger bubbles indicate the higher frequency and vice versa.

3.2. Indian Pines Simulation Experiments

The results of uncertainty assessment for the Indian Pines dataset using DNN and RF were similar to those for the Salinas dataset. For both DNN and RF, classification uncertainty was reduced for larger training samples while the OA values of classification increased. However, these phenomena were less obvious for RF compared with those for DNN (Figure 7). In addition, the improvement of OA values with an increase in training sample size was more distinctive than that for the Salinas dataset. Remarkably, for every corresponding sample size, DNN was not only the more accurate algorithm but also displayed fewer pixels with high uncertainty values. The mode of correct/incorrect classified pixels demonstrated almost the same pattern for both algorithms while there were fewer misclassified pixels within the results of DNN algorithm.

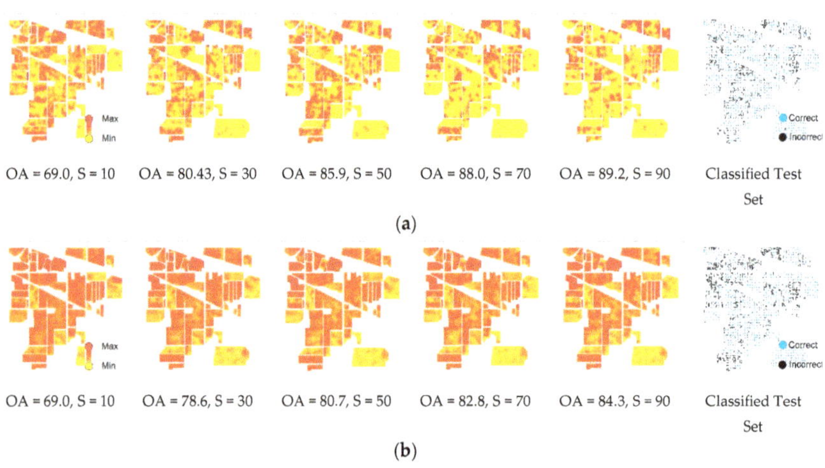

Figure 7. Results of uncertainty assessment for DNN (**a**) and RF (**b**) using different portions of training sample (S, in %) and mode of correct/incorrect classified test data for the Indian Pines dataset. The estimated overall accuracy (OA, in %) of the whole classification scheme is also demonstrated for each training sample.

The higher accuracy of DNN elevates the quality of implemented uncertainty assessment for locating correct/incorrect classifications for this dataset. Nonetheless, to quantify the difference in the quality of uncertainty the assessment between the two algorithms, the RMSE values were estimated for every training sample. The RMSE values also confirmed the superiority of DNN for the majority of training sample sizes in a way that less uncertainty was estimated for correct classified pixels while incorrect classified pixels were identified by more levels of uncertainty. However, the same as the Salinas dataset, RMSE values derived from five consecutive simulation runs of RF are more consistent. This can be easily observed by comparing the difference between the minimum and maximum RMSE values for each sample size that is observable in Figure 8. Obviously, DNN is coupled with more variation between minimum and maximum RMSE values for almost all different sample sizes.

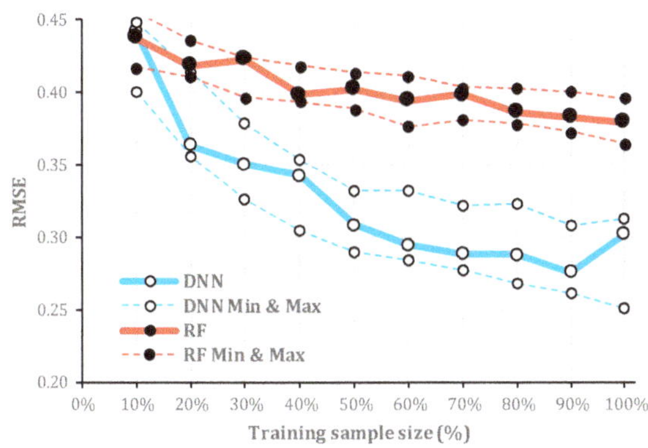

Figure 8. The estimated RMSE values of uncertainty assessment for test datasets (*y*-axis) where the algorithm is trained with different portions of training sample (*x*-axis) of Indian Pines dataset. Dashed lines represent the minimum and maximum RMSE values for each sample size achieved in five consecutive simulation runs.

Finally, the correspondence between mean class uncertainty (i.e., entropy) and class accuracy of Indian pine dataset is demonstrated in Figure 9 for 50% of training sample size using both DNN and RF algorithms. Similar to Salinas dataset results, the achieved results of Indian Pines demonstrated a negative relationship between uncertainty and accuracy for the majority of class labels.

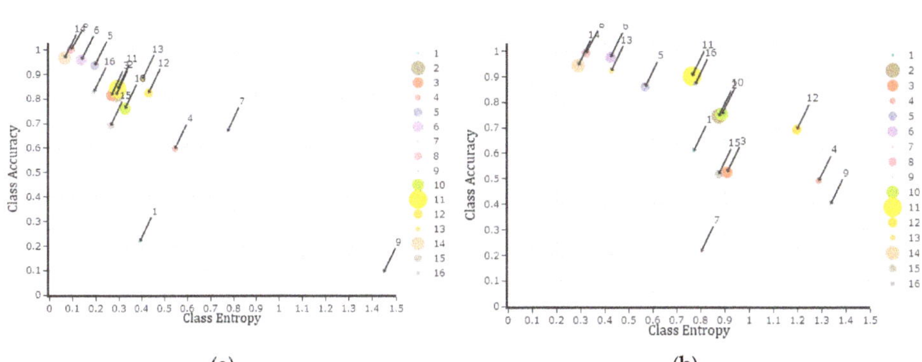

Figure 9. Class entropy/uncertainty (*x*-axis) versus class accuracy (*y*-axis) plots of Indian Pines dataset using DNN (**a**, left) and RF (**b**, right) algorithms observed by applying 50% of training sample size. The bubble sizes represent the frequency of land use class labels while bigger bubbles indicate the higher frequency and vice versa.

4. Discussion

4.1. Comparing the Quality of Uncertainty Assessment Based on RMSE

With reference to the fact that both DNN and RF algorithms may achieve an OA above 70%, even for the minimum portion of training sample size (i.e., 10%), it was expected one algorithm may perform a better uncertainty assessment if it successfully limits the high-uncertainty areas to the spatial vicinity of incorrectly classified pixels while highlighting the remaining areas as low uncertainty. This is regardless of achieved OA, although the RMSE values derived from five consecutive runs of each algorithm indicate that results of uncertainty assessment using RF is more consistent compared with DNN. Nonetheless, comparing the results of uncertainty assessment, for both utilised datasets and every corresponding sample sizes, demonstrates that areas of high uncertainty values were less abundant within the results of DNN algorithm compared with that for RF algorithm (Figures 4, 5 and 7). This may be due to the fact that DNN is optimized to reduce the difference between the predicted distribution and the true data generating distribution by minimizing the cross-entropy of the two probability distributions [64,65]. Therefore, the uncertainty assessment derived from DNN algorithm was superior to RF combined with better OA for these two datasets. However, more studies using different datasets are still required for generalizing the results.

4.2. Quality of Uncertainty Assessment for Different Sample Sizes

For both algorithms and both datasets, larger training samples were found to be more beneficial for uncertainty assessment. The RMSE of uncertainty estimates, which was applied as a goodness of fit to assess the quality of uncertainty maps, decreased from the initial (10%) to final (100%) training sample sizes (Figures 5 and 8). However, this improvement was more obvious for DNN compared with that for RF. This may be due to different formulations of RF and DNN algorithms, which are affecting the performance of the two algorithms for uncertainty assessment. Usually, the training sample size has a crucial role in classification accuracy [66]; thus, it will also affect the uncertainty assessment process. The increased training sample size will typically increase the performance of an algorithm from random sampling [67,68], but not all algorithms will be improved at the same level with a larger sample size. Although RF can also benefit from a larger training sample by extracting more binary rules [69], DNN may achieve a better performance. For DNN, the ratio of uncertainty assessment improvement followed by larger training sample size and more accurate classification depends on the abundance of contextual information per-pixel in the target dataset [70]. As many extensive experimental results confirm the excellent performance of the deep learning-based algorithms matched with rich spectral and contextual information [71], our study suggests this is also beneficial to increase the training sample to achieve a better uncertainty assessment result.

4.3. Uncertainty vs. Accuracy

The existing uncertainties at different stages of the classification procedure influence classification accuracy [66,72]. Therefore, understanding the relationships between the classification uncertainty and accuracy is the key successful contribution to an estimate of accuracy for image classification. Although a low uncertainty classification instance is accompanied with high accuracy, some exceptions may apply to the high uncertainty, which usually belongs to low accuracy estimates. Thus, incorrect predicted class labels are usually located inside high-uncertainty areas with very few exceptions within low-uncertainty regions while correct classified pixel overlay the low uncertainty areas (Figures 4 and 7). In this research, for both applied datasets, the existing correspondence between uncertainty and accuracy was better identified using the DNN algorithm. Having said that, in our study, DNN is demonstrated more potential in uncertainty assessment for hyperspectral image classification. Following accurate classifications combined with minimising high-uncertainty areas, DNN not only offers a lower rate of RMSE but also offers a higher contrast between low and high uncertainty areas.

At a wider scale, considering mean class uncertainty against the class accuracy of test data, it was revealed that usually a lower uncertainty value of a class is followed by a higher accuracy (Figures 6 and 9). In other words, as low uncertainty indicates the probabilities of potential class labels for a pixel are not equal (i.e., unimodal distribution). This simply specifies that based on the available distribution of potential labels and their probability values (Figure 2), defined by either deep learning neurons or tree votes, usually one of the potential class labels (i.e., 16 labels for each applied datasets) has a significant preference to be selected as the estimated label. Accordingly, the concentration of low uncertainty values corresponding to every pixel of the desired class label is anticipated by an acceptable accuracy of classification. In terms of higher values of mean uncertainty for a class, the class accuracy will be reduced due to the abundance of high uncertainty estimates within that class.

5. Conclusions

Due to the weaknesses of the traditional approaches of map accuracy assessment based on a confusion matrix, many uncertainty assessment approaches are being developed as accuracy estimates. In terms of supervised methods, we compared DNN with RF, where an estimate of accuracy is defined by the entropy of all potential probabilities/votes toward different class labels for a pixel, as an uncertainty measure. In this research, entropy was applied to encode the measure of uncertainty, which is applicable to any dataset including hyperspectral image datasets. Considering the results of uncertainty assessment, for both Salinas and Indian Pines datasets, DNN outperformed RF for the purpose of uncertainty assessment. However, the superiority of DNN algorithm was more obvious when applying the Indian Pines dataset, as well as larger training sample sizes. This was due to less-abundant high uncertainty values throughout the classified dataset compared with RF for every corresponding training sample size while having a comparable or better OA. Nonetheless, the achieved uncertainty maps of DNN can facilitate the application of hyperspectral image classification products by alerting map users about the spatial variation of classification uncertainty over the entire mapped region as an estimate of accuracy.

Author Contributions: M.S.R. Designed the methodological framework, processed the data and prepared the manuscript draft. J.A., A.L. and B.A.B. edited the methodological framework and then extensively reviewed the manuscript.

Funding: This research was jointly funded by University of Tasmania and Commonwealth Scientific and Industrial Research Organization (CSIRO), grant number RT109121. BAB was funded by Deakin University.

Acknowledgments: This research was supported by Commonwealth Scientific and Industrial Research Organization (CSIRO) Australian Sustainable Agriculture Scholarship (ASAS) as a top-up scholarship to Majid Shadman, a PhD scholar in the School of Land and Food at the University of Tasmania (RT109121). We thank three anonymous reviewers for their suggestions in improving the manuscript.

Conflicts of Interest: The authors declare no conflict of interest.

References

1. Xie, Y.; Sha, Z.; Yu, M. Remote sensing imagery in vegetation mapping: A review. *J. Plant Ecol.* **2008**, *1*, 9–23. [CrossRef]
2. Roodposhti, M.S.; Safarrad, T.; Shahabi, H. Drought sensitivity mapping using two one-class support vector machine algorithms. *Atmos. Res.* **2017**, *193*, 73–82. [CrossRef]
3. Dutta, R.; Das, A.; Aryal, J. Big data integration shows australian bush-fire frequency is increasing significantly. *R. Soc. Open Sci.* **2016**, *3*, 150241. [CrossRef]
4. Xiao, J.; Shen, Y.; Ge, J.; Tateishi, R.; Tang, C.; Liang, Y.; Huang, Z. Evaluating urban expansion and land use change in shijiazhuang, china, by using gis and remote sensing. *Landsc. Urban Plan.* **2006**, *75*, 69–80. [CrossRef]
5. Weng, Q.; Quattrochi, D.; Gamba, P.E. *Urban Remote Sensing*; CRC Press: Boca Raton, FL, USA, 2018.
6. Congalton, R.G.; Green, K. *Assessing the Accuracy of Remotely Sensed Data: Principles and Practices*; CRC Press: Boca Raton, FL, USA, 2008.

7. Congalton, R.G. A review of assessing the accuracy of classifications of remotely sensed data. *Remote Sens. Environ.* **1991**, *37*, 35–46. [CrossRef]
8. Khatami, R.; Mountrakis, G.; Stehman, S.V. Mapping per-pixel predicted accuracy of classified remote sensing images. *Remote Sens. Environ.* **2017**, *191*, 156–167. [CrossRef]
9. Cohen, J. Weighted kappa: Nominal scale agreement provision for scaled disagreement or partial credit. *Psychol. Bull.* **1968**, *70*, 213. [CrossRef]
10. Richards, J.A.; Jia, X. *Remote Sensing Digital Image Analysis: An Introduction*; Springer Inc.: New York, NY, USA, 1999; p. 363.
11. Ye, S.; Pontius, R.G.; Rakshit, R. A review of accuracy assessment for object-based image analysis: From per-pixel to per-polygon approaches. *ISPRS J. Photogramm. Remote Sens.* **2018**, *141*, 137–147. [CrossRef]
12. Foody, G.M. Status of land cover classification accuracy assessment. *Remote Sens. Environ.* **2002**, *80*, 185–201. [CrossRef]
13. Comber, A.; Fisher, P.; Brunsdon, C.; Khmag, A. Spatial analysis of remote sensing image classification accuracy. *Remote Sens. Environ.* **2012**, *127*, 237–246. [CrossRef]
14. Yu, Q.; Gong, P.; Tian, Y.Q.; Pu, R.; Yang, J. Factors affecting spatial variation of classification uncertainty in an image object-based vegetation mapping. *Photogramm. Eng. Remote Sens.* **2008**, *74*, 1007–1018. [CrossRef]
15. Burnicki, A.C. Modeling the probability of misclassification in a map of land cover change. *Photogramm. Eng. Remote Sens.* **2011**, *77*, 39–49. [CrossRef]
16. Tsutsumida, N.; Comber, A.J. Measures of spatio-temporal accuracy for time series land cover data. *Int. J. Appl. Earth Obs. Geoinf.* **2015**, *41*, 46–55. [CrossRef]
17. Stehman, S.V. Selecting and interpreting measures of thematic classification accuracy. *Remote Sens. Environ.* **1997**, *62*, 77–89. [CrossRef]
18. Legleiter, C.J.; Goodchild, M.F. Alternative representations of in-stream habitat: Classification using remote sensing, hydraulic modeling, and fuzzy logic. *Int. J. Geogr. Inf. Sci.* **2005**, *19*, 29–50. [CrossRef]
19. Loosvelt, L.; Peters, J.; Skriver, H.; Lievens, H.; Van Coillie, F.M.; De Baets, B.; Verhoest, N.E. Random forests as a tool for estimating uncertainty at pixel-level in sar image classification. *Int. J. Appl. Earth Obs. Geoinf.* **2012**, *19*, 173–184. [CrossRef]
20. Dehghan, H.; Ghassemian, H. Measurement of uncertainty by the entropy: Application to the classification of mss data. *Int. J. Remote Sens.* **2006**, *27*, 4005–4014. [CrossRef]
21. Giacco, F.; Thiel, C.; Pugliese, L.; Scarpetta, S.; Marinaro, M. Uncertainty analysis for the classification of multispectral satellite images using svms and soms. *IEEE Trans. Geosci. Remote Sens.* **2010**, *48*, 3769–3779. [CrossRef]
22. Prasad, M.G.; Arora, M.K. A simple measure of confidence for fuzzy land-cover classification from remote-sensing data. *Int. J. Remote Sens.* **2014**, *35*, 8122–8137. [CrossRef]
23. Wang, Q.; Shi, W. Unsupervised classification based on fuzzy c-means with uncertainty analysis. *Remote Sens. Lett.* **2013**, *4*, 1087–1096. [CrossRef]
24. Foody, G.M.; Campbell, N.; Trodd, N.; Wood, T. Derivation and applications of probabilistic measures of class membership from the maximum-likelihood classification. *Photogramm. Eng. Remote Sens.* **1992**, *58*, 1335–1341.
25. Brown, K.; Foody, G.; Atkinson, P. Estimating per-pixel thematic uncertainty in remote sensing classifications. *Int. J. Remote Sens.* **2009**, *30*, 209–229. [CrossRef]
26. McIver, D.K.; Friedl, M.A. Estimating pixel-scale land cover classification confidence using nonparametric machine learning methods. *IEEE Trans. Geosci. Remote Sens.* **2001**, *39*, 1959–1968. [CrossRef]
27. Breiman, L. Random forests. *Mach. Learn.* **2001**, *45*, 5–32. [CrossRef]
28. Breiman, L. Bagging predictors. *Mach. Learn.* **1996**, *24*, 123–140. [CrossRef]
29. Naidoo, L.; Cho, M.A.; Mathieu, R.; Asner, G. Classification of savanna tree species, in the greater kruger national park region, by integrating hyperspectral and lidar data in a random forest data mining environment. *ISPRS J. Photogramm. Remote Sens.* **2012**, *69*, 167–179. [CrossRef]
30. Chan, J.C.-W.; Paelinckx, D. Evaluation of random forest and adaboost tree-based ensemble classification and spectral band selection for ecotope mapping using airborne hyperspectral imagery. *Remote Sens. Environ.* **2008**, *112*, 2999–3011. [CrossRef]

31. Crawford, M.M.; Ham, J.; Chen, Y.; Ghosh, J. Random forests of binary hierarchical classifiers for analysis of hyperspectral data. In Proceedings of the 2003 IEEE Workshop on Advances in Techniques for Analysis of Remotely Sensed Data, Greenbelt, MD, USA, 27–28 October 2003; pp. 337–345.
32. Lawrence, R.L.; Wood, S.D.; Sheley, R.L. Mapping invasive plants using hyperspectral imagery and breiman cutler classifications (randomforest). *Remote Sens. Environ.* **2006**, *100*, 356–362. [CrossRef]
33. Abdel-Rahman, E.M.; Makori, D.M.; Landmann, T.; Piiroinen, R.; Gasim, S.; Pellikka, P.; Raina, S.K. The utility of aisa eagle hyperspectral data and random forest classifier for flower mapping. *Remote Sens.* **2015**, *7*, 13298–13318. [CrossRef]
34. Coulston, J.W.; Blinn, C.E.; Thomas, V.A.; Wynne, R.H. Approximating prediction uncertainty for random forest regression models. *Photogramm. Eng. Remote Sens.* **2016**, *82*, 189–197. [CrossRef]
35. Ließ, M.; Glaser, B.; Huwe, B. Uncertainty in the spatial prediction of soil texture: Comparison of regression tree and random forest models. *Geoderma* **2012**, *170*, 70–79. [CrossRef]
36. Loosvelt, L.; De Baets, B.; Pauwels, V.R.; Verhoest, N.E. Assessing hydrologic prediction uncertainty resulting from soft land cover classification. *J. Hydrol.* **2014**, *517*, 411–424. [CrossRef]
37. Caruana, R.; Karampatziakis, N.; Yessenalina, A. An empirical evaluation of supervised learning in high dimensions. In Proceedings of the 25th International Conference on Machine Learning, Helsinki, Finland, 5–9 July 2008; ACM: New York, NY, USA, 2008; pp. 96–103.
38. Krizhevsky, A.; Sutskever, I.; Hinton, G.E. Imagenet classification with deep convolutional neural networks. In Proceedings of the Advances in Neural Information Processing Systems, Lake Tahoe, NV, USA, 3–8 December 2012; pp. 1097–1105.
39. Hinton, G.E.; Salakhutdinov, R.R. Reducing the dimensionality of data with neural networks. *Science* **2006**, *313*, 504–507. [CrossRef] [PubMed]
40. Chen, Y.; Lin, Z.; Zhao, X.; Wang, G.; Gu, Y. Deep learning-based classification of hyperspectral data. *IEEE J. Sel. Top. Appl. Earth Obs. Remote Sens.* **2014**, *7*, 2094–2107. [CrossRef]
41. Adep, R.N.; Vijayan, A.P.; Shetty, A.; Ramesh, H. Performance evaluation of hyperspectral classification algorithms on aviris mineral data. *Perspect. Sci.* **2016**, *8*, 722–726. [CrossRef]
42. Acquarelli, J.; Marchiori, E.; Buydens, L.; Tran, T.; van Laarhoven, T. Convolutional neural networks and data augmentation for spectral-spatial classification of hyperspectral images. *arXiv* **2017**, arXiv:1711.05512.
43. Pan, B.; Shi, Z.; Xu, X. Mugnet: Deep learning for hyperspectral image classification using limited samples. *ISPRS J. Photogramm. Remote Sens.* **2018**, *145*, 108–119. [CrossRef]
44. Liu, P.; Zhang, H.; Eom, K.B. Active deep learning for classification of hyperspectral images. *IEEE J. Sel. Top. Appl. Earth Obs. Remote Sens.* **2017**, *10*, 712–724. [CrossRef]
45. Wei, W.; Zhang, J.; Zhang, L.; Tian, C.; Zhang, Y. Deep cube-pair network for hyperspectral imagery classification. *Remote Sens.* **2018**, *10*, 783. [CrossRef]
46. Tao, Y.; Xu, M.; Lu, Z.; Zhong, Y. Densenet-based depth-width double reinforced deep learning neural network for high-resolution remote sensing image per-pixel classification. *Remote Sens.* **2018**, *10*, 779. [CrossRef]
47. R Core Team. R: A Language and Environment for Statistical Computing. Available online: https://www.scirp.org/(S(351jmbntvnsjt1aadkposzje))/reference/ReferencesPapers.aspx?ReferenceID=2144573 (accessed on 31 October 2017).
48. Candel, A.; Parmar, V.; LeDell, E.; Arora, A. *Deep Learning with H2O*; H2O.ai Inc.: Mountain View, CA, USA, 2016.
49. Liaw, A.; Wiener, M. Classification and regression by randomforest. *R News* **2002**, *2*, 18–22.
50. Hausser, J.; Strimmer, K.; Strimmer, M.K. Package 'Entropy'. Available online: http://strimmerlab.org/software/entropy/ (accessed on 19 February 2015).
51. Shannon, C.E. A mathematical theory of communication. *ACM SIGMOBILE Mob. Comput. Commun. Rev.* **2001**, *5*, 3–55. [CrossRef]
52. Hinton, G.; Deng, L.; Yu, D.; Dahl, G.E.; Mohamed, A.-R.; Jaitly, N.; Senior, A.; Vanhoucke, V.; Nguyen, P.; Sainath, T.N. Deep neural networks for acoustic modeling in speech recognition: The shared views of four research groups. *IEEE Signal Process. Mag.* **2012**, *29*, 82–97. [CrossRef]
53. Li, L.; Zhao, Y.; Jiang, D.; Zhang, Y.; Wang, F.; Gonzalez, I.; Valentin, E.; Sahli, H. Hybrid deep neural network–hidden markov model (dnn-hmm) based speech emotion recognition. In Proceedings of the 2013 Humaine Association Conference on Affective Computing and Intelligent Interaction (ACII), Geneva, Switzerland, 2–5 September 2013; pp. 312–317.

54. LeCun, Y.; Bengio, Y.; Hinton, G. Deep learning. *Nature* **2015**, *521*, 436–444. [CrossRef]
55. Rumelhart, D.E.; Hinton, G.E.; Williams, R.J. Learning representations by back-propagating errors. *Cogn. Model.* **1988**, *5*, 1. [CrossRef]
56. Ham, J.; Chen, Y.; Crawford, M.M.; Ghosh, J. Investigation of the random forest framework for classification of hyperspectral data. *IEEE Trans. Geosci. Remote Sens.* **2005**, *43*, 492–501. [CrossRef]
57. Belgiu, M.; Drăguţ, L. Random forest in remote sensing: A review of applications and future directions. *ISPRS J. Photogramm. Remote Sens.* **2016**, *114*, 24–31. [CrossRef]
58. Mahapatra, D. Analyzing training information from random forests for improved image segmentation. *IEEE Trans. Image Process.* **2014**, *23*, 1504–1512. [CrossRef]
59. Yu, S.; Jia, S.; Xu, C. Convolutional neural networks for hyperspectral image classification. *Neurocomputing* **2017**, *219*, 88–98. [CrossRef]
60. Kianisarkaleh, A.; Ghassemian, H. Nonparametric feature extraction for classification of hyperspectral images with limited training samples. *ISPRS J. Photogramm. Remote Sens.* **2016**, *119*, 64–78. [CrossRef]
61. Luo, F.; Huang, H.; Duan, Y.; Liu, J.; Liao, Y. Local geometric structure feature for dimensionality reduction of hyperspectral imagery. *Remote Sens.* **2017**, *9*, 790. [CrossRef]
62. Li, L.; Ge, H.; Gao, J. A spectral-spatial kernel-based method for hyperspectral imagery classification. *Adv. Space Res.* **2017**, *59*, 954–967. [CrossRef]
63. Yang, C.; Tan, Y.; Bruzzone, L.; Lu, L.; Guan, R. Discriminative feature metric learning in the affinity propagation model for band selection in hyperspectral images. *Remote Sens.* **2017**, *9*, 782. [CrossRef]
64. Guo, C.; Pleiss, G.; Sun, Y.; Weinberger, K.Q. On calibration of modern neural networks. *arXiv* **2017**, arXiv:1706.04599.
65. Zhu, D.; Yao, H.; Jiang, B.; Yu, P. Negative log likelihood ratio loss for deep neural network classification. *arXiv* **2018**, arXiv:1804.10690.
66. Lu, D.; Weng, Q. A survey of image classification methods and techniques for improving classification performance. *Int. J. Remote Sens.* **2007**, *28*, 823–870. [CrossRef]
67. Foody, G.M.; Mathur, A. The use of small training sets containing mixed pixels for accurate hard image classification: Training on mixed spectral responses for classification by a svm. *Remote Sens. Environ.* **2006**, *103*, 179–189. [CrossRef]
68. Pal, M.; Foody, G.M. Feature selection for classification of hyperspectral data by svm. *IEEE Trans. Geosci. Remote Sens.* **2010**, *48*, 2297–2307. [CrossRef]
69. Li, C.; Wang, J.; Wang, L.; Hu, L.; Gong, P. Comparison of classification algorithms and training sample sizes in urban land classification with landsat thematic mapper imagery. *Remote Sens.* **2014**, *6*, 964–983. [CrossRef]
70. Heydari, S.S.; Mountrakis, G. Effect of classifier selection, reference sample size, reference class distribution and scene heterogeneity in per-pixel classification accuracy using 26 landsat sites. *Remote Sens. Environ.* **2018**, *204*, 648–658. [CrossRef]
71. Zhang, L.; Zhang, L.; Du, B. Deep learning for remote sensing data: A technical tutorial on the state of the art. *IEEE Geosci. Remote Sens. Mag.* **2016**, *4*, 22–40. [CrossRef]
72. Canters, F. Evaluating the uncertainty of area estimates derived from fuuy land-cover classification. *Photogramm. Eng. Remote Sens.* **1997**, *63*, 403–414.

entropy

MDPI

Article

Reconstruction of PET Images Using Cross-Entropy and Field of Experts

Jose Mejia [1], Alberto Ochoa [2] and Boris Mederos [3,*]

1 Department of Electrical and Computation Engineering, Universidad Autónoma de Ciudad Juárez, Ciudad Juárez 32310, Mexico; jose.mejia@uacj.mx
2 Department of Industrial and Systems, Universidad Autónoma de Ciudad Juárez, Ciudad Juárez 32310, Mexico; alberto.ochoa@uacj.mx
3 Department of Physics and Mathematics, Universidad Autónoma de Ciudad Juárez, Ciudad Juárez 32310, Mexico
* Correspondence: boris.mederos@uacj.mx

Received: 17 December 2018; Accepted: 14 January 2019; Published: 18 January 2019

Abstract: The reconstruction of positron emission tomography data is a difficult task, particularly at low count rates because Poisson noise has a significant influence on the statistical uncertainty of positron emission tomography (PET) measurements. Prior information is frequently used to improve image quality. In this paper, we propose the use of a field of experts to model a priori structure and capture anatomical spatial dependencies of the PET images to address the problems of noise and low count data, which make the reconstruction of the image difficult. We reconstruct PET images by using a modified MXE algorithm, which minimizes a objective function with the cross-entropy as a fidelity term, while the field of expert model is incorporated as a regularizing term. Comparisons with the expectation maximization algorithm and a iterative method with a prior penalizing relative differences showed that the proposed method can lead to accurate estimation of the image, especially with acquisitions at low count rate.

Keywords: positron emission tomography; reconstruction; field of experts

1. Introduction

Positron emission tomography is an imaging technology that provides quantitative studies to detect, diagnose, and monitor treatment of different diseases such as hypernated myocardium, cancer, and many others [1,2].

The scan process begins by administering a radioactive substance, the radiotracer, to the patient. The substance is absorbed mainly by the target organs or tissues. The positron emission tomography (PET) data are then obtained by detecting the radiotracer distribution within the body. These acquired data are then processed by reconstruction algorithms to obtain the final image, which is presented to the medical research personnel [2,3].

Quality of the PET image depends of several factors: physical factors such as positron range, non collinearity and spurious events; hardware related factors such as crystal type and size and response time of the electronics; and the software or reconstruction algorithm used to estimate the final image [3].

In this paper, we are interested in improving the quality of the PET images by using algorithms to reconstruct the image. Reconstruction algorithms can be broadly classified into analytic and iterative methods. Iterative methods are popular in PET due to their robustness and ability to incorporate prior data and noise statistics. Iterative methods are mainly based on the maximum likelihood expectation maximization estimator or the least squares model. However, when excessive noise is present in the acquired data, such as in low count acquisitions, most iterative methods have difficulties in obtaining an accurate estimation of the data and regularization techniques are required to stabilize the solution [4].

In this paper a novel approach to reconstruct PET images is presented. Our approach is based on a regularized expectation maximization (EM) algorithm. Here, we propose to regularize teh problem by using a cross-entropy fidelity term and field of experts (FoE) priors [5], which are capable of capturing richer spatial statistics through patches extracted from a dictionary of images. Therefore, we expect that the FoE prior helps to recover the anatomical structure and capture anatomical spatial dependencies of the PET images during the reconstruction process.

The rest of the paper is organized as follows. In Section 2, the proposed algorithm is presented. In Section 3, experiments and results are shown. Finally, conclusions are provided in Section 4.

2. Methodology

2.1. ML-EM Algorithm

The Maximum Likelihood Expectation Maximization (ML-EM) algorithm takes into account the Poisson-based likelihood distribution of the data.

The PET scanner detectors count pairs of events in coincidence throughout the entire ring. These counts are subsequently processed to form the final image. Here, we represent the counts registered by the I detectors of the scanner by the vector $y = [y_1, y_2, y_3, ..., y_I]$. Each element y_i of y is modeled as independent random variables and Poisson distributed with expectation $\bar{y}_i$ given by

$$\bar{y}_i = E[y_i] = \sum_{j=1}^{J} x_j a_{i,j} \tag{1}$$

where x_i represents the radionuclide activity within the scanned subject and a pixel in the reconstructed image of size J pixels. Here, the image will be ordered as a vector $x = [x_1, x_2, x_3, ..., x_J]$. The $a_{i,j}$s are elements of the system matrix A. The probability $P(y|x)$ of observing y_i is a likelihood of the unknown emissions at pixel x_j, thus

$$L(x) = p(y|x) = \prod_{i=1}^{I} e^{-\bar{y}_i} \frac{\bar{y}_i^{y_i}}{y_i!} \tag{2}$$

The log-likelihood is obtained by combining Equations (1) and (2) as

$$\ln(L(x)) = -\sum_{i=1}^{I} [\sum_{j=1}^{J} x_j a_{i,j} + y_i \ln(\sum_{j=1}^{J} x_j a_{i,j}) + \ln(y_i!)] \tag{3}$$

The application of maximum-likelihood estimation techniques in Equation (3) leads the EM iterative scheme for the update of the ith pixel at iteration $(n+1)$ as follows:

$$x_j^{(n+1)} = \frac{x_j^{(n)}}{\sum_i a_{i,j}} \sum_i \frac{a_{i,j} y_i}{\sum_k a_{i,k} x_k^{(n)}} \tag{4}$$

This method has been widely used in PET reconstruction as it produces images with better quality than other techniques. However, it is affected when data are acquired at low-count rates, producing noisy images.

2.2. Cross-Entropy

One of the trade offs of the reconstruction algorithms is to minimize the difference between the measured and reconstructed data, while at the same time maintaining the final result without nuisances such as noise. The term that reflects the degree of similarity is often called fidelity term. In this paper, we adopt as fidelity term the cross entropy, which in information theory measures the resemblance between two probability distributions. This approach is taken in [6], where the authors reconstructed the measured data by minimizing a weighted sum of two cross-entropy terms.

The cross-entropy or Kullback–Leiber distance between Ax and y is defined as

$$J_0(x) = D(y, Ax) = \sum_i (y_i \ln y_i - y_i \ln(Ax)_i - y_i + (Ax)_i) \tag{5}$$

It is known that $D(y, Ax)$ is strictly convex and then a sufficient condition for a global minimum is that $\frac{\partial J_0(x)}{\partial x_j} = 0$, due to

$$\frac{\partial J_0(x)}{\partial x_j} = \frac{\partial D(y, Ax)}{\partial x_j} = -\frac{a_{i,j} y_i}{q_i^{(n)}} + a_{i,j}, \tag{6}$$

and then the sufficient condition is written as

$$-\frac{a_{i,j} y_i}{q_i^{(n)}} + a_{i,j} = 0. \tag{7}$$

It is possible to develop an optimization method based on the EM scheme. In [6], it has been shown that the minimization of Equation (5) with respect to x is equivalent to the maximization of the log-likelihood function in the ML estimate of Equation (4). Thus, using the EM algorithm to minimize Equation (5), and using $q_i^{(n)} = \sum_k a_{i,k} x_k^{(n)}$ and Equation (6), we have

$$x_j^{(n+1)} = \frac{x_j^{(n)}}{\sum_i a_{i,j} y_i} \sum_i \frac{a_{i,j} y_i}{q_i^{(n)}} \tag{8}$$

$$x_j^{(n+1)} = \frac{x_j^{(n)}}{\sum_i a_{i,j}} \sum_i \frac{a_{i,j}(y_i + q_i^{(n)} - q_i^{(n)})}{q_i^{(n)}} \tag{9}$$

$$x_j^{(n+1)} = x_j^{(n)} - \frac{x_j^{(n)}}{\sum_i a_{i,j}} \sum_i \left(-\frac{a_{i,j} y_i}{q_i^{(n)}} + a_{i,j} \right) \tag{10}$$

Now, using Equation (6), we obtain

$$x_j^{(n+1)} = x_j^{(n)} - \frac{x_j^{(n)}}{\sum_i a_{i,j}} \frac{\partial J_0(x^{(n)})}{\partial x_j} \tag{11}$$

This scheme is termed as MXE1 in [6] and has been used in several algorithms for reconstruct PET images [7,8]. It worth to remarking that Equation (11) can be seem as a gradient descent step of the cross-entropy function J_0 with variable step-size $\frac{x_j^{(n)}}{\sum_i a_{i,j}}$. Additionally, in this context, this idea can accommodate more complex objective functions, as will be shown in Section 2.4.

2.3. Field of Experts Model

The Field-of-expert scheme [5,9] models high-dimensional probability distributions by taking the product of several distributions (the experts). This provides a framework for learning image priors from sets of experts. Each expert distribution takes into account certain image structure learned using a database of images. In this manner, a prior captures richer spatial statistics present in the image. The modeled priors represent clique potential functions on a Markov random field (MRF) [10], and the experts are modeled as t-distributions, defined on each clique, with parameters learned by using the contrastive divergence method of Carreira-Perpinan and Hinton [11]. Field of expert have been used in several image processing algorithms and applications, such as segmentation [12], painting [13], and compressed sensing based restoration [14], among others.

In this paper, we propose to model the prior distribution for PET images by using a field of experts. Each expert is represented as a distribution modeled by linear filters J. Thus, the prior function is defined as

$$P(x,\Theta) = \frac{1}{Z(\Theta)} \prod_{k=1}^{K} \prod_{i=1}^{N} \phi(J_i^T x_{(k)}, \alpha_i) \tag{12}$$

where $\Theta = \theta_1, ..., \theta_N$ is a set of learned parameters $\theta_i = \{\alpha_i, J_i\}$, α_i is a parameter of expert i, $Z(\Theta)$ is the normalizing function, and $\phi(\cdot,\cdot)$ represents the experts, and are given by the student t-experts

$$\phi(J_i^T x, \alpha_i) = (1 + \frac{1}{2}(J_i^T x)^2)^{-\alpha_i}) \tag{13}$$

Note that α is similar to the degrees of freedom [15].

Several techniques have been developed to take into account the similarity between the anatomical and functional images to define priors to improve the reconstruction process [16,17].

In this work, to capture anatomical spatial dependencies of the PET images, we use the FoE scheme. To this end, we trained 5×5 filters to model priors that adapt to the images and underlaying anatomy. The training data consisted of 1000 images of real and simulated PET images. We also included a set of 50 images of studies of computer tomography and magnetic resonance without PET to provide the filter with more detailed anatomical characteristics. The obtained filter are shown in Figure 1.

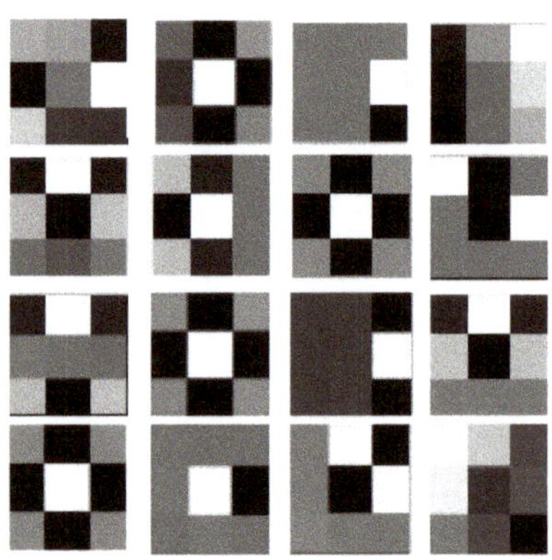

Figure 1. The 5×5 filters obtained by training the oroduct-of-experts model on positron emission tomography (PET) images database. The colors in each frame are proportional to the magnitude of the filter coefficient, using a gray scale.

In this way, we could incorporate into the reconstruction filters adapted to structure of the PET images and anatomical information, in the form of organ and lesion boundaries, derived from CT and MR.

2.4. Proposed Objective Function

In this section, we present our proposed function:

$$J_{FoE}(x) = D(y, Ax) + \beta P(x, \Theta) \tag{14}$$

where D is the cross entropy and P is a prior based in the FoE framework. To derive an optimization method, we examine the derivative of Equation (14).

$$\begin{aligned}\frac{\partial J_{FoE}(x)}{\partial x_j} &= \frac{\partial D(y,Ax)}{\partial x_j} + \beta\frac{\partial P(x,\Theta)}{\partial x_j} \\ &= \sum_i \left[\frac{-y_i a_{i,j}}{(Ax)_i} + a_{i,j}\right] + \beta \sum_k^N J_-^{(k)} * \phi'(J^{(k)} * x; \alpha_i),\end{aligned} \tag{15}$$

where $J^{(k)}$ is the convolutional filter corresponding to J_k and $J_-^{(k)}$ is obtained by mirroring $J^{(k)}$ around the center (for more details, see [5]).

Then, we can generalize the idea of (MXE1 iterative scheme) to the proposed regularization of the cross-entropy (Equation (14))

$$x_j^{(n+1)} = x_j^{(n)} - \frac{x_j^{(n)}}{\sum_i a_{i,j}} \frac{\partial J_{FoE}(x^{(n)})}{\partial x_j} \tag{16}$$

This iterative scheme corresponds to a gradient descent step of the proposed functional J_{FoE} with variable step-size $\frac{x_j^{(n)}}{\sum_i a_{i,j}}$, as mentioned in Section 2.2.

Unfortunately, MXE1 offer no guarantee that the algorithm will preserve non-negativity constraints. This can be remedied by using the line search algorithm LINU described in [18] by setting all negative elements in the new iteration to 0. For implementation of the algorithm, we used a fixed iteration scheme, with 27 iterations, and a $\beta = 0.5$.

3. Results

In this section, we present acquired and simulated datasets of PET images to show how the proposed algorithm deals with noise and structures and borders. We also offer comparisons with the EM algorithm and the algorithm proposed in [19], which uses a concave prior penalizing relative differences (CP) between neighbors. This algorithm was modified in [20] (GE healthcare white paper). Here, we set the parameters of the CP algorithm as $\beta = 0.01$, $\gamma = 0.1$, and $\lambda = 0.97$. The EM was used with 12 iterations.

The simulated images were generated using Simset (a Simulation System for Emission Tomography) software (version 2.9.2, provided by the Division of Nuclear Medicine, University of Washington, Seattle, WA, USA) [21]. We used a simulated 3D PET detector with two axial rings consisting of an aluminum front cover followed by a single layer of 3.5 cm of BGO. From the generated volumetric data, 2D sinograms were taken.

3.1. Simulated Data

In this experiment, the efficacy of the reconstruction under low count conditions was evaluated. We used the reconstructed images from each method to generate selected surfaces to show graphically how close the reconstructed image was to the ground truth.

To this end, we designed a cylindrical software phantom of polymethyl methacrylate with five rows of holes (water-filled cylindrical inserts) of 2, 3, 4, and 5 mm in diameter. Each hole was filled with activity of 1:8 with respect to background. Figure 2 shows the phantom, a simulation of 30 M counts and its reconstruction using expectation maximization (EM).

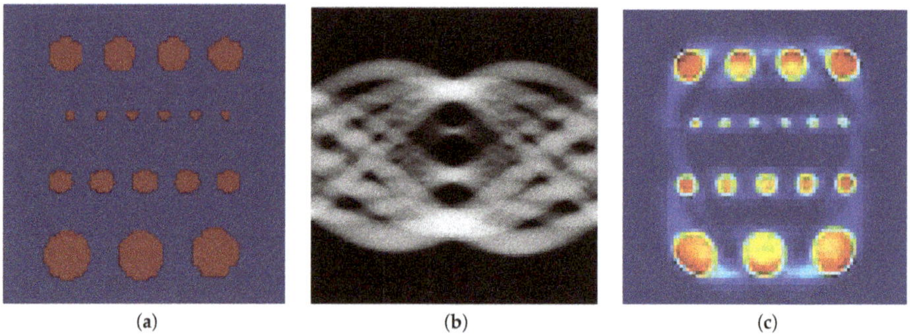

Figure 2. Cylindrical software phantom: (**a**) ground truth; (**b**) simulated sinogram data at 30 M counts; and (**c**) reconstruction with expectation maximization (EM).

We were interested in low count reconstructions, since this means less radiation exposure for the patient. Similar to higher counts data, all methods evaluated had practically equal quality in the reconstruction. Thus, we ran a PET scan simulation at 5 M count using the Simset software. Figure 3a shows the simulated sinogram data at lower count and Figure 3b–d shows the reconstruction using EM, CP, and the proposed method, respectively. It can be seen that the image reconstructed with the proposed method had better definition of the rods.

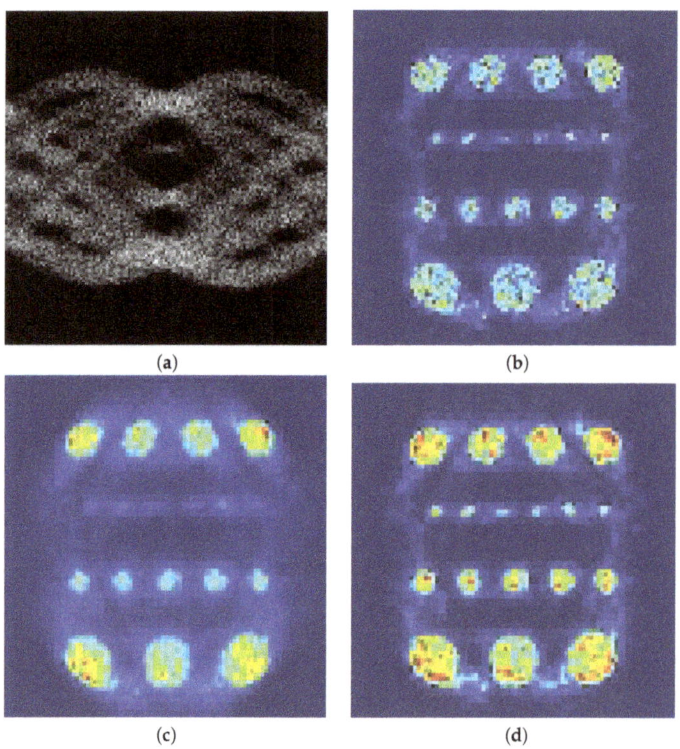

Figure 3. Cylindrical software phantom: (**a**) Input sinogram; (**b**) low count reconstruction with EM; (**c**) low count reconstruction with CP; and (**d**) low count reconstruction with the proposed method.

Figure 4 depicts the surfaces for each row of the holes. The proposed method had a smoother surface than EM, without losing contrast. In addition, contrast for the larger cylinders was slightly better at some points in the surface with CP, and achieved a maximum for the 5 mm surface.

In our next experiment, we used the Digimouse anatomical atlas dataset described in [22] to design a software mouse phantom. The mouse phantom was used to have a more realistic assessment of the structures found on practical data. The PET scans of the mouse were simulated using Simset software (version, publisher, city, state abbreviation if USA or Canada, country).

We simulated two realizations with different number of counts: 30 M and 5 M counts. The 30 M counts realization was taken as a ground truth. Figure 5 presents two different slices of the phantom, the first column shows the 30 M realization reconstructed with EM, while the second, third, and fourth columns show the 5 M realization reconstructed with EM, CP and the proposed method, respectively.

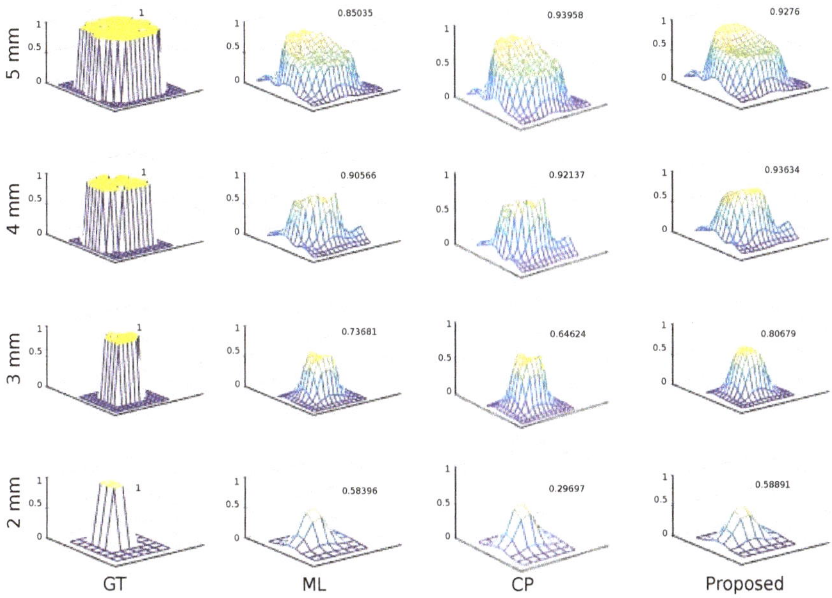

Figure 4. Profiles of the different methods with the cylindrical software phantom. Each row shows the same hole, and each column the same method. The maximum of each surface is indicated next to it.

To quantitatively evaluate image quality in the Digimouse, a channelized Hoteling observer [23,24] was used in the context of lesion detectability. The area under the curve (AUC) was used as a figure of merit. The task performed by the observer was the detection of a lesion with known location. To this end, we used 25 simulations with lesion and 25 without lesion. The lesion has an activity of 1:5 with respect to background, in 10 of the 25 phantoms with lesion, and 1:3 with respect to background in the rest. The images were reconstructed with each method. These images were fed to the observer to analyze its output. Figure 6 shows the AUC attained by each method. As can be seen, our method obtained more AUC than EM and CP. Based on this result, the simulated observer was able to better detect the lesion in the images reconstructed by the proposed method.

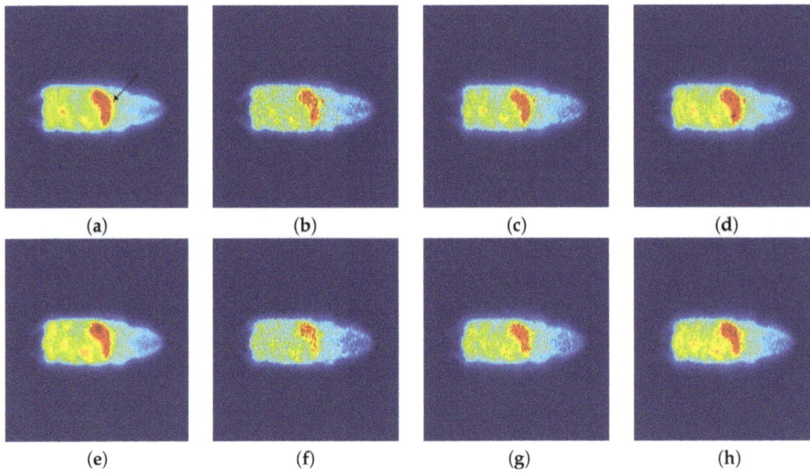

Figure 5. Slices of the Digimouse software phantom: ground truth (**a**,**e**); low count reconstruction with EM (**b**,**f**); low count reconstruction with CP (**c**,**g**); and (**d**,**h**) low count reconstruction with the proposed method. In (**a**), the arrow indicates a lesion.

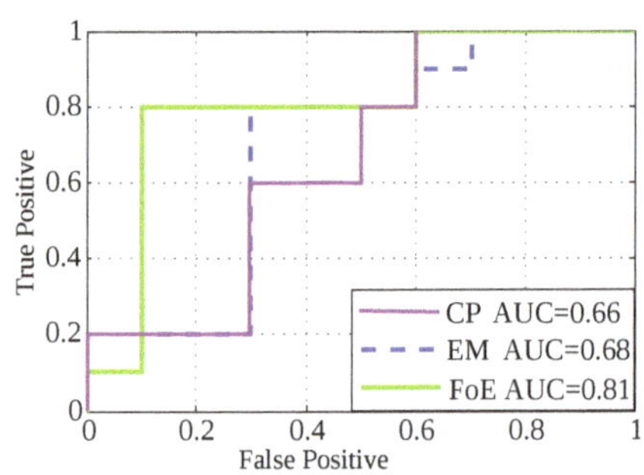

Figure 6. ROC analysis evaluated on the Digimouse phantom.

In the next experiment, we evaluated the performance with measured data. We used data from http:\\web.eecs.umich.edu\~fessler\, which is from a subject who was scanned on a CTI ECAT PET scanner. The raw sinograms were acquired at 160 radial samples and 192 angular samples; data were pre-corrected for delayed coincidences [25]. Figure 7a–c shows the reconstructions of the data using EM, CP, and the proposed method, respectively.

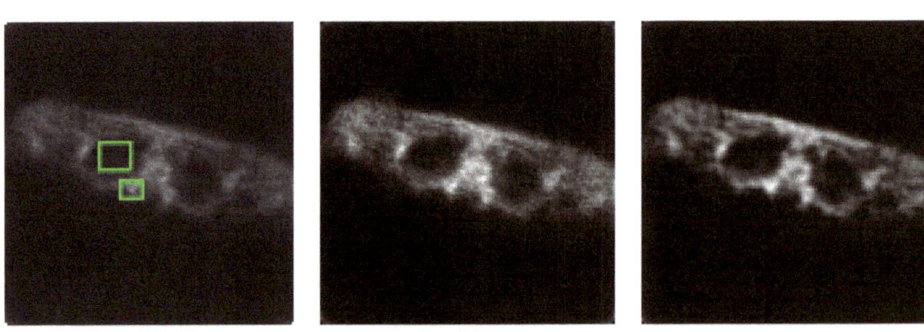

Figure 7. Reconstruction of measured data with: (**a**) EM; (**b**) CP; and (**c**) the proposed method.

Table 1 shows the results of applying the contrast resolution (CR) metric [18] between the two rectangles depicted in Figure 7a. Our method has better CR than the other methods evaluated, and lower noise.

Table 1. Contrast resolution measures.

Method	CR
EM	0.577
CP	0.541
Proposed	0.695

4. Conclusions

In this paper, a novel reconstruction method for PET images is presented based on a cross-entropy fidelity term. We propose to regularize the ill-posed problem by using field of experts priors. In this way, we can incorporate into the reconstruction process prior distributions from adapted filters to structure PET images and anatomical information.

The experimental results show that the proposed method led to a better reconstruction performance than EM and CP. In an experiment with a phantom with rods of different sizes, we found an improvement in the recovering of the pixels of the smallest rods, showing that the proposed method could perform better in the reconstruction of small structures such as lesions. This was also observed in a second experiment with a phantom with lesions, where a ROC analysis of a observer detecting lesions was simulated and the proposed method outperformed EM and CP methods. In the experiment with measured data, our method obtained more contrast resolution, and visually the image of the proposed method had more defined edges than EM and CP. As future work, we plan to implement the algorithm with listmode data and take into account the additional information of time of flight, as well as construct a database of more images including PET/CT and PET/MRI.

Author Contributions: All authors have worked equally in this manuscript. All authors have read and approved the final manuscript.

Funding: This research received no external funding

Conflicts of Interest: The authors declare no conflict of interest.

References

1. Wahl, R.L. *Principles and Practice of Positron Emission Tomography*; Lippincott Williams & Wilkins: Philadelphia, PA, USA, 2002.
2. Lynch, T.B. *PET/CT in Clinical Practice*; Springer: New York, NY, USA, 2007.
3. Bendriem, B.; Townsend, D.W. *The Theory and Practice of 3D PET*; Springer: New York, NY, USA, 2013; Volume 32.
4. Teng, Y.; Zhang, T. Three penalized EM-type algorithms for PET image reconstruction. *Comput. Biol. Med.* **2012**, *42*, 714–723. [CrossRef] [PubMed]
5. Roth, S.; Black, M.J. Fields of experts: A framework for learning image priors. In Proceedings of the CVPR 2005 IEEE Computer Society Conference on Computer Vision and Pattern Recognition, San Diego, CA, USA, 20–25 June 2005; Volume 2, pp. 860–867.
6. Ardekani, B.A.; Braun, M.; Hutton, B.F.; Kanno, I.; Iida, H. Minimum cross-entropy reconstruction of PET images using prior anatomical information. *Phys. Med. Biol.* **1996**, *41*, 2497–2517. [CrossRef] [PubMed]
7. Zhu, H.; Shu, H.; Luo, L.; Zhou, J. Modified minimum cross-entropy algorithm for pet image reconstruction using total variation regularization. In Proceedings of the IEEE International Workshop on Biomedical Circuits and Systems, Singapore, 1–3 December 2004.
8. Zhu, H.; Zhou, J.; Shu, H.; Luo, L. A edge-preserving minimum cross-entropy algorithm for PET image reconstruction using multiphase level set method. In Proceedings of the IEEE International Conference on Acoustics, Speech, and Signal Processing (ICASSP'05), Philadelphia, PA, USA, 23 March 2005; Volume 2, pp. 469–472.

9. Roth, S.; Black, M.J. Fields of experts. *Int. J. Comput. Vis.* **2009**, *82*, 205. [CrossRef]
10. Gu, W.; Lv, Z.; Hao, M. Change detection method for remote sensing images based on an improved Markov random field. *Multimedia Tools Appl.* **2017**, *76*, 17719–17734. [CrossRef]
11. Carreira-Perpinan, M.A.; Hinton, G.E. On contrastive divergence learning. In Proceedings of the Aistats, Bridgetown, Barbados, 6–8 January 2005; Citeseer: University Park, PA, USA, 2005; Volume 10, pp. 33–40.
12. Mahapatra, D.; Buhmann, J.M. A field of experts model for optic cup and disc segmentation from retinal fundus images. In Proceedings of the 2015 IEEE 12th International Symposium on Biomedical Imaging (ISBI), New York, NY, USA, 16–19 April 2015; pp. 218–221.
13. Li, Y.; Wu, H.; Li, L.; Yang, Y.; Zhang, C.; Bie, R. Improved field of experts model for image completion. *Optik-Int. J. Light Electr. Opt.* **2017**, *142*, 174–182. [CrossRef]
14. Lan, X.; Barner, K. Field of experts: Optimal structured bayesian compressed sensing. In Proceedings of the 2017 IEEE Global Conference on Signal and Information Processing (GlobalSIP), Montreal, QC, Canada, 14–16 November 2017; pp. 1130–1134.
15. Roth, S.; Adviser-Black, M. *High-Order Markov Random Fields for Low-Level Vision*; Citeseer: University Park, PA, USA, 2007; Volume 68.
16. Tang, J.; Rahmim, A. Anatomy assisted PET image reconstruction incorporating multi-resolution joint entropy. *Phys. Med. Biol.* **2014**, *60*, 31. [CrossRef] [PubMed]
17. Baete, K.; Nuyts, J.; Van Paesschen, W.; Suetens, P.; Dupont, P. Anatomical-based FDG-PET reconstruction for the detection of hypo-metabolic regions in epilepsy. *IEEE Trans. Med. Imaging* **2004**, *23*, 510–519. [CrossRef] [PubMed]
18. Kaufman, L. Implementing and accelerating the EM algorithm for positron emission tomography. *IEEE Trans. Med. Imaging* **1987**, *6*, 37–51. [CrossRef] [PubMed]
19. Nuyts, J.; Beque, D.; Dupont, P.; Mortelmans, L. A concave prior penalizing relative differences for maximum-a-posteriori reconstruction in emission tomography. *IEEE Trans. Nucl. Sci.* **2002**, *49*, 56–60. [CrossRef]
20. Ross, S.Q. *Clear*; White Paper; GE Healthcare: Uppsala, Sweden, 2014; pp. 1–9.
21. Lewellen, T.; Harrison, R.; Vannoy, S. *The Simset Program*; University of Washington: Seattle, WA, USA, 1998.
22. Stout, D.; Chow, P.; Silverman, R.; Leahy, R.M.; Lewis, X.; Gambhir, S.; Chatziioannou, A. Creating a whole body digital mouse atlas with PET, CT and cryosection images. *Mol. Imaging Biol.* **2002**, *4*, S27.
23. Van de Sompel, D.; Brady, M.; Boone, J. Task-based performance analysis of FBP, SART and ML for digital breast tomosynthesis using signal CNR and Channelised Hotelling Observers. *Med. Image Anal.* **2011**, *15*, 53–70. [CrossRef] [PubMed]
24. Racine, D.; Ryckx, N.; Ba, A.; Becce, F.; Viry, A.; Verdun, F.R.; Schmidt, S. Task-based quantification of image quality using a model observer in abdominal CT: A multicentre study. *Eur. Radiol.* **2018**, *28*, 5203–5210. [CrossRef] [PubMed]
25. Fessler, J.A.; Ficaro, E.P.; Clinthorne, N.H.; Lange, K. Grouped-coordinate ascent algorithms for penalized-likelihood transmission image reconstruction. *IEEE Trans. Med. Imaging* **1997**, *16*, 166–175. [CrossRef] [PubMed]

entropy

MDPI

Article

Objective 3D Printed Surface Quality Assessment Based on Entropy of Depth Maps

Jarosław Fastowicz [1], Marek Grudziński [2], Mateusz Tecław [1] and Krzysztof Okarma [1,*]

[1] Faculty of Electrical Engineering, West Pomeranian University of Technology, Szczecin, 70-313 Szczecin, Poland; jfastowicz@zut.edu.pl (J.F.); mateusz.teclaw@zut.edu.pl (M.T.)

[2] Faculty of Mechanical Engineering and Mechatronics, West Pomeranian University of Technology, Szczecin, 70-310 Szczecin, Poland; mgrudzinski@zut.edu.pl

* Correspondence: okarma@zut.edu.pl

Received: 20 December 2018; Accepted: 18 January 2019; Published: 21 January 2019

Abstract: A rapid development and growing popularity of additive manufacturing technology leads to new challenging tasks allowing not only a reliable monitoring of the progress of the 3D printing process but also the quality of the printed objects. The automatic objective assessment of the surface quality of the 3D printed objects proposed in the paper, which is based on the analysis of depth maps, allows for determining the quality of surfaces during printing for the devices equipped with the built-in 3D scanners. In the case of detected low quality, some corrections can be made or the printing process may be aborted to save the filament, time and energy. The application of the entropy analysis of the 3D scans allows evaluating the surface regularity independently on the color of the filament in contrast to many other possible methods based on the analysis of visible light images. The results obtained using the proposed approach are encouraging and further combination of the proposed approach with camera-based methods might be possible as well.

Keywords: additive manufacturing; 3D prints; 3D scanning; image entropy; depth maps; surface quality assessment; machine vision; image analysis

1. Introduction

One of the most dynamically growing technologies in the era of Industry 4.0 is undoubtedly the additive manufacturing widely known as 3D printing. The applicability of these technologies covers many areas of modern technology, industry, medicine, clothing, preservation of cultural heritage and even food production. The great variety of affordable 3D printers, which may be assembled at home with open source software, causes growing popularity of the cheapest devices utilizing the Fused Deposition Modeling (FDM) technology. Their principle of operation (initially preserved by the patent expired in 2009) is based on the melting process of plastic material (filament) and moving the extruder according to the defined tool path. The dripping filament hardens immediately and forms the visible layers which should adhere to each other. Some other solutions, however much less popular and more expensive, include selective laser sintering, stereolithography and inkjet printing.

The most typical materials used in FDM printing are Polyactic Acid (PLA) and Acrylonitrile Butadiene Styrene (ABS). Both types of filaments have different properties and the most popular thermoplastic polymer for low-cost 3D printers is the PLA filament. Although it is translucent in its natural form, it may be dyed using various colors. This fully biodegradable material can also be used for food packaging purposes but is less durable, more fragile and sensitive to heat in comparison to ABS. Therefore, its melting point is lower (around 150 °C) as the ABS filaments require typically about 200 °C. ABS filaments are more abrasion resistant and lightweight. They can be used to create low cost medical prostheses and has good mechanical properties. However, there are some concerns related to its potential toxicity, especially regarding fumes emitted during printing [1,2].

The final quality of 3D printed objects depends on various factors. Some of them are related to the quality of materials from which the printing device is made as well as the accuracy of its construction. The other elements influencing the quality of 3D prints might be the quality of the filament, temperature and other environmental conditions as well as improper printing parameters (e.g., filament's delivery speed, wrong configuration of the stepper motors, inappropriate melting temperature for the specified type of filament, etc.). The most typical surface distortions may be related to under-filling (dry printing), over-filling or the presence of cracks without adequate adhesion of layers.

One of the future challenges is related to the detection of the internal imperfections of 3D printed objects. The application of X-ray tomography and ultrasonic imaging for the detection of embedded defects and altered printing orientation was examined by Zeltmann [3]. The presence of some internal distortions may also be detected using electromagnetic methods, e.g., terahertz non-destructive testing [4]. Nevertheless, the applicability of these technologies for on-line printing monitoring is very limited in opposition to cameras and reasonably small 3D scanners, which are mounted in some 3D printing devices. Therefore, in this paper, we focus on the assessment of the outer surfaces based on the regularity of visible layer patterns produced by the FDM printers.

Since the process of additive manufacturing is usually time-consuming and the production of a complicated object may take several hours or more, an obviously desired solution is the on-line monitoring of the progress of 3D printing. In such systems, the detection of low printing quality may lead to aborting the printing to save the filament, time and energy or—for smaller imperfections—performing some corrections for previously manufactured layers. However, such a decision is often dependent on hardware capabilities [5].

Some early ideas of the visual monitoring of the 3D printing progress have been related to the use of process signatures applied for fused deposition of ceramics [6,7] and detection of places with missing filament by monitoring the top surface of the manufactured object based on the process dynamics models [8]. Some predefined types of distortions, such as part jams and feeder jams, may be detected using the method proposed by Szkilnyk [9], although this approach is devoted mainly to machine vision fault detection of automated assembly machines. Similar methods for fault detection, which make use of Gaussian Mixed Models (GMMs), blob analysis, optical flow and running average, were presented recently by Chauhan and Surgenor [10,11].

Another exemplary defect detection system for plastic products, namely anaesthetic respiratory masks, based on computer vision, is presented in [12]. Recently, an interesting approach to optimization of tool paths during the 3D printing process was suggested by Fok et al. [13]. Not only does it save the filament and avoid the presence of visible strings, but it also significantly improves the visual quality of manufactured products. The additional advantage of this approach is the minimization of the time spent on traversing transitions.

Some interesting works related to 3D printing issues were published by Jeremy Straub, including the first machine vision system making it possible to detect of the lack of filament, based on five cameras and Raspberry Pi units [14]. Unfortunately, the proposed solution is very sensitive to camera motions as well as the changes of lighting conditions and additionally requires many stops during the printing process. Some other related papers describe alignment issues [15], cybersecurity problems [16,17] and human error prevention [18].

Some other recent attempts to the monitoring of the 3D printing process include the use of neural networks for the 3D inkjet printing of the electronic products [19], matching the reference images of the models of the 3D parts with their manufactured equivalents [20] and the 3D image correlation comparison with the CAD model using two cameras [21].

A recent paper written by Scime [22] presents the system for in-situ monitoring and analysis of powder bed images for the laser powder bed fusion machine. Nevertheless, some consequential anomalies that may be detected require previous training using the machine learning algorithms. As stated by the authors, the analysis of each layer requires approximately 4 s on a single 4.00 GHz i7-4770 processor, thus the method is relatively slow. Another application of machine learning for quality monitoring of 3D prints, based on the trained SVM model, was presented recently by Delli and Chang [23]. The main limitations of this approach are related to the use of top view camera and necessary stops of the printing process for image acquisition during manufacturing.

One of the successful attempts to the use of 3D scanning for multi-material 3D printing monitoring has been presented in the MultiFab platform [24]. The applied 3D scanner module is based on Michelson's interferometer and full-field optical coherence tomography (OCT). Since the authors of the MultiFab system assume the lack of spatial details and significant textures, the direct application of structured light scanning is not possible. Nevertheless, for the low cost FDM based 3D printers, the visibility of each layer of the filament can be assumed for the side view cameras and the applicability of 3D scanners is worth investigating. To verify the usefulness of structured light 3D scanners, which have been successfully applied in many other applications such as CNC machines [25] and visual control of the loading cranes [26], some experiments were conducted for a dataset of several 3D printed flat surfaces subjected to 3D scanning followed by conversion of the obtained STL models into depth maps.

Since the main motivation of our research was the development of an objective method useful for automatic evaluation of 3D printed surfaces, one of the contributions of the paper is the verification of the usefulness of depth maps for this purpose. The novelty of the proposed approach is related to the use of the 3D scanning technology, providing the depth information, together with the development of the original entropy based surface quality assessment method, to increase the color independence of the obtained experimental results.

The rest of the paper consists of the short overview of the previous attempts to automatic quality assessment of the 3D printed surfaces, description of the proposed approach, presentation and discussion of obtained results and conclusions.

2. Methods of Quality Evaluation of 3D Printed Surfaces

Considering the assumptions related to the visibility of texture patterns caused by the use of low cost FDM devices and the application of the side view camera, the first approach to automatic quality assessment of 3D printed surfaces has been the use of texture analysis methods. For this purpose, a method based on the calculation of Haralick features determined from the Gray-Level Co-occurrence Matrix (GLCM) has been used [27]. Since the GLCM, also called the second-order histogram, represents the statistical information related to the neighbouring spatial relations between individual greyscale levels, some regularities may be expected, especially considering the vertical neighbourhoods. In fact, the periodicity of some features of the GLCMs calculated on the assumption that various offsets between the analyzed pixels may be observed. Further experiments made it possible to distinguish the scanned images from their equivalents captured by a camera and properly classify them into high and low quality samples with the use of homogeneity, independently on the image type [28]. Nevertheless, the application of this method requires many computations due to the necessity of analyzing tens of GLCMs and the obtained results have been verified for a limited number of PLA samples.

Another possible approach is based on the application of some image quality assessment (IQA) methods [29]. As the direct application of the most widely known full-reference IQA metrics, e.g., Structural Similarity (SSIM) proposed by Wang and Bovik [30], is not possible due to the lack of reference images, the self-similarity of the 3D printed surfaces is utilized. After the division of images into fragments, the local mutual similarity indexes are calculated to detect the low quality samples due to the presence of high local differences. Nevertheless, the method requires the additional matching of image fragments further examined together with the application of the Monte Carlo method to speed up the computations [31].

The first attempt to the use of image entropy for the assessment of the 3D printed flat surfaces is presented in a conference paper [32], where the direct application of image entropy is shown to be an appropriate method for the classification of high and low quality samples for a specified color of the filament. The method assumes that a perfect surface of the 3D printed sample observed from a side view camera should have only regular visible line patterns representing individual layers. For such regular patterns, the image entropy should be significantly lower than the values obtained for lower quality samples contaminated by various distortions. Nevertheless, the values obtained for the whole images, as well as the local entropy values calculated for the fragments of images, were strongly dependent on the color of the filament and the color to greyscale conversion. Therefore, to overcome this limitation, further experiments described in the present paper were made for the depth maps obtained from the 3D scans of the printed samples.

3. Proposed Method and Its Experimental Verification

To propose a reliable method of objective surface quality assessment, which would be independent on filament's color and further verify its prediction accuracy, several dozen flat samples were printed using three different devices illustrated in Figure 1. Most were obtained using various colors of ABS filaments (108), however 18 of them were manufactured using various PLA materials. Most of the high quality prints were obtained using da Vinci Pro device and the same or very similar settings were also used for the other two printers.

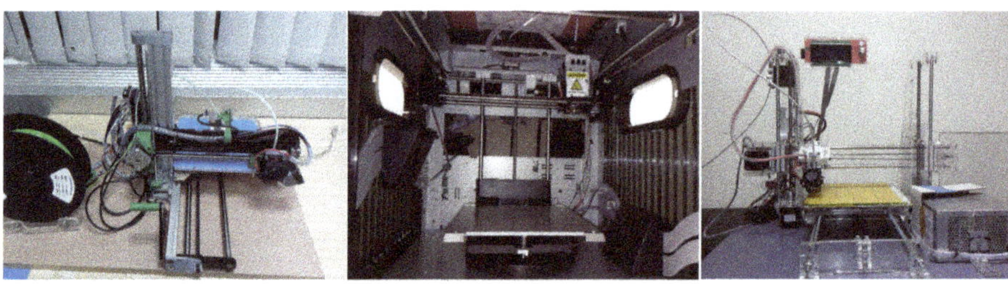

(a) (b) (c)

Figure 1. Three devices used for preparation of samples used in experiments: (**a**) RepRap Pro Ormerod 2; (**b**) da Vinci 1.0 Pro 3-in-1 (view of inner parts); and (**c**) Prusa i3.

Some of the samples were intentionally distorted by forcing "dry printing" and some changes of filament's delivery speed or temperature as well as parameters of stepper motors. Some of such obtained distortions, including the presence of cracks in some of the ABS samples, are illustrated in Figure 2. The only high quality sample (Figure 2a) is an illustration of properties of the surfaces that were subjectively assessed as representing high quality, whereas the sample in Figure 2d was evaluated as moderately high quality due to the presence of a single crack and minor visible distortions. The remaining seven images are examples of low quality surfaces.

All the manufactured samples were assessed by the members of our team as high, moderately high, moderately low or low quality, independently for each side, and therefore each sample was analyzed and assessed twice (in fact, for most of the samples, the assessment results of both sides were the same). Since the calculations of entropy values for each color channel, as well as using various color to greyscale conversion methods for such obtained images, did not lead to satisfactory results, the decision about the use of 3D scanning technology was made.

The entropy values, considered as statistical measure of randomness, expected to be relatively low for the high quality samples with visible regular patterns, were calculated according to the well known formula:

$$E = -\sum p \cdot \log_2 p \tag{1}$$

where p contains the normalized histogram counts obtained for greyscale images.

Figure 2. Exemplary photos of 3D printed flat surfaces: high quality sample No. 24 (**a**); and the illustration of various distortions obtained for lower quality 3D prints (**b**–**i**).

The reason for poor classification results based on the entropy analysis of photos was related to the strong dependence of entropy on the color and brightness of the analyzed samples. Although the results obtained for each color separately were in accordance with expectations, the color independent quality assessment of the 3D printed surfaces was troublesome. Another problem, influencing the results, also obtained using the flatbed scanner instead of a camera, as shown in earlier experiments [28], was related to a partial transparency of some of the filaments, particularly visible for some PLA materials.

To ensure the independence of the entropy values on the color of filament, it can be noticed that any distortions of the structure of consecutive layers should also be noticeable in the depth maps which may be obtained using stereovision camera pairs or 3D scanning technology. As laser 3D scanning technology is time-consuming and requires precise control of the moving laser light source, we focused on the use of fringe pattern based 3D scanning.

After the first attempts made with the use of previously developed 3D scanning system [25,26], in further experiments the more precise ATOS device manufactured by GOM company was used for the verification of the idea. During the experiments, precise fringe patterns were projected onto the scanned surface of the object and captured by two cameras based on the stereo camera principle allowing to scan the reflective surfaces more precisely (to several micrometers). However, it is worth noticing that projected fringe patterns should not be parallel to the visible layers and therefore the perpendicular patterns were used during the 3D scanning. The obtained cloud of points can be stored in the popular STL format, as shown in Figure 3, illustrating the exemplary scan of four 3D printed samples mounted in the dedicated mounting holders.

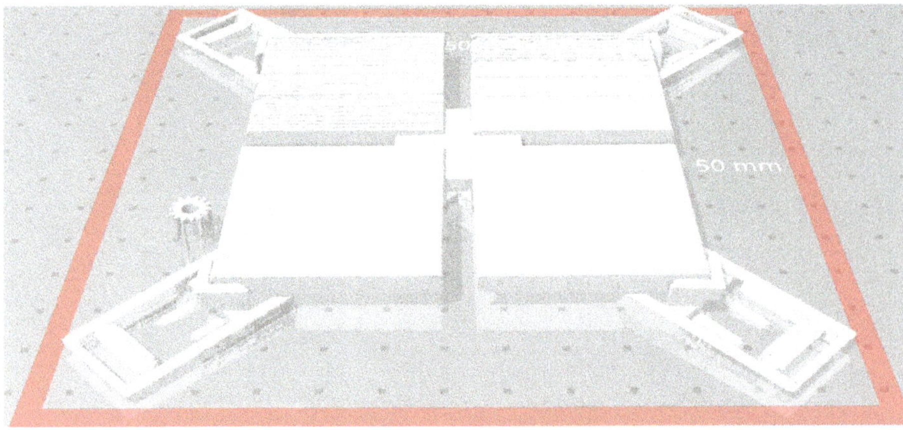

Figure 3. Illustration of the exemplary obtained STL model of four 3D scanned samples with visible mounting elements.

Such obtained STL files were converted into normalized depth maps stored as 1928 × 1928 pixels 16-bit greyscale images, representing the surfaces of the 3D printed and scanned samples. Some exemplary depth maps together with photos of respective samples obtained for green, red, brown and white lower quality 3D prints are presented in Figure 4, whereas similar examples obtained for high quality surfaces are illustrated in Figure 5.

The most relevant preprocessing step was the application of the Contrast Limited Adaptive Histogram Equalization (CLAHE) [33] for balancing the distribution of brightness on the images representing the depth maps. These differences, similar to the effect on non-uniform lighting, originate from the placement of the scanned samples that were not always perfectly parallel to each other. Consequently, after scanning four samples at once, as illustrated in Figure 3, some of them may be represented by various depth ranges equivalent to different brightness values of depth map images. Some differences for the opposite corners of the scanned sample can be observed as well. It can also be noticed for both depth maps obtained for the left images shown in Figure 4 (i.e., green and brown samples).

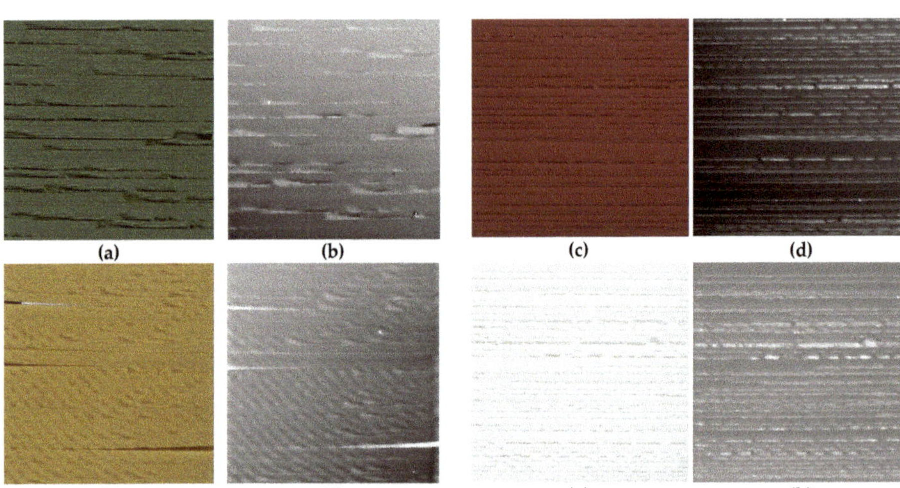

Figure 4. Exemplary lower quality samples: (**a**,**c**,**e**,**g**); and their obtained depth maps: (**b**,**d**,**f**,**h**).

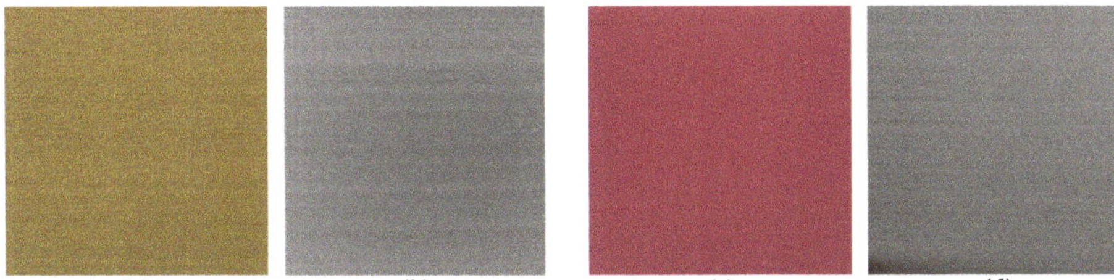

(a) **(b)** **(c)** **(d)**

Figure 5. Exemplary high quality samples: (**a**,**c**); and their obtained depth maps: (**b**,**d**).

Since the direct application of entropy calculated for the depth maps not always leads to satisfactory results, an approach utilizing the variance of the local entropy values was used, which was previously proposed for images of 3D printed samples obtained using flatbed scanner [34]. Nevertheless, in the previously proposed solution, due to some problems with color independence, the application of the combined metric based on on the average local entropy and its variance was used. Additionally, such a metric has been calculated for the hue component in HSV color space as well as the average of the RGB channels. Finally, the color independent metric was obtained and verified for 18 PLA samples. However, its application for the 126 samples used in our experiments did not lead to satisfactory results.

To overcome the problem of non-uniform brightness of the obtained depth maps, apart from the use of the CLAHE algorithm, the local entropy values as well as their variance were utilized. After extensive experimentation, the best results were obtained assuming the division of the depth map images into 64 blocks forming the array of 8×8 subimages. For each of these regions, the local entropy value E_{local} was calculated in addition to the overall entropy calculated for the whole depth map image, according to Equation (1).

The computations were supplemented with the variance of the local entropy Var_E and the final quality metric is proposed as:

$$Q = \log[E_{global} \cdot avg(E_{local}) \cdot {Var_E}^{0.25}] \quad (2)$$

where $avg(E_{local})$ denotes the mean of the 64 local entropy values.

4. Results and Discussion

The initial results achieved for the direct computation of the global entropy for the depth map images are presented in Figure 6 where all samples are presented in groups according to their colors and type of filament (with the first 18 samples being manufactured using PLA filaments). The samples containing cracks are marked with unfilled symbols and each color of the mark represents roughly the color of the respective filament. High quality samples are tagged with more round symbols, whereas lower quality ones are marked with more polygonal shapes. The same convention is also used in Figures 7–9.

All plots presented in Figures 6–9 should be interpreted considering the high values of all presented quality metrics as equivalent to low quality. Since the presence of local distortions causes the increase of entropy as well as its variance, higher values of the metric may be expected, which can be interpreted as a measure of the amount of distortions. Nevertheless, values close to zero might be expected only for perfectly flat surfaces which do not contain any patterns.

Analyzing the distribution of the global entropy values presented in Figure 6, it can be noted that the appropriate classification of the samples according to their quality is impossible. Even for the first four white samples the entropy values are somehow mixed, similar to most of the other colors. A much better situation takes place after the application of the CLAHE algorithm, leading to the results shown in Figure 7 where the separating line can be proposed at 6.4. Nevertheless, some of the samples are still incorrectly classified, e.g., Nos. 1 (white), 26 (brown), and 40 (salmon pink color); many moderately low quality pink samples (Nos. 43–48); two dark green samples; or some of the dark red and blue samples.

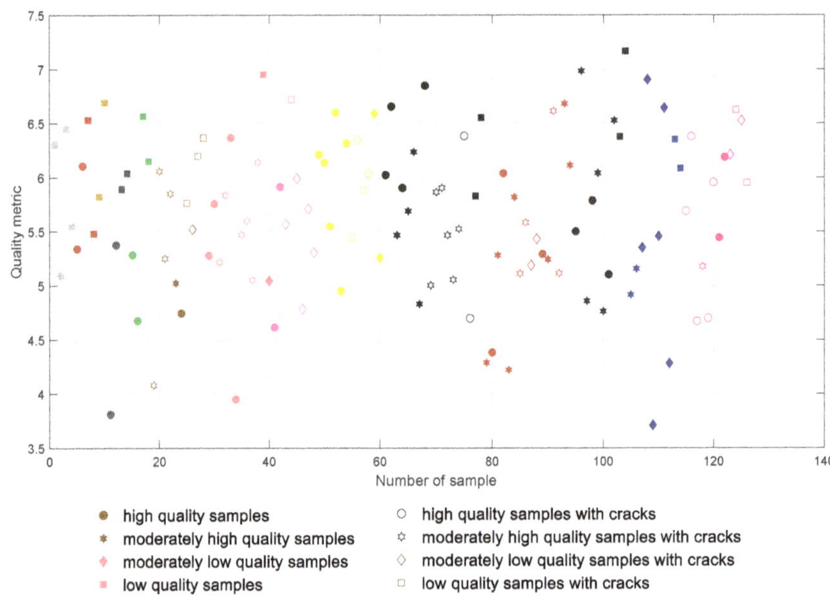

Figure 6. Results obtained for the entropy calculated for depth maps without preprocessing.

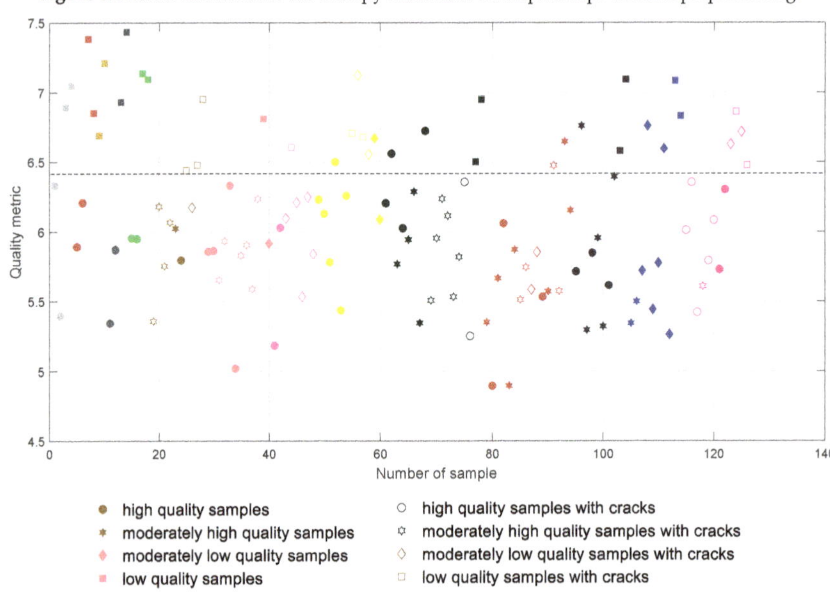

Figure 7. Results obtained for the entropy calculated for depth maps after the CLAHE application.

Although the direct application of the proposed metric does not ensure a satisfactory classification, as shown in Figure 8, its use for the depth map images subjected to CLAHE based preprocessing leads to the promising results illustrated in Figure 9. Applying the proposed solution, most of the samples can be correctly classified, although some individual samples seem to be troublesome, i.e., Nos. 1, 26, 46, 48, 87 and 88, as well as four of the blue samples. Nevertheless, incorrect classification takes place mainly for some of the moderately high or moderately low quality samples and in some cases differences between the visual quality of such surfaces are quite small, as illustrated in Figure 10.

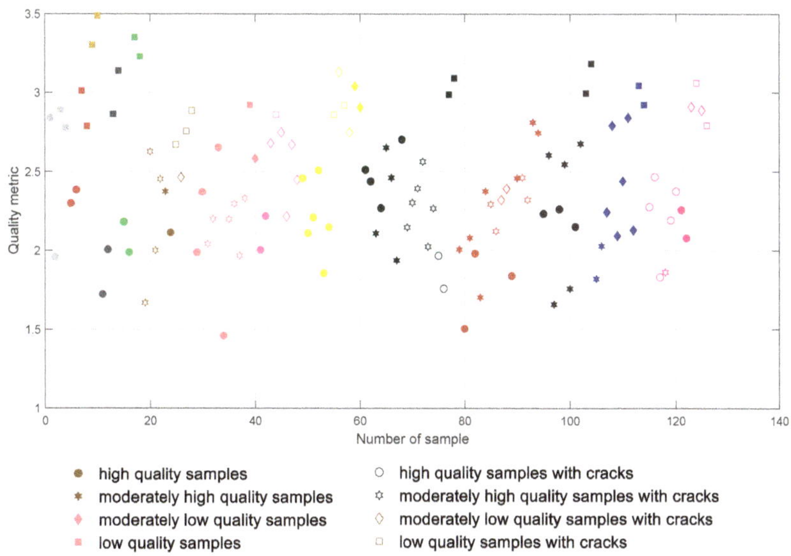

Figure 8. Results obtained for the proposed metric calculated for depth maps without preprocessing.

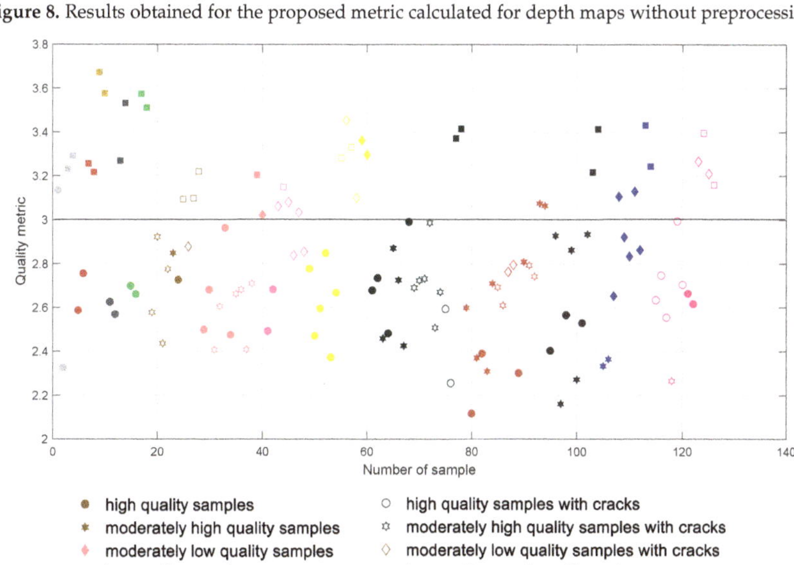

Figure 9. Results obtained for the proposed metric calculated for depth maps after the CLAHE application.

To illustrate the advantages of the proposed method numerically, some typical classification metrics were calculated for the proposed method. Assuming the classification of the samples into two classes—high quality (positives) and low quality (negatives)—true positives (TP) can be defined as high and moderately high quality samples classified as "good". On the other hand, true negatives

(TN) are properly classified distorted surfaces (low and moderately low quality). Consequently, false negatives (FN) and false positives (FP) may be assigned to improperly classified samples. Therefore, some of the most popular classification metrics, namely F-Measure (F1-score), specificity, sensitivity and accuracy, may be calculated as:

$$\mathrm{FM} = \frac{2 \cdot \mathrm{TP}}{2 \cdot \mathrm{TP} + \mathrm{FP} + \mathrm{FN}}, \tag{3}$$

$$\mathrm{Specificity} = \frac{\mathrm{TN}}{\mathrm{FP} + \mathrm{TN}}, \tag{4}$$

$$\mathrm{Sensitivity} = \frac{\mathrm{TP}}{\mathrm{TP} + \mathrm{FN}}, \tag{5}$$

and

$$\mathrm{Accuracy} = \frac{\mathrm{TP} + \mathrm{TN}}{\mathrm{TP} + \mathrm{FP} + \mathrm{FN} + \mathrm{TN}}. \tag{6}$$

Their values obtained for the results illustrated in Figures 6–9 are shown in Table 1.

Table 1. The values of classification metrics obtained using the proposed method in comparison to some other approaches presented in the paper.

	Number of Samples				Classification Metrics			
Method	**TP**	**TN**	**FP**	**FN**	**F-Measure**	**Specificity**	**Sensitivity**	**Accuracy**
Global Entropy (Figure 6)	65	20	26	15	0.760	0.435	0.813	0.675
Entropy + CLAHE (Figure 7)	74	32	14	6	0.881	0.696	0.925	0.841
Q without CLAHE (Figure 8)	75	36	10	5	0.909	0.783	0.938	0.881
Q with CLAHE (Figure 9)	**77**	**37**	**9**	**3**	**0.928**	**0.804**	**0.963**	**0.905**

The reason of some issues for the proper automatic quality assessment of blue prints is related to the specific type of distortions that are present on their surfaces. Since in these cases the over-extruding of the filament may be observed, resulting in filling the groove between the consecutive layers rather than causing the presence of holes, the expected entropy values should decrease. Therefore, the proposed method cannot be efficiently applied for such specific type of distortions. However, in many applications, the inability to recognize individual layers could be considered as an advantage rather than a problem. An illustration of the surface of one such sample is presented in Figure 10. As can be observed, the visual quality of the samples in the top row is slightly lower than the bottom ones despite their identical subjective assessment.

Figure 10. Exemplary moderate quality samples being problematic for the proposed method.

5. Conclusions

The automatic objective quality assessment method of 3D printed surfaces based on the analysis of their depth maps acquired using a fringe patterns based 3D scanner provides encouraging results. The validity and usefulness of the proposed approach was confirmed by the F-Measure exceeding 92% and classification accuracy over 90% for 126 testing samples. In contrast to some machine learning or neural network based solutions which might be potentially used, the presented approach does not require any training process. Potential issues resulting from some imperfections related to therelative position and orientation of the scanned samples according to the 3D scanner can be solved by the application of adaptive histogram equalization using the CLAHE method leading to the correction of the brightness of the depth map images.

The proposed method may be used for further actions such as aborting the printing or correction of the printed surface. The time necessary for printing a single layer—depending on its height—is typically several times longer than the analysis of a single image. Even considering the additional time necessary for 3D scanning using the fringe pattern approach, the total processing time would be short enough to take appropriate action. In practical applications, there is no need to evaluate the quality of the highest layer immediately, especially assuming the location of the camera providing a side view, and therefore some (reasonably short) delays are not critical. In some applications, even without the analysis of the depth maps, the use of a single camera may be useful for monitoring and preventing, e.g., fire caused by heated filament. Nevertheless, these issues have been considered as separate problems, appropriate for further research, and therefore are not analyzed in the paper in details.

Our planned future work will concentrate on the verification of possibilities of data fusion, considering depth maps and photos of the printed surfaces, towards even better classification accuracy. Another direction of our further research is the idea of hybrid quality metrics combining some previously proposed methods.

Author Contributions: J.F. and M.T. worked under the supervision of K.O., J.F. prepared the 3D printed samples, J.F. and M.G. prepared and calibrated the 3D scanning environment and scanned the samples. M.T. extracted and prepared the depth maps for analysis. J.F. and K.O. designed the methodology, proposed and implemented the algorithm and performed the calculations. K.O. wrote the final version of the paper.

Funding: This research received no external funding.

Conflicts of Interest: The authors declare no conflict of interest.

Abbreviations

The following abbreviations are used in this manuscript:

ABS	Acrylonitrile Butadiene Styrene
CAD	Computer Aided Design
CLAHE	Contrast Limited Adaptive Histogram Equalization
CNC	Computerized Numerical Control
FDM	Fused Deposition Modeling
GLCM	Gray-Level Co-occurrence Matrix
IQA	image quality assessment
OCT	Optical Coherence Tomography
SSIM	Structural Similarity
STL	Stereolithography file format
SVM	Support Vector Machines
PLA	Polyactic Acid

References

1. Stephens, B.; Azimi, P.; Orch, Z.E.; Ramos, T. Ultrafine particle emissions from desktop 3D printers. *Atmos. Environ.* **2013**, *79*, 334–339. [CrossRef]

2. Azimi, P.; Zhao, D.; Pouzet, C.; Crain, N.E.; Stephens, B. Emissions of Ultrafine Particles and Volatile Organic Compounds from Commercially Available Desktop Three-Dimensional Printers with Multiple Filaments. *Environ. Sci. Technol.* **2016**, *50*, 1260–1268. [CrossRef] [PubMed]
3. Zeltmann, S.E.; Gupta, N.; Tsoutsos, N.G.; Maniatakos, M.; Rajendran, J.; Karri, R. Manufacturing and Security Challenges in 3D Printing. *JOM* **2016**, *68*, 1872–1881. [CrossRef]
4. Busch, S.F.; Weidenbach, M.; Fey, M.; Schäfer, F.; Probst, T.; Koch, M. Optical Properties of 3D Printable Plastics in the THz Regime and their Application for 3D Printed THz Optics. *J. Infrared Millim. Terahertz Waves* **2014**, *35*, 993–997. [CrossRef]
5. Straub, J. Automated testing and quality assurance of 3D printing/3D printed hardware: Assessment for quality assurance and cybersecurity purposes. In Proceedings of the 2016 IEEE AUTOTESTCON, Anaheim, CA, USA, 12–15 September 2016; pp. 1–5.
6. Fang, T.; Jafari, M.A.; Bakhadyrov, I.; Safari, A.; Danforth, S.; Langrana, N. Online defect detection in layered manufacturing using process signature. In Proceedings of the IEEE International Conference on Systems, Man and Cybernetics, San Diego, CA, USA, 14 October 1998; Volume 5, pp. 4373–4378.
7. Fang, T.; Jafari, M.A.; Danforth, S.C.; Safari, A. Signature analysis and defect detection in layered manufacturing of ceramic sensors and actuators. *Mach. Vis. Appl.* **2003**, *15*, 63–75. [CrossRef]
8. Cheng, Y.; Jafari, M.A. Vision-Based Online Process Control in Manufacturing Applications. *IEEE Trans. Autom. Sci. Eng.* **2008**, *5*, 140–153. [CrossRef]
9. Szkilnyk, G.; Hughes, K.; Surgenor, B. Vision Based Fault Detection of Automated Assembly Equipment. In Proceedings of the ASME/IEEE International Conference on Mechatronic and Embedded Systems and Applications, Parts A and B, Washington, DC, USA, 28–31 August 2011; Volume 3, pp. 691–697.
10. Chauhan, V.; Surgenor, B. A Comparative Study of Machine Vision Based Methods for Fault Detection in an Automated Assembly Machine. *Procedia Manuf.* **2015**, *1*, 416–428. [CrossRef]
11. Chauhan, V.; Surgenor, B. Fault detection and classification in automated assembly machines using machine vision. *Int. J. Adv. Manuf. Technol.* **2017**, *90*, 2491–2512. [CrossRef]
12. Laucka, A.; Andriukaitis, D. Research of the Defects in Anesthetic Masks. *Radioengineering* **2015**, *24*, 1033–1043. [CrossRef]
13. Fok, K.Y.; Cheng, C.; Ganganath, N.; Iu, H.; Tse, C.K. An ACO-Based Tool-Path Optimizer for 3D Printing Applications. *IEEE Trans. Ind. Inform.* **2018**. [CrossRef]
14. Straub, J. Initial Work on the Characterization of Additive Manufacturing (3D Printing) Using Software Image Analysis. *Machines* **2015**, *3*, 55–71. [CrossRef]
15. Straub, J. Alignment issues, correlation techniques and their assessment for a visible light imaging-based 3D printer quality control system. In Proceedings of the SPIE Proceedings—Image Sensing Technologies: Materials, Devices, Systems, and Applications III, Baltimore, MD, USA, 26 May 2016; Volume 9854.
16. Straub, J. Identifying positioning-based attacks against 3D printed objects and the 3D printing process. In Proceedings of the SPIE—Pattern Recognition and Tracking XXVII, Anaheim, CA, USA, 16 June 2017; Volume 10203.
17. Straub, J. 3D printing cybersecurity: detecting and preventing attacks that seek to weaken a printed object by changing fill level. In Proceedings of the SPIE—Dimensional Optical Metrology and Inspection for Practical Applications VI, Anaheim, CA, USA, 9 June 2017; Volume 10220.
18. Straub, J. Physical security and cyber security issues and human error prevention for 3D printed objects: Detecting the use of an incorrect printing material. In Proceedings of the SPIE—Dimensional Optical Metrology and Inspection for Practical Applications VI, Anaheim, CA, USA, 15 June 2017; Volume 10220.
19. Tourloukis, G.; Stoyanov, S.; Tilford, T.; Bailey, C. Data driven approach to quality assessment of 3D printed electronic products. In Proceedings of the 38th International Spring Seminar on Electronics Technology (ISSE), Eger, Hungary, 6–10 May 2015; pp. 300–305.
20. Makagonov, N.G.; Blinova, E.M.; Bezukladnikov, I.I. Development of visual inspection systems for 3D printing. In Proceedings of the 2017 IEEE Conference of Russian Young Researchers in Electrical and Electronic Engineering (EIConRus), St. Petersburg, Russia, 1–3 February 2017; pp. 1463–1465.
21. Holzmond, O.; Li, X. In situ real time defect detection of 3D printed parts. *Addit. Manuf.* **2017**, *17*, 135–142. [CrossRef]
22. Scime, L.; Beuth, J. Anomaly detection and classification in a laser powder bed additive manufacturing process using a trained computer vision algorithm. *Addit. Manuf.* **2018**, *19*, 114–126. [CrossRef]

23. Delli, U.; Chang, S. Automated Process Monitoring in 3D Printing Using Supervised Machine Learning. *Procedia Manuf.* **2018**, *26*, 865–870.
24. Sitthi-Amorn, P.; Ramos, J.E.; Wangy, Y.; Kwan, J.; Lan, J.; Wang, W.; Matusik, W. MultiFab: A Machine Vision Assisted Platform for Multi-material 3D Printing. *ACM Trans. Graph.* **2015**, *34*, 129. [CrossRef]
25. Pajor, M.; Grudziński, M. Intelligent Machine Tool – Vision Based 3D Scanning System for Positioning of the Workpiece. *Solid State Phenom.* **2015**, *220–221*, 497–503. [CrossRef]
26. Pajor, M.; Grudziński, M.; Marchewka, Ł. Stereovision system for motion tracking and position error compensation of loading crane. *AIP Conf. Proc.* **2018**, *2029*, 020050.
27. Okarma, K.; Fastowicz, J. No-Reference Quality Assessment of 3D Prints Based on the GLCM Analysis. In Proceedings of the 2016 21st International Conference on Methods and Models in Automation and Robotics (MMAR), Międzyzdroje, Poland, 29 August–1 September 2016; pp. 788–793.
28. Fastowicz, J.; Okarma, K. Texture Based Quality Assessment of 3D Prints for Different Lighting Conditions. In Proceedings of the Computer Vision and Graphics: International Conference, ICCVG 2016, Warsaw, Poland, 19–21 September 2016; Chmielewski, L.J., Datta, A., Kozera, R., Wojciechowski, K., Eds.; Springer International Publishing: Cham, Switzerland, 2016; Volume 9972, pp. 17–28.
29. Okarma, K.; Fastowicz, J.; Tecław, M. Application of Structural Similarity Based Metrics for Quality Assessment of 3D Prints. In Proceedings of the Computer Vision and Graphics: International Conference, ICCVG 2016, Warsaw, Poland, 19–21 September 2016; Chmielewski, L.J., Datta, A., Kozera, R., Wojciechowski, K., Eds.; Springer International Publishing: Cham, Switzerland, 2016; hlLNCS, Volume 9972, pp. 244–252.
30. Wang, Z.; Bovik, A.; Sheikh, H.; Simoncelli, E. Image quality assessment: From error measurement to Structural Similarity. *IEEE Trans. Image Proc.* **2004**, *13*, 600–612. [CrossRef]
31. Fastowicz, J.; Bąk, D.; Mazurek, P.; Okarma, K. Estimation of Geometrical Deformations of 3D Prints Using Local Cross-Correlation and Monte Carlo Sampling. In *Image Processing and Communications Challenges 9: 9th International Conference, IP&C 2017 Bydgoszcz, Poland, September 2017*; AISC; Choraś, M., Choraś, R.S., Eds.; Springer International Publishing: Cham, Switzerland, 2018; Volume 681, pp. 67–74.
32. Fastowicz, J.; Okarma, K. Entropy Based Surface Quality Assessment of 3D Prints. In *Artificial Intelligence Trends in Intelligent Systems: Proceedings of the 6th Computer Science On-Line Conference 2017 (CSOC2017), Vol 1*; AISC; Silhavy, R., Senkerik, R., Kominkova Oplatkova, Z., Prokopova, Z., Silhavy, P., Eds.; Springer International Publishing: Cham, Switzerland, 2017; Volume 573, pp. 404–413.
33. Zuiderveld, K. Contrast Limited Adaptive Histogram Equalization. In *Graphics Gems IV*; Heckbert, P.S., Ed.; Academic Press Professional, Inc.: San Diego, CA, USA, 1994; pp. 474–485.
34. Okarma, K.; Fastowicz, J. Color Independent Quality Assessment of 3D Printed Surfaces Based on Image Entropy. In *Proceedings of the 10th International Conference on Computer Recognition Systems CORES 2017*; Kurzynski, M., Wozniak, M., Burduk, R., Eds.; Springer International Publishing: Cham, Switzerland, 2018; Volume 578, pp. 308–315.

entropy

MDPI

Article

Nonrigid Medical Image Registration Using an Information Theoretic Measure Based on Arimoto Entropy with Gradient Distributions

Bicao Li [1], Huazhong Shu [2], Zhoufeng Liu [1], Zhuhong Shao [3], Chunlei Li [1,*], Min Huang [4] and Jie Huang [1]

1 School of Electronic and Information Engineering, Zhongyuan University of Technology, Zhengzhou 450007, China; lbc@zut.edu.cn (B.L.); liuzhoufeng@zut.edu.cn (Z.L.); 3983@zut.edu.cn (J.H.)
2 Laboratory of Image Science and Technology, School of Computer Science and Engineering, Southeast University, Nanjing 210096, China; shu.list@seu.edu.cn
3 College of Information Engineering, Capital Normal University, Beijing 100048, China; zhshao@cnu.edu.cn
4 School of Computer and Communication Engineering, Zhengzhou University of Light Industry, Zhengzhou 450002, China; huangmin1998@126.com
* Correspondence: 5540@zut.edu.cn

Received: 12 December 2018; Accepted: 14 February 2019; Published: 18 February 2019

Abstract: This paper introduces a new nonrigid registration approach for medical images applying an information theoretic measure based on Arimoto entropy with gradient distributions. A normalized dissimilarity measure based on Arimoto entropy is presented, which is employed to measure the independence between two images. In addition, a regularization term is integrated into the cost function to obtain the smooth elastic deformation. To take the spatial information between voxels into account, the distance of gradient distributions is constructed. The goal of nonrigid alignment is to find the optimal solution of a cost function including a dissimilarity measure, a regularization term, and a distance term between the gradient distributions of two images to be registered, which would achieve a minimum value when two misaligned images are perfectly registered using limited-memory Broyden–Fletcher–Goldfarb–Shanno (L-BFGS) optimization scheme. To evaluate the test results of our presented algorithm in non-rigid medical image registration, experiments on simulated three-dimension (3D) brain magnetic resonance imaging (MR) images, real 3D thoracic computed tomography (CT) volumes and 3D cardiac CT volumes were carried out on *elastix* package. Comparison studies including mutual information (MI) and the approach without considering spatial information were conducted. These results demonstrate a slight improvement in accuracy of non-rigid registration.

Keywords: Arimoto entropy; free-form deformations; normalized divergence measure; gradient distributions; nonextensive entropy; non-rigid registration

1. Introduction

Volume registration is an essential task of image processing, especially in medical field, such as aiding diagnosis, surgical applications, and image-guided radiation therapy [1,2]. The images to be registered are generally obtained at different times and from different imaging sensors, namely, multi-modality imaging. Different medical imaging modalities could provide various and complementary information. For instance, CT and MRI display the anatomic structures of an organ, while positron emission tomography (PET) and single photon emission computed tomography (SPECT) provide the functional and metabolic information. Therefore, in clinic, these multi-modal volumes are often registered and fused together, in this way, much complementary information derived from different modalities are supplied with the physicians to improve the diagnosis accuracy and assessment efficiency of lesion progression.

Image registration is related by the process to find the optimal mapping function between two images to be aligned [3]. In recent years, the registration algorithms using the similarity based on information theory have been attracted more and more attention in medical image registration, among which maximization of MI was early reported for registration of medical images from different modalities by Collignon et al. [4], Maes et al. [5], Wells et al. [6]. Studholme et al. [7] studied a normalized similarity measure called normalized MI (NMI) to tackle the problem of changing field of view (FOV). MI and NMI are both estimated by the probability distributions of images to be registered. Besides, the concept of cumulative probability distributions was introduced into image registration, and cumulative residual entropy (CRE) was investigated in [8]. Additionally, the relations between CRE and Shannon entropy were ulteriorly researched in [9], and a comparison with MI was reported in [10]. In this new measure, cumulative density functions (CDF) instead of probability density functions (PDF) were adopted to calculate values of the similarity measure, which illustrates a good robustness to noise.

The aforementioned measures based on information theory—such as MI, NMI, and CRE—are constructed by Shannon entropy. Nonetheless, the additivity of Shannon entropy signifies that it is an extensive entropy. However, Antolin et al. [11] pointed out that the extensive entropy does not consider the correlation of two variables. Consequently, they presented a similarity measure exploiting Tsallis entropy. Subsequently, this new divergence was employed to construct a non-rigid registration model for medical images [12,13].

Illuminated by the reference [11], a divergence measure based on Arimoto entropy was presented called the Jensen–Arimoto divergence (JAD) [14]. In [15], some properties, such as the concavity of Arimoto entropy and the boundedness of JAD, have been further investigated. However, the registration method based on JAD does not take the spatial information between voxels into account, which is of significance to medical image registration. This paper aims to present a novel nonrigid registration method of medical images adopting a normalized measure based on Arimoto entropy and gradient distributions. Firstly, the properties of Arimoto entropy and JAD are analyzed, and a distance of gradient distributions is constructed. Secondly, a nonrigid deformation model is chosen and the registration process is formulated by an optimization procedure. In the sequel, the continuous probability distributions are estimated using Parzen window method applying B-splines and the analytical gradient of objective function can be obtained. Finally, the L-BFGS optimization [16] is adopted to obtain the optimal deformation parameters. To assess the performance of our registration framework for medical images, several groups of non-rigid experiments on simulated volumes and real 3D data are implemented.

Our contributions are twofold. Firstly, the related measures based on Shannon entropy do not consider the correlation. Therefore, we present a normalized measure based on Arimoto entropy, a non-extensive entropy, as the dissimilarity measure. Secondly, in the existing measures, such as MI, NMI and JAD, the intensity values are directly exploited to calculate the similarity measure, while the spatial information has been not considered. To take the spatial information between voxels into account, a distance term of gradient distributions is constructed and incorporated into the objective function.

The rest of this work is arranged as follows. In Section 2, the knowledge of information theory was firstly reviewed, and then introduce the Arimoto entropy and JAD measure, constructing gradient distribution distance. We formulate the model of nonrigid registration and give the detailed description of the registration method in Section 3 adopting a normalized measure based on Arimoto entropy with gradient distribution. Section 4 demonstrates the nonrigid test results on 3D MR volumes and real 3D clinical datasets, with the compared results to other registration algorithms illustrated. Finally, we provide the conclusions and perspectives in Section 5.

2. Preliminaries

For this section, we briefly review the theoretical concept of information theory, and then introduce the Arimoto entropy and JAD, along with studying their properties. In addition, the gradient distributions of reference image and float image are constructed and a distance of them is derived.

2.1. Shannon Entropy and Mutual Information

For an arbitrary random variable $X(x_1, x_2, \ldots, x_N)$, with its probability distributions $p(x_1, x_2, \ldots, x_N)$, the Shannon entropy of X is used to measure the amount of average uncertainty included in this random variable,

$$H(X) = -\sum_{i=1}^{N} p(x_i) \log p(x_i) \tag{1}$$

which is also employed to measure the account of information provided by this random variable. Then, considering another random variable Y, $H(X \mid Y)$ is remarked as the conditional entropy of X when Y is known. The reduction in uncertainty due to Y is called the MI. MI of X and Y is defined by

$$I(X,Y) = H(X) - H(X|Y) = \sum_{x}\sum_{y} p(x,y) \log \frac{p(x,y)}{p(x)\cdot p(y)} \tag{2}$$

where $p(x)$ and $p(y)$ denote marginal probability of two random variables, as well as $p(x, y)$ being the joint distribution of them. MI is applied as a measure of the dependence between X and Y. MI is symmetric in X and Y and always nonnegative [17].

2.2. Arimoto Entropy

Arimoto [18] introduced a generalized form of Shannon entropy, Arimoto entropy is defined by

$$A_\alpha(X) = \frac{\alpha}{\alpha-1}\left[1 - \left(\sum_{i=1}^{N} p_i^\alpha\right)^{\frac{1}{\alpha}}\right] \qquad \alpha > 0, \alpha \neq 1 \tag{3}$$

Boekee et al. [19] investigated some significant properties of Arimoto entropy, here, we only exhibit several useful properties as follows.

Non-negativity:

$$A_\alpha(X) \geq 0 \qquad \alpha > 0, \alpha \neq 1 \tag{4}$$

Pseudo-additivity:

$$A_\alpha(X,Y) = A_\alpha(X) + A_\alpha(Y) - \frac{\alpha-1}{\alpha} A_\alpha(X) A_\alpha(Y) \qquad \alpha > 0, \alpha \neq 1 \tag{5}$$

Concavity:

$$A_\alpha(tX_1 + (1-t)X_2) \geq tA_\alpha(X_1) + (1-t)A_\alpha(X_2) \qquad t \in (0,1), \alpha > 0, \alpha \neq 1 \tag{6}$$

Symmetry:

$$A_\alpha(\cdots, p_i, \cdots p_j \cdots) = A_\alpha(\cdots, p_j, \cdots p_i \cdots) \tag{7}$$

Upper bound:

$$A_\alpha(p_1, p_2, \cdots, p_N) \leq A_\alpha\left(\frac{1}{N}, \frac{1}{N}, \cdots, \frac{1}{N}\right) = \frac{\alpha}{\alpha-1}\left[1 - N^{\frac{1-\alpha}{\alpha}}\right] \tag{8}$$

See [19] for the detailed proof of these properties. Property one ensures that Arimoto entropy is non-negative. Its pseudo-addivity illustrates that Arimoto entropy accounts for a non-extensive entropy. The parameter α in (3) accounts for the degree of non-extensivity.

2.3. Jensen Arimoto Divergence

For a random variable X, and $P(p_1, p_2, \ldots, p_N)$ is probability distributions on X. JAD is defined to be [15]

$$JA_\alpha(p_1, p_2, \ldots, p_N) = A_\alpha\left(\sum_{i=1}^{N} \omega_i p_i\right) - \sum_{i=1}^{N} \omega_i A_\alpha(p_i) \qquad \alpha > 0,\ \alpha \neq 1 \tag{9}$$

where $A_\alpha(\cdot)$ denotes Arimoto entropy and ω_i is a weighted vector to constrain $\omega_i \geq 0$ and $\sum_{i=1}^{N} \omega_i = 1$. In the following, we review the properties of JAD.

Proposition 1. *JAD has these properties of non-negativity, symmetry. Also, JAD is identical to 0 when and only when all of the probability distributions are the same as each other. The proof has been reported in [15].*

Li et al. pointed that a distance must fulfill four requirements [20]. JAD does not meet the triangle inequality, so JAD is not a true distance metric. Nonetheless, JAD can still be adopted to measure the disparity among the probability distributions of two random variables.

Proposition 2. *The JA divergence has the maximum when $p_1, p_2, \ldots, p_N$ are degenerate distributions, where $p_i = \delta_{ij} = 1$ when i = j and 0 otherwise. The proof had been provided in [21], and the maximum equals to $A_\alpha(\omega)$.*

According to Proposition 1 and 2, it is obviously observed that JAD is bounded, $0 \leq JA_\alpha(p_1, p_2, \ldots, p_N) \leq A_\alpha(\omega)$.

2.4. Gradient Distributions Distance

Given the reference image R and float image F, ∇F, and ∇R represent the gradients of F and R, along with $q(\nabla F)$ and $p(\nabla R)$ representing the gradient distributions of F and R. Kullback–Leibler divergence (KLD) can be used to calculate the distance of $q(\nabla F)$ and $p(\nabla R)$ as

$$KLD(q||p) = \sum_{\mathrm{x}} q(\nabla F(\mathrm{x})) \log \frac{q(\nabla F(\mathrm{x}))}{p(\nabla R(\mathrm{x}))} \tag{10}$$

where x denotes any point of image gradient, $\mathrm{x} = [x, y, z]^{\mathrm{T}}$. KLD is also known as relative entropy, measuring the diversity of two probability distributions. In (10), we use the convention that $0 \cdot \log(0/0) = 0$. In other word, when R and F are completely registered, KLD between gradient distributions defined in (10) achieves the minimum value. Considering the spatial transformation T_μ, μ denoted by the parameters of transformation model, the gradient distribution distance of F and the transformed F is given by

$$KLD(\nabla F(T_\mu(\mathrm{x}))||\nabla R(\mathrm{x})) = \sum_{d}\sum_{\mathrm{x}} q(\nabla F_d(T_\mu(\mathrm{x}))) \log \frac{q(\nabla F_d(T_\mu(\mathrm{x})))}{p(\nabla R_d(\mathrm{x}))} \tag{11}$$

where $\nabla F_d(T_\mu(\mathrm{x}))$ and $\nabla R_d(\mathrm{x})$ represent the gradients of transformed float image and reference image, and d is the dimension of the images to be registered. Subsequently, the Parzen method is employed to estimate the gradient distributions.

Denote $\beta^{(0)}$ and $\beta^{(3)}$ by zero-order and three-order B-spline, respectively. In the process of image registration, the transformation parameters μ does not affect the reference image, the gradient of reference image is also constant in registration process. Consequently, the gradient of reference image can be calculated before registration to improve the implementation efficiency. In addition, a zero-order

B-spline is exploited to estimate the gradient distributions $\nabla R_d(\mathrm{x})$ of reference image $R(\mathrm{x})$, with the probability density function of $\nabla R_d(\mathrm{x})$ defined by

$$p(\nabla R_d(\mathrm{x})) = \widetilde{p}_d(r_i) = \frac{1}{V}\sum_{\mathrm{x}\in\Omega} \beta^{(0)}\left(r_i - \frac{\nabla R_d(\mathrm{x}) - \nabla R_d^0}{\Delta b_R}\right) \tag{12}$$

where Ω is volume domain to estimate probabilities. V denotes the number of voxels of Ω domain, as well as d being the image dimension. r_i represents intensity levels of $\nabla R_d(\mathrm{x})$, and '∇' is the gradient operator. The three-order B-spline is adopted to compute the gradient distributions $\nabla F_d(T_\mu(\mathrm{x}))$ of transformed float image, with the probability density function of $\nabla F_d(T_\mu(\mathrm{x}))$ shown as

$$q(\nabla F_d(T_\mu(\mathrm{x}))) = \widetilde{q}_d(f_j) = \frac{1}{V}\sum_{x\in\Omega} \beta^{(3)}\left(f_j - \frac{\nabla F_d(T_\mu(\mathrm{x})) - \nabla F_d^0}{\Delta b_F}\right) \tag{13}$$

In Equations (12) and (13), Δb_R and Δb_F are the widths of bins. ∇R_d^0 and ∇F_d^0 is the minimum values in two images, and f_j represents intensity levels of $\nabla F_d(T_\mu(\mathrm{x}))$. Substituting (12) and (13) into (11), we can obtain the following formula.

$$\begin{aligned} KLD(\nabla F(T_\mu(\mathrm{x}))||\nabla R(\mathrm{x})) = & \sum_d\sum_{\mathrm{x}} \widetilde{q}_d(f_j)\log\frac{\widetilde{q}_d(f_j)}{\widetilde{p}_d(r_i)} \\ & = \sum_d\left[H(\widetilde{q}_d) - \sum_{\mathrm{x}}\widetilde{q}_d(f_j)\log\widetilde{p}_d(r_i)\right] \end{aligned} \tag{14}$$

where H is the Shannon entropy. In this paper, KLD of gradient distributions will be applied as a distance term in medical image registration, regularizing that the gradient distribution of float image is similar to the gradient distribution of reference image.

3. Description of Proposed Nonrigid Registration Method

The details of the registration method using a normalized information theoretic measure based on Arimoto entropy with gradient distributions is described. Firstly, an appropriate transformation model needs to be selected, where the registration criteria (objective function or cost function) and optimization algorithm is applied to optimize the criteria. Finally, the images to be registered are aligned using an optimal solution obtained by the optimization scheme. Figure 1 displayed the block diagram of our nonrigid registration framework.

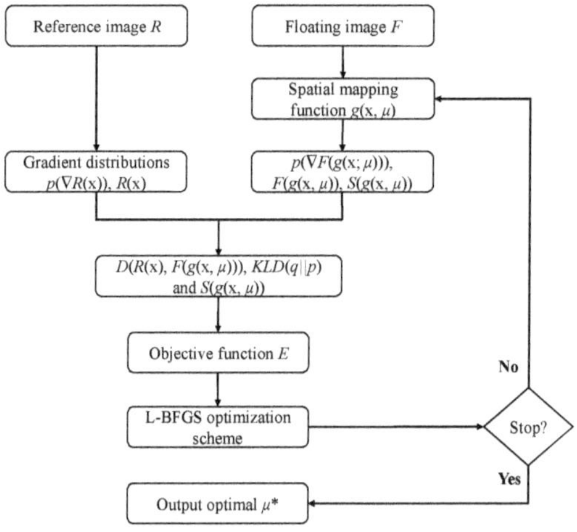

Figure 1. Block diagram of our registration algorithm.

3.1. Formulation

Figure 2a,b depict the corresponding planes in two 3D MR images, which account for T1-weighted MR and T2-weighted MR. Non-rigid registration is formulated as the process of searching for the optimal spatial deformation function of reference image R and float image F as

$$\mathrm{y} = g(\mathrm{x};\mu) \tag{15}$$

where $g(\mathrm{x};\mu)$ is the deformation function, μ denotes the vector of deformation parameters, with x and y being the coordinates of arbitrary point in R and F, respectively. We can formulate the non-rigid registration of F to R as

$$T^* = \underset{\mu}{\mathrm{argmin}} D(F(\mathrm{x}) \circ g(\mathrm{x};\mu), R(\mathrm{x})) = \underset{\mu}{\mathrm{argmin}} D(F(g(\mathrm{x};\mu)), R(\mathrm{x})) \tag{16}$$

where D represents a dissimilarity measure that can achieves its minimum in registration of $R(\mathrm{x})$ and $F(g(\mathrm{x};\mu))$.

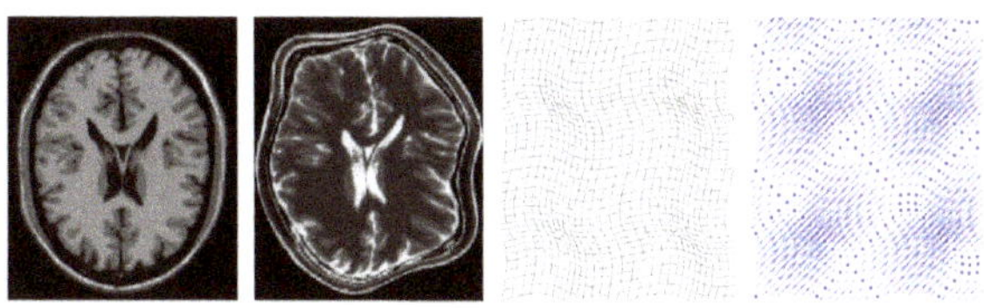

(a) (b) (c) (d)

Figure 2. (**a**) MR T1 image; (**b**) MR T2 image; (**c**) deformation field; (**d**) deformation vector.

However, the process of image registration is an ill-posed issue, and a penalty term need to be incorporated to obtain a smooth transformation. Considering the regularization term, the cost function E is expressed by

$$E = D(F(g(\mathrm{x};\mu)), R(\mathrm{x})) + \lambda S(g(\mathrm{x};\mu)) \tag{17}$$

3.2. Transformation Model

Clinically, there exist some large deformations between medical images. Therefore, a nonrigid transformation is generally employed to deal with organ or tissue deformation. Figure 2c,d display the deformation fields and vectors of the nonrigid transformation between Figure 2a,b. The free-form deformations (FFD) model can preferably described local deformations between medical images. Therefore, cubic B-splines are employed to constructed FFD model [22] to simulate this elastic deformation.

In a 3D image, Φ is a mesh with the size of $n_x \times n_y \times n_z$, as well as the control points $[\omega_i, \omega_j, \omega_k]^T$, and δ is the size of spacing. Then, 3D deformation function of $\mathrm{x} = [x, y, z]^T$ could be defined as

$$g(\mathrm{x};\mu) = \sum_{ijk} \mu_{ijk} \beta^{(3)}\left(\frac{x-\omega_i}{\delta}\right) \beta^{(3)}\left(\frac{y-\omega_j}{\delta}\right) \beta^{(3)}\left(\frac{z-\omega_k}{\delta}\right) \tag{18}$$

where μ_{ijk} is the vector of deformation coefficients, and $\beta^{(3)}$ is the third-order B spine function.

3.3. Registration Criteria

The dissimilarity measure D is the most significant part of objective functions, which is used to measure the difference between two images to be registered. When D achieve the minimum value, the similarity of two volumes is maximized, resulting in the two volumes completely registered. We have employed JAD as similarity measure to register medical images in the presence of non-rigid transformation, in which a negative sign is assigned to the JAD to construct the dissimilarity measure [15]. In this paper, according to *Proposition 2* in Section 2.3, a normalized dissimilarity measure based on JAD is introduced. Its definition is given as

$$D(F(g(x;\mu)), R(x)) = 1 - \frac{JA_\alpha(F(g(x;\mu)), R(x))}{A_\alpha(\omega)} \tag{19}$$

In [15], JAD is expressed by

$$\begin{aligned} JA_\alpha(F(g(x;\mu)), R(x)) &= \tfrac{\alpha}{1-\alpha}\left\{ \left[\sum_{j=1}^{M}\left[\sum_{i=1}^{M} p(r_i)p(f_j|r_i)\right]^\alpha\right]^{\frac{1}{\alpha}} - \sum_{i=1}^{M} p(r_i)\left[\sum_{j=1}^{M} p(f_j|r_i)^\alpha\right]^{\frac{1}{\alpha}} \right\} \\ &= \tfrac{\alpha}{1-\alpha}\left\{ \left[\sum_{j=1}^{M} p(f_j)^\alpha\right]^{\frac{1}{\alpha}} - \sum_{i=1}^{M}\left[\sum_{j=1}^{M} p(r_i, f_j)^\alpha\right]^{\frac{1}{\alpha}} \right\} \end{aligned} \tag{20}$$

where $f = (f_1, f_2, \ldots, f_M)$ and $r = (r_1, r_2, \ldots, r_M)$ are the intensity values in $F(g(x;\mu))$ and $R(x)$. Also, M is bins number. Also, $p(f_j \mid r_i)$ is the conditional probability. JAD and $A_\alpha(\omega)$ are substituted into (19). In consequence, the dissimilarity measure D in (19) is rewritten as

$$D(F(g(x;\mu)), R(x)) = 1 - \left\{ \sum_{i=1}^{M}\left[\sum_{j=1}^{M} p(r_i, f_j)^\alpha\right]^{\frac{1}{\alpha}} - \left[\sum_{j=1}^{M} p(f_j)^\alpha\right]^{\frac{1}{\alpha}} \right\} / \left(1 - M^{\frac{1-\alpha}{\alpha}}\right) \tag{21}$$

where M is the number of bins. To address the problem of nonrigid registration, the smooth deformation needs to be acquired by regularizing the deformation model. Incorporated with the regularization term, the objective function E is rewritten by

$$E(F(g(x;\mu)), R(x)) = D(F(g(x;\mu)), R(x)) + \lambda S(g(x;\mu)) \tag{22}$$

where D is the dissimilarity measure defined in (21), and S is the regularization term, with its expression given as follows.

$$\begin{aligned} S(g(x;\mu)) &= \tfrac{1}{V}\int_0^X\int_0^Y\int_0^Z \left[\left(\tfrac{\partial^2 g(x;\mu)}{\partial x^2}\right)^2 + \left(\tfrac{\partial^2 g(x;\mu)}{\partial y^2}\right)^2 + \left(\tfrac{\partial^2 g(x;\mu)}{\partial z^2}\right)^2 \right. \\ &\left. +2\left(\tfrac{\partial^2 g(x;\mu)}{\partial x \partial y}\right)^2 + 2\left(\tfrac{\partial^2 g(x;\mu)}{\partial x \partial z}\right)^2 + 2\left(\tfrac{\partial^2 g(x;\mu)}{\partial y \partial z}\right)^2 \right] dxdydz, \end{aligned} \tag{23}$$

However, objective function shown in (22) does not take into account the spatial information between voxels. To deal with the issue, the distance between two gradient distributions $q(\nabla F(g(x;\mu)))$ and $p(\nabla R(x))$ displayed in (14) is introduced to (22). As a result, the nonrigid registration process is expressed by

$$\begin{aligned} \mu^* &= \underset{\mu}{\operatorname{argmin}} E(F(g(x;\mu)), R(x)) \\ &= \underset{\mu}{\operatorname{argmin}}\{D(F(g(x;\mu)), R(x)) + \lambda_1 S(g(x;\mu)) + \lambda_2 KLD(\nabla F(g(x;\mu))||\nabla R(x))\} \end{aligned} \tag{24}$$

where KLD represents gradient distribution distance, as well as λ_1 and λ_2 being weight parameters, balancing the tradeoff among a dissimilarity measure D, a regularization S, and a distance term KLD.

3.4. Optimization

Newton–Raphson algorithm has been widely exploited, in which second-order derivatives can show better convergence [23] compared with these strategies based on first-order gradient. L-BFGS [24] does not calculate second-order information. Thus, a high computation efficiency can be achieved. A second-order Taylor approximation [16] of E with respect to μ is given as

$$E(\mu + \Delta\mu) \approx E(\mu) + \Delta\mu^T \cdot \nabla E(\mu) + \frac{1}{2}\Delta\mu^T \cdot \nabla^2 E(\mu) \cdot \Delta\mu, \tag{25}$$

where $\Delta\mu$ is the increment of μ, ∇ is gradient operation. The deformation parameter μ of the L-BFGS optimization algorithm is updated as

$$\mu^{(k+1)} = \mu^{(k)} - \left(H^{(k)}\right)^{-1} \cdot \nabla E(\mu^{(k)}), \tag{26}$$

In the sequel, the derivative of objective function E with respect to μ need to computed.

$$\frac{\partial E}{\partial \mu} = \left[\frac{\partial E}{\partial \mu_1}, \frac{\partial E}{\partial \mu_2}, \ldots, \frac{\partial E}{\partial \mu_n}\right], \tag{27}$$

The pseudo code of our registration approach is displayed in Algorithm 1.

To solve the optimization process, we need calculate the analytical gradient of the objective function E. Traditionally, the probability distributions expressed in (21) was not continuous. Hence, the continuous probability density function (pdf) needs to be estimated by Parzen-window method. The continuous marginal and joint pdfs of two images to be registered have been calculated [15]. The continuous expression of gradient distribution distance has been also provided in Section 2.4. Equalization (18) is substituted into (23), the continuity of smoothness term is acquired. Consequently, the objective function E is continuous and its analytical derivative with respect to μ can be calculated.

Algorithm 1. Nonrigid medical image registration with gradient distributions

Input: Reference image R, floating image F
Output: Optimal deformation parameters μ^*
Set λ_1, λ_2, NMAX, α, M, N, δ, ε
Compute the gradient of R, denote as $\nabla R(\mathrm{x})$ and gradient distributions $p(\nabla R(\mathrm{x}))$
Initialize deformation parameters $\mu^{(0)}$, iteration $k = 0$, $F(g(\mathrm{x}; \mu^{(0)})) = F$, $E(\mu^{(0)}) = 0$
While $|E(\mu^{(k+1)}) - E(\mu^{(k)})| >$ threshold ε or $k <=$NMAX
Obtain the deformed float image $F(g(\mathrm{x}; \mu^{(k+1)}))$ and the regularization $S(g(\mathrm{x}, \mu^{(k+1)}))$
Compute $\nabla F(g(\mathrm{x}; \mu^{(k+1)}))$ and gradient distributions $q(\nabla F(g(\mathrm{x}; \mu^{(k+1)})))$
Estimate the dissimilarity measure D and gradient distributions distance KLD
Calculate objective function $E(\mu^{(k+1)}) = D(R(\mathrm{x}), F(g(\mathrm{x}; \mu^{(k+1)}))) + KLD(q^{(k+1)} || p) + S(g(\mathrm{x}, \mu^{(k+1)}))$
$\mu^{(k+1)} = \mu^{(k)} - (H^{(k)})^{-1} \cdot \nabla E(\mu^{(k)})$
$k = k + 1$
end

Derivative of the Objective Function

The objective function defined in (24) includes dissimilarity measure D, a regularization term S, and a distance term KLD. The derivative of D is deduced as

$$\frac{d[D(F(g(\mathrm{x};\mu)), R(\mathrm{x}))]}{d\mu} = -\frac{1}{A_\alpha(\omega)} \frac{d[JA_\alpha(F(g(\mathrm{x};\mu)), R(\mathrm{x}))]}{d\mu} \tag{28}$$

According to (20), we obtain

$$\frac{d[D(F(g(\mathrm{x};\mu)), R(\mathrm{x}))]}{d\mu} = \frac{1}{\left(1 - M^{\frac{1-\alpha}{\alpha}}\right)} \cdot \left\{\sum_i \sum_j Y \frac{\partial \tilde{p}(r_i, f_j)}{\partial \mu}\right\} \tag{29}$$

$$Y = \left(\sum_j \widetilde{p}(f_j)^{\alpha}\right)^{\frac{1}{\alpha}-1} \widetilde{p}(f_j)^{\alpha-1} - \left(\sum_j \widetilde{p}(f_j|r_i)^{\alpha}\right)^{\frac{1}{\alpha}-1} \widetilde{p}(f_j|r_i)^{\alpha-1} \tag{30}$$

where $\partial\widetilde{p}(f_j, r_i)/\partial\mu$ represents derivative of estimated joint probability. The derivative of $\widetilde{p(f_j, r_i; \mu)}$ is calculated by

$$\begin{aligned} \frac{\partial \widetilde{p(f_j,r_i;\mu)}}{\partial\mu} = & -\frac{1}{N\cdot\Delta b_F}\sum_{x\in\Omega}\beta^{(0)}\left(r_i - \frac{R(x)-R^0}{\Delta b_R}\right) \\ & \times\beta'^{(3)}\left(f_j - \frac{F(g(x;\mu))-F^0}{\Delta b_F}\right) \times \left(\frac{\partial F(s)}{\partial s}|_{s=g(x;\mu)}\right) \times \frac{\partial(g(x;\mu))}{\partial\mu} \end{aligned} \tag{31}$$

with $\beta^{(0)}$ and $\beta'^{(3)}$ being the zero-order B-Splines and the derivative of the three-order B-Splines, respectively. R^0 and F^0 are the minimal intensities in $R(\mathrm{x})$ and $F(g(\mathrm{x};\mu))$, as well as $\partial F(t)/\partial t$ being the gradient of the deformed float image $F(g(\mathrm{x};\mu))$, $\partial(g(x;\mu))/\partial\mu$ can be estimated by FFD model.

To obtain the derivative of the penalty term S, we rewrite (23) as

$$S(g(\mathrm{x};\mu)) = \frac{1}{V}\sum_{\mathrm{x}}\sum_{i,j=1}^{3}\left(\frac{\partial^2 g(\mathrm{x};\mu)}{\partial x_i \partial x_j}\right)^2 \tag{32}$$

where x represents the points in image region, and V denotes the number of pixels. The derivative of S has been provided by Staring and Klein [25],

$$\frac{\partial S(g(\mathrm{x};\mu))}{\partial\mu} = \frac{1}{V}\sum_{\mathrm{x}}\sum_{i,j=1}^{3} 2\left(\frac{\partial^2 g(\mathrm{x};\mu)}{\partial x_i \partial x_j}\right)\frac{\partial}{\partial\mu}\frac{\partial^2 g(\mathrm{x};\mu)}{\partial x_i \partial x_j} \tag{33}$$

where $\partial^2 g(\mathrm{x};\mu)/\partial x_i \partial x_j$ denotes Hessian matrix of deformation function $g(\mathrm{x};\mu)$, $\partial(\partial^2 T/\partial x_i \partial x_j)/\partial\mu$ is the Jacobi of Hessian matrix.

Next, we calculate the derivative of KLD,

$$\frac{d[KLD(\nabla F(g(\mathrm{x};\mu))||\nabla R(\mathrm{x}))]}{d\mu} = \sum_d\sum_{\mathrm{x}}\left(1 + \log\frac{\widetilde{q}_d(f_j)}{\widetilde{p}_d(r_i)}\right)\frac{\partial\widetilde{q}_d(f_j)}{\partial\mu} \tag{34}$$

where $\widetilde{p}_d(r_i)$ and $\widetilde{q}_d(f_j)$ represent the gradient distributions of the transforms float image and reference image, respectively. The derivative of $\widetilde{q}_d(f_j)$ is calculated as

$$\begin{aligned} \frac{\partial\widetilde{q}_d(f_j)}{\partial\mu} = -\frac{1}{V\cdot\Delta b_F} & \sum_{x\in\Omega}\cdot\beta'^{(3)}\left(f_j - \frac{\nabla F_d(g(\mathrm{x};\mu))-\nabla F_d^0}{\Delta b_F}\right) \\ & \cdot\left(\nabla^2 F_d(s)\big|_{s=g(x;\mu)}\right)\cdot\frac{\partial(g(x;\mu))}{\partial\mu} \end{aligned} \tag{35}$$

where $\beta'^{(3)}$ is derivative of three-order B-spline, and $\nabla^2 F_d(t)$ represents the second-order gradient of $F(g(\mathrm{x};\mu))$, as well as $\partial(g(\mathrm{x};\mu))/\partial\mu$ being the derivative of deformation function $g(\mathrm{x};\mu)$ with respect to the parameter μ. Substituting (12), (13), and (35) into (34), we can obtain the derivative of gradient distribution distance. In the terms of (29), (33), and (34), the derivative of E will be easily calculated.

4. Experiments and Results

To evaluate the registration method using the normalized JAD with gradient distribution (NJAD-GD), we designed several groups of tests and performed on simulated and real 3D data, respectively. In Section 4.1, the experimental data is depicted, including simulated and real medical images. The non-rigid registration of simulated MR volumes is performed in Section 4.2. The tests on real 3D thoracic CT images and 3D cardiac data are implemented, and the experimental results are shown in Sections 4.3 and 4.4, respectively. Our nonrigid registration algorithm employing JAD and gradient distributions was implemented in the *elastix* package [26].

4.1. Experimental Data

In this paper, simulated brain MR volumes, thoracic CT volumes and real 3D cardiac CT images were exploited as experimental data. The detailed descriptions of brain MR and 3D thoracic CT images have been reported in [15]. Additionally, non-rigid tests were also performed on twelve 4D cardiac CT sequences acquired from twelve patients. Each of 4D CT sequence consists of 10 3D cardiac CT images, which were obtained from one whole cardiac cycle of one patient. These CT images have 256 × 256 pixels along axial direction. Figure 3 exhibits 10 3D cardiac CT volumes of one 4D CT sequence. It is obviously observed that some elastic deformations are existed between 10 images.

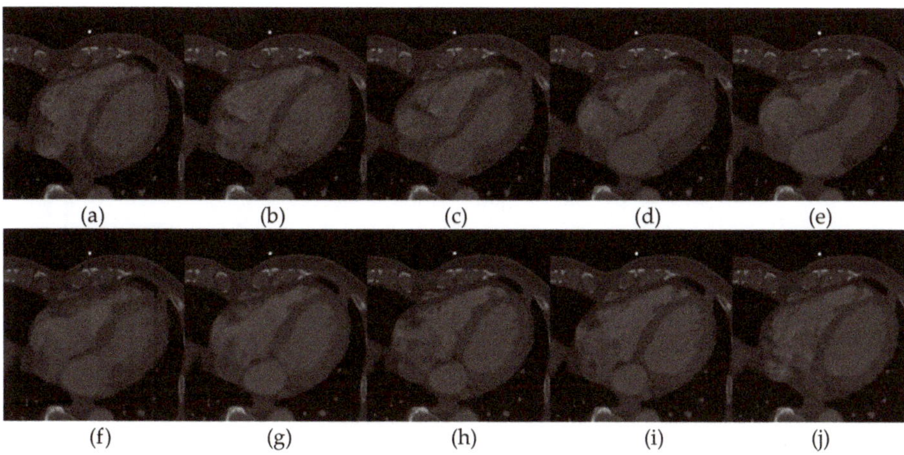

Figure 3. The axis slice of 10 3D cardiac CT images in one 4D sequence. (**a–j**) represent the 10 frames acquired from one whole cardiac cycle of one patient.

4.2. Nonrigid Registration of Simulated Brain Images

Simulated brain volumes were firstly used to design the elastic alignment experiments. Furthermore, we employ a multiresolution hierarchical strategy with three levels to carry out these non-rigid tests. Also, a comparison with JAD without gradient distribution and MI is also reported.

We selected 60 warping indexes (see parameters m of the warping function in [15]), which were yielded randomly from the interval [1,7]. Consequently, 60 float images were produced based on the 60 deformations for each pair of test volumes and 540 nonrigid trials of three pairs of brain MR volumes for NJAD-GD, JAD, and MI algorithms in total.

To assess quantitatively test results of these trails, we exploit registration error as the evaluation standard. Here, the registration error is defined as the difference of true values that can be calculated by warping indexes and the obtained values by optimization strategy. In the registration trails of brain images, the involved parameters are set as follows: the nonextensive parameter $\alpha = 1.5$, bins $M = 16$, the number of random samples $N = 2000$, $\delta = 20 \times 20 \times 20$. The weighting parameters $\lambda_1 = 0.005$ and $\lambda_2 = 0.001$ can provide a good tradeoff among three terms: D, S, and gradient distribution term KLD in the objective function E.

Figure 4 shows the test results of all 540 non-rigid registrations. From Figure 4, the NJAD-GD registration algorithm could result in the lower errors of three pairs of test volumes compared to other two approaches.

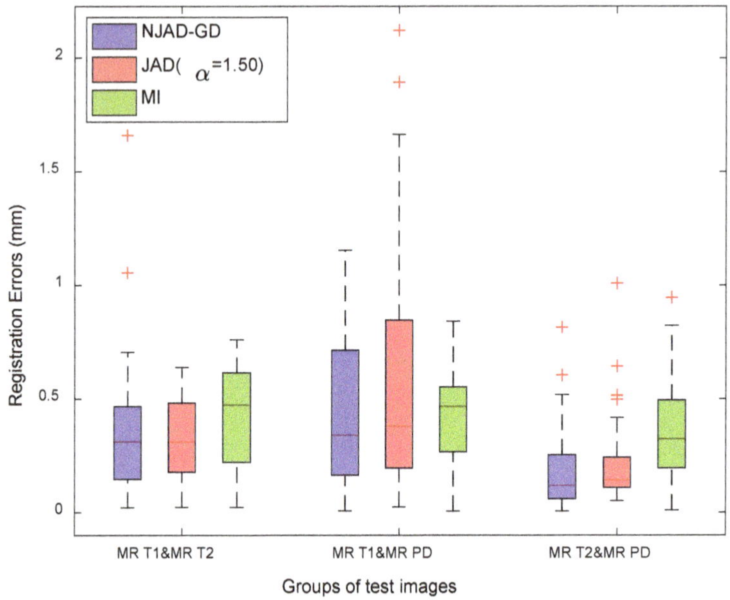

Figure 4. The registration results of the simulated 3D brain MR T1 & MR T2, MR T1 & MR PD, and MR T2 & MR PD volumes using three algorithms. The red color crosses for each box represents these outliers.

4.3. Experiments of 3D Thoracic CT Images

3D thoracic CT volumes were chosen as the test images to carry out non-rigid registrations. These volumes consist of four 4D sequences, and each of them includes 10 3D volumes. A three-level implementation scheme was still employed to decrease registration accuracy and improve computation efficiency.

We denote the 10 volumes from each 4D sequence by T00-T90, in which the maximal inhalation and maximum exhalation are included, with indicated by T10 and T60, respectively. Then, we designed the following experiments: in each 4D sequence, the T60 frame is applied as reference image and the residual nine frames are chosen as float image, leading to nine non-rigid tests. Hence, 36 trails of elastic alignments were yielded in total for four 4D CT sequences. We also compared the results adopting NJAD-GD and JAD without considering spatial information. Finally, 72 elastic registration tests were conducted for two methods. In order to quantify the test errors, target registration error (TRE) and Hausdorff distance meansure (HDM) were calculated. HDM is a widely-used measure to calculate the distance of two clouds of points. In the 3D thoracic CT registration, the manually marked landmarks can be applied to calculate HDM of two images.

Figures 5 and 6 demonstrate the registration results of 72 tests, along with TREs before registration and after alignment. Figure 7 illustrates the box-and-whisker plots of HDM values of four 4D CT sequences. It is observed from these results that the registration errors applying NJAD-GD algorithm are less than these obtained by the method based on JAD without gradient distribution.

In implementation of experiments, the nonextensive parameter $\alpha = 1.5$, bins $M = 16$, the number of random samples $N = 8000$, the spacing of mesh points $\delta = 20 \times 20 \times 20$. The weighting parameters $\lambda_1 = 0.005$ and $\lambda_2 = 0.001$.

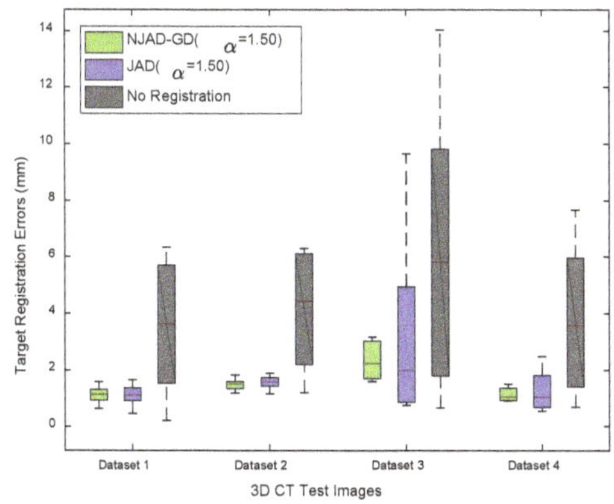

Figure 5. The TREs obtained when employing NJAD-GD algorithm, the registration method based on JAD without gradient distribution.

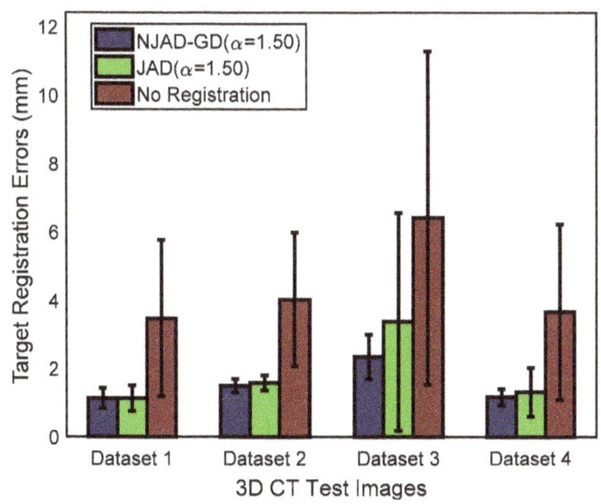

Figure 6. Statistics of TREs before registration and after alignment exploiting the NJAD-GD, JAD methods.

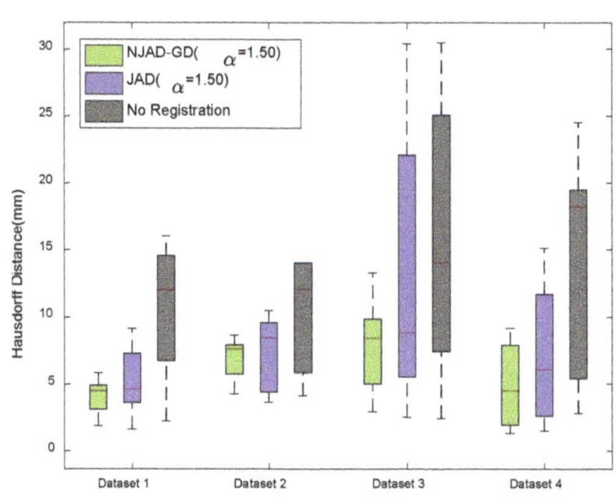

Figure 7. HDMs obtained when employing NJAD-GD algorithm, the registration method based on JAD without gradient distribution.

4.4. Registration of 3D Cardiac CT Image

The cardiac CT data consists of 12 groups of 4D image sequence, and each of which includes 10 3D images acquired from one whole cardiac cycle. One cardiac cycle consists of the phase of systole and phase of diastole. In each 4D CT sequence, two 3D images with the maximum deformation were employed as the test images, and 12 nonrigid registration experiments were carried out adopting NJAD-GD approach. Figure 8 illustrates the checkboard of 12 examples adopting our non-rigid framework ($\alpha = 1.50$) for 12 4D CT sequences. As it can be seen, the registration algorithm based on gradient distribution demonstrates the accuracy results.

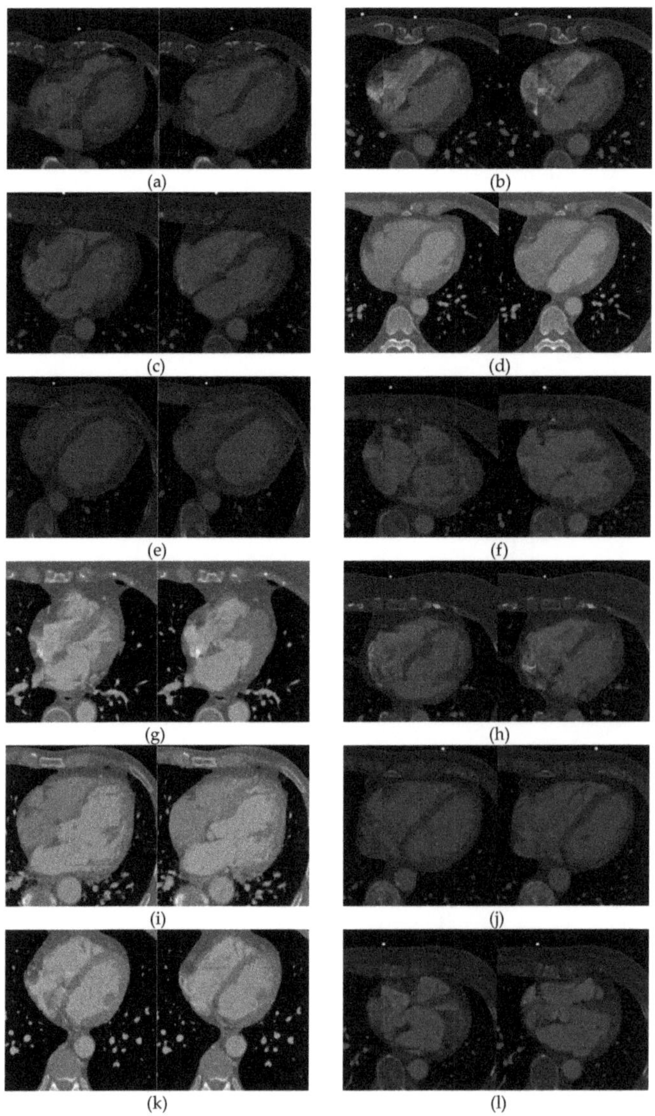

Figure 8. Registration results of 12 groups of 3D cardiac images. (**a**–**l**) display the test results of patient 1 to 12, respectively. In each group, left image represents the checkboard before registration, and the right accounts for the result after registration.

In these tests, a multiresolution scheme with three levels was also exploited to implement these nonrigid registrations. Due to the large deformation, the number of random samples N and the spacing of mesh points δ were set to 10,000 and $10 \times 10 \times 10$, with other parameters being as follows: bins M = 16, the weighting parameters $\lambda_1 = 0.005$ and $\lambda_2 = 0.001$, the maximum number of iterations of the limited memory BFGS scheme NMAX = 200.

5. Conclusions

In this work, we review the definition and properties of Arimoto entropy, with an information measure based on Arimoto entropy, called JAD. The gradient distributions of reference image and float image are constructed and a distance between them is derived. Additionally, a normalized dissimilarity measure based on JAD was presented. A nonrigid registration method exploiting the normalized measure with gradient distributions is proposed.

Arimoto entropy is regarded as a generalized form of the classical Shannon entropy. In the aforementioned section, it is proofed that the JAD measure is equal to MI when α tends to 1. We adopted FFDs as the parameter space for non-rigid registration, along with objective function E including three elements: the normalized JAD as the dissimilarity measure, a regularization to acquire the smooth deformation and a distance term of the gradient distributions.

Author Contributions: B.L. implemented the algorithm, validated the experiments, analyzed the data, and wrote the manuscript. B.L., H.S., Z.S., J.H., and M.H. investigated the project. B.L., H.S. and C.L. conceived and revised the manuscript. B.L., Z.L., Z.S., and C.L. acquired the funding. All authors have read and approved the final manuscript.

Funding: This research was supported by the National Natural Science Foundation of China (no. U1804157, no. 61772576, no. 61601311), the Key Natural Science Foundation of Henan Province (no. 162300410338), Science and technology innovation talent project of Education Department of Henan Province (no. 17HASTIT019), the Henan Science Fund for Distinguished Young Scholars (no. 184100510002), Scientific and technological program projects of Henan Province (no. 192102210127, no. 172102210071), Key scientific research projects of Henan Province (no. 19B510011, no. 17A510006), Project of Beijing Excellent Talents (no. 2016000020124G088), Beijing Municipal Education Research Plan Project (no. SQKM201810028018), Program for Interdisciplinary Direction Team in Zhongyuan University of Technology.

Acknowledgments: The authors acknowledge the *BrainWeb:* Simulated Brain Database (http://brainweb.bic.mni.mcgill.ca/brainweb/) and DIR Validation Data (https://www.creatis.insa-lyon.fr/rio/popi-model) for the data used in this paper. Also, the authors would like to thank the editor-in-chief, the handling editors and the anonymous reviewers for their helpful comments and suggestions.

Conflicts of Interest: The authors declare no conflict of interest.

References

1. Sadozye, A.H.; Reed, N. A review of recent developments in image-guided radiation therapy in cervix cancer. *Curr. Oncol. Rep.* **2012**, *14*, 519–526. [CrossRef] [PubMed]
2. Wang, L.; Gao, X.; Zhou, Z.; Wang, X. Evaluation of four similarity measures for 2D/3D registration in image-guided intervention. *J. Med. Imaging Health Inf.* **2014**, *4*, 416–421. [CrossRef]
3. Song, G.; Han, J.; Zhao, Y.; Wang, Z.; Du, H. A Review on Medical Image Registration as an Optimization Problem. *Curr. Med. Imaging Rev.* **2017**, *13*, 274–283. [CrossRef] [PubMed]
4. Collignon, A.; Maes, F.; Delaer, D.; Vandermeulen, D.; Suetens, P.; Marchal, G. Automated multi-modality image registration based on information theory. *Inf. Process. Med. Imaging* **1995**, *3*, 263–274.
5. Maes, F.; Collignon, A.; Vandermeulen, D.; Marchal, G.; Suetens, P. Multimodality image registration by maximization of mutual information. *IEEE Trans. Med. Imaging* **1997**, *16*, 187–198. [CrossRef] [PubMed]
6. Wells, W.M., III; Viola, P.; Atsumi, H.; Nakajima, S.; Kikinis, R. Multi-modal volume registration by maximization of mutual information. *Med. Image Anal.* **1996**, *1*, 35–51. [CrossRef]
7. Studholme, C.; Hill, D.L.G.; Hawkes, D.J. An overlap invariant entropy measure of 3d medical image alignment. *Pattern Recogn.* **1999**, *32*, 71–86. [CrossRef]
8. Wang, F.; Vemuri, B.C.; Rao, M.; Chen, Y. Cumulative Residual Entropy, A New Measure of Information & its Application to Image Alignment. *IEEE Int. Conf. Comput. Vis.* **2003**. [CrossRef]

9. Rao, M.; Chen, Y.; Vemuri, B.C.; Wang, F. Cumulative residual entropy: A new measure of information. *IEEE Trans. Inf. Theory* **2004**, *50*, 1220–1228. [CrossRef]
10. Wang, F.; Vemuri, B.C. Non-rigid multi-modal image registration using cross-cumulative residual entropy. *Int. J. Comput. Vis.* **2007**, *74*, 201–215. [CrossRef]
11. Antolín, J.; LópezRosa, S.; Angulo, J.C.; Esquivel, R.O. Jensen-tsallis divergence and atomic dissimilarity for position and momentum space electron densities. *J. Chem. Phys.* **2010**, *132*, 131. [CrossRef]
12. Mohammed, K.; Hamza, A.B. Nonrigid image registration using an entropic similarity. *IEEE Trans. Inf. Technol. Biomed.* **2011**, *15*, 681–690. [CrossRef]
13. Khader, M.; Hamza, A.B. An information-theoretic method for multimodality medical image registration. *Expert Syst. Appl.* **2012**, *39*, 5548–5556. [CrossRef]
14. Li, B.; Yang, G.; Shu, H.; Coatrieux, J.L. A New Divergence Measure Based on Arimoto Entropy for Medical Image Registration. In Proceedings of the 2014 22nd International Conference on Pattern Recognition, Stockholm, Sweden, 24–28 August 2014. [CrossRef]
15. Li, B.; Yang, G.; Coatrieux, J.L.; Li, B.; Shu, H. 3d nonrigid medical image registration using a new information theoretic measure. *Phys. Med. Biol.* **2015**, *60*, 8767–8790. [CrossRef] [PubMed]
16. Press, W.H.; Teukolsky, S.A.; Vetterling, W.T.; Flannery, B.P. *Numerical Recipes in C*, 3rd ed.; Cambridge Univ. Press: Cambridge, UK, 2007; Chapter 10; pp. 521–526.
17. Cover, T.M. *Elements of Information Theory (Wiley Series in Telecommunications and Signal Processing)*; Wiley-Interscience: New York, NY, USA, 2017.
18. Arimoto, S. Information-theoretical considerations on estimation problems. *Inf. Control* **1971**, *19*, 181–194. [CrossRef]
19. Boekee, D.E.; Van der Lubbe, J.C. The r-norm information measure. *Inf. Control* **1980**, *45*, 136–155. [CrossRef]
20. Li, M.; Chen, X.; Li, X.; Ma, B.; Vitanyi, P.M.B. The similarity metric. *IEEE Trans. Inf. Theory* **2004**, *50*, 3250–3264. [CrossRef]
21. He, Y.; Hamza, A.B.; Krim, H. A generalized divergence measure for robust image registration. *IEEE Trans. Signal Process.* **2003**, *51*, 1211–1220. [CrossRef]
22. Mattes, D.; Haynor, D.R.; Vesselle, H.; Lewellen, T.K.; Eubank, W. PET-CT image registration in the chest using free-form deformations. *IEEE Trans. Med. Imaging* **2003**, *22*, 120–128. [CrossRef]
23. Klein, S.; Staring, M.; Pluim, J.P.W. Evaluation of optimization methods for nonrigid medical image registration using mutual information and b-splines. *IEEE Trans. Image Process.* **2008**, *16*, 2879–2890. [CrossRef]
24. Nocedal, J. Updating quasi-newton matrices with limited storage. *Math. Comput.* **1980**, *35*, 773–782. [CrossRef]
25. Staring, M.; Klein, S. Itk:: Transforms Supporting Spatial Derivatives. Available online: http://hdl.handle.net/10380/3215 (accessed on 8 September 2010).
26. Klein, S.; Staring, M.; Murphy, K.; Viergever, M.A.; Pluim, J.P.W. Elastix: A toolbox for intensity-based medical image registration. *IEEE Trans. Med. Imaging* **2009**, *29*, 196–205. [CrossRef] [PubMed]

entropy

MDPI

Article

Study on Asphalt Pavement Surface Texture Degradation Using 3-D Image Processing Techniques and Entropy Theory

Yinghao Miao [1,2], Jiaqi Wu [1], Yue Hou [1,*], Linbing Wang [2,3], Weixiao Yu [1] and Sudi Wang [1]

1 Beijing Key Laboratory of Traffic Engineering, Beijing University of Technology, 100 Pingleyuan, Chaoyang District, Beijing 100124, China; miaoyinghao@ustb.edu.cn (Y.M.); wujiaqi@emails.bjut.edu.cn (J.W.); yuweixiao@emails.bjut.edu.cn (W.Y.); wangsudi@emails.bjut.edu.cn (S.W.)
2 National Center for Materials Service Safety, University of Science and Technology Beijing, 30 Xueyuan Road, Haidian District, Beijing 100083, China; wangl@vt.edu
3 The Charles E. Via, Jr. Department of Civil & Environmental Engineering, Virginia Polytechnic Institute and State University, Blacksburg, VA 24061, USA
* Correspondence: yuehou@bjut.edu.cn

Received: 4 February 2019; Accepted: 18 February 2019; Published: 21 February 2019

Abstract: Surface texture is a very important factor affecting the anti-skid performance of pavements. In this paper, entropy theory is introduced to study the decay behavior of the three-dimensional macrotexture and microtexture of road surfaces in service based on the field test data collected over more than 2 years. Entropy is found to be feasible for evaluating the three-dimensional macrotexture and microtexture of an asphalt pavement surface. The complexity of the texture increases with the increase of entropy. Under the polishing action of the vehicle load, the entropy of the surface texture decreases gradually. The three-dimensional macrotexture decay characteristics of asphalt pavement surfaces are significantly different for different mixture designs. The macrotexture decay performance of asphalt pavement can be improved by designing appropriate mixtures. Compared with the traditional macrotexture parameter Mean Texture Depth (MTD) index, entropy contains more physical information and has a better correlation with the pavement anti-skid performance index. It has significant advantages in describing the relationship between macrotexture characteristics and the anti-skid performance of asphalt pavement.

Keywords: pavement; macrotexture; 3-D digital imaging; entropy; decay trend

---

1. Introduction

Surface texture is a very important factor affecting the anti-skid performance of pavements [1–3]. Due to the mutual interactions between tire and pavements during driving, the surface texture wears continuously. Some observations show that anti-skid performance decreases under the vehicle load [2,3]. The study of texture and wear characteristics are therefore helpful for civil engineers to better understand the anti-skid performance of pavements. Generally, according to different influences on the anti-skid performance, the road surface texture is divided into the macrotexture (wavelength of 0.5 to 50 mm and peak-to-peak amplitude of 0.2 to 10 mm) and the microtexture (wavelength of 0 to 0.5 mm and peak-to-peak amplitude of 0 to 0.2 mm) [4]. The Mean Texture Depth (MTD) and Mean Profile Depth (MPD) are commonly used in engineering practice to evaluate the macrotexture [5,6]. However, these indexes still need to be improved in terms of reflecting the effects of texture on anti-skid performance [7,8]. Since it is difficult to test microtextrue on the road surface, it is not required to evaluate it in engineering practices, which is mainly controlled in the stage of aggregate selection [9,10].

The development of three-dimensional testing technology provides a new method for the evaluation of pavement surface texture, as indicated in the previous study [7], like the indoor laser profiler [11], X-ray computerized tomography (CT) [12], laser technology [13], optical three-dimensional scanner [14], three-dimensional laser device [15], and four-source photometric stereo technique [16,17]. With the continuous progress of three-dimensional testing technology, many commercial three-dimensional laser scanners have been developed and applied in the measurement of the three-dimensional texture of pavement surfaces [18–21], and the corresponding resolution is gradually improving. At the same time, in order to meet the needs of rapid testing, researchers are also working to develop some on-board three-dimensional testing devices for on-site road surface texture detection [22,23].

With the fast development of three-dimensional texture testing technology for pavement surfaces, the study of fine texture features based on three-dimensional data is also carried out, like in Fourier analysis [12], fractal theory [7] and texture analysis methods in image processing [24,25]. In recent years, Shannon's Entropy theory has been widely used as a powerful tool for image analysis [26–28], being significantly convenient for describing the complexity of texture. It should be noted that pavement macrotexture has been analyzed using fractal theory [7], Co-occurrence Matrix [24], gray tone difference matrix [25], and degradation analysis [29] in previous research. The introduction of entropy theory to this area can still help civil engineers better study the decay behavior of three-dimensional macrotexture and microtexture of road surfaces. There have been lots of studies on the bulk properties of pavements using various numerical and testing approaches [30–33], and the mechanism of the surface properties, e.g., pavement texture is still not fully understood. In this study, the current research progress and methods of 3D texture data acquisition in the field are introduced first. Following this, the feasibility of entropy theory in describing three-dimensional macrotexture and microtexture features is investigated. Third, the entropy of the three-dimensional texture is taken as an index to investigate the decay behavior of the macrotexture and microtexture of pavement surfaces. Finally, the advantages of three-dimensional macrotexture entropy in describing the decay of pavement anti-skid performance compared with a traditional MTD index are presented.

2. Field Data Collection

In order to study the decay characteristics of asphalt pavement surface texture with traffic wear, different types of asphalt surface on several highway and urban roads in Beijing were tested from November 2010 to November 2012. Seven tests were carried out during the period, where six different types of asphalt pavements were covered, including dense asphalt concrete (DAC), stone matrix asphalt (SMA), rubber asphalt concrete (RAC), ultra-thin wearing course (UTWC), micro-surfacing (MS), and open graded friction course (OGFC). The basic information of the test was described in detail in [7]. Due to the influences of pavement maintenance during the test period, the MS and OGFC measurement points did not result in continuous test data, and some DAC measurement points did not include a decay analysis due to the low traffic volume. The detailed decay analysis of the measurement points and traffic volume were presented in [29].

A commercial hand-held 3-D laser scanner (Creaform Inc., Lévis, QC, Canada), based on the laser triangulation technique, is used to collect the three-dimensional macro and micro textures of the pavement surface. The scanner consists of three charge-coupled devices (CCDs) and a cross laser [24]. By collecting the coordinates of a series of points on the surface of the object, the 3-D image of the surface can be obtained. The minimum sampling point spacing is 0.05 mm, and the measuring accuracy is 0.04 mm. The test results can be outputted to a variety of standard 3-D image file formats, such as stl., iges., etc. For more detailed information, refer to [7]. Figure 1 shows the field test photos. For the macrotexture, in the first two tests, the sampling size was 90 mm × 90 mm, and in the last five tests, the sampling size was 190 mm × 190 mm. All macrotexture scanners use a sampling interval of 0.4 mm to edit the 3-D images obtained by scanning. First, the commercial 3-D image software Geomagic Studio (3D Systems, Inc., Research Triangle Park, NC, USA) is used to edit the images, and then the data is transformed into ordered point clouds with an equal spacing distribution in two horizontal directions through the simulation scanning tool. Furthermore, the Fast Fourier Transform is used to filter the part whose wavelength exceeds 50 mm. According to [4], these components are beyond the scope of the macrotexture. The ordered point cloud data with a sampling interval of 0.5 mm in the x and y directions are finally obtained, and used for the macrotexture analysis. The detailed information was described in detail in [7]. For the microtexture, the test method was the same as for the macrotexture, except that a sampling interval of 0.05 mm was used. Based on the scanned three-dimensional data, the filtering process is carried out according to the frequency and wavelength range of the microtexture [4]. Finally, the ordered point cloud data with a size of 5 mm × 5 mm and a sampling interval of 0.05 mm in the x and y directions are obtained for the analysis of the microtexture. One macrotexture and two microtextures are collected at each test point. Figure 2 is the result of typical macrotexture testing after filtering, and Figure 3 is the result of a typical microtexture by measuring and after filtering. For the macrotexture, since only the part whose wavelength is larger than 50 mm is filtered out, this wavelength value is several times the size of the particle exposed on the pavement surface, and the filtered image has not intuitively changed. For the microtexture, because it is necessary to filter out the part whose wavelength is more than 0.5 mm, which is much smaller than the size of the particle exposed on the pavement surface, the filtered microtexture loses the true morphology characteristics of the pavement surface.

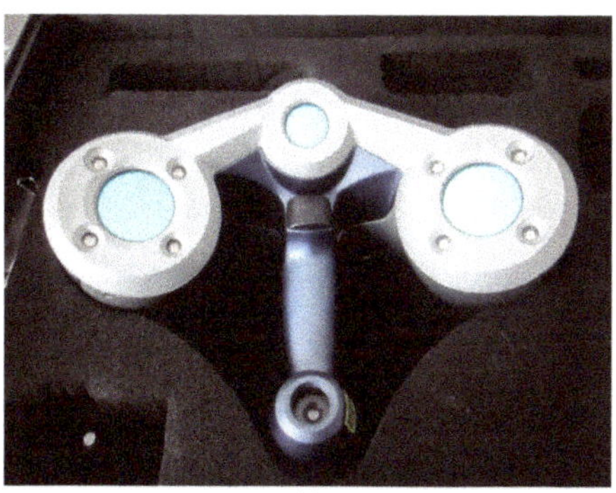

(a)

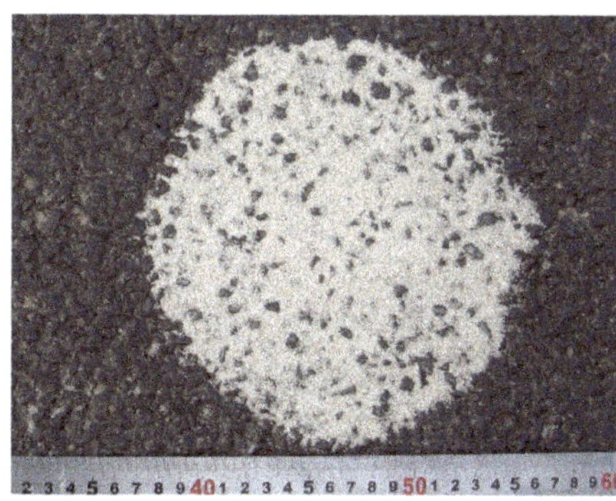

(b)

Figure 1. *Cont.*

(c)

(d)

Figure 1. Field tests: (**a**) the 3-D scanner; (**b**) sand patch test for the mean texture depth (MTD); (**c**) Scanning test; and (**d**) dynamic friction tester (DFT) test.

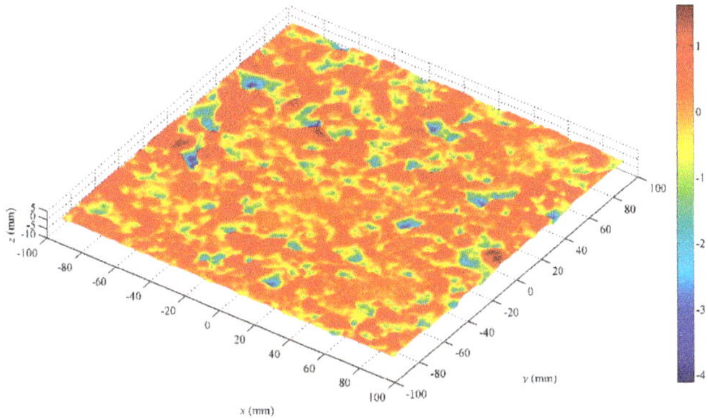

(**a**)

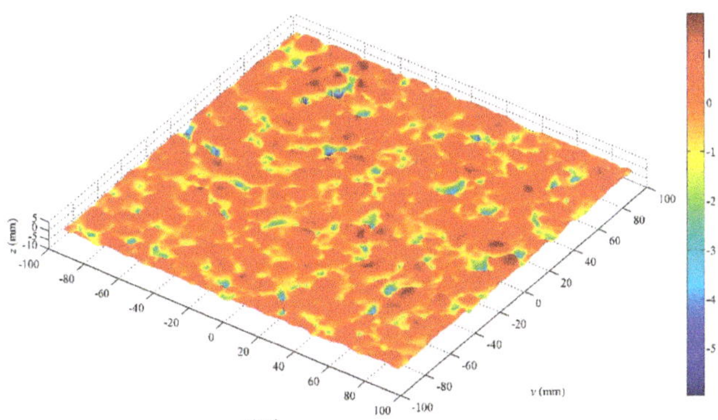

(**b**)

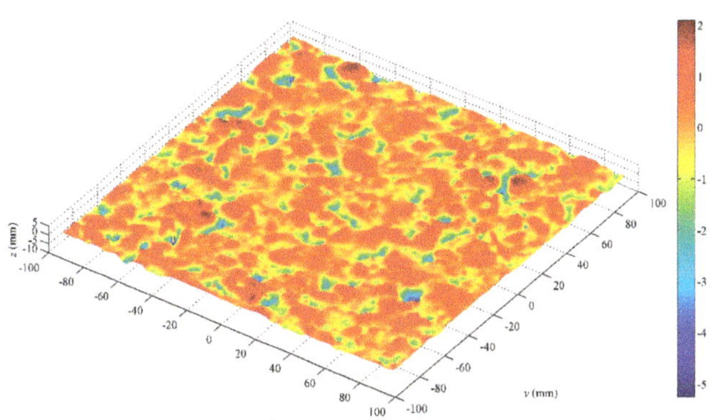

(**c**)

Figure 2. *Cont.*

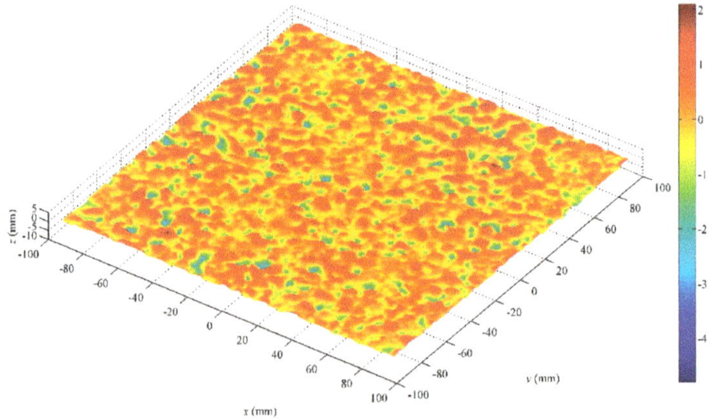

(**d**)

Figure 2. Results of a typical 3-D macrotexture: (**a**) dense asphalt concrete (DAC); (**b**) stone matrix asphalt (SMA); (**c**) rubber asphalt concrete (RAC); and (**d**) ultra-thin wearing course (UTWC).

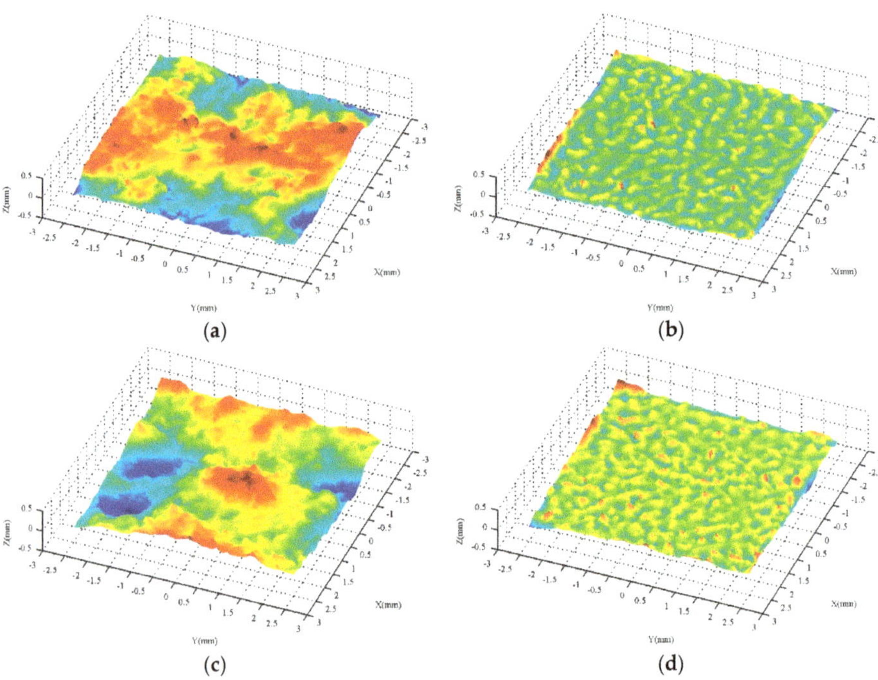

Figure 3. Results of a typical 3-D microtexture: (**a**) DAC, measured; (**b**) DAC, filtered; (**c**) SMA, measured; and (**d**) SMA, filtered.

3. Characterizing Surface Texture of Asphalt Pavement Using Entropy Theory

Denote a grey level image as $I = \{G(x, y), x = 1, 2, \ldots, N_x, y = 1, 2, \ldots, N_y\}$, where $G(x, y)$ is the grey level at (x, y), and N_x and N_y are the pixel numbers along the x and y directions respectively. Following this, the probability of grey level i is

$$p_i = \frac{\sum_{x=1}^{N_x} \sum_{y=1}^{N_y} \delta(G(x,y),i)}{N_x \times N_y} \tag{1}$$

where $\delta(i,j)$ is the Kronecker delta function.

$$\delta(i,j) = \begin{cases} 1 & i = j \\ 0 & i \neq j \end{cases} \tag{2}$$

If an image has the maximum grey level of N_g, the entropy (E) of the image can be defined as [34]

$$E = \sum_{i=1}^{N_g} p_i \log_2\left(\frac{1}{p_i}\right) \tag{3}$$

The 3-D texture measurements should be converted into grey-level images so that they can be characterized by entropy. First, the height range of a given 3-D texture measurement is divided into sections using a given interval. Following this, a corresponding grey-level image is obtained through mapping each height section onto a grey level. Reference [24] describes the conversion techniques in detail. Figures 4 and 5 depict the grey-level images corresponding to the 3-D measurements in Figures 2 and 3, respectively. Finally, the Entropy E can be calculated for each 3-D texture measurement in accordance with Equation (3).

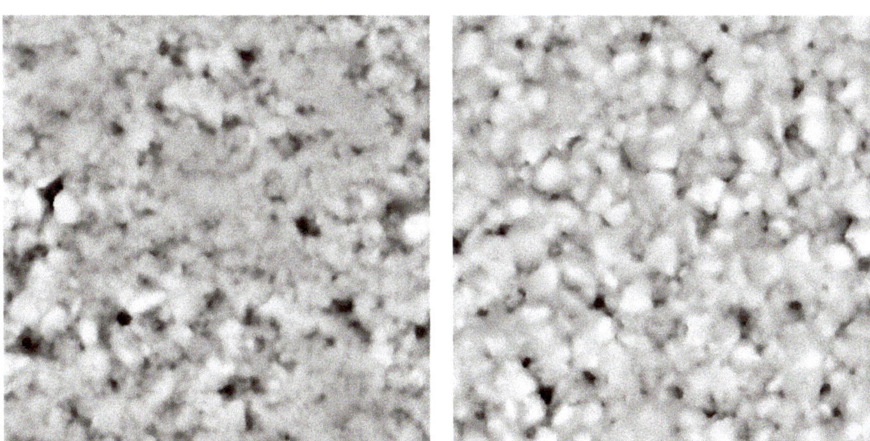

(a) (b)

Figure 4. *Cont.*

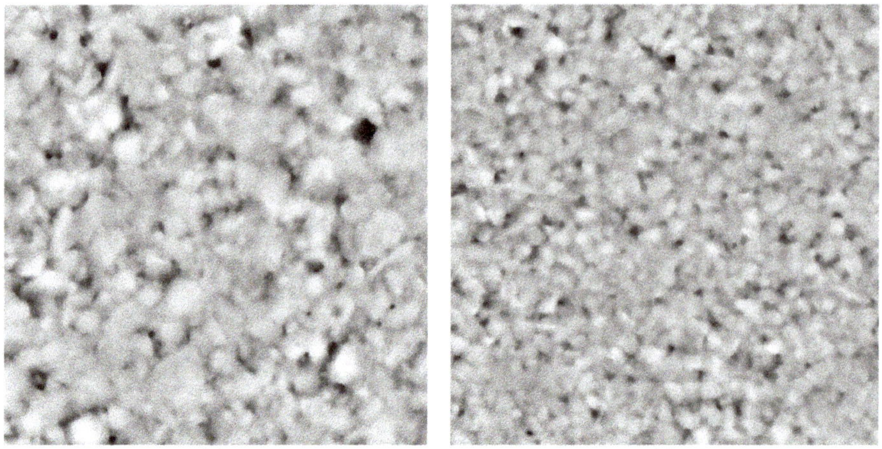

Figure 4. Grey-level images corresponding to the macrotexture shown in Figure 2: (**a**) DAC; (**b**) SMA; (**c**) RAC; and (**d**) UTWC.

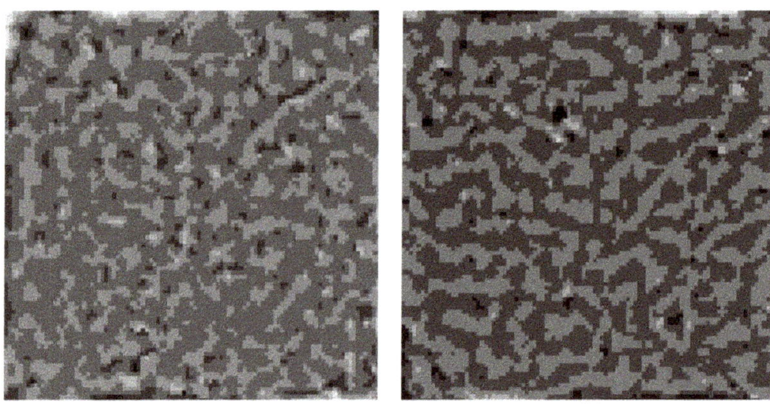

Figure 5. Grey-level images corresponding to the microtexture shown in Figure 3: (**a**) SMA and (**b**) DAC.

Figure 6 presents the entropy distribution of the macrotexture and microtexture of different types of pavements, where D1 and D2 represents the DAC pavement, M represents the MS pavement, O represents the OGFC pavement, R represents the RAC pavement, S represents the SMA pavement, and U1 and U2 represent the UTWC pavements constructed over different years. The detailed information of various pavement parameters is referred to in [29]. As shown in Figure 6a, there are significant differences in the entropy of the macrotexture of different types of pavement surfaces. Among them, the entropy of MS pavement is the smallest and that of OGFC pavement is the largest. It is noted that there is a clear distinction between U1 and U2, in which U1 is the pavement opened in September, 2009, and U2 is the pavement opened in September, 2010; this indicates that entropy can be used to describe the macrotexture decay of the pavement surface. Figure 6a has the same distribution trends as those of previous research [24], indicating that the use of entropy is reasonable and accurate. For the microtexture, the range of the entropy distribution is narrow (Figure 6b), and the difference between different pavements is not very obvious. This may be because the microtexture is mainly affected by mineral aggregates and is less affected by mixture gradation.

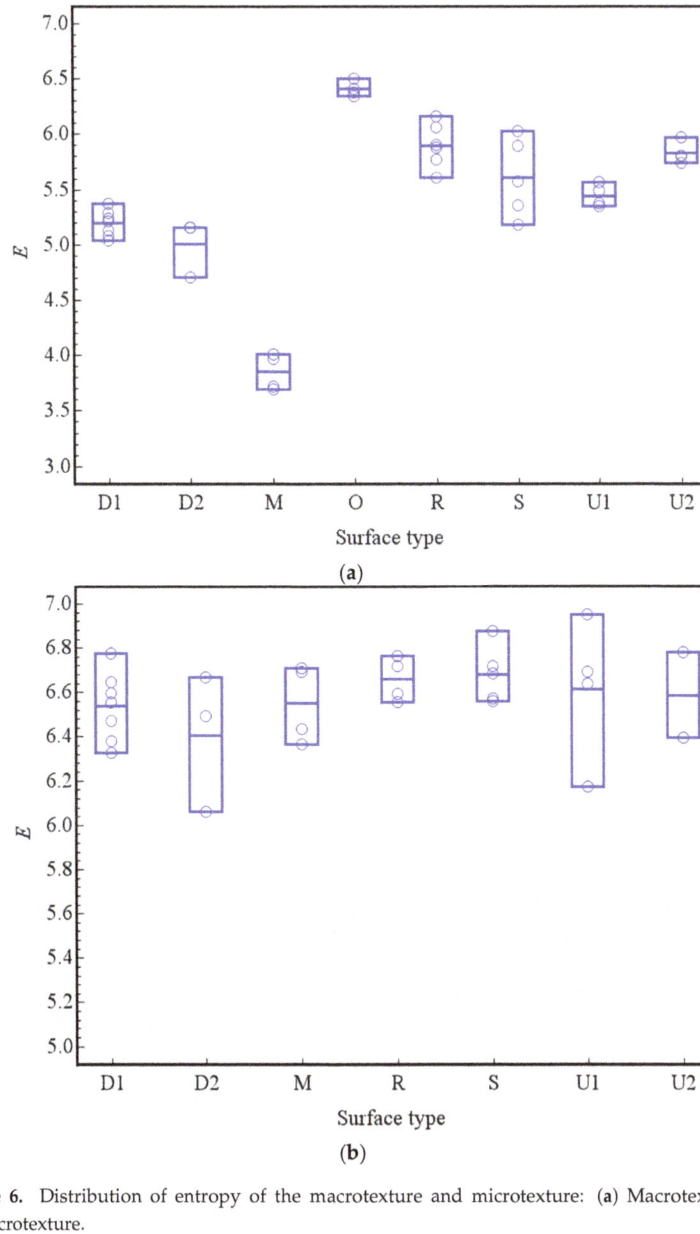

Figure 6. Distribution of entropy of the macrotexture and microtexture: (**a**) Macrotexture and (**b**) Microtexture.

4. Characterization of Macrotexture Degradation with Entropy

4.1. Degradation of Macrotexture Entropy

In order to analyze the decay of the macrotexture of the asphalt pavement by the entropy value, the experimental data is grouped according to the type of pavement and cumulative traffic volume of service. Figure 7 presents the changes of entropy of the macrotextures of DAC, SMA, RAC and UTWC pavement surfaces with cumulative traffic volume by box-and-whisker plots, in which the mark inside the box is the median, the lower and the upper edges of the box are the 1st and 3rd quartiles, respectively, and the "x" are the outliers. Before analyzing the decay trend, it is necessary to note that the aggregate types used in the four types of pavement are not identical. DAC and RAC pavements used one type of aggregate, while SMA and UTWC pavement used another type. In the early service stage of roads, the entropy of the RAC surface macrotexture is the largest amongst the four types of pavement, with an average value of 5.90 (after a 0.22×10^6 standard vehicle passes), followed by the UTWC and SMA pavements, with an average value of 5.83 (after a 0.54×10^6 standard vehicle passes) and 5.61 (after a 0.59×10^6 standard passes), respectively. The entropy of the surface macrotexture of DAC is the smallest, with an average value of 5.20 (after a 0.31×10^6 standard vehicle passes). Despite some fluctuations in data, for the DAC, RAC and UTWC pavements, it is clear that the entropy of the macrotexture of the pavement surface decreases gradually with the increase of the cumulative traffic volume. The DAC pavement decay is the most obvious. After a 2.29×10^6 standard vehicle passes, the average entropy of DAC's macrotexture decays to 4.34, and the average entropy of DAC's macrotexture decays to 8.37% for every 1×10^6 standard vehicle passing. After a 2.20×10^6 standard vehicle passes, the average entropy of RAC's macrotexture decays to 5.48, with an average decay rate of 3.6% for every 1×10^6 standard vehicle passing. After a 7.40×10^6 standard vehicle passes UTWC, the average entropy of the macrotexture decays to 5.40, and the average decay rate is 1.07% for every 1×10^6 standard vehicle passing, respectively. The entropy of the macrotexture of the SMA pavement surface does not obviously decay. After a 4.61×10^6 standard vehicle passes, the average entropy of the macrotexture is still 5.59, which is basically consistent with the mean value of a 0.59×10^6 standard vehicle passing. The analysis using entropy theory obtains similar results compared with our previous research [29]. Pavements with different gradations will have different sizes of aggregates exposed on the surfaces. Due to the difference of wear performance of different size particles, there are different decay trends in the macrotexture of different pavements. On the other hand, the difference of aggregate types should also be noticed.

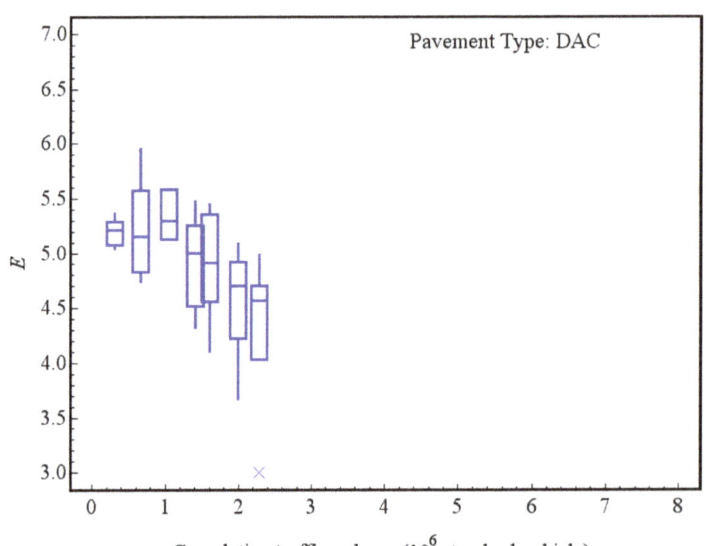

(a)

Figure 7. *Cont.*

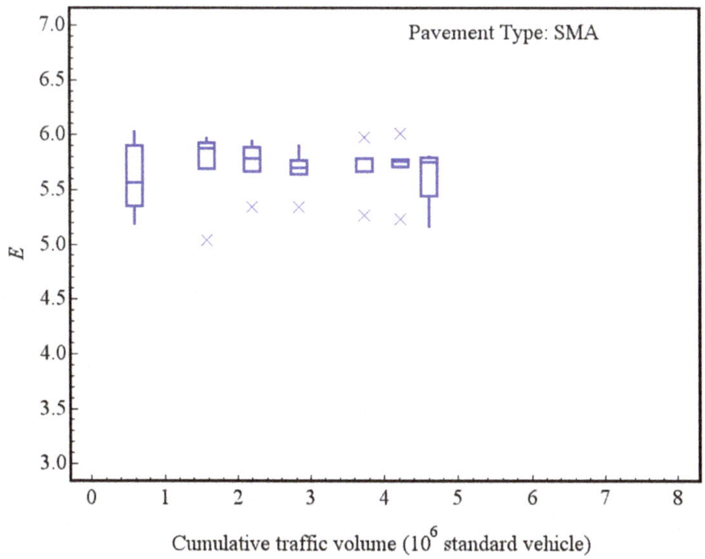

(**b**)

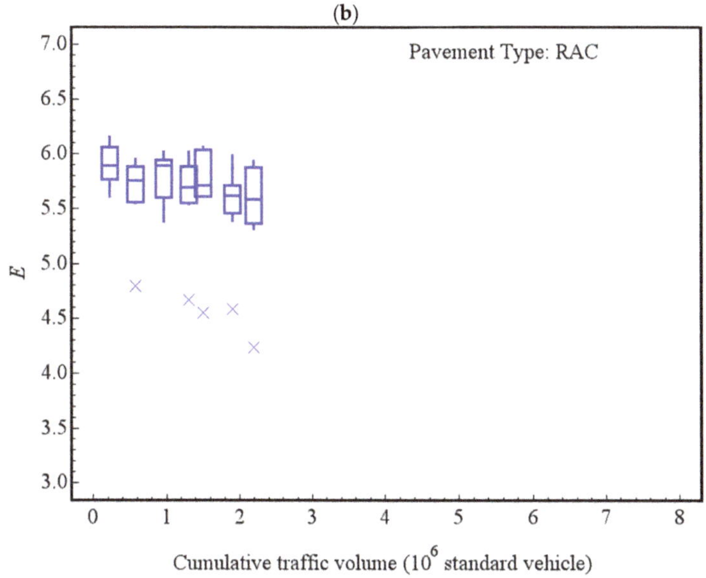

(**c**)

Figure 7. *Cont.*

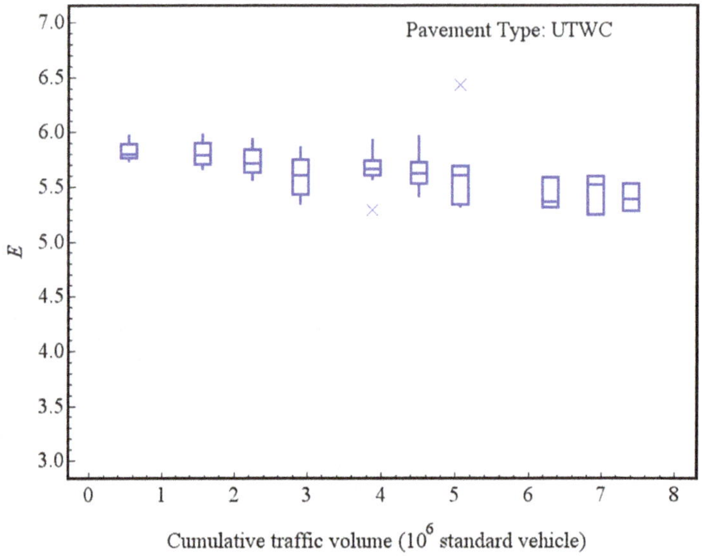

(d)

Figure 7. Change of the macrotexture entropy with the traffic volume: (**a**) DAC; (**b**) SMA; (**c**) RAC; and (**d**) UTWC.

4.2. Changing Trends of Entropy in Macrotexture

In order to quantitatively depict the decay trend of the macrotexture of the pavement surfaces, a logarithmic model (as shown in Equation (4)) is used for a regression analysis of the change trend of the average entropy with the cumulative traffic volume, corresponding to the different cumulative traffic volumes of each pavement. Table 1 gives the least squares analysis results of four kinds of road regression analyses. Table 2 lists the mean square errors (MSEs) and R-squares of regression analyses of four kinds of pavements. The regression parameters of the model are listed in Table 3, as [29].

$$E = a \times \ln(traf) + b \quad (4)$$

where *traf* is the cumulative traffic volume, and *a* and *b* are the fitting coefficients.

Table 1. Least-squares analysis of the model for the mean entropy of the macrotextures. Dense asphalt concrete (DAC); Stone matrix asphalt (SMA); Rubber asphalt concrete (RAC); and Ultra-thin wearing course (UTWC).

Surface Type	Model Sum of Squares	Error Sum of Squares	Corrected Total Sum of Squares	F Value	P > F
DAC	0.4518	0.3416	0.7935	6.61	0.0499
RAC	0.0893	0.0353	0.1246	12.63	0.0163
SMA	0.000343	0.0152	0.0156	0.11	0.7509
UTWC	0.1531	0.0548	0.208	22.35	0.0015

Table 2. The mean square errors (MSEs) and R-squares of the regression model for the mean entropy of the macrotextures.

DAC		RAC		SMA		UTWC	
MSE	R2	MSE	R2	MSE	R2	MSE	R2
0.0683	0.5695	0.0071	0.7167	0.0030	0.0256	0.0068	0.7365

Table 3. Fitting coefficients of the model for the mean entropy of macrotexture.

Surface Type	DAC	RAC	SMA	UTWC
a	−0.3929	−0.1525	0.0105	−0.1609
b	4.9734	5.6611	5.6569	5.8139

As shown in Table 1, for the DAC, RAC, and UTWC pavements, the *p*-value is below 0.05, indicating that the logarithmic model is significant for these three types of pavement. For the SMA pavement, the *p*-value is 0.7509, much higher than 0.05, indicating that the logarithmic model is not significant for the SMA pavement, which is mainly due to the fact that the entropy of the macrotexture of the SMA pavement has not decayed significantly over more than two years of observation. Although the logarithmic model is not significant for the SMA pavement, Table 2 shows that the MSE fitted by the model is only 0.0030. Figure 8 presents the changes of the average entropy of four types of pavement macrotextures fitted by the logarithmic model.

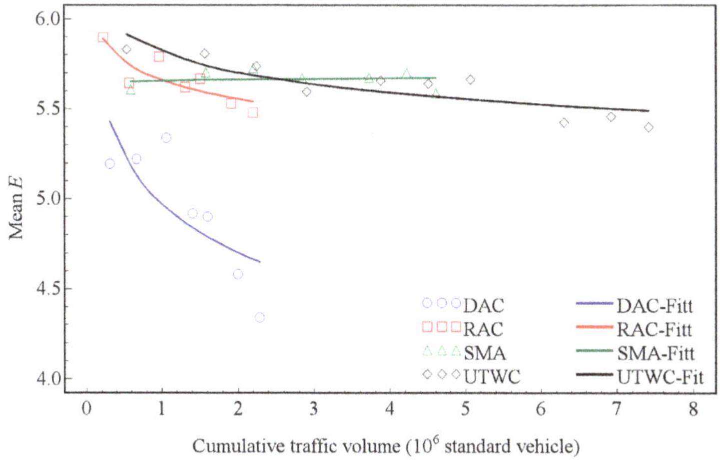

Figure 8. Changing trends of the macrotexture entropy.

According to Equation (4), the coefficient *a* described the decay rate. The smaller the value of *a*, the faster the decay rate. As mentioned above, the DAC and RAC pavements have one same aggregate type, and the SMA and UTWC pavements have another type. Because the aggregate type has a potential impact on the pavement wear, the decay of the macrotexture entropy of the pavement surface should take the difference between different aggregate types into account. From Table 3 and Figure 8, it can be seen that the macrotexture of the DAC pavement decreases fastest. The RAC and UTWC pavements have a similar macrotexture decay trend. The SMA pavement maintains a stable entropy after 4.61×10^6 standard vehicle passes, which should be attributed to the specific gradation of SMA. Note that the trends using the mean entropy are very similar to those from the previous analysis [29], which validates our research accuracy.

4.3. Relationship of Macrotexture Entropy and MTD with DFT60

In the field tests from November 2010 to November 2012, the mean texture depth (MTD) was tested by a sand patch method at each test point, and the friction performance of the pavement was tested by a dynamic friction tester (Nippo Sangyo Co., Ltd., Tokyo, Japan), which is described in detail in [29]. In order to investigate the potential advantages of the entropy theory in describing the texture characteristics of asphalt pavement surfaces, based on the mean values of entropy, MTD and DFT60 for each pavement with different cumulative traffic volumes, the correlation among *E*, MTD and DFT60 are analyzed. The Pearson correlation coefficients are listed in Table 4.

Table 4. Pearson correlation coefficients between the Mean E, Mean texture depth (MTD), and Mean DFT60.

Surface Type	DAC	RAC	SMA	UTWC
Mean E VS Mean MTD	0.6041	0.7139	−0.0932	0.7997
Mean E VS Mean DFT60	0.8283	0.3407	0.7036	0.9169
Mean MTD VS Mean DFT60	0.5323	0.7298	−0.0635	0.8474

It can be seen from Table 4 that for the DAC, RAC, and UTWC pavements, there is a certain correlation between entropy and MTD. However, the Pearson correlation coefficients are below 0.8, and there is no significant correlation between the entropy and MTD of the SMA pavement. This shows that entropy describes some features of the macrotexture of the pavement surface, which the MTD indexes fail to describe. Comparing the correlations between entropy, MTD and DFT60, it is found that the Pearson correlation coefficients of entropy and DFT60 are significantly higher than those of MTD and DFT60 except for RAC. For the DAC, SMA and UTWC pavements, the correlation coefficients of entropy and DFT60 are 0.2960, 0.6401 and 0.0695 higher than the Pearson correlation coefficients of MTD and DFT60, respectively. This shows that entropy has more advantages than MTD in describing the impact of macrotexture on the anti-skid performance of asphalt pavement. Figure 9 plots DFT60 against the entropy *E* and MTD of different types of asphalt pavements, which could more intuitively reflect the advantages of *E*.

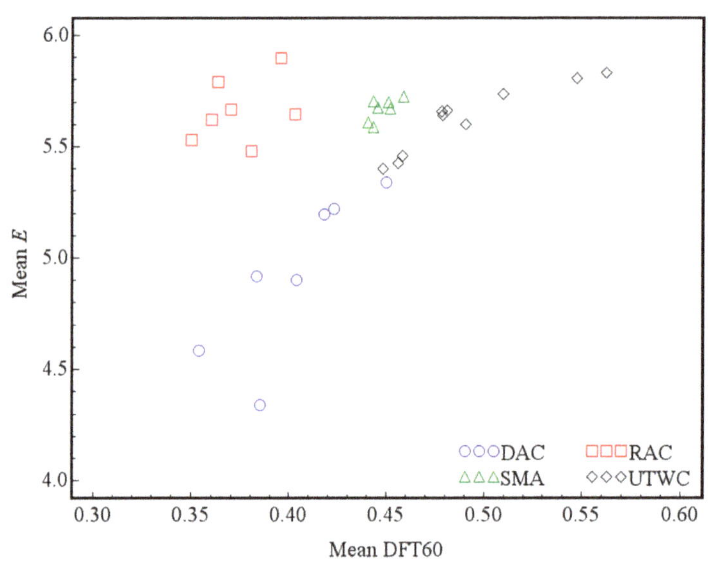

(a)

Figure 9. *Cont.*

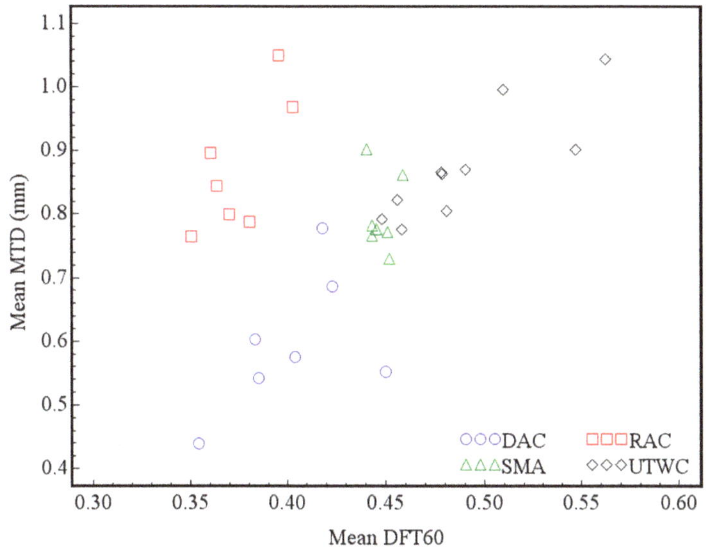

(**b**)

Figure 9. Scatter plots of DFT60 against *E* and MTD: (**a**) DFT60-*E* and (**b**) DFT60-MTD.

5. Characterization of Microtexture Degradation with Entropy

5.1. Degradation of Entropy of Microtexture

In the field test, two microtextures were collected at each test point, and the mean value of the two microtextures' entropy is used as the evaluation basis for the feature evaluation. Similar to the macrotexture, the experimental data is grouped according to the cumulative traffic volume of different pavement types and services. Figure 10 gives the variation of the microtexture entropy of the DAC, SMA, RAC and UTWC pavement surfaces with the cumulative traffic volume by box-and-whisker plots, where the mark inside the box is the median, the lower and the upper edges of the box are the 1st and 3rd quartiles, respectively, and the "x" are the outliers. Note that the aggregate types used in the four types of pavement are not identical. The DAC and RAC pavements used one same type of aggregate, while the SMA and UTWC pavements used another. In the early service stage of roads, the average values of DAC, SMA, RAC and UTWC are 6.54 (after a 0.31×10^6 standard vehicle passes), 6.68 (after a 0.59×10^6 standard vehicle passes), 6.66 (after a 0.22×10^6 standard vehicle passes), and 6.59 (after a 0.54×10^6 standard vehicle passes).

Because the decay of microtexture is mainly caused by the polishing of aggregate particles on the surface of pavements, the decay of microtexture is relatively slow, and the absolute value of entropy decay is much smaller than that of macrotexture. Nevertheless, Figure 10 could identify that the entropy of the four types of pavement surface microtextures gradually decreases with the increase of the traffic volume. After a 2.29×10^6 standard vehicle passes, the average entropy of the DAC texture decays to 6.19, and the average decay rate is 2.71% for every 1×10^6 standard vehicle passing. The average entropy of the microtexture of the RAC pavement decays to 6.27 after a 2.20×10^6 standard vehicle passes, and the average decay rate is 2.97% for every 1×10^6 standard vehicle passing. The average entropy of the SMA pavement texture decays to 5.93 when a 4.61×10^6 standard vehicle passes, with an average decay rate of 2.80% for every 1×10^6 standard vehicle passing. After a 7.40×10^6 standard vehicle passed the UTWC pavement, the average entropy of the microtexture decays to 5.96, and the average decay rate is 1.37% for every 1×10^6 standard vehicle passing. The microtexture decay behavior of different types of pavements is similar.

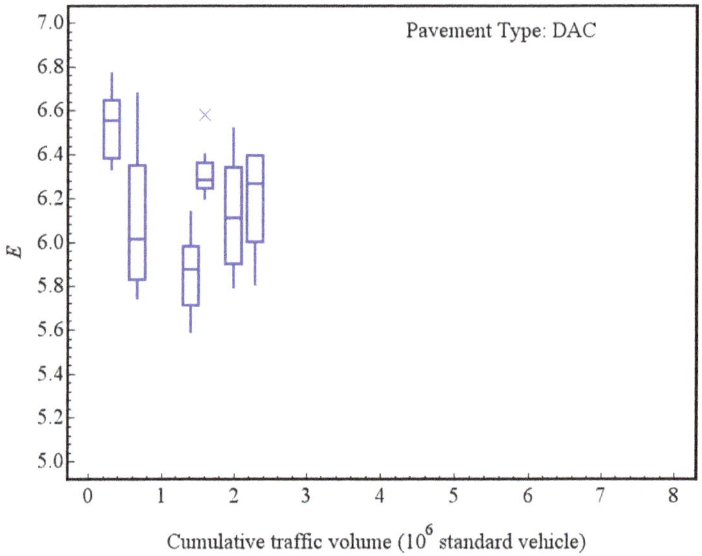

(a)

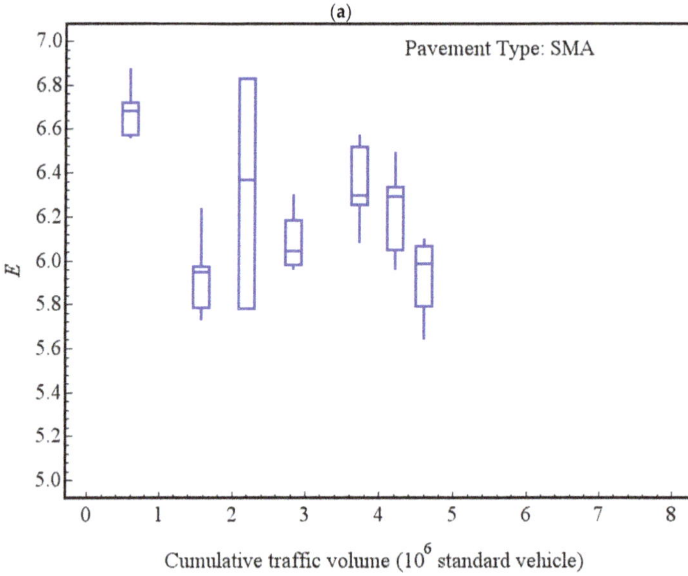

(b)

Figure 10. *Cont.*

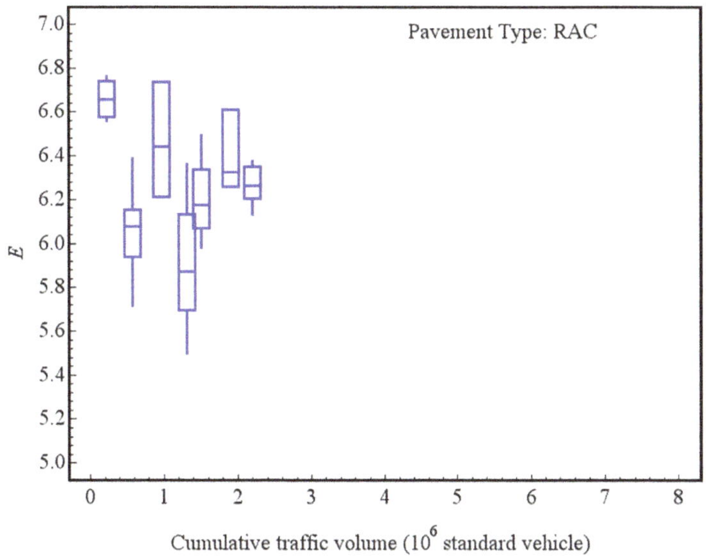

(c)

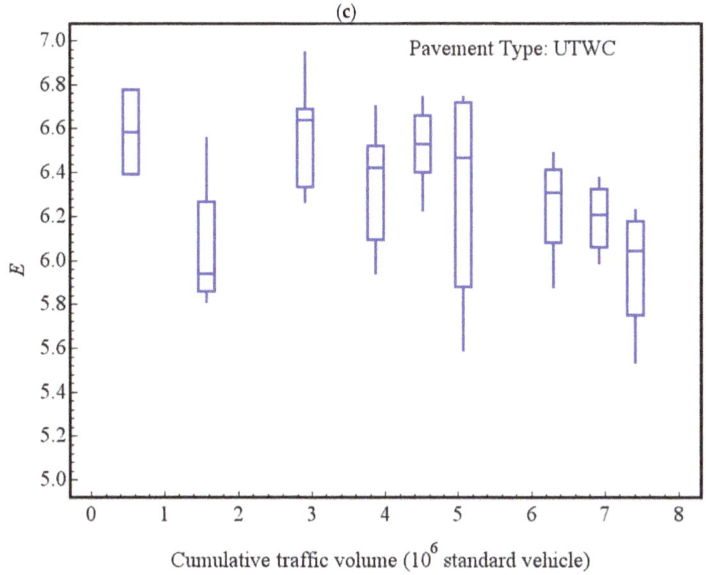

(d)

Figure 10. Change of the microtexture entropy with the traffic volume: (**a**) DAC; (**b**) SMA; (**c**) RAC; and (**d**) UTWC.

5.2. Changing Trends of Entropy of Microtexture

In order to quantitatively depict the decay trend of the microtexture of the pavement surface, based on the average value of entropy corresponding to the different cumulative traffic volume of each pavement, the power function model (as shown in Equation (5)) is used to analyze the trend of the average entropy with the cumulative traffic volume by comparing various models. Table 5 shows the least squares analysis results of four kinds of road regression analyses. Table 6 lists the mean square errors (MSEs) and R-squares of a regression model. The regression parameters of the model are listed in Table 7, as [29].

$$E = a \times traf^{b} \tag{5}$$

where *traf* is the cumulative traffic volume, and *a* and *b* are the fitting coefficients.

Table 5. Least-squares analysis of the model for the mean entropy of the macrotextures.

Surface Type	Model Sum of Squares	Error Sum of Squares	Corrected Total Sum of Squares	F Value	P > F
DAC	229.9	0.1848	0.2590	2487.7	<0.0001
RAC	276.1	0.3141	0.3848	2197.2	<0.0001
SMA	270.9	0.267	0.4222	2536.0	<0.0001
UTWC	358.6	0.2981	0.3765	4210.4	<0.0001

Table 6. The MSEs and R-squares of the regression model for the mean entropy of the microtextures.

DAC		RAC		SMA		UTWC	
MSE	R2	MSE	R2	MSE	R2	MSE	R2
0.0462	0.2864	0.0628	0.1836	0.0534	0.3676	0.0426	0.2082

Table 7. Fitting coefficients of the model for the mean entropy of microtextures.

Surface Type	DAC	RAC	SMA	UTWC
a	6.2081	6.279	6.412	6.4559
b	−0.0257	−0.0216	−0.0357	−0.0183

For the four types of pavements, all the *p*-values below are 0.0001. Table 5 shows that the R-square value is low, and the SMA pavement with the largest R-square value is only 0.3676, indicating that the regression model is not very ideal. This is mainly due to the small decay variation of the microtexture entropy and the fluctuation of the test data. Figure 11 presents the variation curves of the average entropy of the four types of pavement microtextures fitted by power model. Although the regression model is not perfect, the model can still be used for some simple analyses.

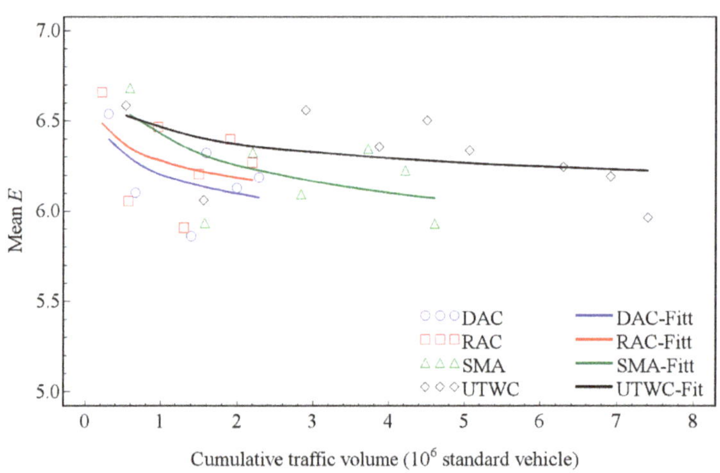

Figure 11. Changing trends of the microtexture entropy.

According to Equation (5), coefficient b describes the decay rate, and the larger the value b, the faster the decay rate. As mentioned above, the DAC and RAC pavements use one same aggregate type and the SMA and UTWC pavements use another type. This difference should be noticed when analyzing the difference of microtexture entropy between the different pavements. According to Table 7 and Figure 11, the SMA pavement surface texture decay rate is the fastest, followed by the DAC and RAC pavements, and the UTWC pavement decay rate is the slowest. The average decay of entropy is less than 2.7% for every 1×10^6 standard vehicle passing, for all four types of pavements.

6. Discussion

According to the definition of pavement texture entropy in Equations (1) to (3), it can be seen that the possible range of entropy of the macro- or micro- textures of the pavement surfaces is $[0, \log_2(N_g)]$. In a plane, if there is only one grey value, the p_i of this grey value is 1, which leads to zero entropy. For the same texture with N_g, when all grey values correspond to the same p_i, the entropy reaches the maximum value [34]. The larger the N_g, the greater the maximum value. For the pavement surface texture, the complexity of the texture increases with the increase of the entropy. Under the polishing action of the vehicle load, the surface of the pavement tends to be smooth, showing the entropy decreasing gradually. Compared with the MTD index, entropy contains more physical information. At the same time, the correlation between macrotexture entropy and DFT60 is significantly higher than for MTD. The entropy has obvious advantages over the traditional MTD index in the macrotexture evaluation of pavement anti-skid performance.

According to the test results in this paper, the difference of microtexture entropy between different pavements is not very significant, and the law of decreasing with traffic volume is not very significant. On one hand, the microtexture is mainly influenced by aggregate particles, and the difference of aggregates between different pavements is not prominent. On the other hand, the polishing process of the aggregate surface is relatively slow, and may require a larger load to show a significant decay trend.

It should be noted that the wavelength of microtextures is less than 0.5 mm according to the division of the texture scale, which requires a very small sampling interval to reflect the real microtexture. Some researchers used a 0.001 mm sampling interval to test and analyze microtextures in laboratory experiments [11]. However, in order to evaluate the microtexture of real pavements, it is necessary to sample the pavement by drilling holes, which causes great damage to the pavement surface. At present, there is no report on pavement field test equipment which can reach a 0.001 mm sampling distance. In this paper, a 3-D scanner with a 0.05 mm sampling distance is used to test and analyze the microtexture of the pavement field. Although the microtexture cannot be fully reflected in this study, the present research can, as an exploration, still provide guidance for future studies.

The valuation of pavement surface texture has important engineering application values. On the one hand, it can directly establish the relationship between the texture of road surfaces and anti-skid performance, which can be used to evaluate anti-skid performance. On the other hand, the pavement surface texture depends on the gradation of the asphalt mixture and the morphological characteristics of aggregates. The pavement surface texture is a bridge connecting the asphalt mixture design and the anti-skid performance of the pavement. The description of texture featured by traditional evaluation indexes provides an insufficient connection between texture evaluation and anti-skid performance, and thus the engineering value of texture evaluation has not been fully revealed, while the widely-used design method of anti-skid performance of asphalt pavement has not been formed. Our research results show that the use of entropy to describe the macrotexture of pavement surfaces contains more physical information, and indicates the relationship between macrotexture and anti-skid performance. On the one hand, with the accumulation of data by the popularized 3-D texture testing method in engineering practice, using entropy as a texture index can improve the evaluation of pavement anti-skid performance. On the other hand, it is feasible to use entropy as a texture index to connect the anti-skid performance of asphalt mixture and pavement and improve the design method of the anti-skid performance of asphalt pavement.

7. Conclusions

In this paper, based on the data of seven field tests on the surface texture and friction characteristics of various types of asphalt pavements over more than two years, the macro-/micro- texture characteristics and decay of 3-D asphalt pavement surfaces were studied using the theory of entropy. Through this research, the following conclusions can be drawn:

(1) The entropy distribution range of the 3-D macrotexture of asphalt pavements is wide, and there are significant differences among different gradation pavement types. There are significant differences in the entropy of the 3-D macrotextures of asphalt pavements with different mixture designs. The difference of 3-D microtextures is not very obvious. Furthermore, the distribution range of macrotexture entropy is wider than that of microtextures. The macrotexture of asphalt pavements is mainly affected by the gradation of mixture, while the microtexture is mainly affected by the surface morphology of aggregates.

(2) There are significant differences in the decay characteristics of 3-D macrotextures of asphalt pavements with different mixture types, which indicates that the decay characteristics of the macrotexture of asphalt pavement surfaces could be significantly improved by choosing appropriate mixture types and optimizing the design.

(3) Compared with the traditional macrotexture parameter MTD, entropy contains more physical information and a better correlation with the pavement anti-skid performance index. It has significant advantages in describing the relationship between macrotexture characteristics and anti-skid performances of asphalt pavements.

(4) This paper attempts to collect the 3-D microtexture of pavement surfaces with a 0.05 mm sampling interval. The decay law of the 3-D microtexture of different types of asphalt pavements is not very significant; this may require a longer observation time and more innovative methods to obtain more detailed microtextures for further studies.

Author Contributions: Conceptualization, Y.M., J.W. and W.Y.; Methodology, Y.H., Y.M. and L.W.; Validation, J.W., Y.H. and Y.M.; Investigation, Y.M., S.W., Y.H. and L.W.; Data Curation, Y.H. and Y.M.; Writing—Original Draft Preparation, J.W. and S.W.; Writing—Review & Editing, Y.H., L.W. and Y.M.; Visualization, W.Y. and J.W.; Supervision, Y.M.; Project Administration, Y.M.; Funding Acquisition, Y.M.

Funding: The research reported in this paper is funded by the National Natural Science Foundation of China (No. 51178013).

Acknowledgments: The authors would like to thank the financial support from National Natural Science Foundation of China.

Conflicts of Interest: The authors declare no conflict of interest.

References

1. Kogbara, R.B.; Masad, E.A.; Kassem, E.; Scarpas, A.; Anupam, K. A state-of-the-art review of parameters influencing measurement and modeling of skid resistance of asphalt pavements. *Constr. Build. Mater.* **2016**, *114*, 602–617. [CrossRef]
2. Henry, J.J. *Evaluation of Pavement Friction of Characteristics*; Transportation Research Board, National Research Council: Washington, DC, USA, 2000.
3. American Association of State Highway and Transportation Officials (AASHTO). *Guide for Pavement Friction*; AASHTO: Washington, DC, USA, 2008.
4. PIARC. Technical Committee Report No 1: Surface Characteristics. In Proceedings of the XVIII World Road Congress, Brussels, Belgium, 13–19 September 1987.
5. Praticò, F.G.; Vaiana, R. A study on the relationship between mean texture depth and mean profile depth of asphalt pavements. *Constr. Build. Mater.* **2015**, *101*, 72–79. [CrossRef]
6. Plati, C.; Pomoni, M.; Stergiou, T. Development of a Mean Profile Depth to Mean Texture Depth Shift Factor for Asphalt Pavements. *Transp. Res. Rec.* **2017**, *2641*, 156–163. [CrossRef]

7. Miao, Y.; Song, P.; Gong, X. Fractal and Multifractal Characteristics of 3D Asphalt Pavement Macrotexture. *J. Mater. Civ. Eng.* **2014**, *26*, 04014033. [CrossRef]
8. Wang, Y.; Yang, Z.; Liu, Y.; Sun, L. The characterisation of three-dimensional texture morphology of pavement for describing pavement sliding resistance. *Road Mater. Pavement Des.* **2018**. [CrossRef]
9. Erichsen, E. Relationship between PSV and in situ friction: A Norwegian case study. *Bull. Eng. Geol. Environ.* **2009**, *68*, 339–343. [CrossRef]
10. Smith, A.B.; Fu, C.N. Correlation of Laboratory and Field Friction Measurements to Optimize Utilization of Bituminous Surface Aggregates in Utah. In Proceedings of the Transportation Research Board—94th Annual Meeting, Washington, DC, USA, 11–15 January 2015. No. 15-4418.
11. Huang, C. *Mathematical Characterization of Road Surface Texture and Its relation to Laboratory Friction Measures*; Michigan Technological University: Houghton, MI, USA, 2002.
12. Abbas, A.; Kutay, M.E.; Azari, H.; Rasmussen, R. Three-dimensional surface texture characterization of portland cement concrete pavements. *Comput.-Aided Civ. Infrastruct. Eng.* **2007**, *22*, 197–209. [CrossRef]
13. Ech, M.; Yotte, S.; Morel, S.; Breysse, D.; Pouteau, B. Laboratory evaluation of pavement macrotexture durability. *Revue Europé-enne de Génie Civil* **2007**, *11*, 643–662. [CrossRef]
14. Wen, J. Study on Evaluating Texture Depth of Asphalt Pavement with Digital Technology. Master's Thesis, Chang'an University, Xi'an, China, 2009.
15. Cackler, E.T.; Ferragut, T.; Harrington, D.S. *Evaluation of U.S. and European Concrete Pavement Noise Reduction Methods*; National Concrete Pavement Technology Center, Iowa State University: Ames, IA, USA, 2006.
16. Gendy, A.E.; Shalaby, A. Mean profile depth of pavement surface macrotexture using photometric stereo techniques. *J. Transp. Eng.* **2007**, *133*, 433–440. [CrossRef]
17. Gendy, A.E.; Shalaby, A.; Saleh, M.; Flintsch, G.W. Stereo-vision applications to reconstruct the 3D texture of pavement surface. *Int. J. Pavement Eng.* **2011**, *12*, 263–273. [CrossRef]
18. Sengoz, B.; Topal, A.; Tanyel, S. Comparison of pavement surface texture determination by sand patch test and 3D laser scanning. *Period. Polytech. Civ. Eng.* **2012**, *56*, 73–78. [CrossRef]
19. Gabriele, B.; Andrea, S.; Fabrizio, G.; Claudio, L. Laser scanning on road pavements: A new approach for characterizing surface texture. *Sensors* **2012**, *12*, 9110–9128. [CrossRef]
20. Čelko, J.; Kováč, M.; Kotek, P. Analysis of the pavement surface texture by 3D scanner. *Transp. Res. Procedia* **2016**, *14*, 2994–3003. [CrossRef]
21. Hu, L.; Yun, D.; Liu, Z.; Du, S.; Zhang, Z.; Bao, Y. Effect of three-dimensional macrotexture characteristics on dynamic frictional coefficient of asphalt pavement surface. *Construct. Build. Mater.* **2016**, *126*, 720–729. [CrossRef]
22. Wang, K.C.P.; Li, L. Pavement surface texture modeling using 1 mm 3D laser images. In Proceedings of the Transportation Systems Workshop 2012, Austin, TX, USA, 5–8 March 2012.
23. Laurent, J.; Hébert, J.F.; Lefebvre, D.; Savard, Y. Using 3D laser profiling sensors for the automated measurement of road surface conditions. *7th RILEM Int. Conf. Crack. Pavements* **2012**, *4*, 157–167.
24. Miao, Y.; Wang, L.; Wang, X.; Gong, X. Characterizing Asphalt Pavement 3-D Macrotexture Using Features of Co-occurrence Matrix. *Int. J. Pavement Res. Technol.* **2015**, *8*, 243–250. [CrossRef]
25. Miao, Y.; Chen, G.; Wang, W.; Gong, X. Application of gray-tone difference matrix-based features of pavement macrotexture in skid resistance evaluation. *J. Southeast Univ. Engl. Ed.* **2015**, *31*, 389–395. [CrossRef]
26. Jernigan, M.E.; D'astous, F. Entropy-based texture analysis in the spatial frequency domain. *IEEE Trans. Pattern Anal. Mach. Intell.* **1984**, *2*, 237–243. [CrossRef]
27. Böhlke, T. Application of the maximum entropy method in texture analysis. *Comput. Mater. Sci.* **2005**, *32*, 276–283. [CrossRef]
28. Sun, Q.; Huang, Y.; Wang, J.; Zhao, S.; Zhang, L.; Tang, W.; Wu, N. Applying CT texture analysis to determine the prognostic value of subsolid nodules detected during low-dose CT screening. *Clin. Radiol.* **2018**. [CrossRef] [PubMed]
29. Miao, Y.; Li, J.; Zheng, X.; Wang, L. Field investigation of skid resistance degradation of asphalt pavement during early service. *Int. J. Pavement Res. Technol.* **2016**, *9*, 313–322. [CrossRef]
30. Yao, H.; You, Z. Effectiveness of Micro-and Nanomaterials in Asphalt Mixtures through Dynamic Modulus and Rutting Tests. *J. Nanomater.* **2016**. [CrossRef]
31. Yao, H.; Dai, Q.; You, Z.; Bick, A.; Wang, M. Modulus simulation of asphalt binder models using Molecular Dynamics (MD) method. *Construct. Build. Mater.* **2018**, *162*, 430–441. [CrossRef]

32. Xu, H.; Xing, C.; Zhang, H.; Li, H.; Tan, Y. Moisture seepage in asphalt mixture using X-ray imaging technology. *Int. J. Heat Mass Transf.* **2019**. [CrossRef]
33. Xu, H.; Guo, W.; Tan, Y. Internal structure evolution of asphalt mixtures during freeze-thaw cycles. *Mater. Des.* **2015**, *86*, 436–446. [CrossRef]
34. MacKay, D.J.C. *Information Theory, Inference, and Learning Algorithms*; Cambridge University Press: Cambridge, UK, 2003.

entropy

MDPI

Article

Entropy and Contrast Enhancement of Infrared Thermal Images Using the Multiscale Top-Hat Transform

Julio César Mello Román [1], José Luis Vázquez Noguera [1,*], Horacio Legal-Ayala [1], Diego P. Pinto-Roa [1], Santiago Gomez-Guerrero [1] and Miguel García Torres [2]

1 Facultad Politécnica, Universidad Nacional de Asunción, San Lorenzo 2160, Paraguay; juliomello@pol.una.py (J.C.M.R.); hlegal@pol.una.py (H.L.-A.); dpinto@pol.una.py (D.P.P.-R.); sgomezpy@gmail.com (S.G.-G.)

2 Division of Computer Science, Universidad Pablo de Olavide, ES-41013 Seville, Spain; mgarciat@upo.es

* Correspondence: jlvazquez@pol.una.py; Tel.: +595-982-652388

Received: 29 December 2018; Accepted: 25 February 2019; Published: 4 March 2019

Abstract: Discrete entropy is used to measure the content of an image, where a higher value indicates an image with richer details. Infrared images are capable of revealing important hidden targets. The disadvantage of this type of image is that their low contrast and level of detail are not consistent with human visual perception. These problems can be caused by variations of the environment or by limitations of the cameras that capture the images. In this work we propose a method that improves the details of infrared images, increasing their entropy, preserving their natural appearance, and enhancing contrast. The proposed method extracts multiple features of brightness and darkness from the infrared image. This is done by means of the multiscale top-hat transform. To improve the infrared image, multiple scales are added to the bright areas and multiple areas of darkness are subtracted. The method was tested with 450 infrared thermal images from a public database. Evaluation of the experimental results shows that the proposed method improves the details of the image by increasing entropy, also preserving natural appearance and enhancing the contrast of infrared thermal images.

Keywords: discrete entropy; infrared images; low contrast; multiscale top-hat transform

1. Introduction

Thermal infrared imaging (TII) is emerging as a powerful and non-invasive tool to accurately evaluate the thermal distribution of a body. TII is based on the physical phenomenon that all bodies above absolute zero emit thermal radiation. The intensity and spectral distribution of emitted radiation depend on the temperature, and its detection allows the creation of a thermal map of temperature distribution. TII uses the thermal radiation to create an image similar to visible light imaging. However, the use of this thermal radiation presents advantages over visible light in extreme situations since it can provide valuable information from an environment independent of the quality of the environmental light source, as is the case in foggy conditions or darkness, where TII can detect the presence of individuals, objects, or animals [1,2]. This feature makes the utilization TII very competitive to traditional methods in different fields as security, engineering, ecology, etc. [1–3].

Despite the advantages of TII, in some scenarios images may present low contrast, as well as low-level and blur details. These issues are due to facts such as limitations of the cameras with which the images are captured, conditions in the environment, etc. Therefore, contrast enhancement techniques may yield higher image details [4,5].

Many algorithms currently exist that enhance the contrast of infrared images. Histogram-based algorithms are widely used to enhance the brightness areas of an infrared image [6–9]. One of the most popular methods is the Histogram Equalization (HE). However, in the process of enhancing an image, HE drastically changes the average brightness of the image, resulting in loss of information and visually deteriorated images [10]. The HE variants cause the same problems, but to a lesser extent. Hence, global histogram-based algorithms cannot improve image entropy [11–15].

Other strategies for improving thermal infrared image are based on mathematical morphology. These are widely used to enhance contrast, improve details and edges, suppress noise, and enhance small targets [4,5,16–22]. However, the technique has some problems associated with the shape and size of the structuring element. In order to solve this problem, proposals have been presented where, in the basic operations of mathematical morphology, two structuring elements of equal sizes and different shapes are used [5,21,23]. Strategies have also been used within multiscale schemes, such as sequential toggle operators, to achieve improvements in infrared images [4,22,24,25].

The top-hat transform is one of the most used operations of mathematical morphology. Image enhancement by top-hat transform consists of adding bright areas and subtracting dark areas from the original image [26–28]. To improve the performance of top-hat transform, it is normally used in a multiscale scheme [29]. The multiscale top-hat transform can extract multiple useful features from the image, which are then used to enhance the infrared image. The multiscale top-hat transform scheme is widely used to make improvements in different types of grayscale images [26,30]. For example, it has been used to enhance retinal images [31], ultrasound images [32], and infrared images [16,22,33]. It has also been used in applications such as visible and infrared image fusion [34–36], image segmentation [37], and detection of small objects [21,22,38].

In the literature, the results obtained by infrared image enhancement algorithms based on multiscale mathematical morphology are generally evaluated using the following metrics: Peak Signal-to-Noise Ratio (PSNR) [21,30], which measures distortion in the improved images; and linear index of fuzziness (γ) measure [4,5,20,33], which quantifies the improvement in blurriness of infrared images. For the results of this work it is also of utmost importance to quantify the richness of the details of the infrared image by means of its entropy [39,40], contrast enhancement [16] to differentiate the objects from their background, and the mean brightness [11], which will tell us if the resulting image maintains its naturalness after the process of enhancement.

In this article we propose a new method based on the multiscale top-hat transform. Two geometrically proportional and flat structuring elements are used in top-hat operations [16]. The method improves the details of infrared images by increasing their global entropy. It also introduces less distortion, preserves natural brightness, and enhances contrast in the resulting thermal infrared images. In the proposed method, first the two structuring elements are selected to improve the performance of the multiscale scheme. It then extracts the light and dark areas of the image on multiple scales, and after that it sums and weighs the light and dark areas obtained. Finally, the infrared thermal image is enhanced by adding the bright regions and subtracting the dark regions.

The contributions of this work are: (1) proposing the top-hat transform by using two structuring elements of different sizes; (2) a new algorithm for improving entropy and contrast in TTI based on the multiscale top-hat transform.

The article is structured as follows: Section 2 presents the preliminary concepts of entropy and contrast, Section 3 presents the proposed method to improve the TII based on the multiscale top-hat transform, Section 4 shows the experimental results, and Section 5 concludes with the main contributions of the work.

2. Entropy and Contrast in Digital Images

TII often presents problems at the time of capture, such as poor details and low contrast. When you want to solve the above problems by means of strategies to improve the image, other types of inconvenience usually appear; for instance, loss of detail and naturalness in the image.

Entropy [39–43] quantifies the information content of the image. It describes how much uncertainty or randomness there is in an image. The more information the image contains, the better its quality. In [44], Wang et al. propose a method based on fractional Fourier entropy map, multilayer perceptron, and Jaya algorithm in multiple sclerosis identification. In [45], Zhang et al. propose a smart detection method for abnormal breasts in digital mammography. In this case, fractional Fourier entropy was employed to extract global features. In [46], Lee et al. investigate a framework for expressing visual information in bits termed visual entropy, based on information theory.

The entropy (E) referred to here is Shannon's entropy. In the field of information theory, entropy, also called entropy of information and Shannon's entropy, measures the uncertainty of a source of information [47]. Shannon's entropy is defined as:

$$E(I) = -\sum_{k=0}^{L-1} p(k)log_2(p(k)), \tag{1}$$

where I is the original image, $p(k)$ is the probability of occurrence of the value k in the image I, and $L = 2^q$ indicates the number of different gray levels. $E(I)$ is a convenient notation for the entropy of an image, and should not be interpreted here as a mathematical expectation since I is not a random variable. It is not difficult to prove that if q is the number of bits representing each pixel in the image, then $E(I) \in [0, q]$; for this work $q = 8$ for infrared thermal images in gray scale.

In Figure 1 we can observe the histogram of an 8-bit image (histogram with uniform distribution). In this case the entropy has maximum value, i.e., the entropy has a value equal to 8. This happens when the probabilities of all possible results are equal. Also, it can be seen that the histogram uses all the available dynamic range, that is to say in the histogram we visualize all the values of intensity in the range $[0, 255]$. Minimal entropy happens when the result is a certainty and its value is zero. In image processing, discrete entropy is a measure of the number of bits required to encode image data [41]. The higher the value of the entropy, the more detailed the image will be.

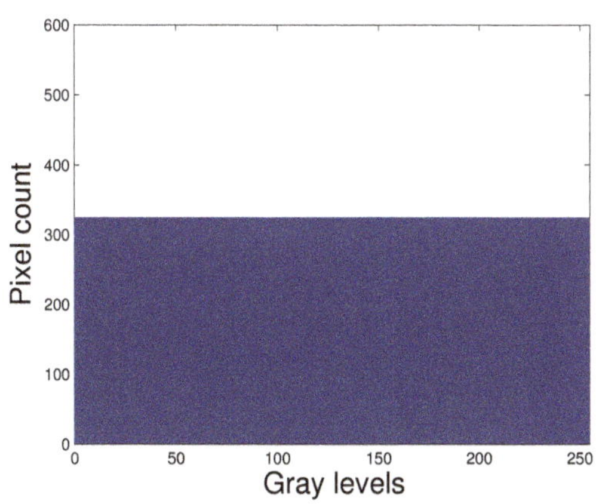

Figure 1. Histogram with uniform distribution.

Contrast is defined as the difference between the light and dark areas of the image. The higher the variance of gray intensities, the higher the contrast. When the difference between the maximum and minimum intensities of an image is very small, the image has low entropy and poor contrast. Niu et al. [48] introduce a contrast enhancement algorithm of tone-preserving entropy maximization. Yoo et al. [10] propose an image enhancement method called MEDHS (Maximum Entropy Distribution based Histogram Specification), which uses the Gaussian distribution to maximize the entropy and preserve the mean brightness.

Unlike the methods mentioned above, in this work we propose a new method based on mathematical morphology. This method increases the global entropy and contrast, improving the details of the TII.

In Figure 2 we can see the infrared thermal image with its associated histogram. Observing the histogram of the image, we can see that it does not effectively use the whole range of available intensity values. This indicates that the image has poor entropy and low contrast. When calculating Shannon's entropy (Equation (1)) we can see that it has a value of $E = 6.008$.

(**a**)

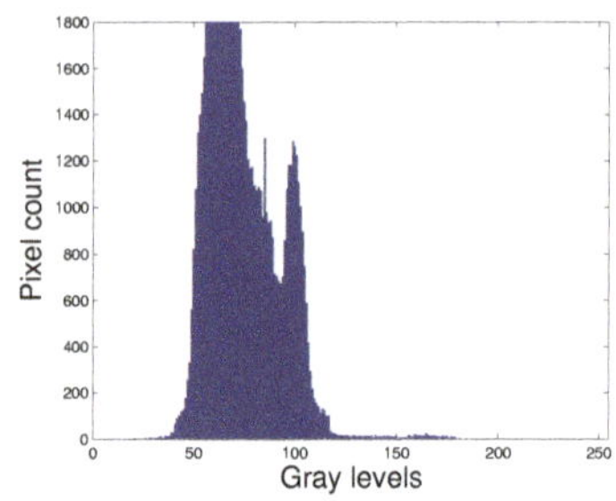

(**b**)

Figure 2. Thermal infrared image. (**a**) Original TII; (**b**) Histogram of TII.

As an example, Figure 3 shows the thermal infrared image (TII) obtained with the HE algorithm and its histogram. The HE method enhanced the contrast of the TII by making it brighter. In the histogram of the improved image we can visualize that the intensities are redistributed towards the available extreme values, leaving many holes. However, the method did not improve Shannon's entropy, obtaining a value of $E = 5.933$, which is less than the entropy of the unprocessed Figure 2a. Visually it is observed in Figure 3a that there is a loss in details, for example it is not possible to differentiate well the horse from the person.

(**a**)

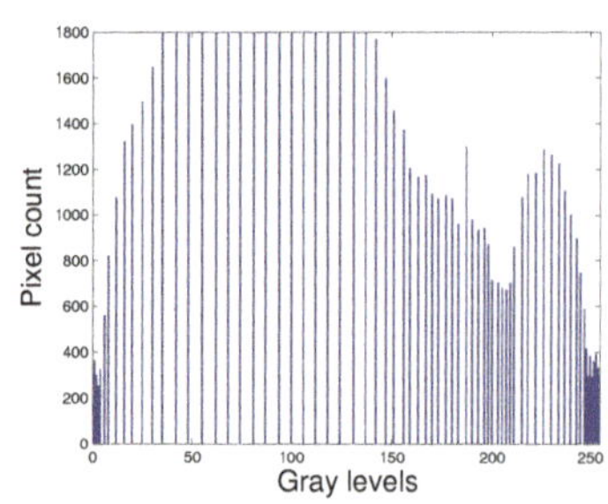

(**b**)

Figure 3. Loss of information with enhanced contrast. (**a**) TII enhanced with HE; (**b**) Histogram of the TII enhanced with HE.

To solve the problem of improving the image without incurring in a loss of the details and the mean brightness of the image, we will make a detailed description of the proposed method based on multiscale mathematical morphology in the following section.

3. Enhancement of Thermal Infrared Images

The top-hat transform is one of the most used operations of mathematical morphology to obtain improvements in the TII [4,5,16,20–22]. Two structuring elements of proportional sizes, equal shapes and planes, will be used to improve the performance of the top-hat transform [16].

3.1. Classic Top-Hat Transform

The top-hat transform is a composite operation of mathematical morphology; it is defined from other morphological operations, namely erosion, dilation, opening, and closing.

The morphological operations of dilation and erosion of $I(u,v)$ for $B(s,t)$, denoted by $(I \oplus B)$ and $(I \ominus B)$, are defined as follows [27,49]:

$$(I \oplus B)(u,v) = \max_{(s,t)\in I} \{I(u+s,v+t) + B(s,t)\}, \tag{2}$$

$$(I \ominus B)(u,v) = \min_{(s,t)\in B} \{I(u+s,v+t) - B(s,t)\}. \tag{3}$$

where I is the original infrared thermal image whose pixels are represented for all (u,v) spatial coordinates and B is the structuring element whose spatial coordinates are represented by (s,t).

The opening $(I \circ B)$ and closing $(I \bullet B)$ morphological operations of $I(u,v)$ for $B(s,t)$ are defined from the dilation and erosion operations as follows [27,49]:

$$I \circ B = (I \ominus B) \oplus B, \tag{4}$$

$$I \bullet B = (I \oplus B) \ominus B. \tag{5}$$

The top-hat transform morphological operation [27] is defined from the morphological opening and closing. White Top-Hat (WTH) is the top-hat transform through opening, Black Top-Hat (BTH) is the top-hat transform through closing. WTH gets the bright areas and BTH gets the dark areas lost in the opening and closing operations. Both transforms are defined as follows:

$$WTH = I - (I \circ B) = I - ((I \ominus B) \oplus B), \tag{6}$$

$$BTH = (I \bullet B) - I = ((I \oplus B) \ominus B) - I. \tag{7}$$

3.2. Modified Top-Hat Transform

The classical top-hat transform is characterised by the use of a single structuring element. This makes its image processing performance inefficient [29]. To improve the performance of the top-hat transform it is proposed to use two structuring elements, whose characteristics will be proportional geometry and flat [16]. The Modified White Top-Hat ($MWTH$) and the Modified Black Top-Hat ($MBTH$) transforms will be used for image improvement within the scheme of multiscale top-hat transform.

Let the structuring elements be G and G' geometrically proportional and flat. Then, the top-hat transform that we will use in the multiscale scheme is defined as follows:

$$MWTH = I - ((I \ominus G) \oplus G'), \tag{8}$$

$$MBTH = ((I \oplus G) \ominus G') - I. \tag{9}$$

Note that if $G = G'$, then Equations (8) and (9) are equal to Equations (6) and (7). Therefore, the classical top-hat transform is a particular case of the modified top-hat transform. In [16], Román et al. show that the modified top-hat transform improves thermal images, enhancing the contrast, preserving the details and introducing less distortion.

3.3. How Entropy is Changed by Top-Hat Transform

The Shannon entropy depends on both (a) the number of distinct values exhibiting a positive frequency, and (b) how uneven the density function is, compared with a discrete uniform distribution.

The top-hat transform, working within a local region of the image, often generates new values of grey, thereby causing a small to moderate increase in the entropy value of the region. This occurs because Equations (8) and (9) can induce one or more new levels of grey when the logic is executed.

When one new level of grey h is added by the algorithm to a region being worked, it replaces another value g at certain spatial position. There are two possibilities: either

- The old value g was unique in the region, with a count of 1, hence it disappears from the region and is replaced by value h. No change in entropy occurs because in the old g bin of the histogram the count of 1 becomes 0, and in the new h bin the count of 0 becomes 1; or
- The old value g existed in $k > 1$ pixels in the region. In this case the count in the g bin decreases to $k - 1$, and the count in the h bin increases to 1. The following Lemma shows that this change in the histogram increases the region's entropy.

Lemma 1. *Consider a rectangular region of m pixels in an image. Let $H(X)$ be the original entropy of the grey scale X in use. Suppose that grey level g appears in k pixels of the original image and grey level h does not appear. Further, suppose that an image transformation replaces grey level g with grey level h at certain pixel of the rectangular region. Then the entropy of the transformed region increases to*

$$H'(X) = H(X) - p_g\left[(1-\varepsilon)\,\log(1-\varepsilon) + \varepsilon\,\log(\varepsilon)\right] \tag{10}$$

where the value of ε is $1/m$, the inverse of the number of pixels in the region.

Proof. Without loss of generality assume 255 levels of grey; thus both g and h are integers in $\{1, ..., 255\}$. As the sum of probabilities before and after the transformation equals 1, the increase in p_h occurs at the expense of a decrease in p_g; that is, p_h increases from 0 to εp_g and p_g decreases to $(1-\varepsilon)p_g$.

For the region under consideration, Equation (1) can be written as

$$H(X) = -\sum_{i=0}^{255} p_i \log(p_i) = -[p_0 \log(p_0) + \cdots + p_{255} \log(p_{255})].$$

After transformation, the probability corresponding to level g is broken down in two: a portion εp_g for newly incorporated level h and a portion $(1-\varepsilon)p_g$ for level g. Thus

$$\begin{aligned}
H'(X) &= [H(X) + p_g \log(p_g)] - \varepsilon p_g \log(\varepsilon p_g) - (1-\varepsilon)p_g \log((1-\varepsilon)p_g) \\
&= [H(X) + p_g \log(p_g)] - \varepsilon p_g(\log(\varepsilon) + \log(p_g)) - (1-\varepsilon)p_g(\log(1-\varepsilon) + \log(p_g)) \\
&= [H(X) + p_g \log(p_g)] - \varepsilon p_g \log(\varepsilon) - \varepsilon p_g \log(p_g) - (1-\varepsilon)p_g \log(1-\varepsilon) - (1-\varepsilon)p_g \log(p_g) \\
&= H(X) - \varepsilon p_g \log(\varepsilon) - (1-\varepsilon)p_g \log(1-\varepsilon) \\
&> H(X).
\end{aligned}$$

The term $-p_g[\varepsilon \log(\varepsilon) + (1-\varepsilon)\log(1-\varepsilon)]$ is a positive value representing the increase in entropy when a new level of grey is incorporated in the region. This completes the proof. □

The smallest possible frequency for any level of grey in a region of m pixels is $1/m$ as mentioned in the Lemma. In practice, ε may be larger than $1/m$; this occurs when more than one pixel is assigned the new grey level h. It is easy to show that $\varepsilon = 0.5$ would yield a maximum increase in entropy at current iteration; however, according to the method proposed below (next subsection), entropy increases incrementally as the algorithm iterates.

3.4. Proposed Method Using Multiscale Top-Hat Transform

The proposed method is based on the multiscale top-hat transform. This method employs two structuring elements in the top-hat transform to improve its performance. The proposed method improves the image in terms of detail, contrast and mean brightness conservation. The infrared image enhancement algorithm initially uses the following parameters: The original image I, the number of iterations n in a range $i \in \{1, 2, \ldots, n\}$, $n > 1$; and two structuring elements G and G'.

Multiple Brightness (MB) and Multiple Darkness (MD) areas will be obtained by top-hat transform as follows:

$$MB_i = I - ((I \ominus G_i) \oplus G_i'), \tag{11}$$

where MB_i is the i-scales of brightness extracted from the image, and G_i and G_i' will grow in each iteration. G' will always be greater than or equal to G.

$$MD_i = ((I \oplus G_i) \ominus G_i') - I, \tag{12}$$

where MD_i are the i-scales of darkness extracted from the image.

The Subtractions of the Neighboring Bright Scales ($SNBS$) are then calculated. This operation is expressed as follows:

$$SNBS_{i-1} = \begin{cases} MB_i - MB_{i-1}, \text{ to } i = 2 \\ MB_i - SNBS_{i-2}, \text{ to } i > 2 \end{cases} \tag{13}$$

where $SNBS_{i-1}$ are the $(i-1)$-differences of the neighboring brightness scales obtained from the image.

Similarly, the Subtractions of the Neighboring Dark Scales ($SNDS$) are calculated. This operation is expressed as follows:

$$SNDS_{i-1} = \begin{cases} MD_i - MD_{i-1}, \text{ to } i = 2 \\ MD_i - SNDS_{i-2}, \text{ to } i > 2 \end{cases} \tag{14}$$

where $SNDS_{i-1}$ are the $(i-1)$-differences of the neighboring dark scales obtained from the image.

The Sum of all the brightness (SMB and $SSNBS$) and darkness (SMD and $SSNDS$) values obtained in the multiscale process are then calculated as follows:

$$SMB = \sum_{i=1}^{n} MB_i, \tag{15}$$

$$SMD = \sum_{i=1}^{n} MD_i, \tag{16}$$

$$SSNBS = \sum_{i=1}^{n-1} SNBS_{i-1}, \tag{17}$$

$$SSNDS = \sum_{i=1}^{n-1} SNDS_{i-1}. \tag{18}$$

Finally, the image enhancement (I_E) will be obtained as follows:

$$I_E = I + \omega \times (SMB + SSNBS) - \omega \times (SND + SSNDS), \tag{19}$$

where $\omega \in [0, 1]$ is a weighting factor or regulator of the bright and dark areas.

The TII enhancement process is described in the following Algorithm 1.

Algorithm 1 Proposed method for TII Enhancement

Input: I, G, G', n, ω
Output: I_E *(Enhanced image)*

Initialization : G, G'

1: **for** $i = 1$ to n **do**

2: *Calculation of top-hat transform.*

$MB_i = I - ((I \ominus G_i) \oplus G_i')$ (Equation (11))

$MD_i = ((I \oplus G_i) \ominus G_i') - I$ (Equation (12))

3: *Calculation of subtractions from neighboring scales, obtained through the top-hat transform. The top-hat is subtracted with the previous difference, from the first subtraction of the first neighboring top-hat.*

$$SNBS_{i-1} = \begin{cases} MB_i - MB_{i-1}, \text{ to } i = 2 \\ MB_i - SNBS_{i-2}, \text{ to } i > 2 \end{cases} \text{(Equation (13))}$$

$$SNDS_{i-1} = \begin{cases} MD_i - MD_{i-1}, \text{ to } i = 2 \\ MD_i - SNDS_{i-2}, \text{ to } i > 2 \end{cases} \text{(Equation (14))}$$

4: **end for**

5: *Calculation of the maximum values of all the multiple scales obtained.*

$SMB = \sum_{i=1}^{n} MB_i$ (Equation (15))

$SMD = \sum_{i=1}^{n} MD_i$ (Equation (16))

$SSNBS = \sum_{i=1}^{n-1} SNBS_{i-1}$ (Equation (17))

$SSNDS = \sum_{i=1}^{n-1} SNDS_{i-1}$ (Equation (18))

6: *TII enhancement calculation.The contrast enhancement calculation consists of adding the results of the multiple bright scales to the original image and subtracting the results of the multiple dark scales.*

$I_E = I + \omega \times (SMB + SSNBS) - \omega \times (SND + SSNDS)$ (Equation (19))

7: **return** I_E

4. Results and Discussion

Experiments were performed by randomly selecting 450 TII of 324 × 256 from a public repository [50]. We analyzed 9 different scenes of 50 images each one. Images were captured with an infrared thermal camera FLIR Tau 320 with a resolution of 324 × 256 pixels. Images in the database are 8-bit and 16-bit. The database has no radiometric data. Tests were performed on the 8-bit images. Figure 4 shows the scenes. The computer used has the following features: Pentium Dual-Core 2.3 GHz processor, RAM 4GB, HD 1TB, and the operating system used was Windows 7.

In order to test the performance of the proposed method, we considered three different experiments:

- In the first part (Section 4.1) we perform a parameter adjustment to find good parameter values that maximize the entropy of the output image after applying the proposed method.
- Then, in the second part (Section 4.2) we analyze the proposed method per iteration and compare its performance with Multiscale Morphological Infrared Image Enhancement (MMIIE) (mathematical morphology-based multiscale approach) [4].
- Finally, in the last part (Section 4.3), we apply the proposed method and compare the results achieved with the proposed techniques with the following competitive methods from the literature: HE, Contrast Limited Adaptive Histogram Equalization (CLAHE) [51], the method of Kun Liang et al. [6] called IRHE2PL for infrared images, and the MMIIE method for infrared images.

(**a**) Scene 1 (**b**) Scene 2 (**c**) Scene 3
(**d**) Scene 4 (**e**) Scene 5 (**f**) Scene 6
(**g**) Scene 7 (**h**) Scene 8 (**i**) Scene 9

Figure 4. Examples of scenes from the database.

4.1. Parameter Tuning

In this section, the goal is to find a good combination of values of the parameters of ω and the number of iterations n. Parameter ω has real values. As we cannot perform tests for all real values, we take a selection criteria for values that we consider representative to get good outcomes for ω and n. The search for more optimal values of these parameters could be approached in future work as an optimization problem. For this experiment we applied the proposed method in the selected dataset. Since we are seeking to optimize the entropy of the resulting image, we use such Equation (1) as evaluation metrics.

The parameter values of the proposed method are presented in Table 1. As shown, we tested different values of the number of iterations n and w. For n, we changed the value from 2 to 10. No larger values were considered because the larger the value is, the more the image becomes distorted. The parameter w was changed in the range of $[0,1]$ in increments of 0.05. A value of 0 gives as result the original image. The initial structuring elements G and G' are squares of 3×3 and 15×15, respectively. In each iteration the two structuring elements will side increase in sizes of two.

Table 2 presents the results obtained. Each column refers to the corresponding iteration n while each row corresponds to a different value of ω. Higher results are highlighted in bold. The highest result is achieved with $n = 8$ and $\omega = 0.35$. It is worth stressing that the entropy increases when the iteration rises its value until the local optima and from there, values start to decrease.

Table 1. Parameter values for the parameter tuning experiment.

Parameter	Value(s)
n	$[2, 10]$
ω	$[0, 1]$
G	3×3
G'	15×15

Table 2. Entropy values of the enhanced thermal infrared imaging (TII) obtained by the proposed method with parameters ω and n.

ω	n								
	2	3	4	5	6	7	8	9	10
0.05	6.5931	6.5962	6.6041	6.6188	6.6396	6.6656	6.6950	6.7271	**6.7607**
0.10	6.5971	6.6120	6.6394	6.6777	6.7224	6.7696	6.8174	6.8658	**6.9129**
0.15	6.6037	6.6315	6.6756	6.7326	6.7900	6.8502	6.9100	6.9653	**7.0099**
0.20	6.6145	6.6577	6.7190	6.7861	6.8548	6.9228	6.9848	7.0326	**7.0593**
0.25	6.6242	6.6821	6.7540	6.8316	6.9081	6.9788	7.0349	**7.0661**	7.0648
0.30	6.6293	6.6970	6.7790	6.8678	6.9498	7.0185	7.0633	**7.0702**	7.0348
0.35	6.6430	6.7217	6.8089	6.9025	6.9851	7.0475	**7.0740**	7.0519	6.9828
0.40	6.6518	6.7420	6.8398	6.9380	7.0181	7.0673	**7.0688**	7.0161	6.9169
0.45	6.6568	6.7545	6.8607	6.9637	7.0394	**7.0735**	7.0505	6.9706	6.8448
0.50	6.6806	6.7872	6.8957	6.9957	7.0596	**7.0726**	7.0225	6.9165	6.7668
0.55	6.6824	6.7928	6.9068	7.0085	**7.0647**	7.0610	6.9900	6.8619	6.6915
0.60	6.6914	6.8113	6.9314	7.0295	**7.0702**	7.0450	6.9516	6.8011	6.6111
0.65	6.6945	6.8201	6.9451	7.0404	**7.0682**	7.0249	6.9112	6.7406	6.5330
0.70	6.7066	6.8408	6.9655	7.0516	**7.0631**	7.0010	6.8676	6.6786	6.4556
0.75	6.7161	6.8588	6.9835	**7.0602**	7.0550	6.9745	6.8216	6.6152	6.3778
0.80	6.7221	6.8707	6.9979	**7.0639**	7.0433	6.9462	6.7747	6.5531	6.3032
0.85	6.7259	6.8791	7.0077	**7.0650**	7.0303	6.9167	6.7289	6.4929	6.2319
0.90	6.7309	6.8906	7.0183	**7.0641**	7.0140	6.8843	6.6802	6.4311	6.1612
0.95	6.7326	6.8959	7.0235	**7.0604**	6.9973	6.8515	6.6320	6.3702	6.0927
1.00	6.7791	6.9368	7.0460	**7.0610**	6.9780	6.8134	6.5783	6.3055	6.0210

In Figure 5 we can see that images of the same scene with similar entropy, but with different configurations of ω and n, get similar visual results. For the other experiments, we select the configuration that has the best average.

(**a**) (**b**) (**c**)

Figure 5. Visual results obtained with the proposed method and the configuration of ω and n. (**a**) Original TII with $E = 6.7893$; (**b**) TII enhancement with proposed method with $\omega = 0.35$, $n = 8$ and $E = 7.4696$; (**c**) TII enhancement with proposed method with $\omega = 0.45$, $n = 7$ and $E = 7.4467$.

4.2. Performance of Proposed Method per Iteration

In this section we compare the results per iteration between proposed method and MMIIE method, using the 450 infrared thermal images. We compare the performance of the entropy (Equation (1)) and with the following metrics:

- The *Standard Deviation* (SD), which quantifies the global contrast of the infrared images, is defined as [16]:

$$SD(I) = \sqrt{\sum_{k=0}^{L-1} (k - A(I))^2 \times p(k)}, \tag{20}$$

where k is the pixel value of the image I, $L-1$ is the maximum gray level, the average intensity of the image is represented by $A(I)$, and $p(k)$ is the probability of occurrence of the value k. If $SD(I_E)$ is greater than $SD(I)$, then there is contrast enhancement.

- The metric adopted to measure the signal-to-noise ratio of an image is the PSNR.

Given the original infrared image I and the infrared image with enhancement I_{EN} where the size of the images is $M \times N$, the PSNR between I and I_{EN} is given by [30]:

$$PSNR(I, I_E) = 10 \times log_{10} \frac{(L-1)^2}{MSE(I, I_E)}. \tag{21}$$

The *Mean Squared Error* (MSE) is defined as:

$$MSE(I, I_E) = \frac{1}{M \times N} \sum_{u=0}^{M-1} \sum_{v=0}^{N-1} (I(u,v) - I_{EN}(u,v))^2. \tag{22}$$

- The *Absolute Mean Brightness Error* (AMBE) [11], which quantifies the conservation of the mean brightness of the processed image, is given by:

$$AMBE(I, I_E) = |A(I) - A(I_E)|, \tag{23}$$

where I and I_E represent the input infrared image and the image enhancement, respectively, $A(I)$ and $A(I_E)$ represent the mean brightness of the input infrared image and the image enhancement. The lower the AMBE value, the better the mean brightness of the image is preserved.

- The linear blur index γ [4] is used to measure the performance of the infrared image enhancement. It is defined as follows:

$$\gamma(I) = \frac{2}{M \times N} \sum_{u=1}^{M} \sum_{v=1}^{N} min\{p_{uv}, (1 - p_{uv})\}, \tag{24}$$

$$p_{uv} = sin[\frac{\pi}{2} \times (1 - \frac{I(u,v)}{L-1})]. \tag{25}$$

where $M \times N$ is the size of the infrared image. $I(u,v)$ is the gray pixel value (u,v). $L-1$ is the maximum gray value of I. The performance of the algorithm is better if the value of γ is small.

Following the recommendations of Bai [4], we set the following parameter values for MMIIE method. The number of iterations n was set to 10, the weights to $w_1 = 0.6$, $w_2 = w_3 = 1.5$, and the initial structuring element B was fixed to 3×3. For the proposed strategy and taking into account the parameter tuning, we fixed w to 0.35 and considered the same parameter values as in the previous experiment; $n = 10$ and G and G' to 3×3 and 15×15 respectively. In each iteration the two structuring elements will increase in sizes of two. The average entropy of the 450 original thermal infrared images is $E = 6.5924$. The proposed method and MMIIE was implemented using ImageJ [49].

In Table 3 it can be seen the results on average on each iteration for the 450 images with the proposed method and MMIIE method. Results in bold refer to the best results by iteration. According to the results obtained we can say that the proposed method outperforms MMIIE method in four of the five evaluation measures on every iteration, while MMIIE method achieves better results with γ. Therefore, on average, the proposed method provides better contrast enhancement and signal-to-noise ratio and higher level of detail. Furthermore, it also preserves better the brightness. On the other hand MMIIEE method provides better blur effect. The computational time of the proposed method is higher than that of the MMIIE method, but the proposed method obtains better results in fewer iterations (lower n).

Table 3. Results achieved with the proposed method and Multiscale Morphological Infrared Image Enhancement (MMIIE) method

n	Proposed Method						MMIIE					
	E	SD	$PSNR$	$AMBE$	γ	Time (ms)	E	SD	$PSNR$	$AMBE$	γ	Time (ms)
2	**6.643**	**40.835**	**40.417**	**0.129**	0.332	2328	6.324	30.247	16.236	38.072	**0.194**	**454**
3	**6.722**	**41.969**	**33.564**	**0.286**	0.320	3752	6.447	32.069	16.176	37.775	**0.193**	**902**
4	**6.809**	**43.632**	**29.256**	**0.466**	0.311	6719	6.441	33.016	16.211	37.528	**0.192**	**1605**
5	**6.902**	**45.854**	**26.023**	**0.714**	0.301	10,629	6.519	34.438	16.204	37.193	**0.193**	**2759**
6	**6.985**	**48.556**	**23.509**	**1.107**	0.293	11,947	6.515	35.486	16.216	36.902	**0.194**	**4770**
7	**7.047**	**51.749**	**21.434**	**1.636**	0.286	16,429	6.570	36.798	16.177	36.580	**0.195**	**7187**
8	**7.074**	**55.353**	**19.693**	**2.299**	0.281	19,979	6.565	37.558	16.194	36.318	**0.196**	**9255**
9	**7.052**	**59.216**	**18.219**	**3.034**	0.275	20,176	6.612	38.453	16.178	36.093	**0.197**	**18,318**
10	**6.983**	**63.111**	**16.980**	**3.830**	0.268	21,527	6.604	39.017	16.201	35.884	**0.197**	**20,003**

Figure 6 presents an example image from the dataset with its corresponding histogram. The original image and its histogram is shown in Figure 6a,b respectively. Figure 6c is the resulting image after applying MMIIE method and Figure 6e when applying the proposed method. In both cases we have selected the iteration in which the entropy is maximum. It can be observed that the proposed method presents a better redistribution of intensity levels in bright areas and has fewer peaks according to its histogram (Figure 6d,f. It therefore leads to a better level of detail and contrast.

(a)

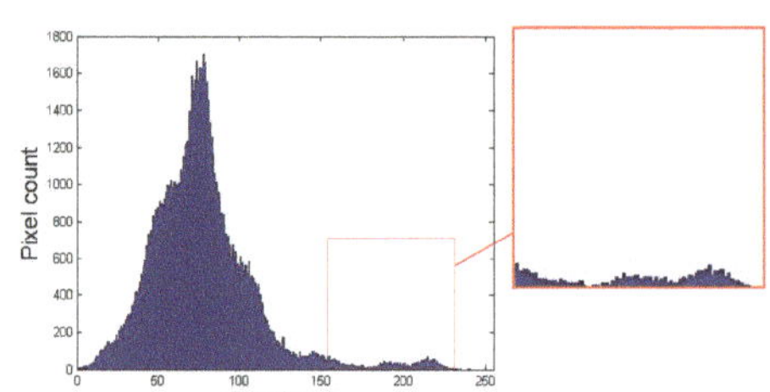

(b)

(c)

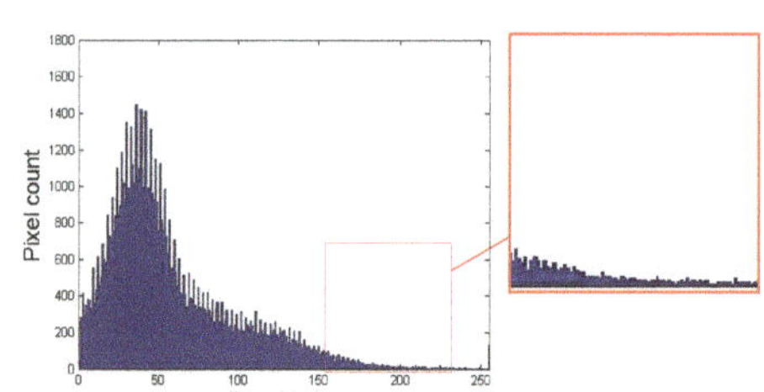

(d)

Figure 6. *Cont.*

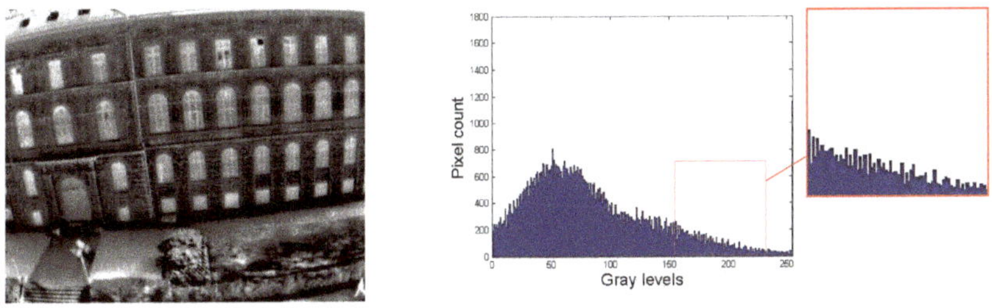

(e) (f)

Figure 6. Example of a thermal infrared imaging (TII) to compare the proposed method and Multiscale Morphological Infrared Image Enhancement (MMIIE) method. (**a**) Original TII with $E = 6.8387$ and $SD = 31.8204$; (**b**) Histogram of the original image; (**c**) TII enhanced with the MMIIE method with $E = 6.8619$, $SD = 42.5731$, and $AMBE = 24.1018$; (**d**) Histogram of the image enhanced with MMIIE; (**e**) TII enhanced with the proposed method with $E = 7.3629$, $SD = 58.8730$, and $AMBE = 2.1848$; (**f**) Histogram of the image enhanced with the proposed method.

4.3. Comparison of the Performance of the Proposed Method with State of the Art Methods

In this section we compare the proposed method with popular methods from the literature. The methods used in this section are HE, CLAHE, IRHE2PL [6], and MMIIE. In this part we analyze two results. First we are interested in knowing the percentage of images that are enhanced compared to the original using each method. Then, we analyze the performance of each method on each different scene.

The parameters of the various methods are as follows. For CLAHE, the method was implemented with the MATLAB program, using its default values. For the IRHE2PL method the parameters are described in [6]. For MMIIE method we set the number of iteration n to 9 and the weights to $w_1 = 0.6$, $w_2 = w_3 = 1.5$. Finally, the initial structuring element B was fixed to 3×3 square. Results for this method use the number of iterations that maximize the entropy. Finally, for the proposed method we selected the best combination found in the first part. In this case $n = 8$, $\omega = 0.35$, and the initial structuring elements G and G' are squares of 3×3 and 15×15, respectively. The HE, IRHE2PL, MMIIE methods and the proposed method were implemented using the ImageJ library [49].

Table 4 shows the percentage of images that have been enhanced in terms of contrast. An image is considered improved if the value of the standard deviation of the processed image is greater than the original image. As we can see, the proposed method enhances the contrast of the 450 TII. HE has a very high value with 98.89%, followed by IRHE2PL method, CLAHE method, and MMIE method with 90.22%, 82.67%, and 47.56%, respectively.

Table 4. Contrast improvement percentage.

Methods	Percentage of Images Improved (%)
HE	98.89%
CLAHE	82.67%
IRHE2PL	90.22%
MMIIE	47.56%
Proposed method	100%

4.3.1. Analysis of Methods by Scenes

The performance of each method in each scene is presented in Table 5. For each scene the first row refers to the original image followed by the different methods used in this study. The two best values for each metric are highlighted in bold. Results of each scene are presented are averaged over all the images that belong to such scene. At the end the average over all images is also presented.

Table 5. Average of the assessments of the 9 scenes obtained by the methods.

	Methods	E	SD	$PSNR$	$AMBE$	γ
Scene 1	I	6.814	32.336	-	-	0.284
	HE	6.596	**73.420**	11.543	48.519	0.406
	CLAHE	**7.557**	50.808	15.984	24.259	0.401
	IRHE2PL	6.814	36.776	**29.603**	**7.065**	0.293
	MMIIE	6.910	43.573	**17.005**	25.392	**0.164**
	Proposed method	**7.418**	**60.136**	16.767	**1.887**	**0.273**
Scene 2	I	7.039	55.330	-	-	0.454
	HE	6.844	**73.364**	20.479	3.101	**0.400**
	CLAHE	**7.500**	53.721	**21.237**	**1.657**	0.488
	IRHE2PL	7.036	58.604	**33.310**	6.802	0.453
	MMIIE	7.038	45.486	13.085	48.345	**0.273**
	Proposed method	**7.601**	**69.425**	18.215	**1.534**	0.419
Scene 3	I	5.945	18.269	-	-	0.477
	HE	5.881	**73.063**	10.789	47.807	0.408
	CLAHE	**6.970**	32.154	19.839	18.275	0.485
	IRHE2PL	5.945	**42.197**	**20.011**	**13.270**	0.326
	MMIIE	6.133	20.900	17.288	32.819	**0.127**
	Proposed method	**6.826**	30.332	**23.205**	**0.136**	**0.273**
Scene 4	I	6.808	41.521	-	-	0.342
	HE	6.642	**73.148**	12.977	41.972	0.407
	CLAHE	**7.482**	48.135	18.560	19.816	0.422
	IRHE2PL	6.808	56.144	**22.617**	**11.306**	0.313
	MMIIE	6.848	35.941	16.106	33.095	**0.149**
	Proposed method	**7.566**	**56.723**	**19.019**	**1.328**	**0.312**
Scene 5	I	7.052	40.839	-	-	0.356
	HE	6.901	**73.319**	14.793	32.298	0.404
	CLAHE	**7.620**	50.630	**17.786**	15.352	0.433
	IRHE2PL	7.048	45.807	**32.631**	**11.444**	**0.307**
	MMIIE	7.025	44.123	15.660	34.956	**0.173**
	Proposed method	**7.505**	**61.669**	17.599	**2.584**	0.317
Scene 6	I	6.272	24.626	-	-	0.152
	HE	6.158	**72.882**	8.091	86.636	0.408
	CLAHE	**7.200**	**42.379**	16.477	31.714	0.263
	IRHE2PL	6.272	37.308	**21.316**	17.733	0.179
	MMIIE	6.035	28.802	**21.602**	**15.389**	**0.048**
	Proposed method	**6.702**	40.545	20.893	**2.284**	**0.114**
Scene 7	I	6.990	67.015	-	-	0.348
	HE	6.783	**73.516**	18.535	19.921	0.398
	CLAHE	**7.548**	66.298	19.828	**7.159**	0.400
	IRHE2PL	6.987	75.173	**33.982**	14.009	**0.295**
	MMIIE	7.125	53.062	12.017	54.459	0.327
	Proposed method	**7.204**	**75.462**	**19.735**	**1.405**	**0.323**
Scene 8	I	6.219	28.522	-	-	0.237
	HE	6.131	**72.805**	8.042	89.033	0.409
	CLAHE	**7.134**	41.543	16.828	31.090	0.309
	IRHE2PL	6.219	**62.031**	11.319	60.173	0.334
	MMIIE	5.952	23.883	**21.792**	**14.567**	**0.077**
	Proposed method	**6.589**	37.486	**23.014**	**2.426**	**0.118**
Scene 9	I	6.191	53.458	-	-	0.448
	HE	6.001	**80.909**	13.975	41.396	**0.329**
	CLAHE	**6.459**	59.638	**19.045**	13.075	0.440
	IRHE2PL	6.188	**67.644**	**30.537**	**10.979**	0.386
	MMIIE	**6.438**	50.305	11.052	65.813	0.433
	Proposed method	6.254	66.401	18.794	**7.105**	**0.378**

Based on the averages of the metrics E, SD, $PSNR$, $AMBE$, and γ we can conclude that:

- E metric: The CLAHE method and the proposed method are the methods that have the best performance in terms of entropy for scenes 1 to 8. However, in scene 9 the CLAHE and MMIIE methods have the best results.
- SD metric: The HE, CLAHE, IRHE2PL methods and the proposed method enhance the contrast of the TII in the 9 scenes. The MMIIE method did not enhance the contrast of scenes 2, 4, 7, 8, and 9. The HE method is the best performing method for all scenes and the proposed method is in second place.
- $PSNR$ metric: The methods that produce the less distortion to TII are the IRHE2PL, the proposed method, and CLAHE.
- $AMBE$ metric: For all scenes, the best method in regards to maintaining the average brightness is the proposed method.
- γ metric: The MMIIE method and the proposed method present the best results in terms of blurring.

In the results of scenes 7 and 9 we can see that the IRHE2PL method does not improve the entropy of the image, but has a higher $PSNR$ than the proposed method. This is because IRHE2PL generates an image very similar to the original image. Figure 7 shows that the TII (scene 9) enhanced with the IRHE2PL method (Figure 7b) is very similar to the original image (Figure 7a), contrary to the proposed method, which improves the entropy and contrast of the image (Figure 7c).

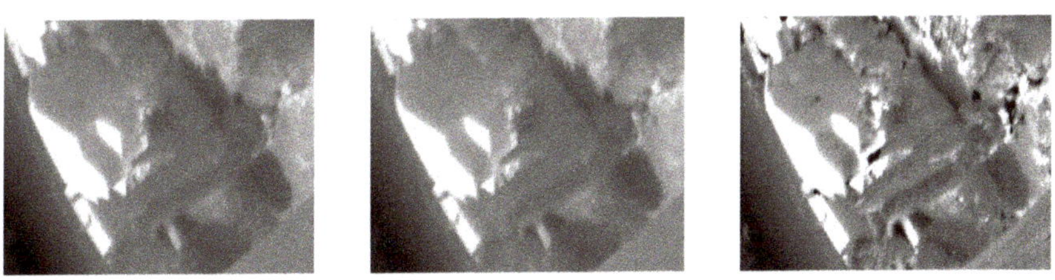

(a) (b) (c)

Figure 7. (**a**) Original TII 449.png, $E = 7.2210$, $SD = 53.1827$; (**b**) TII enhanced with IRHE2PL method, $E = 7.2058$, $SD = 53.3164$, and $PSNR = 48.0940$ and (**c**) TII enhanced with the proposed method, $E = 7.2411$, $SD = 62.8887$, and $PSNR = 20.6578$.

4.3.2. General Analysis of Methods

In general none of the methods presented outperforms the other techniques in all evaluation criteria. In almost all cases, the proposed method is the strategy that achieves the best performance in entropy, contrast, and $AMBE$. Therefore, it is the one that provides better details and keeps a better brightness. HE is the algorithm that yields higher contrast according to its good performance, in all cases, with SD. The CLAHE method performs the best in entropy, so it preserves the details better. The best signal-to-noise ratio is achieved by the IRHE2PL method and, so, it is the one that provides the lowest distortion. Finally, the MMIIE method improves the blur effect in all cases. The proposed method provides the best values in AMBE in almost all cases and in E, SD, $PSNR$, and γ results are very competitive since it achieves high values that are close to the best value in many cases.

Now we can see a couple of examples in Figures 8 and 9. For each image enhancement method, images were selected so that each method achieved the highest entropy value. In Figure 8 represents a patio scene images with Figure 8a the original image. The results of HE is in Figure 8b and it presents an excess of brightness. In Figure 8c the CLAHE method makes a moderate enhancement to the image. However, in Figure 8d the IRHE2PL method does not enhance the contrast of the TII. In Figure 8e the MMIIE method adds distortion to TII. Finally, in Figure 8f the proposed method enhances the contrast and improves the details of the TII. The second example, which represents a person with a horse, is in Figure 9. The analysis is similar to the previous case. Therefore we can emphasize the good performance of the proposed technique.

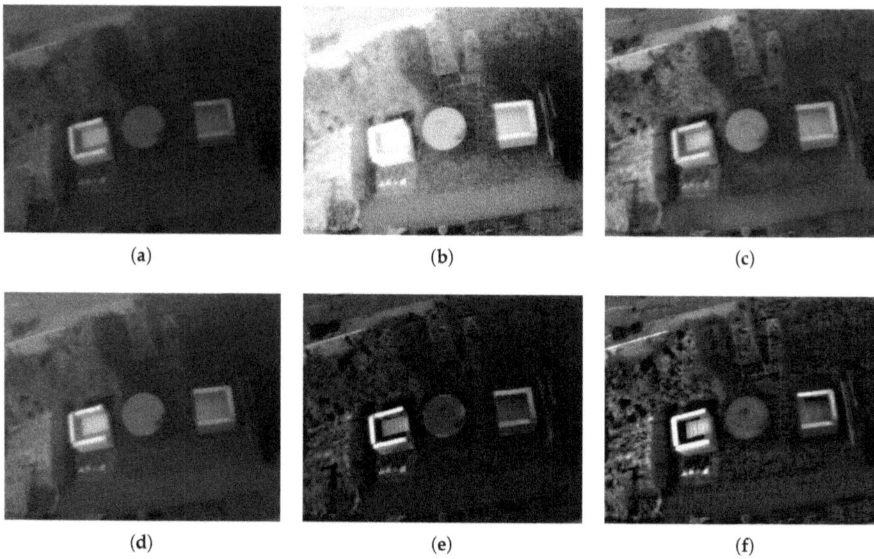

Figure 8. An example of comparison of a TII with a dark background. (**a**) Original TII, (**b**) TII enhanced with HE method, (**c**) TII enhanced with Contrast Limited Adaptive Histogram Equalization (CLAHE) method, (**d**) TII enhanced with IRHE2PL method, (**e**) TII enhanced with MMIIE method, and (**f**) TII enhanced with the proposed method.

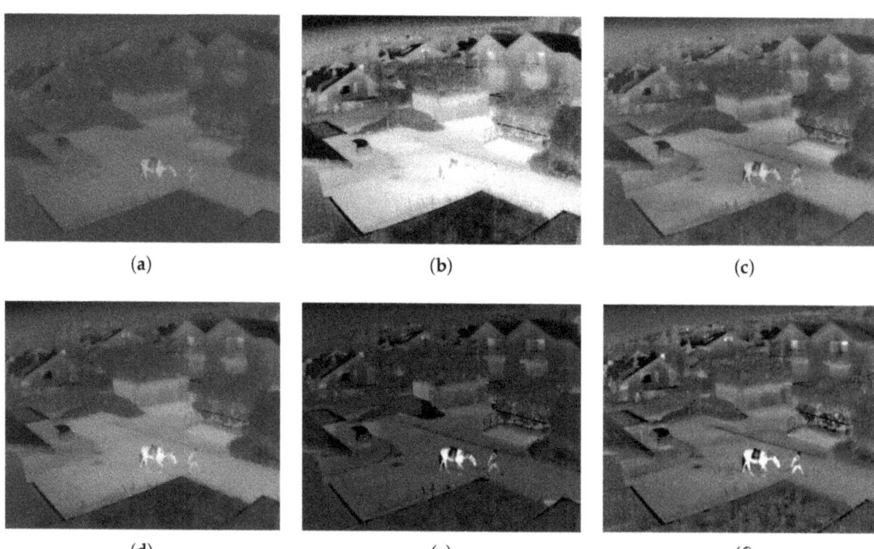

Figure 9. This example shows a TII with a dark background and semi-bright objectives. (**a**) original TII, (**b**) TII enhanced with HE, (**c**) TII enhanced with CLAHE, (**d**) TII enhanced with IRHE2PL, (**e**) TII enhanced with MMIIE, and (**f**) TII enhanced with the proposed method.

Finally, in Figure 10 we analyze the level of detail of the resulting image before and after applying the proposed strategy. Numerical results suggest that the level of details increase with the proposed method and this feature can be visually corroborated. For example, the resulting image in Figure 10b presents an excellent contrast, and a well-defined detail, which makes easier its identification.

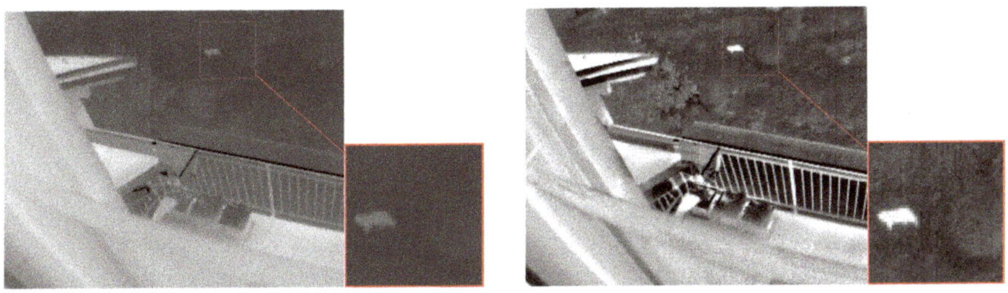

(**a**) (**b**)

Figure 10. TII with improved contrast and detail, (**a**) the Original TII with $E = 6.9334$ and $SD = 58.24$ and (**b**) the TII enhanced with the proposed method with $E = 7.5783$ and $SD = 70.7615$

5. Conclusions

In this work we have introduced an iterative contrast enhancement method for TII. This approach is based on multiscale top-hat transform that improves the entropy of images, which implies an improvement in the level of detail of the resulting image. Furthermore, the proposed method not only improves the entropy but also preserves the brightness and enhances contrast.

The proposed method was compared with state of the art algorithms and has proved to be competitive. It is noteworthy that the proposed method is the only algorithm that improved the original image for all input images in terms of contrast.

Visually, the resulting image after applying the proposed method presents a higher quality than the original image. This result is consistent with the performance of the algorithm. This proposed method could be very useful for infrared thermal image analysis, object recognition, people tracking, and other applications based on infrared thermal images.

Author Contributions: Funding acquisition, H.L.-A. and D.P.P.-R.; Investigation, J.C.M.R.; Methodology, J.C.M.R.; Project administration, J.L.V.N.; Resources, J.L.V.N., H.L.-A. and D.P.P.-R.; Software, J.C.M.R.; Supervision, J.L.V.N., H.L.-A., D.P.P.-R. and M.G.T.; Validation, J.L.V.N. and S.G.-G.; Visualization, J.C.M.R.; Writing—original draft, J.C.M.R. and J.L.V.N.; Writing—review & editing, J.L.V.N., H.L.-A., D.P.P.-R., S.G.-G. and M.G.T.

Funding: This research was funded by CONACYT, Paraguay, grant numbers 14-POS-007 and POS17-53 and by TIN2015-64776-C3-2-R of the Spanish Ministry of Economic and Competitiveness and the European Regional Development Fund (MINECO/FEDER)

Conflicts of Interest: The authors declare no conflict of interest.

References

1. Havens, K.J.; Sharp, E.J. *Thermal Imaging Techniques to Survey and Monitor Animals in the Wild: A Methodology*; Elsevier: Amsterdam, The Netherlands, 2016, doi:10.1016/c2014-0-03312-6.
2. Portmann, J.; Lynen, S.; Chli, M.; Siegwart, R. People detection and tracking from aerial thermal views. In Proceedings of the 2014 IEEE International Conference on Robotics and Automation (ICRA), Hong Kong, China, 31 May–7 June 2014; pp. 1794–1800.
3. Krapels, C.K.; Driggers, C.R.G.; Garcia, C.J.F. Performance of infrared systems in swimmer detection for maritime security. *Opt. Express* **2007**, *15*, 12296–12305. [CrossRef] [PubMed]
4. Bai, X. Morphological infrared image enhancement based on multi-scale sequential toggle operator using opening and closing as primitives. *Infrared Phys. Technol.* **2015**, *68*, 143–151. [CrossRef]
5. Bai, X.; Zhou, F.; Xue, B. Infrared image enhancement through contrast enhancement by using multiscale new top-hat transform. *Infrared Phys. Technol.* **2011**, *54*, 61–69. [CrossRef]

6. Liang, K.; Ma, Y.; Xie, Y.; Zhou, B.; Wang, R. A new adaptive contrast enhancement algorithm for infrared images based on double plateaus histogram equalization. *Infrared Phys. Technol.* **2012**, *55*, 309–315. [CrossRef]
7. Lin, C.L. An approach to adaptive infrared image enhancement for long-range surveillance. *Infrared Phys. Technol.* **2011**, *54*, 84–91. [CrossRef]
8. Lai, R.; Yang, Y.T.; Wang, B.J.; Zhou, H.X. A quantitative measure based infrared image enhancement algorithm using plateau histogram. *Opt. Commun.* **2010**, *283*, 4283–4288. [CrossRef]
9. Li, Y.; Zhang, Y.; Geng, A.; Cao, L.; Chen, J. Infrared image enhancement based on atmospheric scattering model and histogram equalization. *Opt. Laser Technol.* **2016**, *83*, 99–107. [CrossRef]
10. Yoo, J.H.; Ohm, S.Y.; Chung, M.G. Brightness Preservation and Image Enhancement Based on Maximum Entropy Distribution. In *Convergence and Hybrid Information Technology*; Springer: Berlin/Heidelberg, Germany, 2012; pp. 365–372.
11. Aquino-Morínigo, P.B.; Lugo-Solís, F.R.; Pinto-Roa, D.P.; Ayala, H.L.; Noguera, J.L.V. Bi-histogram equalization using two plateau limits. *Signal Image Video Process.* **2017**, *11*, 857–864. [CrossRef]
12. Kim, Y.T. Contrast enhancement using brightness preserving bi-histogram equalization. *IEEE Trans. Consum. Electron.* **1997**, *43*, 1–8. [CrossRef]
13. Ooi, C.H.; Kong, N.S.P.; Ibrahim, H. Bi-histogram equalization with a plateau limit for digital image enhancement. *IEEE Trans. Consum. Electron.* **2009**, *55*, 2072–2080. [CrossRef]
14. Lim, S.H.; Isa, N.A.M.; Ooi, C.H.; Toh, K.K.V. A new histogram equalization method for digital image enhancement and brightness preservation. *Signal Image Video Process.* **2013**, *9*, 675–689. [CrossRef]
15. Pineda, I.A.B.; Caballero, R.D.M.; Silva, J.J.C.; Román, J.C.M.; Noguera, J.L.V. Quadri-histogram equalization using cutoff limits based on the size of each histogram with preservation of average brightness. *Signal Image Video Process.* **2019**. [CrossRef]
16. Román, J.C.M.; Ayala, H.L.; Noguera, J.L.V. Top-Hat transform for enhancement of aerial thermal images. In Proceedings of the 2017 30th SIBGRAPI Conference on Graphics, Patterns and Images (SIBGRAPI), Niteroi, Brazil, 17–20 October 2017; pp. 277–284.
17. Sun, K.; Sang, N. Enhancement of vascular angiogram by multiscale morphology. In Proceedings of the 1st International Conference on Bioinformatics and Biomedical Engineering (ICBBE 2007), Wuhan, China, 6–8 July 2007; pp. 1311–1313.
18. Bai, X. Microscopy mineral image enhancement through center operator construction. *Appl. Opt.* **2015**, *54*, 4678–4688. [CrossRef] [PubMed]
19. Bai, X.; Zhou, F.; Xue, B. Noise-suppressed image enhancement using multiscale top-hat selection transform through region extraction. *Appl. Opt.* **2012**, *51*, 338–347. [CrossRef] [PubMed]
20. Bai, X.; Liu, H. Edge enhanced morphology for infrared image analysis. *Infrared Phys. Technol.* **2017**, *80*, 44–57. [CrossRef]
21. Bai, X.; Zhou, F. Analysis of new top-hat transformation and the application for infrared dim small target detection. *Pattern Recognit.* **2010**, *43*, 2145–2156. [CrossRef]
22. Bai, X.; Zhou, F.; Xue, B. Infrared dim small target enhancement using toggle contrast operator. *Infrared Phys. Technol.* **2012**, *55*, 177–182. [CrossRef]
23. Bai, X.; Zhou, F. Analysis of different modified top-hat transformations based on structuring element construction. *Signal Process.* **2010**, *90*, 2999–3003. [CrossRef]
24. Bai, X.; Zhou, F.; Xue, B. Toggle and top-hat based morphological contrast operators. *Comput. Electr. Eng.* **2012**, *38*, 1196–1204. [CrossRef]
25. Bai, X.; Zhou, F.; Xue, B. Multi-scale toggle operator for constructing image sharpness measure. *Opt. Laser Technol.* **2012**, *44*, 2004–2014. [CrossRef]
26. Mukhopadhyay, S.; Chanda, B. A multiscale morphological approach to local contrast enhancement. *Signal Process.* **2000**, *80*, 685–696. [CrossRef]
27. Gonzalez, R.C.; Woods, R.E.; Eddins, S.L. *Digital Image Processing Using MATLAB*; Pearson-Prentice-Hall: Upper Saddle River, NJ, USA, 2004; Volume 624.
28. Soille, P. *Morphological Image Analysis: Principles and Applications*; Springer: Berlin, Germany, 2004. [CrossRef]
29. Bai, X.; Zhou, F. A unified form of multi-scale top-hat transform based algorithms for image processing. *Optik* **2013**, *124*, 1614–1619. [CrossRef]
30. Bai, X.; Zhou, F.; Xue, B. Image enhancement using multi scale image features extracted by top-hat transform. *Opt. Laser Technol.* **2012**, *44*, 328–336. [CrossRef]

31. Liao, M.; Zhao, Y.Q.; Wang, X.H.; Dai, P.S. Retinal vessel enhancement based on multi-scale top-hat transformation and histogram fitting stretching. *Opt. Laser Technol.* **2014**, *58*, 56–62. [CrossRef]
32. Peng, B.; Wang, Y.; Yang, X. A multiscale morphological approach to local contrast enhancement for ultrasound images. In Proceedings of the 2010 International Conference on Computational and Information Sciences, Chengdu, China, 17–19 December 2010; pp. 1142–1145.
33. Bai, X. Image enhancement through contrast enlargement using the image regions extracted by multiscale top-hat by reconstruction. *Optik* **2013**, *124*, 4421–4424. [CrossRef]
34. Bai, X.; Zhou, F.; Xue, B. Image fusion through local feature extraction by using multi-scale top-hat by reconstruction operators. *Optik* **2013**, *124*, 3198–3203. [CrossRef]
35. Zhao, J.; Zhou, Q.; Chen, Y.; Feng, H.; Xu, Z.; Li, Q. Fusion of visible and infrared images using saliency analysis and detail preserving based image decomposition. *Infrared Phys. Technol.* **2013**, *56*, 93–99. [CrossRef]
36. Zhu, P.; Ding, L.; Ma, X.; Huang, Z. Fusion of infrared polarization and intensity images based on improved toggle operator. *Opt. Laser Technol.* **2018**, *98*, 139–151. [CrossRef]
37. Mukhopadhyay, S.; Chanda, B. Multiscale morphological segmentation of gray-scale images. *IEEE Trans. Image Process.* **2003**, *12*, 533–549. [CrossRef] [PubMed]
38. Ye, B.; Peng, J.x. Small target detection method based on morphology top-hat operator. *J. Image Graph.* **2002**, *7*, 638–642.
39. Tsai, D.Y.; Lee, Y.; Matsuyama, E. Information entropy measure for evaluation of image quality. *J. Dig. Imaging* **2008**, *21*, 338–347. [CrossRef] [PubMed]
40. Singh, K.; Kapoor, R. Image enhancement using exposure based sub image histogram equalization. *Pattern Recognit. Lett.* **2014**, *36*, 10–14. [CrossRef]
41. Ye, Z.; Mohamadian, H.; Pang, S.S.; Iyengar, S. Image contrast enhancement and quantitative measuring of information flow. In Proceedings of the 6th WSEAS International Conference on Information Security and Privacy, Tenerife, Spain, 14–16 December 2007; pp. 172–177.
42. More, L.G.; Brizuela, M.A.; Ayala, H.L.; Pinto-Roa, D.P.; Noguera, J.L.V. Parameter tuning of CLAHE based on multi-objective optimization to achieve different contrast levels in medical images. In Proceedings of the 2015 IEEE International Conference on Image Processing (ICIP), Quebec City, QC, Canada, 27–30 September 2015; pp. 4644–4648.
43. Khellaf, A.; Beghdadi, A.; Dupoisot, H. Entropic contrast enhancement. *IEEE Trans. Med. Imaging* **1991**, *10*, 589–592. [CrossRef] [PubMed]
44. Wang, S.H.; Cheng, H.; Phillips, P.; Zhang, Y.D. Multiple Sclerosis Identification Based on Fractional Fourier Entropy and a Modified Jaya Algorithm. *Entropy* **2018**, *20*, 254. [CrossRef]
45. Zhang, Y.; Wu, X.; Lu, S.; Wang, H.; Phillips, P.; Wang, S. Smart detection on abnormal breasts in digital mammography based on contrast-limited adaptive histogram equalization and chaotic adaptive real-coded biogeography-based optimization. *Simulation* **2016**, *92*, 873–885, doi:10.1177/0037549716667834. [CrossRef]
46. Lee, K.; Lee, S. A new framework for measuring 2D and 3D visual information in terms of entropy. *IEEE Trans. Circuits Syst. Video Technol.* **2016**, *26*, 2015–2027. [CrossRef]
47. Shannon, C.E. A mathematical theory of communication. *Bell Syst. Tech. J.* **1948**, *27*, 379–423. [CrossRef]
48. Niu, Y.; Wu, X.; Shi, G. Image enhancement by entropy maximization and quantization resolution upconversion. *IEEE Trans. Image Process.* **2016**, *25*, 4815–4828. [CrossRef]
49. Burger, W.; Burge, M.J. *Digital Image Processing: An Algorithmic Introduction Using Java*; Springer: London, UK, 2016. [CrossRef]
50. Dataset. Thermal Infrared Dataset. 2014. Available online: https://projects.asl.ethz.ch/datasets/doku.php?id=ir:iricra2014 (accessed on 1 December 2018).
51. Zuiderveld, K. Contrast limited adaptive histogram equalization. In *Graphics Gems, Elsevier*; Elsevier: Amsterdam, The Netherlands, 1994; pp. 474–485. [CrossRef]

entropy

MDPI

Article

Breaking an Image Encryption Algorithm Based on DNA Encoding and Spatiotemporal Chaos

Heping Wen [1,*], **Simin Yu** [1] **and Jinhu Lü** [2]

[1] School of Automation, Guangdong University of Technology, Guangzhou 510006, China; siminyu@163.com
[2] School of Automation Science and Electrical Engineering, State Key Laboratory of Software Development Environment, and Beijing Advanced Innovation Center for Big Data and Brain Computing, Beihang University, Beijing 100191, China; jhlu@iss.ac.cn
* Correspondence: hepingwen@yeah.net

Received: 3 February 2019; Accepted: 26 February 2019; Published: 5 March 2019

Abstract: Recently, an image encryption algorithm based on DNA encoding and spatiotemporal chaos (IEA-DESC) was proposed. In IEA-DESC, pixel diffusion, DNA encoding, DNA-base permutation and DNA decoding are performed successively to generate cipher-images from the plain-images. Some security analyses and simulation results are given to prove that it can withstand various common attacks. However, in this paper, it is found that IEA-DESC has some inherent security defects as follows: (1) the pixel diffusion is invalid for attackers from the perspective of cryptanalysis; (2) the combination of DNA encoding and DNA decoding is equivalent to bitwise complement; (3) the DNA-base permutation is actually a fixed position shuffling operation for quaternary elements, which has been proved to be insecure. In summary, IEA-DESC is essentially a combination of a fixed DNA-base position permutation and bitwise complement. Therefore, IEA-DESC can be equivalently represented as simplified form, and its security solely depends on the equivalent secret key. So the equivalent secret key of IEA-DESC can be recovered using chosen-plaintext attack and chosen-ciphertext attack, respectively. Theoretical analysis and experimental results show that the two attack methods are both effective and efficient.

Keywords: image encryption; DNA encoding; chaotic cryptography; cryptanalysis; image privacy

1. Introduction

With the rapid development of information technologies such as mobile Internet, cloud computing, social networking, and Big Data, the security of multimedia data such as image and video has attracted more and more attention [1–3]. Image is an important part of multimedia data, and its encryption protection techniques are particularly interesting [4,5]. In the past three decades, many novel image encryption schemes based on various methodologies were proposed, such as chaos theory [6], DNA computing [7], cellular automaton [8,9], and quantum information [9,10]. Among them, chaos is the most popular one because it has the unique characteristics of sensitivity to initial values and parameters, ergodicity, and deterministic inherent randomness [11–16], which correspond to the confusion and diffusion properties of encryption [17]. Moreover, DNA computing has the characteristics of high parallelism, large storage capacity, and low energy consumption [7]. Hence, researches on image encryption schemes combined with chaos theory and DNA computing have become a hot topic in recent years [18–20]. Nevertheless, many encryption schemes are actually insecure as a result of their various security defects [21,22]. Therefore, performing cryptanalysis on these existing encryption algorithms is indispensable [21,23,24].

In recent years, with the security analysis and breaking of some existing chaotic image algorithms combining DNA computing and chaos theory [25–30], research interest in cryptanalysis has become increasingly stimulated [31–34]. In 2010, Zhang et al. [25] created an image encryption method using

DNA addition combined with chaotic maps. However, in 2014, Hermassi et al. [26] pointed out that the algorithm in Reference [25] was irreversible and was vulnerable to the chosen-plaintext attack and the known-plaintext attack. In 2015, Zhen et al. [27] proposed an image encryption scheme combining DNA coding and entropy. Nonetheless, in 2016, Su et al. [28] pointed out that the algorithm of Reference [27] was insecure and could be broken using the chosen-plaintext attack. In 2016, Jain et al. [29] proposed a robust DNA chaotic image encryption scheme based on Reference [25] and Reference [26]. Whereas, in 2017, Dou et al. [30] used the chosen-plaintext attack method to break the algorithm proposed in Reference [29]. In addition, Özkaynak et al. [33] and Zhang et al. [34] further concluded that an encryption algorithm may lead to the existence of an equivalent secret key if only a single DNA encoding and operation rule is employed.

Generally speaking, cryptanalysis becomes more difficult as the level of encryption design increases [35]. However, there are still some existing algorithms that can be broken owing to their inherent defects [36]. Moreover, since each encryption algorithm has natural features, the corresponding attack method may also be different. Therefore, it makes sense, even if a similar attack method is used, to reveal the intrinsic characteristics of the different encryption algorithms.

In 2015, an image encryption algorithm based on DNA encoding and spatiotemporal chaos (IEA-DESC) was proposed [37]. In IEA-DESC, pixel diffusion, DNA encoding, DNA-base permutation, and DNA decoding are adopted successively to obtain cipher-images from plain-images. Some security analyses and simulation results are given to prove that it can withstand various common attacks. Despite this, according to the basic criteria of cryptanalysis, some findings in IEA-DESC can be given as follows:

(1) Its pixel diffusion is invalid for attackers.

In IEA-DESC, there is no external secret key during the pixel diffusion phase. According to the cryptographic principle proposed by Kerckhoffs [38], the algorithm is public for attackers. Therefore, its pixel diffusion is essentially useless.

(2) The combination of DNA encoding and DNA decoding can be equivalently simplified.

Although the DNA encoding rule is related to the plain-image, there is a certain relationship between its decoding rule and its encoding rule. This leads to the fact that for any binary bit, the output is the complement of the input after DNA encoding and DNA decoding. Hence, DNA encoding and DNA decoding are a complementary process on the whole.

(3) The sequences for DNA-base permutation are fixed for different plain-images.

During IEA-DESC's DNA-base permutation, the chaos-based sequences for encryption are neither associated with plain-image nor cipher-image. Thus, on the basis of the basic rules of cryptanalysis, under the condition of a given secret key, the encryption sequences are fixed for different plain-images. Once the attackers obtain these sequences, i.e., an equivalent secret key, the DNA-base permutation is deciphered.

On the basis of the above properties, IEA-DESC's pixel diffusion is invalid, and therefore, its security depends only on the DNA domain encryption. Unfortunately, an equivalent secret key exists in the overall DNA domain encryption phase. More specifically, the DNA-based encryption algorithm is essentially a permutation-only process of a quaternary element. Yet, permutation-only encryption algorithms have been analyzed to be insecure [39,40]. Therefore, in this paper, two attack methods for breaking IEA-DESC using the chosen-plaintext attack and chosen-ciphertext attack are proposed, respectively.

The rest of the paper is organized as follows. Section 2 concisely describes IEA-DESC. Section 3 proposes two different attack methods on IEA-DESC. Section 4 presents the experimental simulation results. Section 5 gives some improvement suggestions for the security of chaos-based encryption algorithms. The last section concludes the paper.

2. The Encryption Algorithm under Study

In this section, the DNA coding rules and spatiotemporal chaos used in Reference [37] are introduced, and then the specific steps of IEA-DESC are detailed.

2.1. DNA Coding Rules

A DNA sequence includes four kinds of nucleic acid bases: A, T, C, and G. With respect to these four bases, the total number of coding combinations is $4! = 24$. However, there are only eight kinds of coding combinations because these four bases satisfy the principle of complementary base pairs. More precisely, A and T are complementary to each other, as are C and G. Table 1 shows the eight DNA coding rules.

Table 1. Eight kinds of DNA coding rules.

Rules	1	2	3	4	5	6	7	8
A	00	00	01	01	10	10	11	11
T	11	11	10	10	01	01	00	00
G	01	10	00	11	00	11	01	10
C	10	01	11	00	11	00	10	01

2.2. Spatiotemporal Chaos

Two discrete chaotic maps are used in IEA-DESC [37], one is the logistic map and the other is a spatiotemporal chaos map based on the so-called new chaotic algorithm (NCA) given in Reference [41]. The iterative equation of the Logistic map is represented as

$$x_{n+1} = \mu x_n(1 - x_n), \tag{1}$$

where the state variable $x \in (0, 1)$ and the control parameter $\mu \in (3.57, 4)$. The structure of the functional graph of the Logistic map in a digital computer is quantitatively analyzed in Reference [16].

The spatiotemporal chaos is a dynamic system using discrete time and space, in which the coupled map lattice (CML) is its most common model. The iterative equation of NCA-based CML is modeled by

$$\begin{cases} x_{n+1}(i) = (1-\varepsilon)f(x_n(i)) + \varepsilon\left\{f\left[x_n(i-1)\right] + f\left[x_n(i+1)\right]\right\}/2, \\ f(x) = (1-\beta^{-4}) \cdot \mathrm{ctg}(\alpha/(1+\beta)) \cdot (1+1/\beta)^{\beta} \cdot \mathrm{tg}(\alpha x_n) \cdot (1-x)^{\beta}, \end{cases} \tag{2}$$

where the spatial lattice index $i = 1, 2, \cdots, L$, the time grid index $n = 1, 2, \cdots$, the coupling strength $\varepsilon \in (0, 1)$, the state variable $x_n(i) \in (0, 1)$, and the periodic boundary condition is $x_n(0) = x_n(L)$. The second equation of Equation (2) is the so-called NCA, which is actually an improved logistic map. Given the parameters $\alpha = 1.57, \beta = 3.5, \varepsilon = 0.3$, and $L = 1024$, the system is chaotic, and its attractor is shown in Figure 1.

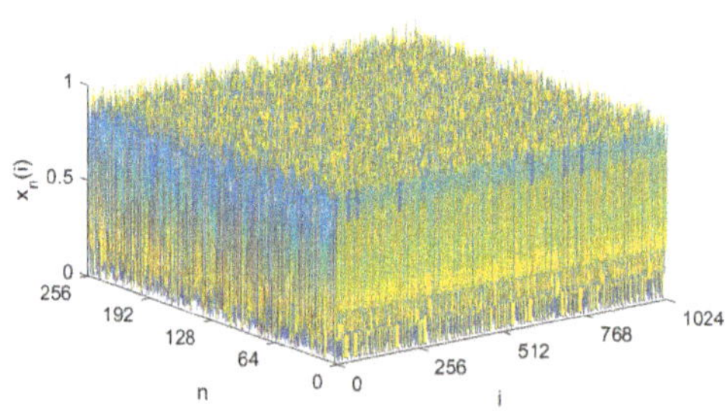

Figure 1. The attractor of the new chaotic algorithm (NCA)-based coupled map lattice (CML).

2.3. Description of IEA-DESC

2.3.1. Secret Key

The secret key of IEA-DESC consists of $x_0, \mu, K_0, N_0, \alpha, \beta, \varepsilon$, and L, where x_0, μ, K_0, and N_0 are the parameters of the logistic map, $\alpha, \beta, \varepsilon$, and L are the parameters of NCA-based CML, and N_0 is the length of discarded sequence for eliminating harmful transient effects.

2.3.2. Encryption Process

The encryption objects of IEA-DESC are 8-bit grayscale images of size $H \times W$ (height $\times$ width). For convenience, the symbolic representation is different without changing the original algorithm. A block diagram of IEA-DESC is shown in Figure 2, where $\boldsymbol{P}, \boldsymbol{P}'$, and $\boldsymbol{C}$ are the plain-image, the diffused image, and the cipher-image, respectively. As can be seen from Figure 2, the encryption process of IEA-DESC includes four phases: pixel diffusion, DNA encoding, DNA-base permutation, and DNA decoding.

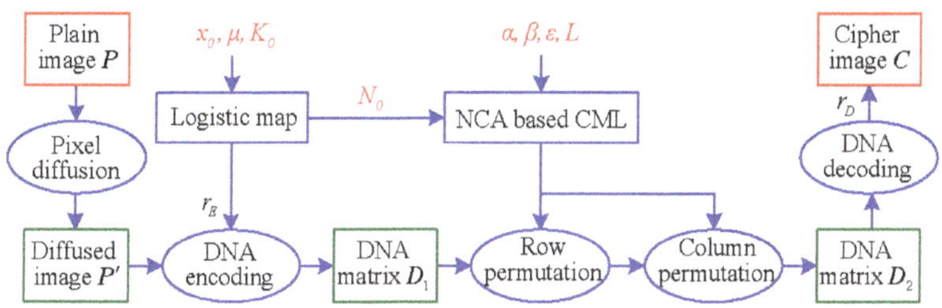

Figure 2. Block diagram of the image encryption algorithm based on DNA encoding and spatiotemporal chaos (IEA-DESC).

The specific descriptions of IEA-DESC are given as follows:

- **Phase 1.** Pixel Diffusion:

  By converting the plain-image $\boldsymbol{P}$ into the corresponding sequence $\{p_1, p_2, \ldots, p_{H\times W}\}$ in raster scanning order, the pixel diffusion equation is defined as

$$\begin{cases} p'_1 = p_1 \oplus p_{H\times W}, \\ p'_{i+1} = p_i \oplus p'_i, \end{cases} \tag{3}$$

  where $i = 1, 2, \ldots, H \times W - 1$, and $\oplus$ represents the bitwise XOR operation. Thus, the diffused image $\boldsymbol{P}'$ of size $H \times W$ is obtained from the diffused sequence $\{p'_1, p'_2, \ldots, p'_{H\times W}\}$.

- **Phase 2.** DNA Encoding:

  Calculating the sum of the plain-image pixels, the K_0-th iteration value x_{K_0} is obtained by Equation (1) under the initial value x_0 and the control parameter μ. The DNA encoding rule r_E, as in Table 1, is further determined by

$$r_E = \lfloor x_{K_0} \times 8 \rfloor + 1, \tag{4}$$

  where $r_E \in [1, 8]$, and $\lfloor a \rfloor$ rounds the element a to the nearest integer toward minus infinity. Then, by the r_E-th encoding rule in Table 1, the diffused image $\boldsymbol{P}'$ of size $H \times W$ is firstly converted into the corresponding binary matrix of size $H \times 8W$, and then encoded as the DNA matrix $\boldsymbol{D}_1$ of size $H \times 4W$.

- **Phase 3.** DNA-Base Permutation:

First, by iterating Equation (1) $N_0 + 4W$ times and then discarding the front N_0 elements under the initial value x_0 and the control parameter μ, a sequence $\{a_1, a_2, \ldots, a_{4W}\}$ of length $4W$ is obtained. Here, the sequence $\{a_1, a_2, \ldots, a_{4W}\}$ is taken as an initial value of the spatiotemporal chaos called NCA-based CML. Thus, by iterating Equation (2) H times under the parameters $\alpha, \beta, \varepsilon$, and L, a real matrix $\boldsymbol{X}$ of size $H \times 4W$ is achieved.

Then, by sorting each row's elements of $\boldsymbol{X}$ in ascending order, the corresponding H row position index sequences are obtained as $\boldsymbol{RI}_i$. Using $\boldsymbol{RI}_i$ to perform permutation for each row on the DNA matrix $\boldsymbol{D}_1$, the corresponding row permuted DNA matrix $\boldsymbol{D}'_1$ is obtained, given by

$$[\boldsymbol{D}'_1]_{i,k} = [\boldsymbol{D}_1]_{i,\boldsymbol{RI}_i(k)}, \tag{5}$$

where $i = 1, 2, \ldots, H$, $k = 1, 2, \ldots, 4W$, $\boldsymbol{RI}_i(k)$ indicates a position index of the k-th element in the i-th row, and $\boldsymbol{RI}_i(k) \in \{1, 2, \ldots, 4W\}$. Similarly, by sorting each column's elements of $\boldsymbol{X}$ in ascending order, the corresponding $4W$ column position index sequences are obtained as $\boldsymbol{CI}_j$. Using $\boldsymbol{CI}_j$ to perform permutation for each column on the DNA matrix $\boldsymbol{D}'_1$, the corresponding column permuted DNA matrix $\boldsymbol{D}_2$ is obtained, represented as

$$[\boldsymbol{D}_2]_{k,j} = [\boldsymbol{D}'_1]_{\boldsymbol{CI}_j(k),j}, \tag{6}$$

where $j = 1, 2, \ldots, 4W$, $k = 1, 2, \ldots, H$, $\boldsymbol{CI}_j(k)$ indicates a position index of the k-th element in the j-th column, and $\boldsymbol{CI}_j(k) \in \{1, 2, \ldots, H\}$.

- **Phase 4.** DNA Decoding:

  Corresponding to Equation (4), the DNA decoding rule r_D is determined as

$$r_D = 9 - r_E, \tag{7}$$

  where $r_D \in [1, 8]$. Thus, by the r_D-th decoding rule in Table 1, the DNA matrix $\boldsymbol{D}_2$ of size $H \times 4W$ is firstly decoded as the corresponding binary matrix of size $H \times 8W$, and then converted into the cipher-image $\boldsymbol{C}$ of size $H \times W$.

2.3.3. Decryption Process

Decryption is the inverse of encryption. First, the cipher-image $\boldsymbol{C}$ is converted into the DNA matrix $\boldsymbol{D}_2$ by the r_D-th encoding rule. Then, the DNA matrix $\boldsymbol{D}_1$ is exacted from the DNA matrix $\boldsymbol{D}_2$ after the anti-permutation. Next, the DNA matrix $\boldsymbol{D}_1$ is decoded as the diffused image $\boldsymbol{P}'$ with the r_E-th decoding rule. Finally, the plain-image $\boldsymbol{P}$ is recovered by anti-diffusion decryption from Equation (3).

3. Cryptanalysis of IEA-DESC

3.1. *Preliminary Analysis of IEA-DESC*

According to modern cryptography principles, encryption algorithms are public and only the secret keys are unknown to attackers [42,43]. More precisely, the security of an algorithm solely depends on its secret key. Four common attack methods for cryptanalysis are shown in Table 2. A secure cryptosystem should be able to resist all types of attacks in Table 2. If a cryptosystem cannot resist anyone of these attacks, one can conclude that the cryptosystem is insecure.

By observing Figure 2, one can divide the encryption process of IEA-DESC into two parts, one is pixel diffusion, and the other is DNA domain encryption. For the pixel diffusion part, there is no secret key involved. Since the algorithm is open from the perspective of cryptanalysis, the diffusion phase of IEA-DESC is essentially invalid for the attacker.

Table 2. Four common attack methods for cryptanalysis.

Attack Methods	Available Resources for Cryptanalysis
Ciphertext-only attack	The attacker only knows the ciphertext.
Known-plaintext attack	The attacker knows any given plaintext, and also knows the corresponding ciphertext.
Chosen-plaintext attack	The attacker can choose the plaintext that would be useful for deciphering, and also knows the corresponding ciphertext.
Chosen-ciphertext attack	The attacker can choose the ciphertext that is useful for deciphering, and also knows the corresponding plaintext.

For this reason, only the DNA domain encryption part is worthy of further discussion here. Following Equations (4) and (7), one knows that there is a definite one-to-one correspondence between DNA encoding rule r_E and DNA decoding rule r_D. Hence, one can list all possible pairs of DNA codec rules, given as

$$(r_E, r_D) \in \{(1,8),(2,7),(3,6),(4,5),(5,4),(6,3),(7,2),(8,1)\}. \tag{8}$$

Accordingly, given the eight pairs of DNA codec rules, within the binary bits before and after DNA coding, there appears a certain regularity [34]. Table 3 shows any 2-bit input and its 2-bit output with the eight pairs of DNA codec rules. As can be seen from Table 3, given any 2-bit input, no matter which DNA encoding rule is taken, the corresponding 2-bit output is the same because $r_E + r_D = 9$ holds. Put explicitly, the 2-bit input and the corresponding 2-bit output are complementary.

Table 3. Any 2-bit input and its 2-bit output with the eight pairs of DNA codec rules.

2-Bit Input	DNA-Base with Encoding Rule r_E								2-Bit Output with Decoding Rule r_D
	1	2	3	4	5	6	7	8	
00	A	A	G	C	G	C	T	T	11
01	G	C	A	A	T	T	C	G	10
10	C	G	T	T	A	A	G	C	01
11	T	T	C	G	C	G	A	A	00

Furthermore, one can see that the chaos-based sequences for DNA-base permutation are fixed for different plain-images under the premise of a given secret key. Indeed, it means that an equivalent secret key exists in IEA-DESC.

On this basis, it is found that IEA-DESC is essentially a combination process of a fixed DNA-base position permutation and bitwise complement. Therefore, a simplified block diagram of IEA-DESC can be illustrated, as is shown in Figure 3, where EK_P is the equivalent secret key. Once EK_P is obtained, IEA-DESC will be broken. Note that the eight different pairs of DNA encoding and decoding, as in Table 3, are equivalent. For simplicity, one sets $r_E = 1$ and $r_D = 8$, i.e., the first DNA encoding rule and the eighth DNA decoding rule are adopted below.

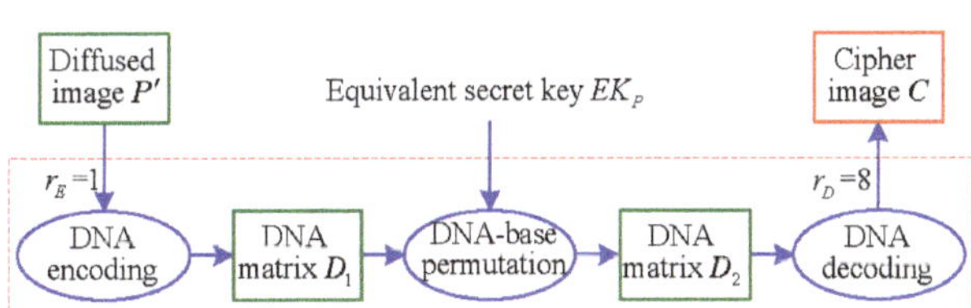

Figure 3. Simplified block diagram of IEA-DESC.

3.2. Analysis of DNA-Base Permutation

To obtain the equivalent secret key EK_P, performing analysis on the DNA-base permutation is significant. Since DNA only has four different bases, the essence of DNA-base permutation is a position shuffling procedure for a quaternary matrix of size $H \times 4W$.

Supposing that one has an input matrix $\boldsymbol{PV}$ of size $H \times 4W$ which satisfies that each element is unequal, and its corresponding one-dimensional sequence in the raster scanning order is $\{0, 1, 2, \ldots, 4HW - 1\}$. Letting $\mathbb{Z}_m$ denote a set $\{0, 1, \cdots, m-1\}$, one has $\boldsymbol{PV} \in \mathbb{Z}_{4HW}$. Obviously, after a position permutation, the output matrix $\boldsymbol{CV}$ corresponding to the input matrix $\boldsymbol{PV}$ also has the feature that all elements are not equal to each other. According to the assumption of the chosen-plaintext attack in Table 2, one can know both $\boldsymbol{PV}$ and $\boldsymbol{CV}$. Thus, the equivalent secret key can be determined by comparing the elements before and after the permutation.

However, since each DNA-base only takes four values, A, G, C, and T, such an input matrix $\boldsymbol{PV}$ does not exist. To cope with the problem, an appropriate transformation is inevitable. Therefore, the specific analysis steps to determine the equivalent secret key EK_P, as in Figure 3, are detailed as follows:

Step 1. Decompose the virtual matrix $\boldsymbol{PV}$ of size $H \times 4W$ into some quaternary matrices of the same size.

The virtual matrix $\boldsymbol{PV}$ is firstly decomposed into the N_C corresponding quaternary matrices $\boldsymbol{PQ}_n (n = 1, 2, \ldots, N_C)$, defined by

$$\boldsymbol{PV} = \sum_{n=1}^{N_C} 4^{n-1} \boldsymbol{PQ}_n = \boldsymbol{PQ}_1 + 4\boldsymbol{PQ}_2 + 4^2 \boldsymbol{PQ}_3 + \ldots + 4^{N_C - 1} \boldsymbol{PQ}_{N_C}, \tag{9}$$

where $\boldsymbol{PV} \in \mathbb{Z}_{4HW}$, $\boldsymbol{PQ}_n \in \mathbb{Z}_4$, and N_C is the minimum amount required to ensure this decomposition method. Referring to Reference [39], generally one has

$$\begin{aligned} N_C &= \lceil \log_4 (H \times 4W) \rceil \\ &= 1 + \lceil \log_4 (HW) \rceil , \end{aligned} \tag{10}$$

where $\lceil a \rceil$ rounds the element a to the nearest integer toward positive infinity.

Step 2. Transform these quaternary matrices into the 8-bit images of size $H \times W$, respectively.

The quaternary matrices $\boldsymbol{PQ}_n (n = 1, 2, \ldots, N_C)$ are transformed into the corresponding decimal matrices using the method whereby every four quaternary elements are combined into a decimal one in order from low to high. For instance, given four quaternary elements are 0, 1, 2, and 3, one gets the corresponding decimal result as 228 because of its combination procedure: $0 \times 1 + 1 \times 4 + 2 \times 4^2 + 3 \times 4^3$. In fact, these decimal matrices are the resulting 8-bit images $\boldsymbol{PI}_n (n = 1, 2, \ldots, N_C)$ of size $H \times W$.

Step 3. Temporarily use the encryption machine to obtain the N_C corresponding cipher-images.

Following Figure 3, the diffused image $\boldsymbol{P}'$ is deemed as an input plain-image. As for the chosen-plaintext attack in Table 2, the input plain-images can be arbitrarily chosen, and the encryption machine can be temporarily used. Therefore, one gets the N_C cipher-images $\boldsymbol{CI}_n (n = 1, 2, \ldots, N_C)$ corresponding to the input plain-images $\boldsymbol{PI}_n (n = 1, 2, \ldots, N_C)$, respectively, after the encryption. Obviously, one has $\boldsymbol{PI}_n \in \mathbb{Z}_{256}$ and $\boldsymbol{CI}_n \in \mathbb{Z}_{256}$.

As shown in Figure 3, it takes three phases from $\boldsymbol{PI}_n$ to $\boldsymbol{CI}_n$: DNA encoding, DNA-base permutation, and DNA decoding. First, the N_C plain-images $\boldsymbol{PI}_n$ are encoded as the corresponding DNA matrices with the first DNA encoding rule of Table 1. Then, the permuted DNA matrices are further obtained after a DNA-base permutation. Finally, the N_C corresponding cipher-images $\boldsymbol{CI}_n$ are decoded from the permuted DNA matrices with the eighth DNA decoding rule.

Step 4. Convert these 8-bit cipher-images into the quaternary matrices of size $H \times 4W$, respectively.

Note that the eighth DNA decoding rule corresponds to the first DNA encoding rule, and the combination process of DNA encoding and DNA decoding is bitwise complement analyzed as in Table 3. Therefore, the complement operation cannot be ignored.

First, similar to the method in Step 2, the N_C 8-bit cipher-images $\boldsymbol{CI}_n(n = 1, 2, \ldots, N_C)$ of size $H \times W$ are converted to the corresponding quaternary matrices $\boldsymbol{CQ}_n(n = 1, 2, \ldots, N_C)$ of size $H \times 4W$. Then, the corresponding complementary quaternary matrices $\overline{\boldsymbol{CQ}}_n(n = 1, 2, \ldots, N_C)$ can be obtained from these quaternary matrices $\boldsymbol{CQ}_n(n = 1, 2, \ldots, N_C)$, respectively. Here, the quaternary complement operation is defined as being subtracted by 3. Specifically, the complements of 0, 1, 2, and 3 are 3, 2, 1, and 0, respectively.

Step 5. Compose the N_C complementary quaternary matrices into a virtual matrix of size $H \times 4W$.

Corresponding to Equation (9), the virtual matrix $\boldsymbol{CV}$ is composed from the N_C complementary quaternary matrices $\overline{\boldsymbol{CQ}}_n(n = 1, 2, \ldots, N_C)$, given as

$$\boldsymbol{CV} = \sum_{n=1}^{N_C} 4^{n-1}\overline{\boldsymbol{CQ}}_n = \overline{\boldsymbol{CQ}}_1 + 4\overline{\boldsymbol{CQ}}_2 + 4^2\overline{\boldsymbol{CQ}}_3 + \ldots + 4^{N_C-1}\overline{\boldsymbol{CQ}}_{N_C}, \tag{11}$$

where $\boldsymbol{CV} \in \mathbb{Z}_{4HW}$, and $\overline{\boldsymbol{CQ}}_n \in \mathbb{Z}_4$.

Step 6. Obtain the equivalent secret key EK_P.

Finally, EK_P is obtained by comparing all the different elements of $\boldsymbol{PV}$ and $\boldsymbol{CV}$.

To better illustrate this analysis process, a simple example is taken. Let a input virtual matrix $\boldsymbol{PV}$ of size 4×4 be

$$\boldsymbol{PV} = \begin{bmatrix} 0 & 1 & 2 & 3 \\ 4 & 5 & 6 & 7 \\ 8 & 9 & 10 & 11 \\ 12 & 13 & 14 & 15 \end{bmatrix}.$$

First, following Steps 1 and 2, one gets two quaternary matrices $\boldsymbol{PQ}_1$ and $\boldsymbol{PQ}_2$, the two 8-bit images $\boldsymbol{PI}_1$ and $\boldsymbol{PI}_2$, and their corresponding DNA matrices as below:

$$\boldsymbol{PQ}_1 = \begin{bmatrix} 0 & 1 & 2 & 3 \\ 0 & 1 & 2 & 3 \\ 0 & 1 & 2 & 3 \\ 0 & 1 & 2 & 3 \end{bmatrix} \rightarrow \boldsymbol{PI}_1 = \begin{bmatrix} 228 \\ 228 \\ 228 \\ 228 \end{bmatrix} \xrightarrow{r_E=1} \begin{bmatrix} A & G & C & T \\ A & G & C & T \\ A & G & C & T \\ A & G & C & T \end{bmatrix},$$

$$\boldsymbol{PQ}_2 = \begin{bmatrix} 0 & 0 & 0 & 0 \\ 1 & 1 & 1 & 1 \\ 2 & 2 & 2 & 2 \\ 3 & 3 & 3 & 3 \end{bmatrix} \rightarrow \boldsymbol{PI}_2 = \begin{bmatrix} 0 \\ 85 \\ 170 \\ 255 \end{bmatrix} \xrightarrow{r_E=1} \begin{bmatrix} A & A & A & A \\ G & G & G & G \\ C & C & C & C \\ T & T & T & T \end{bmatrix}.$$

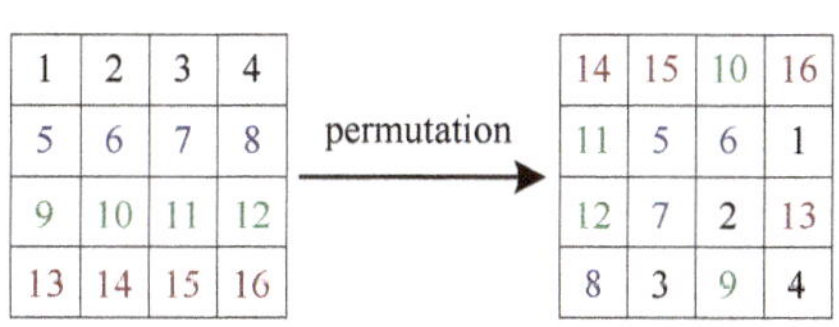

Figure 4. The illustration diagram of a position shuffling for matrices of size 4×4.

Then, as in Steps 3 and 4, supposing that the procedure of DNA-base permutation is given in Figure 4, one obtains the two cipher-images $\boldsymbol{CI}_1$ and $\boldsymbol{CI}_2$, their corresponding DNA matrices, and the two quaternary ones $\boldsymbol{CQ}_1$ and $\boldsymbol{CQ}_2$ via

$$\boldsymbol{CQ}_1 = \begin{bmatrix} 2 & 1 & 2 & 0 \\ 1 & 3 & 2 & 3 \\ 0 & 1 & 2 & 3 \\ 0 & 1 & 3 & 0 \end{bmatrix} \leftarrow \boldsymbol{CI}_1 = \begin{bmatrix} 38 \\ 237 \\ 228 \\ 52 \end{bmatrix} \xrightarrow{r_D=8} \begin{bmatrix} G & C & G & T \\ C & A & G & A \\ T & C & G & A \\ T & C & A & T \end{bmatrix},$$

$$\boldsymbol{CQ}_2 = \begin{bmatrix} 0 & 0 & 1 & 0 \\ 1 & 2 & 2 & 3 \\ 1 & 2 & 3 & 0 \\ 2 & 3 & 1 & 3 \end{bmatrix} \leftarrow \boldsymbol{CI}_2 = \begin{bmatrix} 16 \\ 233 \\ 57 \\ 222 \end{bmatrix} \xrightarrow{r_D=8} \begin{bmatrix} T & T & C & T \\ C & G & G & A \\ C & G & A & T \\ G & A & C & A \end{bmatrix}.$$

Correspondingly, one gets the two complementary quaternary matrices $\overline{\boldsymbol{CQ}}_1$ and $\overline{\boldsymbol{CQ}}_2$ as

$$\overline{\boldsymbol{CQ}}_1 = \begin{bmatrix} 1 & 2 & 1 & 3 \\ 2 & 0 & 1 & 0 \\ 3 & 2 & 1 & 0 \\ 3 & 2 & 0 & 3 \end{bmatrix}, \overline{\boldsymbol{CQ}}_2 = \begin{bmatrix} 3 & 3 & 2 & 3 \\ 2 & 1 & 1 & 0 \\ 2 & 1 & 0 & 3 \\ 1 & 0 & 2 & 0 \end{bmatrix}.$$

Next, as in Step 5, one obtains the corresponding output virtual matrix as

$$\boldsymbol{CV} = \begin{bmatrix} 13 & 14 & 9 & 15 \\ 10 & 4 & 5 & 0 \\ 11 & 6 & 1 & 12 \\ 7 & 2 & 8 & 3 \end{bmatrix}.$$

Finally, the equivalent secret key EK_P is achieved with Step 6.

On the basis of the above discussion, one can conclude that the encryption algorithm given in Figure 3 can be broken just with the equivalent secret key EK_P without knowing any secret key parameter.

3.3. Breaking IEA-DESC Using the Chosen-Plaintext Attack

Following Section 3.2, the diffused image $\boldsymbol{P}'$ is considered as the input of the cryptosystem. However, as shown in Figure 2, the actual input of IEA-DESC is the plain-image $\boldsymbol{P}$ rather than the diffused image $\boldsymbol{P}'$. To accommodate to this change, the input chosen plain-image should be adjusted accordingly.

According to the analysis in Section 3.1, the pixel diffusion part of IEA-DESC is actually useless for attackers. Under the premise that the algorithm is known, there is a certain one-to-one correspondence between the diffused image and the plain-image. Therefore, for 8-bit grayscale images of size $H \times W$, the specific analysis steps for the chosen-plaintext attack are given as follows:

Step 1. Choose some special plain-images.

The N_C 8-bit images $\boldsymbol{PI}_n (n = 1, 2, \ldots, N_C)$ constructed in Section 3.2 are presented as the diffused images $\boldsymbol{P}'_n (n = 1, 2, \ldots, N_C)$, respectively, and then their one-to-one corresponding plain-images $\boldsymbol{P}_n (n = 1, 2, \ldots, N_C)$ are obtained using anti-diffusion decryption, which is defined from Equation (3) as

$$\begin{cases} p_i = p'_{i+1} \oplus p'_i, \\ p_{H\times W} = p_1 \oplus p'_1, \end{cases} \tag{12}$$

where $i = 1, 2, \ldots, H \times W - 1$, $\{p_1, p_2, \ldots, p_{H\times W}\}$ and $\{p'_1, p'_2, \ldots, p'_{H\times W}\}$ are the sequences transformed by the plain-image $\boldsymbol{P}$ and the diffused image $\boldsymbol{P}'$ in the raster scanning order, respectively.

Step 2. Temporarily use the encryption machine to get the corresponding cipher-images.

On the basis of the condition of the chosen-plaintext attack, the corresponding N_C cipher-images $\boldsymbol{C}_n(n = 1, 2, \ldots, N_C)$ are obtained from the N_C plain-images $\boldsymbol{P}_n(n = 1, 2, \ldots, N_C)$ by temporarily using the encryption machine.

Step 3. Achieve the equivalent DNA-base permutation secret key.

By substituting $\boldsymbol{PI}_n(n = 1, 2, \ldots, N_C)$ and $\boldsymbol{CI}_n(n = 1, 2, \ldots, N_C)$ in Section 3.2 with the diffused images $\boldsymbol{P}'_n(n = 1, 2, \ldots, N_C)$ and the cipher-images $\boldsymbol{C}_n(n = 1, 2, \ldots, N_C)$, respectively, one gets the equivalent DNA-base permutation secret key EK_P with the same method as in Section 3.2.

Step 4. Recover the images with the equivalent secret key.

First, using the equivalent secret key EK_P, the corresponding diffused image can be obtained from a cipher-image. Then, the recovered plain-image is obtained from the diffused images with Equation (12).

Therefore, the chosen-plaintext attack is effective to break IEA-DESC, and its data complexity is $O(N_C) = O(1 + \lceil \log_4(HW) \rceil)$.

3.4. Breaking IEA-DESC Using the Chosen-Ciphertext Attack

Since the encryption structure of Figure 3 is symmetrical, the chosen-ciphertext attack is also available. The specific analysis steps based on the chosen-ciphertext attack are detailed below:

Step 1. Choose some specific cipher-images and temporarily use the decryption machine to get the corresponding plain-images.

Here, the N_C images $\{\boldsymbol{PI}_n\}_{n=1}^{N_C}$ in Secction 3.2 are served as the chosen cipher-images $\{\boldsymbol{C}_n\}_{n=1}^{N_C}$ respectively, and then temporarily use the decryption machine to get the corresponding plain-images $\{\boldsymbol{P}_n\}_{n=1}^{N_C}$.

Step 2. Get the corresponding diffused images.

The one-to-one corresponding N_C diffused images $\boldsymbol{P}'_n$ $(n = 1, 2, \ldots, N_C)$ are obtained from these plain-images $\boldsymbol{P}_n(n = 1, 2, \ldots, N_C)$ using Equation (3).

Step 3. Achieve the equivalent secret key.

The equivalent DNA-base permutation secret key is achieved by using the same method as Step 3 in Section 3.3.

Step 4. Recover images with the equivalent secret key:

This step is also the same as Step 4 in Section 3.3, so it is omitted.

Therefore, the chosen-ciphertext attack is also valid for breaking IEA-DESC, and its data complexity is also $O(1 + \lceil \log_4(HW) \rceil)$.

4. The Experiments for Breaking IEA-DESC

To verify the feasibility of the two proposed attack methods, some experimental simulations were performed based on a personal computer with Matlab R2016a. Similar to those in Reference [37], our experimental images are 8-bit grayscale images "Lenna" and "Peppers" of size 256×256.

4.1. Breaking IEA-DESC by Chosen-Plaintext Attack

The experiment for breaking IEA-DESC was firstly carried out by the chosen-plaintext attack method proposed in Section 3.3. Given $H = 256$ and $W = 256$, one gets $N_C = 1 + \lceil \log_4(H \times W) \rceil = 9$ from Equation (10). Correspondingly, the nine 8-bit images $\boldsymbol{PI}_n$ $(n = 1, 2, \ldots, 9)$ constructed using the method in Section 3.2 are shown in Figure 5a–r.

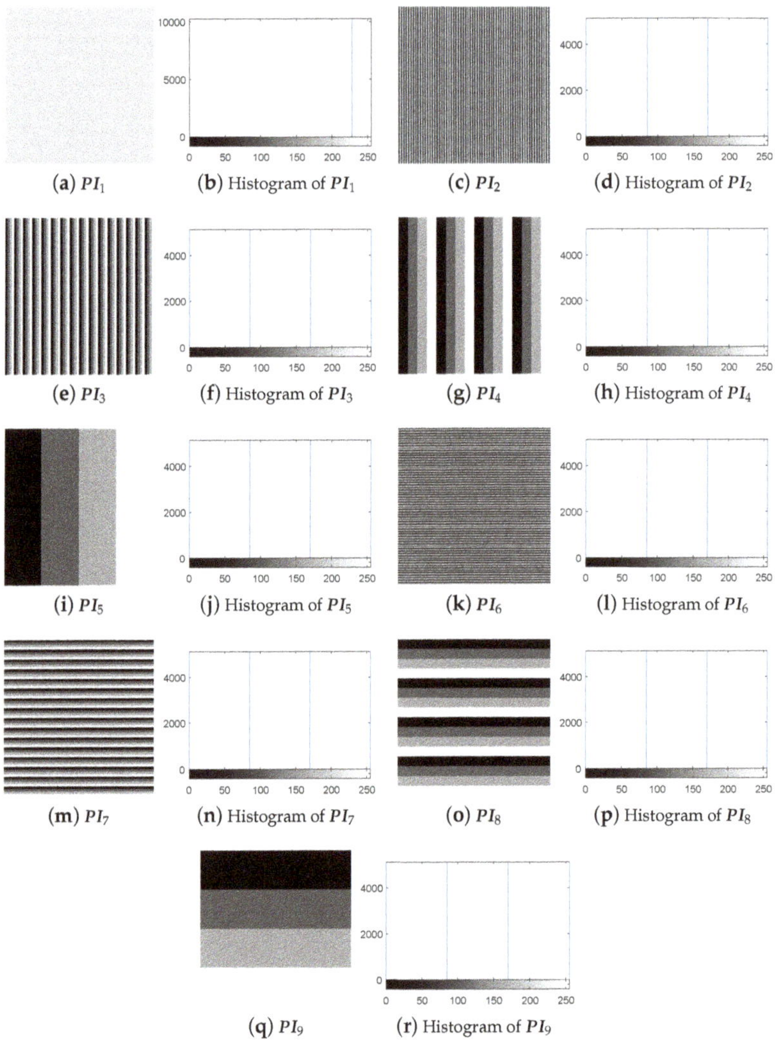

Figure 5. The nine 8-bit special images and their corresponding histograms.

First, following Step 1 in Section 3.3, the nine special images shown in Figure 5 are selected as the diffused images $\boldsymbol{P}'_n(n = 1, 2, \ldots, 9)$, respectively, and then their corresponding plain-images $\boldsymbol{P}_n(n = 1, 2, \ldots, 9)$ are obtained, as shown in Figure 6a–r. Then, according to Step 2 in Section 3.3, the nine corresponding cipher-images $\boldsymbol{C}_n(n = 1, 2, \ldots, 9)$ and their histograms are obtained as shown in Figure 7a–r. Next, using the method in Step 3 in Section 3.3, the equivalent secret key EK_P is obtained using the nine chosen diffused images shown in Figure 5a–r and the nine corresponding cipher-images shown in Figure 7a–r. Finally, the images are recovered using the equivalent secret key EK_P. The attacking results on IEA-DESC with the 8-bit images "Lenna" and "Peppers" are shown in Figures 8a–d and 9a–d, respectively.

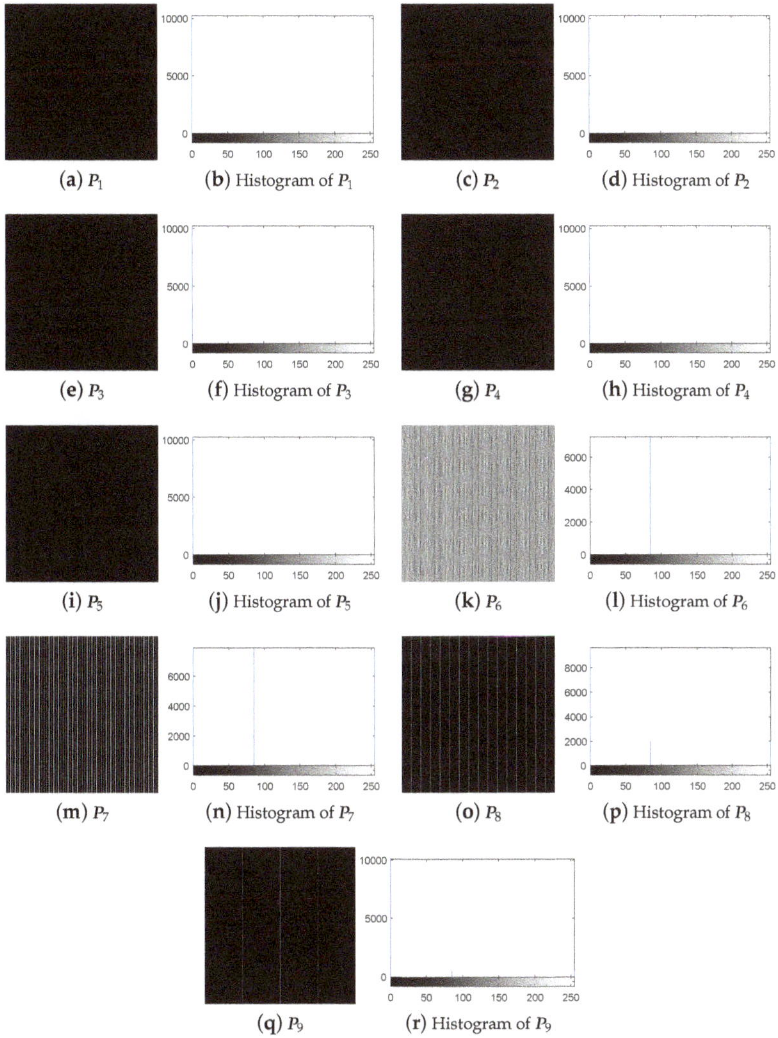

Figure 6. The nine plain-images under chosen-plaintext attack.

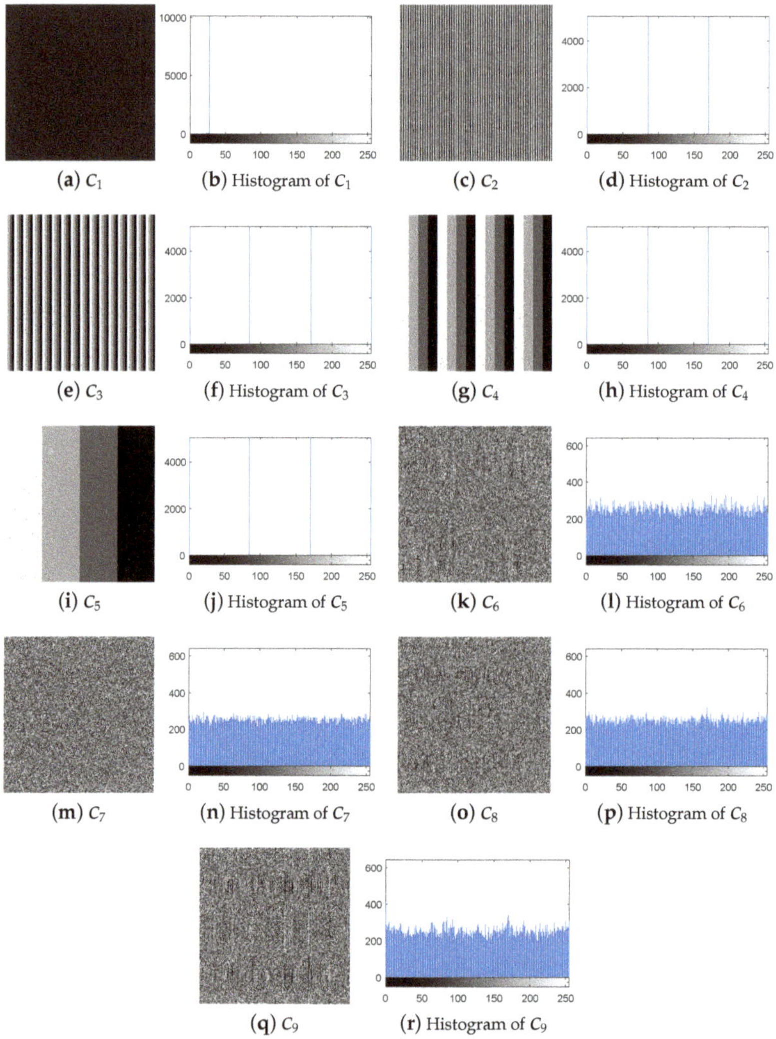

Figure 7. The nine corresponding output cipher-images under chosen-plaintext attack.

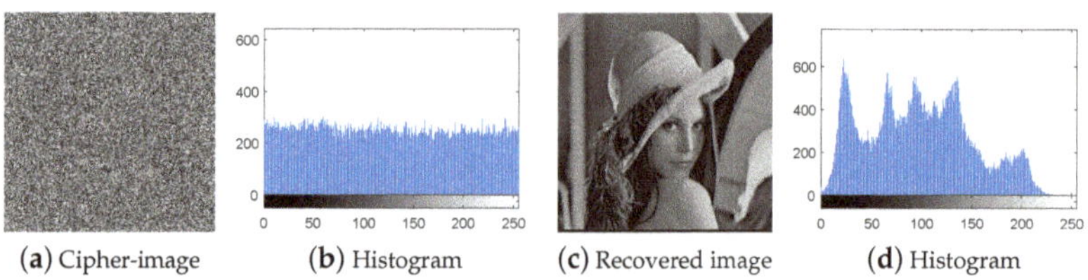

(**a**) Cipher-image
(**b**) Histogram
(**c**) Recovered image
(**d**) Histogram

Figure 8. Attacking result on IEA-DESC with the 8-bit image "Lenna".

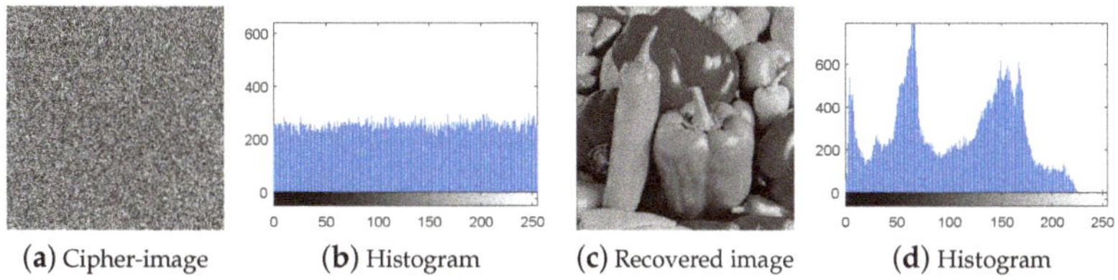

Figure 9. Attacking result on IEA-DESC with the 8-bit image "Peppers".

4.2. Breaking IEA-DESC Using the Chosen-Ciphertext Attack

Accordingly, the experiment for breaking IEA-DESC is accomplished using the chosen-ciphertext attack method proposed in Section 3.4.

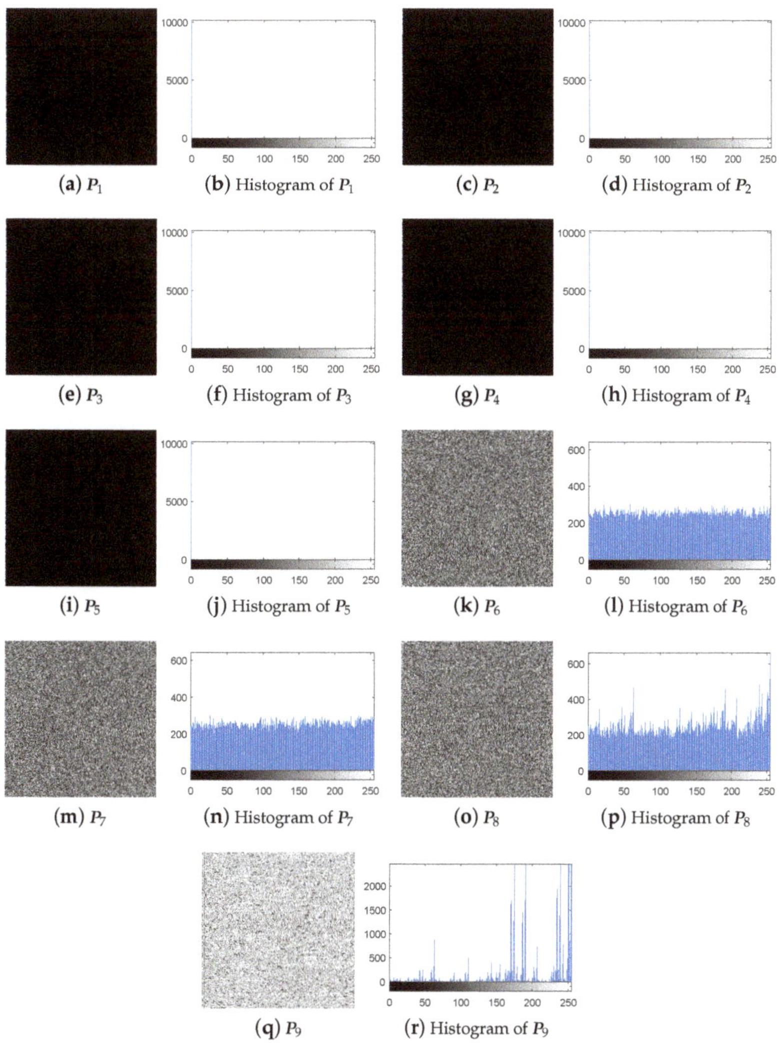

Figure 10. The nine corresponding output plain-images under chosen-ciphertext attack.

First, following Step 1 in Section 3.4, the nine special images shown in Figure 5 are used as the cipher-images $C_n(n = 1, 2, \ldots, 9)$ respectively, and then their corresponding plain-images $P_n(n = 1, 2, \ldots, 9)$ are obtained, as shown in Figure 10a–r. Then, according to Step 2 in Section 3.4, the nine corresponding diffused images $P'_n(n = 1, 2, \ldots, 9)$ and their histograms are obtained as shown in Figure 11a–r, respectively. Next, using the method in Step 3 in Section 3.4, the equivalent secret key EK_P is obtained using the nine chosen cipher-images shown in Figure 5a–r and the nine corresponding diffused images shown in Figure 11a–r. Finally, the images are recovered using the equivalent secret key EK_P. The attacking results on IEA-DESC with the 8-bit images "Lenna" and "Peppers" are also shown in Figures 8a–d and 9a–d, respectively.

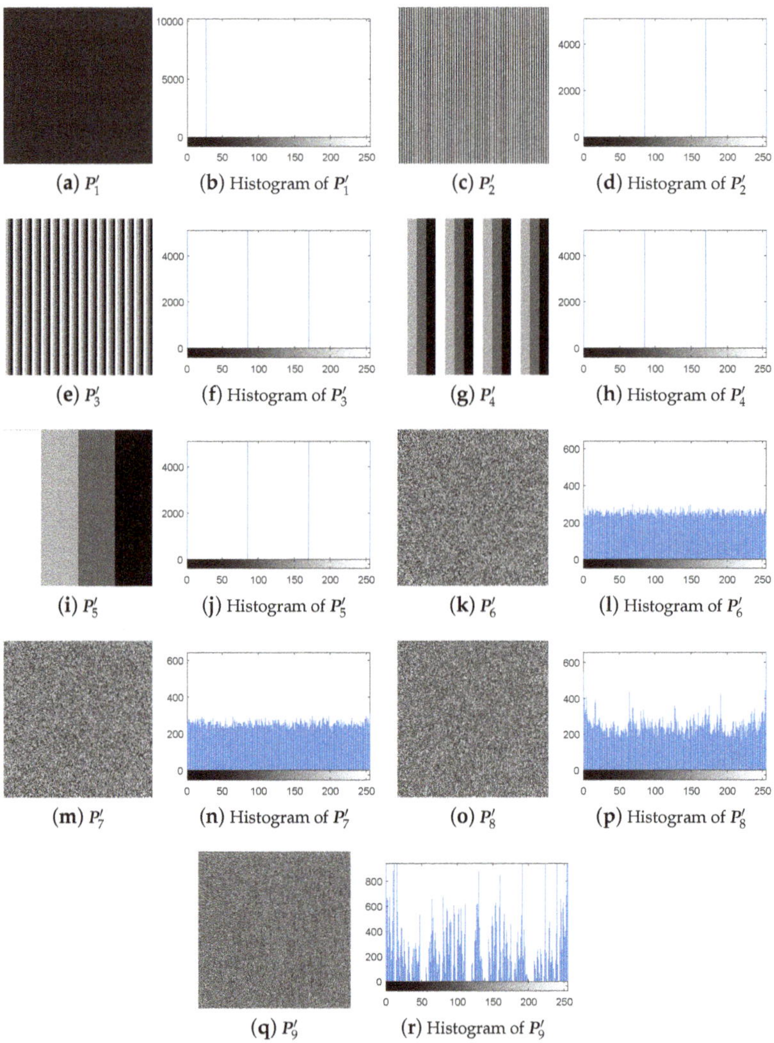

(a) P'_1 **(b)** Histogram of P'_1 **(c)** P'_2 **(d)** Histogram of P'_2 **(e)** P'_3 **(f)** Histogram of P'_3 **(g)** P'_4 **(h)** Histogram of P'_4 **(i)** P'_5 **(j)** Histogram of P'_5 **(k)** P'_6 **(l)** Histogram of P'_6 **(m)** P'_7 **(n)** Histogram of P'_7 **(o)** P'_8 **(p)** Histogram of P'_8 **(q)** P'_9 **(r)** Histogram of P'_9

Figure 11. The nine corresponding diffused images under chosen-ciphertext attack.

4.3. Attack Complexity

In terms of attack complexity, the running times of the chosen-plaintext attack method and the chosen-ciphertext attack method are about 2.1165 s and 2.0785 s, respectively. Moreover, given 8-bit images of size 256 × 256, the data complexity of the two attack methods required for breaking IEA-DESC are both $O(9)$. Therefore, the experimental results verify that the two attack methods are both effective and efficient.

5. Suggestions for Improvement

On the basis of the analysis above, IDE-DESC can neither resist against chosen-plaintext attacks nor chosen-ciphertext attacks because of its inherent security defects. In fact, some other chaos-based ciphers also have similar vulnerabilities as mentioned in Reference [36]. To deal with these problems, some suggestions for improvement to enhance the security are given below:

(1) Checking the validity of each encryption component is significant.

The diffusion part of IEA-DESC is invalid for the attacker because it does not involve any secret key parameter. In fact, it does not contribute to security, but increases the computational complexity of the algorithm. Therefore, the designed algorithms should be scrutinized from the perspective of cryptanalysis to ensure the validity of each encryption component.

(2) Exploiting some novel permutation mechanisms to enhance the security.

Like other permutation-only encryption algorithms, DNA-base permutation only changes the position but does not change the value of each element. The only difference is that the element is quaternary. For permutation-only algorithms, many studies have proved that they are insecure [39,40,44]. To fulfil this demand, exploiting some novel permutation mechanisms is worthwhile.

(3) Avoiding the existence of an equivalent secret key in the algorithm.

The encryption process of the algorithm should be associated with the characteristics of the plain-image or cipher-image [2]. Otherwise, the encryption process for different input images is completely identical, which may lead to the existence of an equivalent secret key. Once the equivalent secret key is obtained by an attacker, the encryption algorithm is broken [36].

(4) Appropriately increasing the number of encryption rounds.

In a single-round encryption algorithm, the confusion and diffusion characteristics maybe insufficient [42]. Increasing the number of encryption rounds can effectively improve this problem. Of course, it also requires higher computational complexity [21]. Therefore, ways in which to balance safety and efficiency deserves more research.

6. Conclusions

In this paper, the security of a recent image encryption algorithm called IEA-DESC has been analyzed in detail. It was claimed that some merits of DNA encoding and spatiotemporal chaos are inherited in the algorithm. However, its algorithm structure has several inherent security pitfalls. It was found that IEA-DESC is actually a combined process of DNA-base permutation and bitwise complement from the perspective of cryptanalysis. Therefore, a chosen-plaintext attack and a chosen-ciphertext attack were proposed to recover the equivalent secret key of IEA-DESC, respectively. Both theoretical analysis and experimental results are provided to support effectiveness and efficiency of two attack methods for breaking IEA-DESC. The reported results would help the designers of DNA-based cryptography pay more attention to importance of the essential structure of an encryption scheme, instead of the elegance of the underlying theory.

Author Contributions: Methodology, H.W.; Software, H.W.; Validation, S.Y. and J.L.; Supervision, S.Y.; Project Administration, S.Y. and J.L.; Funding Acquisition, J.L.

Funding: This research was funded by the National Key Research and Development Program of China (No. 2016YFB0800401), and the National Natural Science Foundation of China (No. 61532020, 61671161).

Conflicts of Interest: The authors declare no conflict of interest.

References

1. Özkaynak, F. Brief review on application of nonlinear dynamics in image encryption. *Nonlinear Dyn.* **2018**, *92*, 305–313. [CrossRef]
2. Wen, H.; Yu, S.; Lü, J. Encryption algorithm based on hadoop and non-degenerate high-dimensional discrete hyperchaotic system. *Acta Phys. Sin.* **2017**, *66*, 230503. [CrossRef]
3. Chen, S.; Yu, S.; Lü, J.; Chen, G.; He, J. Design and FPGA-based realization of a chaotic secure video communication system. *IEEE Trans. Circuits Syst. Video Technol.* **2018**, *28*, 2359–2371. [CrossRef]
4. Fridrich, J. Symmetric ciphers based on two-dimensional chaotic maps. *Int. J. Bifurc. Chaos* **1998**, *8*, 1259–1284. [CrossRef]
5. Chen, G.; Mao, Y.; Chui, C.K. A symmetric image encryption scheme based on 3D chaotic cat maps. *Chaos Solitons Fractals* **2004**, *21*, 749–761. [CrossRef]
6. Lorenz, E.N. Deterministic nonperiodic flow. *J. Atmos. Sci.* **1963**, *20*, 130–141. [CrossRef]
7. Gehani, A.; LaBean, T.; Reif, J. DNA-based cryptography. In *Aspects of Molecular Computing*; Lecture Notes in Computer Science; Springer: Berlin/Heidelberg, Germany, 2003; Volume 2950, pp. 167–188.
8. Abdo, A.; Lian, S.; Ismail, I.; Amin, M.; Diab, H. A cryptosystem based on elementary cellular automata. *Commun. Nonlinear Sci. Numer. Simul.* **2013**, *18*, 136–147. [CrossRef]
9. Yang, Y.G.; Tian, J.; Lei, H.; Zhou, Y.H.; Shi, W.M. Novel quantum image encryption using one-dimensional quantum cellular automata. *Inf. Sci.* **2016**, *345*, 257–270. [CrossRef]
10. Akhshani, A.; Akhavan, A.; Lim, S.C.; Hassan, Z. An image encryption scheme based on quantum logistic map. *Commun. Nonlinear Sci. Numer. Simul.* **2012**, *17*, 4653–4661. [CrossRef]
11. Askar, S.S.; Karawia, A.; Al-Khedhairi, A.; Al-Ammar, F.S. An algorithm of image encryption using logistic and two-dimensional chaotic economic maps. *Entropy* **2019**, *21*, 44. [CrossRef]
12. Karawia, A. Encryption algorithm of multiple-image using mixed image elements and two dimensional chaotic economic map. *Entropy* **2018**, *20*, 801. [CrossRef]
13. Liu, H.; Wang, X. Color image encryption using spatial bit-level permutation and high-dimension chaotic system. *Opt. Commun.* **2011**, *284*, 3895–3903. [CrossRef]
14. He, S.; Sun, K.; Wang, H. Complexity analysis and DSP implementation of the fractional-order lorenz hyperchaotic system. *Entropy* **2015**, *17*, 8299–8311. [CrossRef]
15. Chen, C.; Sun, K.; Peng, Y.; Alamodi, A.O. A novel control method to counteract the dynamical degradation of a digital chaotic sequence. *Eur. Phys. J. Plus* **2019**, *134*. [CrossRef]
16. Li, C.; Feng, B.; Li, S.; Kurths, J.; Chen, G. Dynamic analysis of digital chaotic maps via state-mapping networks. *IEEE Trans. Circuits Syst. I Regul. Pap.* **2019**, *66*. [CrossRef]
17. Shannon, C.E. Communication theory of secrecy systems. *Bell Syst. Tech. J.* **1949**, *28*, 656–715. [CrossRef]
18. Chai, X.; Gan, Z.; Yang, K.; Chen, Y.; Liu, X. An image encryption algorithm based on the memristive hyperchaotic system, cellular automata and DNA sequence operations. *Signal Process. Image Commun.* **2017**, *52*, 6–19. [CrossRef]
19. Zhang, L.; Sun, K.; Liu, W.; He, S. A novel color image encryption scheme using fractional-order hyperchaotic system and DNA sequence operations. *Chin. Phys. B* **2017**, *26*, 100504. [CrossRef]
20. Chai, X.; Fu, X.; Gan, Z.; Lu, Y.; Chen, Y. A color image cryptosystem based on dynamic DNA encryption and chaos. *Signal Process.* **2019**, *155*, 44–62. [CrossRef]
21. Li, C.; Lin, D.; Feng, B.; Lü, J.; Hao, F. Cryptanalysis of a chaotic image encryption algorithm based on information entropy. *IEEE Access* **2018**, *6*, 75834–75842. doi:10.1109/ACCESS.2018.2883690. [CrossRef]
22. Lin, Z.; Yu, S.; Feng, X.; Lü, J. Cryptanalysis of a chaotic stream cipher and its improved scheme. *Int. J. Bifurc. Chaos* **2018**, *28*, 1850086. [CrossRef]
23. Zhu, C.; Wang, G.; Sun, K. Improved cryptanalysis and enhancements of an image encryption scheme using combined 1D chaotic maps. *Entropy* **2018**, *20*, 843. [CrossRef]

24. Li, C.; Liu, Y.; Zhang, L.Y.; Wong, K.W. Cryptanalyzing a class of image encryption schemes based on Chinese Remainder Theorem. *Signal Process. Image Commun.* **2014**, *29*, 914–920. [CrossRef]
25. Zhang, Q.; Guo, L.; Wei, X. Image encryption using DNA addition combining with chaotic maps. *Math. Comput. Model.* **2010**, *52*, 2028–2035. [CrossRef]
26. Hermassi, H.; Belazi, A.; Rhouma, R.; Belghith, S.M. Security analysis of an image encryption algorithm based on a DNA addition combining with chaotic maps. *Multimed. Tools Appl.* **2014**, *72*, 2211–2224. [CrossRef]
27. Zhen, P.; Zhao, G.; Min, L.; Jin, X. Chaos-based image encryption scheme combining DNA coding and entropy. *Multimed. Tools Appl.* **2016**, *75*, 6303–6319. [CrossRef]
28. Su, X.; Li, W.; Hu, H. Cryptanalysis of a chaos-based image encryption scheme combining DNA coding and entropy. *Multimed. Tools Appl.* **2017**, *76*, 14021–14033. [CrossRef]
29. Jain, A.; Rajpal, N. A robust image encryption algorithm resistant to attacks using DNA and chaotic logistic maps. *Multimed. Tools Appl.* **2016**, *75*, 5455–5472. [CrossRef]
30. Dou, Y.; Liu, X.; Fan, H.; Li, M. Cryptanalysis of a DNA and chaos based image encryption algorithm. *Optik-Int. J. Light Electron Opt.* **2017**, *145*, 456–464. [CrossRef]
31. Sun, S. A novel hyperchaotic image encryption scheme based on DNA encoding, pixel-level scrambling and bit-level scrambling. *IEEE Photonics J.* **2018**, *10*, 1–14. [CrossRef]
32. Feng, W.; He, Y. Cryptanalysis and improvement of the hyper-chaotic image encryption scheme based on DNA encoding and scrambling. *IEEE Photonics J.* **2018**, *10*, 1–15. [CrossRef]
33. Özkaynak, F.; Yavuz, S. Analysis and improvement of a novel image fusion encryption algorithm based on DNA sequence operation and hyper-chaotic system. *Nonlinear Dyn.* **2014**, *78*, 1311–1320. [CrossRef]
34. Zhang, Y.; Xiao, D.; Wen, W.; Wong, K.W. On the security of symmetric ciphers based on DNA coding. *Inf. Sci.* **2014**, *289*, 254–261. [CrossRef]
35. Li, C.; Lin, D.; Lü, J. Cryptanalyzing an Image-Scrambling Encryption Algorithm of Pixel Bits. *IEEE Multimed.* **2017**, *24*, 64–71. [CrossRef]
36. Li, C.; Lin, D.; Lü, J.; Hao, F. Cryptanalyzing an image encryption algorithm based on autoblocking and electrocardiography. *IEEE Multimed.* **2018**, *25*, 46–56. [CrossRef]
37. Song, C.; Qiao, Y. A novel image encryption algorithm based on DNA encoding and spatiotemporal chaos. *Entropy* **2015**, *17*, 6954–6968. [CrossRef]
38. Schneier, B. *Applied Cryptography—Protocols, Algorithms, and Souce Code in C*, 2nd ed.; John Wiley & Sons, Inc.: New York, NY, USA, 1996.
39. Li, C.; Lo, K.T. Optimal quantitative cryptanalysis of permutation-only multimedia ciphers against plaintext attacks. *Signal Process.* **2011**, *91*, 949–954. [CrossRef]
40. Jolfaei, A.; Wu, X.W.; Muthukkumarasamy, V. On the security of permutation-only image encryption schemes. *IEEE Trans. Inf. Forensics Secur.* **2016**, *11*, 235–246. [CrossRef]
41. Gao, H.; Zhang, Y.; Liang, S.; Li, D. A new chaotic algorithm for image encryption. *Chaos Solitons Fractals* **2006**, *29*, 393–399. [CrossRef]
42. Alvarez, G.; Li, S. Some basic cryptographic requirements for chaos-based cryptosystems. *Int. J. Bifurc. Chaos* **2006**, *16*, 2129–2151. [CrossRef]
43. Xie, E.Y.; Li, C.; Yu, S.; Lü, J. On the cryptanalysis of Fridrich's chaotic image encryption scheme. *Signal Process.* **2017**, *132*, 150–154. [CrossRef]
44. Li, C. Cracking a hierarchical chaotic image encryption algorithm based on permutation. *Signal Process.* **2016**, *118*, 203–210. [CrossRef]

entropy

MDPI

Article

On Structural Entropy and Spatial Filling Factor Analysis of Colonoscopy Pictures

Szilvia Nagy [1,*], **Brigita Sziová** [1] **and János Pipek** [2]

[1] Széchenyi István University, Egyetem tér 1, H-9026 Gyor, Hungary; szi.brigitta@sze.hu
[2] Budapest University of Technology and Economics, Budafoki út 8, H-1111 Budapest, Hungary; pipek@phy.bme.hu
* Correspondence: nagysz@sze.hu; Tel.: +36-96-613-773

Received: 31 December 2018; Accepted: 27 February 2019; Published: 6 March 2019

Abstract: Colonoscopy is the standard device for diagnosing colorectal cancer, which develops from little lesions on the bowel wall called polyps. The Rényi entropies-based structural entropy and spatial filling factor are two scale- and resolution-independent quantities that characterize the shape of a probability distribution with the help of characteristic curves of the structural entropy–spatial filling factor map. This alternative definition of structural entropy is easy to calculate, independent of the image resolution, and does not require the calculation of neighbor statistics, unlike the other graph-based structural entropies.The distant goal of this study was to help computer aided diagnosis in finding colorectal polyps by making the Rényi entropy based structural entropy more understood. The direct goal was to determine characteristic curves that can differentiate between polyps and other structure on the picture. After analyzing the distribution of colonoscopy picture color channels, the typical structures were modeled with simple geometrical functions and the structural entropy–spatial filling factor characteristic curves were determined for these model structures for various parameter sets. A colonoscopy image analying method, i.e., the line- or column-wise scanning of the picture, was also tested, with satisfactory matching of the characteristic curve and the image.

Keywords: computer aided diagnostics; colonoscopy; Rényi entropies; structural entropy; spatial filling factor

1. Introduction

Colorectal cancer develops from colorectal polyps. The detection of the colorectal polyps is mostly carried out by special endoscopes, called colonoscopes [1,2]. These devices possess not only the image acquiring equipment with light source, but also forceps, needle, laser scalpel, or loop instrument for removing polyps or tissue samples for biopsy. Beside normal, white light pictures, some of the endoscopes can take narrow band images (NBI), which emphasize the blood vessels and the shadows, as can be seen in [3–5], which help find unusual vein patterns that are typical in the case of malignant polyps. In many cases, indigo carmine [4] or other food dyes of bluish hue can be sprayed beside the usual cleansing water to make the pits and valleys of the bowel wall more visible (chromoendoscopy). In addition, magnifying endoscopy is becoming more and more common to detect the fine scale patterns of the surface [4]. Virtual endoscopy [6,7] is a computer tomography based alternative for the optical endoscopy. Capsule endoscopy was developed with the goal of decreasing the discomfort of the patients. It is a small capsule with two cameras and light sources at both ends. It can be swallowed, and travels through the bowel [8]. Unfortunately, it is generally less effective in finding polyps than the classical endoscopy, and cannot perform any operations, as it is a passive device.

Although colonoscopy is considered to be the most effective way of cancer screening [1], it still has a non-negligible miss rate [9], which has not decreased much, even though the equipment has improved over time [10]. A colorectal polyp can be missed for the following reasons. First, if the polyps are small, then they are much harder to find [9], even though curvelet-based methods exist that can improve the diagnosis probability [11]. In addition, usually, the better quality the picture is, the lower the risk of missing a small polyp. Moreover, although before endoscopy sections, the bowel is theoretically purified, impurities often occur [12], mostly in the form of yellowish liquid or solid pieces. These impurities can be removed by spraying water on the given surface segment from the endoscope. The next reason is that the bowel has continuously moving, shiny, pink walls, which sometimes (despite the inflation) fold over polyps. The last factor is fatigue: after a couple of seconds of watching the screen of the endoscopes, the eyes and brain of the gastroenterologist gets used to the environment, and after a longer time of watching they often get tired; this is true not only in the case of inexperienced medical staff [9,10]. This last factor was the reason for thinking about developing a computer aided diagnosis protocol. Computer aided diagnosis is not meant to substitute the human medical expert, only to draw their attentions to certain points, in a way that does not disturb other aspects of the diagnosis.

Image processing tools can improve the diagnosis performance; however, colonoscopes are not developed for machine processing, no matter if it is a conventional or capsule version. Usually, live video signal arrives from the endoscope, which has quite low bit rate, hence the small resolution and/or large compression ratio of the pictures. This means that, even though the video signal seems to be of good or at least acceptable quality for the human eye, the individual pictures have large distortion and many compression artifacts. Since the bowel walls move continuously, usually the pictures are blurred. Impurities are often present, and, even if they are removed by water, this water together with the native liquids makes the bowel walls shiny, and the thus arising reflections make image processing more complicated. Even though chromoendoscopy or NBI can make the color spectrum a bit more stretched, in a common endoscopy picture, the colors are mostly only shades of pink.

In medicine, to determine whether a lesion is benign or potentially malignant, a sample is taken from the tissue for biopsy. As fatigue might also influence the working efficiency of a pathologist, there are possibilities to introduce computer aided diagnosis to this point, too. This branch is investigated by multiple research groups. Kayser and his group applied the neighborhood relations based structural entropy for this purpose [13]; dos Santos and his group used sample entropy [14]; Ribeiro and his fellow researchers proposed curvelet and fractal analysis combined with Haralick structure descriptors and various classification methods [15]; Chaddad and his coworkers combined multiple texture characterizing entropy-related quantities and studied their multiresolution behavior [16,17]; and Wang trained neural networks for the picture components derived for both of the applied stains [18]. However, even though these methods might work well for tissue samples or other illnesses [19], they are unfortunately not applicable directly to endoscopy pictures.

Another image classification task related to colorectal polyps or cancer arose after magnifying endoscopy appeared [20]. Kudo and his coworkers, based on biopsy results, found that the pit pattern of the polyp surface can classify the polyps without performing the biopsy, or removing the benign polyps [21]. The classification of polyps based on the Kudo classes [22] is the second branch of image processing based computer aided diagnosis for colonoscopy [23]. However, the task for finding the polyps is also of great interest.

The colorectal polyps can have a wide range of shapes, from lesions depressed into the bowel wall through flat and slightly protruding sessile to pedunculated polyps with expressed stalk. Generally, the roundish polyps—either sessile or pedunculated—are the target for automatic image processing methods. The MICCAI Endoscopic Vision Challenge made several databases and methods available [24–26]. Its results are summarized in [27]. The methods collected in [27] include evolutionary algorithms, neural networks, and shape- and lighting-based classical similarity recognition algorithms [25]. For the detailed description of the methods and the results achieved by the research groups that participated in the MICCAI Endoscopic Vision Challenge, we refer to [27]. There are also methods for computer aided classification of lesions from capsule endoscopy images [28,29], but most of the studies work with conventional endoscopes, as they are more widespread.

A fuzzy classification scheme based on the method detailed in [30,31] and summarized in [32] also appeared [33,34]. This proposal uses edge density and statistical parameters, such as the mean, standard deviation or the Rényi entropy based structural entropy, for determining whether a segment of the colonoscopy picture contains polyp.

Entropies are often used for image analysis [19,35,36]. There are two approaches for determining the structural entropy. First, a graph theory based definition was given in [37]. Later, independently a Rényi entropy based structural entropy was also introduced. Although this later structural entropy is first applied in electron structure analysis in [38], its use in image processing is first presented in [39] for characterizing microstructures of the metal electrode materials on semiconductor surfaces. The idea to use it in medical image processing are presented in in [33]. The results are good, well above 90% hit rate for some types of pictures (especially the ones where either the color or the pattern is very different for the polyp and the background, or the polyp has strong, visible contour), while for some other types (e.g., when the polyps are lit too strongly from the side, or when there are image distortions on the pictures due to low resolution or some dark impurities), the miss rate is around 50%, which is of course unacceptable. The false positive rate is always very low, except for some extremely impure cases.

The aim of this paper was not to provide or improve a method for classification of images, but to study a tool that can be used in classification algorithms as one of the parameters related to the shape of the pixel intensity distribution of the picture.

The Rényi entropy based structural entropy is a very simple quantity, easy to calculate, and, together with the spatial filling factor, gives visible information about the shape of the studied distribution. In the following, we discuss some of the most important properties of the roundish colorectal polyps and model their pixel intensity distributions and structural entropy behavior to get characteristic curves that help to understand the reason of differences between image segments with and without polyps and image types.

The remainder of this article is organized as follows. The properties of structural entropy and its use in image analysis are summarized in Section 2. Next, in Section 3, the distribution of bowel picture segments with and without polyps are studied and model structures for reproducing certain aspects of the pictures are introduced, which are required to generate structural entropy–spatial filling factor characteristic curves for the picture segments with polyps, as presented in Section 4. In this section, the dependence of the structural entropy–spatial filling factor curves of both the model surfaces and the real images are studied according to several parameters. Finally, the conclusion is summarized in Section 5. The characteristic curves are collected in Appendices A and B.

2. The Rényi Entropy-Based Structural Entropy

One of the first attempts to describe information was by Hartley, who used a number of yes/no questions to identify an element of a set of possible strings as the information revealed by identifying the string [40]. A couple of years earlier, Nyquist used a very similar formula [41]: both used the logarithm of the number of possible choices to define information content. Shannon wrote his article about the theory of communication in 1948 [42]. He first defined entropy in the information content sense, referring to statistical mechanics and the Gibbs entropy [43] when introducing this quantity. In addition, in quantum mechanics, the entropy of a density distribution was introduced [44], as it was based on the notations used by von Neumann [45]; later, it was named after him. In both cases, the entropy of a probability distribution $\{p_1, p_2, \ldots, p_N\}$ is defined as

$$S(p_1, p_2, \ldots, p_N) = K \sum_{i=1}^{N} p_i \log_a \frac{1}{p_i}. \tag{1}$$

where, both the constant K and the basis of the logarithm are freely chosen; however, both in quantum mechanics and in information theory, the constant is generally selected to be one. The basis of the logarithm is 2 in the case of the information theory applications (in this case, the unit of the entropy is Shannon or simply bit), and e in the case of physics.

The entropies used in image processing, as well as the quantities originated from the sample or structural entropy usually define the probability distribution corresponding to an image in a rather complex way. Stantchev based the probabilities on the number of connections of a given node in a graph. The entropy from [13,46] calculates probabilities from distances between neighboring vertexes and connections weights; Humeau-Heurtier and her co-workers generalized the sample entropy, which uses probabilities consisting of the ratio of the number of cases when two sample vectors froming a series have sufficiently small distance, and a similar number for their shortened versions [36,47–49]. All these methods introduce quite complicated concepts, which are scale-dependent. Of course, scale dependency gives valuable information on the structure, such as in the case of the multiscale entropy [36].

However, there is another concept for characterizing the structure the shape of a picture by entropies. Images have native distributions, their pixel intensities, which can be easily normalized to fulfill the conditions for being a probability distribution, i.e.,

$$\sum_{i=1}^{N} p_i = 1 \tag{2}$$

$$p_i \geq 0 \quad \text{for } i = 1,2,\ldots,N, \tag{3}$$

if the already non-negative pixel intensities I_i are divided by their sum as $p_i = I_i / \sum_i I_i$.

In electron structure calculations, instead of the probabilities, the electron density is used; it is also normalized similarly to the probabilities. Although the electron density is usually a continuous function, it can be approximated, or modeled as a distribution over a regular grid, thus the similarity of the electron states and picture pixel intensities can be seen. For measuring how localized is an electron state of a solid, a participation ratio, or delocalization measure, was introduced [50,51] the following way,

$$D = \frac{1}{\sum_{i=1}^{N} p_i^2}. \tag{4}$$

This quantity tells approximately the number of the higher probability grid points, i.e., the number of grid points the electron density extends to, or, in the case of the pictures, it is the number of the light pixels.

If in the entropy we substitute the real probability distribution by a step distribution that extends to the D pixels, we exclude the shape information and keep only the information related to the extension of the distribution. The entropy thus becomes the extension entropy [38]

$$S_{ext} = \log D. \tag{5}$$

This means that, if we subtract S_{ext} from the total Shannon entropy, the remaining part has the information about the shape or the structure of the distribution. Structural entropy was introduced as

$$S_{str} = S_1 - S_{ext} = S_1 - \log D. \tag{6}$$

In Ref. [38], Equation (6) uses natural logarithm, and we apply this convention (even though any basis for the logarithm could be used).

Using Shannon's entropy definition, Alfréd Rényi generalized [52,53] the quantity characterizing the amount of information based on Faddeev's postulates [54]. His zeroth entropy was Hartley's information content; the first one was Shannon's entropy; and the next members of this series are

$$S_n = \frac{1}{1-n} \log \sum_{i=1}^{N} p_i^n. \tag{7}$$

If we study the extension entropy in Equation (5), knowing the Rényi entropy series in Equation (7), we can find that the second Rényi entropy

$$S_2 = \frac{1}{1-2} \log \sum_{i=1}^{N} p_i^2 = \log \frac{1}{\sum_{i=1}^{N} p_i^2}. \tag{8}$$

is the extension entropy itself [55].

Pipek and Varga introduced another quantity that describes the structure of the distribution. If the participation ratio D is compared to the total number of grid points (pixels), i.e.,

$$q = \frac{D}{N} \tag{9}$$

is defined, we receive the so-called spatial filling factor, which is a quantity between 0 and 1. Pipek and Varga [38] showed that, if for a distribution of a given shape its structural entropy S_{str} is plotted versus its spatial filling factor q, then the point is along a curve that is characteristic for the shape of the distribution. Each type of shape, e.g. Gaussian, exponential or power law distribution has its separate characteristic curve (which is of course different for one-, two-, or three-dimensional distributions). Moreover, in [55], the relation

$$\log q = \log \frac{D}{N} = \log D - \log N = S_2 - S_0 \tag{10}$$

is also derived, and it is usual to plot the $S_{str}(\ln q)$ curves, instead of the originally proposed structural entropy–filling factor plots. Some characteristic curves for the two-dimensional exponential, Gaussian and second-order power law distributions are shown in Figure 1. In addition, the theoretical limit of the structural entropy

$$S_{str} \leq -\ln q \tag{11}$$

is plotted. For the proof of this formula, $S_{str} \geq 0$, the completeness of the allowed domain, as well as the shape of the characteristic curves, we refer to the appendices of [38].

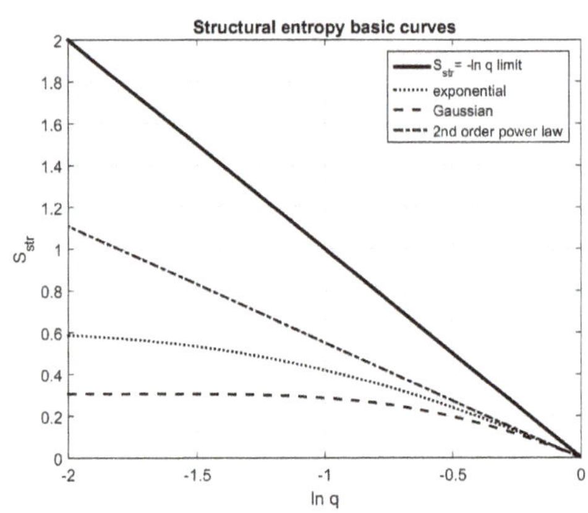

Figure 1. A structural entropy $\ln q$ plot showing the limiting line $S_{str} = -\ln q$ in thick continuous line, and the trend lines for the second-order over law, exponential and Gaussian type distributions for two-dimensional case.

The Rényi entropy based structural entropy and the filling factor is introduced in scanning electron microscope image characterization in [39], and for determining superstructures within a nanostructure in [56]. For characterizing surfaces of electrodes, Bonyár and his coworkers used the structural entropy based localization factor with success [57,58]. Based on these results, we surmised that also colorectal polyps can be identified using their structural entropy versus filling factor plots. The classification of a distribution needs characteristic curves, to which the structural entropy and filling factor point of the distribution can be related. To make the Rényi entropy-based structural entropy applicable for characterizing images or image segments of colonoscopy origin, we need to find possible structures present in such an image, as well as their characteristic lines on the $S_{str}(\ln q)$ map. The purpose of this study was to determine if there are differences between characteristic curves of images with and without polyps, and if there is a way they might be used for distinguishing the two types of images.

3. Results: Simplifying and Modeling the Structures Present on Colonoscopy Images

3.1. Across Real Pictures

We used the database of Etis Larib from the MICCAI Endoscopy Vision Challenge [24] for this study, as their pictures have very high resolution (1225 by 966 pixels), only small black frame, and only very few compression artifacts. The three color channels of two selected images are plotted in Figure 2. (The first one belongs to the well, but not extremely well classifiable group in [33,34], the second to the not too badly classifiable group.) It can be seen that the different color channels emphasize different features of the image: the veins are visible in the green color channel, the shadows can be seen in the blue and red channels and yellowish liquids show in the blue channel.

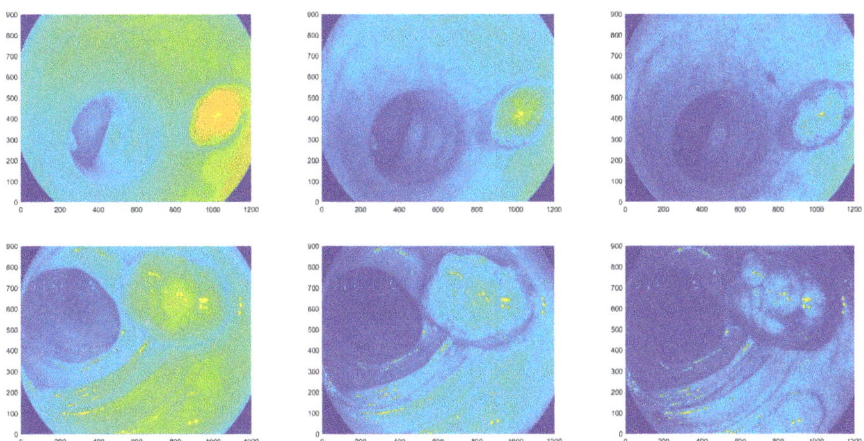

Figure 2. The red, green and blue color channels of pictures 83.tif and 114.tif from database ETIS Larib [24].

In these pictures, the elementary structures seem to be waves and sphere or ellipsoid segments. To understand the behavior of the structural entropy of the different image segments, structural entropy versus filling factor plots of waves with straight or curved wave fronts, as well as of hemispheres are determined and the characteristic lines are given for these structures as a first step. As in these pictures the sphere segments are sitting on the wavy background, the next step would be to plot structural entropy characteristic curves of these superposed structures. To determine the more detailed structure around the polyps, we prepared cross section cuts through the polyps both in both dimensions. Some examples are shown in Figure 3.

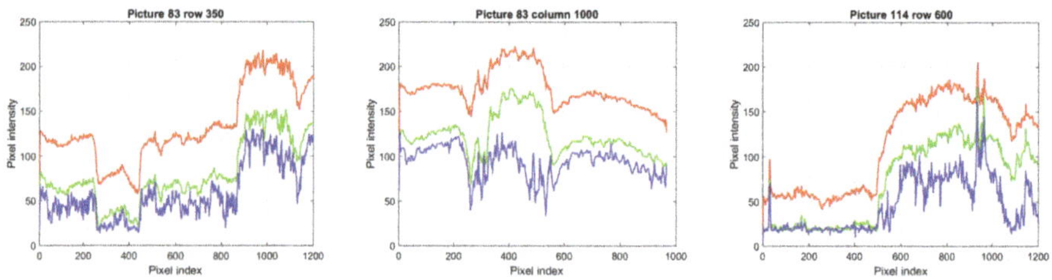

Figure 3. Cross section plots through polyps of Figure 2. The red, green and blue color channels are plotted with red, green and blue, respectively.

According to the cross sections, the environment of the polyps can be modeled as if a hard hemisphere would be pressed into an elastic surface, i.e., almost all the polyps had some kind of ditches around them, similar to the function shown in Figure 4. This is of course the shadow around the polyp. After testing some functions to reproduce this structures, we found that, if we subtract Gaussian function of the same standard deviation as the radius of the sphere, the behavior is rather well modeled.

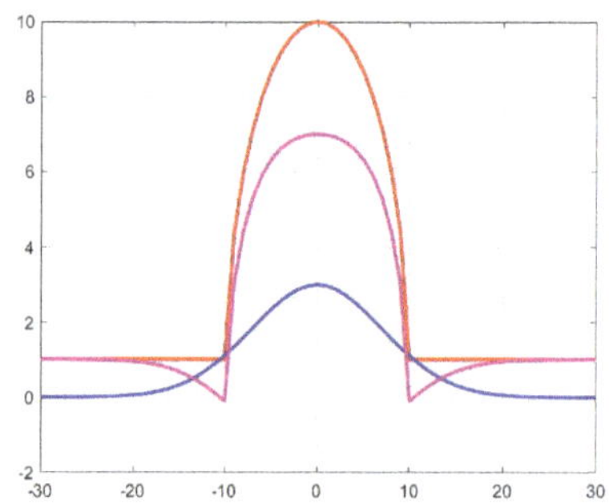

Figure 4. A semicircle (red line), a Gaussian function (blue line) of the same parameters $R_{circle} = \sigma_{Gaussian}$ and their difference (magenta line), which is reminiscent of the shape around the polyp in Figure 3. This kind of function can be used as a simple model the pixel intensity around the polyps.

3.2. Model Structures for Waves of the Bowel Wall

The bowel wall without polyp forms waves. As a first step, these waves can be modeled as sinusoidal function with straight wave profile over a grid of size $N \times N$, as can be seen in Figure 5. A sinusoidal distribution

$$f(x,y) = A \cdot \sin\left(\frac{2\pi}{T}x + \varphi\right) \tag{12}$$

has three parameters: the amplitude A, the wavelength T (frequency $1/T$) and the phase φ. In our case, as the distributions are normalized to be a probability distribution, changing the amplitude is out of the question, thus the remaining parameters are frequency and phase shift. An offset or DC term can also be introduced as a parameter to study, and the angle of the wave front, is the distribution, thus becomes:

$$f_W(x,y) = \sin\left(\frac{2\pi}{T_x}x + \frac{2\pi}{T_y}y + \varphi\right) + B + 0.5, \tag{13}$$

with B being the offset, and T_x and T_y the two components of the wavelength. A default offset 0.5 was introduced to fulfill Equation (3). The distribution is normalized according to Equation (2) before calculating the structural entropies in all the cases, thus the normalization step is not mentioned in the further models.

As in most cases only part of a whole period is visible in the studied image segments, the parameter set was selected to cover the cases when the period of the wave is between 0.1 and 10 times the size of the tile size. There is no point going below 0.1 as the surface is practically a plane with gradient of $2\pi/T$. The offset was studied to be between 10^{-10} and 10^{-1}. The zero offset was not used, as in the $p_i = 0$ case (Equation (1)) is not computable by machine; of course, its limit can be derived by l'Hospital's rule, however for running time reasons the conditional branching for calculating the $p_i = 0$ entropies was not implemented. The parameter set for φ was from 0° to 180°, and the T_y/T_x ratio from 0.1 to 10.

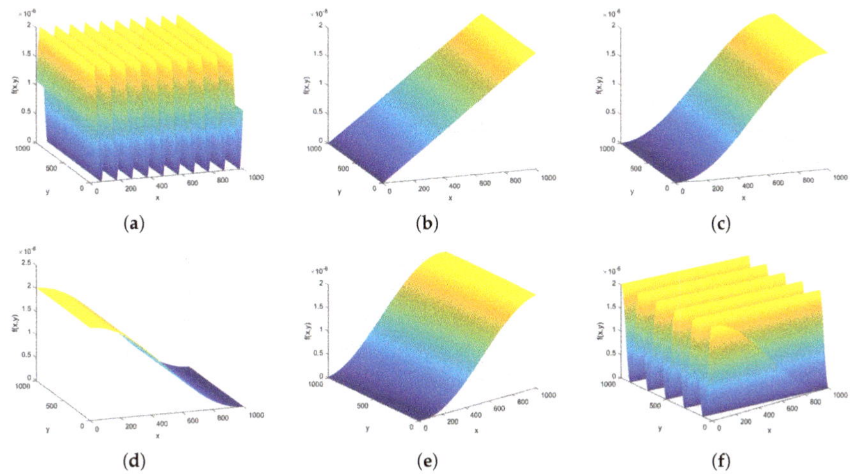

Figure 5. Pictures of straight waves over 1000 by 1000 grid. The limits of the parameter sets are plotted: (**a**,**b**) the largest wavelength and the smallest wavelength used; (**c**,**d**) the minimum and the maximum of the phase shifts; and (**e**,**f**) the minimum and maximum wavefront direction angles.

As the bowel is a tube, and in perspective the waves of the wall might seem to be concentric, the circular, or elliptic waves are also of interest. The distribution for these waves is modeled as

$$f_C(x,y) = \sin\left(\sqrt{\left(\frac{2\pi}{T_x}(x-x_0)\right)^2 + \left(\frac{2\pi}{T_y}(y-y_0)\right)^2} + \varphi\right) + 0.5 + B. \tag{14}$$

The six parameters are the two wavelengths T_x and T_y, the two coordinates of the center x_0 and y_0, the phase φ and the offset B. The studied parameter domains for the ratio T_x/T_y are from 1/5 to 5, for the center coordinates (x_0, y_0) from the center of the tile, i.e., from $(0,0)$ to $(N,0)$, and for the phase φ from 0° to 180°, as can be seen in Figure 6.

Tilted Waves

As in the pictures the further parts of the bowel are darker, waves with a tilt were also studied. In this case, instead of the constant offset B, a plane with a slight slope was also applied. The direction of the slope was perpendicular to the wave front, as mostly the wave fronts are perpendicular to the bowel axis. Plane tilt was given to the waves with straight front and conical tilt to the circular fronts. The parameter was the ratio of the wavelength and the gradient, which was between 2^0 and 2^{-6}.

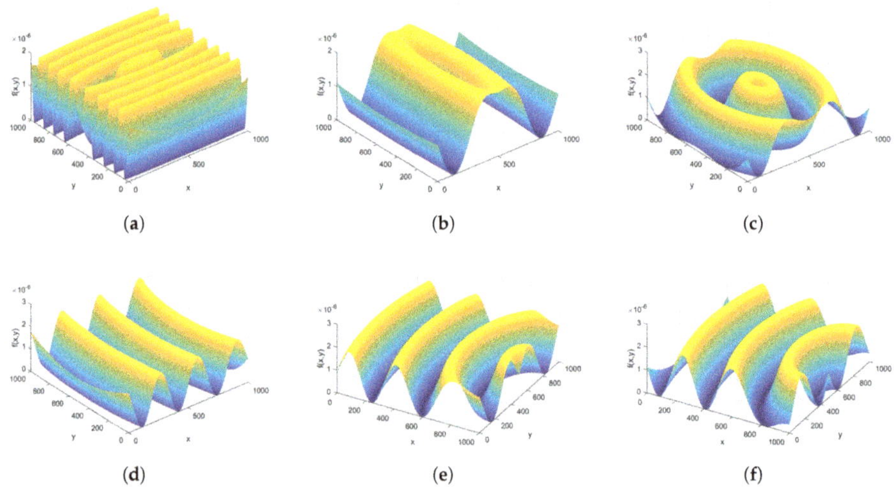

Figure 6. Picture of elliptical waves with minimum and maximum wavelength ratio in (**a**,**b**), minimum and maximum off-center positions in (**c**,**d**), and minimum and maximum phases in (**e**,**f**)

3.3. Model Structures for the Polyps and their Shadows

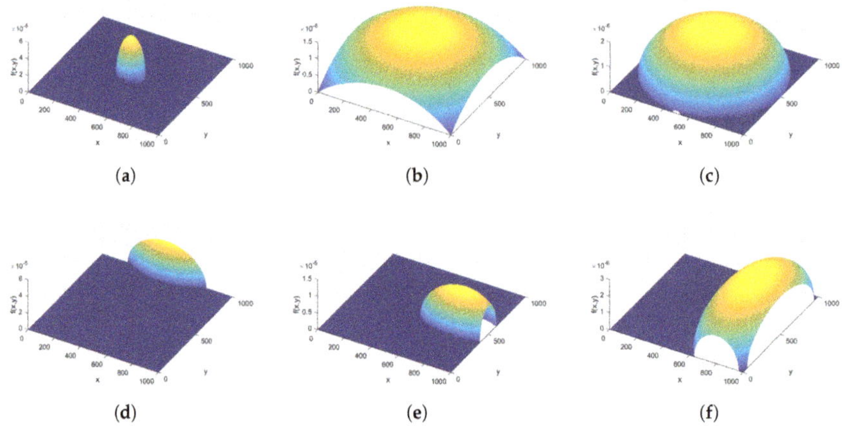

Figure 7. Picture of hemispheres. In (**a**,**b**) the radius is at the two extremum of the parameter set, in (**c**,**d**) the offset, and in (**e**,**f**) the deformation toward an ellipsoid.

The polyp can be quite well modeled as hemispheres, ellipsoid or sphere segments. The studied distribution was

$$f_S(x,y) = \max\left(\sqrt{R^2 - \left(\left(\frac{(x-r_x)}{R_x}\right)^2 + \left(\frac{(y-r_y)}{R_y}\right)^2\right)}, B\right), \tag{15}$$

with R being the radius of a sphere, R_x and R_y the parameters distorting the hemisphere to a half ellipsoid, (r_x, r_y) the coordinates of the center of the object, and B the background height around the ellipsoid or sphere segment, which was usually set to 10^{-10}. The analysis went on in two directions: first the size and the position of a hemisphere was varied, and then the positions remained at the center and at the edge of the picture, but the shape was distorted to ellipsoid. The the distributions corresponding to the limits of the parameter sets are given in Figure 7.

Model Structures for the Shadows around the Polyps

Only Gaussian functions were used for generating the valley representing the shadow around the spheres; however, distributions of type

$$f_G(x,y) = \exp\left(-\sqrt{\left(\frac{(x-r_x)^2}{\sigma_x^2}+\frac{(y-r_y)^2}{\sigma_y^2}\right)^{\alpha}}\right), \tag{16}$$

were also studied. Here, the center was always set as the same position as the center of the half ellipsoid, and the variances σ_x and σ_y as the same as the radial parameters of the ellipsoid. As functions that use higher power α in Equation (16) have wider and flatter central part and quicker decrease, they were also tested for reproducing the shadows around the hemispheres, with less distortion in the spheres. However promising this idea was, the results were usually less similar to the real polyps than the $\alpha = 2$ case, as can be seen in Figure 8.

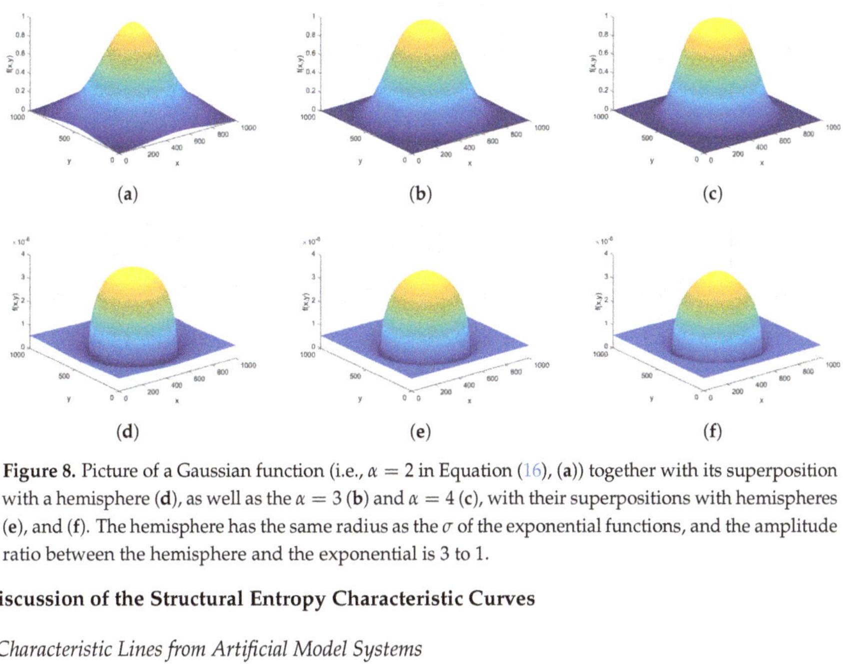

Figure 8. Picture of a Gaussian function (i.e., $\alpha = 2$ in Equation (16), (**a**)) together with its superposition with a hemisphere (**d**), as well as the $\alpha = 3$ (**b**) and $\alpha = 4$ (**c**), with their superpositions with hemispheres (**e**), and (**f**). The hemisphere has the same radius as the σ of the exponential functions, and the amplitude ratio between the hemisphere and the exponential is 3 to 1.

4. Discussion of the Structural Entropy Characteristic Curves

4.1. Characteristic Lines from Artificial Model Systems

After deciding the possible models and their parameter sets, the structural entropy versus spatial filling factor plots were studied. Two parameters were changed in one plot series: the first given as the third axis of the plot, and the second as the color and marker of the plotted points. Even though the parameter sets consist of discrete values, the points corresponding to the second parameter value and varying along the first parameter were connected as a guide of the eye. Most of the characteristic curves are presented in Appendices A and B. The reason for this is manifold. First, with only the title containing the information about the model type, it is easier to see the result. A similar statement is true for the text about the characteristic curves in this section: not breaking the text with images helps keep the focus. Second, there are many parameter combinations that do not seem to be very important at this point, and their results can be summarized in one sentence. Third, usually three or four plots are given for one result, which is too many; however, the 3D plots with parameter–filling factor–structural entropy axes are usually interesting not only from one point of view, but from the three projections and one perspective plot as well.

4.2. Dependence on the Image or Tile Size

The dependence on the tile size can be excluded as a parameter if all the other parameters are given in the relation to the tile size N. The only exception is the offset parameter B. In the cases when B was used only for technical reasons, namely for treating the $p_i = 0$ cases without having to use if–then conditional branchings in the program, the offset should be small enough to be negligible compared to the rest of the intensity values. To set a suitable default offset, the tile size dependence of the structural entropy and the spatial filling factor of all our model surfaces were determined. In all cases, we found that, between the realistic limits of $N = 20$ (smallest applied tile size in the case of lower resolution images in [33]) and $N = 1000$ (magnitude of the full image size of database [24]), the tile size–$\ln q$–S_{str} curves are practically the same if B is smaller than 10^{-5}.

An example (of waves with straight wavefront) is plotted in Figure 9. This serves as a demonstration of how the three projections of the 3D plot look similar. It can also be seen in Figure 9 that, although above the tile size $N = 200$ the values $\ln q$ and S_{str} are almost completely independent of the tile size (even of the tile size to wavelength ratio), in the region of smaller N, the tile size plays not negligible role, thus we can conclude that using larger tile sizes in the evaluation process makes the results more stable and reliable. However, fulfilling the condition of using at least 200 by 200 pixel sized tiles is not always possible, especially if the images are of 384×288 size, such as in the case of the CVC Clinic database [27].

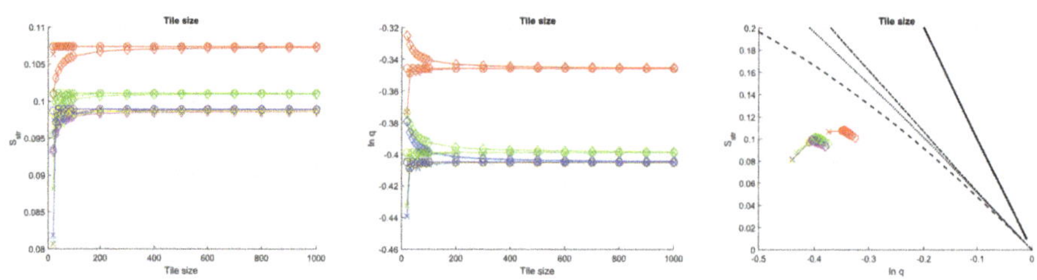

Figure 9. The structural entropy S_{str} and the logarithm of the spatial filling factor q versus the tile size for straight waves. The offsets are the following: red, 0.1; green, 0.01; blue, 0.001; cyan, 0.0001; magenta, 0.00001; yellow, 0.000001. The last three are barely distinguishable. The wavelength to tile size ratios are the following: $\diamond$: 2, $\circ$: 1, $\times$: 0.5.

4.3. Dependence on the Wavelength Compared to the Image Size

If the wavelength T of the wave in the picture segment is chosen as the first parameter, it can be either larger than the tile size N, or at most one third of it, thus the tile size to wavelength ratio was selected to be between $1/4$ and 3. In the case of N to T being 1:4, the resulting wave distribution starts to resemble a plane; this was the reason the parameter space was extended to 1:10 limit, as in that case the limiting behavior could also be studied. In addition, the other limit was extended compared to the realistic case to see whether there is a limiting behavior in the small wavelength domain as well.

The offsets can also play important role if they are larger than 0.001, thus, as a first step, the second parameter was selected to be B. The characteristic curves for both the waves and the limiting planes are given in Figures A1 and A2 in the Appendix, both for straight and for circular wavefronts. Figure A1 gives the three projections of the $S_{str}(\ln q, T/N)$ plots, while Figure A2 shows 3D perspective. It can be seen that in the large $T : N$ domain the curves follow their corresponding limits' characteristics (i.e., the planes for the straight wave front and the cones for the circular wavefront). The small $T : N$ ratio part of the curve oscillates around a value with decreasing amplitude in both $\ln q$ and S_{str}, resulting in ribbon-bow-like, eight-shaped loops in the $S_{str}(\ln q)$ plots.

The second parameter can also be the phase. With a very small offset $B = 10^{-10}$; the characteristic curves can be seen in Figures A3 and A4. It is clearly visible that phase influences the structural entropy, and spatial filling factor values, especially in the lower $T : N$ domains.

Another possibility for the other parameter beside the wavelength is the tilt slope to wavelength ratio. The results are summarized in Figures A5 and A6 in the Appendix. It can be seen that, for wavelengths larger than the tile size, the tilt does not have real influence, however, in the small wavelength direction, the characteristic curve oscillates much more vehemently if tilt is present than in the tiltless cases, and also some points with very high curvature—turning points—arise in the case of the waves with tilt.

4.4. Dependence on the Phase and Center-Shift

In the case of the colonoscopy image categorization, besides dividing the pictures into fixed tiles, applying sliding tiles and analyzing the characteristics of the arising $S_{str}(q)$ or $S_{str}(\ln q)$ point set is another option. This can be represented as a phase scanning in the case of the waves with straight wavefront, and as moving the center in the case of the circular waves.

For the studies of the phase, the fourth dimension can be either the wavelength, or the offset (constant or linear). For all three cases (i.e., wavelength, offset, and tilt), only the 3D plots are given in Figure A7 of Appendix A. The non-varying parameters were set the following way. The offset in the first image, where the wavelength varied, was set to be negligible (10^{-10}). The wavelength in the second column, where the offset varied, was set to $2N$, as for wavelength values smaller than N neither the structural entropy nor the spatial filling factor had dependence on wavelength, in the case of the straight wavefront, and very simple sinusoid wave-like characteristic curves arise in the case of the circular wavefronts. In both cases, if the wavelength is larger than the tile size, the loops formed on the $S_{str}(\ln q)$ plot are turned back at a point, resulting in hook-like lines, which seem to have derivative singularities, or at least very rapid variation in their gradients. As can be seen in the plot with varying offset, if the offset becomes negligible, this turning point becomes a simple inflection on the characteristic line. In the case of the linear offset (tilt) of the third column, the hook-like behavior becomes rather loop-like.

However, as both the offset and the tilt can easily be removed from a picture by image processing means (the offset by a counter-offset, i.e., by setting the minimum of the pixel intensities as 0, whereas the tilt by removing a mean-filtered version of sufficiently large filter size from the image), it is more advisable to remove these unnecessary information sources from the picture.

Moving along a diameter of a circular wave results in the characteristic lines given in Figures A8 and A9. Both the $\ln q$ and the S_{str} curves are periodic at the higher center shift domains, and, similar to the straight waves, they have hook-like characteristics, if the wavelength is larger than the tile size. In the case of smaller wavelengths, the oscillations are of much smaller magnitude.

If the center is moved in the other direction as well, the upper hook becomes more and more asymmetric, and a shift also appears, as is demonstrated in Figures A10 and A11 of Appendix A.

As in the case of scanning a row of a picture the tile size is usually smaller than the wavelength, the large wavelength curves are of greater interest from the point of view of polyp detection. In addition, as the center of the elliptical waves are generally in the more distant parts of the image, i.e., practically never in the same frame as the polyp, the offsets larger than the tile size are of more interest. In these cases, as can be seen in Figures A10 and A11, the straight waves model the behavior of the circular waves very well.

The dependence directions of the straight waves and the axis ratio of the elliptic waves can be seen in Figure A12, and in its 3D version in Figure A13. We can conclude that the direction does not influence the characteristic curves of straight waves if the wavelength is below the tile size. The hook-like characteristic curves with smaller or larger asymmetry remain for both the straight and the circular waves, and for the elliptical waves the ratio of the axes becomes negligible if the center is shifted out of the tile.

To summarize this subsection, scanning the picture with a moderately large window along a line or column can be of greater interest from structure detection point of view. In this case, for larger distance of the center of the elliptical waves, they behave similarly to the waves with straight wavefront: periodic, hook-like characteristic curves are usual, which an be distorted by other parameters.

4.5. Hemispheres

In the case of the sphere or ellipsoid segments, the parameters we selected are the radius to tile size relation, the ratio of the axes of the ellipsoid and the center shift. The characteristic curves can be seen in Figures A14 and A15 of Appendix B. It can be seen that the hemispheres or half ellipsoids have very low structural entropy because, in the case of a sphere with radius to tile size ratio 0.3, a very large part of the picture is completely flat and dark, with 0 entropy (and thus 0 structural entropy). In addition, the radius and axis ratios influence only the spatial filling factor; the structural entropy does not change as long as the whole ellipsoid is within the tile.

If the shadow part of the picture is also included into the model, i.e., the Gaussian like functions (Equation (16)) are subtracted from the hemispheres, the structural entropy of course becomes much larger, as the part of the image with zero pixel intensity becomes very small. The results are given in Figures A16 and A17 of Appendix B. The characteristic curves of the Gaussian-like structures and their negative counterpart are also given, but only in the 3D plot form.

The Gaussian distributions are on their theoretical characteristic line for that central region, where the $S_{str}(\ln q)$ points are constant, and deviate from their theoretical value if significant part of the distribution is outside of the tile (the deviation starts to be visible at the shift of about $5R$ and in very small variation cases, if the shift is larger than about eight times the radius, the structural entropy's deviation starts to grow, and then its value sinks to the origin of the plot.

The structural entropy plots of the hemispheres with shadow have big loops if their radius is small, and hooks start to form with the increasing of the size of the polyp model. The Gaussian-like structures with higher power α were also tested, however, their result did not differ much from the Gaussian case, only the loop area became a little bit smaller, as the power increased.

In addition, the depth of the shadow, or the shadow to ellipsoid height ratio, is interesting. If the shadow is much deeper than the polyp, we arrive at the distant part of the bowel, the tunnel, which almost always has a darkening part and a turning, which often appears as hemisphere or similar object in the cross section of the distribution. The results are plotted in the Appendix in Figures A18 and A19. The pictures show that, as the shadow deepens, the hooks at the sides of the hemisphere decrease, moving inward, toward the point with 0 center shift. In addition, more smaller hooks appear in the inner domain.

Tilt is important in this case, too. The effects of introducing and increasing tilt are shown in Figures A20 and A21 of Appendix B. It is clearly visible that the distance of the received $S_{str}(\ln q)$ points decrease from the origin, and other little hooks emerge in the middle region of the plot. This is of course not always this visible: if the radius is too small, the hooks disappear here as well, such as in the case of the blue curve in Figure A16. This means that, if the polyp is much smaller than the window used for scanning, it behaves completely differently from the ones with radii more similar to the tile size.

The components of the hemisphere with Gaussian shadow were also studied and the effect of the tilt to their properties are given in Figures A22–A27. In the case of the hemispheres, the tilt increases the structural entropy and decreases the spatial filling factor, thus elevating the hemisphere's curve from the $\ln q$ axis. In the case of the Gaussians, the magnitude decreases with increasing tilt. In those center shift values, where the sphere dominates in the tile (i.e., when the tile center is around the center of the sphere), the movement toward the origin is less than those parts that contain picture domains with 0 value.

4.6. Superpositions

The superpositions of the semi-ellipsoids and the waves have rather complex behavior, depending on which component is dominant according to the magnitude and size. If the wavelength is larger than the tile size, and the hemisphere diameter is smaller, then the setup is very similar to a roundish polyp largely protruding into the bowel volume. If the wavelength is smaller, the arising picture is similar to those flatter polyps, which are sitting at the bends of the bowel wall, making these bends only slightly thicker at a given region.

Characteristic curves for such superstructures can be seen in Figures A28 and A29 of Appendix B.

The resulting characteristic curves are also of two types: for the larger wavelengths, the periodic behavior dominates, hooks similar to the ones in Figure A8 appear, and the sphere segment and its shadow causes only slight asymmetries. If the wavelength is smaller that the tile, the two components can decrease each other's structural entropy and filling factor.

The Rényi entropy based structural entropy and the spatial filling factor is able to distinguish parts of a superstructure, if they are multiplied and not added. In the case of a multiplicative superstructure, the S_{str} and $\ln q$ values of the components are simply added together. Unfortunately, in the case of additive superstructures, the S_{str} and $\ln q$ values of the component structures can only be detected, if one of the structures is dominant. It might also be possible to detect components of the superstructures, if they are of different characteristic lengths, and wavelet analysis or other filtering method is used to separate the different characteristic lengths [59].

4.7. Summary of the Artificial Surface Characteristic Curve Properties

For a better visibility, we summarize the previous results in Table 1, concentrating on how other parameters influence the center-shift curves.

Table 1. The effect of the increasing of various parameters to the characteristic curves for the waves and the hemispheres in the case of off-center shift being the first parameter.

Parameter	Wave	Sphere with Shadow
General shape	periodic loops	2 hooks, M-shape
Radius	-	different position
Wavelength	different period	-
Offset	increased magnitudes	Decreased magnitudes
Tilt	broken symmetry of the shapes	smaller curves
Height ratio	-	decreased hook size
-	-	more hooks
Direction	shallower hooks	no systematic effect
Tile size	no effect for large tiles	no effect for large tiles

4.8. Typical Characteristic Curves of Real Images

In the case of real images, instead of fixed tiles, we applied the sliding tile method suggested in Section 4.4. In Figure 10, the two cuts in Figure 3 are scanned with tiles of size 50 by 50. The characteristic curves of these cuts are very similar to the ones given in Figure A16, however, as the distant, dark part of the bowel is also similar to the hemisphere with shadow model profile in some cases (see pixels 200-600 in picture 83, row 350), such occasions may cause misinterpretation of the $S_{str}(\ln q, i)$ curves, and thus false positive categorization.

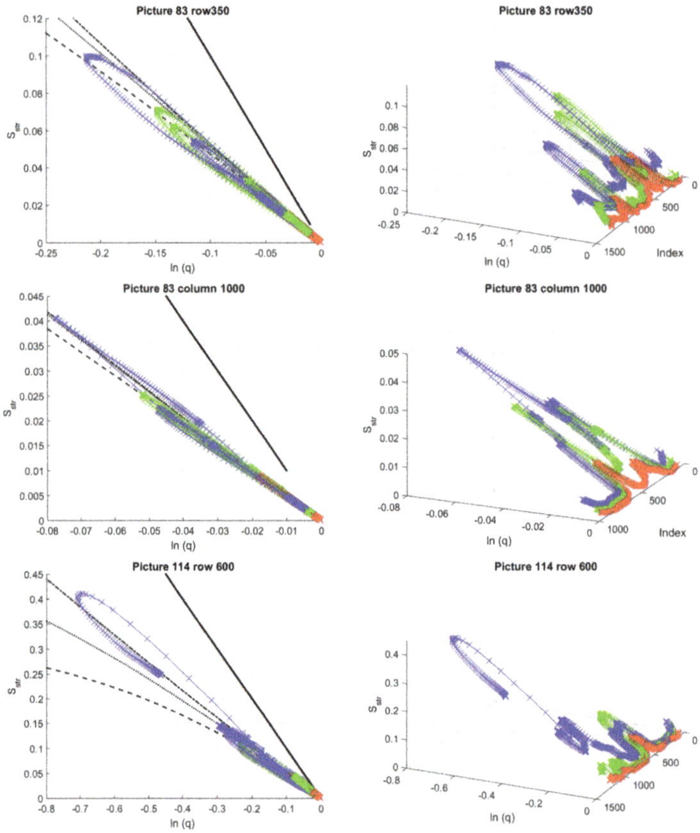

Figure 10. Structural entropy across real polyps. The center of the sliding tile is at the curves plotted in Figure 3. Fixed tile size of 50 by 50 pixels. The polyps are located approximately between pixel indices 900 and 1100 for picture 83, row 350, between 300 and 500 for the same picture's 1000th column, and between 600 and 1100 for picture 114, row 600.

In Figure 11, two scans without polyp are given as an example, one with clearly distinguishable waves, and the other across a polyp-like appearing curvature of the bowel, where the shadows are much larger, and the spherical characteristics are much weaker than in real polyps. These polyp mimicking parts with much more emphasized shadows generally have larger loops than the real polyps. The waves do not have such expressed, curvy hook-like behavior as can be seen in Figures A10 and A11.

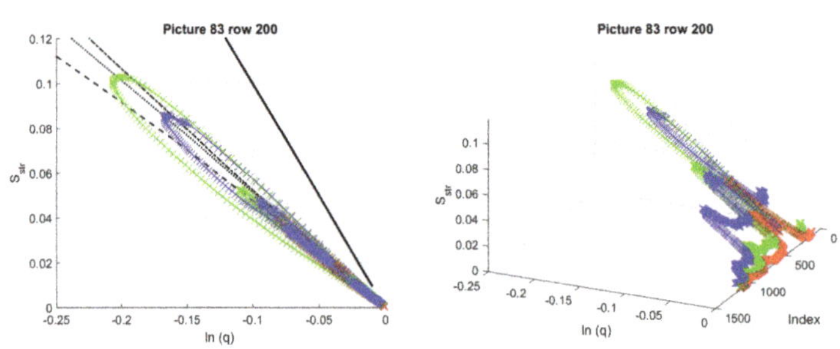

Figure 11. *Cont.*

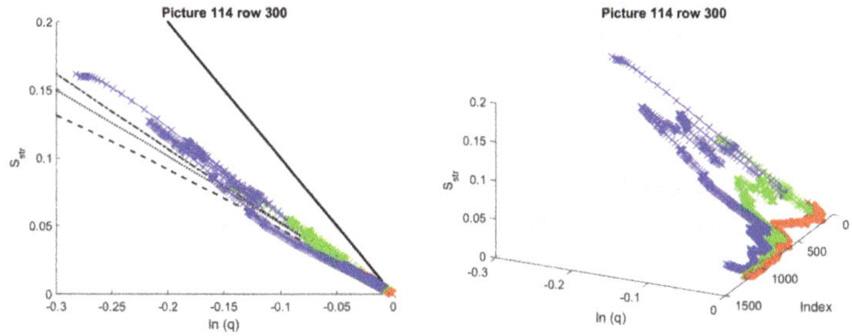

Figure 11. Structural entropy across pictures in scans without polyps. The center of the sliding tile is at the rows 200 and 300 of the pictures in Figure 2. Fixed tile size of 50 by 50 pixels. The dark part mimicking polyp is approximately located between pixel indices 300 and 700 in picture 83, row 200, while the waves are between pixels 200 and 800 in picture 114, row 300.

4.9. Real Picture versus Model: The Applicability

As an example, part of a real picture was studied. As we suggest removing the offset and the tilts from the image by shifting the 0 level and applying a larger scale mean filtered version of the image, first we show their effects on the structural entropy–filling factor plots. We used the same 1000th column of image 83 from database [24], as in the previous section. The tile size remained 50 by 50.

The image preprocessing algorithm consists of only the following steps: reflection removing, histogram stretching and removing of the mean-filtered background pattern [60]. As the image size is around 1000 in both directions, we applied filter sizes of 100 by 100 and 200 by 200. The pictures, the cross section cuts and the structural entropy–filling factor–scanning window center position plots are given in Figure 12. The average diameter of the polyp is also around 200–250 pixels, thus the background generated with the smaller filter size suppressed the sphere-like characteristics of the polyp, as can be seen in Figure 12a,b. The symmetrizing effect of removing the tilt from the background can also be seen, even with such rough background subtracting algorithm.

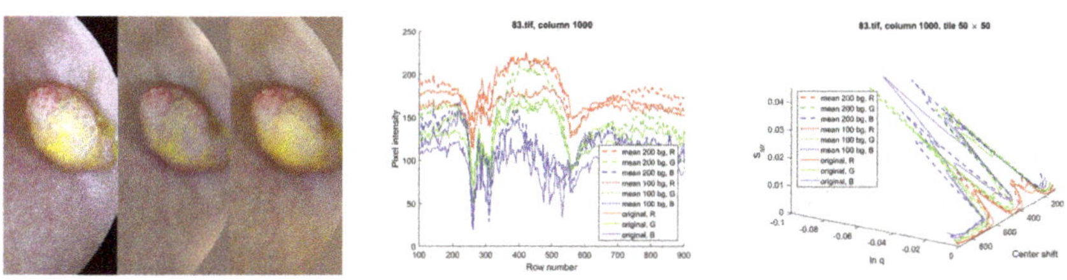

(a) (b) (c)

Figure 12. (**a**) A cut from the not processed, and background subtracted versions of picture 83.tif from database of [24]. The second picture segment's background is generated by a 100 × 100 sized mean filter, the third slices by a 200 × 200 sized one. (**b**) The cross sections at the studied row before and after background subtraction. (**c**) Structural entropy across the 1000th column of the original picture, approximately at the middle of the cuts. Tiles of 50 by 50 pixels.

As the image preprocessing method was applied to the whole picture, not only to the shown segment, there is still an offset in the color channels, about 150 in the red channel, 80 in the green, and 50 in the blue. We used a simplified model to demonstrate that the characteristic curves of model systems are similar to those of the real images, even though many aspects of the real picture, such as the fine patterns, the details of the background, or the yellowish spot that causes a depression in the middle of the polyp in the blue channel of the original picture, are neglected.

In the model system, a hemisphere was used with Gaussian, or higher-order exponential shadow, and flat, constant offset. The offset values were chosen according to the picture color channels' offsets.

The polyp diameter was selected to be 120 pixels, and the height of the hemisphere to be 80 for the red, 120 for the green and 100 for the red channel. The shadows were Gaussian in both the red and the green picture parts, and third-order exponential for the blue part. The depth of the shadows were adjusted to be 30, 60 and 80 for the R, G, and B channels, respectively.

The resulting characteristic curves for the three color channels can be seen in Figure 13. From the $S_str(\ln q)$ plot, one might conclude that the models fit the real image very well, however, from the 3D curves, it can be seen that the fine structure of the structural entropy and filling factor around the shadow-polyp transition is not too well represented. As the fine-scale behavior of the pixel intensity distribution is not studied, these deviations may be attributed to the smaller sized patterns, however, this aspects needs further investigation. As the fine-scale pattern is useful in the case of pit pattern based classification, we decided not to study this problem in this article.

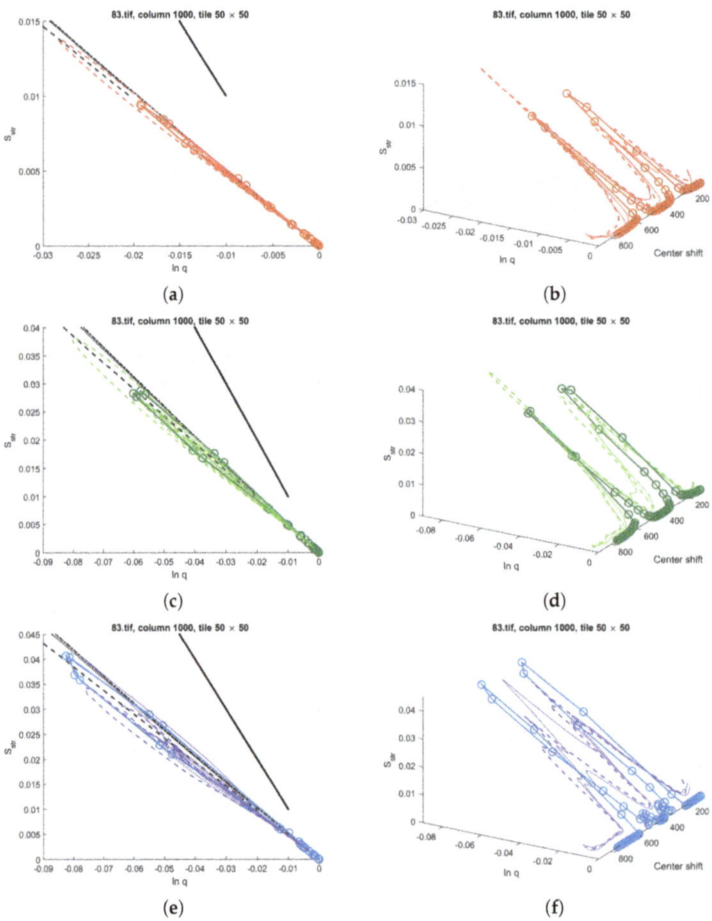

Figure 13. Rényi entropy based structural entropy–spatial filling factor–scanning window center curves for picture 83.tif from database of [24] in continuous lines. The preprocessed-image segments are given in Figure 12 in dashed and dotted lines. The structural entropy and filling factor values of the model system of hemisphere with Gaussian shadow and constant offset model systems. (**a**,**b**): red channel, (**c**,**d**): green channel, (**e**,**f**): blue channel. The shadows in the red and green channels are modeled with Gaussian functions of $\alpha = 2$, while for the blue channel, a higher, $\alpha = 3$rd order polynomial was used in the exponential function.

5. Conclusions

The shape of a probability distribution can be characterized by Rényi entropy based quantities, which are called spatial filling factor and structural entropy. Although the name is similar to the graph neighborhood relation based structural entropy that appeared earlier in the literature, these quantities of the same name are significantly different.

The Rényi entropy based structural entropy uses the native probability distribution of an image, i.e., its pixel intensity distribution, simply normalized in a manner that it would form a probability distribution. This structural entropy is from one point of view more complex than the graph theory based one, as it uses generalized, Rényi entropy differences instead of Shannon entropy. The main advantage of the method, however, is its simpleness. The probability distribution used in the entropies is straightforward, easy to generate, and does not need topological knowledge and neighborhood statistics, which might change with the resolution. These Rényi entropy differences can possibly be used as input parameters for fuzzy, support vector machine, or other metaheuristical or learning algorithms.

The application of the Rényi entropy based structure parameters requires plotting the structural entropy as the function of the filling factor, and comparing the result with existing characteristic curves. This process is easy to be visualized, but might be rather hard to understand and apply. Simply using the two quantities as input parameters for classification methods might loose a lot of information, which lies in the position on the $S_{str}(q)$ map related to characteristic lines. If this information is also to be included into the analysis, previous knowledge about the possible shapes are necessary to know which characteristic lines should be used as references, as these characteristic lines might overlap. This is the main disadvantage of the method.

Characteristic curves of simple distributions such as the Gaussian or exponentially decreasing probability distributions were known for a long time; however, distributions related to structures present on colonoscopy images were never mapped before this study. Here, besides roundish colorectal polyps, different types of waves were also investigated using a rather broad set of possible parameters. The collection of the characteristic curves in the appendices could be used as references or extended and refined if the application deems it necessary.

For some real images, some aspects can be found by using the characteristic structures listed in this contribution. However, we did not pay attention to the fine-scale behavior, as they are not as important in the process of finding a polyp. Superpositions of different types of distributions are rather complicated, if additive and not multiplicative superstructures are studied. We suggest using wavelet-analysis or other, scale sensitive methods to separate the components of an additive superstructure.

To summarize the other findings, the following can be suggested for using structural entropy in image classification methods, especially in colorectal polyp searching cases. Instead of static tiling of the images, and using structural entropy and filling factor as two parameters of the image segment classification—although they provide valuable information about the shape of the distribution—it is more advisable to use sliding tiles and study the thus arising curves on the $S_{str}(\ln q)$ plots. We also suggest removing tilts and offsets from the image segments using simple image processing tools. We demonstrated that simple background subtraction techniques can change the characteristic curves very much without introducing extra information or losing valuable information.

Regarding the characteristic curves, we could conclude the following. The directions of the patterns do not influence the types of the characteristic curves. The size of the tiles also do not influence the results, provided that sufficiently large tile sizes are used. We suggest using larger tile sizes to achieve more stable results. In addition, for larger wavelengths, in realistic cases, the elliptical waves produce similar characteristic curves to those of the waves with straight wavefront, and they are both suitable to model the waves on the bowel walls.

Author Contributions: Conceptualization, J.P.; methodology, J.P.; software, S.N.; validation, S.N.; resources, B.S.; and writing—original draft preparation, S.N. and B.S; correction, S.N. and J.P.

Funding: This research was funded by the New National Excellence Programme of Hungary grant number ÚNKP-18-3.

Conflicts of Interest: The authors declare no conflict of interest.

Appendix A. Characteristic Curves for Waves

Appendix A.1. Waves, Wavelength to Tile Size Ratio, Offset, Limiting Behaviors

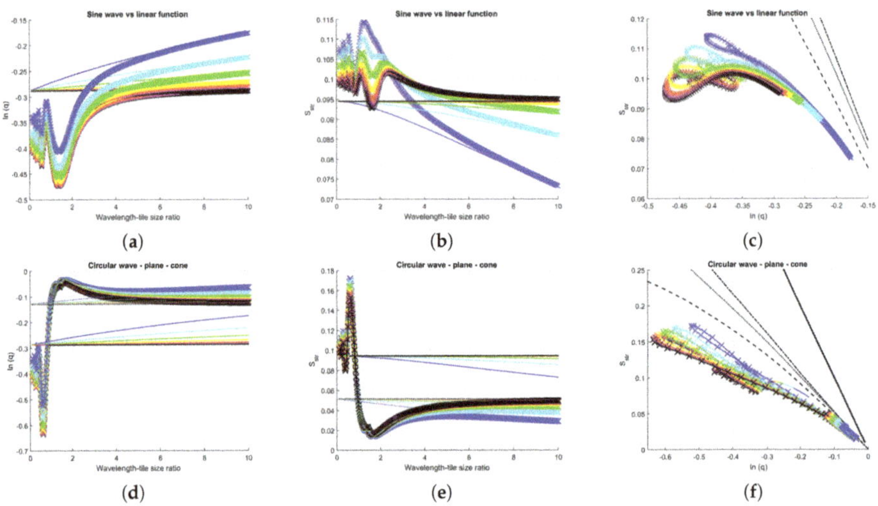

Figure A1. Characteristic curves of waves with straight (**a–c**) and circular (**d–f**) wavefronts, if the parameters are the wavelength to tile size ratio and the constant offset. The colors denote the following constant offsets: $B = 0.1$, blue; $B = 0.05$, cyan; $B = 0.025$, green; $B = 0.0125$, yellow; $B = 0.00625$, red; $B = 0.003125$, magenta; $B = 0.0015625$, black. As a reference, the characteristic curves for planes with the same gradient as the $x \to 0$ gradient of the waves are plotted with solid lines without markers, whereas the cones corresponding to the initial slope of the circular waves are plotted in dotted lines.

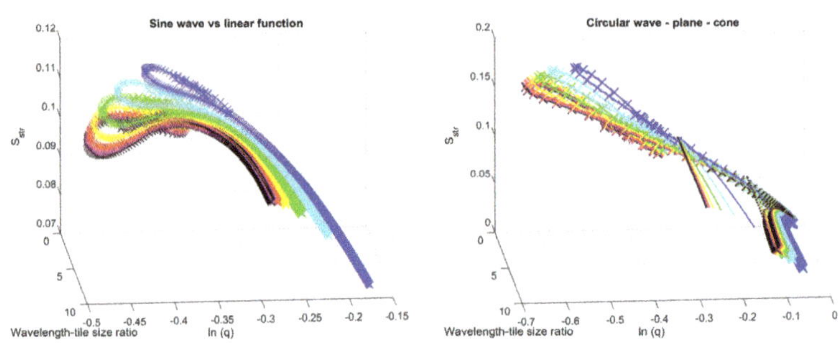

Figure A2. 3D view of Figure A1.

Appendix A.2. Waves, Wavelength to Tile Size, Phase

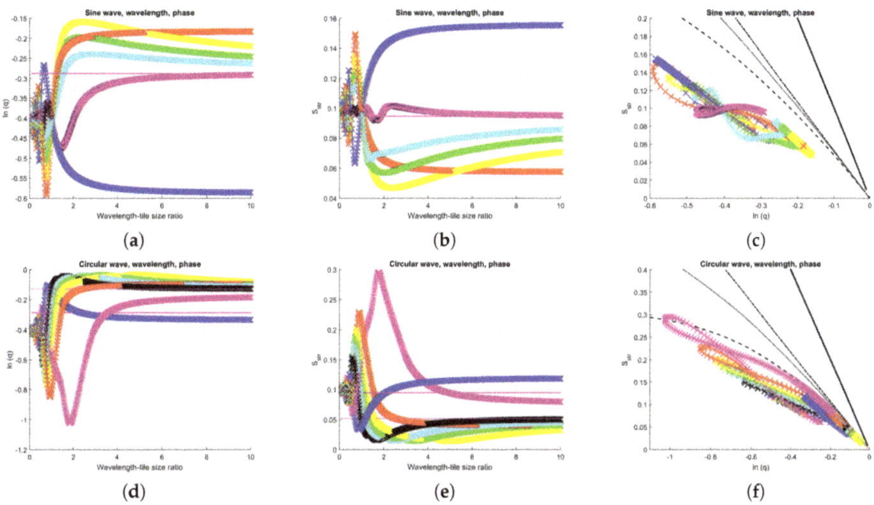

Figure A3. Characteristic curves of waves with straight (**a**–**c**) and circular (**d**–**f**) wavefronts, if the parameters are the wavelength to tile size ratio and the phase φ. The colors denote the following: $\varphi = 0°$, black; $\varphi = 30°$, cyan; $\varphi = 45°$, green; $\varphi = 60°$, yellow; $\varphi = 90°$, red; $\varphi = 180°$, magenta; $\varphi = 270°$, blue. As a reference, the characteristic curves for planes with the same gradient as the $x \to 0$ gradient of the waves are plotted with solid lines without markers, whereas the cones corresponding to the initial slope of the circular waves are plotted in dotted lines.

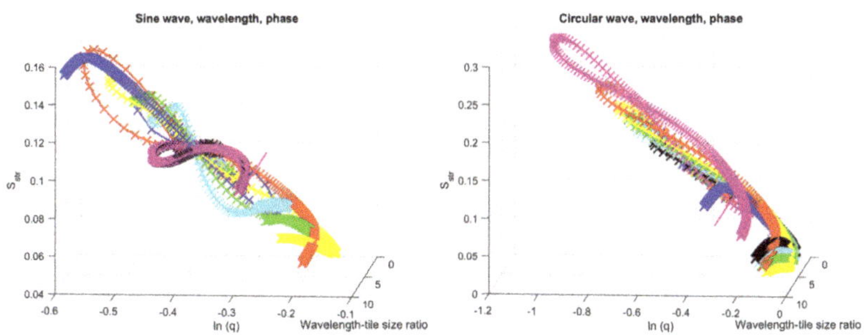

Figure A4. 3D view of Figure A3.

Appendix A.3. Waves, Wavelength to Tile Size, Tilt Gradient

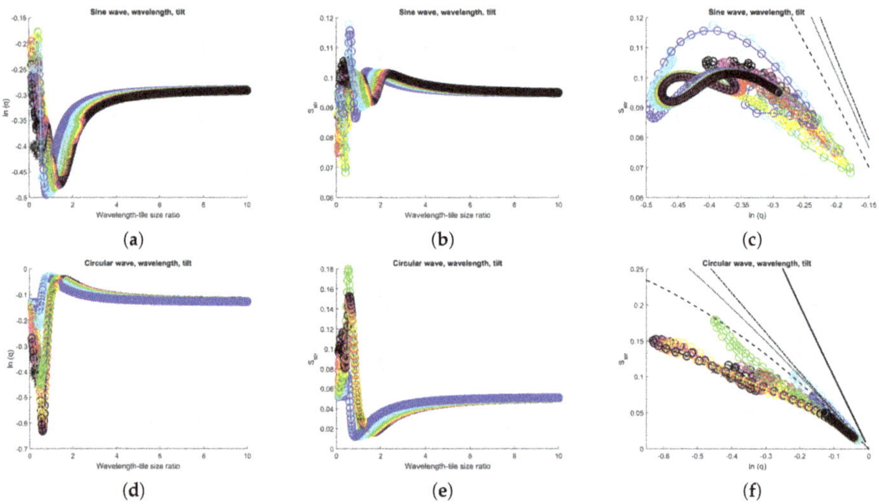

Figure A5. Characteristic curves of waves with straight (**a–c**) and circular (**d–f**) wavefronts, if the parameter is the wavelength to tile size ratio and the tilt gradient. The colors denote the following linear offset slopes to wavelength ratios: $B/T = 1$, blue; $B/T = 0.5$, cyan; $B/T = 0.25$, green; $B/T = 0.125$, yellow; $B/T = 0.0625$, red; $B/T = 0.03125$, magenta; $B/T = 0.015625$, black. As a reference, the characteristic curves waves without tilt are plotted in thinner lines with different $\times$ markers.

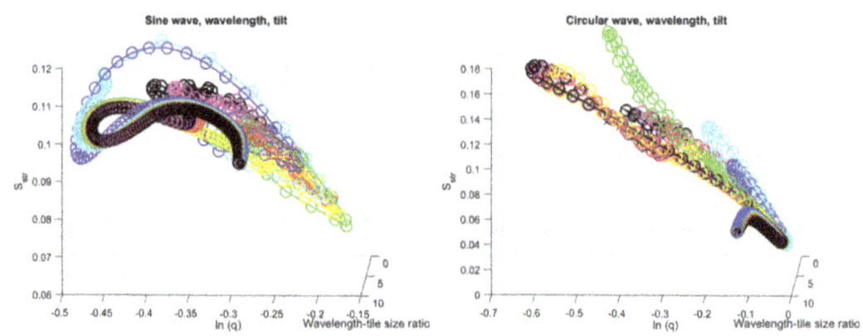

Figure A6. 3D view of Figure A5.

Appendix A.4. Waves, Phase and Wavelenth, Offset, Tilt

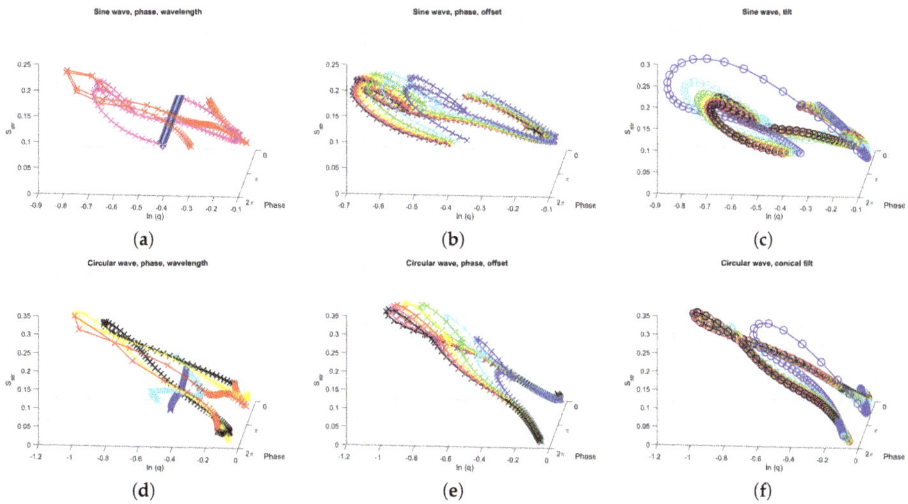

Figure A7. Characteristic curves of waves with straight (**a–c**) and circular (**d–f**) wavefronts, if the parameter is phase. The other parameters are the following. (**a**,**d**): wavelength. The colors denote the following wavelength to tile size ratio: $T/N = 0.1$, blue; $T/N = 0.5$, cyan; $T/N = 1$, black; $T/N = 2$, yellow; $T/N = 10$, red. (**b**,**e**): constant offset. The colors mean the following: $B = 0.1$, blue; $B = 0.05$, cyan; $B = 0.025$, green; $B = 0.0125$, yellow; $B = 0.00625$, red; $B = 0.003125$, magenta; $B = 0.0015625$, black. (**c**,**f**): linear offset. The colors mean the following: $B/T = 1$, blue; $B/T = 0.5$, cyan; $B/T = 0.25$, green; $B/T = 0.125$, yellow; $B/T = 0.0625$, red; $B/T = 0.03125$, magenta; $B/T = 0.015625$, black. The markers $\times$ denote the zero-tilt limit, while the circles show the results for the tilted waves.

Appendix A.5. Wave, Center Shift (One Direction), Wavelength

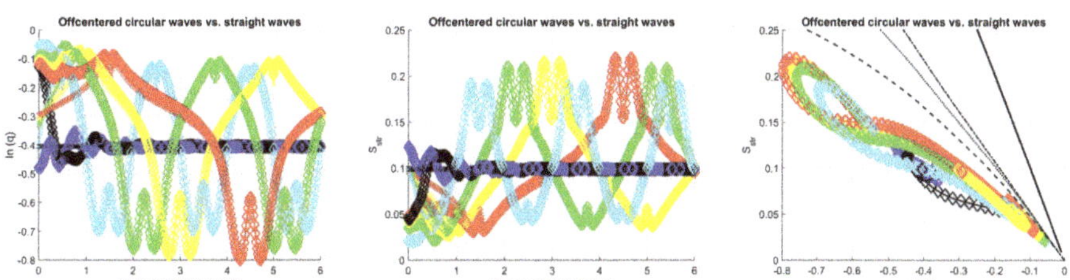

Figure A8. Characteristic curves of circular wavefronts, if the parameter is center shift x_0 and wavelength to tile size ratio. In the first row, the other coordinate of the center shift was set to $y_0 = 0$ and the colors denote the following wavelength to tile size ratios: $T/N = 0.5$, blue; $T/N = 1$, black; $T/N = 2$, cyan; $T/N = 3$, green; $T/N = 4$, yellow; $T/N = 6$, red. The markers are $\diamond$. As a reference, the characteristic curves of straight waves of same parameters were plotted with thinner lines and smaller markers of shape $\times$ (the phase had to be set to π to match the curves). As a reference, the characteristic curve of a straight wave with the same parameters was plotted with thinner black line and smaller markers of shape $\times$ (the phase was again π).

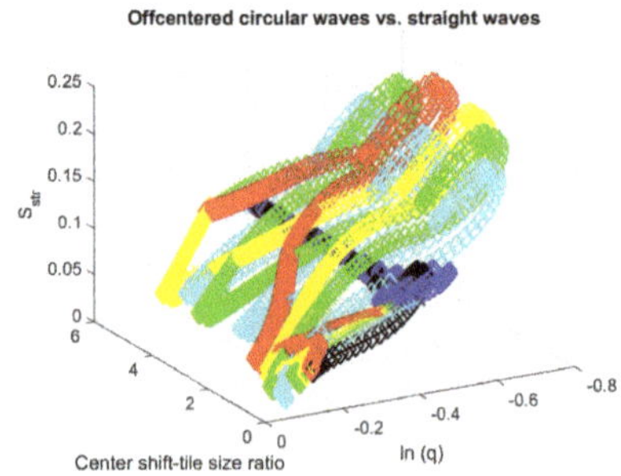

Figure A9. 3D view of Figure A8.

Appendix A.6. Wave, Center Shift, both Directions

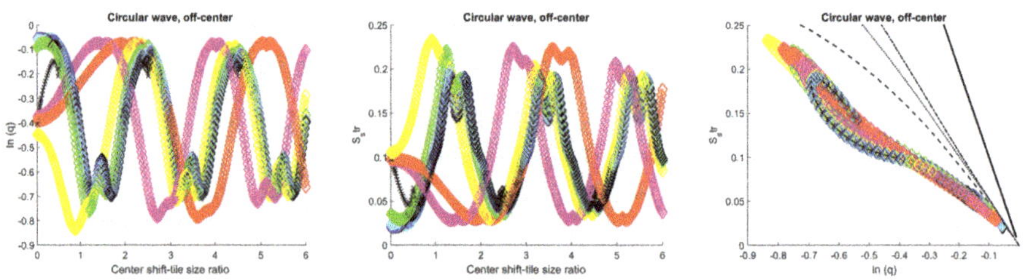

Figure A10. Characteristic curves of circular wavefronts, if the parameter is center shift x_0 and center shift in the other direction. The wavelength to tile size ratio was set to $T/N = 2$, and the colors mean the following other coordinate of the center shift to tile size ratios, $y_0 = 0$ and the colors denote the following: $y_0/N = 0$, black; $y_0/N = 1/8$, blue; $y_0/N = 1/4$, cyan; $y_0/N = 1/2$, green; $y_0/N = 1$, yellow; $y_0/N = 2$, magenta; $y_0/N = 4$, red. The markers are $\diamond$. As a reference, the characteristic curve of a straight wave with the same parameters was plotted with thinner black line and smaller markers of shape $\times$ (the phase was again π).

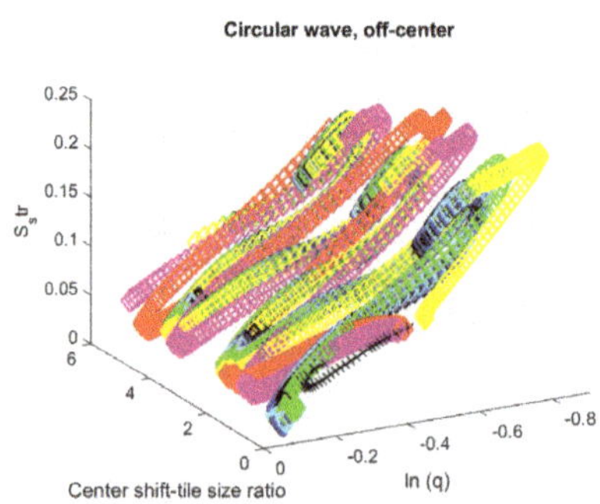

Figure A11. 3D view of Figure A10.

Appendix A.7. Waves, Center Shift, Direction

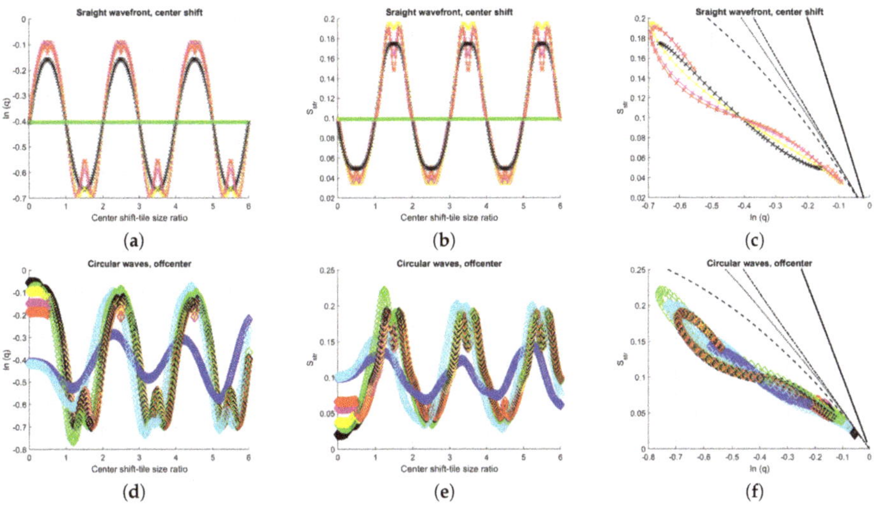

Figure A12. Characteristic curves of straight (**a**–**c**) and circular (**d**–**f**) wavefronts, if the parameters are the center shift and the direction of the wavefront. The colors denote the following ratios: $T_y/T_x = 1/8$, blue; $T_y/T_x = 1/4$, cyan; $T_y/T_x = 1/2$, green; $T_y/T_x = 1$, black; $T_y/T_x = 2$, yellow; $T_y/T_x = 4$, magenta; $T_y/T_x = 8$, red.

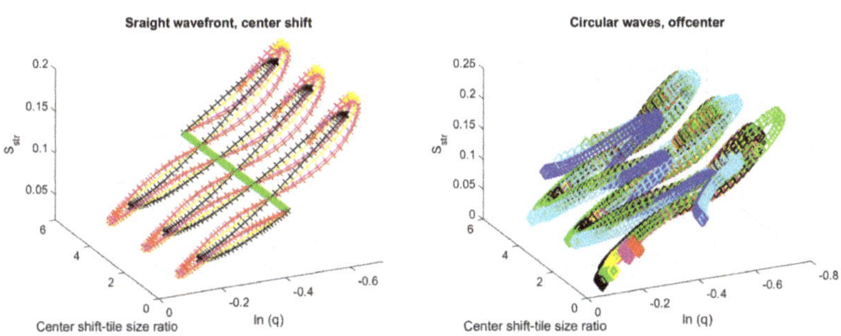

Figure A13. 3D view of Figure A12.

Appendix B. Characteristic Curves for Semi-Ellipsoids, Hemispheres

Appendix B.1. Semi-Ellipsoid, Center Shift, Axis Ratio

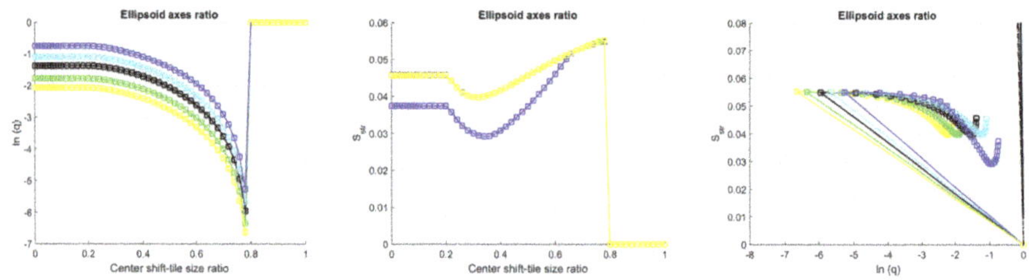

Figure A14. Characteristic curves of hemispheres, if the parameter is center shift x_0. The other coordinate of the center shift was set to $y_0 = 0$. The colors denote the following axes ratios: $R_y/R_x = 0.5$, blue; $R_y/R_x = 0.75$, cyan; $R_y/R_x = 1$, black; $R_y/R_x = 1.5$, green; $R_y/R_x = 2$, yellow. R was set to $0.3N$, while R_y was set to 1.

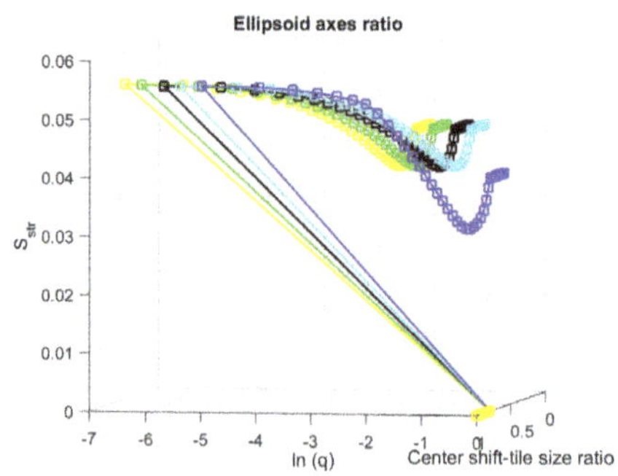

Figure A15. 3D view of Figure A14.

Appendix B.2. Semi-Ellipsoids with Gaussian Subtraction (Shadow), Center Shift, Radius

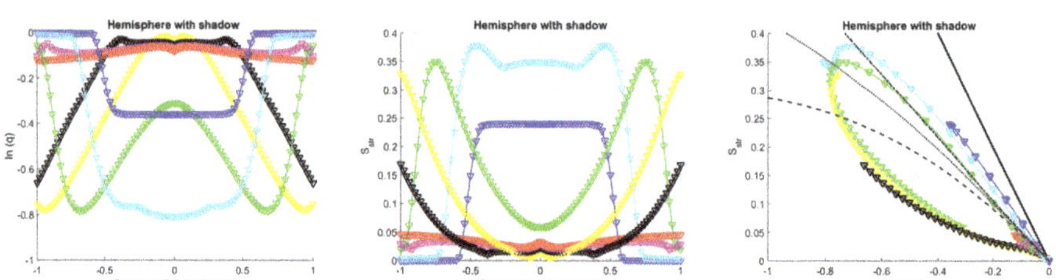

Figure A16. Characteristic curves of hemispheres with Gaussian shadows, if the parameter is center shift x_0 and radius to tile size ratio. The other coordinate of the offset was set to $y_0 = 0$. The colors denote the following radius to tile size ratios: $R/N = 0.1$, blue; $R/N = 0.25$, cyan; $R/N = 0.5$, green; $R/N = 0.75$, yellow; $R/N = 1$, black; $R/N = 1.5$, magenta; $R/N = 2$, red.

Appendix B.3. Components of the Semi-Ellipsoids with Gaussian Shadow, Center Shift, Radius

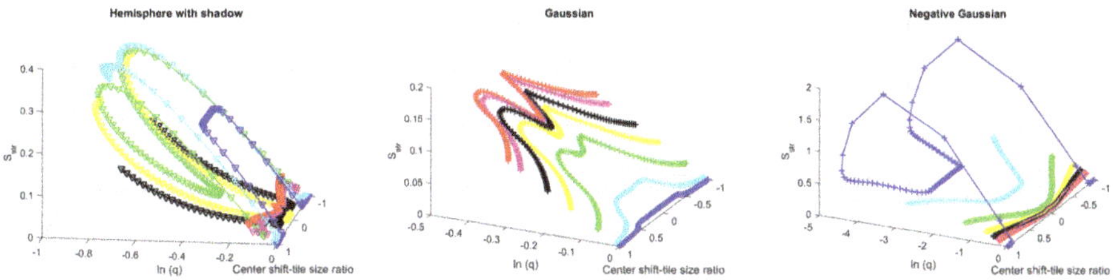

(a) (b) (c)

Figure A17. (**a**) 3D view of Figure A16; (**b**) Gaussian functions; and (**c**) negative Gaussian functions. The variances of the Gaussians are the same as the square of radii of the hemispheres. The color notation is the same as in Figure A16. The ratio of the Gaussian's height and the radius is 1:3.

Appendix B.4. Semi Ellipsoids with Gaussian Shadow, Center Shift, Ratio of the Two Components: Shadow Depth

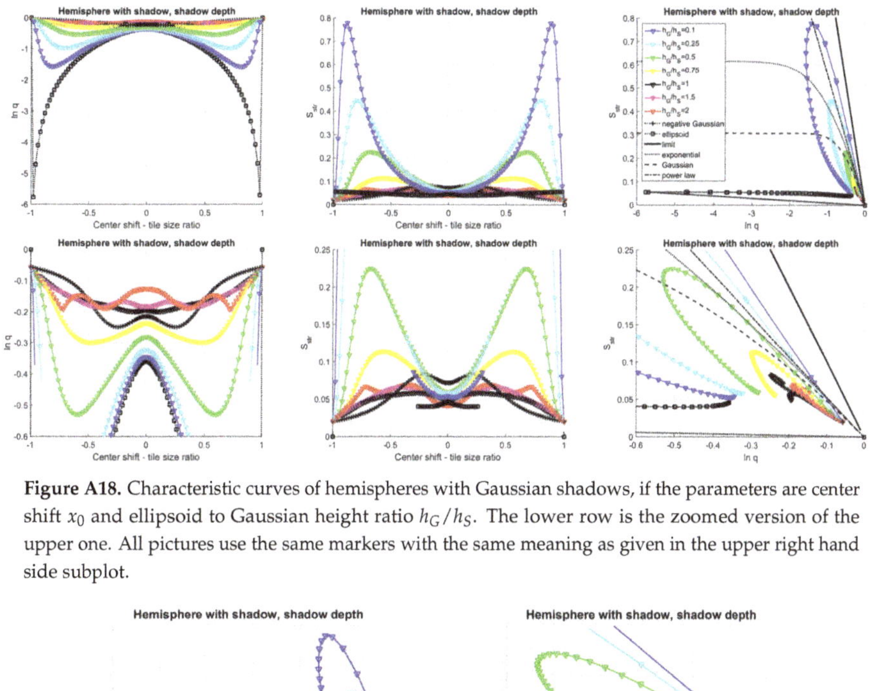

Figure A18. Characteristic curves of hemispheres with Gaussian shadows, if the parameters are center shift x_0 and ellipsoid to Gaussian height ratio h_G/h_S. The lower row is the zoomed version of the upper one. All pictures use the same markers with the same meaning as given in the upper right hand side subplot.

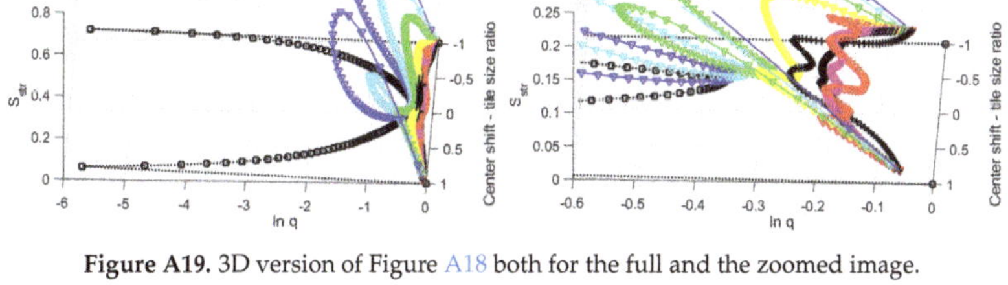

Figure A19. 3D version of Figure A18 both for the full and the zoomed image.

Appendix B.5. Semi-Ellipsoids with Gaussian Shadow, Center Shift, Tilt

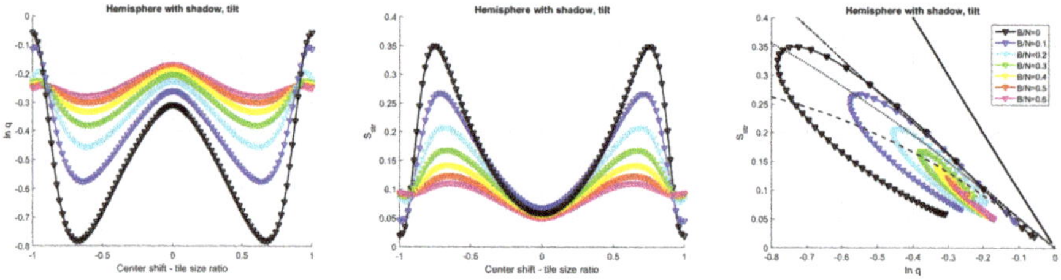

Figure A20. Characteristic curves of hemispheres with Gaussian shadows, if the parameters are center shift x_0 and tilt gradient B.

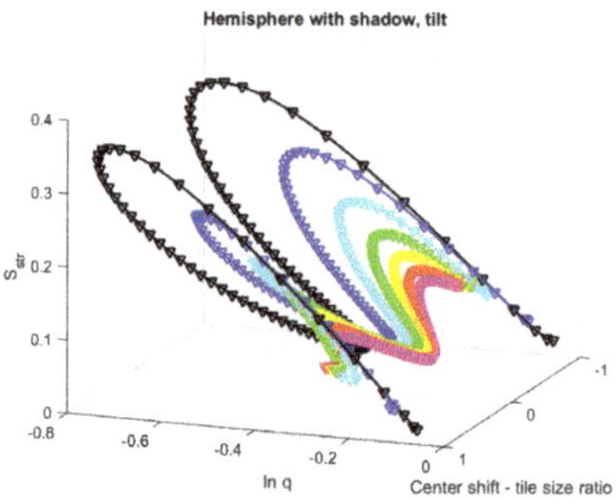

Figure A21. 3D version of Figure A20.

Appendix B.6. Semiellipsoids, Center Shift, Tilt

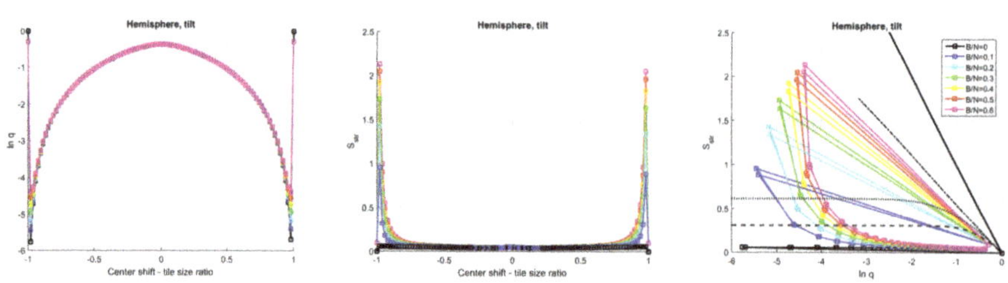

Figure A22. Characteristic curves of hemispheres, if the parameters are center shift x_0 and tilt gradient B.

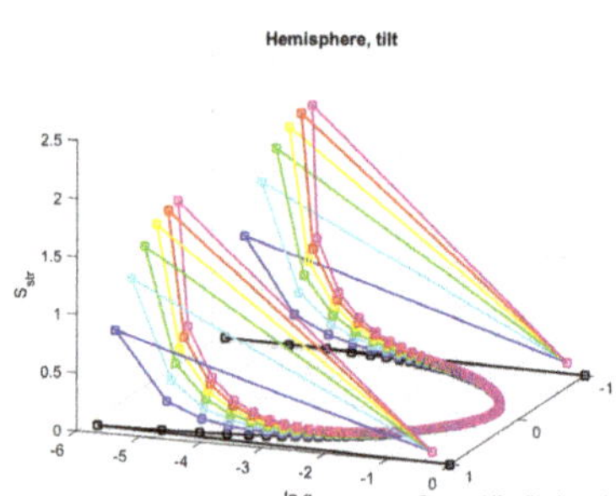

Figure A23. 3D version of Figure A22.

Appendix B.7. Gaussian Distribution, Center Shift, Tilt

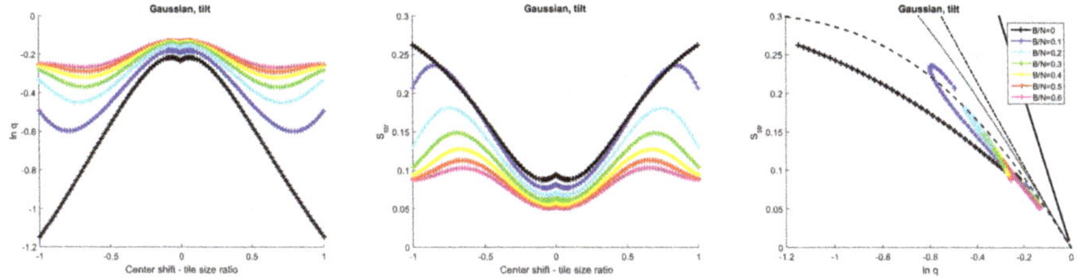

Figure A24. Characteristic curves of Gaussian distribution segments, if the parameters are center shift x_0 and tilt gradient B.

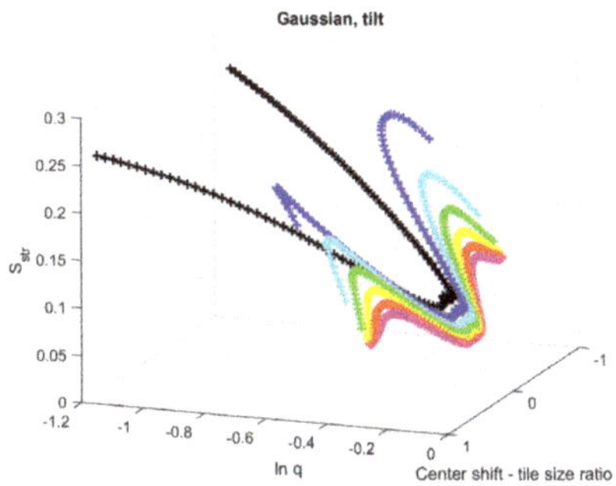

Figure A25. 3D version of Figure A24.

Appendix B.8. the Negative Gaussian Shadow, Center Shift, Tilt

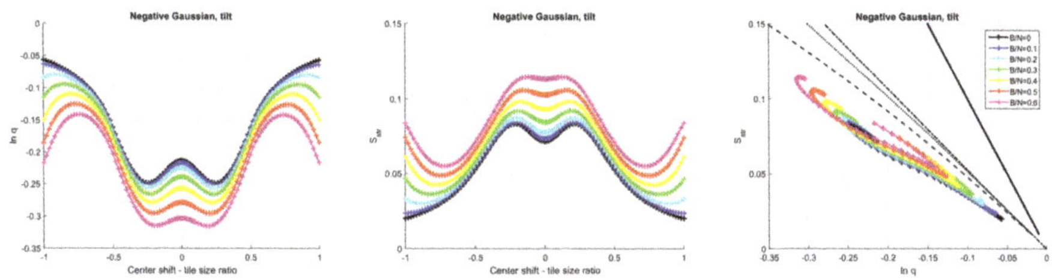

Figure A26. Characteristic curves of the negative Gaussian shadows, if the parameters are center shift x_0 and tilt gradient B.

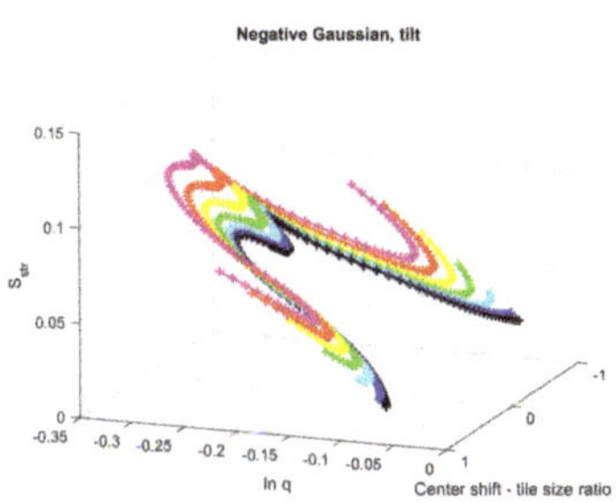

Figure A27. 3D version of Figure A26.

Appendix B.9. Superposition of Wave and Hemisphere with Gaussian Shadow

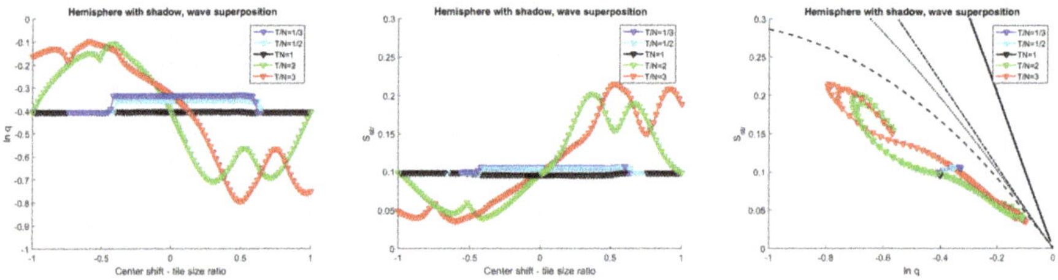

Figure A28. Hemisphere with gaussian shadow superposed to a wave of wavelength T with picture segment size $N \times N$. The radius was set to $0.1N$. The small wavelength curves are similar to polyps sitting on a bending of the bowel wall making it slightly thicker. The large wavelength lines are similar to roundish polyps sitting on slightly curved bowel walls.

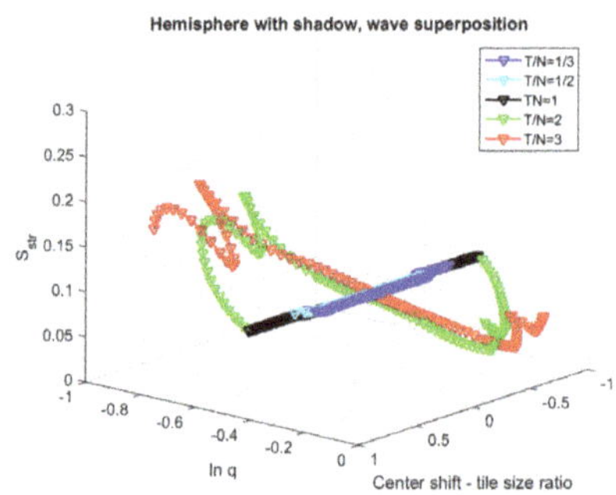

Figure A29. 3D version of Figure A28.

References

1. Burnand B.; Harris J.K.; Wietlisbach, V.; Froelich, F.; Vader, J.P.; Gonvers, J.J. Use, appropriateness, and diagnostic yield of screening colonoscopy: An international observational study (EPAGE). *Gastrointest Endosc.* **2006**, *63*, 1018–1026. [CrossRef] [PubMed]
2. Levin, B.; Lieberman, D.A.; McFarland, B.; Smith, R.A.; Brooks, D.; Andrews, K.S.; Dash, C.; Giardiello, F.M.; Glick, S.; Levin, T.R.; et al. Screening and surveillance of the early detection of colorectal cancer and adenomatous polyps, 2008: A joint guideline from the American Cancer Society, the US Multi-Society Task Force on Colorectal Cancer, and the American College of Radiology. *Gastroenterology* **2008**, *134*, 1570–1595. [CrossRef]
3. Gono, K.; Obi, T.; Yamaguchi, M.; Machida, H.; Sano, Y.; Yoshida, S.; Hamamoto, Y.; Endo, T. Appearance of enhanced tissue features in narrow-band endoscopic imaging. *J. Biomed. Opt.* **2004**, *9*, 568–577. [CrossRef] [PubMed]
4. Tischendorf, J.J.W.; Wasmuth, H.E.; Koch, A.; Hecker, H.; Trautwein, C.; Winograd, R. Value of magnifying chromoendoscopy and narrow band imaging (NBI) in classifying colorectal polyps: A prospective controlled study. *Endoscopy* **2007**, *39* 1092–1096. [CrossRef] [PubMed]
5. Stehle, T.; Auer, R.; Gross, S.; Behrens, A.; Wulff, J.; Aach, T.; Winograd, R.; Tautwein, C.; Tischendorf, J. Classification of colon polyps in NBI endoscopy using vascularization features. In Proceedings of the Medical Imaging 2009: Computer-Aided Diagnosis (SPIE 7260), Lake Buena Vista (Orlando Area), FL, USA, 7–12 February 2009; Volume 7260. [CrossRef]
6. Pickhardt, P.J.; Choi, J.R.; Hwang, I.; Butler, J.A.; Puckett, M.L.; Hildebrandt, H.A.; Wong, R.K.; Nugent, P.A.; Mysliwiec, P.A.; Schindler, W.R. Computed Tomographic Virtual Colonoscopy to Screen for Colorectal Neoplasia in Asymptomatic Adults. *N. Eng. J. Med.* **2003**, *349*, 2191–2200. [CrossRef] [PubMed]

7. Dachman, A.H. Diagnostic performance of virtual colonoscopy. *Abdom. Imaging* **2002**, *27*, 260–267. [CrossRef] [PubMed]
8. Van Gossum, A.; Munoz-Navas, M.; Fernandez-Urien, I.; Carretero, C.; Gay, G.; Delvaux, M.; Lapalus, M.G.; Ponchon, T.; Neuhaus, H.; Philipper, M.; et al. Capsule Endoscopy versus Colonoscopy for the Detection of Polyps and Cancer. *N. Engl. J. Med.* **2009**, *361*, 264–270. [CrossRef]
9. Rex, D.K.; Cutler, C.S.; Lemmel, G.T.; Rahmani, E.Y.; Clark, D.M.; Helper, D.J.; Lehman, G.A.; Mark, D.G. Colonoscopic miss rates of adenomas determined by back-to-back colonoscopies. *Gastroenterology* **1997**, *112*, 24–28. [CrossRef]
10. Pullens, H.J.M; Leenders, M.; Schipper, M.E.I.; van Oijen, M.G.H.; Siersema, P.D. No Decrease in the Rate of Early or Missed Colorectal Cancers After Colonoscopy With Polypectomy Over a 10-Year Period: A Population-Based Analysis. *Clin. Gastroenterol. Hepatol.* **2015**, *13*, 140–147. [CrossRef]
11. Liu, G.; Yan, G.; Kuang, S.; Wang, Y. Detection of small bowel tumor based on multi-scale curvelet analysis and fractal technology in capsule endoscopy. *Comput. Biol. Med.* **2016**, *70*, 131–138. [CrossRef]
12. Clark, B.T.; Protiva, P.; Nagar, A.; Imaeda, A.; Ciarleglio, M.M.; Deng, Y.H.; Laine, L. Quantification of Adequate Bowel Preparation for Screening or Surveillance Colonoscopy in Men. *Gastroenterology* **2016**, *150*, 396–405. [CrossRef] [PubMed]
13. Kayser, G.; Görtler, J.; Weis, C.A.; Borkenfeld, S.; Kayser, K. The application of structural entropy in tissue based diagnosis. *Diagnos. Pathol.* **2017**, *3*, 251. [CrossRef]
14. dos Santos, L.F.S.; Neves, L.A.; Rozendo, G.B.; Ribeiro, M.G.; do Nascimento, M.Z.; Azevedo Tosta, T.A. Multidimensional and fuzzy sample entropy (SampEnMF) for quantifying H&E histological images of colorectal cancer. *Comput. Biol. Med.* **2018**, *103*, 148–160. [CrossRef]
15. Ribeiro, M.G.; Neves, L.A.; do Nascimento, M.Z.; Roberto, G.F.; Martins, A.S.; Tosta, T.A.A. Classification of colorectal cancer based on the association of multidimensional and multiresolution features. *Expert Sys. Appl.* **2019**, *120*, 262–278. [CrossRef]
16. Chaddad, A.; Daniel, P.; Niazi, T. Radiomics Evaluation of Histological Heterogeneity Using Multiscale Textures Derived From 3D Wavelet Transformation of Multispectral Images. *Front. Oncol.* **2018**, *8*, 96. [CrossRef] [PubMed]
17. Chaddad, A.; Tanougast, C. Texture Analysis of Abnormal Cell Images for Predicting the Continuum of Colorectal Cancer. *Anal. Cell. Pathol.* **2017**, *2017*, 8428102. [CrossRef] [PubMed]
18. Wang, C.; Shi, J.; Zhang, Q.; Ying, S. Histopathological image classification with bilinear convolutional neural networks. In Proceedings of the 39th Annual International Conference of the IEEE Engineering in Medicine and Biology Society (EMBC), Seogwipo, Korea, 11–15 July 2017; pp. 4050–4053. [CrossRef]
19. Chaddad, A.; Desrosiers, C.; Niazi, T.M. Deep radiomic analysis of MRI related to Alzheimer's disease. *IEEE Access* **2018**, *6*, 58213–58221. [CrossRef]
20. Buchner, A.M.; Shahid, M.W.; Heckman, M.G.; McNeilm, R.B.; Cleveland, P.; Gill, K.R.; Schore, A.; Ghabril, M.; Raimondo, M.; Gross, S.A.; et al. High-Definition Colonoscopy Detects Colorectal Polyps at a Higher Rate Than Standard White-Light Colonoscopy. *Clin. Gastroenterol. Hepatol.* **2009**, *8*, 364–370. [CrossRef]
21. Kudo, S.; Hirota, S.; Nakajima, T.; Hosobe, S.; Kusaka, H.; Kobayashi, T.; Himori, M.; Yagyuu, A. Colorectal tumours and pit pattern. *J. Clin. Pathol.* **1994**, *47*, 880–885. [CrossRef]
22. Kudo, S.; Tamura, S.; Nakajima, T.; Yamano, H.; Kusaka, H.; Watanabe, H. Diagnosis of colorectal tumorous lesions by magnifying endoscopy. *Gastrointest. Endosc.* **1996**, *44*, 8–14. [CrossRef]
23. Rácz, I.; Horváth, A.; Szalai, M.; Spindler, Sz.; Kiss, Gy.; Regoczi, H.; Horváth, Z. Digital Image Processing Software for Predicting the Histology of Small Colorectal Polyps by Using Narrow-Band Imaging Magnifying Colonoscopy. *Gastrointest. Endosc.* **2015**, *81*, AB259. [CrossRef]
24. Silva, J.S.; Histace, A.; Romain, O.; Dray, X.; Grando, B. Towards embedded detection of polyps in WCE images for early diagnosis of colorectal cancer. *Int. J. Comput. Assist. Radiol. Surg.* **2014**, *9*, 283–293. [CrossRef] [PubMed]
25. Bernal, J.; Sánchez, F.J.; Vilariño, F. Towards Automatic Polyp Detection with a Polyp Appearance Model. *Pattern Recognit.* **2012**, *45*, 3166–3182. [CrossRef]
26. Bernal, J.; Sánchez, F.J.; Fernández-Esparrach, G.; Gil, D.; Rodrígez, C.; Vilariño, F. WM-DOVA maps for accurate polyp highlighting in colonoscopy: Validation vs. saliency maps from physicians. *Comput. Med. Imaging Graph.* **2015**, *43*, 99–111. [CrossRef]

27. Bernal, J.; Tajkbaksh, N.; Sánchez, F.J.; Matuszewski, B.; Chen, H.; Yu, L.; Angermann, Q.; Romain, O.; Rustad, B.; Balasingham, I.; et al. Comparative Validation of Polyp Detection Methods in Video Colonoscopy: Results from the MICCAI 2015 Endoscopic Vision Challenge. *IEEE Trans. Med. Imaging* **2017**, *36*, 1231–1249. [CrossRef]
28. Charisis, V.S.; Hadjileontiadis, L.J. Potential of hybrid adaptive filtering in inflammatory lesion detection from capsule endoscopy images. *World J. Gastroenterol.* **2016**, *22*, 8641–8657. [CrossRef] [PubMed]
29. Charfi, S.; El Ansari, M. Computer-aided diagnosis system for colon abnormalities detection in wireless capsule endoscopy images. *Multimed. Tools Appl.* **2018**, *77*, 4047–4064. [CrossRef]
30. Lilik, F.; Botzheim, J. Fuzzy based Prequalification Methods for EoSHDSL Technology. *Acta. Tech. Jaurinensis* **2011**, *4*, 135–144.
31. Lilik, F.; Kóczy, L.T. Performance Evaluation of Wire Pairs in Telecommunications Networks by Fuzzy and Evolutionary Models. In Proceedings of the IEEE Africon, Pointe-Aux-Piments, Mauritius, 9–12 September 2013; pp. 712–716. [CrossRef]
32. Lilik, F.; Nagy, Sz.; Kóczy, L.T. Improved Method for Predicting the Performance of the Physical Links in Telecommunications Access Networks. *Complexity* **2018**, *2018*. [CrossRef]
33. Nagy, Sz.; Lilik, F.; Kóczy, L.T. Entropy based fuzzy classification and detection aid for colorectal polyps. In Proceedings of the IEEE Africon, Cape Town, South Africa, 18–20 September 2017; pp. 78–82. [CrossRef]
34. Nagy, Sz.; Sziová, B.; Kóczy, L.T. The effect of image feature qualifiers on fuzzy colorectal polyp detection schemes using KH interpolation - towards hierarchical fuzzy classification of coloscopic still images. In Proceedings of the FuzzIEEE, Rio de Janeiro, Brazil, 8–13 July 2018; pp. 1–7. [CrossRef]
35. Lin, Y.-H.; Liao, Y.-Y.; Yeh, C.-K.; Yang, K.-C.; Tsui, P.-H. Ultrasound Entropy Imaging of Nonalcoholic Fatty Liver Disease: Association with Metabolic Syndrome. *Entropy* **2018**, *20*, 893. [CrossRef]
36. Humeau-Heurtier, A. The multiscale entropy algorithm and its variants: A review. *Entropy* **2015**, *17*, 3110–3123. [CrossRef]
37. Stantchev, I. Structural Entropy: A New Approach for Systems Structure's Analysis. In *Cybernetics and Systems '86*; Trappl, R., Ed.; Springer: Dordrecht, The Netherlands, 1986; pp. 139–186._19. [CrossRef]
38. Pipek, J.; Varga, I. Universal classification scheme for the spatial-localization properties of one-particle states in finite, *d*-dimensional systems. *Phys. Rev. A* **1992**, *46*, 3148–3163. [CrossRef] [PubMed]
39. Mojzes, I.; Dominkovics, C.; Harsányi, G.; Nagy, Sz.; Pipek, J.; Dobos, L. Heat treatment parameters effecting the fractal dimensions of AuGe metallization on GaAs. *Appl. Phys. Lett.* **2007**, *91*, 073107. [CrossRef]
40. Hartley, R.V.L. Transmission of Information. *Bell Syst. Tech. J.* **1928**, *7*, 535–563. [CrossRef]
41. Nyquist, H. Certain Factors Affecting Telegraph Speed. *Bell Syst. Tech. J.* **1924**, *3*, 324–346. [CrossRef]
42. Shannon, C.E. A mathematic theory of communication. *Bell Syst. Tech. J.* **1948**, *27*, 379–423. [CrossRef]
43. Gibbs, J.W. *Elementary Principles in Statistical Mechanics, Developed with Especial Reference to the Rational Foundation of Thermodynamics*; Charles Scribner's Sons: New York, NY, USA, 1902; p. 179.
44. Von Neumann, J. Thermodynamik quantenmechanischer Gesamtheiten. *Nachrichr. Ges. Wiss. Gött. Math.-Phys. Kl.* **1927**, *102*, 273–291. (In German)
45. von Neumann, J. *Mathematische Grundlagen der Quantenmechanik*; Springer: Berlin, Germany, 1932; p. 26. (In German)
46. Craciunescu, T.; Murari, A.; Gelfusa, M. Improving Entropy Estimates of Complex Network Topology for the Characterization of Coupling in Dynamical Systems. *Entropy* **2018**, *20*, 891. [CrossRef]
47. Humeau-Heurtier, A.; Omoto, A.C.M.; Silva, L.E. Bi-dimensional multiscale entropy: Relation with discrete Fourier transform and biomedical application. *Comput. Biol. Med.* **2018**, *100*, 36–40. [CrossRef] [PubMed]
48. Silva, L.E.V.; Senra Filho, A.C.S.; Fazan, V.P.S.; Felipe, J.C.; Murta Junior, L.O. Two-dimensional sample entropy: Assessing image texture through irregularity. *Biomed. Phys. Eng. Express* **2016**, *2*, 045002. [CrossRef]
49. Azami, H.; Escudero, J.; Humeau-Heurtier, A. Bidimensional Distribution Entropy to Analyze the Irregularity of Small-Sized Textures. *IEEE Signal. Proc. Lett.* **2017**, *24*, 1338–1342. [CrossRef]
50. Bell, R.J.; Dean, P. Atomic vibrations in vitreous silica. *Discuss. Faraday. Soc.* **1970**, *50*, 55–61. [CrossRef]
51. Pipek, J. Localization measure and maximum delocalization in molecular systems. *Int. J. Quantum Chem.* **1989**, *36*, 487–501. [CrossRef]
52. Rényi, A. On measures of information and entropy. In Proceedings of the fourth Berkeley Symposium on Mathematics, Statistics and Probability, Berkeley, CA, USA, 20 June–30 July 1960; pp. 547–561.

53. Amigó, J.M. , Balogh, S.G.; Hernández, S. A Brief Review of Generalized Entropies. *Entropy* **2018**, *20*, 813. [CrossRef]
54. Faddeev, D.K. Zum Begriff der Entropie Einer endlichen Wahrscheinlichkeitsschenmas. In *Arbeiten zu Informationstheorie*; Deutschen Verlag der Wissenschaften: Berlin, Germany, 1957; pp. 85–90. (In German)
55. Varga, I.; Pipek, J. Rényi entropies characterizing the shape and the extension of the phase space representation of quantum wave functions in disordered systems. *Phys. Rev. E* **2003**, *68*, 026202. [CrossRef] [PubMed]
56. Molnár, L.M.; Nagy, S.; Mojzes, I. Structural entropy in detecting background patterns of AFM images. *Vacuum* **2010**, *84*, 179–183. [CrossRef]
57. Bonyár, A.; Molnár, L.M.; Harsányi, G. Localization factor: A new parameter for the quantitative characterization of surface structure with atomic force microscopy (AFM). *Micron* **2012**, *43*, 305–310. [CrossRef]
58. Bonyár, A. AFM characterization of the shape of surface structures with localization factor. *Micron* **2016**, *87*, 1–9. [CrossRef]
59. Nagy, Sz.; Fehér, A. Topology analysis of scanning microscope images with structural entropy and discrete wavelet transform. In Proceedings of the 18th International Conference on Systems, Signals and Image Processing IWSIP2011, Sarajevo, Bosnia Herzegovina, 16–18 June 2011; pp. 101–104.
60. Nagy, Sz.; Sziová, B.; Solecki, L. The effect of background and outlier subtraction on the structural entropy of two-dimensional measured data. *Int. J. Reason. Intell. Syst.* **2019**, submitted for publication.

entropy

MDPI

Article

Primality, Fractality, and Image Analysis

Emanuel Guariglia [1,2]

[1] Department of Mathematics and Applications "R. Caccioppoli", University of Naples Federico II, 80126 Naples, Italy; emanuel.guariglia@gmail.com

[2] School of Economics, Management and Statistics, University of Bologna, 40126 Bologna, Italy

Received: 22 February 2019; Accepted: 18 March 2019; Published: 21 March 2019

Abstract: This paper deals with the hidden structure of prime numbers. Previous numerical studies have already indicated a fractal-like behavior of prime-indexed primes. The construction of binary images enables us to generalize this result. In fact, two-integer sequences can easily be converted into a two-color image. In particular, the resulting method shows that both the coprimality condition and Ramanujan primes resemble the Minkowski island and Cantor set, respectively. Furthermore, the comparison between prime-indexed primes and Ramanujan primes is introduced and discussed. Thus the Cantor set covers a relevant role in the fractal-like description of prime numbers. The results confirm the feasibility of the method based on binary images. The link between fractal sets and chaotic dynamical systems may allow the characterization of the Hénon map only in terms of prime numbers.

Keywords: binary image; Cantor set; Hénon map; Minkowski island; prime-indexed primes; Ramanujan primes

MSC: Primary: 11N05; 62H35; Secondary: 28A80; 37D45

---

1. Introduction

The distribution of prime numbers has played an important role in mathematics from the beginning. Nevertheless, the structure of prime numbers has represented a big challenge for generations of mathematicians. In particular, algebraic and analytic features related to the distribution of prime numbers can entail several theoretical problems (the Riemann hypothesis, Goldbach's conjecture, etc.). However, it is not our purpose to investigate the unsolved problems involving prime numbers [1,2].

Prime numbers belong to number theory, that is, pure mathematics. Nevertheless, they can be linked with almost all modern scientific applications, such as those in quantum cryptography [3], biology [4,5], medicine [6], and dynamical systems [7]. Quite recently, considerable attention has been paid to the link between the distribution of prime numbers, fractal geometry, and chaos theory [8–11]. In particular, the fractal nature of prime-indexed primes (PIPs) has been discussed in [12], where the author has numerically shown the quasi-self similar fractality of finite-differenced PIP sequences. The main result arising from this paper is that order of prime numbers and scale of fractal sets play the same role. In [13] the existence of a link between some prime sequences and the Cantor set is shown. The author investigated Cantor primes, that is, prime numbers p such that $1/p$ belongs to the Cantor set, providing characterization results. The importance and complexity of the prime distribution led to the development of several techniques for the resolution of unsolved problems. In modern number theory many subsets of the prime numbers (twin primes, Ramanujan primes, Chen primes, etc.) play a relevant role. Thus the fractal nature of prime numbers can be partially determined, showing the fractality of these subsets. The main difficulty in carrying out this technique is that the set of prime numbers cannot currently be partitioned into these subsets. However, the gap between prime numbers seems to be the key point for understanding the structure of prime distribution. In fact, in [14] the author has investigated the possibility of expressing gaps between consecutive primes in terms of the prime counting function. In addition, the same author has shown that the distribution of prime numbers has a powerlike behavior, that is, a long-range dependence. This suggests some kind of self-similarity in prime distribution.

The existence of patchiness in prime distribution can be studied by gaps between consecutive primes. The results of Wolf [11,14] indicate that these patches can be expressed in terms of the prime counting function, which allows a self-similar approximation [15]. Investigation into the nature of these gaps may reveal the hidden structure of prime numbers. For instance, the randomness in prime distribution is linked to the Sierpinski gasket in [10]. All these partial results indicate that the behavior of prime numbers resembles the recursive law of a fractal set. Thus this work is intended to motivate further investigation into the chaoticness of prime numbers.

The main purpose of this paper is to shine new light on the hidden structure of the prime distribution through an investigation in fractal geometry. In particular, the key research question is whether or not the coprimality condition, PIPs, and Ramanujan primes have fractal-like behavior. The approach adopted for this study is based on image analysis. In fact, the construction of a Boolean matrix enables us to define the correspondent correlation matrix, which provides the binary image sought. The results resemble well-known fractal patterns. More precisely, the coprimality condition and Ramanujan primes exhibit behavior based on the Minkowski island and Cantor dust, respectively. Furthermore, PIPs and Ramanujan primes are compared and the results discussed. Cattani and Ciancio [16] showed that the distribution of PIPs is similar to the Cantor dust. The Hénon map, which is related to a Cantor-like set, provides an application in chaotic dynamical systems. Thus the Cantor set seems to play a noteworthy and relevant role in the distribution of prime numbers.

The remainder of the paper is organized as follows. Section 2 presents some preliminaries on PIPs, Ramanujan primes, and Rényi dimension. Section 3 outlines the main results of the binary image linked to both the coprimality condition and Ramanujan primes. Furthermore, the comparison between k-order PIPs and Ramanujan primes is presented and discussed. In Section 4, the Hénon map provides an application in chaotic dynamical systems. Finally, Section 5 summaries the results and concludes with a suggestion for further investigation.

2. Preliminaries

Throughout this paper and unless otherwise specified, k and n will indicate elements of $\mathbb{N} = \{1, 2, 3, \ldots\}$. Likewise, α and x will always denote real numbers. The greatest common divisor of two integers a and b is denoted by (a, b). In addition, a and b are called coprime (or relatively prime) whenever $(a, b) = 1$. The cardinality of a set A will be denoted by $\# A$. For the convenience of the reader and for brevity the preliminaries are reported below without proofs [8,17,18], thus making our exposition self-contained.

2.1. Prime Numbers, PIPs, and Ramanujan Primes

Let D_n be the set given by

$$\mathrm{D}_n := \{k \in \mathbb{N} : k \mid n\}, \quad n \geq 1 .$$

It follows immediately that the elements of D_n are simply all the positive divisors of n. Any natural number $p > 1$ is called prime if $\# \mathrm{D}_p = 2$. For simplicity of notation, let p_i $(i \geq 1)$ stand for the ith prime number. Thus the set of prime numbers is defined as follows:

$$\mathbb{P} := \{p_i : i \in \mathbb{N}\} .$$

Therefore $\mathbb{P} = \{2, 3, 5, 7, 11, \ldots\}$. Euclid showed that $\# \mathbb{P} = \infty$ although the distribution of prime numbers within $\mathbb{N}$ is still an open problem. The main result is a (statistical) property called the prime number theorem. This involves the prime counting function, $\pi : \mathbb{R} \to \mathbb{N}$, defined by

$$\pi(x) = \#\mathbb{P}_x\,, \quad \mathbb{P}_x := \{p_i \in \mathbb{P} : p_i \leq x\}\,,$$

that is, the number of primes less than or equal to a given real number x. Note that $\mathbb{P}_x \subseteq \mathbb{P}$. Gauss and Legendre conjectured that $\pi(x)$ asymptotically tends to $x/\log x$:

$$\pi(x) \sim \frac{x}{\log x} \qquad \text{as } x \longrightarrow +\infty\,. \tag{1}$$

Independently, Hadamard and de la Vallée Poussin definitely showed approximation (1), hence the prime number theorem states that

$$\lim_{x\to\infty} \frac{\pi(x)}{x/\log x} = 1\,.$$

Both proofs are based on nonvanishing of the Riemann ζ function on the line $\sigma = 1$, that is $\zeta(1+it) \neq 0$ for any real t. The introduction of the logarithmic integral Li leads to a better approximation given by

$$\mathrm{Li}(x) \sim \pi(x) \qquad \text{as } x \longrightarrow +\infty\,,$$

which is the current version of the prime number theorem (see [19] for more details).

As already mentioned in Section 1, the focus of this paper is the joint investigation on two subsets of $\mathbb{P}$ (PIPs and Ramanujan primes). According to [8,20] the sequence of PIPs, that is the sequence of primes with a prime index, is the subset of $\mathbb{P}$ given by

$$\mathbb{P}_1 := \{p_{p_i} \in \mathbb{P} : p_i \in \mathbb{P}\}\,.$$

Note that the subset $\mathbb{P}_1$ can be iteratively generalized as follows:

$$\mathbb{P}_2 := \{p_{p_{p_i}} \in \mathbb{P} : p_{p_i} \in \mathbb{P}\}\,,$$

and so on. This entails that for each integer $k \geq 1$, the set $\mathbb{P}_k$ of the k-order PIPs is immediately built. According to the definition of PIPs it follows that $\mathbb{P}_0 = \mathbb{P}$. Therefore $\mathbb{P}_k$ for $k = 0, 1, 2$ is given by

$$\begin{aligned}
\mathbb{P} &= \{2,3,5,7,11,13,17,19,23,29,\ldots\}\,,\\
\mathbb{P}_1 &= \{3,5,11,17,31,41,59,67,83,109,\ldots\}\,,\\
\mathbb{P}_2 &= \{5,11,31,59,127,179,277,331,431,599,\ldots\}\,.
\end{aligned}$$

The definition of PIPs entails $\mathbb{P}_{k+1} \subset \mathbb{P}_k$ for any nonnegative integer k. This is consistent with the following approximation

$$\pi^2(x) \sim \frac{x}{\log^2 x} \qquad \text{as } x \longrightarrow +\infty\,, \quad \pi^2(x) := \pi(\pi(x))\,, \tag{2}$$

shown by Broughan and Barnett [8]. Note that $\pi^2(x)$ is the number of PIPs not greater than a given real x. For any integer $n \geq 1$, the nth Ramanujan prime is defined as the smallest positive integer R_n such that

$$x \geq R_n \Longrightarrow \pi(x) - \pi\left(\frac{x}{2}\right) \geq n\,. \tag{3}$$

Condition (3) ensures the existence of at least n prime numbers in the interval $\left(\frac{x}{2}, x\right]$ whenever $x \geq R_n$. The minimality entails both that R_n is always a prime number and that the interval $\left(\frac{R_n}{2}, R_n\right]$ contains exactly n primes [17]. The properties of the gamma function allowed Ramanujan to compute the first five values of R_n, showing that

$$\pi(x) - \pi\left(\frac{x}{2}\right) \geq 1,2,3,4,5,\ldots, \quad \text{whenever} \quad x \geq 2,11,17,29,41,\ldots, \quad \text{respectively}.$$

This implies the existence of R_n. Hence the sequence of Ramanujan primes is the subset of $\mathbb{P}$ given by

$$\mathbb{P}_R = \{2,11,17,29,41,47,59,67,71,97,\ldots\}.$$

The infinitude of PIPs follows directly from approximation (2). A similar approximation shows the infinitude of Ramanujan primes [18].

2.2. Fractality and Rényi Dimension

In the last decades, fractal geometry, which plays an important role in modern mathematics, has attracted much attention from research teams. In recent years, there has been a growing interest in the link between fractal geometry and number theory. Fractal sets can often be defined in number theoretic terms. There are different parameters which can used to characterize any fractal set. In particular, fractal dimension and lacunarity [9,21] allow us to investigate the fractal nature of prime sequences.

Recent research on fractal geometry has often been focused on dynamical systems. In fact, previous studies indicate the close link between fractal sets, strange attractors, and entropy [22]. The literature on these topics shows a variety of approaches. In particular, the joint approach has attracted significant attention from many researchers, becoming very popular. The chaoticity can be described by many parameters. Each of them is an index related to the complexity of the system. In particular, the fractal dimension is a noninteger number which describes the irregular shape of fractal sets. There are several definitions of fractal dimension. Investigation along these lines exceeds the scope of this paper. It suffices for our purposes to assume that the fractal dimension coincides with the box-counting dimension. This involves no loss of generality. Let A be a non-empty bounded subset of $\mathbb{R}^n$ and $N_{\delta(A)}$ the smallest number of δ-boxes needed to cover A. The box-counting dimension of A is given [21,23] by

$$\mathrm{Dim}_B(A) := \lim_{\delta\to 0} \frac{\log N_\delta(A)}{\log(1/\delta)}, \tag{4}$$

whenever the limit (4) exists. Definition (4) entails that the box-counting dimension of A is the minimum number of δ-boxes covering the set A on a uniform grid. The fractal dimension remains the most important parameter in fractal modeling. The box-counting dimension is strongly linked with the concept of entropy. In fact, let X be a discrete alphabet random variable and let $N = N(\delta)$ be the total number of δ-boxes with $p_i > 0$. The Rényi dimension, a parameter which describes the multifractality [23,24], is given by

$$D_\alpha(X) := \frac{1}{\alpha - 1} \lim_{\delta\to 0} \frac{\log_b \sum_{i=1}^{N} p_i^\alpha}{\log_b \delta}, \quad \alpha \geq 0, \tag{5}$$

reduces to the box-counting dimension for $\alpha = 0$. It follows immediately that

$$D_\alpha = -\lim_{\delta\to 0} \frac{H_\alpha}{\log_b \delta},$$

where H_α is the Rényi entropy. For a fuller treatment see [24] and the references given there.

Fractal sets are mainly characterized by self-similarity and space-filling properties. The concept of lacunarity is related to this second property. In particular, the lacunarity takes into account the texture of a fractal set and describes the distribution of the gaps (lacunae) within it. High values of lacunarity indicate a large size distribution of gaps, that is, a high degree of gappiness. Therefore, the lacunarity draws attention to the homogeneity of a set [9]. High levels of homogeneity correspond to low values of lacunarity, hence the lacunarity provides a valuable idea of the heterogeneity associated with fractal sets. This parameter can be used to distinguish fractal sets with similar fractal dimensions. In other words, fractal dimension and lacunarity play the same role in fractal geometry that mean and variance play in statistics. Nevertheless, lacunarity has been gaining importance in recent years as independent tool to deal with spatial patterns and is often computed by a gliding box algorithm [25].

3. Binary Image and Primality

First and foremost, our investigation into the hidden nature of $\mathbb{P}_1$ and $\mathbb{P}_R$ may be carried out by considering them as data sets. The major advantage of this approach is that each data set can be described in terms of entropy. In particular, the concept of information entropy introduced by Shannon [24] allows us to compute the entropy of data sets. The information entropy of a discrete alphabet random variable X defined on the probability space $(\Omega, \mathcal{B}, P)$ is given by

$$H(X) := -\sum_{i=1}^{N} p_i \log_b p_i \,, \quad p_i = P(X = x_i) \,.$$

In the previous definition of information entropy, the alphabet size is N and the most common logarithmic bases are $b = 2$ and $b = e$. This approach turned out to be inappropriate, as well as a multifractal analysis based on the Rényi dimension (5). In particular, the value of the information entropy does not give us any further information into the hidden structure of prime numbers. Recently, Cattani and Ciancio proposed a new approach to the investigation into the fractality of PIPs [16]. This paper outlines an empiric study on the similarity between the binary image of the PIP (or prime) distribution and Cantor dust. The proposed method is dot-plot based and provides the binary image of these distributions. We now apply this technique to the coprimality condition.

Let $S_1 = \{i_1, i_2, \ldots, i_k\}$, $S_2 = \{j_1, j_2, \ldots, j_k\}$ be two k-length ordered sequences of natural numbers. The indicator function for coprimality is defined by the binary map

$$f : S_1 \times S_2 \to \mathrm{U} \,, \tag{6}$$

such that U is a $k \times k$ matrix, called the correlation matrix for coprimately and defined as $\mathrm{U}(k \times k) = \{f(i_m, i_n)\} := \{f_{mn}\}$ by

$$f_{mn} := \begin{cases} 1\,, & (i_m, i_n) = 1\,, \\ 0\,, & \text{otherwise}\,. \end{cases}$$

Note that $(i_m, i_n) \in S_1 \times S_2$ and $1 \le m, n \le k$. The correlation matrix U for the coprimality condition is shown in Table 1. Clearly, whenever $S_1 = S_2$, the matrix U is simply the autocorrelation of a k-length sequence. The indicator function allows us to identify hidden symmetries between two given sequences. In particular, any element of U can be mapped into black and white pixels as follows:

$$\begin{cases} 1 \longrightarrow \text{black pixel}\,, \\ 0 \longrightarrow \text{white pixel}\,. \end{cases} \tag{7}$$

The result of the mapping (7) is a two-dimensional image $I_{k \times k}$, often called a binary image. In image analysis, any image is a set of pixels adjacent to each other. The binary image can be drawn on a white or black support. The white support holds the mapping (7), where foreground pixels are printed in black and background pixels in white (and conversely on a dark support). The composition of the maps f and (7) given by

$$S(k) \times S(k) \xrightarrow{f} \mathrm{U}(k \times k) \xrightarrow{(7)} I_{k \times k} \,,$$

allows the investigation of the coprimality by fractal geometry. In fact, comparison of Table 1 and Figure 1 (top) shows clearly the presence of the Minkowski island, a variant of the Minkowski curve built on a square. For more details on the Minkowski curve we refer the reader to [26] and the references given there. The lack of some symmetries in Figure 1 (top) might be interpreted as the introduction of randomness in the Minkowski island. Thus the coprimality condition among natural numbers shows a behavior similar to a well-known fractal set. A full discussion requires analytic techniques and lies beyond the scope of this paper.

The statement discussed above is confirmed for higher values. In fact, Figure 1 (bottom) shows the coprimality condition for $1 \le k \le 200$. The behavior is still based on the Minkowski island and is in accordance with theory. In fact, the binary image is symmetric with respect to the main diagonal. This comes from the symmetry of the coprimality condition, that is $(m, n) = (n, m)$ for any $m, n \in \mathbb{N}$.

Table 1. Correlation matrix for coprimality $U(k \times k)$ with $1 \le k \le 15$.

⋮	⋮	⋮	⋮	⋮	⋮	⋮	⋮	⋮	⋮	⋮	⋮	⋮	⋮	⋮	⋮	⋰
15	1	1	0	1	0	0	1	1	0	0	1	0	1	1	0	...
14	1	0	1	0	1	0	0	0	1	0	1	0	1	0	1	...
13	1	1	1	1	1	1	1	1	1	1	1	1	0	1	1	...
12	1	0	0	0	1	0	1	0	0	0	1	0	1	0	0	...
11	1	1	1	1	1	1	1	1	1	1	0	1	1	1	1	...
10	1	0	1	0	0	0	1	0	1	0	1	0	1	0	0	...
9	1	1	0	1	1	0	1	1	0	1	1	0	1	1	0	...
8	1	0	1	0	1	0	1	0	1	0	1	0	1	0	1	...
7	1	1	1	1	1	1	0	1	1	1	1	1	1	0	1	...
6	1	0	0	0	1	0	1	0	0	0	1	0	1	0	0	...
5	1	1	1	1	0	1	1	1	1	0	1	1	1	1	0	...
4	1	0	1	0	1	0	1	0	1	0	1	0	1	0	1	...
3	1	1	0	1	1	0	1	1	0	1	1	0	1	1	0	...
2	1	0	1	0	1	0	1	0	1	0	1	0	1	0	1	...
1	1	1	1	1	1	1	1	1	1	1	1	1	1	1	1	...
f_{mn}	1	2	3	4	5	6	7	8	9	10	11	12	13	14	15	...

3.1. PIPs and Ramanujan Primes

In view of all that has been outlined in [16], the PIP distribution seems similar to the Cantor dust. The technique based on the binary map (6) can also be applied to the distribution of Ramanujan primes. The correlation matrix U_R can easily be computed by definition (3). This give rise to a Boolean table like Table 1. Clearly, the matrix U_R is nonzero whenever both the components are Ramanujan primes. According to the notation introduced in definition (6) it is

$$f_{mn} = \begin{cases} 1, & (i_m, i_n) \in \mathbb{P}_R^2, \\ 0, & \text{otherwise}. \end{cases}$$

Note that $S_1 = S_2 = \mathbb{P}_R$ and $1 \le m, n \le k$. As with the correlation matrix for coprimality, there is no loss of generality in assuming $S_1 = S_2$. Thus the mapping (7) provides the binary image of the Ramanujan primes among the natural numbers, as shown in Figure 2. Like the PIP distribution [16], the distribution of Ramanujan primes looks like the Cantor dust except for some white zones. This lack of symmetry might simply be seen as the presence of randomness in the Cantor dust. The same problem also occurs in the PIP distribution [16]. Let us now define a Cantor prime as any prime number p such that $1/p$ belongs to the (middle-third) Cantor set. Salas has given a characterization of the Cantor set in terms of prime numbers. In particular, he has shown that any prime number $p > 3$ if a Cantor prime if and only if it satisfies an equation which involves both p and the order of 3 modulo p (see [13] for more details). Accordingly, fractal geometry seems to play a relevant and hidden role in the distribution of prime numbers. However, more study of the issue is required. Further research into the fractal structure of $\mathbb{P}_1$ and $\mathbb{P}_R$ should be focused on the main geometric representations of $\mathbb{P}$ (sieve of Eratosthenes, Ulam prime spiral, etc.).

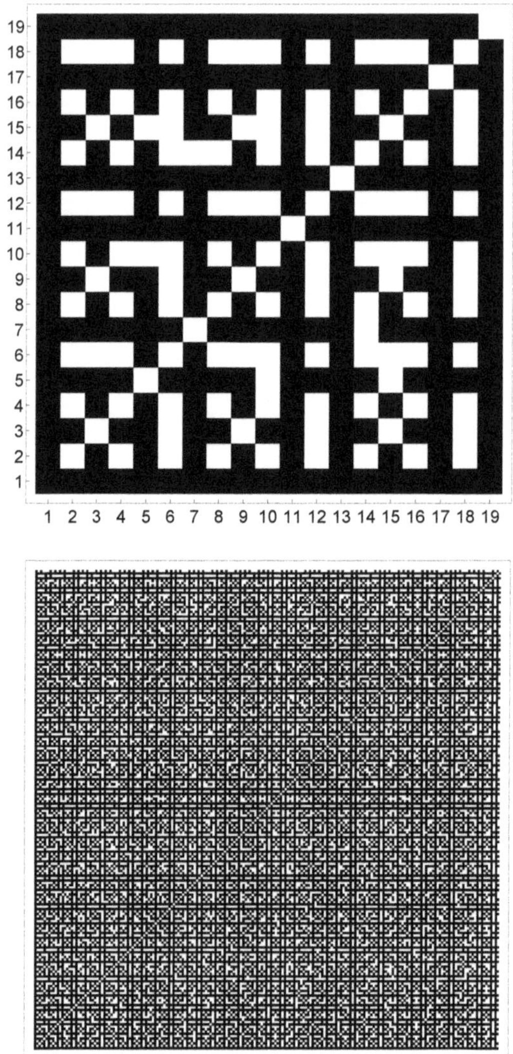

Figure 1. Binary image for the correlation matrix $U(k \times k)$ of Table 1 with $1 \leq k \leq 19$ (**top**) and $1 \leq k \leq 200$ (**bottom**).

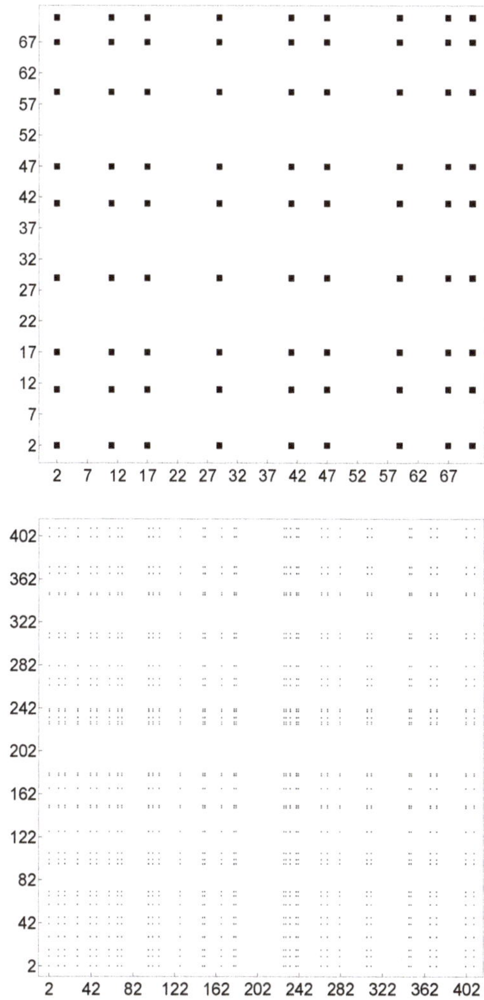

Figure 2. Binary image for the distribution of Ramanujan primes R_i among the natural numbers with $2 \leq R_i \leq 71$ (**top**) and $2 \leq R_i \leq 409$ (**bottom**).

3.2. *Asymptoticity and k-Order PIPs*

We conclude this section by showing and discussing the recent asymptotic results on PIPs and Ramanujan primes. As mentioned in Section 2, the density of PIPs is $\mathcal{O}\left(x/\log^2 x\right)$. This property is an equivalent of the prime number theorem for the PIP distribution. Broughan and Barnett [8] announced that the same holds for the k-order PIPs. Similarly, the results given by Bayless et al. in [20] suggest that

$$\pi^k(x) \sim \frac{x}{\log^k x} \qquad \text{as } x \longrightarrow +\infty\,, \tag{8}$$

where $\pi^k(x)$ denotes the number of k-order PIPs less than or equal to a given real number x. In addition to this, the counting function π^k allows explicit bounds. In particular, for any $k \geq 2$ there exists a computable $x_0(k)$ such that

$$\frac{x}{\log^k x}\left(1+\frac{1}{\log x}\right)^k\left(1+\frac{\log\log x}{\log x}\right)^{k-1} < \pi^k(x) < \frac{x}{\log^k x}\left(1+\frac{1.5}{\log x}\right)^k\left(1+\frac{1.5\log\log x}{\log x}\right)^{k-1}, \quad \forall x \geq x_0(k).$$

The detailed proof can be found in [20]. Note that these bounds are certainly not the best possible. Nevertheless, the finding of better bounds seems difficult from any perspective and related studies are still lacking. Figure 3 shows the behavior of the k-order PIP counting function for $k = 1,2,3$. According to the asymptotic equivalence (8), the family of distributions is strictly decreasing with the order k. Thus the natural question arises regarding the asymptotic comparison between k-order PIPs and Ramanujan primes. Let us denote the counting function of the Ramanujan primes with π_R. Note that the main difficulty in carrying out this issue is given by the asymptotic behavior of π_R. Sondow [17] and Shevelev [18] showed that

$$\pi_R(x) \sim \frac{x}{2\log x} \sim \frac{\pi(x)}{2} \quad \text{as } x \longrightarrow +\infty. \tag{9}$$

Thus the comparison of asymptotic equivalences (8) and (9) assures us that the distribution of Ramanujan primes goes to infinity faster than the k-order PIP distribution. In addition to illustrating the asymptotic behavior, the comparison between k-order PIPs and Ramanujan primes for finite values of x is of interest. Fix a PIP h and consider the subset of $\mathbb{P}_1$ given by $H = \{3,5,11,\ldots,h-1,h\}$ of cardinality n. Figure 4 suggests that the PIP distribution increases more slowly than that of Ramanujan primes, except the values $n \leq 8$. All simulations have confirmed this assertion up to $n = 1500$. Clearly, this is only an empirical clue and far from being a conclusive proof. Summarizing, we may formulate the following

Conjecture 1. *The distribution of Ramanujan primes increases faster than the k-order PIP distribution except for the first eight elements.*

The main difficulty with any eventual proof of Conjecture 1 arises from the lack of results on both PIPs and Ramanujan primes. Note that it has to be based on analytic techniques and will probably appear in a forthcoming publication.

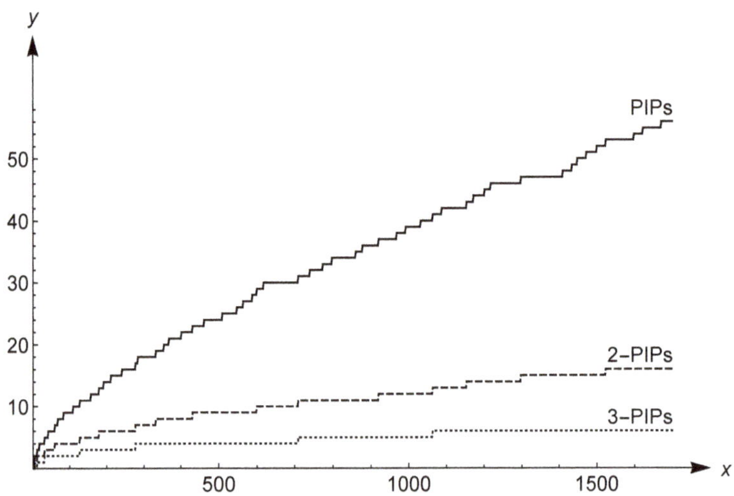

Figure 3. The behavior of k-order PIPs by the counting function π^k for $k = 1,2,3$.

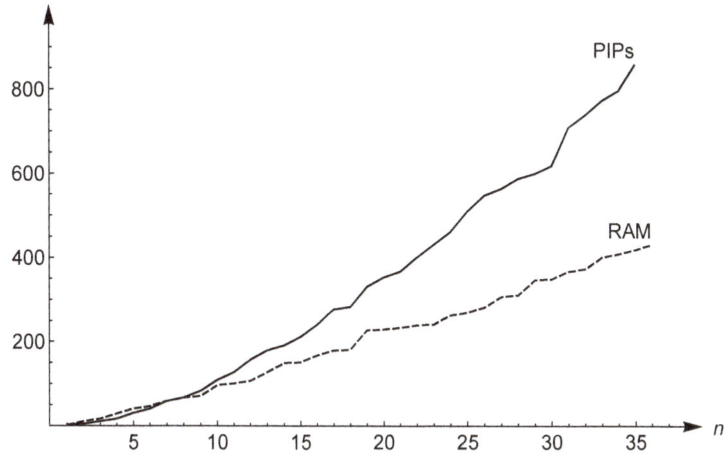

Figure 4. Comparison between PIPs and Ramanujan primes (RAM) in terms of cardinality. The first $n = 35$ elements are depicted for both distributions.

4. An Application in Dynamical Systems: The Hénon Map

The results given in Section 3 provide an application in dynamical systems. In the literature, the link between dynamical systems and fractal geometry is widely investigated and discussed (see for instance [27] (Ch. 13)). According to Section 3, several studies show the central role of the Cantor set not only in fractal geometry but also in pure and applied mathematics. Likewise, the Hénon map H plays an important role in the theory of dynamical systems. In particular, it is defined as the family of maps $H : \mathbb{R}^2 \to \mathbb{R}^2$ of the form

$$H(x,y) = \left(1 + y - ax^2, bx\right) ,$$

where a and b are two real parameters such that $a > 0$ and $|b| < 1$ [28]. Each detailed analysis of H involves several issues and exceeds the scope of this paper. The major drawback occurs in the variation of a and b, which may entail sudden changes in behavior (bifurcations). The Hénon map shows periodicity, mixing, and sensitivity to initial conditions. Hénon dealt with the chaotic behavior of H for $a = 1.4$ and $b = 0.3$, called canonical parameter values for the Hénon map. This is illustrated in Figure 5. A multifractal study of H based on the Rényi dimension and the Lyapunov exponent was recently presented in [29]. For $b = 0.3$ the simulation results indicate the convergence speed of orbits to the attractor for $a \in [1, 1.4]$. Messano et al. [30] showed that all the aperiodic orbits can be removed from the dynamics of H, thus making the new dynamical systems globally convergent.

Among all the dynamical systems, the Hénon map has a close link with the Cantor set. The map H can of course be decomposed into an area-preserving bend, a contraction, and a reflection. The contraction suggests the presence of a fractal-like set. In fact, several simulations indicate that the Hénon map is locally the product of a line segment and a Cantor-like set F [27] (p. 196). Numerical results have clearly shown that the box-counting dimension of the attractor is about 1.26 for the canonical parameter values [29]. Summarizing, the simulation results mean simply that there are smooth bijections ϕ such that

$$\phi : [0,1] \times F \to B \,, \tag{10}$$

where B are small neighbourhoods of the attractor. The results given in Section 3 suggest that F, being a Cantor-like set, may be characterized by a suitable subset of prime numbers. The proof of this conjecture exceeds the scope of this paper. Nevertheless, let us now mention two important consequences. First, note that the role of prime numbers in the theory of chaotic dynamical systems might not be limited to only the description of periodic orbits, becoming a noteworthy parameter of their chaoticity. Second, and perhaps more importantly, the Cantor set and its generalizations are likely to play an increasingly relevant role in number theory, according to recent results [13,15,16].

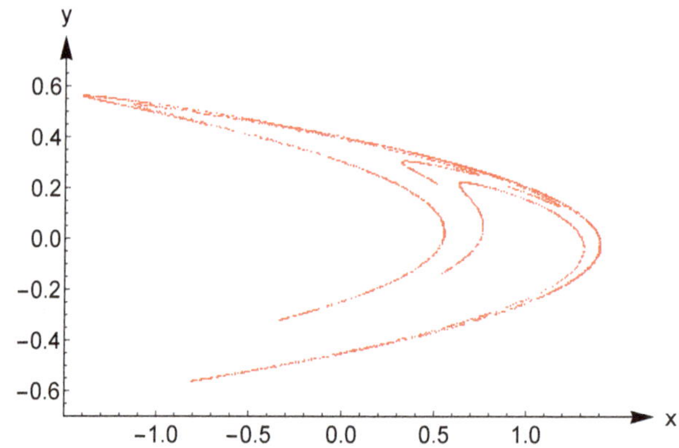

Figure 5. Iterative plot for the Hénon map with $a = 1.2$, $b = 0.4$, and 3000 iterations.

5. Conclusions

This paper has investigated the link between fractal geometry and prime numbers. In particular, the study set out to find further confirmation of the fractal nature of prime numbers, as suggested in [10,13,15,16]. The proposed method is based on image analysis. The paper deals with the application of the binary image to three different topics (coprimality, PIPs, and Ramanujan primes), which play a central role in the theory of prime numbers. In all these cases the results, showing a fractal-like behavior, bear out the fractality of prime numbers. In fact, the coprimality condition depicts a binary image based on the Minkowski island. Likewise, the distribution of Ramanujan primes resembles the Cantor dust. In both the above cases, the lack of some symmetries might indicate the presence of randomness. Thus both the binary images may be seen as random fractals. Recently, Cattani and Ciancio [16] have shown that the PIP distribution looks like the Cantor dust. Accordingly, the Cantor set seems to play a relevant role in the hidden structure of prime numbers. In addition, the comparison between k-order PIPs and Ramanujan primes is discussed and the main conclusions are drawn.

The fractal-like behavior of prime numbers enables us to provide an application in chaotic dynamical systems. Being locally given by the smooth bijections (10), that is, by a Cantor-like set, the Hénon map may be described in terms of prime numbers. Therefore the chaoticity of several dynamical systems could be characterized by a suitable subset of $\mathbb{P}$. Further research should be done to build the binary image of other prime subsets (Mersenne primes, twin primes, etc.). Clearly, several other questions remain to be addressed. In particular, the major drawback of this approach is that all these numerical results will require a theoretical proof.

Funding: The author has not received funds for covering the costs to publish in open access.

Conflicts of Interest: The author declares no conflict of interest.

References

1. Guy, R. *Unsolved Problems in Number Theory*; Springer: New York, NY, USA, 2010. [CrossRef]
2. Nash, J.F., Jr.; Rassias, M.T. (Eds.) *Open Problems in Mathematics*; Springer: New York, NY, USA, 2016. [CrossRef]
3. Bennet, C.H.; Brassard, G. Quantum cryptography: Public key distribution and coin tossing. *Theor. Comput. Sci.* **2014**, *560*, 7–11. [CrossRef] [CrossRef]

4. Goles, E.; Schulz, O.; Markus, M. Prime number selection of cycles in a predator-prey model. *Complexity* **2001**, *6*, 33–38. [CrossRef] [CrossRef]
5. Yan, J.F.; Yan, A.K.; Yan, B.C. Prime numbers and the amino acid code: Analogy in coding properties. *J. Theor. Biol.* **1991**, *151*, 333–341. [CrossRef] [CrossRef]
6. Bershadskii, A. Hidden Periodicity and Chaos in the Sequence of Prime Numbers. *Adv. Math. Phys.* **2011**, *2011*, 519178. [CrossRef] [CrossRef]
7. van Zyl, B.P.; Hutchinson, D.A.W. Riemann zeros, prime numbers, and fractal potentials. *Phys. Rev. E* **2003**, *67*, 066211. [CrossRef] [CrossRef] [PubMed]
8. Broughan, K.A.; Barnett, A.R. On the subsequence of primes having prime subscripts. *J. Integer Seq.* **2009**, *12*, 1–12. [CrossRef]
9. Allain, C.; Cloitre, M. Characterizing the lacunarity of random and deterministic fractal sets. *Phys. Rev. A* **1991**, *44*, 3552–3558. [CrossRef] [CrossRef]
10. Ares, S.; Castro, M. Hidden structure in the randomness of the prime number sequence? *Physica A* **2006**, *360*, 285–296. [CrossRef] [CrossRef]
11. Wolf, M. $1/f$ noise in the distribution of prime numbers. *Physica A* **1997**, *241*, 493–499. [CrossRef] [CrossRef]
12. Batchko, R.G. A prime fractal and global quasi-self-similar structure in the distribution of prime-indexed primes. *arXiv* **2014**, arXiv:1405.2900v2. [CrossRef]
13. Salas, C. Base-3 repunit primes and the Cantor set. *Gen. Math.* **2011**, *19*, 103–107. [CrossRef]
14. Wolf, M. Some heuristics on the gaps between consecutive primes. *arXiv* **2011**, arXiv:1102.0481. [CrossRef]
15. Vartziotis, D.; Wipper, J. The Fractal Nature of an Approximate Prime Counting Function. *Fractal Fract.* **2017**, *1*, 10. [CrossRef] [CrossRef]
16. Cattani, C.; Ciancio, A. On the fractal distribution of primes and prime-indexed primes by the binary image analysis. *Physica A* **460**, 222–229. [CrossRef]
17. Sondow, J. Ramanujan Primes and Bertrand's postulate. *Am. Math. Mon.* **2009**, *116*, 630–635. [CrossRef] [CrossRef]
18. Shevelev, V. Ramanujan and Labos Primes, Their Generalizations, and Classifications of Primes. *J. Integer Seq.* **2012**, *15*, 1–15.
19. Newman, D.J. Simple analytic proof of the prime number theorem. *Am. Math. Mon.* **1980**, *87*, 693–696. [CrossRef] [CrossRef]
20. Bayless, J.; Klyve, D.; Oliveira e Silva, T. New bounds and computations on prime-indexed primes. *Integers* **2013**, *13*, 1–21. [CrossRef]
21. Hutchinson, J.E. Fractals and self similarity. *Indiana Univ. Math. J.* **1981**, *30*, 713–747. [CrossRef] [CrossRef]
22. Hentschel, H.G.E.; Procaccia, I. The infinite number of generalized dimensions of fractals and strange attractors. *Physica D* **1983**, *8*, 435–444. [CrossRef] [CrossRef]
23. Guariglia, E. Harmonic Sierpinski Gasket and Applications. *Entropy* **2018**, *20*, 714. [CrossRef] [CrossRef]
24. Guariglia, E. Entropy and Fractal Antennas. *Entropy* **2016**, *18*, 84. [CrossRef] [CrossRef]
25. Plotnick, R.E.; Gardner, R.H.; Hargrove, W.W.; Prestegaard, K.; Perlmutter, M. Lacunarity analysis: A general technique for the analysis of spatial patterns. *Phys. Rev. E* **1996**, *53*, 5461–5468. [CrossRef] [CrossRef]
26. Kubacki , R.; Czyżewski, M.; Laskowski, D. Minkowski Island and Crossbar Fractal Microstrip Antennas for Broadband Applications. *Appl. Sci.* **2018**, *8*, 334. [CrossRef] [CrossRef]
27. Falconer, K.J. *Fractal Geometry: Mathematical Foundations and Applications*; John Wiley & Sons: Hoboken, NJ, USA, 2003. [CrossRef]
28. Hénon, M. A two-dimensional mapping with a strange attractor. *Commun. Math. Phys.* **1976**, *50*, 69–77. [CrossRef] [CrossRef]
29. Tirnakli, U. Two-dimensional maps at the edge of chaos: Numerical results for the Henon map. *Phys. Rev. E* **2002**, *66*, 066212. [CrossRef] [CrossRef] [PubMed]
30. Di Lena, G.; Franco, D.; Martelli, M.; Messano, B. From Chaos to Global Convergence. *Mediterr. J. Math.* **2011**, *8*, 473–489. [CrossRef] [CrossRef]

entropy

MDPI

Article

Kapur's Entropy for Color Image Segmentation Based on a Hybrid Whale Optimization Algorithm

Chunbo Lang and Heming Jia *

College of Mechanical and Electrical Engineering, Northeast Forestry University, Harbin 150040, China; langchunbo@nefu.edu.cn

* Correspondence: jiaheming@nefu.edu.cn

Received: 14 March 2019; Accepted: 21 March 2019; Published: 23 March 2019

Abstract: In this paper, a new hybrid whale optimization algorithm (WOA) called WOA-DE is proposed to better balance the exploitation and exploration phases of optimization. Differential evolution (DE) is adopted as a local search strategy with the purpose of enhancing exploitation capability. The WOA-DE algorithm is then utilized to solve the problem of multilevel color image segmentation that can be considered as a challenging optimization task. Kapur's entropy is used to obtain an efficient image segmentation method. In order to evaluate the performance of proposed algorithm, different images are selected for experiments, including natural images, satellite images and magnetic resonance (MR) images. The experimental results are compared with state-of-the-art meta-heuristic algorithms as well as conventional approaches. Several performance measures have been used such as average fitness values, standard deviation (STD), peak signal to noise ratio (PSNR), structural similarity index (SSIM), feature similarity index (FSIM), Wilcoxon's rank sum test, and Friedman test. The experimental results indicate that the WOA-DE algorithm is superior to the other meta-heuristic algorithms. In addition, to show the effectiveness of the proposed technique, the Otsu method is used for comparison.

Keywords: Kapur's entropy; color image segmentation; whale optimization algorithm; differential evolution; hybrid algorithm; Otsu method

---

1. Introduction

Image segmentation is a fundamental and key technique in image processing, computer vision, and pattern recognition, the purpose of which is to partition a given image into specific regions with unique characteristics and then extract the objects of interest [1–4]. Hence, the segmentation technique to be adopted determines the performance of higher level systems that introduced above [5]. At present, the main techniques of image segmentation include edge-based technique, region-based technique, neural network-based technique, wavelet transform-based technique, and threshold-based technique [6–10]. Among the available techniques, threshold-based technique (thresholding) is the most popular one that many scholars have done much work in this domain.

More specifically, the thresholding technique determines the segmentation thresholds by optimizing some criteria, such as maximum between-class variance and various entropy criteria [11]. In 1985, Kapur et al. maximized the histogram entropy of segmented classes to obtain the optimal threshold values, which is known as Kapur's entropy technique [12]. This thresholding technique is adopted extensively and show remarkable performance in many image segmentation problems. However, when dealing with complex image segmentation problem, the high threshold operation will increase the computational complexity of the algorithm significantly. Thus, scholars introduce various meta-heuristic algorithms into this domain with the view of reducing computational complexity and improving segmentation accuracy. Shen et al. [13] proposed a modified flower pollination algorithm (MFPA)-based technique for segmenting both real-life images and remote sensing images. The experimental results show that the MFPA algorithm gives higher values in terms of PSNR and SSIM, which is suitable for high dimensional complex image segmentation. In 2016, Kapur's entropy thresholding technique was adopted by Sambandam and Jayaraman [14] for multilevel medical image thresholding. The proposed technique was then optimized by dragonfly optimization (DFO) with the purpose of reducing computational complexity. It can be seen from the results that the proposed algorithm can efficiently explore the search space and obtain the optimal thresholds. In 2017, Khairuzzaman and Chaudhury [5] proposed a grey wolf optimizer (GWO)-based technique for multilevel image thresholding. Kapur's entropy and Otsu methods are used to determine the segmentation thresholds. Experimental results show that the GWO-based technique using both Kapur's entropy and Otsu thresholding techniques performs better than particle swarm optimization (PSO) and bacterial foraging optimization (BFO)-based methods. Besides, there are still many other meta-heuristic algorithms have been successfully applied to multilevel image thresholding, such as artificial bee colony (ABC) [15], firefly algorithm (FA) [16], cuckoo search (CS) [17], wind driven optimization (WDO) [18], krill herd optimization (KHO) [19], moth-flame optimization (MFO) [20], etc. It is well known that the overwhelming majority of images in practical engineering problems are color images, which are often complex and contain a lot of information, whereas, most of the techniques above are used to segment the grayscale images rather than color images. This phenomenon motivated us to introduce an efficient technique to satisfy the practical requirements.

The whale optimization algorithm (WOA) is a novel meta-heuristic algorithm that simulates the behavior of humpback whales in nature [21]. There are mainly three foraging behaviors, namely encircling prey, bubble-net attacking, and search for prey. WOA is a simple and powerful algorithm that has attracted wide attention from scholars recently [22]. In 2018, Xiong et al. [23] used a WOA algorithm to extract the parameters of solar photovoltaic (PV) models. Compared to the conventional as well as recently-developed methods, the proposed algorithm can determine the parameters more accurately. Sun et al. [24] proposed a modified whale optimization algorithm (MWOA) for solving large-scale global optimization (LSGO) problems. Twenty-five benchmark test functions with various dimensions were utilized to verify the performance. The experimental results indicated that the proposed algorithm is superior to other state-of-the-art optimization algorithms in terms of accuracy and stability. In 2017, Mafarja and Mirjalili [25] introduced two hybridization models of WOA and simulated annealing (SA) and then applied the proposed methods to feature selection domain. The SA was adopted to enhance the exploitation capability. It can be observed that the proposed hybrid algorithm outperformed other wrapper-based algorithms in classification accuracy, which is suitable for the current optimization task [25]. To sum up, these promising results motivate us to introduce the WOA algorithm into color image segmentation domain.

It is worth mentioning that color image multilevel thresholding operations need to determine the thresholds of every color component (red, green, and blue), while a meta-heuristic algorithm with strong optimizing capacity can improve the accuracy of image segmentation, as it can obtain appropriate thresholds [26]. Therefore, an improved whale optimization algorithm is proposed which is known as WOA-DE. In the proposed algorithm, differential evolution (DE) is served as local search technique to enhance the exploitation ability. What's more, introducing DE operator improves the situation that the traditional WOA is easy to fall into local optimum in the later iteration. In order to obtain an efficient and universal segmentation method, the performance of WOA-DE using Kapur's entropy is investigated. A series of experiments are conducted on both natural images and satellite images. All experimental results are compared with state-of-the-art algorithms as well as conventional methods. It can be observed from the results that the WOA-DE based methods outperform other meta-heuristic based methods in terms of average fitness values, standard deviation (STD), peak signal to noise ratio (PSNR), structural similarity index (SSIM), feature similarity index (FSIM), and the Wilcoxon's rank sum test as well as the Friedman test. The goal of this paper is as follows:

1. Obtain an efficient segmentation technique for multilevel color image thresholding task.

2. Improve the optimizing capability of WOA to determine the optimal thresholds.
3. Investigate the adaptability of WOA-DE based techniques in the field of natural, satellite, and MR image segmentation.
4. Evaluate the performance of proposed technique from various aspects.

The structure of this paper is presented as follows: Section 2 gives the definition of Kapur's entropy thresholding technique. Section 3 introduces a brief review of the WOA algorithm. The description of the DE algorithm is presented in Section 4. In Section 5, the proposed WOA-DE-based multilevel color image thresholding technique is described in details. Experiments and discussion can be found in Section 6. Finally, Section 7 presents the conclusions and future work directions.

2. Multilevel Thresholding

The image threshold methods can be summarized into two categories: bi-level thresholding methods and multilevel thresholding methods. Bi-level thresholding methods involve one threshold value which partitions the image into two classes: foreground and background, however if the image is quite complex and contains various objects, the bi-level thresholding method is not very effective [27–30]. Therefore, multilevel thresholding methods are used extensively for image segmentation [31–33]. In this paper, a famous multilevel thresholding technique is used to determine the threshold values, namely, Kapur's entropy. A brief formulation of this technique is given in the following subsections. In addition, the RGB image has three basic color components of red, green, and blue, so these thresholding techniques are executed three times to determine the optimal threshold values of each color component [16].

Kapur's Entropy

Kapur's method is also an unsupervised automatic thresholding technique, which selects the optimum thresholds based on the entropy of segmented classes [12]. Assuming that $[th_1, th_2, \ldots, th_n]$ represents the thresholds combination which divided the image into various classes. Then the object function of Kapur's method can be defined as:

$$H(th_1, th_2, \ldots, th_n) = H_0 + H_1 + \ldots + H_n \tag{1}$$

where:

$$H_0 = -\sum_{j=0}^{th_1-1} \frac{p_j}{\omega_0} \ln \frac{p_j}{\omega_0}, \ \omega_0 = \sum_{j=0}^{th_1-1} p_j \tag{2}$$

$$H_1 = -\sum_{j=th_1}^{th_2-1} \frac{p_j}{\omega_1} \ln \frac{p_j}{\omega_1}, \ \omega_1 = \sum_{j=th_1}^{th_2-1} p_j \tag{3}$$

$$H_n = -\sum_{j=th_n}^{L-1} \frac{p_j}{\omega_n} \ln \frac{p_j}{\omega_n}, \ \omega_n = \sum_{j=th_n}^{L-1} p_j \tag{4}$$

$H_0, H_1, \ldots, H_n$ denote the entropies of distinct classes, $\omega_0, \omega_1, \ldots, \omega_n$ are the probability of each class.

In order to obtain the optimal threshold values, the fitness function in Equation (5) is maximized:

$$f_{Kapur}(th_1, th_2, \ldots, th_n) = \text{argmax}\{H(th_1, th_2, \ldots, th_n)\} \tag{5}$$

It is worth noting that the computational complexity of the thresholding technique above will result in exponential growth as the number of thresholds increase. Under such circumstances, Kapur's entropy method is not very effective for multilevel thresholding. Therefore, the WOA-DE-based method using Kapur's entropy is proposed to improve the accuracy and computation speed of thresholding techniques. The ultimate goal of proposed method is to determine the optimal threshold values by maximizing the objective function given in Equation (1).

3. Whale Optimization Algorithm

The whale optimization algorithm, which was proposed by Mirjalili and Lewis in 2016, is inspired by the foraging behavior of humpback whales in nature [21]. Humpback whales tend to create spiral bubbles, and then swim to the prey along the trajectory of bubbles (see Figure 1) [25]. The encircling prey and bubble-net attacking behaviors represent the exploitation phase of optimization. The other phase of optimization namely exploration is represented by the search for prey behavior. It is worth noting that the position vector of search agent is defined in a d-dimensional space, where d denotes the number of decision variables of an optimization problem. Thus, the population X of n search agents can be represented by a $(n \times d)$-dimensional matrix, which is shown in Equation (6):

$$X = \begin{bmatrix} x_{1,1} & x_{1,2} & \cdots & x_{1,d} \\ x_{2,1} & x_{2,2} & \cdots & x_{2,d} \\ \vdots & \vdots & \cdots & \vdots \\ x_{n,1} & x_{n,2} & \cdots & x_{n,d} \end{bmatrix} \tag{6}$$

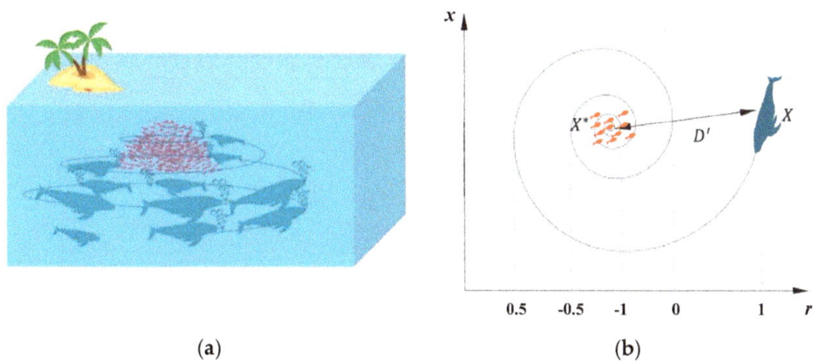

Figure 1. Bubble-net feeding behavior of humpback whale (**a**) and the position update model (**b**).

3.1. Exploitation Phase (Encircling Prey and Bubble-Net Attacking Method)

In the process of hunting, the humpback whales first encircle the prey, which can be represented as follows:

$$D = |C \cdot X^*(t) - X(t)| \tag{7}$$

$$X(t+1) = X^*(t) - A \cdot D \tag{8}$$

where X^* represents the best solution obtained so far, X denotes the position vector, t is the current iteration, $|\ |$ is the absolute value, $\cdot$ is an element-by-element multiplication, A and C are two essential parameters that can be evaluated by:

$$A = 2a \cdot r - a \tag{9}$$

$$C = 2 \cdot r \tag{10}$$

where r is a random number in the range of [0,1] and a is a constant that will decrease linearly from 2 to 0 within the whole iterative process (both exploration and exploitation). It can be observed from Equation (8) that search agents can update their position $X(t)$ according the best solution X^*. The parameters A and C determine the distance between the updated position $X(t+1)$ and the optimal position X^*.

The bubble-net attacking behavior can be mathematically represented by the following equation:

$$D' = |X^*(t) - X(t)| \tag{11}$$

$$X(t+1) = D' \cdot e^{br} \cdot \cos(2\pi r) + X^*(t) \tag{12}$$

where $D\prime$ shows the distance between the current search agent position and the optimal position, b is a constant that determine the shape of a logarithmic spiral, and r is a random number in the range of [−1,1]. In order to transform these two mechanisms (encircling prey and bubble-net attacking method) of exploitation phase, assume that each mechanism will be executed with 50% probability. Thus, the mathematical model of the entire exploitation phase can be expressed as:

$$X(t+1) = \begin{cases} X^*(t) - A \cdot D & if\ p < 0.5 \\ D' \cdot e^{br} \cdot \cos(2\pi r) + X^*(t) & if\ p \geq 0.5 \end{cases} \tag{13}$$

where p is a random number in the range of [0,1].

3.2. *Exploration Phase (Search for Prey)*

In order to enhance the exploration capability of algorithm, a global search strategy is utilized. The search agents update their position according to a random agent in the population rather than the best solution obtained so far. It is worth mentioning that the absolute value of A determines the phase of optimization to be selected, namely the exploration and exploitation phases. Thus, the search for prey behavior can be mathematically represented as follows:

$$D = |C \cdot X_{rand}(t) - X(t)| \tag{14}$$

$$X(t+1) = X_{rand}(t) - A \cdot D \tag{15}$$

where X_{rand} denotes a random individual in the current population.

Pseudo code of traditional whale optimization algorithm based multilevel thresholding has been given in Algorithm 1.

Algorithm 1 Pseudo code of whale optimization algorithm based multilevel thresholding

```
Initialize the position of whales X_i.
Initialize the best search agent X*.
  WHILE t < Maximum number of iterations
    FOR i = 1:n
      Calculate the objective value of each search agent by using the Equation (1) for Kapur's entropy.
      Update the best search agent X*.
      Update a, A, C, r, and p
      IF1 p < 0.5
        IF2 |A| < 1
          Update the position of search agent using Equations (7) and (8).
        ELSE
          Update the position of search agent using Equations (14) and (15).
        END IF2
      ELSE
        Update the position of search agent using Equations (11) and (12).
      END IF1
      Correct the position of the current search agent if it is beyond the border.
    END FOR
  END WHILE
Return X*, which represents the optimal threshold values of segmentation.
```

4. Differential Evolution

Differential evolution (DE) algorithm is a simple and powerful algorithm for solving optimization problems [34–36]. Basically, the DE algorithm contains two significant parameters, namely mutation scaling factor denoted by *SF* and crossover probability denoted by *CR* [37]. For the standard DE algorithm, the mutation, crossover, and selection operators can be summarized as follows [38]:

4.1. Mutation Operation

The mutation operation of DE algorithm is defined as follows:

$$m_i^{g+1} = x_{r1}^g + SF \times \left(x_{r2}^g - x_{r3}^g\right) \tag{16}$$

where m_i^{g+1} represents the mutant individual in the $(g + 1)$-th generation. x_{r1}^g, x_{r2}^g, and x_{r3}^g are different individuals from the population. In other words, r_1, r_2, and r_3 cannot be equal. *SF* is a constant that indicates the mutation scaling factor.

4.2. Crossover Operation

In the process of crossover, the trial individual c_i^{g+1} is selected from the current individual x_i^g or the mutant individual m_i^{g+1} on account of enhancing the diversity of population. The crossover operation of DE algorithm is described as:

$$c_i^{g+1} = \begin{cases} m_i^{g+1} & if\ rand \leq CR \\ x_i^g & if\ rand > CR \end{cases} \tag{17}$$

where *rand* represents a random value which is in the range [0,1]. *CR* is a constant that shows the crossover probability.

4.3. Selection Operation

After the process of selection, the individual of next generation x_i^{g+1} is selected according to the comparison of fitness value between the trail individual c_i^{g+1} and the target individual x_i^g. For a problem to be minimized, the selection operation of DE algorithm can be summarized as follows:

$$x_i^{g+1} = \begin{cases} c_i^{g+1} & if\ f\left(c_i^{g+1}\right) < f\left(x_i^g\right) \\ x_i^g & otherwise \end{cases} \tag{18}$$

where *f* denotes the fitness function value of a given problem.

5. The Proposed Method

In this section, a detailed introduction of the WOA-DE-based method is given, and the algorithm will be used to obtain the optimal threshold values for image segmentation. A hybrid of the WOA and DE algorithms is introduced to balance the two essential phases of optimization, namely exploration and exploitation. The flowchart of WOA-DE for finding the optimal threshold values is shown in Figure 2.

It is worth mentioning that a better balance between exploration and exploitation plays an important role in improving the optimization ability of algorithm. Therefore, an efficient hybrid strategy is introduced to balance and improve these two phases. On the one hand, the WOA algorithm has strong ability to explore the solution space and is used as global search technique. On the other hand, the DE algorithm is adopted as local search technique, which can increase the precision of solutions.

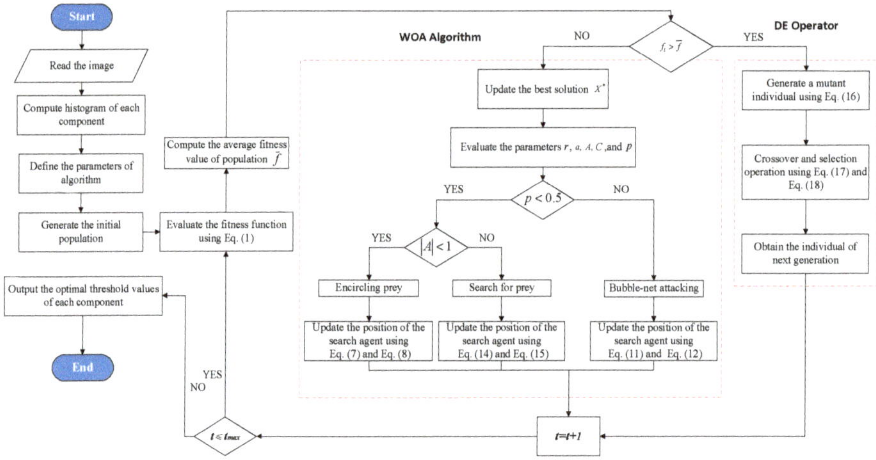

Figure 2. Framework of the WOA-DE based method.

In addition, the purpose of introducing DE operator is not only to enhance the local search ability of the algorithm, but also to overcome the drawback that WOA algorithms easily fall into local optima in the late iterations. As described above, the random variable A will change in the range $[-2,2]$ as a decreases progressively. If the value larger than 1 or less than -1, Equation (15) will be adopted to enhance the exploration capability of the algorithm. On the contrary, Equation (8) will be adopted as local search strategy when the value in the range $[-1,1]$. In order to more intuitively reflect the change of random variable A during the whole iterative process, a relevant schematic diagram is presented in Figure 3. It can be observed from the figure that the value of random variable A is fixed in the interval of $[-1,1]$ after 250 iterations. This means that the global search strategy has no chance to be adopted after half of the iterative process, even if the current best solution may not the global optimum. Therefore, the traditional WOA algorithm will fall into the local optimum, resulting in an unsatisfactory solution accuracy. Especially for complex multi-dimensional optimization problems, such as multilevel color image segmentation, traditional WOA algorithms cannot handle them. On the contrary, DE operators can scale the difference between any two search agents in the population, which makes the particles jump out of the current search area. In Equation (16), $\left(x_{r2}^{g} - x_{r3}^{g}\right)$ can be considered as the difference between two individuals, and SF is the scaling factor. The latter term in Equation (16) "$SF \times \left(x_{r2}^{g} - x_{r3}^{g}\right)$" is crucial to the mutation operator. For the exploration stage, particles tend to be very far apart, and there is a big difference between the individuals. Scaling this big difference can enhance the diversity of population. For the exploitation stage, particles tend to be close together, scaling a small difference makes the algorithm effectively optimize in a small range, improving the accuracy of the solution and avoiding local optimum.

In this paper, the average fitness value of the population is computed in the iterative process to evaluate the quality of each particle. The proposed hybrid model enables particles with better quality to exploit the current promising area to ensure the convergence speed, while the particles with poor quality can explore the unknown area to prevent local optimization. Although the global search strategy of traditional WOA algorithm will not be adopted in the later iteration, the introduced DE operator can effectively overcome this shortcoming, as discussed above. Exactly speaking, if $f_i > \bar{f}$, the DE algorithm will be used to update the solution x_i^g using Equations (16)–(18). However, if $f_i \leq \bar{f}$, then the current solution will be updated using Equations (8), (12), or (15). In addition, a series of experiments are conducted in the following section to verify the advantages of WOA-DE algorithm from various aspects.

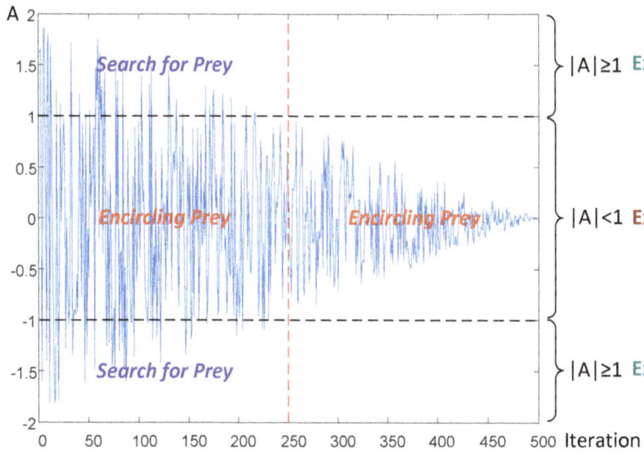

$$X(t+1)=\begin{cases} X^{*}(t)-A\cdot D & if\ |A|<1 \\ X_{rand}(t)-A\cdot D & if\ |A|\geq 1 \end{cases}$$

Figure 3. Schematic diagram of the change in random variable A.

6. Experiments and Results

6.1. Experimental Setup

In this paper, Kapur's entropy thresholding technique is utilized to determine the optimal threshold values for image segmentation. The performance of our WOA-DE-based method is evaluated on fourteen images. Among them, five images are natural images from the Berkeley segmentation database [39], five images are satellite images from [40], and four images are brain magnetic resonance images (MRI) from [41]. Besides, all the images and their corresponding histogram images are shown in Figure 4. Both state-of-the-art and conventional methods, such as the traditional WOA [21], salp swarm algorithm (SSA) [42], sine cosine algorithm (SCA) [43], ant lion optimizer (ALO) [44], harmony search optimization (HSO) [45], bat algorithm (BA) [46], particle swarm optimization (PSO) [47,48], betaDE (BDE) [49], and improved differential search algorithm (IDSA) [50] are used to validate the superiority of proposed algorithm, whose parametric settings are presented in Table 1, except for the population size N set to 30 and the number of iterations $t_{\max}$ set to 500 for fair comparison. The experiments are carried out through the simulation in "Matlab2017" (The MathWorks Inc., Natick, MA, USA) and implemented on a computer equipped with the Microsoft Windows 10 operating system and 8 GB memory space.

Table 1. Parameters of the algorithms.

No.	Algorithm	Parameter Setting	Year	Reference
1	WOA-DE	$CR = 0.9$(crossover rate), $SF = 0.5$(scaling factor)	—	—
2	WOA OA	$a \in [0,2]$	2016	[21]
3	SSA	$c_1 \in [0,2]$	2017	[42]
4	SCA	$r_1 \in [0,2]$	2016	[43]
5	ALO	$\omega \in [2,6]$(constant)	2015	[44]
6	HSO	$HMCR = 0.9, PAR = 0.3$(pitch adjusting rate)	2001	[45]
7	BA	$r_i \in [0,1]$(rate of pluse emission), $A_i \in [1,2]$(loudness value)	2015	[46]
8	PSO	$c_1 = c_2 = 2, w \in [0.4, 0.9], v_{max} = 25.5$	1995	[47]
9	BDE	$a \in [0,1]$(beta distribution parameter)	2018	[49]
10	IDSA	—	2018	[50]

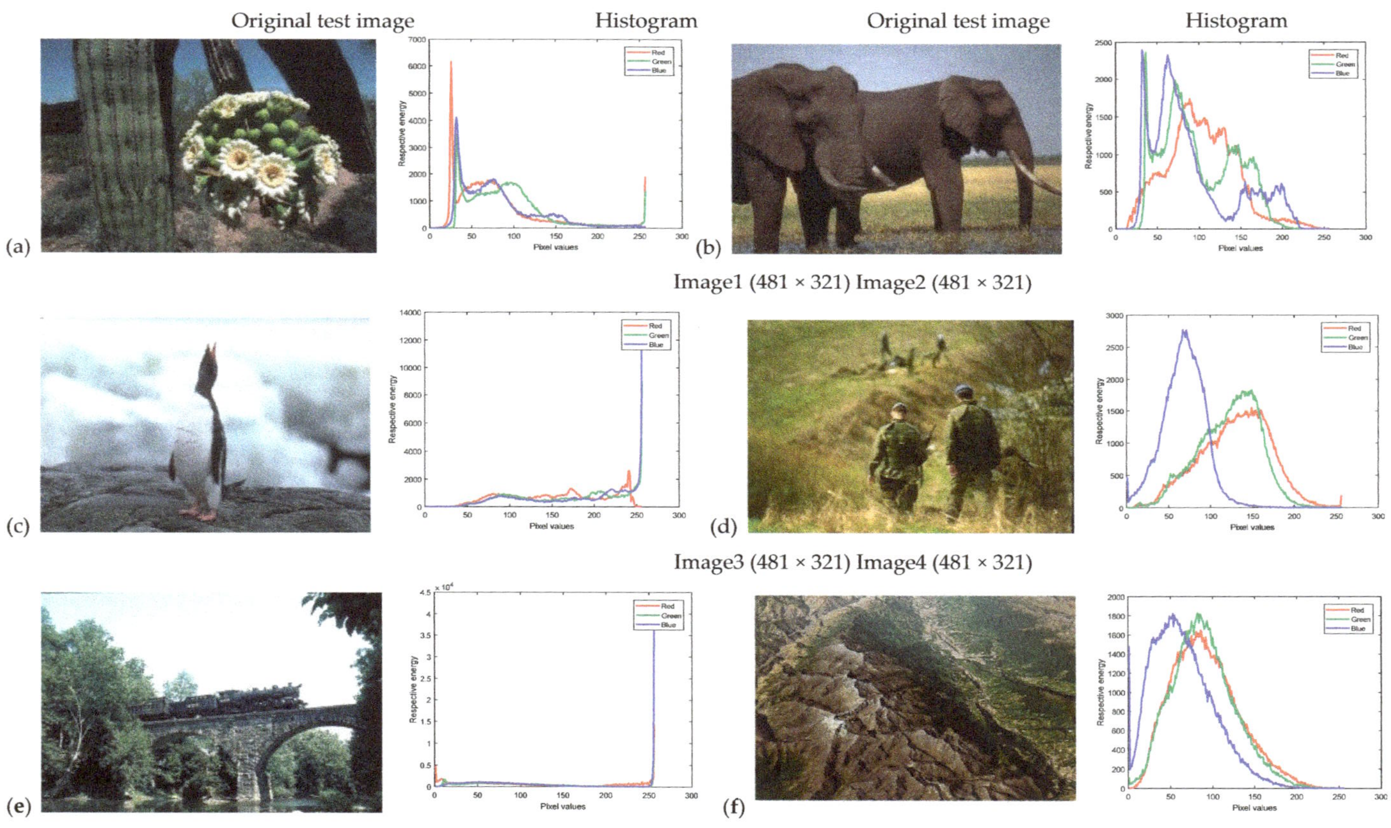

Image1 (481 × 321) Image2 (481 × 321)

Image3 (481 × 321) Image4 (481 × 321)

Image5 (481 × 321) Image6 (481 × 321)

Figure 4. *Cont.*

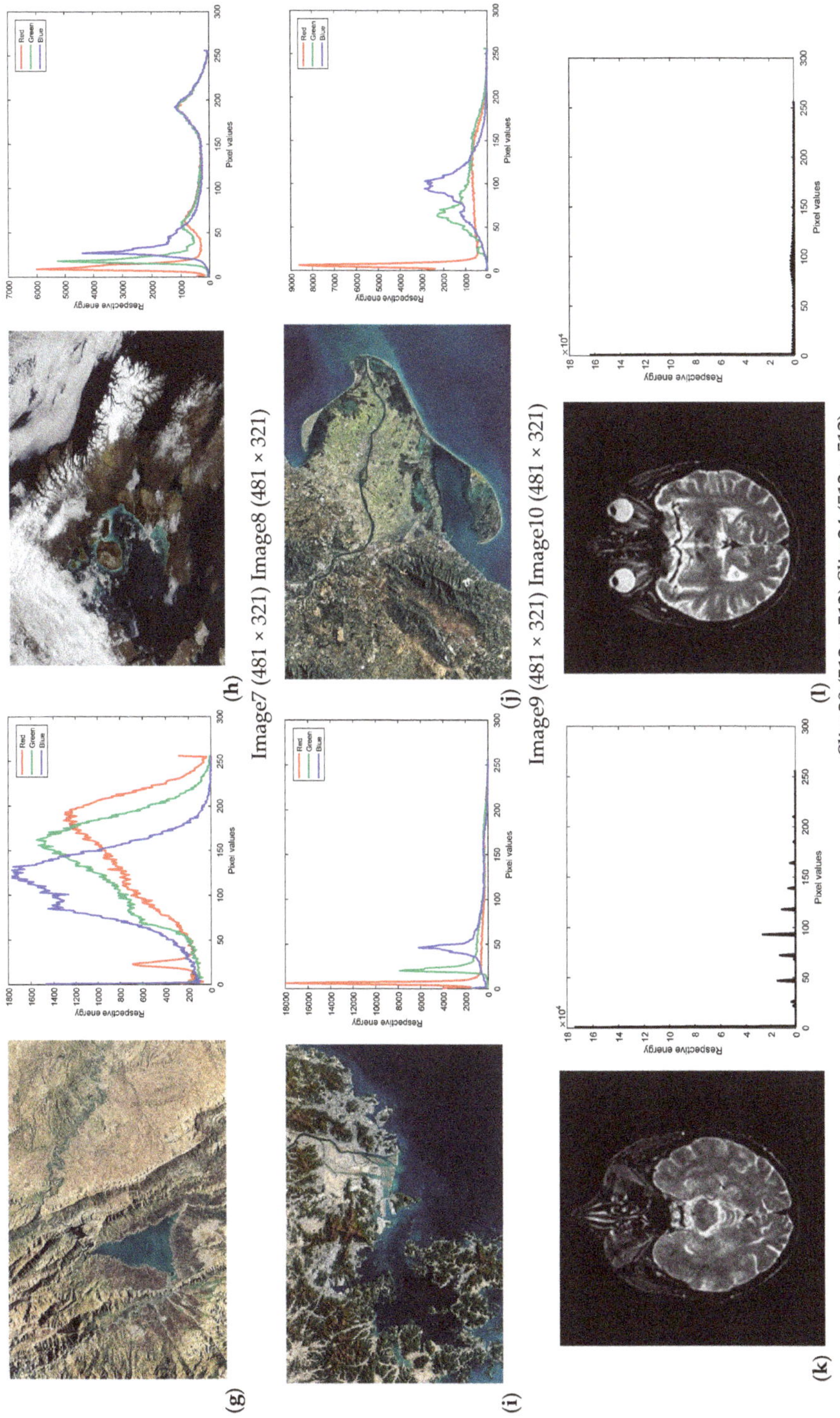

(g) (h) Image7 (481 × 321) Image8 (481 × 321)

(i) (j) Image9 (481 × 321) Image10 (481 × 321)

(k) (l) Slice20 (512 × 512) Slice24 (512 × 512)

Figure 4. *Cont.*

(m)

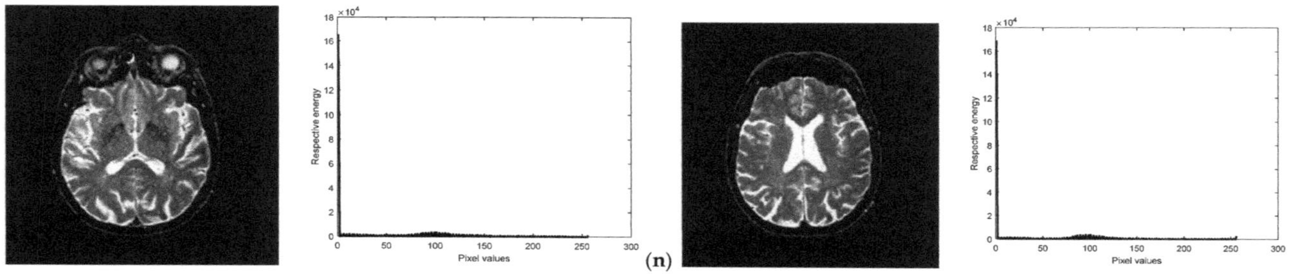

Slice28 (512 × 512) Slice32 (512 × 512)

Figure 4. Original test images and the corresponding histograms.

6.2. Objective Function Measure

As discussed above, Kapur's entropy is used to determine the segmentation thresholds. The segmented images of "Image2" and "Image10" obtained by WOA-DE using Kapur's entropy method with different threshold levels are given in Figures 5 and 6, respectively. Due to the stochastic nature of meta-heuristic algorithms, the experiments are conducted over 30 runs. Then the average objective values of "Image1" and "Image6" are presented in Table 2. It can be seen from the table that the WOA-DE based method gives the best values in general.

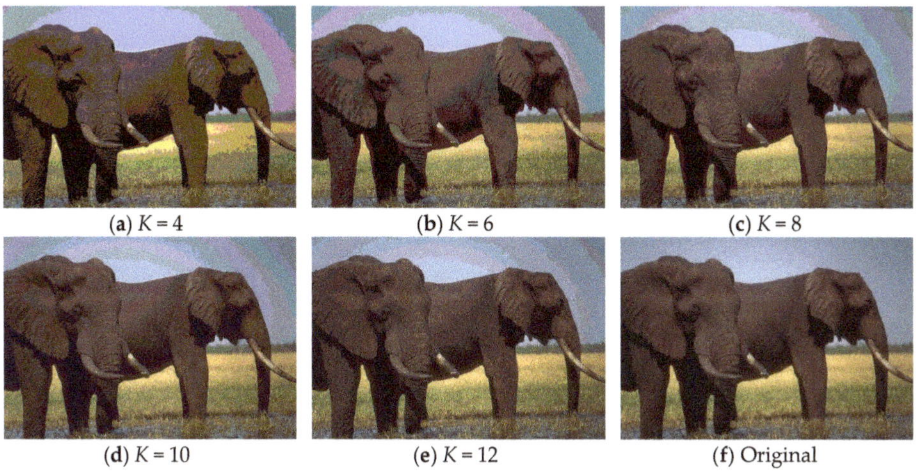

Figure 5. The segmented results of "Image2" at different threshold levels obtained by WOA-DE-Kapur.

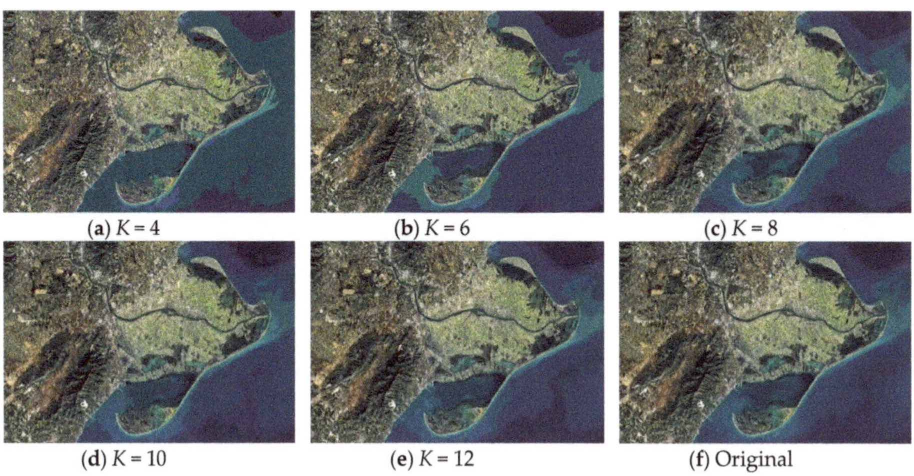

Figure 6. The segmented results of "Image10" at different threshold levels obtained by WOA-DE-Kapur.

The entropy of an image reflects its average information content [51]. Therefore, higher value of Kapur's entropy indicates more information in the image. It can be observed from Table 2 that the objective function value of each algorithm increases with the number of threshold values. This promising result shows that high-quality image with more information is obtained when the threshold level is high (such as $K = 10$ and 12).

Table 2. The average fitness values and STD values obtained by all algorithms.

Measures	Image	*K*	WOA-DE	WOA	SSA	SCA	ALO	HSO	BA	PSO	BDE	IDSA
Mean	Image1	4	18.5843	18.5843	18.5843	18.5632	18.5843	18.5761	18.5818	18.5842	18.5843	18.5843
		6	23.8418	23.73	23.8408	23.479	23.8417	23.755	23.8085	23.8412	23.8115	23.8383
		8	28.5094	28.4605	28.5051	27.8225	28.4627	28.385	27.7631	28.4991	28.5118	28.5139
		10	32.8462	32.8432	32.8325	31.3685	32.8443	32.6682	32.0858	32.7519	32.8455	32.7269
		12	36.8534	36.7164	36.7269	34.5881	36.7313	36.6221	34.7641	36.696	36.7642	36.7764
	Image6	4	18.4839	18.4784	18.4817	18.4434	18.4836	18.4778	18.4745	18.4839	18.4816	18.4836
		6	24.0059	23.9988	23.9994	23.765	24.005	23.9687	23.9225	24.0015	24.0051	24.0059
		8	28.937	28.8743	28.8836	27.9973	28.9272	28.8508	28.2696	28.9293	28.9342	28.9196
		10	33.3483	33.3009	33.1851	31.8768	33.2743	33.0867	31.6321	33.3197	33.3079	33.2562
		12	37.3674	37.271	37.1046	35.8876	37.3246	36.8644	35.1068	37.1813	37.3553	37.2569
STD	Image1	4	0	2.66×10^{-5}	2.66×10^{-5}	4.39×10^{-3}	5.83×10^{-5}	3.19×10^{-3}	2.17×10^{-3}	2.68×10^{-5}	2.47	1.58×10^{-1}
		6	3.25×10^{-5}	1.61×10^{-4}	9.58×10^{-4}	7.45×10^{-2}	2.95×10^{-4}	1.97×10^{-2}	5.39×10^{-2}	4.33×10^{-4}	8.41×10^{-1}	1.59
		8	4.32×10^{-4}	2.74×10^{-2}	8.89×10^{-3}	1.33×10^{-1}	3.24×10^{-2}	3.38×10^{-2}	1.82×10^{-1}	7.30×10^{-3}	1.3	9.98×10^{-1}
		10	3.38×10^{-3}	5.36×10^{-3}	3.90×10^{-2}	2.67×10^{-1}	4.91×10^{-2}	3.24×10^{-2}	1.51×10^{-1}	3.34×10^{-2}	7.76×10^{-1}	7.27×10^{-1}
		12	1.83×10^{-2}	3.25×10^{-2}	7.48×10^{-2}	2.30×10^{-1}	6.12×10^{-2}	5.02×10^{-2}	6.41×10^{-1}	7.35×10^{-2}	6.66×10^{-1}	6.65×10^{-1}
	Image6	4	3.91×10^{-3}	7.91×10^{-3}	4.81×10^{-3}	1.24×10^{-2}	8.29×10^{-1}	5.49×10^{-3}	4.60×10^{-3}	3.93×10^{-3}	6.96×10^{-3}	2.92×10^{-1}
		6	1.68×10^{-2}	3.81×10^{-3}	1.94×10^{-2}	6.82×10^{-2}	4.20×10^{-3}	3.58×10^{-2}	2.96×10^{-2}	5.03×10^{-3}	4.66	1.75
		8	1.57×10^{-2}	2.64×10^{-2}	1.64×10^{-2}	1.60×10^{-1}	2.48×10^{-2}	4.37×10^{-2}	3.94×10^{-1}	6.14×10^{-2}	3.88	9.82×10^{-1}
		10	2.15×10^{-2}	4.26×10^{-2}	3.64×10^{-2}	1.40×10^{-1}	5.49×10^{-2}	2.50×10^{-2}	4.03×10^{-1}	3.55×10^{-2}	2.28	8.98×10^{-1}
		12	1.53×10^{-2}	3.02×10^{-2}	9.90×10^{-2}	2.85×10^{-1}	2.52×10^{-2}	6.41×10^{-2}	3.37×10^{-1}	3.43×10^{-2}	1.67	1.25

6.3. Stability Analysis

Standard deviation (STD): a value indicates the dispersion of sample data and it is mathematically represented as:

$$STD = \sqrt{\frac{1}{n-1}\sum_{i=1}^{n}\left(f_i - \overline{f}\right)^2} \tag{19}$$

where n is the sample size, f_i is the fitness value of the i-th individual, and $\overline{f}$ indicates the average value of the sample.

In order to verify the stability of proposed algorithm, the STD indicator is also used. A lower value of STD indicates better stability. The STD values of "Image1" and "Image6" obtained by all algorithms are presented in Table 2. From the table it is found that WOA-DE based method gives lower values as compared to other algorithms, which shows the better consistency and stability of proposed algorithm.

6.4. Peak Signal to Noise Ratio (PSNR)

Peak signal to noise ratio (PSNR): an index which is used to evaluate the similarity of the processed image against the original image [13]:

$$PSNR = 10\log_{10}\left(\frac{255^2}{MSE}\right) \tag{20}$$

MSE represents the mean squared error and is calculated as:

$$MSE = \frac{1}{MN}\sum_{i=1}^{M}\sum_{j=1}^{N}[I(i,j) - K(i,j)]^2 \tag{21}$$

where $I(i, j)$ and $K(i, j)$ denote the gray level of the original image and the segmented image in the i-th row and j-th column, respectively. M and N denote the number of rows and columns in the image matrix, respectively. A higher value of PSNR indicates a better quality segmented image.

Table 3 shows the PSNR values of "Image2" and "Image7" obtained by all algorithms and Kapur's entropy method. According to the table, the WOA-DE-based method gives the highest values in 9 out of 10 cases using Kapur's entropy. When the threshold level is small, all algorithms give similar result, while the obtained values become different as the number of thresholds increases, and the proposed method can present the best result in most cases. This phenomenon indicates that WOA-DE-based method can determine the appropriate thresholds and then present high-quality segmented image that are more similar to the original image. Figure 7 shows the visual comparison of all available methods at different threshold levels. The results of proposed method are represented as "black" lines and "square" data points.

Table 3. The PSNR, SSIM, and FSIM values obtained by all algorithms under different threshold levels.

Measures	Image	K	WOA-DE	WOA	SSA	SCA	ALO	HSO	BA	PSO	BDE	IDSA
PSNR	Image2	4	18.6558	18.6558	18.6558	18.6533	18.6558	18.5722	18.4352	18.6558	18.6452	18.6558
		6	22.2481	20.8588	21.3402	21.5799	21.3148	20.861	20.2596	21.7136	20.8588	20.9995
		8	24.8821	23.1744	23.6373	23.4877	24.1624	24.5724	23.1158	23.372	23.5863	23.5837
		10	27.9116	25.3956	25.9502	27.0446	27.7051	25.3211	25.1289	25.3938	25.87	25.9861
		12	29.8805	29.8395	29.6719	28.6767	29.4309	26.3023	29.2218	29.34	29.0663	29.2001
	Image7	4	23.2367	22.947	22.9286	22.9305	23.0442	22.9765	22.923	22.982	22.982	22.9122
		6	26.6481	26.5553	26.5205	26.5953	26.656	26.4685	26.5963	26.6156	26.527	26.5732
		8	29.1886	29.1004	28.9405	27.8606	29.132	29.063	27.9378	29.0088	29.1151	29.0763
		10	31.2154	30.9433	31.1186	29.6665	30.9579	30.64	28.3997	30.9199	31.0374	31.169
		12	32.7203	32.6538	31.7022	30.101	32.6566	31.6295	28.988	32.7035	32.6774	32.6603
SSIM	Image2	4	0.5266	0.5266	0.5266	0.5253	0.5186	0.5212	0.5212	0.5266	0.5242	0.5266
		6	0.652	0.6103	0.617	0.6105	0.6192	0.6379	0.5864	0.6332	0.6103	0.6197
		8	0.7361	0.6944	0.7052	0.6976	0.7281	0.7224	0.7094	0.6978	0.7004	0.6963
		10	0.8064	0.7551	0.7608	0.7701	0.7996	0.7534	0.7247	0.7594	0.7705	0.7733
		12	0.8505	0.8484	0.8432	0.7859	0.8411	0.8463	0.8182	0.8367	0.8483	0.8269
	Image7	4	0.8494	0.8414	0.8407	0.8419	0.8456	0.8405	0.8382	0.8416	0.8416	0.8399
		6	0.9136	0.9097	0.907	0.9086	0.9115	0.9091	0.9098	0.9087	0.9091	0.9079
		8	0.9422	0.9409	0.9392	0.9254	0.9415	0.9412	0.9241	0.9404	0.9416	0.9418
		10	0.9648	0.9608	0.9603	0.946	0.9587	0.9552	0.9255	0.9624	0.9633	0.9637
		12	0.9726	0.9715	0.9624	0.9543	0.9724	0.963	0.9338	0.9718	0.9724	0.9725
FSIM	Image2	4	0.7151	0.7151	0.7151	0.7149	0.7151	0.7117	0.7115	0.7151	0.7142	0.7151
		6	0.7921	0.7707	0.7723	0.7708	0.7711	0.7876	0.7577	0.7799	0.7707	0.7866
		8	0.8435	0.8257	0.8289	0.8246	0.8426	0.8313	0.8093	0.824	0.8256	0.8198
		10	0.8745	0.8617	0.8582	0.8423	0.8738	0.8616	0.8153	0.864	0.8686	0.8682
		12	0.9041	0.9036	0.9005	0.8987	0.9007	0.8978	0.8531	0.8806	0.9022	0.8915
	Image7	4	0.9012	0.8964	0.8959	0.8965	0.8991	0.8968	0.8953	0.8972	0.8972	0.8956
		6	0.9469	0.9445	0.946	0.9458	0.9467	0.9449	0.9468	0.9465	0.9453	0.9464
		8	0.9671	0.9666	0.9658	0.9591	0.9665	0.9659	0.9559	0.9658	0.9666	0.966
		10	0.9773	0.9758	0.9752	0.971	0.9759	0.9754	0.9574	0.9765	0.9761	0.9763
		12	0.9834	0.9824	0.9799	0.9729	0.9829	0.9802	0.9664	0.9828	0.9825	0.9826

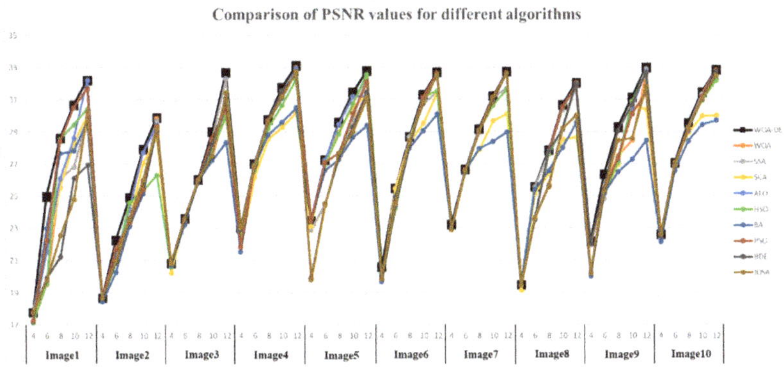

Figure 7. Comparison of PSNR values for different algorithms using Kapur's entropy at 4, 6, 8, 10, and 12 levels.

6.5. Structural Similarity Index (SSIM)

Structural similarity index (SSIM) [52,53]: a measure of the similarity between the original image and the segmented image, which takes various factors such as brightness, contrast, and structural similarity into account:

$$SSIM(x,y) = \frac{(2\mu_x\mu_y + c_1)(2\sigma_{xy} + c_2)}{\left(\mu_x^2 + \mu_y^2 + c_1\right)\left(\sigma_x^2 + \sigma_y^2 + c_2\right)} \tag{22}$$

where μ_x and μ_y denote the mean intensities of the original image and the segmented image respectively. σ_x^2 and σ_y^2 are the standard deviation of the original image and the segmented image respectively. σ_{xy} denotes the covariance between the original image and the segmented image. c_1 and c_2 are constants. The value of SSIM is in the range [0,1], and a higher value shows better performance.

The SSIM values obtained by all algorithms are given in Table 3 and Figure 8, respectively. It can be seen from the table that the WOA-DE-based method gives competitive results again compared with other methods in terms of SSIM indicator. The values obtained by all algorithms increase with the number of thresholds, which indicates that the segmented image is more similar to the original image in terms of brightness, contrast, and structural similarity. The experimental results in this section verify the remarkable performance of the proposed algorithm from another perspective.

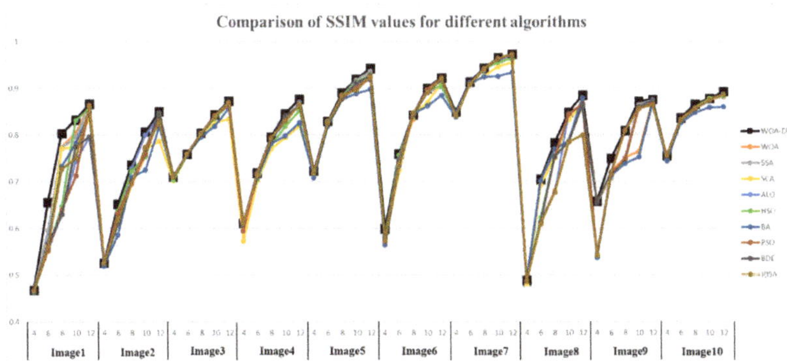

Figure 8. Comparison of SSIM values for different algorithms using Kapur's entropy at 4, 6, 8, 10, and 12 levels.

6.6. Feature Similarity Index (FSIM)

Feature similarity index (FSIM) [54,55]: another measure of the image quality through evaluating the feature similarity between the original image and the segmented image:

$$FSIM = \frac{\sum_{x \in \Omega} S_L(x) \times PC_m(x)}{\sum_{x \in \Omega} PC_m(x)} \tag{23}$$

where Ω represents the whole image pixel domain. $S_L(x)$ is a similarity score. $PC_m(x)$ denotes the phase consistency measure, which is defined as:

$$PC_m(x) = \max(PC_1(x), PC_2(x)) \tag{24}$$

where $PC_1(x)$ and $PC_2(x)$ represent the phase consistency of two blocks, respectively:

$$S_L(x) = [S_{PC}(x)]^{\alpha} \cdot [S_G(x)]^{\beta} \tag{25}$$

$$S_{PC}(x) = \frac{2PC_1(x) \times PC_2(x) + T_1}{PC_1^2(x) \times PC_2^2(x) + T_1} \tag{26}$$

$$S_G(x) = \frac{2G_1(x) \times G_2(x) + T_2}{G_1^2(x) \times G_2^2(x) + T_2} \tag{27}$$

$S_{PC}(x)$ denotes the similarity measure of phase consistency. $S_G(x)$ denotes the gradient magnitude of two regions $G_1(x)$ and $G_2(x)$. α, β, T_1, and T_2 are all constants. The value of FSIM is also in the range [0,1], and a higher value shows better segmented image quality.

On comparing the FSIM values, which are given in Table 3 and Figure 9, it can be observed that WOA-DE-based method again outperforms the other methods. The feature similarity between the original image and the segmented image is considered in this experiment to verify the quality of segmented image comprehensively. The relevant results indicate that the proposed method has a strong feature preserving ability as compared to other methods.

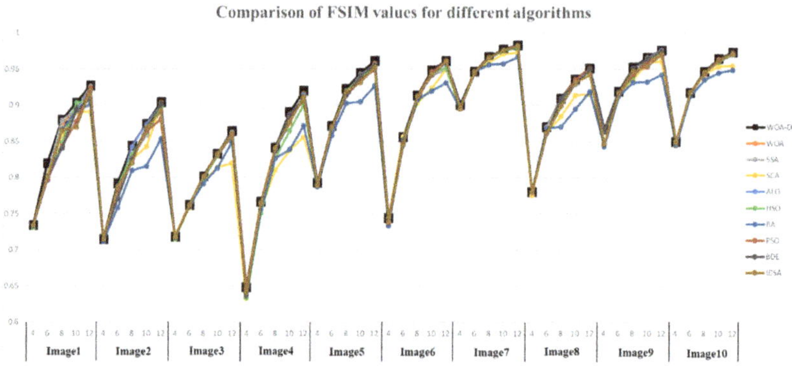

Figure 9. Comparison of FSIM values for different algorithms using Kapur's entropy at 4, 6, 8, 10, and 12 levels.

6.7. Convergence Performance

In this section, the convergence performance of all algorithms is evaluated and discussed in details. In order to reflect the performance of WOA-DE more intuitively, the convergence curves of Kapur's entropy function (for $K = 12$) are shown in Figure 10. Four different images are selected for testing, namely "Image1", "Image4", "Image7", and "Image10". It can be found that the proposed algorithm outperforms other algorithms in general. In other words, the WOA-DE-based method gives higher position curves using Kapur's entropy technique.

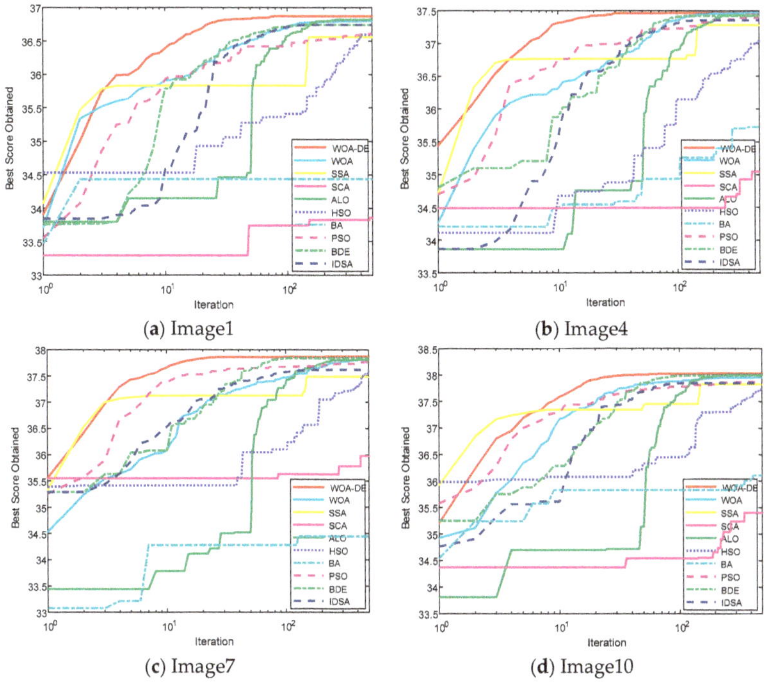

(a) Image1 (b) Image4 (c) Image7 (d) Image10

Figure 10. The convergence curves for fitness function using Kapur's entropy method at 12 levels thresholding.

As discussed above, the main drawbacks of the standard WOA are premature convergence and unbalanced exploration-exploitation, which are clearly reflected in the curves. For example, under the circumstance of "Image1" segmentation, the objective function value of WOA is almost never updated after 100 iterations, while the optimal value obtained is not the best. This phenomenon illustrates the premature convergence shortcoming of WOA. However, the proposed WOA-DE algorithm gives the highest objective function value under the premise of ensuring the convergence speed. In fact, the remarkable performance of the proposed algorithm is not only reflected in the segmentation task of "Image1", but also in other images. The experimental results in this section indicate that WOA-DE algorithm can better balance the exploration and exploitation, and the complex image segmentation tasks are also competent.

6.8. Computation Time

The average CPU time of different algorithms considering all cases is given in Table 4. It can be found from the table that HSO is the fastest among available methods, but the segmentation accuracy discussed above is not ideal. The standard WOA algorithm gives competitive results in some cases, and the proposed algorithm namely WOA-DE is slightly slower than the standard WOA. The reason for this phenomenon is the premature convergence of HSO algorithm, which cannot well balance exploration and exploitation. On the contrary, the WOA-DE algorithm combines the advantages of both WOA and DE, which determine the most appropriate threshold value, despite not being the fastest. To sum up, WOA-DE is a high-performance hybrid algorithm that improves segmentation precision while maintaining runtime.

Table 4. The average computation time (s) considering all images under different threshold levels.

K	WOA-DE	WOA	SSA	SCA	ALO	HSO	BA	PSO	BDE	IDSA
4	1.40087	1.047	1.49062	1.49438	7.8046	1.03739	1.97122	1.70887	2.21335	1.41216
6	1.55259	1.14527	1.63902	1.62171	9.66773	1.10338	2.08452	1.88491	2.40389	1.5397
8	1.67041	1.18449	1.72857	1.6764	12.09074	1.18478	2.31804	1.99103	2.48257	1.56885
10	1.74287	1.24446	1.79294	1.86849	15.31865	1.23933	2.36836	2.13532	2.58595	1.67435
12	1.88104	1.39335	1.95442	1.98369	17.19651	1.30339	2.513	2.23487	2.74791	1.70745

6.9. Statistical Analysis

In this section, a non-parametric statistical test known as "Wilcoxon's rank sum test" is used to evaluate the significant difference between algorithms [56]. The experiments are conducted 30 runs at significance level 5%. All experimental data obtained based on Kapur's entropy are used for testing. The alternative hypothesis (H_1) assumes that there is a significant difference between the two algorithms being compared. The null hypothesis H_0 considers that there is no significant difference between the algorithms. The results of the statistical experiments are given in Table 5.

Table 5. Wilcoxon's rank sum test results.

Comparison	*p*-Value
WOA-DE versus WOA	2.3197×10^{-4}
WOA-DE versus SSA	9.0193×10^{-8}
WOA-DE versus SCA	6.8546×10^{-7}
WOA-DE versus ALO	4.2264×10^{-10}
WOA-DE versus HSO	7.6791×10^{-7}
WOA-DE versus BA	3.2115×10^{-9}
WOA-DE versus PSO	7.6473×10^{-8}
WOA-DE versus BDE	4.5474×10^{-5}
WOA-DE versus IDSA	7.0546×10^{-4}

It can be observed from the table that the *p*-values acquired are far less than 0.05. This promising result indicates that H_0 can be rejected in all cases and there is a significant difference between the proposed algorithm and other methods.

6.10. Comparison of Otsu and Kapur's Entropy Methods

In order to obtain a simple and powerful technique for color image segmentation, an experiment of comparison between Otsu and Kapur's entropy thresholding techniques based on WOA-DE is conducted in this section. More details of Otsu thresholding technique can be found in [11].

The PSNR, SSIM, and FSIM values obtained by WOA-DE-based method are given in Table 6. It can be seen that WOA-DE-based method using Kapur's entropy gives higher values than using Otsu technique in general for PSNR values. However, the Otsu-based technique performs better when comparing SSIM values. Considering the FSIM indicator, these two thresholding techniques are equal. Precisely speaking, on comparing the PSNR values, the Otsu technique presents better results in 11 out of 50 cases (10 images and five thresholds), whereas, Kapur's entropy technique gives better results in 39 out of 50 cases. Considering other two indicators, the Kapur's entropy technique outperforms in 21 cases for SSIM and 25 cases for FSIM, while the Otsu technique outperforms in 29 cases for SSIM and 25 cases for FSIM. To sum up, the WOA-DE-based method through Otsu gives better results in 65 out of 150 cases (10 images, five thresholds, and three performance measures) and the WOA-DE-based method through Kapur's entropy gives satisfactory results in 85 cases. To some extent, these satisfactory results prove that WOA-DE-based method using Kapur's entropy is superior to the method using Otsu. However, as the no free lunch (NFL) theorem goes, there is no technique that can handle all image segmentation tasks [57]. Thus, the WOA-DE algorithm based on different thresholding techniques has potential in the field of color image segmentation, which may exhibit superior performance in some engineering problems that have not been solved so far.

Table 6. Comparison of Kapur's entropy and Otsu methods based on WOA-DE algorithm.

Images	K	PSNR		SSIM		FSIM	
		Otsu	Kapur	Otsu	Kapur	Otsu	Kapur
Image1	4	20.3428	17.7781	0.5798	0.4681	0.7771	0.734
	6	22.6702	24.977	0.6815	0.6559	0.8458	0.8197
	8	24.0516	28.6092	0.7446	0.8033	0.9122	0.8798
	10	25.2164	30.6687	0.7898	0.833	0.9225	0.9039
	12	26.1897	32.2054	0.8059	0.8672	0.926	0.9271
Image2	4	18.459	18.6558	0.608	0.5266	0.7582	0.7151
	6	20.9182	22.2481	0.7095	0.652	0.8245	0.7921
	8	24.5622	24.8821	0.8164	0.7361	0.8684	0.8435
	10	25.7585	27.9116	0.8421	0.8064	0.8878	0.8745
	12	28.4144	29.8805	0.8964	0.8505	0.917	0.9041
Image3	4	17.5776	20.8247	0.6971	0.7109	0.6972	0.7182
	6	22.7555	23.6059	0.7431	0.7592	0.7469	0.7619
	8	27.8967	26.0132	0.7948	0.8036	0.795	0.8017
	10	29.5405	29.0184	0.8341	0.8433	0.8322	0.8329
	12	31.6891	32.6886	0.8633	0.8729	0.8619	0.8646
Image4	4	19.0015	23.013	0.6151	0.612	0.7012	0.6484
	6	24.4296	26.9872	0.7631	0.7188	0.8129	0.7665
	8	27.9781	29.7682	0.8434	0.7953	0.8793	0.8415
	10	32.0713	31.7603	0.8888	0.8456	0.9193	0.8907
	12	33.9227	33.096	0.9194	0.8767	0.9431	0.9203
Image5	4	23.4509	23.4495	0.8082	0.7231	0.8469	0.7925
	6	27.2396	27.2417	0.8948	0.8286	0.9142	0.8716
	8	29.6073	29.5852	0.926	0.8903	0.9415	0.9229
	10	31.5024	31.4704	0.9345	0.919	0.9585	0.9452
	12	32.9105	32.782	0.9457	0.9429	0.9672	0.9614
Image6	4	19.2192	20.597	0.6626	0.5995	0.7712	0.7443
	6	23.4934	25.4883	0.8061	0.7587	0.8673	0.8562
	8	27.6467	28.6898	0.8732	0.8427	0.9192	0.9136
	10	29.7289	31.292	0.9104	0.9002	0.9416	0.9489
	12	32.0406	32.7058	0.9384	0.9227	0.9599	0.9615
Image7	4	18.9474	23.2367	0.7898	0.8494	0.848	0.9012
	6	23.6742	26.6481	0.8938	0.9136	0.9198	0.9469
	8	26.8294	29.1886	0.9383	0.9422	0.9513	0.9671
	10	30.559	31.2154	0.9626	0.9648	0.9728	0.9773
	12	32.9021	32.7203	0.9781	0.9726	0.9828	0.9834
Image8	4	20.3695	19.4801	0.5372	0.4881	0.786	0.7807
	6	23.4982	25.5717	0.6365	0.7043	0.8643	0.8705
	8	25.5399	27.8173	0.7326	0.7823	0.9007	0.9102
	10	27.2326	30.6727	0.8174	0.8479	0.9228	0.9361
	12	30.4945	32.0442	0.8483	0.8849	0.943	0.9514
Image9	4	20.5858	22.1696	0.6759	0.6581	0.8498	0.8671
	6	25.1403	26.3449	0.7465	0.7492	0.9174	0.9197
	8	28.672	29.2954	0.7938	0.8082	0.9476	0.9524
	10	30.9026	31.126	0.8711	0.8716	0.9664	0.9661
	12	32.5855	32.9878	0.9012	0.8757	0.9761	0.9764
Image10	4	20.2121	22.6128	0.7399	0.7551	0.8312	0.8499
	6	24.9168	27.0397	0.8128	0.8355	0.9075	0.9179
	8	29.1254	29.5441	0.8865	0.8649	0.9503	0.947
	10	30.9532	31.447	0.9196	0.8774	0.9641	0.9645
	12	32.5129	32.8351	0.9284	0.8923	0.9729	0.9734
Rank		2(11)	1(39)	2(29)	1(21)	1(25)	1(25)

6.11. Robustness Testing on Noisy Images

In order to further investigate the performance of proposed algorithm, an experiment is conducted on two famous benchmark test images with various noise levels. "Lena" and "Peppers" images are used in this section (see Figure 11), which can be obtained from [58]. The mean value is fixed in this experiment, and the level of Gaussian noise is adjusted by setting the variance as 0.00625, 0.0125, 0.025, 0.05, and 0.1, respectively. The experiment is carried out at 12 threshold level, in which case the difference between algorithms is the most obvious. The relevant results are presented in Figures 12–15. It can be observed from the results that the value of performance measures and quality of segmented image decrease with the increase of noise level, and the WOA-DE-Kapur outperforms other methods using Kapur entropy. The promising results indicate that the proposed technique has strong robustness, which can be competent for complex image segmentation tasks with noise.

(**a**) "Lena" image (**b**) "Peppers" image

Figure 11. Original "Lena" and "Peppers" images from Berkeley Segmentation Dataset.

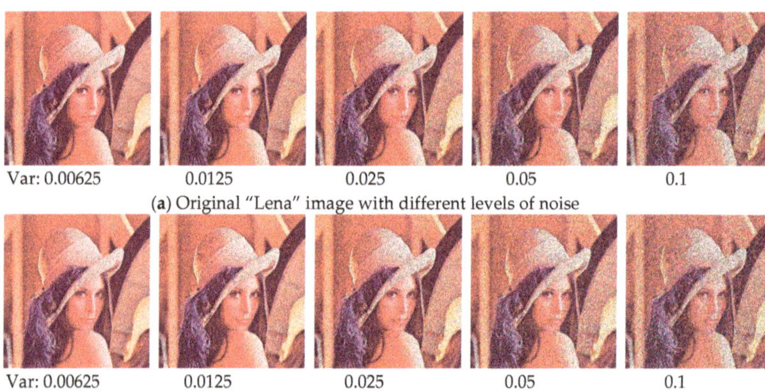

Var: 0.00625 0.0125 0.025 0.05 0.1

(**a**) Original "Lena" image with different levels of noise

Var: 0.00625 0.0125 0.025 0.05 0.1

(**b**) Segmented "Lena" image with different levels of noise

Figure 12. The original "Lena" image and the corresponding segmented results under various noise levels.

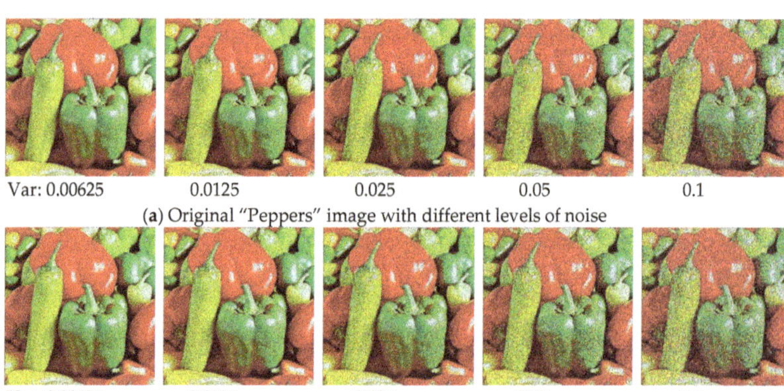

Var: 0.00625 0.0125 0.025 0.05 0.1

(**a**) Original "Peppers" image with different levels of noise

Var: 0.00625 0.0125 0.025 0.05 0.1

(**b**) Segmented "Peppers" image with different levels of noise

Figure 13. The original "Peppers" image and the corresponding segmented results under various noise levels.

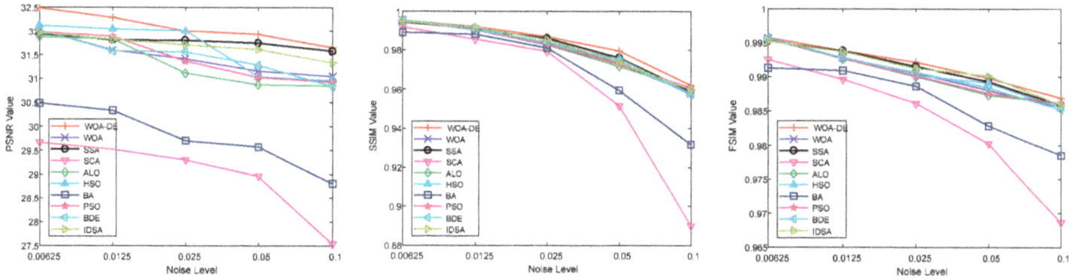

(**a**) PSNR (**b**) SSIM (**c**) FSIM

Figure 14. The value of various performance measures over "Lena" image with different levels of noise.

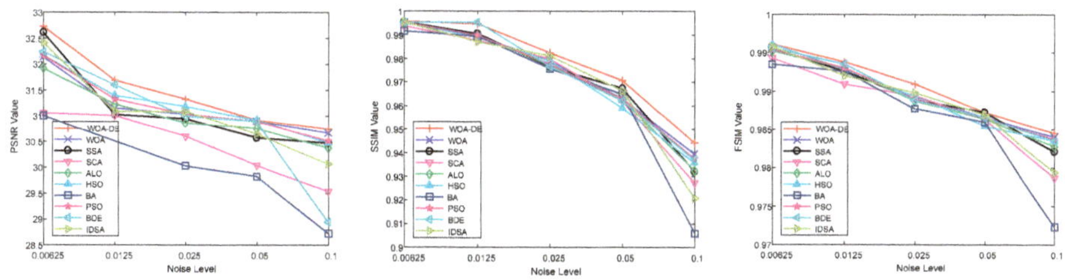

(**a**) PSNR (**b**) SSIM (**c**) FSIM

Figure 15. The value of various performance measures over "Peppers" image with different levels of noise.

6.12. Application in MR Image

In this section, the WOA-DE-Kapur-based multilevel thresholding technique is applied to the field of MR image segmentation. The purpose of this experiment is to investigate whether the proposed algorithm is capable of producing high quality segmented MR images. Two other threshold-based MR image segmentation techniques are used for comparison, namely the crow search algorithm-based method using minimum cross entropy thresholding (CSA-MCET) [59] and adaptive bacterial foraging algorithm-based method using Otsu (ABF-Otsu) [60]. The combination of thresholds (K = 2, 3, 4, and 5) selected is the same as that used by above two algorithms in their corresponding articles. Besides, the parameter values are set according to the original literature, except for the population size N set to 30 and the number of iterations $t_{\max}$ set to 500 for fair comparison. All experiments are performed 30 times to eliminate errors.

The experimental results are shown in three tables. Table 7 presents the optimal thresholds and PSNR values, Table 8 gives the SSIM and FSIM values, and Table 9 indicates the segmented images obtained by all methods. It can be found from these results that WOA-DE-Kapur method can determine more accurate thresholds compared to other methods. For quantitative analysis, the values of performance measures obtained by proposed method is higher, which indicate the better quality of segmented image. For visual analysis, WOA-DE-Kapur method gives more informative segmented MR images, and the details of image become more prominent as the number of thresholds increases.

Since the experiments of three methods are the same, it is necessary to carry out relevant statistical tests. In this section, Friedman test [61] and Wilcoxon's rank sum test [56] are used as non-parametric statistical test to evaluate the performance of these methods considering 5% as significant level. Null hypothesis (H_0) in Friedman test states equality of medians between the algorithms, and the alternative hypothesis (H_1) indicates the difference. A more detailed description of Friedman test can be found in literature [62]. The results of the relevant statistical tests can be observed in Tables 10 and 11. Table 10 presents the average rank and p-value of all algorithms at different threshold levels. As can be found, ABF-Otsu obtains the first rank for K = 3, and WOA-DE-Kapur provides the first rank in other cases.

In other words, the proposed technique gives the best result in general. The p-value for all threshold levels is very small indicating the significant difference among available methods. Table 11 gives the result of Wilcoxon's rank sum test. It can be observed that the p-value is less than 0.05 in most cases, which verifies the remarkable performance of WOA-DE-Kapur technique in a statistical and meaningful way.

Table 7. Comparison of Optimal threshold and PSNR value obtained by WOA-DE-Kapur, ABF-Otsu, and CSA-MCET.

Images	K	Optimal Threshold Value			PSNR		
		WOA-DE-Kapur	ABF-Otsu	CSA-MCET	WOA-DE-Kapur	ABF-Otsu	CSA-MCET
Slice20	2	94 167	28 97	13 84	16.8586	16.524	15.9746
	3	9 118 219	29 87 151	18 64 134	23.9008	23.1061	22.4605
	4	8 29 129 210	7 53 100 153	16 64 98 147	24.6228	25.4972	24.3967
	5	16 36 94 171 211	21 54 98 156 190	3 40 61 113 150	30.4912	27.3411	28.5034
Slice24	2	111 182	48 145	19 118	19.7345	21.0839	20.8004
	3	34 117 182	40 108 172	7 56 136	23.4428	22.9913	23.5030
	4	17 73 129 193	23 70 118 182	6 50 101 161	26.7848	26.2061	24.7095
	5	14 70 115 165 210	20 63 102 143 196	4 27 66 111 170	28.9204	28.3318	25.3871
Slice28	2	114 179	52 151	20 121	19.6991	18.6884	19.1865
	3	20 81 156	46 110 175	7 56 139	24.8983	24.3616	23.7032
	4	22 78 137 192	27 76 126 187	6 48 103 161	26.9455	27.0419	25.8075
	5	13 72 117 157 203	23 68 109 149 203	6 36 74 115 174	29.6822	29.1884	28.1382
Slice32	2	115 175	53 159	20 137	23.3496	22.888	22.6576
	3	16 76 143	50 120 189	8 54 148	24.711	23.2735	25.9537
	4	16 74 131 186	21 70 122 191	7 52 107 172	27.5852	27.958	27.947
	5	18 71 118 162 205	19 63 105 147 206	3 28 67 116 180	29.7914	28.6183	29.598
Rank		—	—	—	1(10)	2(4)	3(2)

Table 8. Comparison of SSIM and FSIM value obtained by WOA-DE-Kapur, ABF-Otsu, and CSA-MCET.

Images	K	SSIM			FSIM		
		WOA-DE-Kapur	ABF-Otsu	CSA-MCET	WOA-DE-Kapur	ABF-Otsu	CSA-MCET
Slice20	2	0.7923	0.7726	0.7882	0.8743	0.8565	0.8421
	3	0.8784	0.8061	0.8811	0.9411	0.9305	0.9594
	4	0.9225	0.8408	0.9208	0.9608	0.9614	0.9599
	5	0.9435	0.8862	0.9249	0.9882	0.9674	0.9723
Slice24	2	0.6809	0.7886	0.7865	0.7772	0.8178	0.8117
	3	0.8391	0.8318	0.8343	0.8686	0.8660	0.8394
	4	0.8791	0.8770	0.8742	0.9081	0.9026	0.8944
	5	0.9015	0.8959	0.8997	0.9277	0.9253	0.9099
Slice28	2	0.7832	0.7678	0.7792	0.813	0.8394	0.8274
	3	0.8365	0.8238	0.8275	0.8849	0.8846	0.8585
	4	0.8672	0.8687	0.8691	0.9084	0.9136	0.9156
	5	0.8993	0.8937	0.9010	0.9371	0.9355	0.9366
Slice32	2	0.8123	0.7973	0.7862	0.8617	0.8388	0.8589
	3	0.8465	0.832	0.8513	0.8864	0.8943	0.9009
	4	0.8794	0.8824	0.8784	0.9199	0.9271	0.9275
	5	0.9023	0.8705	0.8991	0.9477	0.9237	0.9347
Rank		1(10)	3(2)	2(4)	1(9)	3(3)	2(4)

Table 9. The segmented MRI for different algorithms at 2, 3, 4, and 5 levels.

K	WOA-DE-Kapur	ABF-Otsu	CSA-MCET	WOA-DE-Kapur	ABF-Otsu	CSA-MCET
	Slice20			Slice24		
2	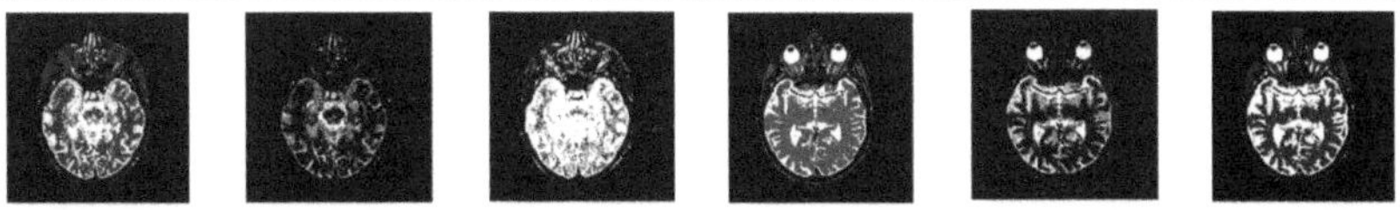					
3	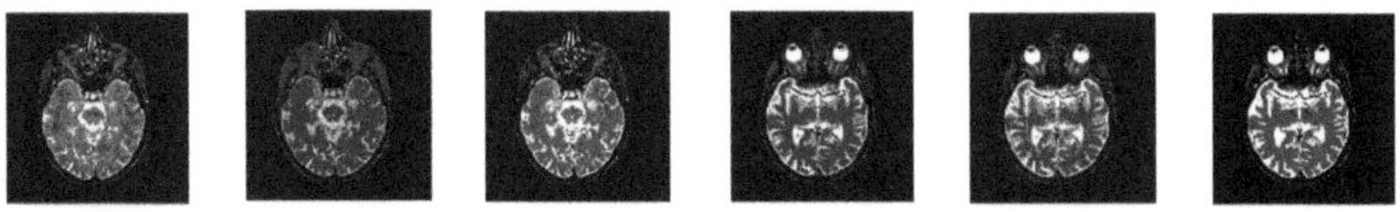					
4	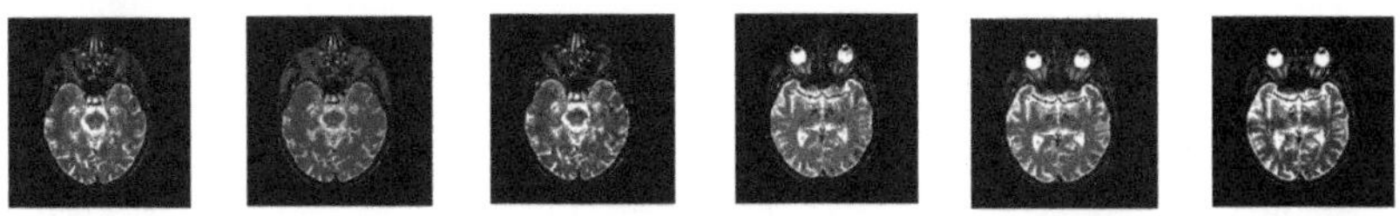					
5	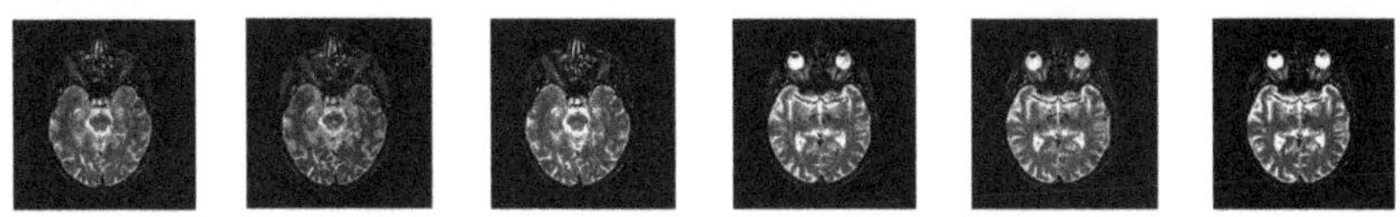					
K	Slice28			Slice32		
2	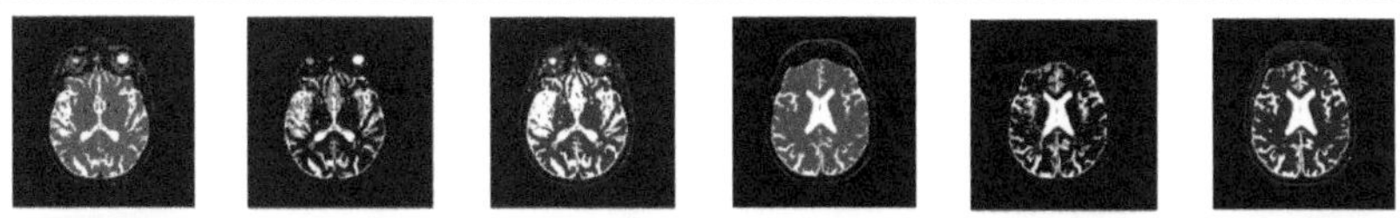					
3	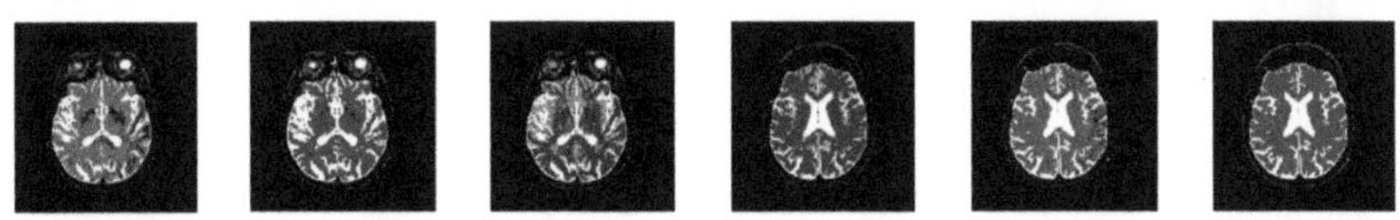					
4	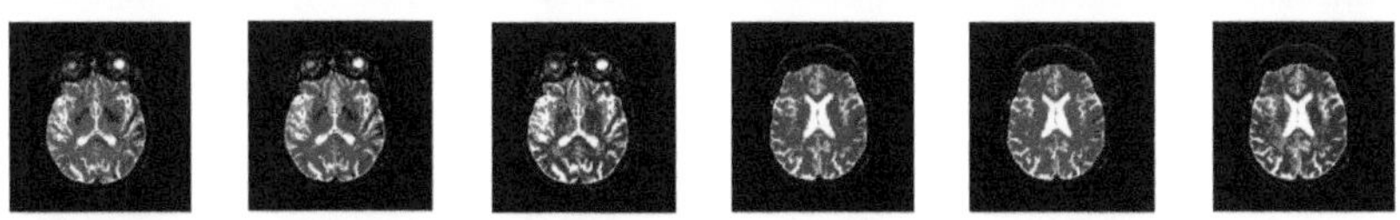					
5	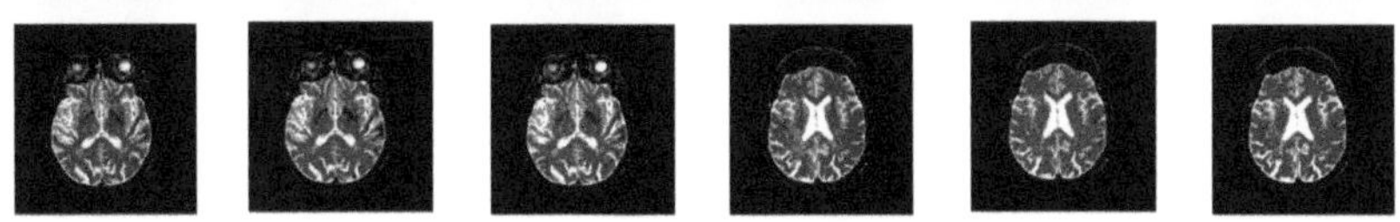					

Table 10. Friedman test for WOA-DE-Kapur, ABF-Otsu, and CSA-MCET on MR images.

K	Average Rank			*p*-Value
	WOA-DE-Kapur	ABF-Otsu	CSA-MCET	
2	1.6667	2.0000	2.3333	2.2619×10^{-7}
3	1.5833	2.5833	1.8333	1.1603×10^{-8}
4	2.0000	1.6667	2.3333	7.2217×10^{-9}
5	1.0833	2.7500	2.1667	5.3467×10^{-9}

Table 11. Wilcoxon's rank sum test for WOA-DE-Kapur, ABF-Otsu, and CSA-MCET on MR images.

K	WOA-DE-Kapur vs. ABF-Otsu		WOA-DE-Kapur vs. CSA-MCET	
	p-Value	*h*	*p*-Value	*h*
2	< 0.05	1	< 0.05	1
3	< 0.05	1	0.0926	0
4	< 0.05	1	< 0.05	1
5	< 0.05	1	< 0.05	1

7. Conclusions

In order to obtain an efficient technique for color image segmentation, an improved WOA-based method is introduced in this paper, which is known as WOA-DE. In the proposed algorithm, DE is adopted as a local search strategy with the purpose of enhancing exploitation capability. Compared to the traditional WOA, the WOA-DE algorithm can effectively avoid falling into a local optimum and prevent the loss of population diversity in the later iterations. A series of experiments have been conducted on various color images including natural images and satellite images. Seven meta-heuristic algorithms are utilized for comparison. The experimental results indicate that the proposed techniques outperform other methods in terms of average fitness values, standard deviation (STD), peak signal to noise ratio (PSNR), structural similarity index (SSIM), and feature similarity index (FSIM) as well as the Wilcoxon's rank sum test. In addition, to give more convincing and reliable results, another thresholding technique namely Otsu is adopted for testing. The experimental results indicate that WOA-DE-based technique through Kapur's entropy gives better results than using the Otsu technique in most cases. However, there is no technique that can handle all image segmentation tasks. Thus, it is necessary to introduce more and better techniques to meet the requirements of different image segmentation problems and this is also the motivation for our future research. The performance of some novel meta-heuristic algorithms will be evaluated in this domain, such as salp swarm algorithm, spotted hyena optimizer, emperor penguin optimizer, etc.

Author Contributions: C.L. and H.J. contributed to the idea of this paper; C.L. performed the experiments; C.L. wrote the paper; C.L. and H.J. contributed to the revision of this paper.

Funding: This research received no external funding.

Acknowledgments: The authors would like to thank the anonymous reviewers for their constructive comments and suggestions.

Conflicts of Interest: The authors declare no conflict of interest.

References

1. Qian, P.; Zhao, K.; Jiang, Y.; Su, K.; Deng, Z.; Wang, S.; Muzic, R.F. Knowledge-leveraged transfer fuzzy C-Means for texture image segmentation with self-adaptive cluster prototype matching. *Knowl.-Based Syst.* **2017**, *130*, 33–50. [CrossRef] [PubMed]
2. Robert, M.; Carsten, M.; Olivier, E.; Jochen, P.; Noordhoek, N.J.; Aravinda, T.; Reddy, V.Y.; Chan, R.C.; Jürgen, W. Automatic Segmentation of Rotational X-Ray Images for Anatomic Intra-Procedural Surface Generation in Atrial Fibrillation Ablation Procedures. *IEEE Trans. Med. Imaging* **2010**, *29*, 260–272.
3. Lee, S.H.; Koo, H.I.; Cho, N.I. Image segmentation algorithms based on the machine learning of features. *Pattern Recognit. Lett.* **2010**, *31*, 2325–2336.
4. Ye, J.; Fu, G.; Poudel, U.P. High-accuracy edge detection with Blurred Edge Model. *Image Vis. Comput.* **2005**, *23*, 453–467.
5. Khairuzzaman, A.K.M.; Chaudhury, S. Multilevel thresholding using grey wolf optimizer for image segmentation. *Expert Syst. Appl.* **2017**, *86*, 64–76. [CrossRef]
6. Chen, W.; Yue, H.; Wang, J.; Wu, X. An improved edge detection algorithm for depth map inpainting. *Opt. Lasers Eng.* **2014**, *55*, 69–77. [CrossRef]

7. Liu, K.; Guo, L.; Li, H.; Chen, J. Fusion of Infrared and Visible Light Images Based on Region Segmentation. *Chin. J. Aeronaut.* **2009**, *22*, 75–80.
8. Fu, X.; Liu, T.; Xiong, Z.; Smaill, B.H.; Stiles, M.K.; Zhao, J. Segmentation of histological images and fibrosis identification with a convolutional neural network. *Comput. Biol. Med.* **2018**, *98*, 147–158. [CrossRef] [PubMed]
9. Demirhan, A.; Törü, M.; Güler, İ. Segmentation of Tumor and Edema Along With Healthy Tissues of Brain Using Wavelets and Neural Networks. *IEEE J. Biomed. Health Inf.* **2015**, *19*, 1451–1458. [CrossRef] [PubMed]
10. Ouadfel, S.; Taleb-Ahmed, A. Social spiders optimization and flower pollination algorithm for multilevel image thresholding: A performance study. *Expert Syst. Appl.* **2016**, *55*, 566–584. [CrossRef]
11. Otsu, N. A threshold selection method from gray-level histograms. *IEEE Trans. Syst. Man Cybern.* **1979**, *9*, 62–66. [CrossRef]
12. Kapura, J.N.; Sahoob, P.K.; Wongc, A.K.C. A new method for gray-level picture thresholding using the entropy of the histogram. *Comput. Vis. Graph. Image Proc.* **1985**, *29*, 273–285. [CrossRef]
13. Shen, L.; Fan, C.; Huang, X. Multi-Level Image Thresholding Using Modified Flower Pollination Algorithm. *IEEE Access* **2018**, *6*, 30508–30519. [CrossRef]
14. Sambandam, R.K.; Jayaraman, S. Self-adaptive dragonfly based optimal thresholding for multilevel segmentation of digital images. *J. King Saud Univ. Comput. Inf. Sci.* **2018**, *30*, 449–461. [CrossRef]
15. Gao, H.; Fu, Z.; Pun, C.; Hu, H.; Lan, R. A multi-level thresholding image segmentation based on an improved artificial bee colony algorithm. *Comput. Electr. Eng.* **2018**, *70*, 931–938. [CrossRef]
16. He, L.; Huang, S. Modified firefly algorithm based multilevel thresholding for color image segmentation. *Neurocomputing* **2017**, *240*, 152–174. [CrossRef]
17. Pare, S.; Kumar, A.; Bajaj, V.; Singh, G.K. An efficient method for multilevel color image thresholding using cuckoo search algorithm based on minimum cross entropy. *Appl. Soft Comput.* **2017**, *61*, 570–592. [CrossRef]
18. Kotte, S.; Pullakura, R.K.; Injeti, S.K. Optimal multilevel thresholding selection for brain MRI image segmentation based on adaptive wind driven optimization. *Measurement* **2018**, *130*, 340–361. [CrossRef]
19. Beevi, S.; Nair, M.S.; Bindu, G.R. Automatic segmentation of cell nuclei using Krill Herd optimization based multi-thresholding and Localized Active Contour Model. *Biocybern. Biomed. Eng.* **2016**, *36*, 584–596. [CrossRef]
20. Aziz, M.A.E.; Ewees, A.A.; Hassanien, A.E. Whale Optimization Algorithm and Moth-Flame Optimization for multilevel thresholding image segmentation. *Expert Syst. Appl.* **2017**, *83*, 242–256. [CrossRef]
21. Mirjalili, S.; Lewis, A. The Whale Optimization Algorithm. *Adv. Eng. Softw.* **2016**, *95*, 51–67. [CrossRef]
22. Oliva, D.; Aziz, M.A.E.; Hassanien, A.E. Parameter estimation of photovoltaic cells using an improved chaotic whale optimization algorithm. *Appl. Energy* **2017**, *200*, 141–154. [CrossRef]
23. Xiong, G.; Zhang, J.; Shi, D.; He, Y. Parameter extraction of solar photovoltaic models using an improved whale optimization algorithm. *Energy Convers. Manag.* **2018**, *174*, 388–405. [CrossRef]
24. Sun, Y.; Wang, X.; Chen, Y.; Liu, Z. A modified whale optimization algorithm for large-scale global optimization problems. *Expert Syst. Appl.* **2018**, *114*, 563–577. [CrossRef]
25. Mafarja, M.M.; Mirjalili, S. Hybrid Whale Optimization Algorithm with simulated annealing for feature selection. *Neurocomputing* **2017**, *260*, 302–312. [CrossRef]
26. Pare, S.; Kumar, A.; Bajaj, V.; Singh, G.K. A multilevel color image segmentation technique based on cuckoo search algorithm and energy curve. *Appl. Soft Comput.* **2016**, *47*, 76–102. [CrossRef]
27. Díaz-Cortés, M.; Ortega-Sánchez, N.; Hinojosa, S.; Oliva, D.; Cuevas, E.; Rojas, R.; Demin, A. A multi-level thresholding method for breast thermograms analysis using Dragonfly algorithm. *Infrared Phys. Technol.* **2018**, *93*, 346–361. [CrossRef]
28. Ewees, A.A.; Elaziz, M.A.; Oliva, D. Image segmentation via multilevel thresholding using hybrid optimization algorithms. *J. Electron. Imaging* **2018**, *27*, 1. [CrossRef]
29. Bhandari, A.K.; Kumar, A.; Singh, G.K. Tsallis entropy based multilevel thresholding for colored satellite image segmentation using evolutionary algorithms. *Expert Syst. Appl.* **2015**, *42*, 8707–8730. [CrossRef]
30. Sathya, P.D.; Kayalvizhi, R. Modified bacterial foraging algorithm based multilevel thresholding for image segmentation. *Eng. Appl. Artif. Intell.* **2011**, *42*, 595–615. [CrossRef]
31. Manikandan, S.; Ramar, K.; Iruthayarajan, M.W.; Srinivasagan, K.G. Multilevel thresholding for segmentation of medical brain images using real coded genetic algorithm. *Measurement* **2014**, *47*, 558–568. [CrossRef]

32. Bhandari, A.K.; Singh, V.K.; Kumar, A.; Singh, G.K. Cuckoo search algorithm and wind driven optimization based study of satellite image segmentation for multilevel thresholding using Kapur's entropy. *Expert Syst. Appl.* **2014**, *41*, 3538–3560. [CrossRef]
33. Pare, S.; Bhandari, A.K.; Kumar, A.; Singh, G.K. A new technique for multilevel color image thresholding based on modified fuzzy entropy and Lévy flight firefly algorithm. *Comput. Electr. Eng.* **2018**, *70*, 476–495. [CrossRef]
34. Ibrahim, R.A.; Elaziz, M.A.; Lu, S. Chaotic opposition-based grey-wolf optimization algorithm based on differential evolution and disruption operator for global optimization. *Expert Syst. Appl.* **2018**, *108*, 1–27. [CrossRef]
35. Zorlu, H. Optimization of weighted myriad filters with differential evolution algorithm. *AEU Int. J. Electron. Commun.* **2017**, *77*, 1–9. [CrossRef]
36. Lin, Q.; Zhu, Q.; Huang, P.; Chen, J.; Ming, Z.; Yu, J. A novel hybrid multi-objective immune algorithm with adaptive differential evolution. *Comput. Oper. Res.* **2015**, *62*, 95–111. [CrossRef]
37. Jadon, S.S.; Tiwari, R.; Sharma, H.; Bansal, J.C. Hybrid Artificial Bee Colony algorithm with Differential Evolution. *Appl. Soft Comput.* **2017**, *58*, 11–24. [CrossRef]
38. Yüzgeç, U.; Eser, M. Chaotic based differential evolution algorithm for optimization of baker's yeast drying process. *Egypt. Inf. J.* **2018**, *19*, 151–163. [CrossRef]
39. The Berkeley Segmentation Dataset and Benchmark. Available online: https://www2.eecs.berkeley.edu/Research/Projects/CS/vision/grouping/segbench/ (accessed on 15 June 2018).
40. Landsat Imagery Courtesy of NASA Goddard Space Flight Center and U.S. Geological Survey. Available online: https://landsat.visibleearth.nasa.gov/index.php?&p=1 (accessed on 17 October 2018).
41. Harvard Medical School. Available online: http://www.med.harvard.edu/AANLIB/ (accessed on 22 December 2018).
42. Mirjalili, S.; Gandomi, A.H.; Mirjalili, S.Z.; Saremi, S.; Faris, H.; Mirjalili, S.M. Salp Swarm Algorithm: A bio-inspired optimizer for engineering design problems. *Adv. Eng. Softw.* **2017**, *114*, 163–191. [CrossRef]
43. Mirjalili, S. SCA: A Sine Cosine Algorithm for solving optimization problems. *Knowl.-Based Syst.* **2016**, *96*, 120–133. [CrossRef]
44. Mirjalili, S. The Ant Lion Optimizer. *Adv. Eng. Softw.* **2015**, *83*, 80–98. [CrossRef]
45. Geem, Z.W.; Kim, J.H.; Loganathan, G.V. A new heuristic optimization algorithm: Harmony search. *Simulation* **2001**, *76*, 60–68. [CrossRef]
46. Ye, Z.; Wang, M.; Liu, W.; Chen, S. Fuzzy entropy based optimal thresholding using bat algorithm. *Appl. Soft Comput.* **2015**, *31*, 381–395. [CrossRef]
47. Kennedy, J.; Eberhart, R.C. Particle swarm optimization. In Proceedings of the IEEE International Conference on Neural Networks, Perth, Australia, 27 November–1 December 1995; Volume 4, pp. 1942–1948.
48. Li, Y.; Bai, X.; Jiao, L.; Xue, Y. Partitioned-cooperative quantum-behaved particle swarm optimization based on multilevel thresholding applied to medical image segmentation. *Appl. Soft Comput.* **2017**, *56*, 345–356. [CrossRef]
49. Bhandari, A.K. A novel beta differential evolution algorithm-based fast multilevel thresholding for color image segmentation. *Neural Comput. Appl.* **2018**, 1–31. [CrossRef]
50. Kotte, S.; Kumar, P.R.; Injeti, S.K. An efficient approach for optimal multilevel thresholding selection for gray scale images based on improved differential search algorithm. *Ain Shams Eng. J.* **2018**, *9*, 1043–1067. [CrossRef]
51. de Albuquerque, M.P.; Esquef, I.A.; Mello, A.R.G.; de Albuquerque, M.P. Image thresholding using Tsallis entropy. *Pattern Recognit. Lett.* **2004**, *25*, 1059–1065.
52. John, J.; Nair, M.S.; Kumar, P.R.A.; Wilscy, M. A novel approach for detection and delineation of cell nuclei using feature similarity index measure. *Biocybern. Biomed. Eng.* **2016**, *36*, 76–88. [CrossRef]
53. Wang, Z.; Bovik, A.C.; Sheikh, H.R.; Simoncelli, E.P. Image quality assessment: From error visibility to structural similarity. *IEEE Trans. Image Process.* **2004**, *13*, 600–612. [PubMed]
54. Pare, S.; Bhandari, A.K.; Kumar, A.; Singh, G.K. An optimal color image multilevel thresholding technique using grey-level co-occurrence matrix. *Expert Syst. Appl.* **2017**, *87*, 335–362. [CrossRef]
55. Zhang, L.; Zhang, L.; Mou, X.; Zhang, D. FSIM: A Feature Similarity Index for Image Quality Assessment. *IEEE Trans. Image Process.* **2011**, *20*, 2378–2386. [CrossRef] [PubMed]

56. Frank, W. Individual Comparisons of Grouped Data by Ranking Methods. *J. Econ. Entomol.* **1946**, *39*, 269–270.
57. Wolpert, D.H.; Macready, W.G. No free lunch theorems for optimization. *Evolut. Comput. IEEE Trans.* **1997**, *1*, 67–82.
58. The USC-SIPI Image Database. Available online: http://sipi.usc.edu/database/ (accessed on 7 December 2018).
59. Oliva, D.; Hinojosa, S.; Cuevas, E.; Pajares, G.; Avalos, O.; Gálvez, J. Cross entropy based thresholding for magnetic resonance brain images using Crow Search Algorithm. *Expert Syst. Appl.* **2017**, *79*, 164–180. [CrossRef]
60. Sathya, P.D.; Kayalvizhi, R. Optimal segmentation of brain MRI based on adaptive bacterial foraging algorithm. *Neurocomputing* **2011**, *74*, 2299–2313. [CrossRef]
61. Friedman, M. The use of ranks to avoid the assumption of normality implicit in the analysis of variance. *J. Am. Stat. Assoc.* **1937**, *32*, 676–701. [CrossRef]
62. Derrac, J.; García, S.; Molina, D.; Herrera, F. A practical tutorial on the use of nonparametric statistical tests as a methodology for comparing evolutionary and swarm intelligence algorithms. *Swarm Evol. Comput.* **2011**, *1*, 3–18. [CrossRef]

entropy

MDPI

Article

Image Encryption Based on Pixel-Level Diffusion with Dynamic Filtering and DNA-Level Permutation with 3D Latin Cubes

Taiyong Li [1,*], Jiayi Shi [1], Xinsheng Li [2,*], Jiang Wu [1] and Fan Pan [3]

1 School of Economic Information Engineering, Southwestern University of Finance and Economics, Chengdu 611130, China; 218081202002@smail.swufe.edu.cn (J.S.); wuj_t@swufe.edu.cn (J.W.)
2 College of Computer Science, Sichuan University, Chengdu 610064, China
3 College of Electronics and Information Engineering, Sichuan University, Chengdu 610064, China; panfan@scu.edu.cn
* Correspondence: litaiyong@gmail.com (T.L.); lixinsheng@scu.edu.cn (X.L.)

Received: 12 March 2019; Accepted: 21 March 2019; Published: 24 March 2019

Abstract: Image encryption is one of the essential tasks in image security. In this paper, we propose a novel approach that integrates a hyperchaotic system, pixel-level Dynamic Filtering, DNA computing, and operations on 3D Latin Cubes, namely DFDLC, for image encryption. Specifically, the approach consists of five stages: (1) a newly proposed 5D hyperchaotic system with two positive Lyapunov exponents is applied to generate a pseudorandom sequence; (2) for each pixel in an image, a filtering operation with different templates called dynamic filtering is conducted to diffuse the image; (3) DNA encoding is applied to the diffused image and then the DNA-level image is transformed into several 3D DNA-level cubes; (4) Latin cube is operated on each DNA-level cube; and (5) all the DNA cubes are integrated and decoded to a 2D cipher image. Extensive experiments are conducted on public testing images, and the results show that the proposed DFDLC can achieve state-of-the-art results in terms of several evaluation criteria.

Keywords: image encryption; dynamic filtering; DNA computing; 3D Latin cube; permutation; diffusion

1. Introduction

As one of the most important information carriers, hundreds of millions of images are generated, stored, and transmitted every day. How to ensure image security has become a very hot topic of research in recent years. Image encryption is one of the most important image security methods. Encryption algorithms can be roughly classified into two categories: symmetric key (private key) and asymmetric key (public key) algorithms. The former uses the same key for both encryption and decryption, while the latter uses a key for encryption and another key for decryption. Typical private key algorithms include data encryption standard (DES), international data encryption algorithm (IDEA), advanced encryption standard (AES) and so on. Rivest-Shamir-Adleman (RSA) and Elliptic-curve cryptography (ECC) are among the very popular public key algorithms. The symmetric key algorithms are fast, efficient, but difficult to manage keys, while the asymmetric encryption algorithms are slow but have higher security [1,2]. Due to the inherent characteristics of images such as strong correlation, high redundancy and bulky data capacity, the above mentioned encryption algorithms are usually not suitable for direct applications in images. To address this issue, a variety of image encryption algorithms have been proposed in recent years [3–6].

There are many kinds of operations for the purpose of image encryption, such as shuffling, permutation, rotation, substitution, confusion, diffusion, transposition, and so on [7]. Among the operations, diffusion and permutation are very popular ones because they can achieve good results and are easy to implement. The diffusion is to change the values of the pixels in images, while the permutation aims at changing the positions of the pixels. Some practical image encryption algorithms are capable of handling diffusion and permutation jointly. Due to the characteristics of ergodicity, pseudorandomness, unpredictability, and extreme sensitivity to initial values and parameters, chaos-based image encryption has become increasingly popular in recent years. The main idea of chaos-based image encryption is to conduct diffusion and/or permutation according to the pseudorandom sequences generated from chaotic systems [8–13]. Very recently, Flores-Vergara et al. have implemented a chaotic cryptosystem on embedded systems with multiprocessors. The NIST statistical test and the security analysis have confirmed the proposed cryptosystem is very secure and robust for image encryption [14]. Wang et al. used a spatiotemporal chaotic system to generate a pseudorandom sequence, and then used the sequence to conduct permutation and diffusion simultaneously [15]. Pareek et al. employed two chaotic logistic maps and eight different types of operations to encrypt the pixels of images, and the experimental results demonstrated the proposed scheme was real-time, efficient and secure [16]. Hua et al. put forward a new 2D Logistic-Sine-coupling map that has more complex behavior, better ergodicity, and larger chaotic range than some other 2D chaotic maps, for image encryption scheme. The experiments showed that the proposed scheme had better security performance than several state-of-the-art encryption approaches [17]. Sahari and Boukemara proposed a novel 3D chaotic map by integrating the piecewise and logistic maps for color image encryption, the experimental results showed the efficiency and safety of the proposed scheme [18]. Zhou et al. proposed a novel image encryption scheme by combining quantum 3D Arnold transform and quantum XOR operations with scaled 3D Zhongtang chaotic system [19]. Low-dimensional chaotic systems have the advantages of simple forms, only a few parameters, and easy implementation. However, such properties may make it easy to estimate the orbits and the initial parameters of the low-dimensional chaotic systems and hence the security of encryption is limited.

In a dynamical system, the Lyapunov exponent (LE) is used to measure the rate of separation of infinitesimally close trajectories [20]. If a chaotic system has at least two positive LEs, the system is said to be hyperchaotic. The image encryption algorithms with hyperchaotic systems have been demonstrated more secure [2,6,21–26]. Chai et al. used a 4D memristive hyperchaotic system to encrypt 4 compound bit planes recombined from the 24-bit planes of components R, G, and B [27]. Li et al. proposed a quantum image compression-encryption approach with quantum cosine transform and a 5D hyperchaotic system, and the experiments demonstrated that the proposed compression-encryption approach outperformed some classical image encryption approaches [28]. Zhou et al. used a 5D hyperchaotic system for quantum color image encryption. Some researchers also applied 6D or 7D hyperchaotic systems to generate hyperchaotic sequences for image encryption [6,29].

Like other tasks in signal processing, image encryption can also be conducted in both spatial or transform domain [30–35]. The encryption in spatial domain is very direct, which changes the values and/or the positions of pixels. To improve the efficiency or the effectiveness of image encryption, sometime the algorithms can be conducted on higher-level data (blocks of pixels) or lower-level data (DNA-level data and bit-level data) [36–38]. Generally speaking, for the same processing power, the lower the data level, the more pixels will be involved in encryption. Therefore, the encryption processing lower-level data usually achieves better performance of encryption [6]. In the field of image encryption, the introduction of transform domain is for the purpose of compressing images. Typical transform methods include discrete cosine transform (DCT) [39–43], Fourier transform [44–46], and wavelet transform [47–50]. With these transform methods, the image encryption can focus on the high-energy parts of the images only and discard some low-energy (zero coefficients) parts. Then the image can be recovered by decryption and corresponding reverse transform.

Some recent progress has improved the performance of image encryption. Regarding diffusion, Hua and Zhou introduced filtering, a very popular technique in image processing, into image encryption. The authors make the filtering reversible by setting the right-bottom point of the filter to "1", and they proposed an image encryption algorithm using block-based scrambling and image filtering (BSIF) with a fixed filter for all pixels [51]. Very recently, Hua et al. have extended image encryption with Josephus scrambling and filtering diffusion, where the filter is a 2×2 square with fixed values [52]. Li et al. used a 1×3 or 3×1 filter with dynamically variable values decided by a 7D hyperchaotic system for filtering (so-called dynamic filtering), and bit cuboid operations, namely DFBC, for image encryption, and the experiments demonstrated the DFBC could achieve state-of-the-art results [6]. As far as permutation is concerned, in theory, any reversible position transform can be used for image encryption. Latin squares are such popular transforms which help to achieve good results of permutation [53–55]. Xu et al. extended the use of Latin squares in image encryption, and they treated the pixel-level image as a 3D bit matrix and then conducted operations of Latin cubes on the 3D matrix, and the experimental results showed that the proposed image encryption achieves both a desirable level of security and high efficiency [56].

Motivated by the diffusion with filtering and the permutation with Latin cubes, in this paper, we propose a novel approach that integrats a hyperchaotic system, Dynamic Filtering, DNA computing, and Latin Cubes, termed as DFDLC, for image encryption. Specifically, the DFDLC consists of five stages: (1) A 5D hyperchaotic system with 2 positive LEs is applied to generate the chaotic sequences for subsequent diffusion and permutation. (2) Filters with variable values are generated from the chaotic sequences, and filtering is conducted on each pixel of the image with a different filter. That is to say, the value of each pixel is changed by a different filter. This is called pixel-level diffusion with dynamic filtering. (3) The 2D pixel plane is transformed into several DNA cubes via DNA encoding rules determined by the chaotic sequence. (4) For each DNA cube, we generate a Latin cube with the same size and then change the position of each element in the DNA cube via the Latin cube. This operation is called DNA-level permutation with 3D Latin cubes. (5) All the DNA cubes are integrated and decoded to a 2D pixel image. The main contributions of this paper are three-aspect: (1) We propose a novel image encryption using a newly found 5D hyperchaotic system. (2) Pixel-level dynamic filtering and DNA-level permutation with Latin cubes are used to improve the performance of image encryption. (3) Extensive experiments on several public images show that the DFDLC is promising for image encryption.

The remainder of this paper is structured as the following. A brief description of a 5D hyperchaotic system with two positive LEs, filtering, DNA computing and Latin square is given in Section 2. In Section 3, a novel image encryption algorithm with dynamic filtering and Latin cube transformation, namely DFDLC, is proposed in detail. Experimental results are reported and analyzed in Section 4. Finally, the paper is concluded in Section 5.

2. Preliminaries

2.1. Hyperchaotic Systems

As one of the most popular chaotic systems, the Lorenz chaotic system and its extensions are very popular in image encryption. Most recently, Wang et al. have found a new 5D autonomous hyperchaotic system with 2 positive LEs by adding feedback controllers to the Lorenz system, formulated as Equation (1) [57]:

$$\begin{cases} \dot{x_1} = x_2 \\ \dot{x_2} = -x_2 + ax_1 + bx_1^3 + cx_1x_5 \\ \dot{x_3} = x_4 \\ \dot{x_4} = -x_4 + dx_3 + ex_3^3 + fx_3x_5 \\ \dot{x_5} = -gx_5 + hx_1^2 + ix_3^2 \end{cases} \tag{1}$$

where $x_j(j = 1, 2, \cdots, 5)$ are state variables, and $(a, b, c, d, e, f, g, h, i)$ are constant parameters. There are several numerical methods to solve this system, such as Forward Euler (FE), 4th order Runge-Kutta (RK) and newly proposed trigonometric polynomials [58]. In this paper, we use the 4th order RK method with a step size of $h = 0.001$ to solve the hyperchaotic system. When the parameters $(a, b, c, d, e, f, g, h, i) = (4, -1, -1, 2, -1, 2, 0.0, 6, -1)$ and initial values $(x_1^0, x_2^0, x_3^0, x_4^0, x_5^0) = (1.618, 3.14, 2.718, 4.6692, 0.618) \times 10^{-2}$, the attractors of the 5D hyperchaotic system are shown in Figure 1.

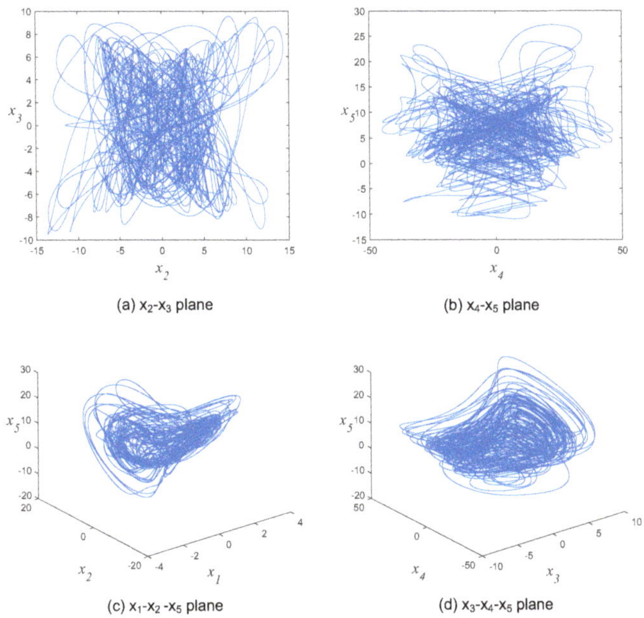

Figure 1. The attractors of the 5D hyperchaotic system.

2.2. Filtering

Filtering, also termed as convolution, is a very popular operation in the field of image processing, which can be applied to denoising, smoothing, and sharpening images by changing the values of pixels. Typically, the operation of filtering is to do convolution between a mask, also known as a kernel/filter/window, and an image. The values of pixels in an image are changed and hence it seems that filtering can be used for diffusion directly. However, since traditional filtering cannot be reversible, the cipher image with such diffusion cannot be recovered. To cope with this issue, Hua and Zhou set the right-bottom point of the filter to "1" and then align this point to the processed pixel in the image for convolution, and they proposed a novel image encryption algorithm with block-based scrambling and such image filtering (BSIF) [51]. However, the BSIF used a fixed filter for all pixel when doing convolution, limiting the encryption performance. An ideal scheme should use a variable/dynamic filter for convolution with each pixel.

2.3. DNA Computing

DNA computing, invented by Leonard Adleman, is a type of parallel computing technique that the information is expressed by four nucleic acids, i.e., adenine (A), cytosine (C), guanine (G), and thymine (T) [59]. The key factors of DNA for encryption are encoding and decoding rules, and algebraic operations for DNA sequences. Like 0 and 1 are complementary pairs in binary, 00 (0) and 11 (3), and 01 (1) and 10 (2) are also complementary pairs in DNA computing. Although there are $4! = 24$ combinations in total for DNA encoding, there are only 8 kinds of DNA bases are capable of meeting the DNA complementary rules, as listed in Table 1. With the encoding rule, an 8-bit pixel in grayscale image can be expressed by 4 letters. For example, following Rule 5 and Rule 8 in Table 1 , the decimal gray-level 156 ('10011100' in binary) can be transformed into a 4-letter DNA sequence 'TAGC' and 'ATCG', respectively. It can be seen that for a fixed binary sequence, different rules lead to totally different DNA sequences.

Table 1. Encoding and decoding rules of DNA.

RULE	Rule 1	Rule 2	Rule 3	Rule 4	Rule 5	Rule 6	Rule 7	Rule 8
00	A	T	T	A	C	G	C	G
01	C	G	C	G	A	A	T	T
10	G	C	G	C	T	T	A	A
11	T	A	A	T	G	C	G	C

In image encryption, several algebraic operations, such as addition (++), subtraction (--) and exclusive OR (XOR, ⊗⊗), as listed in Tables 2–4, can be used to change the values of nucleic acids [2].

Table 2. Addition (++) operation.

++	A	C	G	T
A	C	A	T	G
C	A	C	G	T
G	T	G	C	A
T	G	T	A	C

Table 3. Subtraction (--) operation.

--	A	C	G	T
A	C	G	T	A
C	A	C	G	T
G	T	A	C	G
T	G	T	A	C

Table 4. XOR (⊗⊗) operation.

⊗⊗	A	C	G	T
A	A	C	G	T
C	C	A	T	G
G	G	T	A	C
T	T	G	C	A

2.4. Latin Square

A Latin square of order N is an $N \times N$ matrix which includes a set S with N different symbol elements, and each symbol shows only once in each row and each column [53]. For instance, L is a Latin square of order N, i and j represent the row and column index of an element in L respectively, and S_k is the $k-$th element in set S. We can draw a formula as follows:

$$f(i,j,k) = \begin{cases} 1, L(i,j) = S_k \\ 0, otherwise \end{cases} \quad (2)$$

Given $S = \{0, 1, \cdots, N-1\}$, Figure 2 shows an example of Latin square of order 4.

$$\begin{bmatrix} 0 & 1 & 2 & 3 \\ 2 & 3 & 0 & 1 \\ 1 & 0 & 3 & 2 \\ 3 & 2 & 1 & 0 \end{bmatrix}$$

Figure 2. An example of Latin square of order 4.

3. The Proposed Image Encryption Approach

3.1. Hyperchaotic Sequence Generation

In this paper, we used the 5D hyperchaotic system described in Section 2.1 to generate the hyperchaotic sequence for encryption. Specifically, the generating process has three steps:

Step 1: The sequences generated by the first N_0 iterations are discarded to eliminate the adverse effects.

Step 2: The 5D hyperchaotic system continues to iterate to generate sequences long enough for image encryption. In the $j-$th iteration, we can obtain five state values denoted as $s^j = \{x_1^j, x_2^j, \cdots, x_5^j\}$.

Step 3: When the iteration completes, a hyperchaotic sequence S can be obtained by contacting all the $s^j (j = 1, 2, \cdots, N)$ as

$$\begin{aligned} S = \{s^1, s^2, \cdots, s^N\} &= \{x_1^1, x_2^1, \cdots, x_5^1, \cdots, x_1^N, x_2^N, \cdots, x_5^N\} \\ &= \{s_1, s_2, s_3, \cdots, s_{5N-2}, s_{5N-1}, s_{5N}\}. \end{aligned} \tag{3}$$

The real value sequence S is further mapped to an integral sequence as Equation (4):

$$k_i = mod(\left\lfloor mod((|k_i| - \lfloor |k_i| \rfloor) \times 10^{15}), 10^8) \right\rfloor, 256), \tag{4}$$

where mod, $|\cdot|$ and $\lfloor \cdot \rfloor$ denote the operations of modulo, absolute value, and flooring, respectively [2,6].

3.2. Dynamic Filtering

The modified filtering can be applied to image encryption, according to the very recent work BSIF by Hua and Zhou [51]. However, the BSIF does convolution on each pixel in an image with a fixed kernel generated from a random sequence. Li et al. used a 1×3 or 3×1 variable kernel to convolute each pixel in an image, that is to say, the kernels associated with each pixel for convolution are different, so-called dynamic filtering [6]. The experimental results demonstrated the effectiveness of dynamic filtering. A reasonable assumption is that a dynamic kernel with larger size (e.g., 3×3 or 5×5) will lead to better encryption. An example of dynamic filtering with two 3×3 filters is shown in Figure 3, where the 3×3 red kernel and the 3×3 blue kernel are conducted on the pixels of 34 and 178 in the plain image, and the results of dynamic filtering will be 140 and 214 in the cipher image, respectively. We can see that with dynamic filtering, the values of pixels in the plain image are changed, and this procedure can be reversible [51]. Therefore, we can use this operation for diffusion.

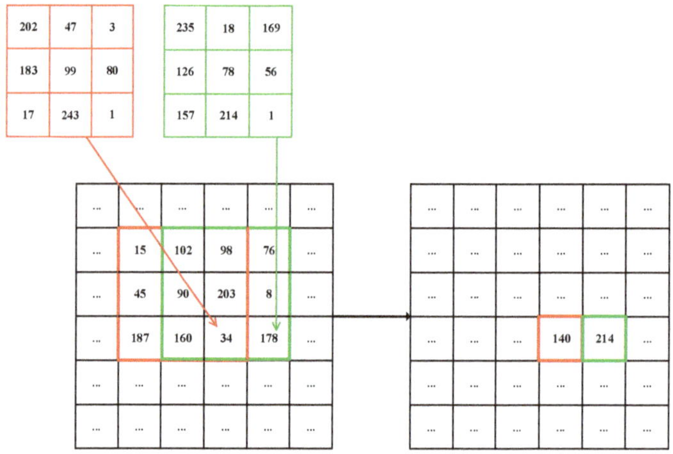

Figure 3. An example of dynamic filtering.

3.3. Image to Cubes

Since 3D Latin cube transformation can be conducted on cubes only, the image for encryption must be reshaped to one or several cubes. The pseudocode of such transformation algorithm (I2C) can be described as below.

Step 1: Given an image with size $h \times w \times d$, where h, w, and d represent the height, width, and depth, respectively, calculate the number of the pixels $N = h \times w \times d$.
Step 2: Let $L = \sqrt[3]{N}$, if L is an integer, jump to Step 3, else jump to Step 4.
Step 3: Get a cube with size $L \times L \times L$, return.
Step 4: Define $K = 2^n, n \in \mathbb{N}$, find the biggest K that meets $K \leq L$; then we get a cube with size $K \times K \times K$.
Step 5: Update $N = N - K^3$, if $N = 0$, return; else jump to Step 2.

For instance, a DNA-level image with size $512 \times 512 \times 4$ can be transformed into 4 cubes with size $64 \times 64 \times 64$, while a DNA-level image with size $256 \times 256 \times 4$ can be transformed into 8 $32 \times 32 \times 32$ cubes. Unlike the previous work that can only encrypts images of specified sizes [56], the proposed DFDLC can handle images of any sizes with such transformation.

Accordingly, one or several cubes can be merged into a plain image with the reverse procedure of the I2C.

3.4. 3D Latin Cube

Latin cube is a generalized version of the Latin square. A Latin cube of order N is an $N \times N \times N$ cube which includes a set S with N different symbol elements, and each symbol occurs only once in each row, each column, and each file [56]. Given a chaotic sequence $x = \{x_0, x_1, \cdots, x_{q^n-1}\}$ (q is a prime and q^n is the order of the Latin cubes to generate), we can sort the sequence by ascending to get an index sequence $y = \{y_0, y_1, \cdots, y_{q^n-1}\}$ and then construct a finite field F_{q^n} on y via redefining "+" and "×". With three distinct nonzero elements p_1, p_2 and p_3 in F_{q^n}, the element of $L_t(i, j, s)$ can be obtained by Equation (5):

$$L_t(i, j, s) = y_s + p_t \times y_j + p_t^2 \times y_i, \tag{5}$$

where $t = \{1, 2, 3\}$ is the index of the Latin cube, and "+" and "×" are the addition and multiplication in F_{q^n}, respectively [56]. Figure 4 shows three Latin cubes of order 3 on the set $S = \{0, 1, 2\}$, named as L_1, L_2, and L_3. When we superimpose the same position of three Latin cubes on the set S, if each combination occurs only once, we can say these three Latin cubes are orthogonal. For example, when we combine the three Latin cubes L_1, L_2 and L_3, each of the 27 combinations $000, 001, 002, \cdots, 222$ occurs only once, so they are orthogonal. By combining L_1, L_2 and L_3, we can get a new cube K shown in Figure 5. Then a spatial permutation is obtained: $(0,0,0) \to (0,0,0), (0,1,0) \to (1,1,2), (0,2,0) \to (2,2,1), \cdots, (2,2,2) \to (0,2,1)$, i.e., the element in the left position is transferred to the right position. More generally, $K_s(i,j) = (L_1(i,j,s), L_2(i,j,s), L_3(i,j,s))$, where s is the index of K (or L), and i and j are the indices of the row and the column, respectively.

$$L_1 \begin{bmatrix} 0 & 1 & 2 \\ 1 & 2 & 0 \\ 2 & 0 & 1 \end{bmatrix} \begin{bmatrix} 1 & 2 & 0 \\ 2 & 0 & 1 \\ 0 & 1 & 2 \end{bmatrix} \begin{bmatrix} 2 & 0 & 1 \\ 0 & 1 & 2 \\ 1 & 2 & 0 \end{bmatrix}$$

$$L_2 \begin{bmatrix} 0 & 1 & 2 \\ 2 & 0 & 1 \\ 1 & 2 & 0 \end{bmatrix} \begin{bmatrix} 1 & 2 & 0 \\ 0 & 1 & 2 \\ 2 & 0 & 1 \end{bmatrix} \begin{bmatrix} 2 & 0 & 1 \\ 1 & 2 & 0 \\ 0 & 1 & 2 \end{bmatrix}$$

$$L_3 \begin{bmatrix} 0 & 2 & 1 \\ 2 & 1 & 0 \\ 1 & 0 & 2 \end{bmatrix} \begin{bmatrix} 1 & 0 & 2 \\ 0 & 2 & 1 \\ 2 & 1 & 0 \end{bmatrix} \begin{bmatrix} 2 & 1 & 0 \\ 1 & 0 & 2 \\ 0 & 2 & 1 \end{bmatrix}$$

Figure 4. Three examples of Latin cube of order 3.

$$K_1 \begin{bmatrix} (0,0,0) & (1,1,2) & (2,2,1) \\ (1,2,2) & (2,0,1) & (0,1,0) \\ (2,1,1) & (0,2,0) & (1,0,2) \end{bmatrix}$$

$$K_2 \begin{bmatrix} (1,1,1) & (2,2,0) & (0,0,2) \\ (2,0,0) & (0,1,2) & (1,2,1) \\ (0,2,2) & (1,0,1) & (2,1,0) \end{bmatrix}$$

$$K_3 \begin{bmatrix} (2,2,2) & (0,0,1) & (1,1,0) \\ (0,1,1) & (1,2,0) & (2,0,2) \\ (1,0,0) & (2,1,2) & (0,2,1) \end{bmatrix}$$

Figure 5. A new cube K constructed by L_1, L_2, L_3. K_1, K_2 and K_3 are the 1st, 2nd and 3rd squares of K respectively.

3.5. DFDLC: The Proposed Image Encryption Approach with Dynamic Filtering and Latin Cubes

The DFDLC is conducted on pixel-level diffusion and DNA-level permutation. Specifically, regarding pixel-level diffusion, we mainly apply dynamic filtering on each pixel in a plain 2D image. We also used the ciphertext diffusion in crisscross pattern (CDCP) to improve the diffusion results [60]. For DNA-level permutation, we mainly use Latin cube to change the position of each nucleic acid. In addition, a kind of global DNA permutation similar to the global bit permutation is adopted for DNA permutation [6]. The proposed DFDLC is illustrated in Figure 6. With the hyperchaotic sequence generated by the 5D chaotic system, the main steps of the DFDLC are as the following: hyperchaotic sequence generation, pixel-level diffusion, pixel-to-DNA transformation, DNA permutation, and DNA-to-pixel transformation.

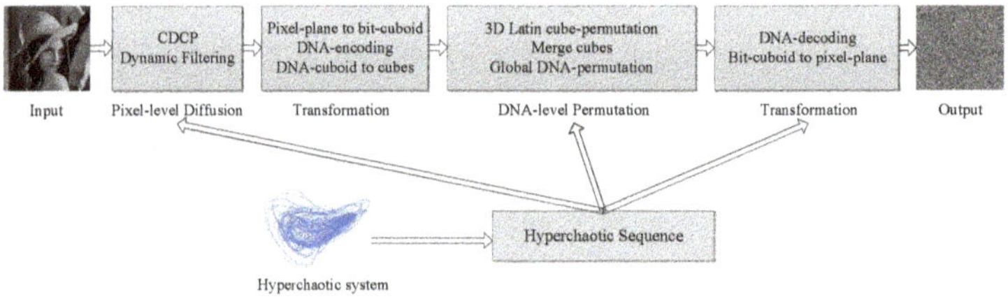

Figure 6. The framework of the proposed DFDLC.

The details of the DFDLC are described as follows:

Step 1: Given the keys, generate a hyperchaotic sequence with Equations (1), (3) and (4).
Step 2: Conduct CDCP with pixels of the image. This operation expands a little change in one pixel of the plain image to very large changes in a variety of pixels of the cipher image.
Step 3: Dynamic filtering on the image. For each pixel, firstly, generate a 3×3 kernel with the hyperchaotic sequence and set the right-bottom grid to 1. Secondly, do convolution with the kernel and corresponding sub-region of the image associated with the pixel. Thirdly, use the result of the convolution as the new value of the pixel in the cipher image.
Step 4: Transform the pixel image to a DNA image. For each pixel, use an encoding rule decided by the hyperchaotic sequence to encode one pixel into a string with 4 nucleic acids. The DNA encoding rule (Rule N) can be formulated as: $N = 1 + mod(x, 8)$, where x is a corresponding value in the hyperchaotic sequence regarding the pixel.
Step 5: Transform the DNA image into one or several cubes using I2C.
Step 6: Conduct DNA-level Latin cube permutation. For each DNA-level cube, generate a Latin cube and then change the position of each nucleic acid according to the Latin cube. In addition, the DNA XOR operation is conducted on the DNA-level cube with a generated DNA cube from the hyperchaotic sequence.
Step 7: Integrate all the DNA-level cubes into a DNA image.
Step 8: Conduct global DNA permutation as described in [2].
Step 9: Decode the DNA image into a pixel image. For each nucleic acid, the DNA encoding rule is decided as the encoding rule in Step 6. The pixel image is the cipher image.

The proposed DFDLC consists of five stages: hyperchaotic sequence generation (Step 1), pixel-level diffusion (Step 2-3), a transformation from a plane image to cubes (Step 4-5), DNA-level Latin cube permutation (Step 6-8) and a transformation from cubes to a plane image (Step 9). The keys of the DFDLC are Step 3 and Step 6, i.e., pixel-level diffusion with dynamic filtering and DNA-level permutation with 3D Latin cubes, respectively. Although the main objective of Latin cubes in the DFDLC is for permuting the DNA, it also results in diffusion because the change of the position of DNA can change the corresponding value of the pixel naturally [6].

The cipher image can be easily decrypted by the inverse steps as listed above.

4. Experimental Results

4.1. Experimental Settings

To validate the performance of the proposed DFDLC, we compare it with some state-of-the-art image encryption schemes, such as the image encryption with a fractional-order hyperchaotic system and DNA computing (FOHCDNA) [2], the hyperchaotic and DNA sequence-based method (HC-DNA) [61], CDCP [60], BSIF [51] and DFBC [6]. We set the parameters for the DFDLC as following. For the 5D hyperchaotic system, we set $(x_1^0, x_2^0, x_3^0, x_4^0, x_5^0) = (1.618, 3.14, 2.718, 4.6692, 0.618) \times 10^{-2}$ and

1000 as the initial values and the preiterating times, respectively. For the compared methods, we use the parameters as set by the corresponding references.

We used ten publicly accessed images for validating the proposed DFDLC, and the details of the images are listed in Table 5.

Table 5. Testing images.

Image	Size ($h \times w$)	Image	Size ($h \times w$)
Lena	512 × 512	Cameraman	512 × 512
Barbara	512 × 512	Mandril	512 × 512
Bw	512 × 512	Pirate	512 × 512
Couple	512 × 512	Finger	512 × 512
Peppers	512 × 512	Houses	512 × 512

The experiments were conducted using MATLAB 2016b (MathWorks, Natick, MA, USA) on a 64-bit Windows 7 Ultimate (Microsoft, Redmond, WA, USA) with 32 GB memory and a 3.6 GHz I7 CPU.

4.2. Security Key Analysis

A feasible image encryption algorithm should have a large enough key space and extreme sensitivity to the key to resist brute force attacks. In this subsection, we will analyze the key space and the sensitivity of the security key.

4.2.1. Key Space

The key space is the set of all possible security keys that can be used in a system of image encryption. It was reported that the size of a key space larger than 2^{100} can provide enough security [62]. Basically, the 5 initial values of the 5D hyperchaotic systems, i.e., $(x_1^0, x_2^0, x_3^0, x_4^0, x_5^0)$ for Equation (1), can be constructed as the security keys. If each initial value has the same precision of 10^{-15}, the DFDLC has a key space with size of $10^{15*5} = 10^{75} \approx 2^{249}$, which is much larger than 2^{100}. Therefore, the DFDLC can resist all types of brute-force attacks from current computers. Besides, the distinct nonzero elements in the finite filed for Latin cubes can be used as security keys to improve the key space.

4.2.2. Sensitivity to Security Key

An ideal image encryption approach should be sensitive enough to the security key, that is to say, a very little change in the security keys will lead to a completely different decrypted image.

We use two groups of slightly different keys to validate the sensitivity to the security keys of the proposed DFDLC. The first group keys are the initial values of the hyperchaotic system, i.e., $g_1 = (x_1^0, x_2^0, x_3^0, x_4^0, x_5^0) = (1.618, 3.14, 2.718, 4.6692, 0.618) \times 10^{-2}$, while the second groups are almost the same as the first group except $x_1^0 = 0.0168 + 10^{-15}$, i.e., $g_2 = (x_1^0 + 10^{-13}, x_2^0, x_3^0, x_4^0, x_5^0) = (1.618 + 10^{-13}, 3.14, 2.718, 4.6692, 0.618) \times 10^{-2}$. We apply g_1 and g_2 to decrypt the first five images in Table 5, and the results are shown in Figure 7. It is clear that even the security keys are changed very little such as 10^{-15}, the cipher images cannot be recovered correctly, demonstrating the high sensitivity to security keys of the proposed DFDLC [6].

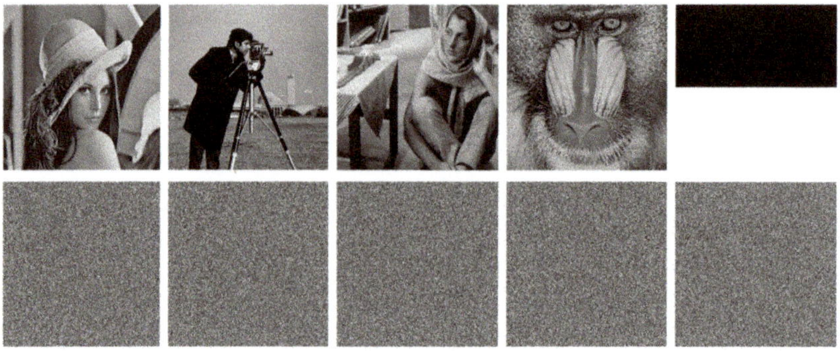

Figure 7. Decrypted images of Lena, Cameraman, Barbara, Mandril and Bw with security keys g_1 and g_2. The first and the second row is with g_1 and g_2, respectively.

4.3. *Statistical Analysis*

Statistical analysis, including histogram analysis, entropy analysis, and correlation analysis are essential for a cryptosystem. An ideal image encryption algorithm should have the ability to resist kinds of statistical attacks.

4.3.1. Histogram Analysis

Histogram describes the distribution of pixels for an image. The histogram of a natural image usually shows an irregular (unevenly distributed) shape. A good image encryption approach should change the irregular shape of a plain image as evenly distributed as possible, leading to a completely random-like cipher image. Regarding evaluating the image encryption approach with histogram, the more uniform the histogram is, the better the encryption approach is [2]. The histograms of the plain images and the corresponding cipher images are shown in Figure 8.

It can be seen that the histograms of the plain images except Bw look like mountains, including peaks and valleys. However, the histograms of their corresponding cipher images are so flat that they are very close to uniform distributions. It is worth pointing out that regarding the image Bw, it has only two values of grayscale level, i.e., 0 and 255, and its histogram looks like two needles. However, the histogram of its cipher image is still very uniform similar to histograms of other cipher images. Although the plain images are very different, the histograms of their corresponding cipher images are so uniform and so close that it looks like that each grayscale level appears about 1000 times in all cipher images. This characteristic of cipher images can be easily found in the last column in Figure 8. The experiments indicate that the proposed DFDLC can obtain very uniform histograms for different types of images and hence it can resist histogram attacks very well.

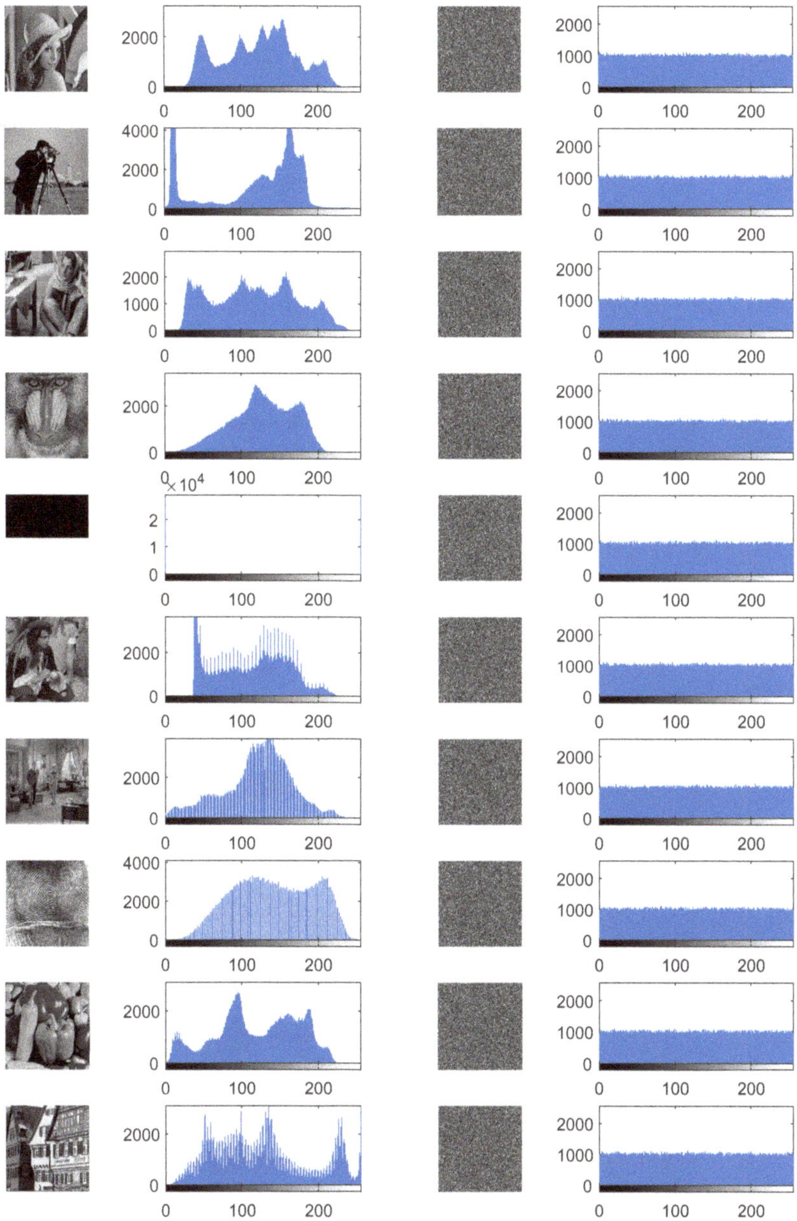

Figure 8. Histograms of the plain images and their corresponding cipher images. The first and the second columns are the plain images and their corresponding histograms, respectively. The third and the fourth columns are the cipher images and their corresponding histograms, respectively.

4.3.2. Information Entropy

Information entropy (IE), originally proposed by Shannon, is one of the key measures to quantify the degree of uncertainty (randomness) of a given system in information theory [63]. It can be applied to measure the randomness of an image encryption system. Given an 8-bit grayscale level that has $2^8 = 256$ possible pixel values, i.e., $0, 1, \cdots, 255$, the IE can be formulated as Equation (6)

$$\mathrm{IE}(I) = -\sum_{i=0}^{255} p(I_i) log_2 p(I_i), \tag{6}$$

where $p(I_i)$ is the probability of the $i-$th gray value I_i appears in an image I. For a cipher image, when each gray value I_i appears with equal probability, i.e., $\frac{1}{256}$, the IE obtains the maximum 8. Therefore, an ideal image encryption approach should have an IE close to 8.

The IEs of the test images and corresponding cipher images with the DFDLC and the compared approaches are shown in Table 6. It can be seen that the testing natural images in this experiment have close IEs around 7, while the image of Bw has the lowest entropy 1, showing that the distribution of pixel values is irregular, as indicated by their histograms in Figure 8. It can be seen that the IEs of all cipher images are very close to the ideal value 8. Specifically, all encryption approaches except for HCDNA achieve very stable IEs, i.e., $7.9992 \sim 7.9994$, which are also very close to 8, indicating that these approaches are secure enough to resist entropy attacks. Although the IEs achieved by the HCDNA are slightly worse than those by the other approaches, they are still very close to the ideal value except that the IE of Bw by HCDNA is as low as 7.9158. Among the approaches, the BSIF obtains the highest IEs with 6 out of 10 cases, followed by DFDLC, FHDNA and DFBC, which all achieve the highest IEs 4 out of 10 times. However, the HCDNA achieves the highest IE only once. The experimental results demonstrate that the DFDLC are advantageous over or comparable to other approaches in terms of IE.

As mentioned above, the IEs reflect the randomness of the grayscale values in an image. The IEs achieved by DFDLC are very close to 8, indicating that the pixel values are distributed very uniformly, as the histograms shown in the last column in Figure 8. Therefore, the results of histograms are consistent with the analysis of IEs, confirming that the proposed DFDLC has good statistical properties in terms of image encryption.

Table 6. The IEs of the testing images.

Image	**Input**	**Cipher Images**					
		DFDLC	**FHDNA** [2]	**HCDNA** [61]	**CDCP** [60]	**IC-BSIF** [51]	**DFBC** [6]
Lena	7.4455	7.9993	7.9993	**7.9994**	7.9993	**7.9994**	**7.9994**
Cameraman	7.0480	7.9992	**7.9993**	7.9981	**7.9993**	**7.9993**	7.9992
Barbara	7.6321	7.9993	**7.9994**	7.9993	7.9992	7.9993	7.9993
Mandril	7.2925	**7.9994**	7.9992	7.9992	7.9993	7.9993	7.9993
Bw	1.0000	**7.9993**	7.9992	7.9158	7.9992	**7.9993**	**7.9993**
Pirate	7.2367	**7.9994**	7.9993	7.9988	7.9993	**7.9994**	7.9993
Couple	7.0572	**7.9993**	7.9992	7.9992	**7.9993**	7.9992	**7.9993**
Finger	6.7279	7.9993	**7.9994**	7.9990	7.9992	**7.9994**	7.9993
Peppers	7.5925	7.9993	**7.9994**	7.9991	7.9993	7.9993	**7.9994**
Houses	7.6548	7.9992	7.9993	7.9993	**7.9994**	**7.9994**	7.9993

4.3.3. Correlation Analysis

Natural images usually show high correlation, that is, neighboring pixels have very close grayscale levels. When an image is permutated, the neighboring pixels will be randomly distributed in the whole image and hence the high correlation in plain image is broken. An ideal image encryption approach should decrease the correlation to zero in the cipher image. One of the popular ways to measure the correlation in images is the correlation coefficient γ defined as Equation (7) [6,64]

$$\begin{aligned} E(x) &= \frac{1}{M}\sum_{i=1}^{M} x_i, \\ S(x) &= \frac{1}{M}\sum_{i=1}^{M} (x_i - E(x))^2, \\ cov(x,y) &= \frac{1}{M}\sum_{i=1}^{M} (x_i - E(x))(y_i - E(y)), \\ \gamma &= \frac{cov(x,y)}{\sqrt{S(x)S(y)}}, \end{aligned} \tag{7}$$

where x and y are grayscale levels of two adjacent pixels in an image, and M denotes the number of pairs of involved pixels, and $E(x)$, $S(x)$ and $cov(x,y)$ are the expectation of x, the standard deviation of x and the covariance of x and y, respectively.

To analyze the correlation, we firstly use all the pairs of adjacent pixels from each plain image and the corresponding cipher image in the horizontal direction, the vertical direction, and the diagonal direction to compute the correlation coefficients, denoted by γ_h, γ_v and γ_d, respectively. The results are shown in Table 7. We can see that the correlation coefficients of all plain images in all directions are very high, especially the γ_h of the image Bw equals to the maximum value of 1. However, all the correlation coefficients of the encrypted images decrease to close to zero, showing that the high correlation in plain images is broken. Regarding the encryption approaches, each outperforms others in several cases, indicating they are comparable in terms of reducing the correlation in images. If we consider the range of the γ achieved by the approaches, we can see that the ranges by DFDLC, FHDNA, HCDNA, CDCP, BSIF and DFBC are [−0.0023,0.0030], [−0.0049, 0.0057], [−0.0032, 0.0038], [−0.0032, 0.0028], [−0.0032, 0.0034] and [−0.0029, 0.0027], respectively. Accordingly, the interval widths of γ by the approaches are $0.0053, 0.0106, 0.0070, 0.0060, 0.0066$ and 0.0056. Among the interval widths, the DFDLC achieves the narrowest one, indicating that the DFDLC is the most stable approach in terms of γ.

Then, we randomly select 4000 pairs of horizontally adjacent pixels from each plain image and its corresponding cipher image to plot the distribution maps of the grayscale levels of the adjacent pixels, as shown in Figure 9. It can be seen that the correlation of natural images is so strong that the grayscale levels of the adjacent pixels are concentrated near the diagonal line. The figure of the plain Bw is a special case because its distribution has only two possible combinations, i.e., $(0,0)$ and $(255,255)$. The strong correlation of all the plain images is thoroughly destroyed by the proposed DFDLC so that the grayscale levels of adjacent pixels are evenly distributed over the entire plane. It further demonstrates that the DFDLC has good performance regarding correlation.

Table 7. The correlation coefficients γ of the testing images.

Image	γ	Input	Cipher Images					
			DFDLC	FHDNA [2]	HCDNA [61]	CDCP [60]	IC-BSIF [51]	DFBC [6]
	γ_h	0.9691	0.0023	**0.0000**	−0.0015	−0.0004	−0.0032	0.0002
Lena	γ_v	0.9841	**0.0009**	−0.0022	−0.0020	0.0028	0.0013	0.0010
	γ_d	0.9639	0.0008	**0.0004**	0.0024	0.0016	−0.0009	0.0006
	γ_h	0.9830	0.0011	0.0013	0.0004	**−0.0001**	−0.0015	−0.0008
Cameraman	γ_v	0.9887	0.0009	0.0033	**0.0003**	0.0019	0.0010	−0.0013
	γ_d	0.9746	−0.0002	**−0.0000**	−0.0013	0.0010	−0.0012	−0.0002
	γ_h	0.8940	−0.0003	−0.0022	0.0010	−0.0026	**−0.0002**	0.0027
Barbara	γ_v	0.9572	0.0030	**−0.0002**	0.0004	0.0006	−0.0004	−0.0029
	γ_d	0.8942	−0.0029	**−0.0000**	−0.0009	0.0005	0.0010	−0.0005
	γ_h	0.9322	0.0022	0.0016	−0.0007	0.0012	0.0026	**−0.0006**
Mandril	γ_v	0.9100	0.0005	0.0035	**−0.0001**	0.0009	**−0.0001**	−0.0018
	γ_d	0.8647	−0.0023	−0.0025	−0.0017	−0.0004	**0.0001**	0.0016
	γ_h	1.0000	0.0019	0.0006	0.0004	−0.0004	0.0003	**0.0000**
Bw	γ_v	0.9922	−0.0006	0.0009	0.0013	**0.0001**	−0.0005	−0.0002
	γ_d	0.9961	−0.0012	−0.0012	**−0.0002**	0.0005	**0.0002**	−0.0016
	γ_h	0.9593	**−0.0000**	0.0015	−0.0023	−0.0012	−0.0026	−0.0012
Pirate	γ_v	0.9675	0.0009	0.0057	**−0.0000**	−0.0008	−0.0006	0.0013
	γ_d	0.9432	0.0015	**0.0001**	0.0011	0.0006	0.0005	0.0005
	γ_h	0.9451	0.0012	0.0013	0.0014	**−0.0001**	−0.0006	−0.0009
Couple	γ_v	0.9514	0.0025	−0.0026	0.0008	**0.0001**	0.0023	0.0022
	γ_d	0.9116	0.0017	−0.0011	−0.0007	**0.0005**	−0.0008	−0.0024
	γ_h	0.9343	**−0.0001**	0.0002	0.0007	−0.0023	0.0004	−0.0025
Finger	γ_v	0.9168	**0.0002**	−0.0025	0.0029	−0.0032	−0.0009	0.0004
	γ_d	0.8664	0.0017	**0.0005**	−0.0022	−0.0010	0.0030	−0.0006
	γ_h	0.9733	0.0003	−0.0045	**0.0000**	−0.0003	−0.0031	0.0008
Peppers	γ_v	0.9763	−0.0010	−0.0049	−0.0005	**0.0003**	−0.0010	**−0.0003**
	γ_d	0.9650	0.0011	−0.0012	**−0.0005**	−0.0025	0.0017	−0.0010
	γ_h	0.9077	0.0020	0.0006	0.0004	0.0026	**0.0001**	−0.0002
Houses	γ_v	0.9173	0.0015	0.0004	−0.0032	**0.0002**	0.0017	0.0006
	γ_d	0.8439	0.0020	0.0021	0.0038	−0.0011	0.0034	**0.0002**
Range		[0.8439,1.000]	[−0.0023,0.0030]	[−0.0049, 0.0057]	[−0.0032, 0.0038]	[−0.0032, 0.0028]	[−0.0032, 0.0034]	[−0.0029, 0.0027]
Interval Width		0.1561	**0.0053**	0.0106	0.0070	0.0060	0.0066	0.0056

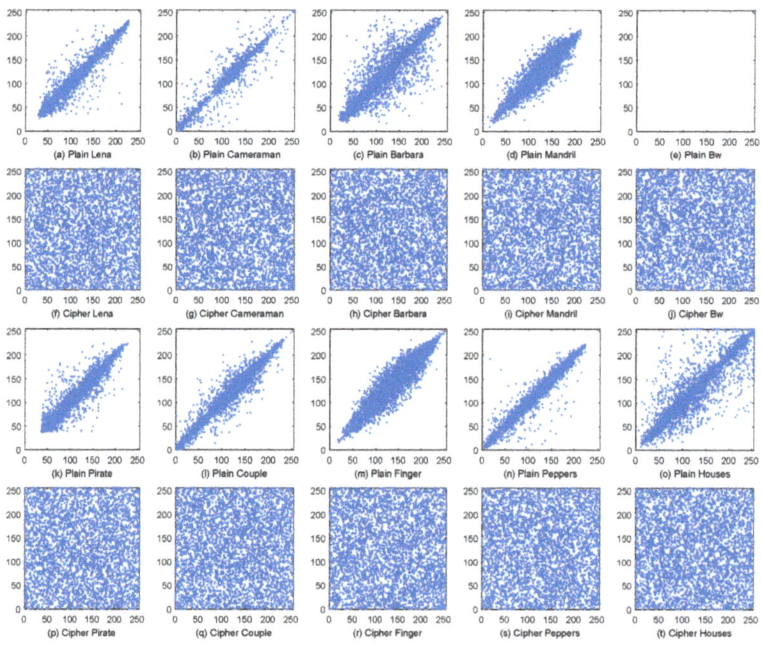

Figure 9. The adjacent-pixel distribution maps of the plain images and the corresponding cipher images in horizontal direction.

4.4. Analysis of Resisting Differential Attacks

Differential attack is to study how a tiny change in a plain image can affect the corresponding cipher image. A good encryption approach should have the ability to resist differential attacks, that is to say, any small changes (even if changing a bit) in a plain image will result in a completely different cipher image. Two of the most popular indices to quantify the performance of resisting differential attacks in image encryption are the number of pixels change rate (NPCR) and the unified average changing intensity (UACI), as defined by Equations (8) and (9), respectively [65]

$$\text{NPCR} = \frac{1}{W \times H} \sum_{i=1}^{W} \sum_{j=1}^{H} d_{ij} \times 100\%, \tag{8}$$

$$\text{UACI} = \frac{1}{255 \times W \times H} \sum_{i=1}^{W} \sum_{j=1}^{H} \left| C_{ij}^{1} - C_{ij}^{2} \right| \times 100\%, \tag{9}$$

where W and H denote the width and the height of the cipher images respectively, C^1 and C^2 are two cipher images, and d_{ij} is defined as Equation (10)

$$d_{ij} = \begin{cases} 0, & C_{ij}^{1} = C_{ij}^{2}, \\ 1, & C_{ij}^{1} \neq C_{ij}^{2}. \end{cases} \tag{10}$$

As far as the two indices are concerned, the NPCR focuses on the variation ratio of two cipher images whose plain images are slightly changed while the UACI defines the mean intensity of the two cipher images. Wu et al. proposed a threshold and a range for NPCR and UACI respectively to evaluate if an encryption approach can pass the differential attack test for a given specified size image at a significance level α. Specifically, for a 512×512 8-bit grayscale image, if the NPCR score is bigger than the threshold $\mathcal{N}_{0.05}^{*} = 99.5893\%$, it passes the NPCR test at $\alpha = 0.05$. In addition, if the UACI score falls into the interval $\left(\mathcal{U}_{0.05}^{*l}, \mathcal{U}_{0.05}^{*u}\right) = (33.3730\%, 33.5541\%)$, it is said to pass the UACI test at $\alpha = 0.05$ [65].

We add 1 to the value of a randomly selected pixel to compute one score of the NPCR and the UACI. The computation is repeated 10 times and then the mean, standard deviation, and times of passing the test of NPCR and UACI are reported in Tables 8 and 9, respectively. The mean scores that pass the NPCR or the UACI tests at a significance level $\alpha = 0.05$ are shown in bold. One can see that both DFDLC and BSIF can pass both tests on all images in terms of the mean scores of NPCR and UACI, while CDCP and DFDC can pass most tests. In contrast, the FHDNA and the HCDNA failed the tests with all images, although the mean scores by the FHDNA are very close to $\mathcal{N}_{0.05}^{*}$ and $\left(\mathcal{U}_{0.05}^{*l}, \mathcal{U}_{0.05}^{*u}\right)$. If we look at the times of passing the NPCR test, both the DFDLC and the BSIF can pass the test in 99 out of $10 \times 10 = 100$ times and they are far superior to other methods. However, regarding times of passing the UACI test, the DFDLC is slightly worse than the BSIF, but it outperforms other methods. The experimental results demonstrate that the proposed DFDLC is capable of resisting differential attacks.

Table 8. The mean / standard deviation / times of passing the test of NPCR (%) of running the schemes 10 times ($\alpha = 0.05$).

Image	DFDLC	FHDNA [2]	HCDNA [61]	CDCP [60]	BSIF [51]	DFBC [6]
Lena	**99.6103**/0.0129/10	99.5814/0.0119/6	43.5948/16.8360/0	**99.6201**/0.2837/5	**99.6166**/0.0109/10	**99.5995**/0.0002/10
Cameraman	**99.6055**/0.0126/10	99.5795/0.0137/4	64.6306/31.1442/0	**99.6146**/0.2372/6	**99.6057**/0.0121/10	**99.6143**/0.0002/10
Barbara	**99.6171**/0.0075/10	99.5842/0.0099/8	37.8473/19.6663/0	**99.6048**/0.2136/6	**99.6165**/0.0144/10	99.5833/0.0002/10
Mandril	**99.6047**/0.0117/10	99.5774/0.0125/3	51.2024/28.3679/0	99.5697/0.2107/4	**99.6070**/0.0117/10	**99.5998**/0.0001/10
Bw	**99.6030**/0.0094/10	99.3196/0.2433/1	47.5142/15.7628/0	**99.6362**/0.2321/5	**99.6180**/0.0142/10	**99.6033**/0.0000/10
Pirate	**99.6176**/0.0133/10	99.5812/0.0127/4	35.8150/27.9995/0	**99.6403**/0.3222/7	**99.6116**/0.0116/10	99.5751/0.0002/0
Couple	**99.6089**/0.0133/10	99.5779/0.0076/5	58.1698/27.5116/0	99.5718/0.1939/3	**99.6079**/0.0112/10	99.5586/0.0001/0
Finger	**99.6097**/0.0158/10	99.5792/0.0152/4	60.3329/29.8886/0	**99.5984**/0.1420/6	**99.6171**/0.0096/10	**99.6132**/0.0002/10
Peppers	**99.6099**/0.0154/9	99.5800/0.0119/5	45.6316/38.2206/0	99.5493/0.2509/4	**99.6099**/0.0150/9	**99.6166**/0.0001/10
Houses	**99.6130**/0.0126/10	99.5795/0.0083/6	63.0733/19.2267/0	**99.6039**/0.2053/7	**99.6135**/0.0119/10	**99.6151**/0.0001/10

Table 9. The mean / standard deviation / times of passing the test of UACI (%) of running the schemes 10 times ($\alpha = 0.05$).

Image	DFDLC	FHDNA [2]	HCDNA [61]	CDCP [60]	BSIF [51]	DFBC [6]
Lena	**33.4504**/0.0466/9	33.2700/0.0490/0	18.5974/9.5490/0	**33.5212**/0.0775/6	**33.4714**/0.0339/10	**33.4818**/0.0005/10
Cameraman	**33.4909**/0.0457/9	33.3010/0.0320/0	27.0047/13.9227/0	**33.4222**/0.0658/7	**33.4755**/0.0485/10	**33.4406**/0.0005/10
Barbara	**33.4451**/0.0350/10	33.2533/0.0431/0	13.6480/8.2289/0	**33.4464**/0.1075/7	**33.4722**/0.0476/9	**33.4808**/0.0007/10
Mandril	**33.4704**/0.0334/10	33.2988/0.0336/0	22.2006/14.1993/0	**33.4467**/0.0928/5	**33.4449**/0.0423/10	**33.5136**/0.0003/10
Bw	**33.4334**/0.0471/10	32.0705/1.0272/0	18.5654/6.1072/0	**33.4555**/0.1144/4	**33.4500**/0.0468/10	41.6585/0.0010/0
Pirate	**33.4736**/0.0275/10	33.3021/0.0431/1	14.8888/14.1945/0	**33.4664**/0.0766/8	**33.4644**/0.0328/10	**33.4668**/0.0003/10
Couple	**33.4282**/0.0385/9	33.2796/0.0381/0	17.8782/7.3362/0	**33.4293**/0.1011/9	**33.4632**/0.0439/10	**33.4717**/0.0003/10
Finger	**33.4311**/0.0504/8	33.2907/0.0413/0	26.0775/14.9380/0	**33.4911**/0.1004/6	**33.4856**/0.0399/9	**33.5263**/0.0006/10
Peppers	**33.4618**/0.0432/10	33.2735/0.0347/0	19.8106/18.6275/0	**33.4626**/0.0752/6	**33.4301**/0.0379/10	**33.4525**/0.0009/10
Houses	**33.4634**/0.0358/10	33.3322/0.0273/1	22.1296/7.4892/0	**33.4721**/0.0467/10	**33.4448**/0.0343/10	**33.4545**/0.0004/10

4.5. Discussion

The proposed DFDLC conducts encryption on pixel-level and DNA-level, with dynamic filtering for diffusion and Latin cubes for permutation. From the above analysis, we can see that the DFDLC can resist brute force attacks, statistical attacks as well as differential attacks, and the experiments have also demonstrated that DFDLC is superior or comparable to the compared state-of-the-art image encryption methods. In addition, the proposed I2C allows the DFDLC to handle images with any sizes, making it more practical.

One limitation of the DFDLC is the running time. It takes about 0.84s and 3.15s to encrypt an image of size 256×256 and 512×512 respectively in our experimental environment. The DFDLC is time consuming because the DNA operations (DNA encoding, decoding and algebraic operation) are actually operations on strings. This can be resolved by introducing lookup tables of DNA operations. Another possible way is to use GPU to accelerate DNA operations.

5. Conclusions

Image encryption is one of the core tasks of image security. To improve image security, in this paper, a novel image encryption algorithm that uses a 5D hyperchaotic system with 2 positive LEs, pixel-level dynamic filtering, DNA computing, and 3D Latin cubes, namely DFDLC, is proposed. The novelty of the DFDLC is introducing a new type of dynamic filtering to conduct pixel-level diffusion and permutating images with DNA-level data via Latin cubes. Extensive experiments on ten public test images have indicated that the proposed DFDLC has a large key space, is very sensitive to security keys, has good statistical characteristics, and can resist types of attacks. In the future, we will extend the proposed DFDLC in several aspects. First, we will apply trigonometric polynomials to generate the hyperchaotic sequence for the DFDLC. Second, we will try a variety of shapes of the filters for dynamic filtering. Third, we may use GPU or lookup tables to speed up the encoding and decoding of DNA and corresponding arithmetic operations. Finally, we can apply the DFDLC to color image encryption.

Author Contributions: Investigation, T.L., J.W. and F.P.; Methodology, T.L. and X.L.; Software, T.L., J.S. and X.L.; Supervision, T.L.; Writing—original draft, T.L. and J.S.; Writing—review & editing, T.L. and J.W.

Funding: This research was funded by the Fundamental Research Funds for the Central Universities (Grant No. JBK1902029, No. JBK1802073 and No. JBK170505), the Ministry of Education of Humanities and Social Science Project (Grant No. 19YJAZH047), Sichuan Science and Technology Program (Grant No. 2019YFG0117) and the Scientific Research Fund of Sichuan Provincial Education Department (Grant No. 17ZB0433).

Acknowledgments: This work was supported by the Fundamental Research Funds for the Central Universities (Grant No. JBK1902029, No. JBK1802073 and No. JBK170505), the Ministry of Education of Humanities and Social Science Project (Grant No. 19YJAZH047), Sichuan Science and Technology Program (Grant No. 2019YFG0117) and the Scientific Research Fund of Sichuan Provincial Education Department (Grant No. 17ZB0433).

Conflicts of Interest: The authors declare no conflict of interest.

References

1. Ahmad, J.; Hwang, S.O.; Ali, A. An experimental comparison of chaotic and non-chaotic image encryption schemes. *Wirel. Pers. Commun.* **2015**, *84*, 901–918. [CrossRef]
2. Li, T.; Yang, M.; Wu, J.; Jing, X. A novel image encryption algorithm based on a fractional-order hyperchaotic system and DNA computing. *Complexity* **2017**, *2017*, 9010251. [CrossRef]
3. Abd El-Latif, A.A.; Abd-El-Atty, B.; Talha, M. Robust encryption of quantum medical images. *IEEE Access* **2018**, *6*, 1073–1081. [CrossRef]
4. Guillén-Fernández, O.; Meléndez-Cano, A.; Tlelo-Cuautle, E.; Núñez-Pérez, J.C.; de Jesus Rangel-Magdaleno, J. On the synchronization techniques of chaotic oscillators and their FPGA-based implementation for secure image transmission. *PLoS ONE* **2019**, *14*, e0209618. [CrossRef] [PubMed]

5. Flores-Vergara, A.; García-Guerrero, E.; Inzunza-González, E.; López-Bonilla, O.; Rodríguez-Orozco, E.; Cárdenas-Valdez, J.; Tlelo-Cuautle, E. Implementing a chaotic cryptosystem in a 64-bit embedded system by using multiple-precision arithmetic. *Nonlinear Dyn.* **2019**, 1–20. [CrossRef]
6. Li, X.; Xie, Z.; Wu, J.; Li, T. Image encryption based on dynamic filtering and bit cuboid operations. *Complexity* **2019**, *2019*, 7485621. [CrossRef]
7. Praveenkumar, P.; Thenmozhi, K.; Rayappan, J.B.B.; Amirtharajan, R. Inbuilt Image Encryption and Steganography Security Solutions for Wireless Systems: A Survey. *Res. J. Inf. Tech.* **2017**, *9*, 46–63.
8. Chen, G.; Mao, Y.; Chui, C.K. A symmetric image encryption scheme based on 3D chaotic cat maps. *Chaos Solitons Fractals* **2004**, *21*, 749–761. [CrossRef]
9. Guan, Z.H.; Huang, F.; Guan, W. Chaos-based image encryption algorithm. *Phys. Lett. A* **2005**, *346*, 153–157. [CrossRef]
10. Huang, X. A designed image encryption algorithm based on chaotic systems. *J. Comput. Theor. Nanosci.* **2012**, *9*, 2130–2135. [CrossRef]
11. Wu, Y.; Yang, G.; Jin, H.; Noonan, J.P. Image encryption using the two-dimensional logistic chaotic map. *J. Electron. Imaging* **2012**, *21*, 013014. [CrossRef]
12. Hua, Z.; Zhou, Y. Image encryption using 2D Logistic-adjusted-Sine map. *Inf. Sci.* **2016**, *339*, 237–253. [CrossRef]
13. Liu, X.; Xiao, D.; Xiang, Y. Quantum image encryption using intra and inter bit permutation based on logistic map. *IEEE Access* **2019**, *7*, 6937–6946. [CrossRef]
14. Flores-Vergara, A.; Inzunza-González, E.; García-Guerrero, E.E.; López-Bonilla, O.R.; Rodríguez-Orozco, E.; Hernández-Ontiveros, J.M.; Cárdenas-Valdez, J.R.; Tlelo-Cuautle, E. Implementing a Chaotic Cryptosystem by Performing Parallel Computing on Embedded Systems with Multiprocessors. *Entropy* **2019**, *21*, 268. [CrossRef]
15. Wang, Y.; Wong, K.W.; Liao, X.; Chen, G. A new chaos-based fast image encryption algorithm. *Appl. Soft Comput.* **2011**, *11*, 514–522. [CrossRef]
16. Pareek, N.K.; Patidar, V.; Sud, K.K. Image encryption using chaotic logistic map. *Image Vis. Comput.* **2006**, *24*, 926–934. [CrossRef]
17. Hua, Z.; Jin, F.; Xu, B.; Huang, H. 2D Logistic-Sine-coupling map for image encryption. *Signal Process.* **2018**, *149*, 148–161. [CrossRef]
18. Sahari, M.L.; Boukemara, I. A pseudo-random numbers generator based on a novel 3D chaotic map with an application to color image encryption. *Nonlinear Dyn.* **2018**, *94*, 723–744. [CrossRef]
19. Zhou, N.; Yan, X.; Liang, H.; Tao, X.; Li, G. Multi-image encryption scheme based on quantum 3D Arnold transform and scaled Zhongtang chaotic system. *Quantum Inf. Process.* **2018**, *17*, 338. [CrossRef]
20. Cvitanović, P.; Artuso, R.; Mainieri, R.; Tanner, G.; Vattay, G.; Whelan, N.; Wirzba, A. *Chaos: Classical and Quantum*; Niels Bohr Institute: Copenhagen, Denmark, 2005.
21. Gangadhar, C.; Rao, K.D. Hyperchaos based image encryption. *Int. J. Bifurcation Chaos.* **2009**, *19*, 3833–3839. [CrossRef]
22. Ye, G.; Wong, K.W. An image encryption scheme based on time-delay and hyperchaotic system. *Nonlinear Dyn.* **2013**, *71*, 259–267. [CrossRef]
23. Chai, X.; Gan, Z.; Yang, K.; Chen, Y.; Liu, X. An image encryption algorithm based on the memristive hyperchaotic system, cellular automata and DNA sequence operations. *Signal Process. Image Commun.* **2017**, *52*, 6–19. [CrossRef]
24. Zhang, L.M.; Sun, K.H.; Liu, W.H.; He, S.B. A novel color image encryption scheme using fractional-order hyperchaotic system and DNA sequence operations. *Chin. Phys. B* **2017**, *26*, 100504. [CrossRef]
25. Bouslehi, H.; Seddik, H. Innovative image encryption scheme based on a new rapid hyperchaotic system and random iterative permutation. *Multimed. Tools Appl.* **2018**, *77*, 30841–30863. [CrossRef]
26. Zhou, H.; Wilke, V.S. Research on image selective encryption and compression algorithm under hyperchaotic system. *J. Intell. Fuzzy Syst.* **2018**, *35*, 4329–4337. [CrossRef]
27. Chai, X.; Gan, Z.; Lu, Y.; Zhang, M.; Chen, Y. A novel color image encryption algorithm based on genetic recombination and the four-dimensional memristive hyperchaotic system. *Chin. Phys. B* **2016**, *25*, 100503. [CrossRef]
28. Li, X.; Chen, W.; Wang, Y. Quantum image compression-encryption scheme based on quantum discrete cosine transform. *Int. J. Theor. Phys.* **2018**, *57*, 2904–2919. [CrossRef]

29. Wu, X.; Wang, D.; Kurths, J.; Kan, H. A novel lossless color image encryption scheme using 2D DWT and 6D hyperchaotic system. *Inf. Sci.* **2016**, *349*, 137–153. [CrossRef]
30. Alfalou, A.; Brosseau, C.; Abdallah, N.; Jridi, M. Simultaneous fusion, compression, and encryption of multiple images. *Opt. Express* **2011**, *19*, 24023–24029. [CrossRef]
31. Li, T.; Zhou, M. ECG classification using wavelet packet entropy and random forests. *Entropy* **2016**, *18*, 285. [CrossRef]
32. Annaby, M.H.; Rushdi, M.A.; Nehary, E.A. Color image encryption using random transforms, phase retrieval, chaotic maps, and diffusion. *Opt. Laser Eng.* **2018**, *103*, 9–23. [CrossRef]
33. Li, T.; Hu, Z.; Jia, Y.; Wu, J.; Zhou, Y. Forecasting Crude Oil Prices Using Ensemble Empirical Mode Decomposition and Sparse Bayesian Learning. *Energies* **2018**, *11*, 1882. [CrossRef]
34. Zhou, Y.; Li, T.; Shi, J.; Qian, Z. A CEEMDAN and XGBOOST-based approach to forecast crude oil prices. *Complexity* **2019**, *2019*, 4392785. [CrossRef]
35. Deng, W.; Zhang, S.; Zhao, H.; Yang, X. A novel fault diagnosis method based on integrating empirical wavelet transform and fuzzy entropy for motor bearing. *IEEE Access* **2018**, *6*, 35042–35056. [CrossRef]
36. Zhang, Q.; Wang, Q.; Wei, X. A novel image encryption scheme based on DNA coding and multi-chaotic maps. *Adv. Sci. Lett.* **2010**, *3*, 447–451. [CrossRef]
37. Wu, X.; Kan, H.; Kurths, J. A new color image encryption scheme based on DNA sequences and multiple improved 1D chaotic maps. *Appl. Soft. Comput.* **2015**, *37*, 24–39. [CrossRef]
38. Faragallah, O.S.; Alzain, M.A.; El-Sayed, H.S.; Al-Amri, J.F.; El-Shafai, W.; Afifi, A.; Naeem, E.A.; Soh, B. Block-based optical color image encryption based on double random phase encoding. *IEEE Access* **2019**, *7*, 4184–4194. [CrossRef]
39. Naeem, E.A.; Abd Elnaby, M.M.; Soliman, N.F.; Abbas, A.M.; Faragallah, O.S.; Semary, N.; Hadhoud, M.M.; Alshebeili, S.A.; Abd El-Samie, F.E. Efficient implementation of chaotic image encryption in transform domains. *J. Syst. Softw.* **2014**, *97*, 118–127. [CrossRef]
40. Qian, Z.; Zhang, X.; Ren, Y. JPEG encryption for image rescaling in the encrypted domain. *J. Vis. Commun. Image Represent.* **2015**, *26*, 9–13. [CrossRef]
41. Lima, J.B.; da Silva, E.S.; Campello de Souza, R.M. Cosine transforms over fields of characteristic 2: Fast computation and application to image encryption. *Signal Process. Image Commun.* **2017**, *54*, 130–139. [CrossRef]
42. Wu, J.; Guo, F.; Liang, Y.; Zhou, N. Triple color images encryption algorithm based on scrambling and the reality-preserving fractional discrete cosine transform. *Optik* **2014**, *125*, 4474–4479. [CrossRef]
43. Wu, J.; Zhang, M.; Zhou, N. Image encryption scheme based on random fractional discrete cosine transform and dependent scrambling and diffusion. *J. Mod. Opt.* **2017**, *64*, 334–346. [CrossRef]
44. Chen, B.; Yu, M.; Tian, Y.; Li, L.; Wang, D.; Sun, X. Multiple-parameter fractional quaternion Fourier transform and its application in colour image encryption. *IET Image Process.* **2018**, *12*, 2238–2249. [CrossRef]
45. Liu, X.; Xiao, H.; Li, P.; Zhao, Y. Design and implementation of color image encryption based on qubit rotation about axis. *Chin. J. Electron.* **2018**, *27*, 799–807. [CrossRef]
46. Liansheng, S.; Xiao, Z.; Chongtian, H.; Ailing, T.; Asundi, A.K. Silhouette-free interference-based multiple-image encryption using cascaded fractional Fourier transforms. *Opt. Laser Eng.* **2019**, *113*, 29–37. [CrossRef]
47. Fan, C.; Ding, Q. A novel image encryption scheme based on self-synchronous chaotic stream cipher and wavelet transform. *Entropy* **2018**, *20*, 445. [CrossRef]
48. Lv, X.; Liao, X.; Yang, B. A novel scheme for simultaneous image compression and encryption based on wavelet packet transform and multi-chaotic systems. *Multimed. Tools Appl.* **2018**, *77*, 28633–28663. [CrossRef]
49. Vaish, A.; Kumar, M. Color image encryption using MSVD, DWT and Arnold transform in fractional Fourier domain. *Optik* **2017**, *145*, 273–283. [CrossRef]
50. Raja, S.P. Joint medical image compression-encryption in the cloud using multiscale transform-based image compression encoding techniques. *Sadhana-Acad. Proc. Eng. Sci.* **2019**, *44*, 28. [CrossRef]
51. Hua, Z.; Zhou, Y. Design of image cipher using block-based scrambling and image filtering. *Inf. Sci.* **2017**, *396*, 97–113. [CrossRef]
52. Hua, Z.; Xu, B.; Jin, F.; Huang, H. Image encryption using josephus problem and filtering diffusion. *IEEE Access* **2019**, *7*, 8660–8674. [CrossRef]

53. Wu, Y.; Zhou, Y.; Noonan, J.P.; Agaian, S. Design of image cipher using latin squares. *Inf. Sci.* **2014**, *264*, 317–339. [CrossRef]
54. Panduranga, H.T.; Kumar, S.K.N.; Kiran. Image encryption based on permutation-substitution using chaotic map and Latin Square Image Cipher. *Eur. Phys. J. Spec. Top.* **2014**, *223*, 1663–1677. [CrossRef]
55. Xu, M.; Tian, Z. A novel image encryption algorithm based on self-orthogonal Latin squares. *Optik* **2018**, *171*, 891–903. [CrossRef]
56. Xu, M.; Tian, Z. A novel image cipher based on 3D bit matrix and latin cubes. *Inf. Sci.* **2019**, *478*, 1–14. [CrossRef]
57. Wang, H.; Li, X. A novel hyperchaotic system with infinitely many heteroclinic orbits coined. *Chaos Solitons Fractals* **2018**, *106*, 5–15. [CrossRef]
58. Pano-Azucena, A.; Tlelo-Cuautle, E.; Rodriguez-Gomez, G.; de la Fraga, L. FPGA-based implementation of chaotic oscillators by applying the numerical method based on trigonometric polynomials. *AIP Adv.* **2018**, *8*, 075217. [CrossRef]
59. Adleman, L.M. Molecular computation of solutions to combinatorial problems. *Science* **1994**, *266*, 1021–1024. [CrossRef]
60. Zhu, C.; Hu, Y.; Sun, K. New image encryption algorithm based on hyperchaotic system and ciphertext diffusion in crisscross pattern. *J. Electron. Inf. Tech.* **2012**, *34*, 1735–1743. [CrossRef]
61. Zhan, K.; Wei, D.; Shi, J.; Yu, J. Cross-utilizing hyperchaotic and DNA sequences for image encryption. *J. Electron. Imaging* **2017**, *26*, 013021. [CrossRef]
62. Stinson, D.R. *Cryptography: Theory and Practice*; CRC Press: Boca Raton, FL, USA, 2005.
63. Shannon, C.E. Communication theory of secrecy systems. *Bell Syst. Tech. J.* **1949**, *28*, 656–715. [CrossRef]
64. Wang, Z.; Huang, X.; Li, Y.; Song, X. A new image encryption algorithm based on the fractional-order hyperchaotic Lorenz system. *Chin. Phys. B* **2013**, *22*, 010504. [CrossRef]
65. Wu, Y.; Noonan, J.P.; Agaian, S. NPCR and UACI randomness tests for image encryption. *J. Sel. Areas Telecommun.* **2011**, 31–38.

entropy

MDPI

Article

A Chaotic Electromagnetic Field Optimization Algorithm Based on Fuzzy Entropy for Multilevel Thresholding Color Image Segmentation

Suhang Song, Heming Jia * and Jun Ma

College of Mechanical and Electrical Engineering, Northeast Forestry University, Harbin 150040, China; songsuhang@nefu.edu.cn (S.S.); majun@nefu.edu.cn (J.M.)
* Correspondence: jiaheming@nefu.edu.cn

Received: 1 April 2019; Accepted: 12 April 2019; Published: 15 April 2019

Abstract: Multilevel thresholding segmentation of color images is an important technology in various applications which has received more attention in recent years. The process of determining the optimal threshold values in the case of traditional methods is time-consuming. In order to mitigate the above problem, meta-heuristic algorithms have been employed in this field for searching the optima during the past few years. In this paper, an effective technique of Electromagnetic Field Optimization (EFO) algorithm based on a fuzzy entropy criterion is proposed, and in addition, a novel chaotic strategy is embedded into EFO to develop a new algorithm named CEFO. To evaluate the robustness of the proposed algorithm, other competitive algorithms such as Artificial Bee Colony (ABC), Bat Algorithm (BA), Wind Driven Optimization (WDO), and Bird Swarm Algorithm (BSA) are compared using fuzzy entropy as the fitness function. Furthermore, the proposed segmentation method is also compared with the most widely used approaches of Otsu's variance and Kapur's entropy to verify its segmentation accuracy and efficiency. Experiments are conducted on ten Berkeley benchmark images and the simulation results are presented in terms of peak signal to noise ratio (PSNR), mean structural similarity (MSSIM), feature similarity (FSIM), and computational time (CPU Time) at different threshold levels of 4, 6, 8, and 10 for each test image. A series of experiments can significantly demonstrate the superior performance of the proposed technique, which can deal with multilevel thresholding color image segmentation excellently.

Keywords: fuzzy entropy; electromagnetic field optimization; chaotic strategy; color image segmentation; multilevel thresholding

1. Introduction

Image segmentation is an important technology in image processing, which is a frontier research direction in computer vision, as well as one of the key preprocessing steps in image analysis [1,2]. It has been widely adopted in medicine, agriculture, industrial production, and various other fields. Image segmentation can be defined as the procedure of dividing an image into different regions [3]. In the subsequent research, the relevant regions can be extracted from the segmented image expediently according to specific requirements. Nowadays, the common image segmentation methods include threshold-based, cluster-based, edge-based methods and so on. Thresholding is extensively applied due to its simplicity, efficiency, and robustness. Depending on the number of thresholds, it can be classified as bi-level segmentation and multilevel segmentation [4]. Bi-level thresholding techniques use one threshold to partition an image into two segments; whereas multilevel segmentation determines several thresholds to separate an image into more than two classes. Many thresholding approaches have been proposed by scholars around the world in the past few years, Otsu's (between-class variance criterion) [5,6] technique pushes the thresholding segmentation to an upsurge and inspires the scholars constantly in this field. Then diverse entropy-based criteria have emerged in the thresholding segmentation study, such as maximum entropy (Kapur's) [7], minimum cross entropy [8], fuzzy entropy [9], etc.

Gray-scale image thresholding technology is relatively popular and mature. Compared with the segmentation of gray-scale images, color image segmentation plays a more beneficial role in practical applications, which separates an image into several disjoint and homogenous components based on the information of texture, color or histogram [10]. Color image segmentation is more complex and challenging than gray-scale images. Nevertheless, considering that color images contain more characteristics and they are closer to human visual effects [11], the research of color image segmentation is more meaningful. There will appear some problems when a traditional segmentation method is adopted to segment a color image, for example, the computation is massive and accuracy of segmented images cannot be guaranteed [12,13]. In this paper, fuzzy entropy is one of the research objects with high segmentation accuracy. In the fuzzy entropy thresholding technique, each threshold needs to be determined by three fuzzy parameters. Hence the calculation of thresholds is more accurate, at the same time the process is more complicated and the running time of the program will be longer. With the improvement of the threshold level, the computation of fuzzy entropy will exponentially increase for searching the optimal thresholds and the efficiency of segmentation will gradually decrease [14–16]. In order to enhance the practicability of fuzzy entropy thresholding technique, this paper combines fuzzy entropy thresholding with intelligent optimization algorithms to improve the performance with respect to accuracy and efficiency.

Meta-heuristic algorithms are utilized to obtain the optimal solution of the problem [17]. Generally, they are inspired by nature and try to handle the problems from mimicking ethology, biology or physics [18]. For instance, Bird Swarm Algorithm (BSA) [19], Firefly Algorithm (FA) [20], and Flower Pollination Algorithm (FPA) [21] are inspired by ethology or biology; Electromagnetic Optimization (EMO) [22], Wind Driven Optimization (WDO) [23], and Gravitational Search Algorithm (GSA) [24] are inspired by physics. At present, a number of scholars have coupled the optimization algorithms with the field of image segmentation in the literature. For instance, Sowjanya et al. [25] combined a WDO algorithm with Otsu's method for the segmentation of brain MRI images, it has shown the superior performance in the experiment results. Wasim et al. [26] proposed an improved Bee Algorithm (BA) for multilevel image segmentation, whereby they embedded Levy fight into a Bees Algorithm (the Levy Bees Algorithm, LBA), and the results show that LBA is more stable than BA in this field. Rakoth et al. [27] tried to combine Dragonfly Optimization with Self-Adaptive weight (SADFO) and used SADFO for image segmentation experiments with satisfying results. These references confirm the feasibility of applying optimization algorithms to image thresholding segmentation. However, the above experiments all concentrate on gray-scale images and do not extend the experiments to the analysis of color images. Applying meta-heuristic algorithms to the field of multilevel image segmentation can enhance the convergence speed and efficiency [28]. Therefore, in this paper, Electromagnetic Field Optimization algorithm (EFO) [29] is modified and combined with fuzzy entropy thresholding method to eliminate the complex computation, which is used into the multilevel color image segmentation field for searching the best threshold values.

Electromagnetic Field Optimization is a new meta-heuristic algorithm inspired by the electromagnetic theory developed in physics. EFO algorithm has been applied in several applications, for example, Behnam et al. [30] created a method using EFO for hiding sensitive rules simultaneously, which has fewer lost rules than other well-known algorithms. Bouchekara et al. [31] proposed the optimal coordination of directional overcurrent relays based on EFO, and the results show that EFO is better than other optimization algorithms such as Particle Swarm Optimization (PSO) [32], or the Differential Evolution (DE) algorithm [33], etc. This paper embeds a new chaos strategy into standard EFO algorithm according to the specific problem of color image segmentation named as Chaotic Electromagnetic Field Optimization (CEFO). Employing the CEFO algorithm to optimize the fuzzy parameters which determine the optimal thresholds of an image in fuzzy entropy. To the best of our knowledge, this topic has not been investigated yet. The rest of this paper is organized as follows: in Section 2, the concept of EFO algorithm is elaborated. In Section 3, the chaotic strategy in CEFO algorithm is introduced and explained. In Section 4, the problem definitions and formulas of the Otsu's, Kapur's entropy, and the fuzzy entropy are illustrated. In Section 5, the experimental environment is reported. In Section 6, the experimental results and discussions are provided and analyzed. Finally, a brief conclusion of this paper and future works are drawn in Section 7.

2. Electromagnetic Field Optimization

Electromagnetic Field Optimization is a novel meta-heuristic intelligent algorithm proposed by Hosein in 2016 [29]. In contrast to the swarm-based meta-heuristic algorithms widely inspired by biology, the EFO algorithm is based on the electromagnetic field principle used in physics. In the EFO algorithm, due to the forces of attraction and repulsion in the electromagnetic field, the electromagnetic particle (EMP) keeps away from the worst solution and moves towards the best solution. In the end, all the electromagnetic particles (EMPs) gather around the optimal solution.

A magnetic field is generated around the electrified iron core, which is made of an electromagnet. An electromagnet has only one polarity and it is contingent on the direction of the electric current. Hence, an electromagnet has two characteristics of attraction or repulsion, electromagnets with the different polarity attract each other, and those with identical polarity repel each other. The intensity of attraction is 5-10% higher than repulsion and the ratio between attraction and repulsion is set as golden ratio [29,31], which can promote the algorithm to explore the optimal solution effectively in the search space. The essence of the optimization problem is to find the pole (maximum or minimum) about the objective function and the corresponding fitness in the prescriptive range [34]. Each potential solution of the problem is represented with an electromagnetic particle composed of a group of electromagnets. The electromagnetic field comprises several electromagnetic particles and it can be defined as a space in 1-D (dimension), 2-D, 3-D, or hyperdimensional space [35]. The number of electromagnets of an electromagnetic particle corresponds to variables of the optimization problem, as well as the dimension of the electromagnetic space. Moreover, all electromagnets of one electromagnetic particle have the same polarity. Therefore, an electromagnetic particle has the same polarity with its electromagnets. The set of electromagnetic particles can be considered in a matrix as:

$$EMPs = \begin{bmatrix} P_{1,1} & P_{1,2} & \cdots & P_{1,d} \\ P_{2,1} & P_{2,2} & \cdots & P_{2,d} \\ \vdots & \vdots & \vdots & \vdots \\ P_{n,1} & P_{n,2} & \cdots & P_{n,d} \end{bmatrix} \tag{1}$$

where n is the number of electromagnetic particles and j is the number of variables (dimension).

The mechanism of the EFO algorithm can be described as follows:

Step 1: A certain number of electromagnetic particles are generated randomly in the electromagnetic field, and the fitness of each electromagnetic particle is evaluated by the objective function. Then the electromagnetic particles are sorted on the basis of their fitness.

Step 2: The electromagnetic field is divided into three regions: positive, negative and neutral. Then all electromagnetic particles are classified into these three groups. The first group consists of the best particles with positive polarity. The second group consists of the worst particles with negative polarity. The third group consists of neutral particles which have a little negative polarity almost near zero. And all electromagnetic particles are located in the corresponding electromagnetic regions.

Step 3: In each iteration of the algorithm, a new electromagnetic particle (EMP^{New}) is generated. If the fitness of EMP^{New} is better than the original worst particle, the EMP^{New} will remain and its fitness and polarity will depend on the list of fitness, furthermore, the original worst particle will be eliminated. If else, the will be eliminated directly. This process continues until the algorithm reaches the maximum number of iterations.

The core of the EFO is the method of generating EMP^{New} in each iteration, and each electromagnet in EMP^{New} is shaped separately. The main process can be described as follows: three electromagnetic particles are randomly extracted from three electromagnetic regions (one EMP from each region), and then three electromagnets are randomly extracted from three electromagnetic particles obtained just now (one electromagnet from each EMP). Consequently, there are three electromagnets with different polarities. The neutral electromagnet is attracted and repelled by positive and negative electromagnets. Owing to the intensity of attraction is stronger than repulsion and the neutral electromagnet has a slight negative polarity, the neutral electromagnet moves a distance away from the negative electromagnet and approaches towards the positive electromagnet. In other words, each electromagnet in EMP^{New} is a result of interaction between attraction and repulsion, which is shown in Figure 1.

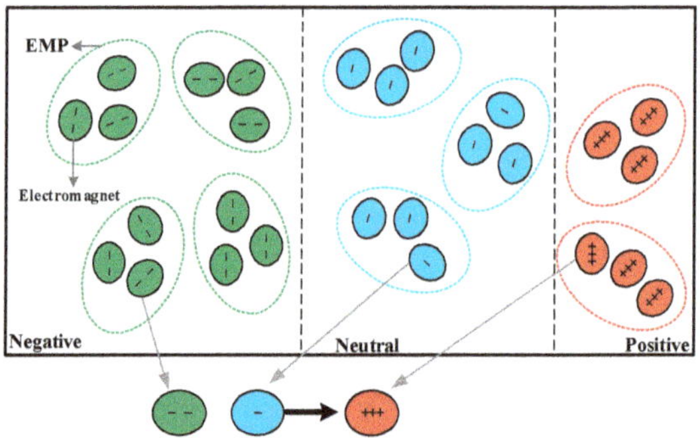

Figure 1. Schematic diagram of the electromagnetic field. The relationship between the electromagnetic particle (EMP) and electromagnet. The method of generating the new electromagnetic particle

Figure 1 shows the process of generating EMP^{New}, in this figure, each electromagnetic particle contains three electromagnets for example, and positive, neutral and negative electromagnets are colored as green, blue and red respectively. In accordance with the above mechanism, three electromagnets of EMP^{New} is selected from nine original electromagnets, which increases randomness and enhances the strength of the optimization algorithm. Establishing a mathematical model to describe the update mechanism of EMP^{New} as below:

$$D_j^{P_j K_j} = EMP_j^{P_j} - EMP_j^{K_j} \tag{2}$$

$$D_j^{N_j K_j} = EMP_j^{N_j} - EMP_j^{K_j} \tag{3}$$

$$EMP_j^{New} = EMP_j^{K_j} + [(\varphi * r) * D_j^{P_j K_j}] - (r * D_j^{N_j K_j}) \tag{4}$$

where j is the number of electromagnets in EMP; $EMP_j^{P_j}$ is the positive electromagnet; $EMP_j^{N_j}$ is the negative electromagnet; $EMP_j^{K_j}$ is the neutral electromagnet; $D_j^{P_j K_j}$ is the distance between positive and neutral electromagnets. $D_j^{N_j K_j}$ is the distance between negative and neutral electromagnets; r is the random value between 0 and 1; φ is the golden ratio of $(\sqrt{5}+1)/2$.

In order to preserve the diversity of particles in the electromagnetic field and reduce the probability of falling into local optima [36], randomness is an indispensable part in EFO algorithm. Therefore, the probability of *Ps_rate* about the new position is determined by the selected electromagnet from a positive field, which accelerates the convergence rate and improves the accuracy of the optimum. Additionally, the probability of *R_rate* is used to replace one electromagnet in EMP^{New} with randomly generated electromagnet within the space. The most important feature of EFO algorithm is the high degree of cooperation among particles. Another pivotal characteristic is high randomization, which avoids obtaining the local optimum. Meanwhile, the application of the golden ratio makes EFO more efficient. All of the above strategies lead EFO to a robust optimization algorithm.

3. Proposed Algorithm

One of the essential points in the EFO algorithm is the degree of chaos about the electromagnetic particles in the electromagnetic field; if the degree of chaos is higher, the search power will be stronger. In the literature, the initial position of electromagnetic particles is processed by a chaotic strategy, which disturbs the distribution of particles and increases the unpredictability of the system.

Chaotic phenomena refer to the external complex behavior in a non-linear deterministic system due to the inherent randomness [37]. Almost all meta-heuristic algorithms need to be initialized randomly, and usually it is achieved by using probability distribution, which can advantageous to replace such randomness with chaotic map [38]. Owing to the dynamic behavior of chaos, chaotic maps have been commonly acknowledged in the field of optimization, which can promote algorithms in exploring optima more effective globally in the search space. Table 1 lists some common chaotic maps, which are expressed by mathematical equations.

Table 1. Chaotic maps.

Name	Chaotic Map
Logistic	$x_{i+1} = ax_i(1 - x_i)$
Sine	$x_{i+1} = \frac{a}{4}\sin(\pi x_i)$
Cubic	$x_{i+1} = ax_i(1 - x_i^2)$
Circle	$x_{i+1} = mod(x_i + b - (\frac{a}{2\pi})\sin(2\pi x_i), 1)$
Iterative	$x_{i+1} = \sin(\frac{a\pi}{x_i})$
Tent	$x_{i+1} = a - (1 - a)\lvert x_i \rvert$

For instance, logistic chaos is widely used because of its simple expression and good performance, and it is shown in Figure 2. As can be seen, the logistic system has missed certain values. In consideration of the multilevel color image segmentation problem, this paper proposes a new chaotic map as follows:

$$x_{n+1} = rand() \times \sin(2\pi x_n) + x_n \tag{5}$$

where $rand()$ is the random value between 0 and 1.

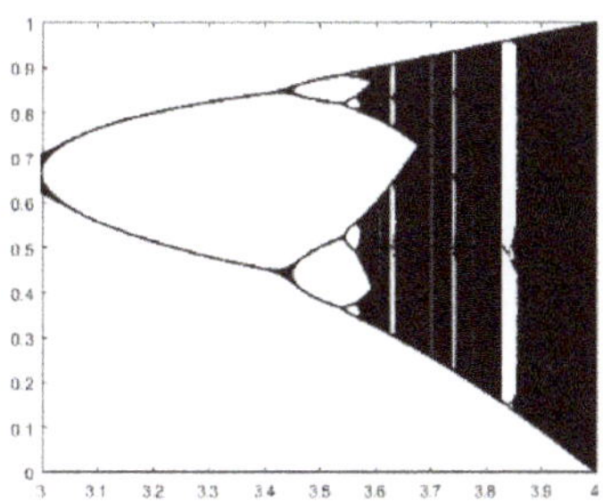

Figure 2. Logistic Chaotic Map.

The new chaotic map is shown in Figure 3 and its distribution is more symmetrical than Logistic chaotic map. Taking advantage of this chaos strategy in EFO, the total performance of the algorithm will be improved and it is known as CEFO. The pseudo-code of the CEFO algorithm is presented in Algorithm 1.

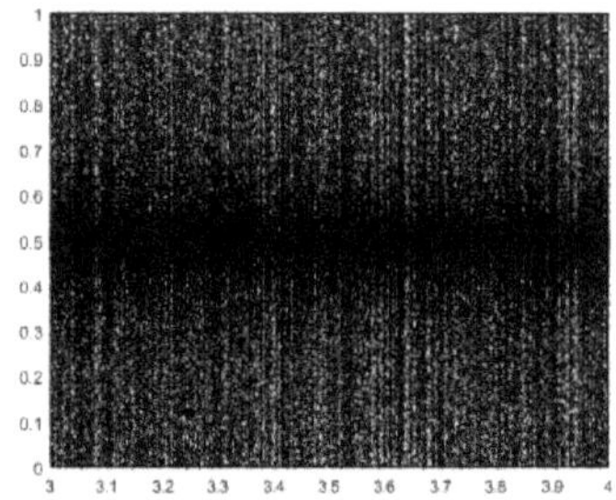

Figure 3. Chaotic Map in this paper.

Algorithm 1. Pseudo-code of CEFO algorithm

```
/* Part 1: Algorithm parameters initialization */
  N_var: The number of electromagnets in each electromagnetic particle.
  N_emp: The number of electromagnetic particles in population.
  Ps_rate: The probability of changing one electromagnet with a random electromagnet.
  R_rate: The probability of selecting electromagnets from the positive field.
  P_field: The portion of particles belonging to positive.
  N_field: The portion of particles belonging to negative.
  min = lower boundary; max = upper boundary
/* Part 2: Main loop of the algorithm */
  for i = 1 to N_emp do
      for j = 1 to N_var do
        position [i, j] = min + rand ( )*(max − min)
      end for
  end for
  Update position by using the chaotic map of Equation (5)
  fitness = function (position)
  while t (current iteration) < max iterations
      Divide the electromagnetic field into three regions
      for i = 1 to N_var do
          if rand (0,1) > Ps_rate
             Generate the EMP^New by Equation (4)
          else
             Generate the EMP^New from positive particles
          end if
          Check if any particle beyond the search space
      end for
      if rand (0,1) < R_rate
          Change one electromagnet of EMP^New randomly
      end if
      Compare the fitness of EMP^New with worst particle
      t = t + 1
  end while
  Output the best particle
```

4. Thresholding Segmentation Methods

The process of multilevel thresholding color image segmentation is to find more than two optimal thresholds to segment three components (red, green, and blue) respectively. In RGB images, each color component consists of P pixels and L number of gray levels. The obtained thresholds are within the range of [0, $L-1$], L is considered as 256 and each gray-level is associated with the histogram representing the frequency of its gray level pixel used by $g(x, y)$.

4.1. Between-Class Variance Thresholding

Between-class variance (Otsu's) [5] thresholding method can be defined as follows:

Assuming that $n-1$ thresholds form the threshold vector $T = [t_1, t_2, \cdots, t_{n-1}]$ to split an image into n classes:

$$\begin{cases} C_1 = \{(x,y)|0 \le g(x,y) \le t_1 - 1\} \\ C_2 = \{(x,y)|t_1 \le g(x,y) \le t_2 - 1\} \\ \vdots \\ C_n = \{(x,y)|t_{n-1} \le g(x,y) \le L - 1\} \end{cases} \tag{6}$$

Constructing image histogram $\{f_0, f_1, \cdots, f_{L-1}\}$, where f_i is the frequency of gray-level i. Then, the probability of gray-level i can be represented as:

$$p_i = \frac{f_i}{\sum_{i=0}^{L-1} f_i}, \sum_{i=0}^{L-1} p_i = 1 \tag{7}$$

For every class C_k, the cumulative probability ω_k and average gray level μ_k in every region can be defined as:

$$\omega_k = \sum_{i \in C_k} p_i, \ \mu_k = \sum_{i \in C_k} \frac{i \cdot p_i}{\omega_k} \tag{8}$$

and Otsu's function can be expressed as:

$$\sigma_B^2 = \sum_{k=0}^{K} \omega_k \cdot (\mu_k - \mu_T)^2, \ \mu_T = \sum_{i=0}^{L-1} i \cdot p_i \tag{9}$$

where μ_k is the average gray intensity of the image.

Therefore, the optimal threshold vector is as follows:

$$T^* = \text{argmax}(\sigma_B^2) \tag{10}$$

4.2. Kapur's Entropy Thresholding

Kapur's entropy method maximizes the entropy value of the segmented histogram such that each separated region has more centralized distribution [39]. Extending Kapur's entropy for multilevel image segmentation problem:

$$\begin{aligned} H_1 &= -\sum_{i=0}^{t_1-1} \frac{p_i}{\omega_1} \ln\frac{p_i}{\omega_1}, \ \omega_1 = \sum_{i=0}^{t_1-1} p_i \\ H_2 &= -\sum_{i=t_1}^{t_2-1} \frac{p_i}{\omega_2} \ln\frac{p_i}{\omega_2}, \ \omega_2 = \sum_{i=t_1}^{t_2-1} p_i \\ H_j &= -\sum_{i=t_{j-1}}^{t_j-1} \frac{p_i}{\omega_j} \ln\frac{p_i}{\omega_j}, \ \omega_j = \sum_{i=t_{j-1}}^{t_j-1} p_i \\ H_n &= -\sum_{i=t_n}^{L-1} \frac{p_i}{\omega_n} \ln\frac{p_i}{\omega_n}, \ \omega_n = \sum_{i=t_n}^{L-1} p_i \end{aligned} \tag{11}$$

where H_j represents the entropy value of j-th region in the image.

There are n thresholds which can be configured as the n dimensional optimization problem. And the optimal threshold vector is obtained analogously by:

$$T^* = \text{argmax}(\sum_{i=0}^{m} H_i) \tag{12}$$

4.3. Fuzzy Entropy Thresholding

In the fuzzy entropy technique, let an original image be $D = \{(i,j)|i = 0,\cdots,M-1; j = 0,\cdots,N-1\}$, where M and N represent the width and height of an image. Supposed that t_1 and t_2 are two thresholds to divide the original image into 3 parts named as E_d, E_m, E_b [10]. E_d consists of pixels of low gray levels; E_m is made of pixels with middle gray levels; E_b is composed of pixels of high gray levels. Usually, using (13) to calculate the image histogram:

$$h_k = \frac{n_k}{M*N} \tag{13}$$

where $k = 0,1,\cdots,255$; n_k is the number of the k-th pixel in D_k; h_k is the histogram of the image at gray-level k, $\sum_{k=0}^{255} h_k = 1$.

Consider $\Pi_3 = \{E_d, E_m, E_b\}$ as an unknown probabilistic partition of D, whose probability distribution can be expressed as:

$$p_d = P(E_d); \quad p_m = P(E_m); \quad p_b = P(E_b) \tag{14}$$

For each $(i,j) \in D$, let:

$$\begin{aligned} D_d &= \{(i,j)|0 \le g(i,j) \le t_1\}, \\ D_m &= \{(i,j)|t_1 \le g(i,j) \le t_2\}, \\ D_b &= \{(i,j)|t_2 \le g(i,j) \le L-1\} \end{aligned} \tag{15}$$

Utilizing μ_d, μ_m, μ_b as the membership functions of E_d, E_m, E_b, which is shown in Figure 4 [40]. There are six fuzzy parameters of u_1, v_1, w_1, u_2, v_2, w_2 in the membership functions, in other words, t_1 and t_2 are determined by these six parameters. According to the above statement, we can have the probability distribution of three regions expressed as:

$$\begin{aligned} p_d &= \sum_{k=0}^{255} p_k \cdot p_{d|k} = \sum_{k=0}^{255} p_k \cdot \mu_d(k) \\ p_m &= \sum_{k=0}^{255} p_k \cdot p_{m|k} = \sum_{k=0}^{255} p_k \cdot \mu_m(k) \\ p_b &= \sum_{k=0}^{255} p_k \cdot p_{b|k} = \sum_{k=0}^{255} p_k \cdot \mu_b(k) \end{aligned} \tag{16}$$

where $p_{d|k}$, $p_{m|k}$, $p_{b|k}$ are the conditional probability of a pixel partitioned into three classes. Moreover, a pixel of k in an image satisfies the constraint of $p_{d|k} + p_{m|k} + p_{b|k} = 1$.

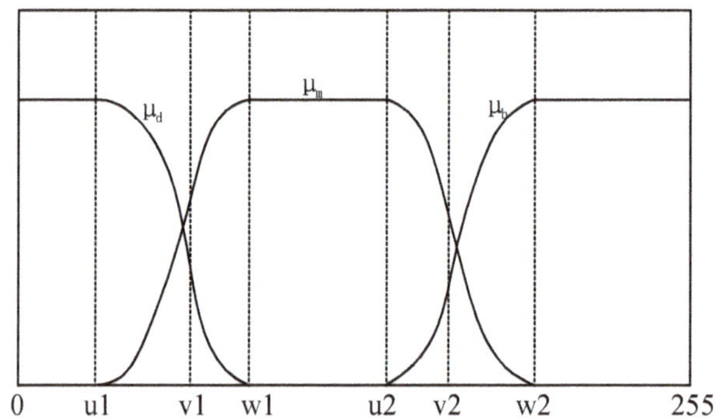

Figure 4. Membership function graph.

The three membership functions have been shown in Figure 4. And these mathematical formulas are defined as follows:

$$\mu_d(k) = \begin{cases} 1 & k \leq u_1 \\ 1 - \frac{(k-u_1)^2}{(w_1-u_1)\cdot(v_1-u_1)} & u_1 \leq k \leq v_1 \\ \frac{(k-w_1)^2}{(w_1-u_1)\cdot(w_1-v_1)} & v_1 \leq k \leq w_1 \\ 0 & k \geq w_1 \end{cases} \tag{17}$$

$$\mu_m(k) = \begin{cases} 0 & k \leq u_1 \\ \frac{(k-u_1)^2}{(w_1-u_1)\cdot(v_1-u_1)} & u_1 \leq k \leq v_1 \\ 1 - \frac{(k-w_1)^2}{(w_1-u_1)\cdot(w_1-v_1)} & v_1 \leq k \leq w_1 \\ 1 & w_1 \leq k \leq u_2 \end{cases} \tag{18}$$

$$\mu_b(k) = \begin{cases} 0 & k \leq u_2 \\ \frac{(k-u_2)^2}{(w_2-u_2)\cdot(v_2-u_2)} & u_2 \leq k \leq v_2 \\ 1 - \frac{(k-w_2)^2}{(w_2-u_2)\cdot(w_{21}-v_2)} & v_2 \leq k \leq w_2 \\ 1 & k \geq w_2 \end{cases} \tag{19}$$

where $u_1, v_1, w_1, u_2, v_2, w_2$ should meet the condition of $0 \leq u_1 < v_1 < w_1 < u_2 < v_2 < w_2 \leq 255$.

Then, the fuzzy entropy of each part is as follows:

$$\begin{aligned} H_d &= -\sum_{k=0}^{255} \frac{p_k \cdot \mu_d(k)}{p_d} \cdot \ln\left(\frac{p_k \cdot \mu_d(k)}{p_d}\right) \\ H_m &= -\sum_{k=0}^{255} \frac{p_k \cdot \mu_m(k)}{p_m} \cdot \ln\left(\frac{p_k \cdot \mu_m(k)}{p_m}\right) \\ H_b &= -\sum_{k=0}^{255} \frac{p_k \cdot \mu_b(k)}{p_b} \cdot \ln\left(\frac{p_k \cdot \mu_b(k)}{p_b}\right) \end{aligned} \tag{20}$$

The whole fuzzy entropy function is defined as:

$$H(u_1, v_1, w_1, u_2, v_2, w_2) = H_d + H_m + H_b \tag{21}$$

Equation (21) is determined by six variables which are called fuzzy parameters. Seeking the optimal group of u_1, v_1, w_1, u_2, v_2, w_2 when (21) reach the maximum value. Therefore, the most applicable threshold can be calculated as:

$$\begin{aligned} \mu_d(t_1) = \mu_m(t_1) = 0.5 \\ \mu_m(t_2) = \mu_b(t_2) = 0.5 \end{aligned} \tag{22}$$

As is shown in Figure 4, according to the above equation, t_1 and t_2 can be defined by (17)–(19), and the result is as follows:

$$\begin{aligned} t_1 &= \begin{cases} u_1 + \sqrt{(w_1 - u_1)\cdot(v_1 - u_1)/2} & (u_1 + w_1)/2 < v_1 < w_1 \\ w_1 - \sqrt{(w_1 - u_1)\cdot(w_1 - v_1)/2} & u_1 < v_1 < (u_1 + w_1)/2 \end{cases} \\ t_2 &= \begin{cases} u_2 + \sqrt{(w_2 - u_2)\cdot(v_2 - u_2)/2} & (u_2 + w_2)/2 < v_2 < w_2 \\ w_2 - \sqrt{(w_1 - u_2)\cdot(w_2 - v_2)/2} & u_2 < v_2 < (u_2 + w_2)/2 \end{cases} \end{aligned} \tag{23}$$

Fuzzy entropy thresholding can meet the requirement from single threshold segmentation to multiple thresholds segmentation, and the optimal threshold vector obtained is more precise. However, each threshold should be determined by three parameters in fuzzy entropy thresholding, and thresholds need to be defined by $3n$ fuzzy parameters [41,42].

With the increase of threshold level gradually, the degree of the computation will be significantly risen, which diminish the speed of the process and the practicability will be reduced. In order to improve the convergence efficiency, it is necessary to use the optimization algorithm for searching the optimal threshold vector. This paper takes advantage of the CEFO to ensure the segmentation accuracy and greatly decrease the execution time. The general flow of fuzzy entropy thresholding based on the CEFO algorithm is presented in Figure 5.

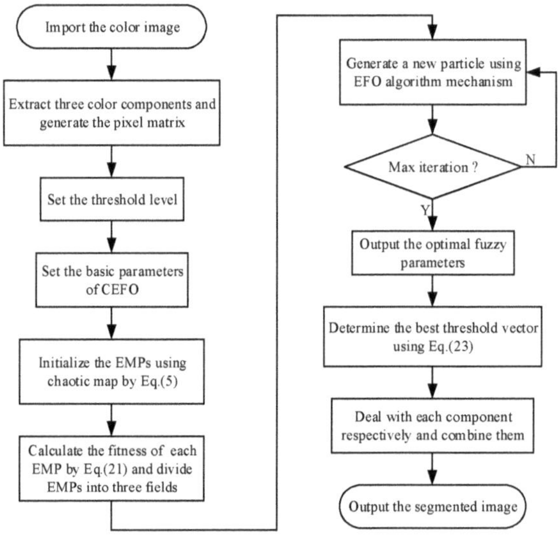

Figure 5. Flow chart of fuzzy entropy thresholding method based on the Chaotic Electromagnetic Field Optimization (CEFO) algorithm.

5. Experimental Environment

In order to verify the superiority of the CEFO algorithm in dealing with the multilevel color image segmentation problem, this section will introduce the description of our benchmark images and then select several other algorithms for comparison. The parameters of each algorithm will be described firstly and a series of quality metrics used to evaluate the quality of segmented images will be calculated at the end.

5.1. Benchmark Images

In this experiment, ten images are chosen from the Berkeley segmentation data set, which is shown in Figure 6. It has presented the histogram of three components about every color image. Among these images, Test 1–3 are animal images; Test 4 and 5 are about human; Test 7 and 8 are landmark buildings; Test 6 and 9 are images related to landscape architecture; Test 10 is the normal scenery image.

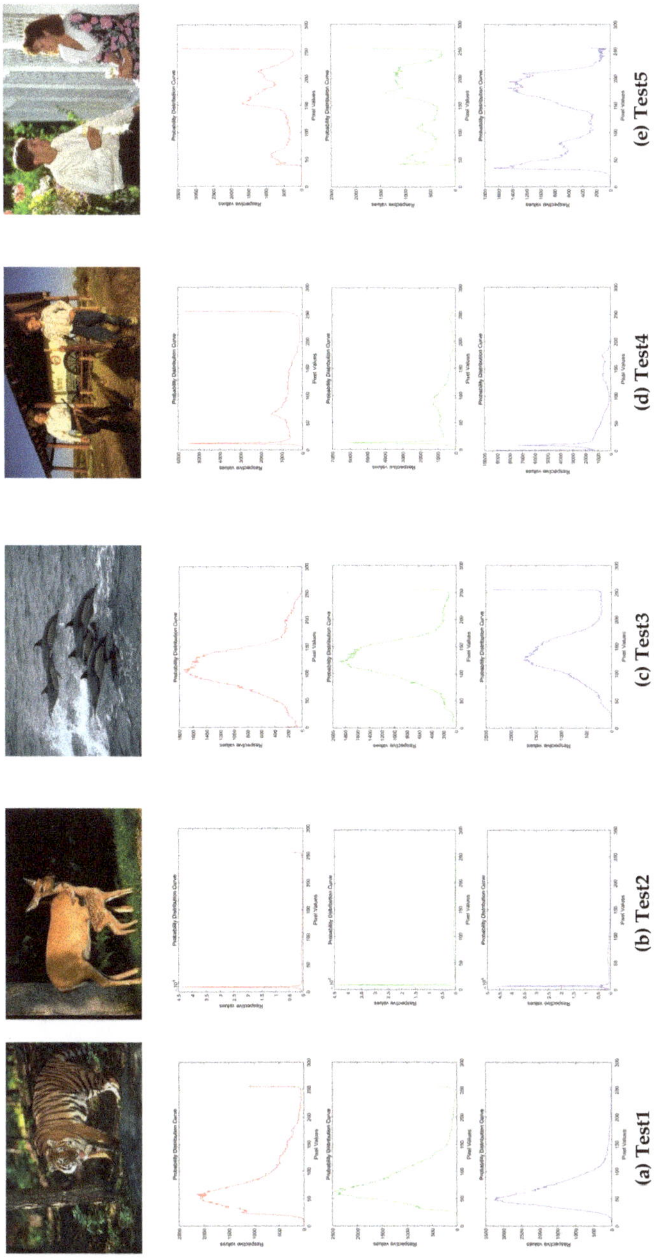

Figure 6. *Cont.*

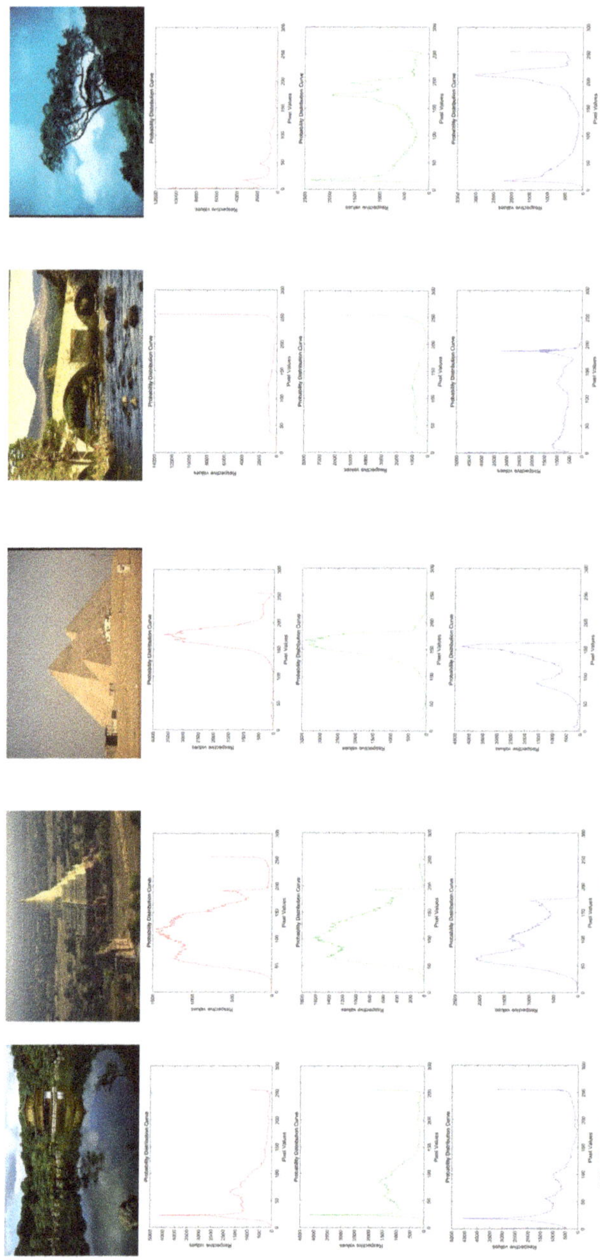

(f) Test6 (g) Test7 (h) Test8 (i) Test9 (j) Test10

Figure 6. Experimental images of Berkeley. Ten classical images and their histograms of three component (red, green, and blue) are exhibited.

5.2. Experimental Settings

When applied to solve the problem of multilevel color image segmentation, different meta-heuristic algorithms have different optimization performances due to their strategies and mathematical formulations [43]. Therefore, it is essential to compare the CEFO algorithm with other different algorithms such as EFO, ABC [44], BA [10], BSA [19], WDO [23]. Among these algorithms, ABC, BA, and BSA are proposed from biology; EFO and WDO are inspired from physics. The number of maximum iterations of each algorithm is set to 500, and the initial population is set to 15, with a total of 30 runs per algorithm, other specific parameters are presented in Table 2.

Table 2. Specific values of parameters used in selected algorithms.

Algorithm	Parameter	Explanation	Value
EFO	P_{field}	The portion of particles belonging to the positive field.	0.1
	N_{field}	The portion of particles belonging to the negative field.	0.45
	Ps_{field}	The probability of selecting electromagnets directly from the positive field.	0.3
	R_{rate}	The probability of changing one electromagnet directly from the positive field.	0.2
ABC	$limit$	The value of the max trial limit.	10
BA	r_i	The rate of pulse emission.	[0, 1]
	A_i	The value of loudness.	[1, 2]
WDO	a	A constant.	0.4
	g	The constant of gravitation.	0.2
	RT	A coefficient.	3
	c	Coriolis coefficient.	0.4
BSA	a_1, a_2	The values of indirect and direct effects on the birds' vigilance behaviors.	2
	c_1, c_2	The values of the cognitive coefficient and social coefficient.	2
	FL	The frequency of birds' flight behaviors.	0.6

All the algorithms are programmed in Matlab R2016a (The Mathworks Inc., Natick, MA, USA) and implemented on a Windows 7 – 64 bit with 8 GB RAM environment.

5.3. Segmented Image Quality Metrics

To evaluate the quality of segmented images under different algorithms at selected threshold levels, four metrics are selected as follows [45,46]:

- Peak Signal to Noise Ratio (PSNR)

The index is used to measure the difference between the original image and the segmented image, and a higher value is gained when the segmented image has a better effect. It can be defined as:

$$\begin{aligned} PSNR(x,y) &= 20 \cdot \log_{10}\left(\frac{255}{\sqrt{MSE}}\right) \\ MSE &= \frac{1}{MN}\sum_{i=0}^{M-1}\sum_{j=0}^{N-1}\left\|x(i,j) - y(i,j)\right\|^2 \end{aligned} \tag{24}$$

where M and N represent the size of the image; x is the original image; y is the segmented image.

- Mean Structural Similarity (MSSIM)

The index evaluates the overall image quality, which is in the range of $[-1,1]$. The higher value of MSSIM is obtained when it represents the segmented image is more similar to the original image. The MSSIM is the average of every component and SSIM can be calculated as:

$$SSIM(x,y) = \frac{(2\mu_x\mu_y + c_1)\,(2\sigma_{xy} + c_2)}{(\mu_x^2 + \mu_y^2 + c_1)\,(\sigma_x^2 + \sigma_y^2 + c_2)} \tag{25}$$

- Feature similarity (FSIM)

The index is in the range of $[0, 1]$, and the segmented image is better when the value is closer to 1. The FSIM can be expressed as:

$$FSIM(x, y) = \frac{\sum_{x \in \Omega} S_L(X) PC_m(x)}{\sum_{x \in \Omega} PC_m(x)} \tag{26}$$

- Computation Time (CPU Time)

The index measures the convergence rate of each algorithm. The algorithm is more efficient when the time is shorter.

6. Results and Discussions

6.1. Comparison of Other Meta-Heuristic Algorithms

Utilizing 6 algorithms based on fuzzy entropy criterion to conduct the experiment on 10 images at the threshold level of 4, 6, 8, and 10 (K = 4, 6, 8, 10). The results of the optimal threshold vector are presented in Tables 3–5 exhibiting each threshold level of three component about every image. And the results of segmented images are presented in Figures 7–16, which take each Test image as a group. Furthermore, the results of the four metrics are shown in Tables 6 and 7.

Table 3. Comparison of optimal threshold values between Chaotic Electromagnetic Field Optimization (CEFO) and Electromagnetic Field Optimization (EFO) at K = 4, 6, 8, 10 based on fuzzy entropy.

Image	K	CEFO			EFO		
		R	G	B	R	G	B
Test 1	4	56 93 153 187	57 88 132 191	59 10 157 197	61 90 144 188	18 74 115 170	58 87 148 202
	6	20 59 92 128 167 199	26 47 81 114 161 203	25 48 76 111 138 195	26 60 95 128 162 198	19 47 80 119 171 210	18 57 85 122 154 212
	8	11 31 58 82 107 132 168 211	12 42 64 92 114 149 181 210	18 50 73 97 143 168 196 223	17 47 71 90 115 138 171 209	12 39 67 93 130 170 205 223	26 56 78 100 113 162 190 216
	10	14 37 55 75 90 111 130 159 182 210	11 41 61 95 122 146 168 185 208 237	14 40 64 80 105 136 161 189 212 231	10 26 56 82 102 133 148 167 192 215	11 32 59 76 90 107 134 166 198 222	24 41 56 72 93 126 156 181 203 229
Test 2	4	10 93 114 171	86 143 194 226	91 110 137 189	80 116 149 183	93 113 146 216	104 124 170 206
	6	18 41 75 107 147 177	87 101 133 153 192 234	54 72 97 121 156 192	73 83 108 139 176 219	94 114 135 162 197 231	16 50 85 129 156 199
	8	15 27 44 59 96 127 171 204	14 22 43 65 117 144 190 222	39 61 82 117 148 176 195 223	7 34 55 78 107 135 166 199	49 65 76 139 168 193 211 232	45 61 82 101 129 154 182 213
	10	32 55 75 108 144 171 195 215 226 239	5 50 65 85 102 121 139 167 197 228	31 47 64 87 108 125 158 179 209 239	4 24 47 67 102 122 147 180 204 223	19 33 53 71 99 119 144 170 195 224	19 36 57 80 108 141 154 177 201 230
Test 3	4	44 92 139 188	31 92 139 188	27 97 138 185	33 80 126 181	37 87 142 206	29 106 154 188
	6	32 62 93 128 173 199	19 53 90 119 162 206	36 71 105 136 183 208	32 76 116 150 177 207	50 69 101 143 190 216	15 46 70 113 153 197
	8	22 43 68 91 112 138 172 208	17 51 73 98 123 148 181 207	11 36 58 88 106 141 170 213	19 54 94 125 141 164 190 214	25 49 72 96 126 156 192 223	15 49 71 95 128 159 187 214
	10	26 39 55 69 86 110 131 154 175 207	42 29 56 79 104 121 141 163 197 219	9 25 49 82 109 132 154 176 199 222	17 31 45 62 93 124 168 194 212 229	14 39 56 76 95 114 148 176 196 224	16 13 34 58 82 115 150 180 199 223
Test 4	4	52 84 127 164	67 102 141 204	49 74 111 155	59 88 129 164	64 105 153 199	61 101 145 203
	6	38 62 97 137 172 195	48 71 111 144 175 213	43 77 109 147 178 235	40 65 89 112 147 184	44 67 105 126 167 211	39 63 98 134 160 190
	8	13 31 54 92 110 142 175 198	35 60 86 114 137 169 200 226	24 41 64 94 129 156 193 224	15 26 46 68 97 134 160 189	18 43 73 100 128 155 190 225	23 57 105 137 169 195 221 237
	10	6 25 43 64 84 106 132 152 172 189	7 27 39 52 75 99 130 157 190 221	15 25 48 78 103 130 157 178 219 240	8 18 31 48 74 100 124 153 181 211	8 23 39 57 80 104 127 157 188 215	10 27 50 87 117 146 171 194 218 234
Test 5	4	52 98 149 198	55 89 164 215	77 125 162 209	87 122 152 190	38 58 106 162	80 120 160 194
	6	22 63 93 131 179 225	16 69 97 125 158 204	15 61 97 131 166 207	23 64 90 140 183 213	26 75 114 134 169 211	20 59 84 132 169 205
	8	16 59 77 94 116 134 162 199	20 58 75 98 129 155 185 214	19 56 77 115 145 173 196 227	14 56 75 104 127 162 191 225	37 37 60 76 111 135 175 214	17 39 66 97 122 171 205 234
	10	13 29 57 77 93 107 130 165 188 214	14 30 62 78 94 126 152 179 205 231	19 44 56 72 89 109 129 159 176 205	11 26 50 72 92 115 149 180 201 224	16 25 48 64 85 106 127 142 175 206	24 41 56 70 96 123 150 177 202 229
Test 6	4	55 97 158 195	64 99 153 190	46 75 131 179	52 100 144 192	56 92 123 178	52 88 149 176
	6	43 60 91 118 165 202	25 38 76 119 165 205	47 69 97 133 155 190	44 75 112 150 176 211	60 87 109 135 167 196	32 51 85 125 172 207
	8	16 42 73 91 129 157 190 221	10 34 52 79 108 136 168 207	23 39 64 86 120 137 172 202	28 45 74 91 120 150 191 225	15 26 55 85 111 141 176 205	23 38 57 87 119 159 183 214
	10	10 32 50 77 99 134 162 192 209 228	13 36 47 69 107 140 164 185 200 223	14 33 60 82 101 130 156 183 208 228	8 30 46 72 99 125 145 166 189 221	10 28 53 72 100 125 146 172 187 206	14 31 51 68 86 115 134 167 190 221
Test 7	4	66 95 139 176	25 66 108 157	23 88 134 220	26 89 132 179	43 89 133 178	30 94 137 208
	6	17 33 80 120 168 205	20 80 108 133 164 189	13 59 84 102 128 151	20 41 60 92 131 180	21 54 86 111 144 183	31 67 101 138 196 223
	8	20 54 79 106 126 146 169 209	13 50 80 102 132 166 197 223	19 40 63 88 119 148 200 228	14 58 80 98 122 146 167 207	18 55 82 105 128 161 193 228	18 36 67 91 128 149 200 220
	10	10 30 41 63 89 110 129 154 178 216	14 54 68 82 106 132 169 197 220 243	15 41 66 85 105 133 151 190 214 233	12 28 59 82 115 146 179 199 215 227	12 31 57 76 92 117 141 168 200 223	10 34 63 95 120 140 157 185 212 232
Test 8	4	48 73 173 212	62 107 141 207	43 79 124 215	22 83 139 199	55 105 149 201	37 102 183 234
	6	38 64 101 141 181 224	32 61 85 126 155 196	38 61 91 126 197 218	40 66 96 151 182 212	40 66 97 140 175 209	34 54 88 117 182 230
	8	15 41 64 85 110 133 175 213	44 34 74 110 135 158 179 203	25 40 69 104 136 182 207 231	31 55 77 93 108 135 178 217	25 51 80 117 155 179 206 227	30 32 60 103 117 141 179 215
	10	18 39 58 87 115 137 172 186 210 230	21 36 63 79 111 129 150 180 208 227	28 47 74 90 116 131 153 178 208 226	22 42 69 84 109 133 158 182 208 227	39 91 103 127 146 159 176 193 208 225	21 35 65 98 119 140 165 184 211 231
Test 9	4	47 74 107 149	55 90 132 167	42 66 117 149	33 67 109 155	53 87 123 165	45 89 155 211
	6	13 43 62 92 126 165	15 50 89 125 160 195	19 45 81 115 151 230	12 42 59 96 127 160	24 33 76 115 143 190	28 48 80 116 149 167
	8	14 36 53 75 92 123 152 173	11 40 55 77 101 126 151 183	26 41 66 91 128 155 200 233	11 38 54 81 115 139 166 187	17 66 86 105 125 145 172 197	28 51 76 104 132 154 172 207
	10	13 23 41 59 82 106 126 158 181 198	11 43 54 72 89 105 134 163 188 213	24 47 71 94 118 140 160 177 208 237	9 41 60 92 116 131 150 167 185 203	13 30 47 67 86 108 131 150 183 212	20 43 71 92 115 155 175 190 226 240
Test 10	4	48 76 132 175	31 69 117 169	53 86 116 200	62 101 132 190	45 70 123 186	52 83 139 193
	6	34 57 90 127 166 196	53 90 116 142 180 214	42 56 104 136 169 220	33 56 87 127 162 198	43 76 118 155 190 225	42 70 100 133 164 226
	8	19 33 62 91 133 165 195 214	20 26 57 90 112 150 195 220	12 38 55 88 113 150 186 229	34 60 98 128 148 170 194 221	17 38 58 87 119 147 187 216	10 23 49 79 121 149 170 189
	10	16 35 54 71 93 119 141 169 196 231	9 26 36 51 72 106 132 166 195 224	10 26 41 61 77 108 134 160 185 213	14 33 56 86 106 129 159 189 212 235	8 29 43 75 99 118 141 168 196 226	10 33 48 69 90 113 144 169 194 230

Table 4. Comparison of optimal threshold values between Artificial Bee Colony (ABC) and Bat Algorithm (BA) at $K = 4, 6, 8, 10$ based on fuzzy entropy.

Image	K	ABC			BA		
		R	G	B	R	G	B
Test 1	4	55 103 154 201	69 104 156 207	40 96 165 219	49 78 150 214	66 92 128 195	78 116 176 213
	6	24 63 109 149 184 220	47 74 111 152 183 218	39 84 126 154 188 220	31 58 81 105 150 195	20 45 85 115 165 212	29 71 118 154 190 230
	8	24 51 81 113 142 177 211 226	20 54 77 108 150 174 200 223	15 47 75 107 143 171 207 237	12 51 80 98 117 155 193 225	13 48 77 109 159 181 205 225	33 63 94 110 143 168 203 242
	10	16 41 59 85 113 142 167 187 211 236	15 43 62 88 112 139 167 189 209 232	13 39 62 83 113 139 165 192 221 236	20 50 82 99 125 149 168 190 211 231	15 41 88 107 129 146 158 177 206 232	31 51 72 84 99 129 162 191 227 242
Test 2	4	63 108 164 216	78 119 159 231	72 98 171 224	117 150 188 275	129 152 182 231	88 113 140 189
	6	49 76 112 149 196 236	52 79 125 159 189 223	56 78 124 148 182 219	81 107 143 173 205 232	83 97 112 130 212 233	83 114 153 178 206 238
	8	38 51 73 110 143 182 221 240	37 56 90 122 143 170 195 222	31 51 79 109 149 175 207 246	35 48 68 89 140 181 210 236	51 85 120 156 182 214 242 250	5 20 45 82 113 149 177 217
	10	22 37 57 85 112 142 167 192 223 237	6 45 60 80 109 133 160 188 218 244	23 40 65 88 115 144 163 188 218 239	64 69 75 81 98 120 155 187 210 230	30 50 70 92 120 147 171 193 214 238	28 58 79 98 135 162 186 206 234 248
Test 3	4	31 103 162 206	60 102 151 209	44 103 157 206	53 99 141 197	53 121 164 203	23 100 143 191
	6	34 68 103 140 190 226	29 66 101 140 186 217	31 67 98 138 186 217	29 66 116 154 185 215	39 72 122 154 181 210	45 86 116 143 173 219
	8	18 51 79 111 138 173 205 232	28 56 84 108 140 168 194 231	20 50 77 103 141 168 199 222	20 43 75 97 136 181 205 236	25 51 80 108 136 177 203 231	12 56 86 110 138 175 202 227
	10	19 37 63 87 112 137 170 195 222 244	22 46 65 92 118 141 165 194 216 237	21 52 71 96 110 136 157 196 214 230	19 40 64 90 117 147 174 206 227 244	36 62 88 105 116 135 162 196 216 239	30 51 74 100 113 135 146 160 184 214
Test 4	4	40 90 136 172	63 118 170 210	63 96 146 208	68 110 141 168	60 112 161 208	49 109 189 234
	6	32 69 114 155 190 219	42 66 111 161 201 237	35 65 104 146 186 235	38 59 116 147 173 197	37 78 135 168 193 223	40 74 108 137 155 226
	8	21 45 78 107 137 169 206 227	32 56 83 120 142 177 212 241	26 49 80 116 156 182 213 233	8 36 61 92 125 197 222 241	26 52 84 112 135 160 198 228	26 58 91 121 146 170 198 228
	10	25 45 60 84 114 140 168 191 217 232	26 43 62 86 108 136 161 191 216 245	22 38 63 85 113 137 159 193 223 241	8 33 64 96 128 156 174 195 217 234	38 63 84 105 129 145 168 192 206 231	25 41 54 67 96 130 152 186 209 233
Test 5	4	34 94 167 222	29 93 159 220	75 118 169 231	33 107 160 196	29 75 134 225	72 106 149 208
	6	26 72 97 138 187 223	21 73 108 140 188 226	20 71 118 153 191 231	59 86 122 163 202 228	19 62 94 144 184 234	52 101 136 176 213 234
	8	16 52 84 117 141 175 202 229	13 62 86 117 143 178 203 228	18 56 77 103 135 172 200 234	27 61 85 117 142 171 201 238	89 124 141 157 171 183 198 225	56 78 98 126 157 167 199 234
	10	18 36 68 99 124 147 167 183 206 228	12 50 68 88 114 144 169 192 219 238	18 41 65 92 111 133 163 190 220 241	17 37 54 72 104 138 160 178 191 231	14 68 86 104 118 141 168 197 216 231	16 43 68 98 128 154 177 198 220 241
Test 6	4	72 116 162 197	59 92 142 177	60 91 143 196	60 121 178 210	55 91 135 174	59 94 128 166
	6	35 54 103 142 184 222	31 66 111 161 192 228	36 74 114 158 197 218	12 59 96 121 156 201	61 110 147 175 204 233	44 74 103 131 190 227
	8	16 45 79 113 151 189 210 231	13 51 85 119 156 182 212 234	21 44 84 120 148 176 206 224	13 33 48 60 119 152 190 236	41 64 98 119 158 183 203 222	11 40 62 125 165 183 213 234
	10	15 42 60 86 114 144 164 194 214 231	11 41 60 81 101 130 155 181 212 233	26 46 64 80 110 131 164 185 201 220	11 47 75 102 121 134 150 178 204 227	14 39 58 84 108 138 161 186 210 235	21 38 51 71 90 115 144 172 191 228
Test 7	4	55 97 147 210	45 96 148 216	41 92 139 220	67 96 125 171	84 112 163 244	24 76 173 219
	6	33 72 111 153 177 219	19 75 110 147 181 227	25 71 109 148 190 227	57 75 92 122 166 212	27 87 118 141 185 234	21 67 100 140 192 235
	8	24 59 79 102 129 157 188 221	29 65 87 113 145 166 195 231	24 56 86 113 147 189 211 234	41 72 92 116 138 154 169 212	22 45 79 104 133 162 200 226	14 45 72 111 154 188 212 244
	10	15 39 68 98 125 142 166 195 220 237	13 40 69 87 110 140 170 202 223 242	17 44 71 89 113 138 158 193 214 234	21 62 94 122 144 163 183 200 217 242	43 80 102 121 140 159 175 196 223 237	20 50 84 111 131 167 180 205 223 239
Test 8	4	55 86 183 220	60 104 153 225	33 85 121 226	32 78 163 226	63 106 148 222	54 97 126 192
	6	38 82 126 154 193 229	44 74 114 152 181 230	35 57 76 105 130 235	46 87 148 181 215 237	53 78 119 144 184 227	27 42 65 90 112 141
	8	14 48 81 105 142 175 212 239	24 55 90 116 146 174 208 236	29 62 80 110 133 172 208 236	28 71 98 124 149 179 204 233	24 54 81 104 142 177 211 236	9 21 42 84 113 132 192 240
	10	11 34 57 78 101 133 167 191 216 234	17 35 64 85 109 134 155 183 210 239	20 40 72 99 116 135 167 186 218 240	16 51 79 103 126 144 169 194 224 239	18 50 66 83 101 121 147 172 207 227	18 47 82 105 127 145 169 194 220 240
Test 9	4	64 110 138 182	43 96 149 194	52 101 145 221	68 104 122 151	35 73 109 182	38 92 144 208
	6	19 62 101 131 157 197	39 74 106 145 183 212	30 65 102 135 166 234	39 63 90 116 141 181	12 51 100 129 161 198	40 54 72 101 129 158
	8	21 56 98 134 154 173 194 214	16 43 67 99 141 175 197 216	22 50 83 115 146 168 205 239	38 58 74 104 128 145 175 198	40 60 105 141 163 187 213 236	15 29 60 96 126 167 210 239
	10	12 42 63 87 114 141 169 197 223 238	15 46 66 86 110 137 160 187 214 236	25 45 60 85 111 136 164 193 227 242	23 66 88 123 142 154 164 179 199 216	14 42 64 84 111 135 159 184 203 226	16 32 48 99 135 158 174 202 227 243
Test 10	4	58 104 164 221	70 110 161 213	77 122 159 215	74 134 165 199	70 109 164 238	78 104 163 193
	6	38 76 125 159 192 229	53 79 111 139 176 218	48 71 101 135 182 221	26 80 137 166 196 233	51 96 118 138 169 209	42 74 121 155 198 229
	8	31 49 76 110 142 169 202 240	32 52 86 117 148 180 206 226	29 52 83 114 146 171 195 234	32 67 94 119 147 174 199 240	47 75 90 103 117 159 187 219	35 67 108 136 162 185 215 243
	10	27 47 69 94 120 143 164 185 212 242	26 43 66 95 117 142 165 192 221 239	22 39 62 89 110 135 157 185 214 238	33 66 88 111 125 143 156 180 201 239	11 29 53 84 114 136 166 193 220 234	13 39 58 105 132 166 183 207 230 247

Table 5. Comparison of optimal threshold values between Wind Driven Optimization (WDO) and Bird Swarm Algorithm (BSA) at $K = 4, 6, 8, 10$ based on fuzzy entropy.

Image	K	WDO			BSA		
		R	G	B	R	G	B
Test 1	4	68 107 150 192	65 103 156 198	77 115 164 196	54 119 190 243	12 57 174 243	24 93 175 238
	6	44 74 109 134 172 205	53 84 120 149 175 205	61 95 131 154 186 218	22 90 100 165 239 255	6 30 78 158 214 239	35 58 93 153 186 227
	8	47 72 101 121 144 168 188 211	48 75 104 129 149 170 192 213	46 68 91 110 131 156 174 212	7 57 108 118 129 158 216 251	9 55 68 97 181 209 241 255	4 32 78 123 152 179 196 247
	10	45 67 93 115 131 142 161 176 194 210	51 65 82 98 114 133 152 171 195 217	25 45 65 80 95 113 139 164 194 223	12 27 46 70 102 136 153 178 198 228	17 37 49 72 95 132 155 194 213 226	12 33 63 99 132 161 189 212 228 243
Test 2	4	123 146 174 202	128 154 183 216	98 131 170 208	7 39 194 232	24 112 145 227	71 105 128 183
	6	100 117 138 160 185 205	109 129 153 171 195 225	87 103 128 164 181 214	17 51 79 129 191 230	3 18 55 104 145 223	26 70 90 142 217 240
	8	78 93 104 118 134 146 169 200	93 111 131 147 165 182 202 228	62 74 92 111 131 154 179 211	17 52 83 123 181 216 230 247	18 73 92 119 152 190 225 252	59 90 102 121 142 198 224 240
	10	70 82 95 107 122 134 150 168 189 209	80 90 101 112 120 130 142 157 170 190	70 83 103 118 137 147 166 182 200 218	1 20 75 79 82 87 118 142 179 211	16 38 73 114 130 163 196 248 253 255	22 73 90 123 137 148 179 205 215 239
Test 3	4	48 104 158 199	68 121 163 196	77 119 161 201	39 94 147 214	42 117 173 201	24 107 208 221
	6	44 80 113 148 176 212	47 77 105 141 180 214	56 83 114 136 168 211	62 96 128 176 199 223	16 42 63 76 122 219	12 34 67 126 183 212
	8	37 67 92 112 135 163 189 217	38 59 82 102 129 156 188 213	46 70 91 117 140 162 190 213	1 37 51 77 115 159 209 237	19 60 102 149 177 203 223 251	20 57 73 95 125 159 188 221
	10	36 69 100 120 142 156 172 190 204 223	30 54 73 96 119 137 158 178 198 222	24 49 67 83 104 120 143 170 192 214	8 77 109 140 164 179 202 207 231 254	1 25 82 98 109 164 178 196 230 255	17 56 80 97 119 146 172 216 229 251
Test 4	4	54 94 132 169	71 117 154 200	78 109 147 207	46 89 145 214	18 95 184 244	36 77 130 213
	6	49 72 101 128 156 183	58 87 109 143 181 212	51 89 115 152 194 223	35 52 85 128 167 193	41 70 93 146 203 231	18 54 89 128 169 221
	8	40 67 94 120 139 156 176 193	48 73 93 118 138 166 196 225	45 76 106 129 153 176 205 229	1 3 30 81 120 181 218 237	26 56 92 133 185 205 224 244	23 42 91 133 163 204 223 236
	10	41 59 83 103 124 141 151 167 185 198	44 64 81 100 115 134 155 175 199 223	43 67 86 103 123 141 159 176 198 227	13 26 41 66 92 119 147 190 226 246	23 54 74 99 123 145 167 190 216 240	15 26 59 95 113 130 156 183 212 238
Test 5	4	86 123 161 198	88 123 165 203	78 118 159 199	8 55 142 203	82 107 159 198	22 57 114 226
	6	71 90 116 154 188 218	78 103 127 157 186 211	65 101 133 162 190 222	42 106 138 203 227 253	11 41 110 152 204 248	75 133 164 194 214 237
	8	33 67 88 114 131 158 183 212	59 76 97 114 135 164 192 222	58 78 97 121 142 168 192 227	11 33 86 110 175 195 211 247	3 63 123 177 197 216 239 255	26 54 85 123 150 178 211 230
	10	57 77 100 116 132 148 166 181 201 223	36 52 70 88 109 128 146 160 183 201	64 87 101 115 128 144 161 179 196 217	32 46 62 94 121 149 171 188 209 238	8 21 50 87 104 138 148 166 192 229	9 36 88 110 123 142 163 182 216 249
Test 6	4	59 104 157 197	69 106 147 193	59 100 130 177	25 127 153 246	16 78 149 224	28 98 172 231
	6	54 81 108 141 175 202	55 83 115 151 177 209	41 71 104 137 163 199	36 100 122 148 178 204	29 87 116 142 163 198	13 28 69 91 137 228
	8	45 74 100 123 146 167 191 215	55 81 105 131 151 171 191 213	38 67 92 122 142 169 186 211	8 52 79 123 158 181 210 243	11 50 109 132 169 204 219 234	20 63 104 130 137 168 207 246
	10	37 53 73 90 110 131 147 170 197 222	47 69 89 105 128 143 162 179 197 219	36 61 83 99 118 137 157 175 199 216	23 42 60 77 108 132 152 182 218 235	6 33 61 108 126 158 180 207 228 245	14 71 105 126 149 173 208 219 230 244
Test 7	4	74 110 145 176	85 107 145 184	75 100 136 182	28 95 224 255	43 70 135 235	46 134 198 247
	6	63 84 114 143 170 201	22 76 100 132 169 221	48 79 107 130 151 191	7 30 49 115 157 194	30 71 107 144 183 242	4 55 137 166 205 251
	8	41 68 97 120 139 165 184 216	43 76 97 119 140 164 181 220	34 57 82 109 129 151 200 233	16 37 65 85 108 141 169 222	13 44 70 118 144 157 177 236	13 57 95 117 145 183 214 230
	10	53 70 89 109 122 141 159 176 195 217	25 44 66 81 103 123 142 158 180 227	55 80 105 123 133 144 156 164 184 234	14 35 57 75 88 120 141 154 173 214	11 38 58 76 100 126 149 172 189 240	17 37 74 100 124 155 175 210 227 242
Test 8	4	69 124 160 207	62 118 152 207	44 92 122 189	48 83 156 213	22 81 125 216	61 98 114 212
	6	50 85 114 157 184 213	54 85 116 149 176 216	38 71 103 128 192 221	6 31 73 108 214 254	49 103 122 161 190 235	14 55 109 163 195 234
	8	41 61 86 111 136 161 188 220	41 67 88 110 130 157 193 224	38 68 91 105 123 140 184 232	21 41 67 76 107 134 163 231	21 47 88 119 159 194 221 242	12 37 67 116 137 160 184 235
	10	34 47 68 88 112 132 157 180 214 236	45 70 89 108 125 137 152 164 178 206	28 42 55 76 104 132 167 194 210 230	15 38 76 102 125 160 178 209 224 243	10 33 46 73 101 138 162 204 227 242	33 47 68 93 114 132 142 153 177 213
Test 9	4	70 104 138 160	71 106 142 176	58 107 149 225	47 86 157 234	43 92 125 219	37 92 147 223
	6	57 82 110 133 156 177	61 92 116 137 163 187	43 74 102 126 155 206	41 93 139 167 213 221	13 79 118 156 166 205	11 71 112 155 175 239
	8	50 72 97 120 138 155 172 190	54 75 93 110 131 151 173 194	37 62 91 121 143 165 205 219	3 19 32 70 111 155 188 227	11 35 76 122 141 181 216 230	23 74 94 113 147 189 214 241
	10	49 64 79 92 105 118 137 158 173 191	50 69 90 108 121 139 155 170 187 204	38 67 85 98 113 129 145 164 196 222	7 35 69 99 115 136 157 184 210 251	5 17 37 91 113 158 202 229 247 252	25 57 90 112 141 155 174 196 216 232
Test 10	4	62 89 136 179	70 103 136 191	70 111 156 191	32 132 173 217	37 90 134 215	70 129 166 192
	6	52 93 129 155 187 226	54 88 116 148 184 216	50 80 111 141 171 207	51 85 123 155 187 218	14 33 61 117 187 237	41 67 108 139 182 219
	8	35 66 89 107 136 164 188 219	49 72 96 117 141 167 198 223	47 67 89 117 143 166 193 224	21 53 63 86 123 185 213 236	19 56 78 112 143 186 238 253	11 44 70 123 154 173 193 215
	10	41 67 90 114 131 153 174 200 215 232	41 62 80 96 109 128 139 157 189 227	41 61 81 102 121 136 158 172 199 232	19 37 65 91 104 134 165 194 223 246	29 50 74 98 114 137 161 190 226 240	6 28 57 105 132 157 180 199 224 239

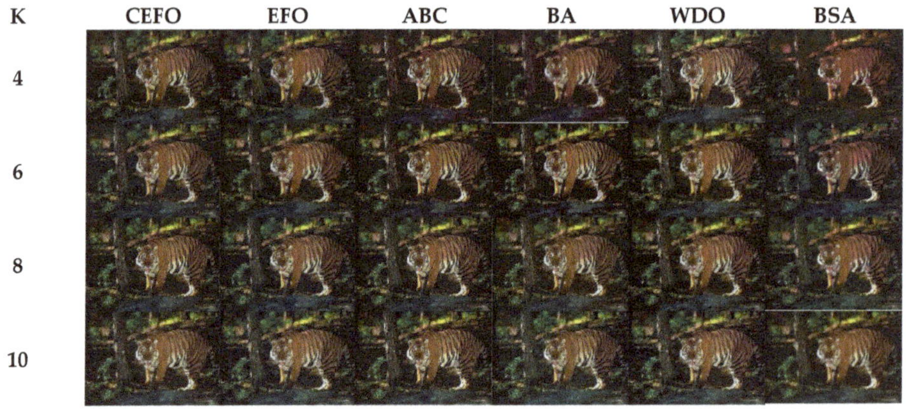

Figure 7. Segmented images of Test 1 at $K = 4, 6, 8, 10$ using selected algorithms based on fuzzy entropy.

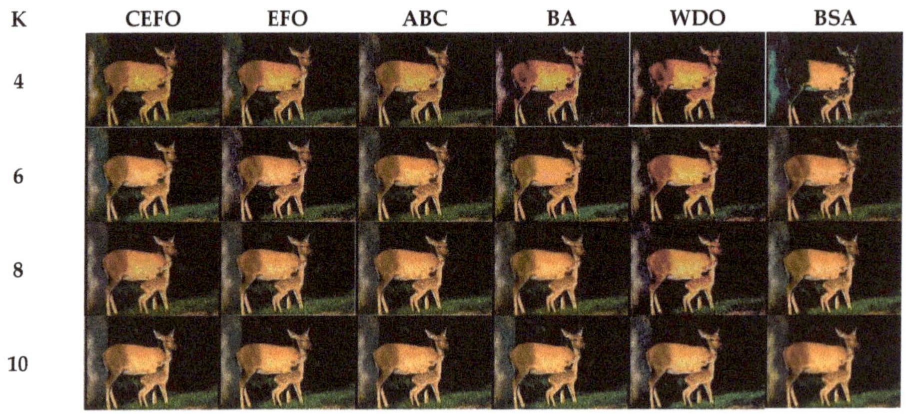

Figure 8. Segmented images of Test 2 at $K = 4, 6, 8, 10$ using selected algorithms based on fuzzy entropy.

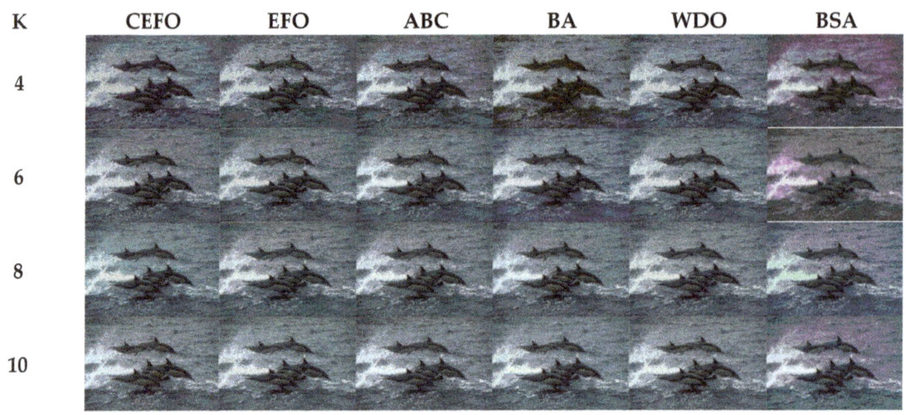

Figure 9. Segmented images of Test 3 at $K = 4, 6, 8, 10$ using selected algorithms based on fuzzy entropy.

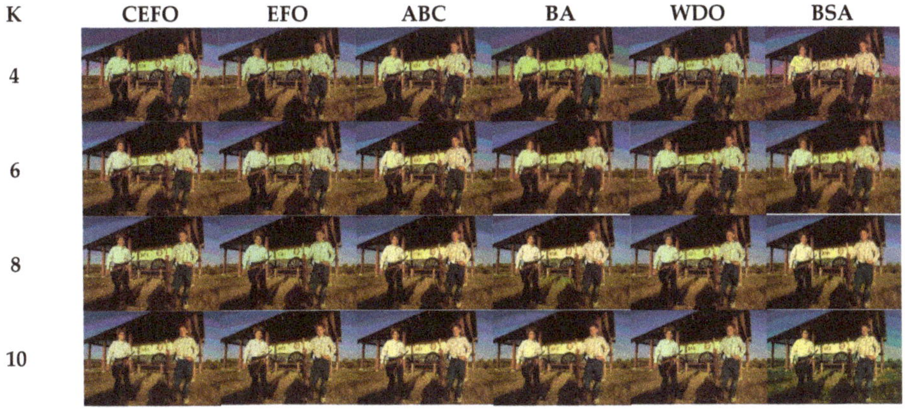

Figure 10. Segmented images of Test 4 at $K = 4, 6, 8, 10$ using selected algorithms based on fuzzy entropy.

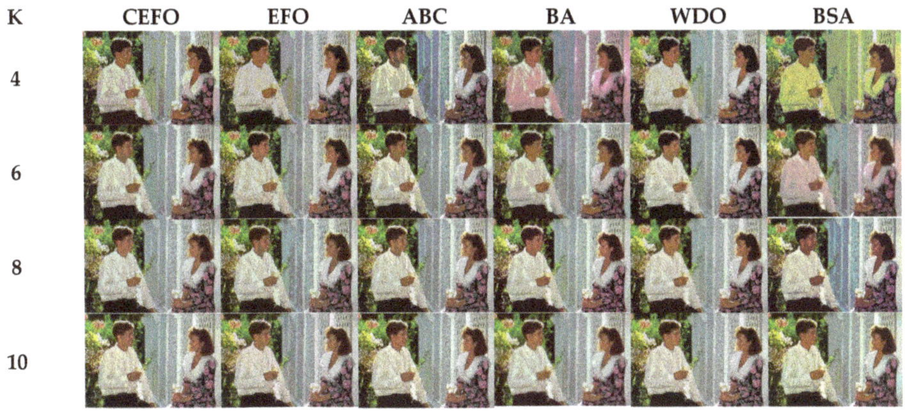

Figure 11. Segmented images of Test 5 at $K = 4, 6, 8, 10$ using selected algorithms based on fuzzy entropy.

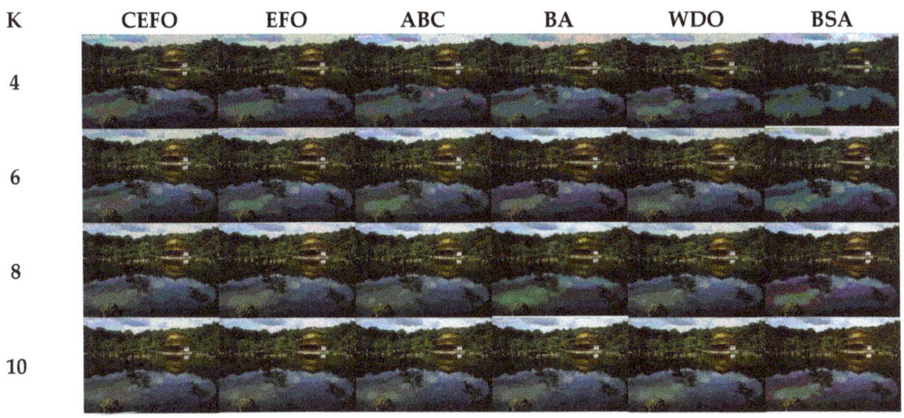

Figure 12. Segmented images of Test 6 at $K = 4, 6, 8, 10$ using selected algorithms based on fuzzy entropy.

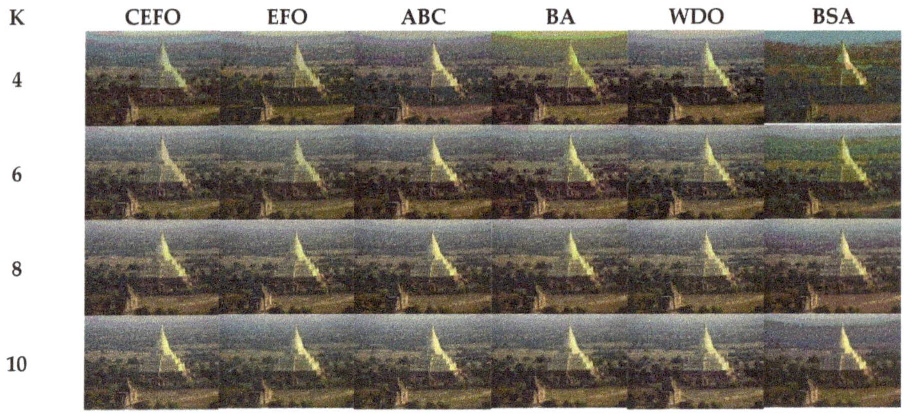

Figure 13. Segmented images of Test 7 at $K = 4, 6, 8, 10$ using selected algorithms based on fuzzy entropy.

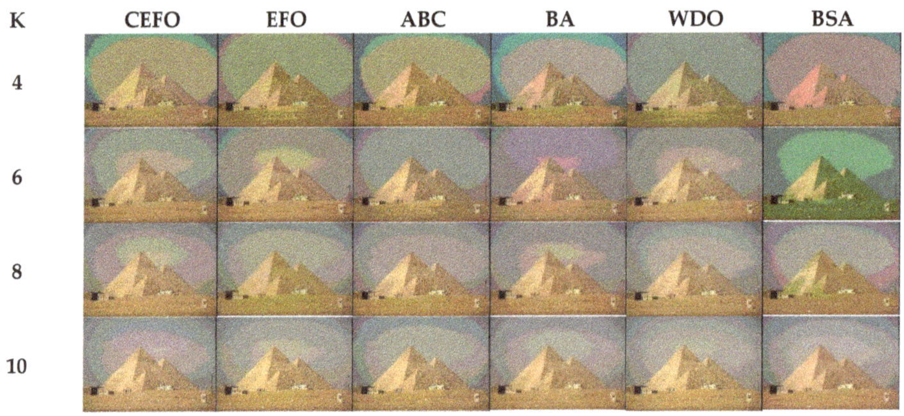

Figure 14. Segmented images of Test 8 at $K = 4, 6, 8, 10$ using selected algorithms based on fuzzy entropy.

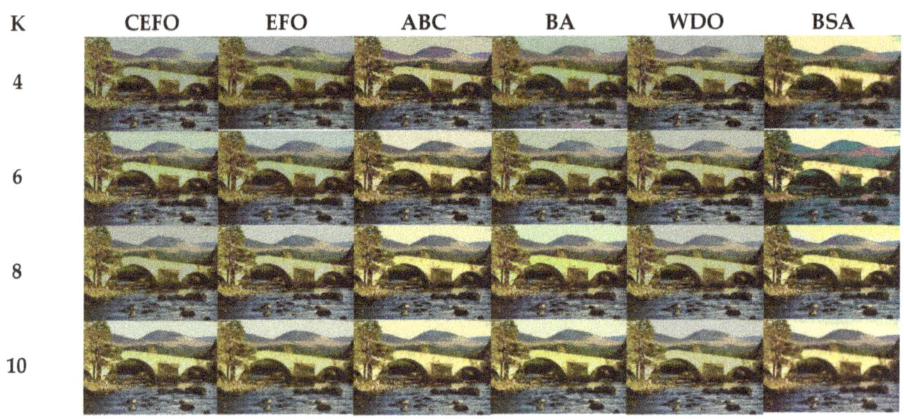

Figure 15. Segmented images of Test 9 at K = 4, 6, 8, 10 using selected algorithms based on fuzzy entropy.

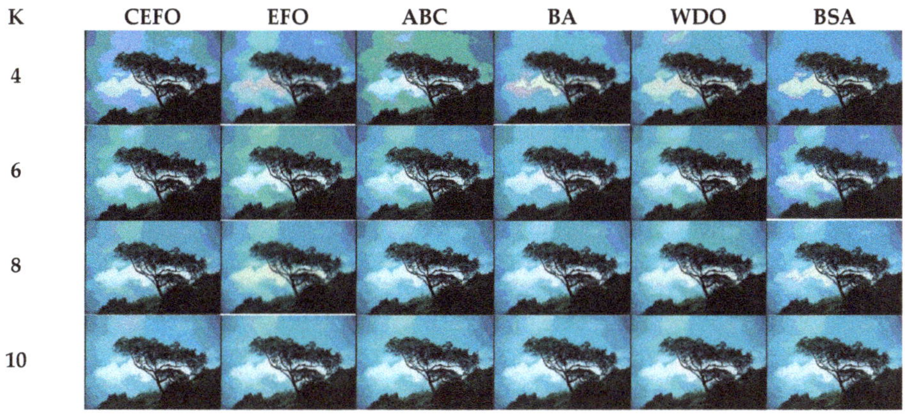

Figure 16. Segmented images of Test 10 at K = 4, 6, 8, 10 using selected algorithms based on fuzzy entropy.

Table 6 and Figures 17 and 18 compare the CPU Time and PSNR values while Table 7 and Figures 19 and 20 compare the MSSIM and FSIM values of the segmented images. As can be seen from these tabulated values, all algorithms have lower values of PSNR, MSSIM, and FSIM at lower threshold levels. With the improvement of the threshold level, the values of PSNR, MSSIM, and FSIM increase gradually. Consequently, it can be clearly known that segmentation performance will be improved as the threshold level increases. However, the time of each algorithm will rise equally on the increasing threshold levels indicating the computation of algorithms is more complex on the higher threshold levels. PSNR, MSSIM, and FSIM are used to measure the similarity and qualify among the segmented images. Higher PSNR, MSSIM, and FSIM demonstrate that segmented images have more excellent segmentation performances.

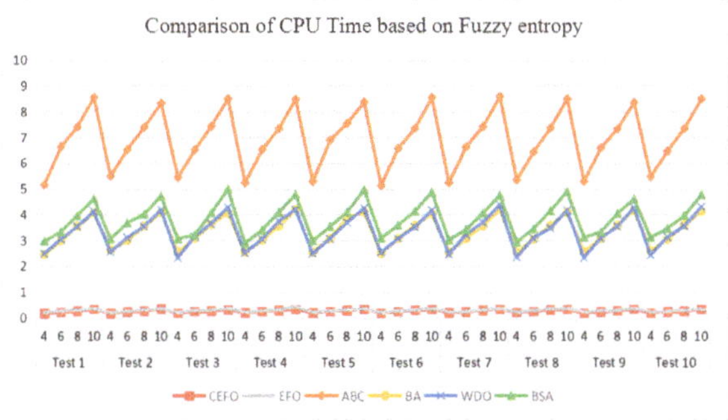

Figure 17. Comparison of Computational Time (CPU Time) based on fuzzy entropy.

Table 6. Comparison of CPU Time (in seconds) and PSNR computed by CEFO, EFO, ABC, BA, WDO, and BSA using fuzzy entropy. The bold numbers are the best values in the relevant index.

Image	K	Computational Time (CPU Time)						Peak Signal to Noise Ratio (PSNR)					
		CEFO	EFO	ABC	BA	WDO	BSA	CEFO	EFO	ABC	BA	WDO	BSA
Test 1	4	**0.17541**	0.20379	5.13759	2.45288	2.49367	2.97138	**19.0869**	18.5304	17.8020	17.3212	15.9576	16.8238
	6	**0.21623**	0.26617	6.63971	2.98102	3.04503	3.33423	**22.4966**	21.0851	20.8826	19.4872	19.2521	19.3995
	8	**0.26219**	0.30416	7.40398	3.54215	3.53203	3.96657	**23.9315**	23.3153	23.0079	23.0902	21.2553	21.9558
	10	**0.33284**	0.35769	8.54834	4.05983	4.12060	4.61008	**25.1603**	24.8085	24.9762	23.5623	23.5174	24.2976
Test 2	4	**0.18152**	0.19745	5.48045	2.62491	2.55021	3.05768	**18.5011**	16.6384	17.9217	14.9956	14.6621	16.8609
	6	**0.23014**	0.25140	6.52806	2.98864	3.11644	3.67576	21.4186	20.8410	**21.4261**	17.9962	16.1975	22.5361
	8	**0.26745**	0.30171	7.39266	3.54545	3.54003	4.00924	**23.4745**	23.5633	24.0574	18.3652	19.0371	22.6106
	10	**0.35415**	0.36493	8.32270	4.07706	4.17821	4.71338	**27.1960**	26.5362	26.5537	24.7670	19.9620	22.7476
Test 3	4	**0.19354**	0.20952	5.42962	2.54004	2.32354	3.05836	17.9266	**18.0706**	17.7559	15.0795	17.8259	16.1882
	6	**0.24406**	0.27724	6.51639	3.07212	3.14887	3.21041	**21.4595**	20.9724	21.0554	20.6465	20.9096	18.7586
	8	**0.26572**	0.30195	7.43254	3.61201	3.64706	4.05836	**23.5257**	22.7557	22.9832	22.7031	23.0729	21.4050
	10	**0.31104**	0.35756	8.49756	4.05537	4.27057	4.97929	**25.1107**	24.1796	24.6243	23.9076	24.8630	21.4820
Test 4	4	**0.20654**	0.22286	5.22364	2.53570	2.51094	2.89167	**18.5791**	18.5562	18.4538	17.8594	18.1050	18.4835
	6	**0.25613**	0.27055	6.52943	3.00267	3.05336	3.40388	**21.9962**	21.5752	21.4728	21.0931	20.4419	21.3494
	8	**0.29565**	0.31734	7.34861	3.53905	3.74083	4.08592	**23.6574**	22.9513	22.5091	22.4950	22.1239	21.8930
	10	**0.34226**	0.43022	8.48234	4.19234	4.20027	4.77867	**24.8212**	24.7081	24.3937	24.5774	23.1124	24.5733
Test 5	4	**0.20017**	0.22525	5.27645	2.53857	2.48718	2.98304	**18.0239**	17.6242	16.8089	16.3870	17.3516	15.8329
	6	**0.25485**	0.26024	6.89415	3.06993	3.06272	3.53148	20.3479	**20.4329**	20.0547	20.2630	19.1668	18.4271
	8	**0.29424**	0.32101	7.55179	3.86221	3.68722	4.09332	22.1661	21.8127	**22.1693**	21.5206	21.5001	19.8220
	10	**0.34480**	0.35174	8.37159	4.09532	4.24325	4.95875	**24.0138**	23.7462	23.8704	23.4936	22.2399	23.5245
Test 6	4	**0.19472**	0.20841	5.12137	2.49369	2.53905	3.08066	**18.1095**	17.6491	17.6509	17.7857	17.7293	16.8704
	6	**0.24263**	0.26745	6.57661	3.07078	3.06447	3.58303	**21.4831**	20.6287	21.0117	20.2254	20.2431	19.4964
	8	**0.30864**	0.32049	7.35641	3.57464	3.50718	4.12885	23.2290	**23.3256**	22.7773	21.8122	21.4135	20.8800
	10	**0.33725**	0.35708	8.54900	4.03751	4.15783	4.86694	**24.7681**	24.4737	24.5259	24.6509	23.1793	22.5603
Test 7	4	**0.20678**	0.23005	5.22556	2.53131	2.44727	3.00952	**18.8739**	18.5039	18.0277	16.4128	16.4863	16.0152
	6	**0.23562**	0.25809	6.62490	3.07199	3.21664	3.41591	**22.1443**	21.6213	20.0762	19.9806	20.8248	18.3119
	8	**0.29807**	0.30283	7.41492	3.56056	3.69153	4.05134	**23.7839**	23.1932	23.2384	22.0415	24.3675	22.0538
	10	**0.34217**	0.36273	8.58824	4.20910	4.36751	4.74359	**24.7433**	24.0335	24.3244	23.0629	23.8627	24.2057
Test 8	4	**0.21450**	0.22378	5.33752	2.52727	2.33784	2.93353	**18.9495**	17.4911	16.6399	16.5774	18.2614	17.6314
	6	**0.25068**	0.28080	6.43523	3.03905	3.08865	3.44368	21.4109	20.5791	**21.8933**	20.7661	22.1747	19.2986
	8	**0.31994**	0.35980	7.36667	3.57058	3.46608	4.14337	**23.0677**	22.4398	23.0619	22.3759	23.4846	21.3875
	10	**0.33725**	0.36888	8.48988	4.08254	4.16608	4.89443	25.4357	25.3720	23.9689	23.7873	**25.4598**	24.9694
Test 9	4	**0.19324**	0.23558	5.28317	2.55034	2.32281	3.11904	16.4568	15.9776	16.5525	16.1450	16.1994	**18.1708**
	6	**0.23789**	0.27384	6.58909	3.06363	3.08147	3.32353	18.4675	18.6749	**19.2639**	18.5698	18.4175	19.1084
	8	**0.27057**	0.29738	7.32034	3.58204	3.53299	4.04083	**21.0098**	20.5312	20.6003	20.3908	19.9597	21.0047
	10	**0.33151**	0.36361	8.34554	4.15394	4.26364	4.60027	22.3666	23.0655	**24.4486**	22.4630	20.8358	22.9661
Test 10	4	**0.19039**	0.23247	5.45731	2.58281	2.42582	3.10718	**18.5817**	18.4860	18.1025	17.9358	18.4517	17.3690
	6	**0.23656**	0.25762	6.46701	3.03982	3.12420	3.45783	**20.8938**	20.8450	20.8187	20.7592	20.5685	19.7354
	8	**0.26390**	0.29360	7.34870	3.59042	3.54711	3.95625	22.6241	22.3346	22.3827	22.2330	**22.6466**	22.6210
	10	**0.33304**	0.34301	8.49122	4.12545	4.29727	4.76364	**25.3640**	24.8916	24.8559	24.7299	23.3085	24.8469

Table 7. Comparison of MSSIM and FSIM computed by CEFO, EFO, ABC, BA, WDO, and BSA using fuzzy entropy. The bold numbers are the best values in the relevant index.

Image	K	Mean Structural Similarity (MSSIM)						Feature Similarity (FSIM)					
		CEFO	EFO	ABC	BA	WDO	BSA	CEFO	EFO	ABC	BA	WDO	BSA
Test 1	4	**0.97368**	0.96312	0.96192	0.94201	0.94017	0.95642	**0.75882**	0.74880	0.74221	0.72940	0.72485	0.69178
	6	**0.98631**	0.98280	0.98244	0.95769	0.97051	0.97305	**0.85039**	0.83938	0.81125	0.76648	0.80694	0.75111
	8	**0.99199**	0.98985	0.98934	0.98929	0.98299	0.98442	**0.88618**	0.87683	0.86915	0.86860	0.85435	0.83095
	10	**0.99361**	0.99251	0.99329	0.98959	0.98878	0.99215	**0.91098**	0.90282	0.90927	0.88384	0.89480	0.89012
Test 2	4	**0.96013**	0.95306	0.95484	0.93035	0.92588	0.95814	**0.72972**	0.68909	0.69247	0.64807	0.64540	0.65013
	6	**0.98459**	0.97495	0.97746	0.93722	0.94664	0.98110	**0.81315**	0.73786	0.74709	0.70236	0.67594	0.73024
	8	**0.99098**	0.98908	0.98343	0.96678	0.97116	0.98765	**0.86210**	0.84695	0.85625	0.71072	0.74300	0.83126
	10	**0.99631**	0.99389	0.99545	0.99194	0.97709	0.98899	**0.92876**	0.92497	0.91465	0.87880	0.76162	0.84512
Test 3	4	**0.97373**	0.97293	0.97077	0.90106	0.96489	0.95882	0.71919	**0.72285**	0.71887	0.68456	0.70871	0.65393
	6	**0.98862**	0.98623	0.98767	0.98523	0.98145	0.97832	**0.83144**	0.81185	0.82559	0.80544	0.82705	0.74121
	8	**0.99319**	0.99143	0.99196	0.99137	0.99274	0.98806	**0.88510**	0.87295	0.87866	0.87292	0.87457	0.84056
	10	**0.99513**	0.99393	0.99447	0.99282	0.99467	0.98472	**0.91471**	0.88683	0.91214	0.90164	0.90996	0.86459
Test 4	4	**0.97504**	0.97382	0.97329	0.97005	0.96999	0.97144	**0.75966**	0.75493	0.75345	0.75329	0.72873	0.73845
	6	**0.98817**	0.98736	0.98634	0.98489	0.98256	0.98795	**0.85825**	0.84736	0.84706	0.82680	0.80402	0.84761
	8	**0.99221**	0.99099	0.99161	0.99025	0.98786	0.98684	**0.89738**	0.88582	0.89347	0.88408	0.84668	0.86167
	10	**0.99485**	0.99452	0.99446	0.99308	0.99011	0.99385	**0.92218**	0.91879	0.91227	0.90467	0.86575	0.91742
Test 5	4	**0.97653**	0.97315	0.96886	0.96022	0.96675	0.96267	**0.76795**	0.76296	0.74913	0.75639	0.75709	0.72777
	6	**0.98616**	0.98513	0.98517	0.98732	0.97710	0.97627	**0.82839**	0.82738	0.81941	0.82174	0.80032	0.78086
	8	**0.99131**	0.98955	0.99027	0.98258	0.98718	0.98158	**0.86282**	0.85410	0.85843	0.85236	0.85559	0.82891
	10	**0.99470**	0.99435	0.99429	0.99253	0.98864	0.99286	**0.89313**	0.88408	0.88847	0.88520	0.88349	0.87589
Test 6	4	**0.97060**	0.96579	0.96406	0.95931	0.96633	0.95680	**0.77366**	0.74182	0.73894	0.75263	0.74499	0.71969
	6	**0.98591**	0.98494	0.98417	0.97846	0.98073	0.97718	**0.83568**	0.82893	0.83366	0.79984	0.81545	0.80549
	8	**0.99198**	0.99165	0.98932	0.98702	0.98448	0.98218	**0.88446**	0.88168	0.86619	0.85645	0.84359	0.82360
	10	**0.99424**	0.99330	0.99346	0.99299	0.98979	0.98715	**0.90343**	0.89737	0.90076	0.89585	0.87933	0.85478
Test 7	4	**0.97596**	0.97279	0.97349	0.95629	0.95168	0.95706	**0.75191**	0.72087	0.73196	0.69886	0.72229	0.62681
	6	**0.98862**	0.98840	0.98061	0.98013	0.98278	0.97028	**0.84983**	0.83905	0.79482	0.79773	0.82547	0.70824
	8	**0.99260**	0.99116	0.99129	0.98974	0.99020	0.98792	**0.88582**	0.87568	0.87649	0.86537	0.87662	0.83777
	10	**0.99407**	0.99259	0.99320	0.99084	0.99053	0.99268	**0.90659**	0.89269	0.90124	0.87211	0.88049	0.89262
Test 8	4	**0.98354**	0.97469	0.96164	0.97004	0.98276	0.97990	**0.79086**	0.76787	0.75329	0.78219	0.77483	0.78629
	6	**0.99111**	0.98909	0.99031	0.98855	0.99055	0.98273	**0.84716**	0.83915	0.83630	0.82678	0.83681	0.79247
	8	**0.99381**	0.99227	0.99103	0.99217	0.99266	0.99066	**0.86616**	0.84681	0.85892	0.84004	0.85703	0.80136
	10	**0.99847**	0.99600	0.99506	0.99363	0.99606	0.99554	**0.88832**	0.88298	0.88288	0.87188	0.87718	0.86633
Test 9	4	**0.97162**	0.97039	0.96942	0.96822	0.96587	0.97086	**0.81290**	0.79545	0.80453	0.78891	0.79135	0.80267
	6	**0.98281**	0.98209	0.98173	0.98096	0.97984	0.97906	**0.86751**	0.85710	0.85210	0.84483	0.85514	0.83386
	8	**0.98634**	0.98826	0.98486	0.98571	0.98601	0.98727	0.88414	0.89537	**0.90678**	0.87892	0.88468	0.87814
	10	**0.99255**	0.99236	0.99222	0.99119	0.98845	0.99031	**0.92326**	0.91715	0.91732	0.91380	0.89638	0.90067
Test 10	4	**0.97860**	0.97680	0.97229	0.96956	0.97424	0.97019	**0.78955**	0.78169	0.74377	0.74445	0.75473	0.78499
	6	**0.98606**	0.95543	0.98550	0.98345	0.98326	0.98265	0.81837	**0.82121**	0.80852	0.80624	0.79444	0.80907
	8	**0.99207**	0.99177	0.99129	0.98994	0.98941	0.99109	**0.86164**	0.86015	0.85335	0.84145	0.83516	0.84721
	10	**0.99532**	0.99500	0.99441	0.99384	0.99105	0.99423	**0.89375**	0.88711	0.87918	0.87236	0.84516	0.88534

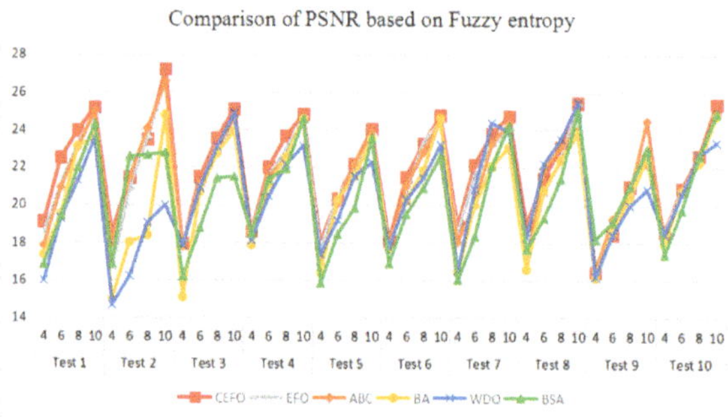

Figure 18. Comparison of Peak Signal to Noise Ratio (PSNR) based on fuzzy entropy.

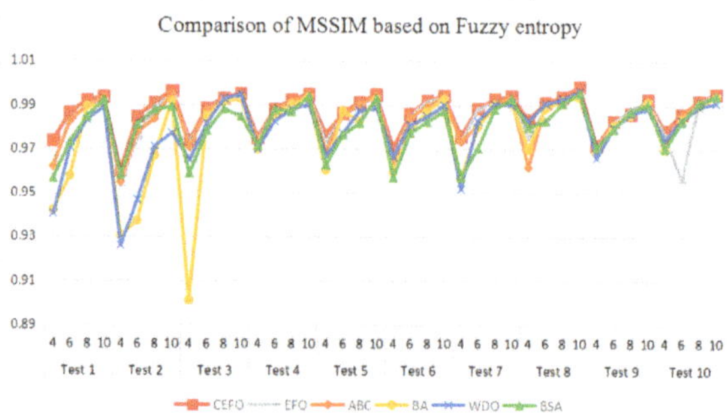

Figure 19. Comparison of Mean Structural Similarity (MSSIM) based on fuzzy entropy.

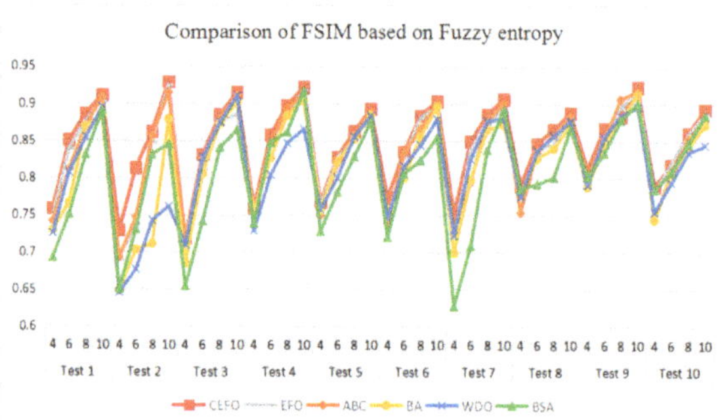

Figure 20. Comparison of Feature Similarity (FSIM) based on fuzzy entropy.

Then, when comparing the differences in CPU time between various algorithms, it can be found that CEFO and EFO are significantly faster than ABC, BA, WDO, and BSA. Moreover, the running time of CEFO has decreased about 12.26% when comparing with EFO, which indicates the modified electromagnetic field optimization algorithm has a faster convergence rate. As for other algorithms, ABC has the longest time of computation due to its slow convergence rate, it needs nearly 30 times as much as CEFO. Afterward, BA, WDO, and BSA have an approximate running time, they are about 15 times longer than CEFO. With the increase of execution time, the practicability of the algorithm will be reduced. For a clearer presentation of convergence speed about these algorithms, the convergence curves are shown in Figure 21.

In terms of PSNR, the chart of all algorithms is shown in Figure 18. It can be seen that CEFO has higher values among these algorithms; ABC has similar values to EFO in some images at a high threshold level. For instance, in Test 1 of K=10, PSNR value of EFO is 24.8085 while ABC is 24.9762, but CEFO is 25.1603. WDO has much lower values in smaller threshold level and BSA has good values in higher threshold level, all in all, PSNR values of BA, WDO, and BSA have different diversification, but they are mediocre on the whole.

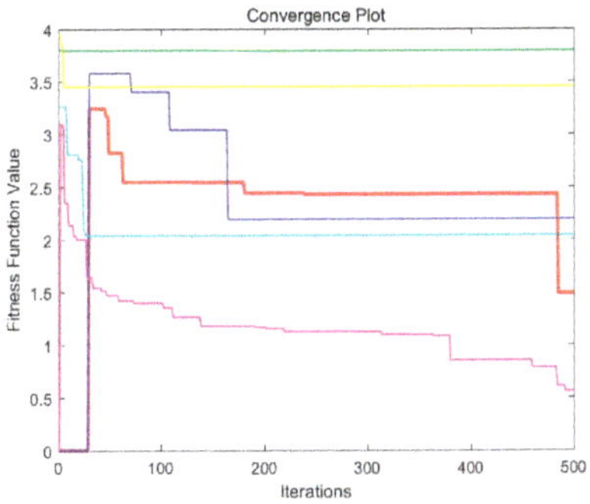

(**a**) Convergence curves of Test 1.

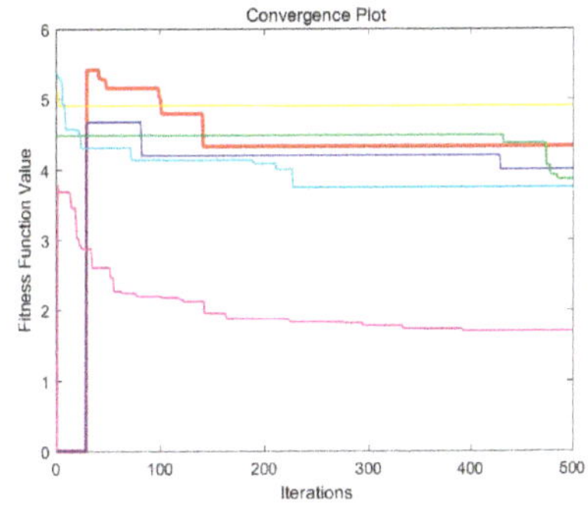

(**b**) Convergence curves of Test 2.

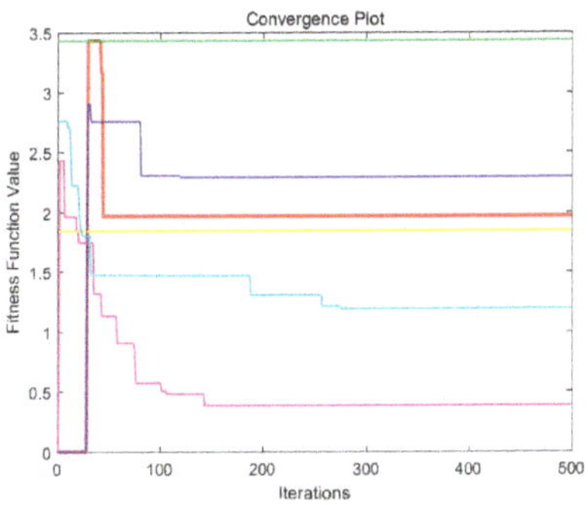

(**c**) Convergence curves of Test 3.

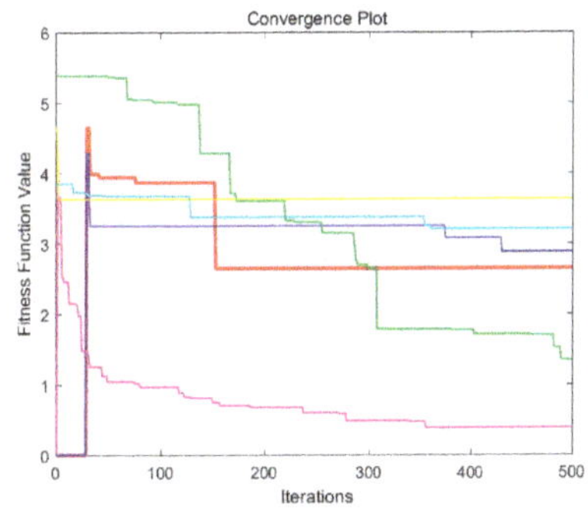

(**d**) Convergence curves of Test 4.

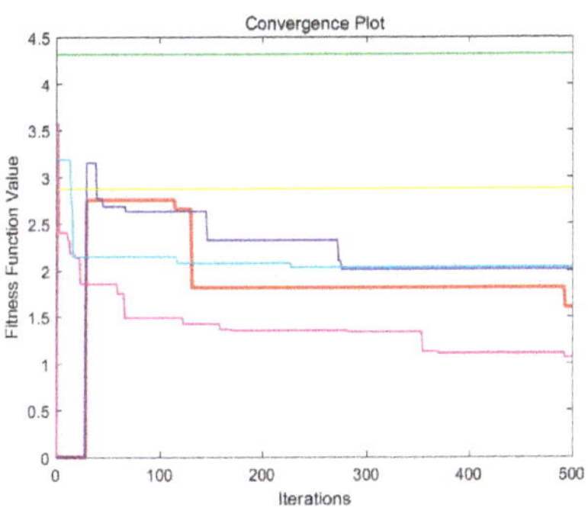

(**e**) Convergence curves of Test 5.

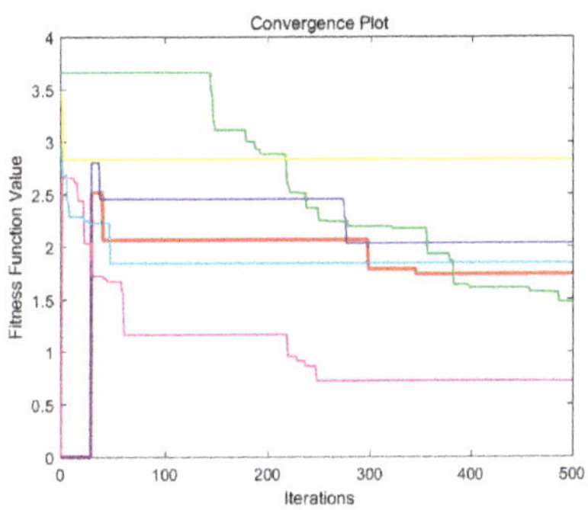

(**f**) Convergence curves of Test 6.

Figure 21. *Cont.*

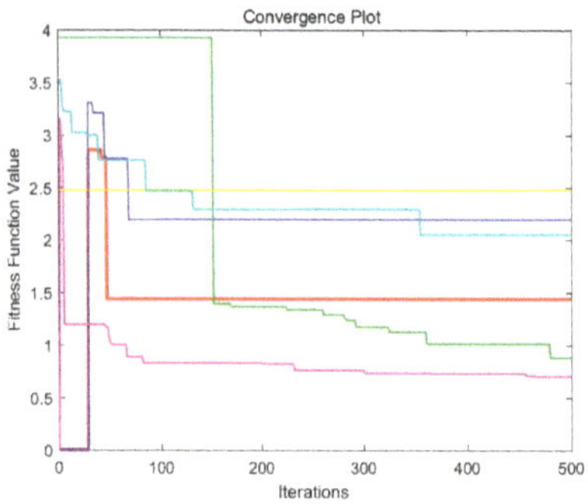

(**g**) Convergence curves of Test 7.

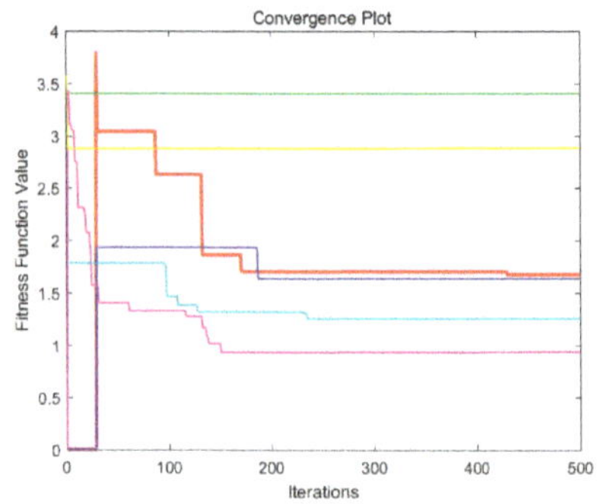

(**h**) Convergence curves of Test 8.

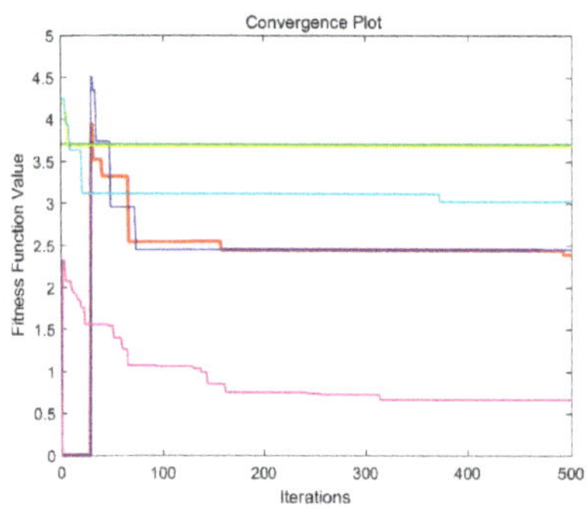

(**i**) Convergence curves of Test 9.

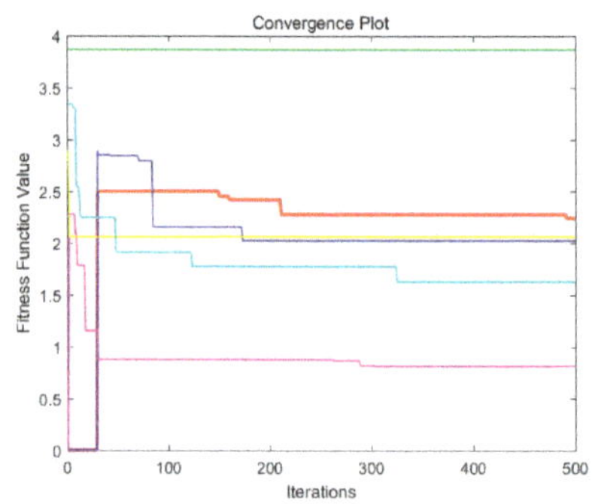

(**j**) Convergence curves of Test 10.

Figure 21. Convergence curves of CEFO, EFO, ABC, BA, WDO, and BSA based on fuzzy entropy at $K = 10$. (Red line represents CEFO; Blue line represents EFO; Cyan line represents ABC; Yellow line represents BA; Magenta line represents WDO; Green ling represents BSA.)

Comparing the results of MSSIM and FSIM in Table 7, FSIM is considered to be more authoritative and application and CEFO also performs better than other algorithms in this index. Although EFO can have a banner performance at $K = 4$, ABC will usually be close to EFO at high levels. BA has better values at some images such as in Test 2, 4, etc. WDO and BSA have higher values than BA at $K = 6$ in some images but they are all lower than CEFO on the whole.

From what has been mentioned above, the CEFO algorithm has superior performance when searching for the optimal threshold vector in multilevel thresholding color image segmentation.

6.2. Comparison of Other Segmentation Methods

In the last experiment, the superiority of CEFO has been verified. And in this experiment, fuzzy entropy has been regarded as the research objective. To show the performance of fuzzy entropy thresholding in multilevel color image segmentation, Otsu's and Kapur's entropy based on color image segmentation are used to be a comparison. Applying the CEFO algorithm to Fuzzy entropy, Otsu's, and Kapur's entropy respectively to segment selected 10 Berkeley images in Figure 6. The threshold level is chosen as = 4, 6, 8, and 10, which is used to obtain the corresponding threshold points for each component of the color image. The results of segmented images are in Figure 22, and the corresponding optimal threshold values are in Table 8. Table 9 compares the performance of different thresholding approaches based on parameters of CPU Time, PSNR, MSSIM, and FSIM.

As can be seen in Table 9 and Figures 23–26, Otsu's thresholding has the fastest speed of execution time, fuzzy entropy and Kapur's entropy are a bit slower than Otsu's. However, three thresholding segmentation methods are all in 0.5 (seconds) at different threshold levels, which can also indicate CEFO algorithm has a fast convergence rate. In terms of PSNR, it is clear that fuzzy entropy thresholding has higher values in general, the ranking of PSNR among these segmentation methods is Fuzzy > Kapur's > Otsu's. As for MSSIM and FSIM, Fuzzy entropy also performs well, which is in advance of two other methods overall. And the ranking of MSSIM and FSIM among three methods is Fuzzy > Kapur's > Otsu's. Therefore, fuzzy entropy is better as compared to others showing CEFO based on the fuzzy entropy technique can be applied in the color image segmentation field excellently.

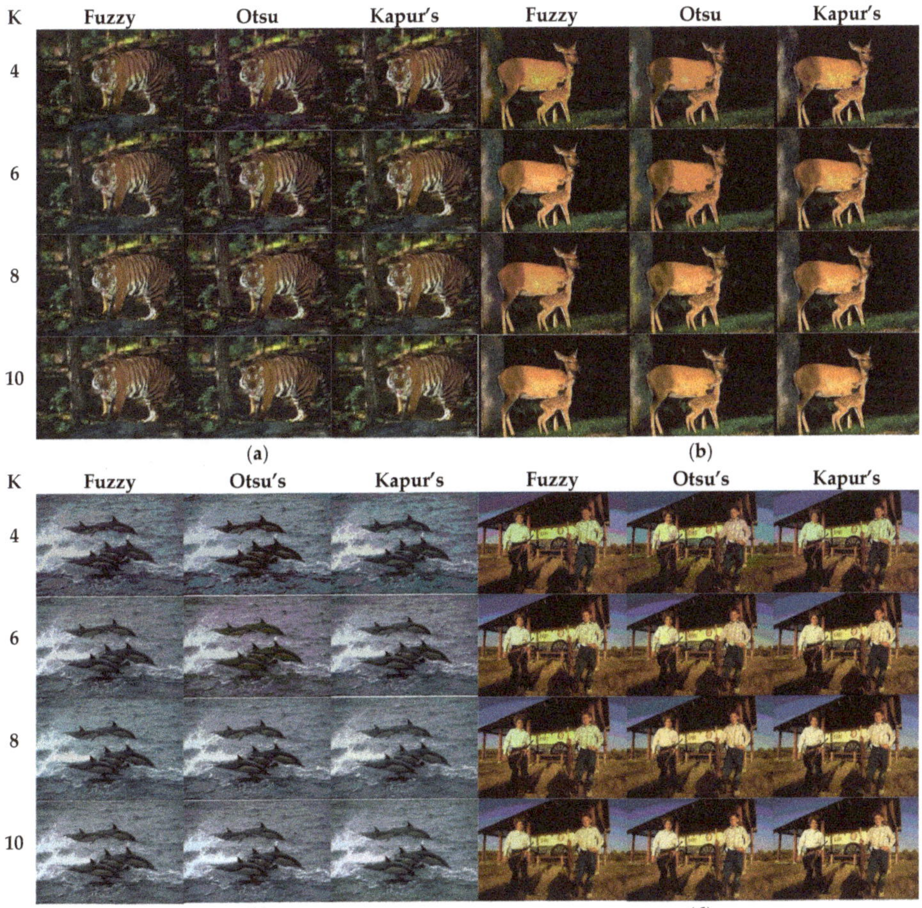

Figure 22. *Cont.*

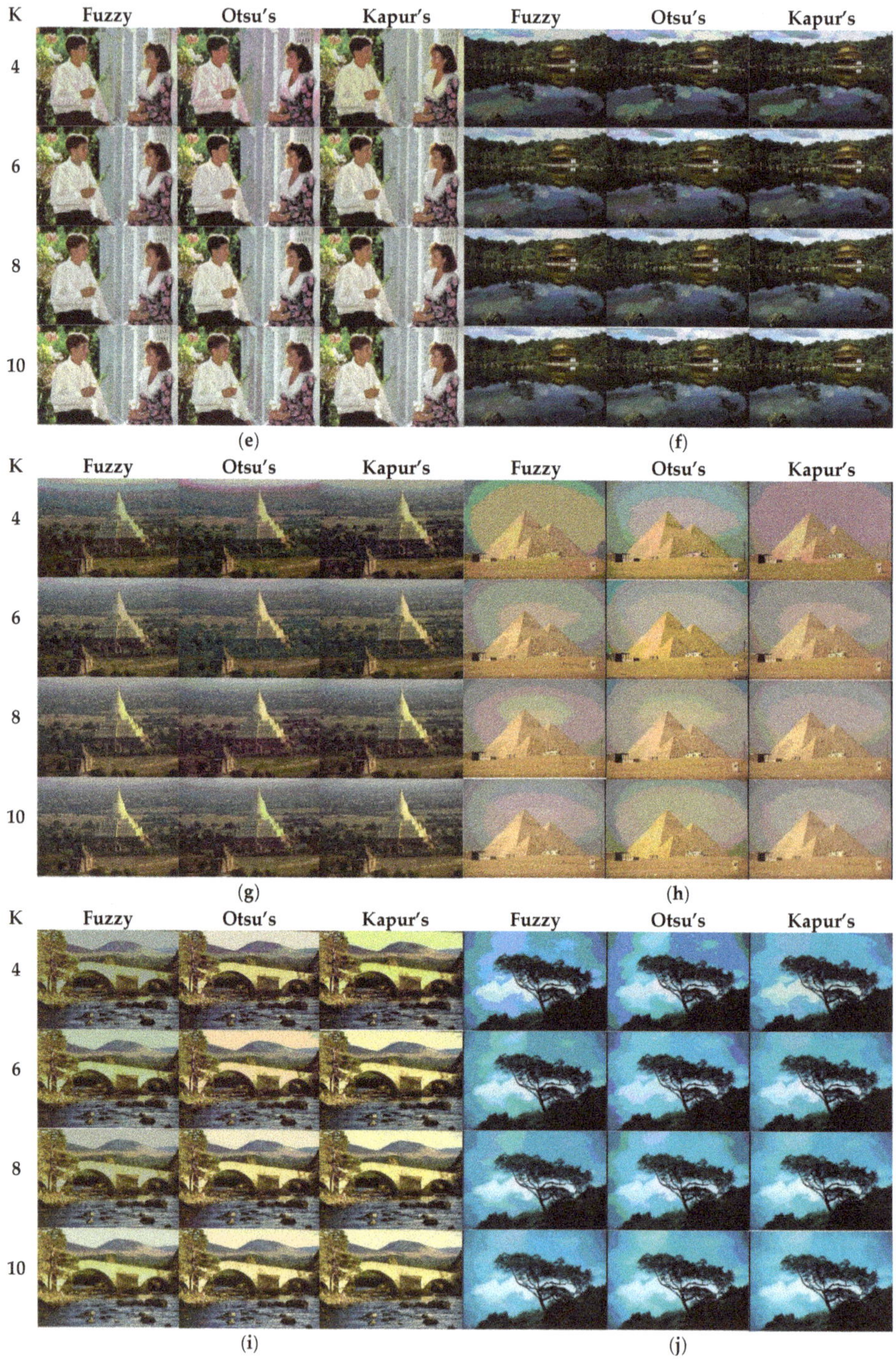

Figure 22. Segmented images at $K = 4, 6, 8, 10$ using CEFO algorithm based on fuzzy entropy, Otsu's and Kapur's entropy.

Table 8. Comparison of optimal threshold values of CEFO at $K = 4, 6, 8$, and 10 using fuzzy entropy, Otsu's and Kapur's entropy.

Image	K	Fuzzy R	Fuzzy G	Fuzzy B	Otsu R	Otsu G	Otsu B	Kapur R	Kapur G	Kapur B
Test 1	4	56 93 153 187	57 88 132 191	59 10 157 197	52 86 130 190	60 80 100 173	50 89 133 223	64 103 145 188	75 117 156 199	67 109 150 194
	6	20 59 92 128 167 199	26 47 81 114 161 203	25 48 76 111 138 195	52 74 105 161 177 217	68 80 101 124 167 190	10 55 85 105 155 212	51 84 116 150 184 212	55 87 117 147 178 212	56 87 115 150 187 221
	8	11 31 58 82 107 132 168 211	12 42 64 92 114 149 181 210	18 50 73 97 143 168 196 223	25 36 69 90 124 152 169 231	42 59 70 79 115 158 202 238	47 59 68 79 107 120 177 197	38 60 81 104 132 158 185 213	49 74 99 122 147 173 199 224	16 56 87 117 143 172 200 227
	10	14 37 55 75 90 111 130 159 182 210	11 41 61 95 122 146 168 185 208 237	14 40 64 80 105 136 161 189 212 231	45 51 67 107 119 129 165 199 222 237	41 70 72 86 91 112 156 192 210 230	35 50 82 100 101 115 139 148 178 197	12 38 58 81 106 133 161 184 203 227	45 61 82 103 124 146 168 189 210 228	17 52 74 94 116 137 163 185 212 238
Test 2	4	10 93 114 171	86 143 194 226	91 110 137 189	39 92 144 206	36 79 116 158	20 45 86 141	76 120 164 208	89 130 171 210	68 111 150 194
	6	18 41 75 107 147 177	87 101 133 153 192 234	54 72 97 121 156 192	22 59 88 121 183 197	29 61 90 129 165 179	18 23 36 64 95 233	20 58 96 135 173 213	19 58 99 138 177 215	41 74 106 137 165 199
	8	15 27 44 59 96 127 171 204	14 22 43 65 117 144 190 222	39 61 82 117 148 176 195 223	26 47 103 127 151 181 222 235	5 27 61 84 102 120 132 170	13 45 55 61 104 140 180 186	16 48 79 109 140 170 200 228	17 52 84 114 147 175 201 228	15 41 67 92 117 143 169 200
	10	32 55 75 108 144 171 195 215 226 239	5 50 65 85 102 121 139 167 197 228	31 47 64 87 108 125 158 179 209 239	18 23 47 56 99 115 143 184 194 237	22 71 76 78 99 108 138 158 196 212	18 43 56 97 129 145 162 234 235 237	13 35 56 81 104 130 155 179 205 230	15 40 65 92 112 134 159 183 206 229	13 37 58 81 102 123 143 163 190 209
Test 3	4	44 92 139 188	31 92 139 188	27 97 138 185	63 113 115 170	93 99 136 192	90 113 150 192	49 98 145 202	25 85 133 192	58 122 157 202
	6	32 62 93 128 173 199	19 53 90 119 162 206	36 71 105 136 183 208	48 78 103 126 153 190	36 71 114 168 185 203	18 30 101 133 159 201	30 70 102 127 166 200	20 65 99 143 181 208	26 80 104 135 168 202
	8	22 43 68 91 112 138 172 208	17 51 73 98 123 148 181 207	11 36 58 88 106 141 170 213	15 68 93 126 130 160 177 225	7 40 95 123 146 172 186 209	18 80 114 137 167 197 206 239	13 33 60 87 125 157 185 214	29 52 73 97 124 161 187 222	12 42 61 86 108 137 167 205
	10	26 39 55 69 86 110 131 154 175 207	42 29 56 79 104 121 141 163 197 219	9 25 49 82 109 132 154 176 199 222	1 37 64 71 99 122 151 176 212 231	13 51 74 75 104 137 170 197 215 246	51 82 106 111 130 168 181 203 233 241	11 32 51 73 100 132 154 173 199 223	14 37 55 75 93 109 123 149 181 212	16 44 66 99 130 150 166 187 208 225
Test 4	4	52 84 127 164	67 102 141 204	49 74 111 155	36 86 170 196	66 107 139 229	41 83 91 146	55 94 133 173	68 124 178 218	40 79 126 203
	6	38 62 97 137 172 195	48 71 111 144 175 213	43 77 109 147 178 235	33 63 93 127 163 215	36 66 112 138 168 231	22 48 83 135 183 206	51 86 121 155 190 224	46 80 116 149 183 219	27 55 83 118 155 203
	8	13 31 54 92 110 142 175 198	35 60 86 114 137 169 200 226	24 41 64 94 129 156 193 224	33 49 87 98 142 193 231 251	6 25 60 85 118 156 173 218	19 37 39 41 47 67 103 150	34 58 87 116 145 173 201 231	35 61 86 112 139 165 192 223	26 54 79 106 129 155 182 204
	10	6 25 43 64 84 106 132 152 172 189	7 27 39 52 75 99 130 157 190 221	15 25 48 78 103 130 157 178 219 240	39 48 73 105 135 152 156 167 197 225	26 39 48 70 106 136 158 191 199 220	9 28 49 66 71 94 131 159 170 186	32 56 78 99 123 145 168 190 211 235	26 48 69 90 112 133 157 180 204 229	21 43 63 80 101 121 145 165 183 204
Test 5	4	52 98 149 198	55 89 164 215	77 125 162 209	95 147 188 198	59 135 142 212	81 116 164 203	74 110 145 194	78 120 160 205	22 88 147 218
	6	22 63 93 131 179 225	16 69 97 125 158 204	15 61 97 131 166 207	92 132 165 192 233 243	78 116 119 138 187 229	68 125 134 158 190 248	66 92 118 145 180 216	67 98 126 160 191 223	22 62 99 142 180 219
	8	16 59 77 94 116 134 162 199	20 58 75 98 129 155 185 214	19 56 77 115 145 173 196 227	89 111 120 148 188 207 240 246	62 96 152 160 186 194 199 215	17 76 100 131 141 169 197 231	65 90 115 140 162 184 204 225	64 88 112 135 158 182 206 227	18 59 89 119 146 172 197 221
	10	13 29 57 77 93 107 130 165 188 214	14 30 62 78 94 126 152 179 205 231	19 44 56 72 89 109 129 159 176 205	36 66 73 82 104 120 126 147 181 210	60 104 127 170 175 181 185 190 206 221	35 58 87 97 135 174 175 189 214 222	54 71 87 106 127 145 168 189 208 225	62 81 100 120 143 163 183 201 218 233	22 52 81 108 133 154 176 197 217 235
Test 6	4	55 97 158 195	64 99 153 190	46 75 131 179	54 104 163 222	69 112 141 214	37 76 121 191	60 109 155 201	66 112 152 199	58 110 156 213
	6	43 60 91 118 165 202	25 38 76 119 165 205	47 69 97 133 155 190	49 60 78 135 146 239	52 64 81 132 190 233	28 57 96 130 201 227	45 79 112 147 181 214	54 91 126 155 185 217	38 78 119 155 188 222
	8	16 42 73 91 129 157 190 221	10 34 52 79 108 136 168 207	23 39 64 86 120 137 172 202	45 50 81 113 133 174 184 197	20 37 51 76 112 141 168 205	3 11 39 75 110 120 187 211	35 60 86 112 139 166 194 222	44 68 91 116 138 163 189 217	32 58 86 112 138 164 196 226
	10	10 32 50 77 99 134 162 192 209 228	13 36 47 69 107 140 164 185 200 223	14 33 60 82 101 130 156 183 208 228	41 49 70 99 111 112 171 191 237 245	42 60 61 85 102 148 186 248 249	13 24 50 87 104 131 137 156 178 229	11 36 59 82 109 135 161 187 209 232	38 58 79 100 119 140 165 189 208 228	15 37 60 86 110 133 158 186 211 232
Test 7	4	66 95 139 176	25 66 108 157	23 88 134 220	85 132 159 183	66 104 171 190	73 104 155 175	75 115 155 193	81 119 157 195	47 88 131 180
	6	17 33 80 120 168 205	20 80 108 133 164 189	13 59 84 102 128 151	77 106 144 188 212 224	51 65 76 110 152 200	45 49 79 90 131 154	63 96 128 160 194 225	47 77 106 134 163 195	17 47 79 112 144 180
	8	20 54 79 106 126 146 169 209	13 50 80 102 132 166 197 223	19 40 63 88 119 148 200 228	48 67 83 123 141 182 224 234	79 100 140 150 165 183 214 254	6 45 49 80 107 131 165 205	38 61 84 110 144 169 195 223	60 88 116 142 168 196 216 237	12 31 56 82 111 133 154 180
	10	10 30 41 63 89 110 129 154 178 216	14 54 68 82 106 132 169 197 220 243	15 41 66 85 105 133 151 190 214 233	38 63 76 101 128 131 157 164 185 186	27 41 83 91 103 119 157 184 219 226	14 55 72 83 86 115 124 157 164 214	41 61 82 101 120 142 167 194 216 237	46 69 92 114 138 158 176 196 216 239	17 42 67 84 111 130 144 161 179 196
Test 8	4	48 73 173 212	62 107 141 207	43 79 124 215	115 174 190 225	92 120 136 156	46 100 139 150	59 116 149 202	83 122 182 218	55 96 130 164
	6	38 64 101 141 181 224	32 61 85 126 155 196	38 61 91 126 197 218	54 146 170 197 219 232	98 119 143 156 167 212	12 42 50 121 142 182	42 81 118 148 179 209	45 83 116 147 182 218	25 49 72 101 131 164
	8	15 41 64 85 110 133 175 213	44 34 74 110 135 158 179 203	25 40 69 104 136 182 207 231	65 72 140 154 178 188 218 239	69 111 141 150 170 218 245	28 29 72 99 116 128 188 200	37 61 90 120 144 171 200 226	46 73 101 126 152 182 204 225	20 43 61 80 99 121 143 164
	10	18 39 58 87 115 137 172 186 210 230	21 36 63 79 111 129 150 180 208 227	28 47 74 90 116 131 153 178 208 226	28 57 59 113 123 163 168 191 213 214	39 70 72 102 128 154 169 183 191 196	9 21 27 77 79 128 168 185 187 197	35 58 80 100 121 142 165 187 207 230	36 47 71 91 116 139 164 183 214 228	20 37 53 70 89 107 126 143 164 182

Table 8. *Cont.*

Image	K	Fuzzy			Otsu			Kapur		
		R	G	B	R	G	B	R	G	B
Test 9	4	47 74 107 149	55 90 132 167	42 66 117 149	40 93 131 211	80 137 195 205	29 81 119 172	76 123 167 212	77 129 180 227	43 90 140 194
	6	13 43 62 92 126 165	15 50 89 125 160 195	19 45 81 115 151 230	57 73 109 164 194 230	64 105 136 166 201 231	15 61 91 114 123 165	62 97 132 167 203 236	62 95 128 162 195 231	31 63 95 126 158 194
	8	14 36 53 75 92 123 152 173	11 40 55 77 101 126 151 183	26 41 66 91 128 155 200 233	45 62 101 143 153 191 200 229	45 92 128 151 156 193 199 217	12 65 78 106 118 135 169 214	55 83 110 138 162 188 213 238	53 80 108 137 164 191 215 238	25 47 70 97 119 142 171 199
	10	13 23 41 59 82 106 126 158 181 198	11 43 54 72 89 105 134 163 188 213	24 47 71 94 118 140 160 177 208 237	30 48 87 101 131 159 164 220 227 253	30 50 59 64 107 112 136 166 189 228	12 20 47 80 93 106 122 132 150 165	47 68 92 115 140 160 182 201 219 239	38 55 74 95 118 142 164 187 212 238	19 36 54 72 91 111 133 156 176 194
Test 10	4	48 76 132 175	31 69 117 169	53 86 116 200	57 112 164 206	63 108 179 202	47 95 172 216	49 91 133 182	63 111 157 206	59 99 148 192
	6	34 57 90 127 166 196	53 90 116 142 180 214	42 56 104 136 169 220	18 53 72 129 133 155	55 85 150 177 181 231	38 75 127 175 203 229	30 67 106 136 177 217	46 79 111 141 170 2070	44 78 114 152 190 225
	8	19 33 62 91 133 165 195 214	20 26 57 90 112 150 195 220	12 38 55 88 113 150 186 229	34 71 116 118 139 141 159 199	15 42 60 103 149 155 170 210	6 45 70 127 139 200 219 248	22 52 79 109 133 161 189 219	39 67 95 124 153 180 206 230	40 66 91 116 142 167 194 224
	10	16 35 54 71 93 119 141 169 196 231	9 26 36 51 72 106 132 166 195 224	10 26 41 61 77 108 134 160 185 213	13 28 49 58 95 105 127 145 151 184	26 65 103 145 169 178 183 193 207 239	19 49 63 107 161 187 191 226 242 249	21 42 67 95 121 148 173 197 217 232	34 55 78 101 122 166 186 207 231	29 47 70 92 117 139 161 183 204 226

Table 9. Comparison of CPU Time, PSNR, MSSIM, and FSIM computed by CEFO at $K = 4, 6, 8, 10$ using fuzzy entropy, Otsu's and Kapur's. The bold numbers are the best values in the relevant index.

Image	K	CPU Time			PSNR			MSSIM			FSIM		
		Fuzzy	Otsu's	Kapur	Fuzzy	Otsu's	Kapur	Fuzzy	Otsu's	Kapur	Fuzzy	Otsu's	Kapur
Test 1	4	0.17541	**0.11919**	0.18065	**19.0869**	16.0378	16.0378	**0.97368**	0.95866	0.94199	**0.75882**	0.75699	0.73718
	6	0.21623	**0.12506**	0.24117	**22.4966**	19.0529	19.0529	**0.98631**	0.96735	0.97092	**0.85039**	0.79878	0.81115
	8	0.26219	**0.13443**	0.25866	**23.9315**	21.3692	21.9768	**0.99199**	0.98135	0.98558	**0.88618**	0.83430	0.86369
	10	0.33284	**0.15629**	0.27287	**25.1603**	22.8071	23.8019	**0.99361**	0.98815	0.99060	**0.91098**	0.85722	0.89851
Test 2	4	0.18152	**0.10603**	0.24967	**18.5011**	18.4665	18.4854	**0.96013**	0.95859	0.95939	**0.72972**	0.72349	0.71672
	6	0.23014	**0.11012**	0.25587	**21.4186**	20.9856	21.2795	**0.98459**	0.98013	0.98049	**0.81315**	0.80687	0.80330
	8	0.26745	**0.12666**	0.26542	**23.4745**	23.2473	23.2436	**0.99098**	0.98926	0.98958	**0.86210**	0.85725	0.86029
	10	0.35415	**0.14591**	0.30354	**27.1960**	24.6605	26.8256	**0.99631**	0.99286	0.99584	**0.92876**	0.87911	0.92489
Test 3	4	0.19354	**0.11462**	0.39732	**17.9266**	17.5340	17.7798	**0.97373**	0.96767	0.97158	**0.71919**	0.70489	0.71365
	6	0.24406	**0.12074**	0.45047	**21.4595**	20.4054	21.3185	**0.98862**	0.98374	0.98793	**0.83144**	0.79214	0.82536
	8	0.26572	**0.13348**	0.48285	**23.5257**	21.5710	22.7384	**0.99319**	0.98726	0.99194	**0.88510**	0.82899	0.87669
	10	0.31104	**0.16254**	0.21154	**25.1107**	23.3835	24.3602	**0.99513**	0.99227	0.99417	0.91471	0.88218	**0.92049**

Table 9. *Cont.*

Image	K	CPU Time			PSNR			MSSIM			FSIM		
		Fuzzy	Otsu's	Kapur	Fuzzy	Otsu's	Kapur	Fuzzy	Otsu's	Kapur	Fuzzy	Otsu's	Kapur
Test 4	4	0.20654	**0.09202**	0.21866	**18.5791**	18.4569	18.4989	**0.97504**	0.94314	0.97497	**0.75966**	0.74368	0.74512
	6	0.25613	**0.12600**	0.22851	21.9962	21.7162	22.0191	**0.98817**	0.98612	0.98724	0.85825	0.84935	**0.85897**
	8	0.29565	**0.15338**	0.24107	**23.6574**	23.0068	23.5891	0.99221	0.99073	**0.99239**	**0.89738**	0.88483	0.89628
	10	0.34226	**0.16649**	0.27016	**24.8212**	24.5519	24.7934	**0.99485**	0.99308	0.99460	**0.92218**	0.91235	0.92176
Test 5	4	0.20017	**0.10569**	0.23254	**18.0239**	16.4797	17.2779	**0.97653**	0.96171	0.97062	**0.76795**	0.74241	0.76655
	6	0.23485	**0.12750**	0.24186	**20.3479**	17.9620	20.0647	**0.98616**	0.97050	0.98300	**0.82839**	0.81379	0.82815
	8	0.29424	**0.15211**	0.27258	**22.1661**	19.6993	21.4806	**0.99131**	0.97874	0.98620	0.86282	0.83396	**0.87859**
	10	0.34480	**0.16173**	0.29775	**24.0138**	21.9181	23.1757	**0.99470**	0.98889	0.99059	**0.89313**	0.85475	0.89169
Test 6	4	0.19472	**0.10286**	0.24964	**18.1095**	17.8512	17.9759	**0.97060**	0.97041	0.96622	**0.77366**	0.76319	0.74541
	6	0.24263	**0.12325**	0.28237	**21.4831**	20.4060	20.6824	**0.98591**	0.98186	0.98165	**0.83568**	0.81530	0.82596
	8	0.30864	**0.15339**	0.29413	**23.2290**	22.1354	22.9859	**0.99198**	0.98855	0.98949	**0.88446**	0.85408	0.87426
	10	0.33725	**0.18392**	0.32098	24.7681	22.8894	**25.0296**	0.99424	0.98975	**0.99350**	**0.90343**	0.86898	0.90198
Test 7	4	0.20678	**0.10402**	0.24912	**18.8739**	17.8704	17.8217	**0.97596**	0.96464	0.96471	**0.75191**	0.72297	0.74535
	6	0.23562	**0.10956**	0.25531	**22.1443**	20.3251	21.6210	**0.98862**	0.97872	0.98609	**0.84983**	0.79339	0.84398
	8	0.29807	**0.12937**	0.27367	**23.7839**	20.7719	23.1949	**0.99260**	0.97916	0.99048	**0.88582**	0.81106	0.87668
	10	0.34217	**0.14226**	0.27555	**24.7433**	23.1843	24.3686	**0.99407**	0.99097	0.99363	0.90659	0.85053	**0.90773**
Test 8	4	0.21450	**0.10617**	0.23422	18.9495	**19.0541**	18.9415	0.98354	0.98371	**0.98401**	0.79086	0.79082	**0.79759**
	6	0.25068	**0.12896**	0.26131	21.4109	21.1411	**21.4439**	**0.99111**	0.98687	0.99032	**0.84716**	0.80299	0.81085
	8	0.31994	**0.14136**	0.28517	23.0677	23.2842	**23.5914**	0.99381	0.99375	**0.99576**	0.86616	0.85771	**0.87685**
	10	0.33725	**0.18727**	0.29926	**25.4357**	22.6056	25.4062	**0.99847**	0.99165	0.99681	0.88832	0.82970	**0.89139**
Test 9	4	0.19324	**0.11092**	0.23595	16.4568	16.3593	**17.1580**	**0.97162**	0.97064	0.96985	**0.81290**	0.80596	0.81157
	6	0.23789	**0.12999**	0.17958	18.4675	18.0109	**19.0843**	0.98281	0.98205	**0.98394**	0.86751	0.86452	**0.89018**
	8	0.27057	**0.14229**	0.22847	21.0098	21.0045	**21.3567**	0.98634	0.98629	**0.98699**	0.88414	0.87084	**0.90497**
	10	0.33151	**0.18128**	0.25452	22.3666	22.3547	**23.1507**	0.99255	0.99251	**0.99428**	0.92326	0.91262	**0.93015**
Test 10	4	0.19039	**0.10466**	0.20511	**18.5817**	17.9607	18.0765	0.97860	0.97268	**0.97934**	**0.78169**	0.76586	0.77935
	6	0.23656	**0.11261**	0.24046	20.8938	20.8925	**21.0027**	**0.98606**	0.98526	0.98586	0.82121	0.82907	**0.83085**
	8	0.26390	**0.12200**	0.24834	22.6241	21.6691	**23.1804**	**0.99207**	0.98882	0.99119	**0.86015**	0.83365	0.85931
	10	0.33304	**0.12677**	0.27179	**25.3640**	23.8567	25.0886	**0.99532**	0.99289	0.99428	**0.88711**	0.88195	0.88555

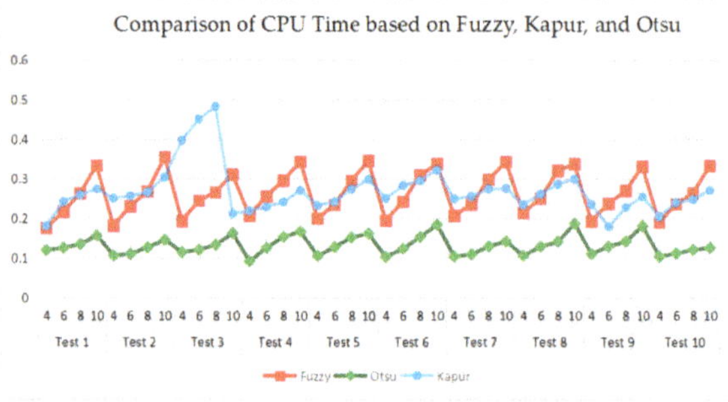

Figure 23. Comparison of CPU Time based on fuzzy entropy.

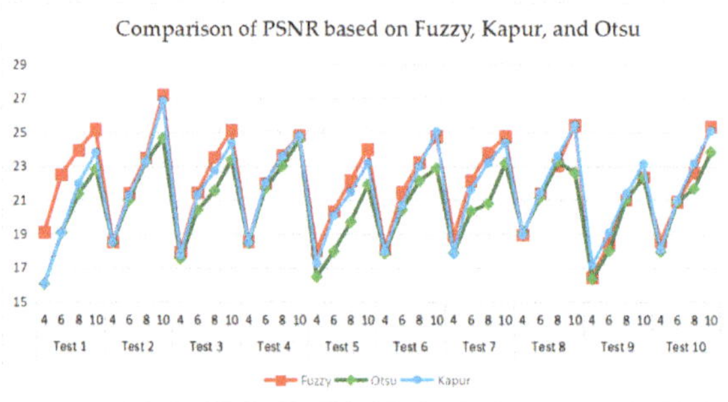

Figure 24. Comparison of PSNR based on fuzzy entropy.

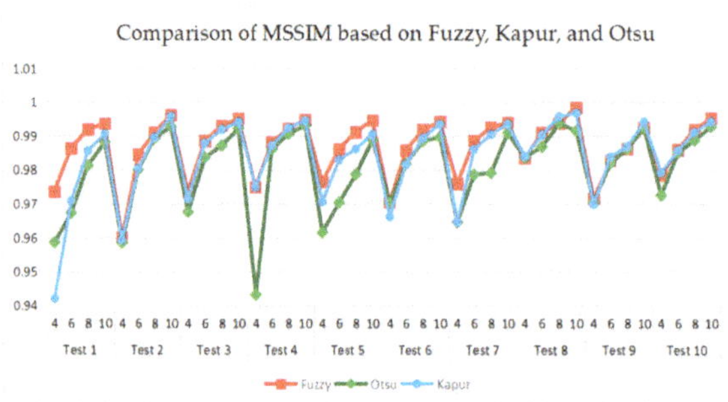

Figure 25. Comparison of MSSIM based on fuzzy entropy.

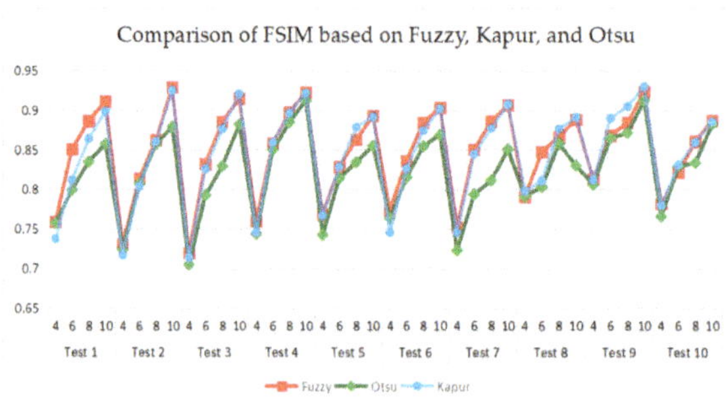

Figure 26. Comparison of FSIM based on fuzzy entropy.

6.3. ANOVA Test

A statistical test known as "the analysis of variance" (ANOVA) has been performed at 5% significance level to evaluate the significant difference between algorithms. In the experiment, CEFO algorithm is regarded as the control group and is compared with EFO, ABC, BA, WDO and BSA algorithms in terms of four measure metrics. The null hypothesis assumes that there is no significant difference between the mean values of 5 selected algorithms, whereas, the alternative hypothesis can be considered as a significant difference between them. Table 10 exhibits the –value of CPU Time, PSNR, MSSIM, and FSIM by the ANOVA test. As can be seen, the -value for CPU Time is less than 0.05, which implies a significant difference between the proposed algorithm and other algorithms and CEFO has a much fast convergence rate. With respect to another three measures, CEFO also has significant difference about BA, WDO and BSA. It can be observed that CEFO algorithm has a better performance.

Table 10. Comparison of *p*-values between CEFO and other algorithms based on Fuzzy entropy.

Dependent Variable	Proposed Algorithm	Algorithms	*p*-Value	Dependent Variable	Proposed Algorithm	Algorithms	*p*-Value
CPU Time	CEFO	EFO	0.038791(*)	**MSSIM**	CEFO	EFO	0.201224
		ABC	3.76E-50(*)			ABC	0.157924
		BA	1.08E-46(*)			BA	0.006606(*)
		WDO	1.23E-43(*)			WDO	0.004560(*)
		BSA	1.95E-47(*)			BSA	0.006092(*)
PSNR	CEFO	EFO	0.457477	**FSIM**	CEFO	EFO	0.364938
		ABC	0.466803			ABC	0.297925
		BA	0.037665(*)			BA	0.034212(*)
		WDO	0.021986(*)			WDO	0.016480(*)
		BSA	0.020720(*)			BSA	0.003728(*)

7. Conclusions and Future Work

In this paper, multilevel thresholding color image segmentation has been considered as an optimization problem in which the fuzzy entropy technique has been presented as the objective function. To achieve efficient segmentation, it is essential for algorithms to search the optimal fuzzy parameters and threshold values. Electromagnetic Field Optimization is a novel meta-heuristic algorithm which use is attempt herein for the first time in this field. Additionally, a new chaotic strategy is proposed and embedded into the EFO algorithm to accelerate the convergence rate and enhance segmentation accuracy. In order to demonstrate the superior performance of the CEFO-based fuzzy entropy technique, a series of experiments have been conducted and results are evaluated in terms of CPU Time, PSNR, MSSIM, and FSIM. On the one hand, the CEFO algorithm is compared with EFO, ABC, BA, WDO, and BSA based on fuzzy entropy for segmenting ten Berkeley benchmark images at different threshold levels ($K = 4$, 6, 8, and 10). The obtained results illustrate the obvious effect of proposed chaotic strategy and CEFO needs less than 0.35 seconds to find the optimal threshold vector which makes it an effective algorithm to handle the above problem. On the other hand, the fuzzy entropy method is compared with Otsu's variance and Kapur's entropy method based on CEFO on the basis of the same experimental environment. The high precision of fuzzy entropy has been validated with four metrics. Although CEFO-fuzzy is not the fastest among the three techniques, its execution time is suitable for practical applications within 0.5 seconds. To sum up, CEFO-based fuzzy entropy is a robust technique in multi-threshold color image segmentation.

In the future, the proposed technique can be applied to solve practical problems such as medical images, satellite images, etc. It is also interesting to modify EFO algorithm in other aspects to improve its performance for higher threshold levels (e.g., $K = 15$ and 20). Furthermore, the merits of CEFO can be investigated using Tsallis entropy, Renyi's entropy, and cross entropy for multilevel thresholding.

Author Contributions: S.S. and H.J. contributed to the idea of this paper; J.M. performed the experiments; S.S. and J.M. wrote the paper; H.J. contributed to the revision of this paper.

Funding: This research received no external funding.

Acknowledgments: The authors would like to thank the anonymous reviewers for their constructive comments and suggestions.

Conflicts of Interest: The authors declare no conflict of interest.

References

1. He, L.F.; Huang, S.W. Modified firefly algorithm based multilevel thresholding for color image segmentation. *Neurocomputing* **2017**, *240*, 152–174. [CrossRef]
2. Mlakar, U.; Potočnik, B.; Brest, J. A hybrid differential evolution for optimal multilevel image thresholding. *Expert Syst. Appl.* **2016**, *65*, 221–232. [CrossRef]
3. Agrawal, S.; Panda, R.; Bhuyan, S.; Panigrahi, B.K. Tsallis entropy based optimal multilevel thresholding using cuckoo search algorithm. *Swarm Evol. Comput.* **2013**, *11*, 16–30. [CrossRef]
4. Pare, S.; Bhandar, A.K.; Kumar, A.; Singh, G.K. An optimal color image multilevel thresholding technique using grey-level co-occurrence matrix. *Expert Syst. Appl.* **2017**, *87*, 335–362. [CrossRef]
5. Otsu, N. A threshold selection method from gray-level histograms. *Automatica* **1975**, *11*, 23–27. [CrossRef]
6. Otsu, N. A threshold selection method from gray level histograms. *IEEE Trans. Syst. Man Cybern.* **1979**, *9*, 62–66. [CrossRef]
7. Kapur, J.N.; Sahoo, P.K.; Wong, A.K. A new method for gray-level picture thresholding using the entropy of the histogram. *Comput. Vis. Graph. Image Process.* **1985**, *29*, 273–285. [CrossRef]
8. Li, C.H.; Lee, C.K. Minimum cross entropy thresholding. *Pattern Recogn.* **1993**, *26*, 617–625. [CrossRef]
9. Jiang, Y.C.; Tang, Y.; Liu, H.; Chen, Z.Z. Entropy on intuitionistic fuzzy sets and on interval-valued fuzzy sets. *Inf. Sci.* **2013**, *240*, 95–114. [CrossRef]
10. Ye, Z.W.; Wang, M.W.; Liu, W.; Chen, S.B. Fuzzy entropy based optimal thresholding using bat algorithm. *Appl. Soft Comput.* **2015**, *31*, 381–395. [CrossRef]
11. Gao, H.; Pun, C.M.; Kwong, S. An efficient image segmentation method based on a hybrid particle swarm algorithm with learning strategy. *Inf. Sci.* **2016**, *369*, 500–521. [CrossRef]
12. Bohat, V.K.; Arya, K.A. A new heuristic for multilevel thresholding of images. *Expert Syst. Appl.* **2019**, *117*, 176–203. [CrossRef]
13. Akay, B. A study on particle swarm optimization and artificial bee colony algorithms for multilevel thresholding. *Appl. Soft Comput.* **2013**, *13*, 3066–3091. [CrossRef]
14. Agarwal, P.; Singh, R.; Kumar, S.; Bhattacharya, M. Social spider algorithm employed multi-level thresholding segmentation approach. *Proc. First Int. Conf. Inf. Commun. Technol. Intell. Syst.* **2016**, *2*, 149–259. [CrossRef]
15. Bhandari, A.K.; Singh, V.K.; Kumar, A.; Singh, G.K. Cuckoo search algorithm and wind driven optimization based study of satellite image segmentation for multilevel thresholding using Kapur's entropy. *Expert Syst. Appl.* **2014**, *41*, 2538–2560. [CrossRef]
16. Sumathi, R.; Venkatesulu, M.; Arjunan, S.P. Extracting tumor in MR brain and breast image with Kapur's entropy based Cuckoo Search Optimization and morphological reconstruction filters. *Biocybern. Biomed. Eng.* **2018**, *38*, 918–930. [CrossRef]
17. Chen, K.; Zhou, Y.F.; Zhang, Z.S.; Dai, M.; Chao, Y.; Shi, J. Multilevel Image Segmentation Based on an Improved Firefly Algorithm. *Math. Probl. Eng.* **2016**, *2016*. [CrossRef]
18. Mohamed, A.E.A.; Ahmed, A.E.; Aboul, E.H. Whale Optimization Algorithm and Moth-Flame Optimization for multilevel thresholding image segmentation. *Expert Syst. Appl.* **2017**, *83*, 242–256. [CrossRef]
19. Wang, X.H.; Deng, Y.M.; Duan, H.B. Edge-based target detection for unmanned aerial vehicles using competitive Bird Swarm Algorithm. *Aerosp. Sci. Technol.* **2018**, *78*, 708–720. [CrossRef]
20. Rajinikanth, V.; Couceiro, M.S. RGB histogram based color image segmentation using firefly algorithm. *Procedia Comput. Sci.* **2015**, *46*, 1449–1457. [CrossRef]
21. Gao, M.L.; Jin, S.; Jun, J. Visual tracking using improved flower pollination algorithm. *Optik* **2018**, *156*, 522–529. [CrossRef]
22. Oliva, D.; Cuevas, E.; Pajares, G.; Zaldivar, D.; Osuna, V. A multilevel thresholding algorithm using electromagnetism optimization. *Neurocomputing* **2014**, *139*, 357–381. [CrossRef]

23. Bayraktar, Z.; Komurcu, M.; Bossard, J.A.; Werner, D.H. The Wind Driven Optimization Technique and its Application in Electromagnetics. *IEEE Trans. Antenn. Propag.* **2013**, *61*, 2745–2757. [CrossRef]
24. Rashedi, E.; Nezamabadi-Pour, H.; Saryazdi, S. GSA: A gravitational search algorithm. *Inf. Sci.* **2009**, *179*, 2232–2248. [CrossRef]
25. Kotte, S.; Pullakura, R.K.; Injeti, S.K. Optimal multilevel thresholding selection for brain MRI image segmentation based on adaptive wind driven optimization. *Measurement* **2018**, *130*, 340–361. [CrossRef]
26. Hussein, W.A.; Sahran, S.; Abdullah, S.N.H.S. A fast scheme for multilevel thresholding based on a modified bees algorithm. *Knowl.-Based Syst.* **2016**, *101*, 114–134. [CrossRef]
27. Sambandam, R.K.; Jayaraman, S. Self-adaptive dragonfly based optimal thresholding for multilevel segmentation of digital images. *Comput. Inform. Sci.* **2018**, *30*, 449–461. [CrossRef]
28. Jia, H.M.; Ma, J.; Song, W.L. Multilevel Thresholding Segmentation for Color Image Using Modified Moth-Flame Optimization. *IEEE Access* **2019**, *7*, 2169–3536. [CrossRef]
29. Abedinpourshotorban, H.; Shamsuddin, S.M.; Beheshti, Z. Electromagnetic field optimization: A physics-inspired metaheuristic optimization algorithm. *Swarm Evol. Comput.* **2016**, *26*, 8–22. [CrossRef]
30. Talebi, B.; Dehkordi, M.N. Sensitive association rules hiding using electromagnetic field optimization algorithm. *Expert Syst. Appl.* **2018**, *114*, 155–172. [CrossRef]
31. Bouchekara, H.R.E.H.; Zellagui, M.; Abido, M.A. Optimal coordination of directional overcurrent relays using a modified electromagnetic field optimization algorithm. *Appl. Soft Comput.* **2017**, *54*, 267–283. [CrossRef]
32. Ghamisi, P.; Couceiro, M.S.; Martins, F.M.; Benediktsson, J.A. Multilevel image segmentation based on fractional-order Darwinian particle swarm optimization. *IEEE Trans. Geosci. Remote Sens.* **2014**, *52*, 2382–2394. [CrossRef]
33. Sarkar, S.; Das, S.; Chaudhuri, S.S. A multilevel color image thresholding scheme based on minimum cross entropy and differential evolution. *Pattern Recogn. Lett.* **2015**, *54*, 27–35. [CrossRef]
34. Mirjalili, S.; Mirjalili, S.M.; Lewis, A. Grey Wolf Optimizer. *Adv. Eng. Softw.* **2014**, *69*, 46–61. [CrossRef]
35. Mirjalili, S. Moth-flame optimization algorithm: A novel nature-inspired heuristic paradigm. *Knowl.-Based Syst.* **2015**, *89*, 228–249. [CrossRef]
36. Cuevas, E.; Zaldivar, D.; Pérez-Cisneros, M. A novel multi-threshold segmentation approach based on differential evolution optimization. *Expert Syst. Appl.* **2010**, *37*, 5265–5271. [CrossRef]
37. Pecora, L.M.; Carroll, T.L. Synchronization of chaotic systems. *Phys. Rev. Lett.* **2015**, *25*. [CrossRef] [PubMed]
38. Kohli, M.; Arora, S. Chaotic grey wolf optimization algorithm for constrained optimization problems. *J. Comput. Des. Eng.* **2017**, *5*, 458–472. [CrossRef]
39. Bhandari, A.K.; Kumar, A.; Singh, G.K. Modified artificial bee colony based computationally efficient multilevel thresholding for satellite image segmentation using Kapur's, Otsu and Tsallis functions. *Expert Syst. Appl.* **2015**, *42*, 1573–1601. [CrossRef]
40. Tao, W.B.; Wen, T.J.; Liu, J. Image segmentation by three-level thresholding based on maximum fuzzy entropy and genetic algorithm. *Pattern Recogn. Lett.* **2003**, *24*, 3069–3078. [CrossRef]
41. Laing, H.N.; Jia, H.M.; Xing, Z.K.; Ma, J.; Peng, X.X. Modified Grasshopper Algorithm-Based Multilevel Thresholding for Color Image Segmentation. *IEEE Access* **2019**, *7*, 2169–3536. [CrossRef]
42. Bhandari, A.K.; Kumar, A.; Singh, G.K. Tsallis entropy based multilevel thresholding for colored satellite image segmentation using evolutionary algorithms. *Expert Syst. Appl.* **2015**, *42*, 8707–8730. [CrossRef]
43. Horng, M.H. A multilevel image thresholding using the honey bee mating optimization. *Appl. Math. Comput.* **2010**, *215*, 3302–3310. [CrossRef]
44. Sağ, T.; Çunkaş, M. Color image segmentation based on multi-objective artificial bee colony optimization. *Appl. Soft Comput.* **2015**, *34*, 389–401. [CrossRef]
45. Wang, Z.; Bovik, A.C.; Sheikh, H.R.; Simoncelli, E.P. Image quality assessment: From error visibility to structural similarity. *IEEE Trans. Image Process.* **2004**, *13*, 600–612. [CrossRef]
46. Sowmya, B.; Rani, B.S. Colour image segmentation using fuzzy clustering techniques and competitive neural network. *Appl. Soft Comput.* **2011**, *11*, 3170–3178. [CrossRef]

entropy

MDPI

Article

A q-Extension of Sigmoid Functions and the Application for Enhancement of Ultrasound Images

Paulo Sergio Rodrigues [1], Guilherme Wachs-Lopes [1], Ricardo Morello Santos [1], Eduardo Coltri [1] and Gilson Antonio Giraldi [2,*]

[1] Computer Science Department, Centro Universitário FEI, São Bernardo do Campo 09850-901, SP, Brazil; psergio@fei.edu.br (P.S.R.); gwachs@fei.edu.br (G.W.-L.); unifrsantos@fei.edu.br (R.M.S.); unifecoltri@fei.edu.br (E.C.)

[2] National Laboratory for Scientific Computing, Petrópolis 25651-075, RJ, Brazil

* Correspondence: gilson@lncc.br; Tel.: +55-24-2233-6088

Received: 14 March 2019; Accepted: 17 April 2019; Published: 23 April 2019

Abstract: This paper proposes the q-sigmoid functions, which are variations of the sigmoid expressions and an analysis of their application to the process of enhancing regions of interest in digital images. These new functions are based on the non-extensive Tsallis statistics, arising in the field of statistical mechanics through the use of q-exponential functions. The potential of q-sigmoids for image processing is demonstrated in tasks of region enhancement in ultrasound images which are highly affected by speckle noise. Before demonstrating the results in real images, we study the asymptotic behavior of these functions and the effect of the obtained expressions when processing synthetic images. In both experiments, the q-sigmoids overcame the original sigmoid functions, as well as two other well-known methods for the enhancement of regions of interest: slicing and histogram equalization. These results show that q-sigmoids can be used as a preprocessing step in pipelines including segmentation as demonstrated for the Otsu algorithm and deep learning approaches for further feature extractions and analyses.

Keywords: contrast enhancement; sigmoid; Tsallis statistics; q-exponential; q-sigmoid; q-Gaussian; ultra-sound images

---

1. Introduction

The use of sigmoid functions has been one of the most prominent strategies in the field of digital image processing, applied not only to highlight characteristics of scenes, objects, or regions of interest but also to segment different image modalities [1,2]. In this paper, we focus on sigmoid functions applied to the enhancement of regions of interest or to the enhancement of global contrast. In this way, the work of Reference [3] proposes an image enhancement method using a modified sigmoid function, where a 3×3 kernel was used as a low-pass convolution filter in the Y component of the $YCbCr$ color system. In the same line, in addition to using the sigmoid function as a filter, Reference [4] proposed the improvement of the support vector regression (SVR) method using a sigmoid kernel for high-resolution imaging without the need for training datasets. In this approach, the best models were chosen using the Bayesian decision theory.

For segmentation, the work of Reference [5] presented a neural network with a nucleus based on a sigmoid function in order to segment cysts in liver tissues. Also, in Reference [6], methodology called group method of data handling (GMHD) for malignant tumors detection in computed tomography is applied. However, considering the low contrast in some medical CT scans, the segmentation of organs, tumors, or other types of components becomes difficult tasks and wrong results can compromise diagnoses.

Such a problem motivates the study that can be found in Reference [7], which proposed a contrast enhancement algorithm and compares its performance with counterpart methods for several modalities of images. On the other hand, the work in Reference [8] presents a sigmoid model based on a cross-correlation algorithm to solve estimation errors in ultrasound images. More recently, saliency models and convolutional neural networks (CNNs) have been used for automatic tumor detection and breast segmentation and in ultrasound images [9,10].

Beyond the medical images field, another application of sigmoid functions is in the area of image fusion for the global enhancement of focus or contrast. Thus, Reference [11] presented a method of merging several partially focused images into different regions for generating a globally focused image using a sigmoid function. Moreover, in Reference [12], it is proposed a fusion algorithm of multifocal imaging based on a spectral comparison. In this algorithm, a sigmoid function was also applied in the fusion process. Also in the same line, the work by Reference [13] presented an image fusion method based on the segmentation of regions with the use of a sigmoid function applied to the method named adaptive multi-strategy fusion rule (AMFR).

In addition to segmentation and fusion of images, another important application of sigmoid functions is for the improvement of stereoscopic images. In this way, Reference [14] presented a visual comfort enhancement approach in 3-D stereoscopic images using nonlinear disparity mapping adaptable to fiducial features.

However, the efficiency of sigmoid functions in the mentioned applications depends mainly on the parameterization involved, which in turn leads the curve fitting over the regions of interest. On the other hand, the flexibility of sigmoid functions, based on traditional exponential and Gaussian functions, is restricted to the standard deviation and mean, imposing limitations due to reduced degrees of freedom. However, in the mid-1980s, the Tsallis entropy approach allowed more general functions to be proposed in the field of non-extensive statistical mechanics, also called q-statistics, which proved to be a generalization of the traditional statistics [15–17]. Since its initial proposal, this theory has been accepted in many areas of applications [18]. In the medical imaging field, in particular, it has been successfully used for feature extraction [19] but mainly for image segmentation [20–23]. Also, it allowed the emergence of the q-exponential functions, of which the parameter $q \in \mathbb{R}^+$, called the non-extensive parameter, can be used for fine tuning the image information.

Thus, in this paper, we introduce the q-exponential as a kernel of sigmoid expression, generating the q-sigmoid functions. Then, we demonstrate the versatility of q-sigmoids in contrast enhancement. The q-sigmoids are applied in the core of the image processing algorithm, and the obtained results are compared with the original sigmoid functions, as well as two other well-known methods for image enhancement: slicing and histogram equalization. We have shown that their superior performance over traditional methods is achieved by adapting the profile of these new functions to the image intensity distributions involved. Moreover, we demonstrate the advantages of q-sigmoid as a preprocessing step of input images in segmentation pipelines that include Otsu thresholding or CNN architectures [24,25]. The U-Net is a known deep network in this field, outperforming prior best methods [10].

The paper is organized as follows. Section 2 focuses on the q-statistics approach and q-exponential functions and summarizes some of their basic properties. The q-sigmoids families are proposed in Section 3, and important properties of such functions for image enhancement are proved in Section 4. The computational results are presented in Section 5. Next, Section 6 discusses some important points that have emerged from this work, the relationship with related topics, and further improvements regarding automatic parametrization. Finally, in Section 7, we conclude the paper, summarizing its main contributions and describing possible future work.

2. The q-Exponential Distribution

In the last decades, Tsallis [16] has proposed the following generalized non-extensive entropic form:

$$S_q = k\frac{1 - \sum_{i=1}^{W} p_i^q}{q-1}, \tag{1}$$

where k is a positive constant, $0 \leq p_i \leq 1$ is a probability distribution, W is the total number of states of the system (grayscale intensities, in the case of digital imaging), and $q \in \mathbb{R}^+$ is called the entropic index, or q-index also. This expression recovers the Shannon entropy in the limit $q \to 1$. The Tsallis entropy, computed by Equation (1), offers a new formalism in which the real parameter q quantifies the level of nonextensivity of a physical systems [16]. In particular, a general principle of maximum entropy (PME) has been considered to find out the distribution p_i to describe such systems. In this PME, the goal is to find the maximum of S_q subjected to

$$\sum_{i=1}^{W} p_i = 1, \tag{2}$$

and

$$\frac{\sum_{i=1}^{W} e_i p_i^q}{\sum_{i=1}^{W} p_i^q} = U_q, \tag{3}$$

where U_q is a known application dependent value and e_i represents the possible states of the system (in image processing, the grayscale intensities). Equation (2) is just a necessary condition for p_i to be probability, and Equation (3) is a generalized expectation value of the e_i (if $q = 1$, we get the usual mean value). The proposed PME can be solved using Lagrange multipliers, and the solution has the form [15,16]

$$p_j = \frac{\left[1 - (1-q)\widetilde{\beta} e_j\right]^{\frac{1}{1-q}}}{\widetilde{Z}_q}, \tag{4}$$

where $\widetilde{\beta}$ and $\widetilde{Z}_q$ are defined by Equations (5) and (6):

$$\widetilde{\beta} = \frac{\beta}{\sum_{j=1}^{W} p_j^q + (1-q)\,\beta U_q}, \tag{5}$$

$$\widetilde{Z}_q = \sum_{j=1}^{W} \left[1 - (1-q)\widetilde{\beta} e_j\right]^{\frac{1}{1-q}}, \tag{6}$$

with β being the Lagrange multiplier associated with the constraint given by Equation (3), and if $q < 1$, then $p_i = 0$ whenever $1 - (1-q)\widetilde{\beta} e_j < 0$ (cutoff condition). Then, Equations (1) and (4) inspire the definition of the q-exponential function [26]:

$$\exp_q(x) = \begin{cases} [1 + (1-q)\,x]^{\frac{1}{1-q}}, & \text{if } 1 + (1-q)\,x > 0, \\ 0, & \text{otherwise.} \end{cases} \tag{7}$$

It can be shown that the traditional exponential function (exp) is given by the limit:

$$\exp(x) = \lim_{q \to 1} \exp_q(x). \tag{8}$$

3. Proposed q-Sigmoid Functions

We built our proposal from the q-exponential function which, in turn, is derived from the PME and Tsallis entropy formalism (Section 2). In this context, we suppose a grayscale image $I : D \to [0, L]$, where $D \subset \mathbb{R}^2$ is the image domain and a specific region in D with an average luminance value β and standard deviation α. With the purpose of contrast enhancement, we consider the sigmoid transformation given by

$$I_1(I;\beta,\alpha,\lambda) = \frac{2}{1+\exp\left(\lambda\left(\frac{|I-\beta|}{\alpha}\right)\right)}. \tag{9}$$

Figure 1 shows a schematic example of the use of parameters β and α in Equation (9). In this figure, we indicate the β value in the input image as well as the range, defined through the parameter α, around the luminance given by β. We can notice that the luminance nearby β was mapped close to 1, while outside the rage $[\beta-\alpha,\beta+\alpha]$, it became darker. In this way, we get an enhancement of the target region, of which the pixel intensities fall in the range $[\beta-\alpha,\beta+\alpha]$ in the transformed field I_1.

On the other hand, we get interesting enhancement effects of values nearby β by using another sigmoid-like function defined as

$$I_2(I;\beta,\alpha,\lambda) = \begin{cases} \frac{1}{1+\exp\left(-\frac{\lambda}{\left(\frac{|I-\beta|}{\alpha}\right)}\right)}, & if \quad I \neq \beta, \\ 1, & otherwise. \end{cases} \tag{10}$$

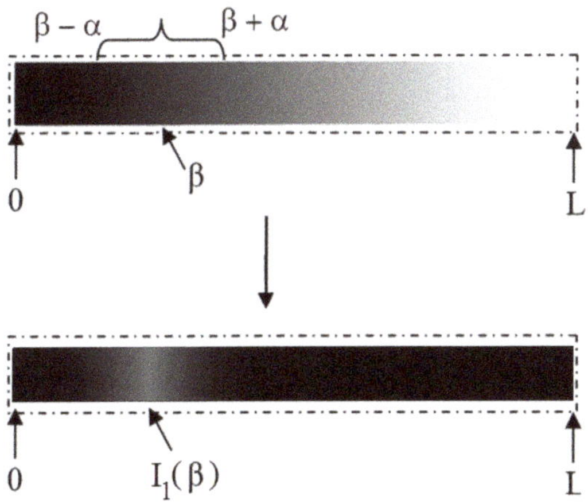

Figure 1. An example of the luminance transformation obtained by Equation (9) with the domain $0 \leq I(x,y) \leq L$ and parameters β and α indicated. The upper row shows the input image while the bottom shows the obtained result with an enhancement of the region nearby $I_1(\beta,\alpha,\lambda)$.

Following the idea of non-extensive systems for natural as well as medical images [22,27–29], in this work, we propose an extended version of Equations (9) and (10), called here as q-sigmoid functions, which are defined based on the q-exponential function, given by Equation (7) in the forms bellow.

- q-Sigmoid for $q < 1$:

$$\tilde{I}_1(I;\beta,\alpha,\lambda,q) = \frac{2}{1+\left[1+\lambda(1-q)\left(\frac{|I-\beta|}{\alpha}\right)\right]^{\frac{1}{1-q}}}, \tag{11}$$

- q-Sigmoid for $q > 1$:

$$\tilde{I}_2(I;\beta,\alpha,\lambda,q) = \begin{cases} \frac{1}{1+[1+\lambda(1-q)F(I)]^{\frac{1}{1-q}}}, & \text{if} \quad I \neq \beta \\ 1 & \text{otherwise} \end{cases}, \tag{12}$$

where

$$F\left(I\right) = -\frac{1}{\left(\frac{|I-\beta|}{\alpha}\right)}.$$

Using the fact that, in the limit, $q \to 1$ non-extensive expressions are reduced to extensive ones, and applying usual limit properties, it is straightforward to show that

$$\lim_{q\to 1} \tilde{I}_1(I;\beta,\alpha,\lambda,q,) = I_1(I;\beta,\alpha,\lambda), \tag{13}$$

and

$$\lim_{q\to 1} \tilde{I}_2(I;\beta,\alpha,\lambda,q) = I_2(I;\beta,\alpha,\lambda), \tag{14}$$

which prove that q-sigmoids are extensions of sigmoid functions.

The idea behind the use of q-sigmoids rather than sigmoid functions is motivated by the fact that q-sigmoids have the extra non-extensive parameter to control the curve's profile. Therefore, one can better customize filters to each class of applications by tuning q besides β, α, and λ parameters. This idea has been used in several fields of applications, mainly in image processing and computer vision applied to medical areas [22,27,29].

It is interesting to compare the profile of sigmoid and q-sigmoid functions under the variation of parameters, such as λ. For this task, we consider some values inside the interval $0 \leq q \leq 3$. Also, we set $I \in [0,255]$, $\beta = 128$, and $\alpha = 30$ and built Figure 2 that allows a comparison of the behavior of Equations (9) and (10) for $\lambda \in \{0.1, 0.5, 1.0, 2.0\}$. Also, Figure 3 shows the profiles of Equations (11) and (12) for $q \in \{0.1, 0.5, 1.5, 3.0\}$ and $\lambda \in \{0.5, 1.0, 2.0\}$.

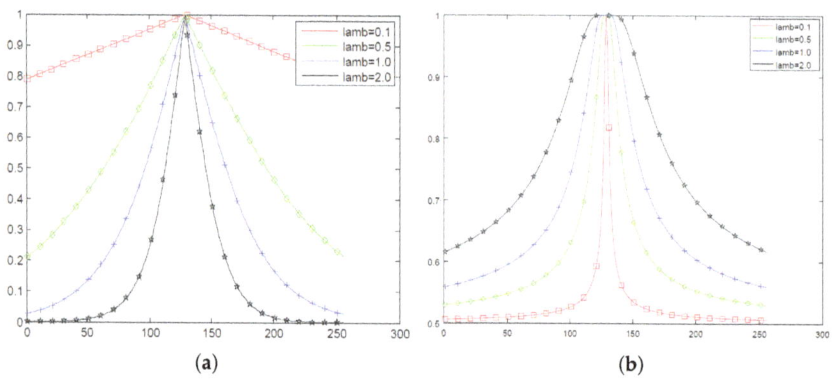

Figure 2. (**a**) The Sigmoid function profile for different λ in Equation (9); (**b**) the modified sigmoid function given by Equation (10).

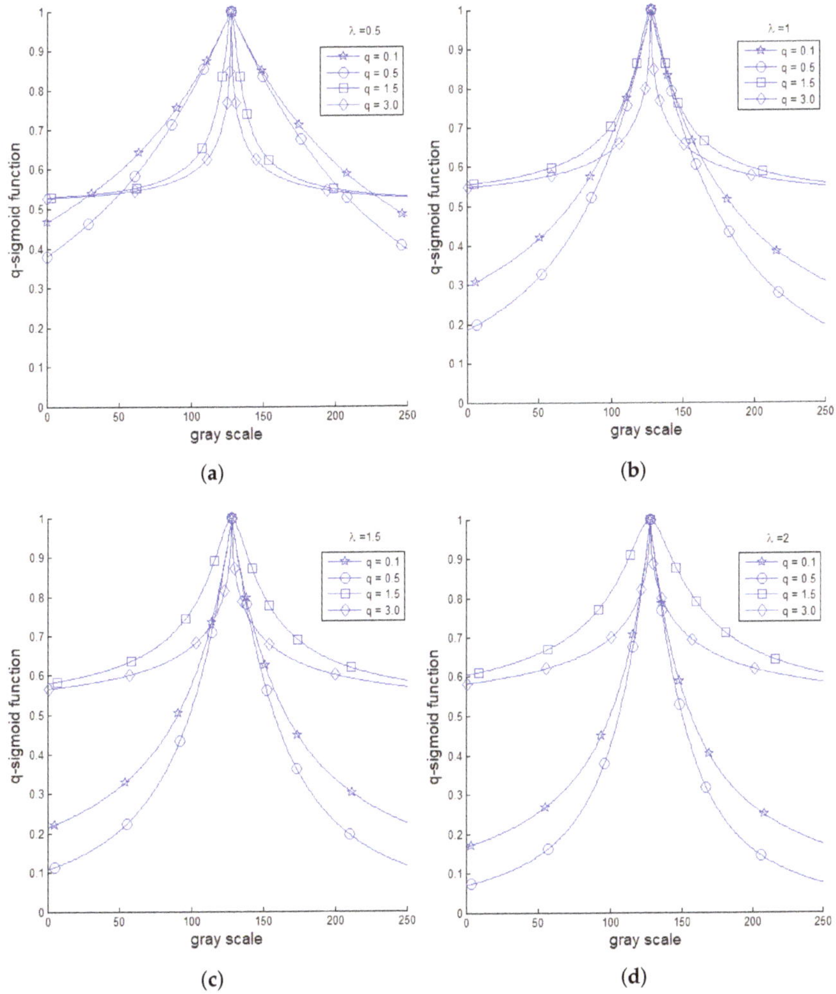

Figure 3. A comparison of q-sigmoid profiles of Equations (11) and (12): (**a**) $\lambda = 0.5$; (**b**) $\lambda = 1.0$; (**c**) $\lambda = 1.5$; and (**d**) $\lambda = 2.0$.

The visual analysis of Figures 2 and 3 allows us to guess that

(i) Both sigmoid and q-sigmoid functions have global maximum at $I = \beta$.

(ii) Equation (12) seems to have an asymptotic behavior for large values of I.

(iii) The slope is given by

$$\lim_{I \to \beta\pm} \frac{df}{dI}, \tag{15}$$

where $f \in \{I_1, I_2, \tilde{I}_1, \tilde{I}_2\}$ could give information about the decay of the corresponding expressions nearby the intensity $I = \beta$.

(iv) The limit value in Equation (15) is highly influenced by λ if $f = I_1$.

(v) We shall analyze Equation (15) to get the way in which the q value influences the decay for $f \in \{\tilde{I}_1, \tilde{I}_2\}$ and analogously for expression I_2 with respect to λ.

(vi) Equation (10) has a null derivative with respect to I at $I = \beta$.

Property (i) is a direct consequence of Equations (9)–(12), and it is true for the whole family of considered sigmoid and q-sigmoid functions. Property (ii) can be verified by observing that

$$\lim_{I\to\infty} \tilde{I}_2(I;\beta,\alpha,q) = 1/2. \tag{16}$$

The other points will be studied in the next section through the derivative of sigmoid and q-sigmoid functions.

4. Analysis of the Derivatives

For simplicity, we will consider the expressions only for $I \geq \beta$. The case $I < \beta$ is analogous with a change of signal only due to the symmetries of the target expressions with respect to $I = \beta$.

- Derivative of sigmoid in Equation (9) for $I > \beta$:

$$\frac{dI_1}{dI} = -2\left(\frac{\lambda}{\alpha}\right)\left(1+\exp\left[\lambda\left(\frac{I-\beta}{\alpha}\right)\right]\right)^{-2}\left(\exp\left[\lambda\left(\frac{I-\beta}{\alpha}\right)\right]\right). \tag{17}$$

- Derivative of sigmoid in Equation (10) for $I > \beta$:

$$\frac{dI_2}{dI} = -\frac{\lambda}{\alpha}\left(\frac{I-\beta}{\alpha}\right)^{-2}\left(1+\exp\left(-\frac{\lambda}{\left(\frac{I-\beta}{\alpha}\right)}\right)\right)^{-1}\exp\left(-\frac{\lambda}{\left(\frac{I-\beta}{\alpha}\right)}\right). \tag{18}$$

- Derivative of q-sigmoid (Equation (11)) if $q < 1$ and $I > \beta$:

$$\frac{d\tilde{I}_1}{dI} = -\lambda\frac{2}{\alpha}\left(1+\left[1+\lambda(1-q)\left(\frac{I-\beta}{\alpha}\right)\right]^{\frac{1}{1-q}}\right)^{-2}\left[1+\lambda(1-q)\left(\frac{I-\beta}{\alpha}\right)\right]^{\frac{q}{1-q}}. \tag{19}$$

- Derivative of q-sigmoid, given by Equation (12), if $q > 1$ and $I > \beta$:

$$\frac{d\tilde{I}_2}{dI} = -\frac{\lambda}{\alpha}\left(1+\left[1+\lambda(1-q)F\left(I\right)\right]^{\frac{1}{1-q}}\right)^{-2}\left[1+\lambda(1-q)F\left(I\right)\right]^{\frac{q}{1-q}}\left(\frac{I-\beta}{\alpha}\right)^{-2}. \tag{20}$$

From Equations (17)–(19), the following properties are straightforwardly verified:

$$\lim_{I\to\beta_+} \frac{dI_1}{dI} = -\frac{\lambda}{2\alpha}; \tag{21}$$

$$\lim_{I\to\beta_+} \frac{dI_2}{dI} = 0. \tag{22}$$

$$\lim_{I\to\beta_+} \frac{d\tilde{I}_1}{dI} = -\frac{\lambda}{2\alpha}. \tag{23}$$

As stated in Property (v) of Section 3, we shall consider the derivative results above to quantify the influence of parameters q and λ. From Equation (21), it becomes obvious that the behavior of Equation (9) is highly influenced by λ, as also noticed in the plots of Figure 2a and pointed out in Property (iv). Equation (22) demonstrates Property (vi) which states that the modified sigmoid Equation (10) is smooth at $I = \beta$ with a null derivative with respect to I. Equation (23) shows that the decay of $\tilde{I}_1$ nearby the intensity $I = \beta$ is almost insensitive to the q variation. To get more information about the decay of $\tilde{I}_1$ in a neighborhood of $I = \beta$, we shall consider the first terms of the Taylor series to approximate Equation (11) as

$$\tilde{I}_1\left(I;\beta,\alpha\right) \approx 1 - \frac{\lambda}{2\alpha}\left(I-\beta\right), \tag{24}$$

Hence, we can compare the first-order approximation in Equation (24) with the q-sigmoid through

$$R\left(I;\beta,\alpha,\lambda,q\right)=\frac{\tilde{I}_1\left(I;\beta,\alpha,\lambda,q\right)}{\left(1-\frac{\lambda}{2\alpha}\left(I-\beta\right)\right)}, \tag{25}$$

We experimentally notice that this expression is monotonically increasing with respect to I. Moreover, for $\lambda=1$, $I=(\beta+\alpha/2)=143$, and $0<q<1$, we get $R\left(I=143;128,30,q\right)<1.07$ which shows that the (almost) linear behavior observed for $q\in\{0.1,0.5\}$ in Figure 2c is a general property of Equation (11) .

An analogous procedure can be performed by Equation (9). However, such an analysis is not so simple when considering Equation (12) because its derivative with respect to I does not have an analytical form for $I=\beta$. To get some guess about Equation (15) for $I>\beta$, we plot in Figure 4 the difference equation:

$$D\left(I;\beta,\alpha,q\right)=\frac{\tilde{I}_2\left(I+\Delta I\right)-\tilde{I}_2\left(I\right)}{\Delta I} \tag{26}$$

for $I=128$, $\beta=128$, $\alpha=30$, $q\in\{1.5,3.0\}$, and $0.0001\leq\Delta I\leq0.1$

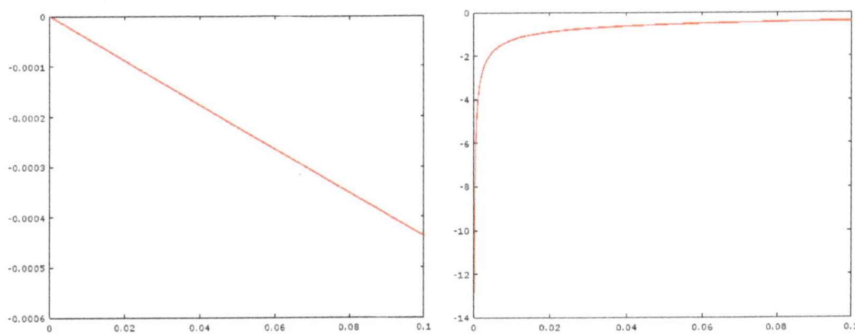

(a) (b)

Figure 4. (**a**) The difference Equation (26) for $I=128$, $\beta=128$, $\alpha=30$, $q=1.5$, and $0.0001\leq\Delta I\leq0.1$; (**b**) The behavior of Equation (26) in $I=128$ with the same parameters, but $q=3.0$.

The plots in Figure 4 indicate that the derivative of Equation (12) depends on the q value, which also agrees with the visual analysis of Figure 2c. From Figure 4a, we could guess that $\lim_{\Delta I\to0_+}D\left(128,\Delta I;128,30,1.5\right)=0$ while $\lim_{\Delta I\to0_+}D\left(128,\Delta I;128,30,1.5\right)\in(-14,-12)$ which implies a dependence of $d\tilde{I}_2/dI$ with respect to parameter q that cannot be explained with the first-order approximation for all ranges $q>1$.

5. Experimental Results

Before starting the computational experiments, it is worthwhile to state the practical consequences of Properties (i)–(vi) and the theoretical analysis of Section 4. Firstly, due to Property (i), we can assure that functions (9)–(12) highlight the pixel intensities nearby $I=\beta$, as pictured in Figure 1. From Equations (21)–(23), we can say that the way the transformed image intensity decays in a neighborhood of $I=\beta$ is linear (respect to I) for Equations (9) and (11) and quadratic (at least) for the modified sigmoid (10). Moreover, in the linear case, it does not depend on the q value, and the linear coefficients are proportional to λ and inversely proportional to α. However, the numerical experiments reported in Figure 4 show that the derivative of Equation (20) undergoes a different behavior nearby $I=\beta$, which depends on the q value besides λ.

In this paper, initially we propose an experimental analysis under artificial data to make easier a quantitative evaluation of the q-sigmoid for image enhancement. Then, we repeat the experiments for real images. The implementations were performed using the MATLAB Release 13 facilities. For the artificial data analysis, we built a 2-D Gaussian distribution so that we can manipulate its parameters (mean μ and standard deviation δ). The Gaussian's amplitude is normalized between $[0, 1]$ so that we can observe its 3-D profile as a grayscale image, creating artificial conditions for the application of the proposed filters.

With this idea at hand, we can use the Gaussian parameters in order to create severe conditions to improve the contrast and to enhance some regions of interest. For this purpose, we defined as a region to be highlighted the one with an intensity in the interval $\mu \pm 3 \times \delta$. Figure 5 shows this setup, where Figure 5a is the original image with a clear region of interest, and Figure 5b is its corresponding perspective visualization.

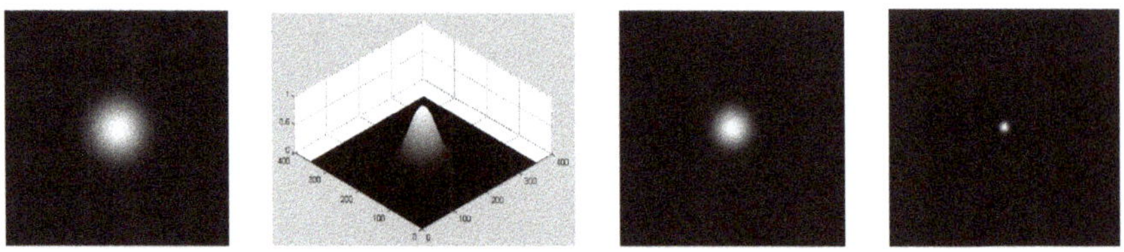

(a) (b) (c) (d)

Figure 5. (**a**) The intensity profile of a normalized 2-D Gaussian distribution, with a mean $\mu = 175$ and a variance $\delta = 30$ in orthogonal view; (**b**) the corresponding 3-D perspective; (**c**) an example of filtering with sigmoid when $\alpha = 0.02$; (**d**) an example of filtering with sigmoid when $\alpha = 0.01$.

5.1. Locally Linear Behavior

In this section, we perform computational experiments using the families given by Equations (9) and (11), of which the derivatives at $I = \beta$ are linear functions with respect to λ (see Equations (21) and (23)). Therefore, we set the parameters $\beta = 1.0$ and $\lambda = 1.0$ and vary the α value for both the sigmoid (Equation (9)) and q-sigmoid (Equation (11)) functions. Under the scenario of Equations (21) and (23), when α is set to small values (e.g., 0.02 or 0.01), the rate of decrease of the transformed image intensity becomes too high. Consequently, the highlighted region in the output field can be very small compared to the original one (e.g., Figure 5c–d).

Now, we keep the value of $\beta = 1.0$ and $\lambda = 1.0$ and set the value of $\alpha = 0.03$ for both the sigmoid and q-sigmoid functions. The q-sigmoid function was parameterized by taking also $q \in \{0.7, 0.8, 0.9, 0.95, 0.98, 0.99, 0.999\}$. We apply the two filters in the image shown in Figure 5a.

The values of q were chosen in order to highlight significant behaviors. For instance, for $q < 0.7$, the function q-sigmoid gets optimal results for all α values studied here (the case $q > 1.0$ is discussed later). Note that the value $q = 1.0$ corresponds to the traditional sigmoid function, since q-sigmoid reduces to sigmoid when q tends to 1.0, as explained in Section 3.

For each value of q, the q-sigmoid filter of Equation (11) was applied on the artificial image of Figure 5a. The same was done for the sigmoid filter in Equation (9). Since each highlighted region may represent a connected set of pixels, a comparison between both results (by sigmoid and q-sigmoid) is accomplished by relating the convex-hull of both enhanced and original areas, according to the following equation:

$$Err = \left| 1 - \frac{A_1}{A_0} \right| \tag{27}$$

where A_1 is the convex-hull area of the q-sigmoid/sigmoid achieved region and A_0 is the ground-truth area defined in a Gaussian grayscale profile image by setting $\mu \pm 3 \times \delta$. When Err achieves zero, it means that q-sigmoid/sigmoid filtering gets the optimal performance.

Table 1 shows the error for each value of q cited above. Figure 6 shows the corresponding visual result when $q = 1.0$. Figure 6a shows the original Gaussian grayscale profile but now with a circle around indicating the convex hull of the ground truth area. Figure 6b is the corresponding filled area, and Figure 6c is the achieved area after sigmoid filtering. The visually imperceptible difference (confirmed in Table 1 for $q = 1.0$) between areas of Figure 6b,c indicates that this was a good performance of filtering approach.

Table 1. *Err* demonstration under a range of q values with $q < 1.0$, $\alpha = 0.03$, $\beta = 1.0$, and $\lambda = 1.0$.

q	0.7	0.8	0.9	0.95	0.98	0.99	1.0
Err	0.0	0.0	0.0	0.00	0.0	0.00	0.0001

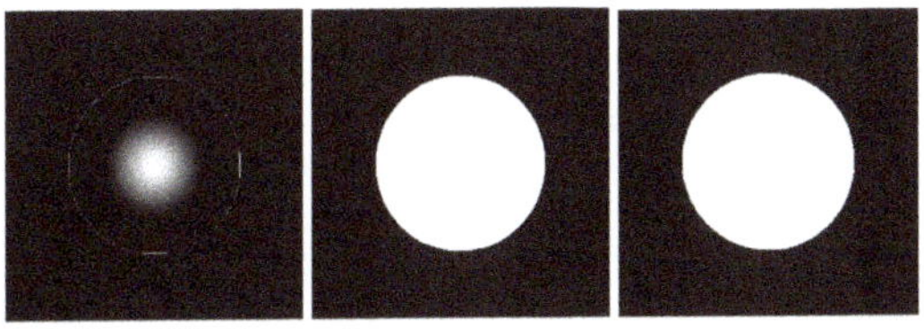

(**a**) ground truth (**b**) convex hull (**c**) $\alpha = 0.03$

Figure 6. (**a**) The intensity profile of a 2-D Gaussian distribution in orthogonal view with the corresponding convex-hull points indicating the ground truth area; (**b**) the corresponding convex-hull filled area; and (**c**) the convex-hull-filled area achieved when $\alpha = 0.03$, $\beta = 1.0$, $\lambda = 1.0$, and $q = 1.0$.

On the other hand, in the more extreme case, we set $\alpha = 0.01$, $\beta = 1.0$, and $\lambda = 1.0$ and apply the q-sigmoid function for all values of q in our set up. Under this new scenario, the value of α is too small, severely hampering the search by the region of interest. To allow the comparison with the results of Table 1 and Figure 6, we report in Table 2 the *Err* measure found for each q value and show in Figure 7 the corresponding visual results.

Table 2. *Err* demonstration under a range of q values around $q = 1.0$ and $\alpha = 0.01$.

q	0.7	0.8	0.9	0.95	0.98	0.99	1.0
Err	0.0	0.0	0.0	0.0052	0.8412	0.8992	0.9266

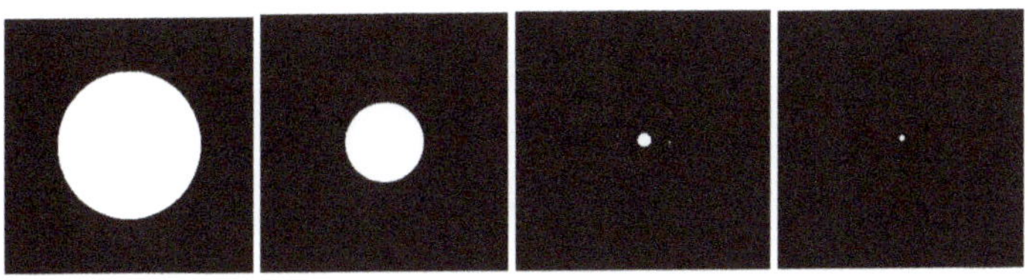

(**a**) $q = 0.95$ (**b**) $q = 0.98$ (**c**) $q = 0.99$ (**d**) $q = 1.0$

Figure 7. (**a**) The convex-hull area achieved when $q = 0.95$; (**b**) the convex-hull area achieved when $q = 0.98$; (**c**) the convex-hull area achieved when $q = 0.99$; and (**d**) the convex-hull area achieved when $q = 1.0$.

Now, according to Table 2, some values of q are more sensitive to the variation of α and present a low performance for the same input of Figure 5a. In Figure 7, we present four convex hulls found for three q-sigmoid filters (when $q \in \{0.95, 0.98, 0.99\}$) and the sigmoid one (when $q = 1.0$). For all other q values, including $q \geq 1.0$, under $\alpha = 0.01$, the q-sigmoid filtering behavior was optimal.

For completeness, we show in Table 3 the same results observed in the Tables 1 and 2 but add the results for other values of $\alpha \in \{0.01, 0.015, 0.02, 0.025, 0.03\}$. As said before, for $\alpha \geq 0.03$, it is no longer extreme cases and both filters achieve the same optimal performances, so these cases are not shown here.

Table 3. *Err* demonstration for $q < 1.0$, $\beta = 1.0$, $\lambda = 1.0$, and some α values.

$Err \backslash q$	0.7	0.8	0.9	0.95	0.98	0.99	1.0
$\alpha = 0.01$	0.0	0.0	0.0	0.0052	0.8412	0.8992	0.9426
$\alpha = 0.015$	0.0	0.0	0.0	0.0	0.3893	0.7298	0.8996
$\alpha = 0.02$	0.0	0.0	0.0	0.0	0.0037	0.0182	0.8173
$\alpha = 0.025$	0.0	0.0	0.0	0.0	0.0000	0.0027	0.5422
$\alpha = 0.03$	0.0	0.0	0.0	0.0	0.0000	0.00	0.001

Finally, Figure 8 shows the overall observed behavior in Table 3. In this figure, we can clearly see that as the α value increases (shown inside the small box), the value of *Err* also decreases for all values of q. However, for each value of α, the value of $q = 1.0$, the traditional sigmoid function presents the worst performance. In this figure, we also show results for $q > 1.0$.

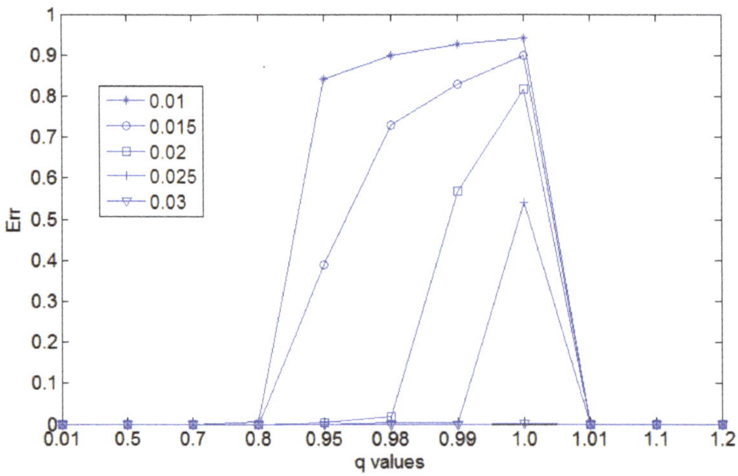

Figure 8. The overview performance under increasing α values. Each curve corresponds to an α value used for an enhancement filtering of the artificial grayscale 3-D Gaussian profile. We also show *Err* for $q > 1.0$.

Clearly, Table 3 and Figure 8 show the superiority of the q-sigmoid filtering over the traditional sigmoid one to highlight images in extreme conditions when the range of values around β are known to be small.

5.2. Generalized Behavior

Now, we shall analyze results using Equation (12) that depends on the q value besides λ. We follow the same methodology of previous sections and apply the proposed enhanced technique over the input image pictured in Figure 5a.

Initially, we set the value of $\beta = 1.0$, $\alpha = 0.03$, and $\lambda = 1.0$ for both sigmoid (Expression (10)) and q-sigmoid functions (Equation (12)) and applied the filters in the image shown in Figure 5a. The q-sigmoid function was applied with values of $q \in \{1.01, 1.1, 1.2\}$. A comparison between both results was accomplished by computing Equation (27). However, the errors for all considered value of q, when $\beta = 1.0$ and $\alpha = 0.03$ were zero.

On the other hand, in the extreme case, we set $\alpha = 0.01$ and applied the q-sigmoid function for all values of $q \in \{1.01, 1.1, 1.2\}$. Under this new scenario, the value of α was too small, severely hampering the search by the region of interest.

Table 4 shows the performance of sigmoid ($q = 1$) and q-sigmoid for increasing values of α and q. Note that Err for values of $q > 1.0$ tends to zero for all α ranges.

Table 4. Err demonstration under a range of q values around $q = 1.0$ and a range of α values.

$Err \backslash q$	1.0	1.001	1.1	1.2
$\alpha = 0.01$	0.0	0.00	0.00	0.0
$\alpha = 0.015$	0.0	0.00	0.00	0.0
$\alpha = 0.02$	0.0	0.00	0.00	0.0
$\alpha = 0.025$	0.0	0.00	0.00	0.0
$\alpha = 0.03$	0.0	0.00	0.00	0.0

5.3. Experiments with Ultrasound Images

The tests in artificial images indicate how q-sigmoid functions behave for image enhancement regarding their parameters. We conclude that q-sigmoids defined by Equations (11)–(12) achieve the better performance in the enhancement tests using Figure 5a. The q-sigmoid results tend to highlight with a higher contrast the region of interest from the background, as shown when comparing the synthetic experiments of Figures 6 and 7. We would like to exploit this property in applications involving medical images.

Besides the influence of noise and low contrast, the morphology of regions of interest is also a factor that hampers the assessment of filters considered to highlight the target information. For instance, ultrasound images as seen in Figure 9 are heavily corrupted by speckle noise, besides having target objects not clearly defined. Under such conditions, the region of interest (lesion) can merge with the background, making difficult further segmentation operations.

Thus, the greatest challenge faced by image enhancing methods automatically isolates the lesion from its background, reducing the interference of noise and artifacts in further operations.

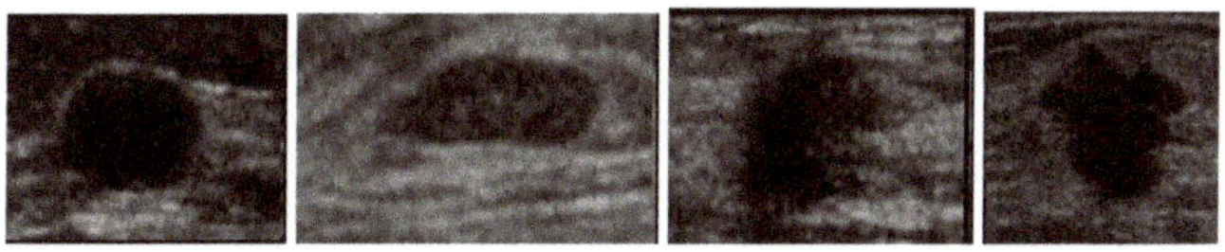

Figure 9. Four examples of breast ultrasound images with lesions in a darker gray scale nearby the center of the images.

A well-known method for image enhancement is the slicing, which consists of mapping the original intensity range within a small track [30]. The most popular mapping is one that generally excludes the lower and higher intensity ranges of the original image. The justification for this slicing method is that the human visual system is much more sensitive to medium intensity ranges than low and high intensities. Besides slicing, histogram equalization also improves the visual contrast of such images, but it does not mean that they facilitate the automatic extraction of the lesion. One reason is because these two popular methods, although they tend to produce more visually pleasing results, also enhance noise, artifacts, and contrast between unwanted areas. In order to compare the slicing and histogram equalization results with the ones obtained with q-sigmoids, we use a statistical measure that quantifies the difference in intensity between two images, in our case, the original image (X) and the output one (Y) obtained by the considered methods. Specifically, we apply the absolute mean brightness error (AMBE), which can be calculated through the absolute difference between the mean intensities of X and Y:

$$AMBE = |E(X) - E(Y)|, \tag{28}$$

where $E(\cdot)$ denotes the statistical mean.

Equation (28) clearly shows that AMBE is designed to detect excessive change in the brightness. In current practice, a lower AMBE implies that the original brightness is better preserved and, hence, that the image Y should yield a better enhancement of the region of interest.

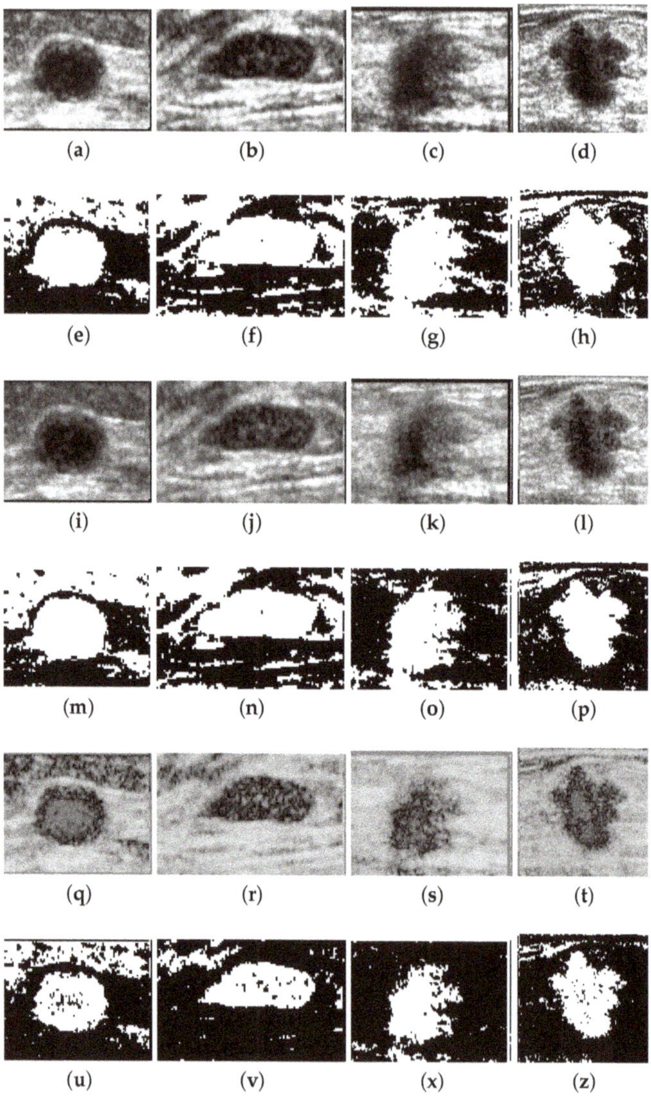

Figure 10. The results of application of enhancing methods over the four images of Figure 9. (**a–d**) The results of histogram equalization. (**e–h**) The segmentation after histogram equalization. (**i–l**) The slicing map results. (**m–p**) The segmentation after slicing. (**q–t**) The results obtained with q-sigmoid when $\beta = 0.15$, $\alpha = 0.03$, $\lambda = 1.0$, and $q = 2.0$ in Equation (12). (**u–z**) The segmentation results after preprocessing with q-sigmoid.

Figure 10 allows a visual comparison between the results of the histogram equalization, slicing, and q-sigmoid given by Equation (12) when applied to the four images in Figure 9. In these examples, it can be noted from Figure 10a–d and Figure 10i–l that the histogram equalization method and slicing-based technique obtain a better visual result than the proposed method (Figure 10q–t). However, the output of q-sigmoid has a much stronger contrast, clearly facilitating the isolation of region of interest from the background as we can check in the images of Figure 10q–t. In Figure 10u–z, we show the segmentation results obtained by preprocessing the original images with q-sigmoid and entering the obtained result in the Otsu thresholding algorithm [24]. If compared with the segmentation results obtained after preprocessing with counterpart techniques (Figure 10e–h and Figure 10m–p), we can visually confirm that the lesion segmentation, in general, is better extracted after a q-sigmoid enhancement. Specifically, Figure 10e–m,u shows equivalent results in the sense that the region of interest is isolated from the surrounding object. In this line, the results in Figure 10v,x favor a q-sigmoid against the counterpart methods (Figure 10f,n,g,o) once there are no bridges between the target object and its neighborhood. The segmentation presented in Figure 10h,p,z can be considered equivalent ones.

In order to obtain a quantitative comparison, we apply the AMBE, formally described by Equation (28). The test database contains 250 ultrasound images of breast cancer, 100 malignant, and 150 benign. The images in Figure 9 are four samples of this database. The images are processed by the considered methods, and Equation (28) was applied for a subsequent analysis.

For the proposed method, we set $\beta = 0.15$, $\alpha = 0.01$ and $\lambda = 1.0$, and we varied the q value in q-sigmoid as $q \in \{0.1, 0.5, 0.999, 1.1, 2.0\}$. For each of the 250 images of the database, the three methods were then applied. Figure 11 allows the comparison of the performance of classical methods (equalization and slicing) and the proposed method with the five q values cited before, given a total of eight comparative instances for each sample. The box-plot of Figure 11 clearly shows that the proposed method outperforms the other two (histogram equalization and slicing-mapping) with any value of q in the q-sigmoid function.

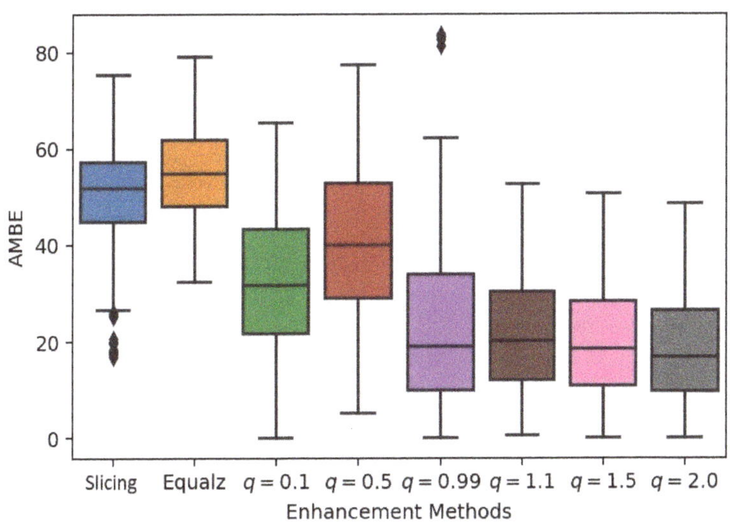

Figure 11. An overview performance under histogram equalization (first left block), slicing (second block), and q-sigmoids (Equations (11) and (12)) with increasing q values (third to eighth block). Each block stands for a mean and standard deviation of absolute mean brightness error (AMBE) measure in the vertical axis.

It can be seen that the best performance was obtained with the value of $q = 2.0$. On the other hand, although for the value of $q = 0.999$ we have obtained a better result than the two other traditional methods, it is the configuration that generates the highest standard deviation.

5.4. *q*-Sigmoid and CNN Segmentation

This section discusses the q-sigmoid efficiency as a preprocessing step in CNN-based image segmentation pipelines. To demonstrate this idea, we used two databases in our experiments. The first one is a 256×256 artificial image database, generated through the Gaussian function shown in Figure 5a. In order to build such a database, we add to the gray scale image in Figure 5a different Gaussian noise levels, generating 100 images with signal to noise ratio (SNR) from $SNR = 0$ to $SNR = 100$. The second grayscale image database is composed by the 250 breast cancer ultrasound images used in Section 5.3. Examples of these images can be seen in Figure 9, where the central part of the images represent the lesion (region of interest). For both databases, we randomly separate 70% images for training and the remaining ones for test, using a cross-validation strategy for CNN training.

The databases were tested on two CNNs with the same architecture: 3 convolutional, 2 max pooling, and 1 transpose layer. The convolutional layers use Relu, and the transpose applies the linear activation functions. The networks are named CNN_A and CNN_U, of which the purpose is to segment the input images (artificial for CNN_A and ultrasound for CNN_U) by separating the region of interest from its background.

The segmentation quality measure used in this work compares the CNNs prediction with a ground-truth. For the artificial database, the ground-truth is the image shown in Figure 5a. In the case of ultrasound images, they were manually segmented by an expertise to build the ground-truth as pictured in Figure 12.

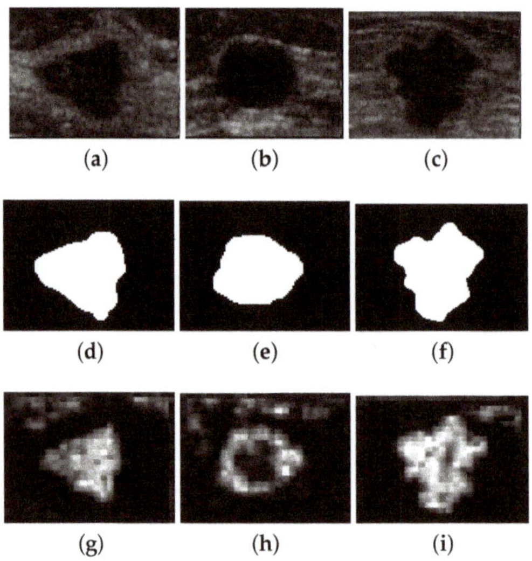

Figure 12. **(a–c)** Original ultrasound images. **(d–f)** The segmentation ground-truth. **(g–i)** The segmentation obtained by CNN after filtering by q-sigmoid when $\beta = 0.15$, $\alpha = 0.03$, $\lambda = 1.0$, and $q = 0.1$ in Equation (12).

Therefore, we fed the trained $CNNs$ with a test image and computed the similarity between its segmentation result (say A) and the ground-truth (say G) by expression:

$$S(A,G) = \frac{1}{2}\left(\frac{\#(A \cap G)}{\#(A \cup G)} + \frac{\#[(G \oplus B - G) \cap \sim A]}{\#(G \oplus B - G)}\right), \tag{29}$$

where # means the number of pixels, $\oplus$ is the dilation operation, B is the canonical 3×3 structuring element [31], and $\sim A$ means the negative of image A. The above expression is motivated by the fact that, in the case of ultrasound images, the CNN_U might extract other objects with similar texture patterns besides the region of interest, as observed in Figure 12g–i. If those regions are disconnected in the segmentation result, it is easy to isolate them in a further process of information extraction. In this case, the ratio $\#\left[(G \oplus B - G) \cap \sim A\right] / (\#(G \oplus B - G))$ is approximately one, indicating that the CNN_U isolates the region of interest. Otherwise, this ratio falls in the interval $[0, 1]$, penalizing the similarity measure because the lesion is connected with other regions in the segmented image.

Therefore, we apply the q-sigmoid filter to each image before entering the CNN and computing Equation (29) to measure the quality of the segmentation obtained. Figure 13a shows the CNN_A efficiency for the segmentation of the synthetic database (black line) and its filtered version (blue line). Figure 13b reports an analogous result for CNN_U.

From Figure 13a,b, we notice that the application of q-sigmoid as a preprocessing step, in general, improves the segmentation results for both CNN_A and CNN_U, although this fact is more evident in the case of artificial images because these images are less challenging than the ultrasound ones to be segmented.

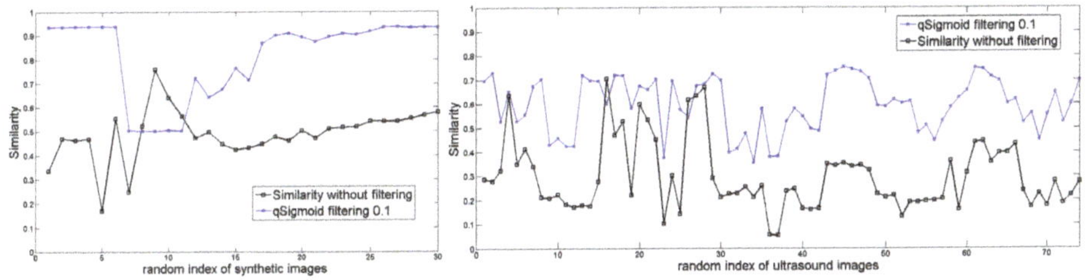

(a) (b)

Figure 13. (**a**) A comparative performance of the CNN_A output without filtering (black line) and with preprocessing by q-sigmoid ($q = 0.1, \beta = 1.0, \alpha = 0.8, \gamma = 1.0$ in Equation (11)), shown by the blue line. The vertical axis is the similarity calculated according to Equation (29), and the horizontal axis lists the synthetic images with different noise levels randomly chosen; (**b**) The analogous result for CNN_U with $q = 0.1, \beta = 0.6, \alpha = 0.2$, and $\gamma = 1.0$ in Equation (11).

6. Discussion

This paper describes the applicability of the proposed q-sigmoid functions, defined by Equations (11) and (12) in breast ultrasound images, which are images highly affected by speckle noise. Moreover, tests were performed with artificial images, under controlled parameterization demonstrating the superiority of q-sigmoids against their classical sigmoid versions.

It should be emphasized that our goal is not to improve global contrast but to highlight a specific region of interest, as in the case of the lesions in the breast ultrasound images of Figure 9. The box-plot of Figure 11 clearly shows that the performance of the proposed q-sigmoid function overcomes the techniques of slicing and equalization in this task. Clearly, there is a decay of the AMBE measure as the q value increases, indicating that, although the ideal value of q has not been pre-calculated, an acceptable range is easily defined, that is, the proposed q-sigmoid function produces a good performance for a large range of q values.

It is also noticed in the synthetic experiments of Section 5.1 the great influence of the values of q in relation to function decay. Thus, although this sensitivity is large, due to the overall good performance for a wide range of q values, this behavior does not affect negatively the final result.

The results of Sections 5.3 and 5.4 indicate that the q-sigmoid preprocessing can simplify segmentation stages if used in pipelines, such as the ones of CAD (Computer Aided Diagnosis) systems. This is the main factor that highlights the proposed functions as promising strategies to be used in computational systems of image analysis even considering recent approaches based on CNNs. In fact, Figures 10 and 13 show some results, indicating that the q-sigmoid can improve the segmentation obtained by Otsu and CNN in this application.

Also, we shall include some comments about the relationship between our work and the fractional logistic function written as

$$D(I) = \frac{D_{max}}{1 + C\exp\left(-\left(k\left(I - I_0\right)\right)^b\right)}, \tag{30}$$

where $D_{max}, C, k, b > 0$ with $0.5 < b < 3.0$ [32,33]. We can demonstrate that the derivative of Equation (30) with respect to I is zero in $I = I_0$ and that this point is a local minimum. For $b = 2$, named a quadratic logistic function in Reference [33], we can verify that $\lim_{I\to+\infty} D(I) = \lim_{I\to-\infty} D(I) = D_{max}$. Moreover, Equation (30) does not have other critical point different from $I = I_0$. Consequently, Equation (30) has the profile given by Figure 14a if $b = 2$. In Figure 14b, we show the profile of $f(I) = 1 - D(I)$, which is a bell-shaped function. Therefore, it resembles the plots in Figure 2b, where we have $dI_2/dI = 0$ if $I = \beta$, as shown by Equations (18) and (22).

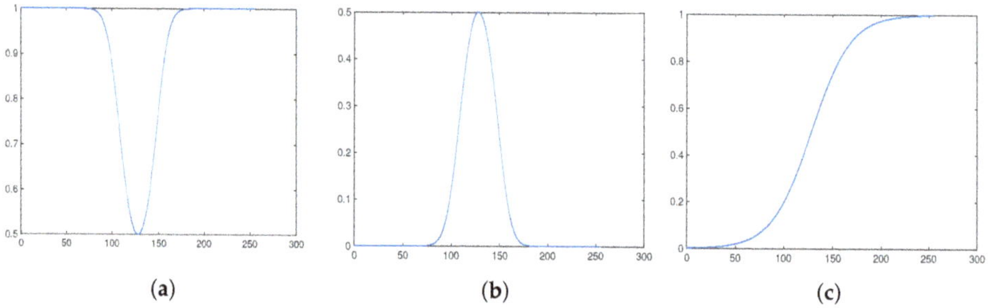

Figure 14. (**a**) The profile of fractional logistic function with $D_{max} = 1$, $C = 1$, $b = 2$, $k = 0.05$, and $I_0 = 128$; (**b**) Negative of the fractional logistic function; (**c**) The S-shaped logistic function obtained by setting $b = 1$ and keeping the other parameters unchanged.

Consequently, a local analysis nearby $I_0 = \beta$ shows an equivalent behavior between $I_2(I)$ in Equation (10) and $f(I) = 1 - D(I)$. Far from $I = \beta$, we can adjust the parameters α and λ such that we obtain $I_2(I) \approx f(t)$. Hence, the quadratic fractional logistic function and the sigmoid in Equation (10) generate (almost) equivalent enhancement results. Consequently, once q-sigmoid functions surpass the sigmoid one, we expect the same conclusion regarding the quadratic fractional logistic function.

On the other hand, if $b = 1$ in Equation (30), we get an S-shaped profile discussed in References [32,33], like the one pictured in Figure 14c. Therefore, in this case, the behavior of the logistic function family is very different from the bell-shaped q-sigmoid functions represented in Figure 3.

However, in another way, from the result of Equation (8), we can write

$$\lim_{q\to 1}\left[1 + (1-q)\left(-(kt)^b\right)\right]^{1/(1-q)} = \exp\left(-(kt)^b\right). \tag{31}$$

Therefore, we can generalize Equation (30) as

$$\widetilde{D}(I) = \frac{D_{max}}{1 + C\left[1 + (1-q)\left(-(kt)^b\right)\right]^{1/(1-q)}}, \tag{32}$$

and study its behavior for $q \neq 1$. We intend to perform such task as future works.

Regarding the parametrization of q-sigmoids, we shall perform some comments once we set the parameters β, α, λ, and q. The parameters β and α define the interval $[\beta - \alpha, \beta + \alpha]$ that we want to highlight in the output image, as pictured in Figure 1. Therefore, these values are set using the knowledge about the regions of interest. In the case of Equation (11) of the material, the parameter λ controls the decay of the image intensity nearby $I = \beta$, as noticed in Equation (24). If we want to sharpen the highlighted region, we must increase it. Otherwise, we shall set it smaller. In the case of images with a gray scale intensity in the interval $[0,1]$, a systematic procedure would be to start with $\lambda = 1$ and then to make it larger or smaller depending on the effect the user wants. The setting of parameter q is more involved, and in general, it is performed by defining a sequence of q-values and analyzing the results to realize the best one. One suggestion is to take some values smaller and larger than $q = 1$ besides the $q = 1$ itself that corresponds to the sigmoid functions. We proceed in this manner to generate Figure 11 of the submitted manuscript.

Moreover, we would like to comment another approach that we shall analyze in further works. We can transform the intensity values of the input image such that the histogram of the output field matches the solution of the PME discussed in Section 2. Formally, we suppose a random variable $u \geq 0$ with a probability density $p_1(u)$ given by the histogram of the input image. Then, inspired by the histogram equalization approaches [34], we transform the variable u in another random variable $v \geq 0$ such that its probability density $p_2(v)$ is given by the solution of the PME. To perform this task, it is just a matter of defining the functions:

$$\int_0^u p_1(x)\,dx = F_1(u), \quad \int_0^v p_2(y)\,dy = F_2(v), \tag{33}$$

and imposing that the variable v must satisfy the constraint $F_2(v) = F_1(u)$, which gives

$$v(u) = F_2^{-1}(F_1(u)). \tag{34}$$

Now, the idea is to solve the problem:

$$\arg\min_{\beta,\alpha,\lambda,q} \left| v(u) - \widetilde{I}(u;\beta,\alpha,\lambda,q) \right| \tag{35}$$

and to seek for parameters $(\beta,\alpha,\lambda, andq)$ such that the q-sigmoid $\widetilde{I}(u,\beta,\alpha,\lambda, andq)$, wherever $\widetilde{I}_1$ or $\widetilde{I}_2$ are, approximates v defined by Equation (34). In this manner, we strengthen the association of the q-sigmoid functions with Tsallis entropy because the random variable v has a probability density $p_2(v)$ that is the solution of the Tsallis PME. On the other hand, we can generate a systematic procedure to set the model parameters in the q-sigmoids.

Moreover, it is important to emphasize the consequences of the q-calculus and q-analysis, which are subproducts of the Tsallis formalism to the signal processing area. In this way, in Reference [35,36], the q-deformation of known functions, introduced in the q-calculus [37], are considered that allow the consideration of some modifications in known functional transforms (wavelets and Gabor transform, for instance) in order to discuss the performance of image processing techniques on noisy objects in the frequency domain. Also, q-Gaussian and difference-of-q-Gaussian kernels have been used as generalizations of the classical counterparts, with outstanding results for noise reduction and high-pass filtering (see Reference [38] and references therein).

7. Conclusions

This paper introduced the q-sigmoid functions as spatial filters for image enhancing in order to isolate the region of interest. The main objective was to highlight a particular region and not visually improve the image through the contrast enhancement. The q-sigmoid functions are generalizations of the well-known sigmoid families, obtained when setting $q = 1$ in Equations (11) and (12).

Tests were carried out with synthetic images when the proposed method outperformed the sigmoid functions under a range of parameter values that leave the scenario in extreme conditions. When the α value is extremely low, the q-sigmoid functions always get better results than the sigmoid family, especially for values of $q \neq 1.0$. The method was also applied to ultrasound breast cancer images, which are severely affected by the speckle noise, low contrast, and subjective boundary. In this case, the proposed method was also compared with two other known methods of contrast enhancement (histogram equalization and slicing) using the AMBE, which allows the measurement of the contrast level of the transformed image against the level of contrast of the original one. The proposed method achieved better results in terms of AMBE than the counterpart ones for $q \in \{0.1, 0.5, 0.999, 1.1, 2.0\}$, suggesting that the q-sigmoids are promising for region enhancement and can be used in CAD systems for this type of image. In particular, the Otsu and CNN segmentation results obtained when preprocessing ultrasound images with q-sigmoids indicate the capabilities of the functions in segmentation pipelines.

As further works, we intend to apply q-sigmoid functions for other image modalities to emphasize the contribution of our proposal as a general technique for image enhancement. Also, we shall compare q-sigmoids with the generalization of fractional logistic functions (Equation (32)) and to implement a scheme to solve problem (Equation (35)).

Author Contributions: P.S.R., G.W.-L., R.M.S. and G.A.G. contributed equally to the conceptualization, methodology, formal analysis, and writing. The author E.C. contributed to the software, validation, and data curation.

Funding: This research was funded by PCI-LNCC.

Acknowledgments: We thank to FAPESP (Foundation for Research of the Sao Paulo State, Brazil), CNPq (National Council for Scientific and Technological Development, Brazil), CAPES (Coordination of High Level Graduation, Brazil), and FEI (Foundation for Ignatian Education, São Paulo, Brazil) for all the support given to this paper.

Conflicts of Interest: The authors declare no conflict of interest.

References

1. Imtiaz, M.S.; Wahid, K.A. Color Enhancement in Endoscopic Images Using Adaptive Sigmoid Function and Space Variant Color Reproduction. *Comp. Math. Methods Med.* **2015**, *2015*, 607407. [CrossRef]
2. Foruzan, A.H.; Chen, Y.W. Improved segmentation of low-contrast lesions using sigmoid edge model. *Int. J. Comput. Assist. Radiol. Surg.* **2015**, *11*, 1267–1283. [CrossRef] [PubMed]
3. Gupta, B.; Agarwal, T.K. New contrast enhancement approach for dark images with non-uniform illumination. *Comput. Electr. Eng.* **2017**. doi:10.1016/j.compeleceng.2017.09.007. [CrossRef]
4. Jebadurai, J.; Peter, J.D. SK-SVR: Sigmoid kernel support vector regression based in-scale single image super-resolution. *Pattern Recognit. Lett.* **2017**, *94*, 144–153. doi:10.1016/j.patrec.2017.04.013. [CrossRef]
5. Lee, C.C.; Shih, C.Y. Classification of Liver Disease from CT Images Using Sigmoid Radial Basis Function Neural Network. In Proceedings of the 2009 WRI World Congress on Computer Science and Information Engineering, Los Angeles, CA, USA, 31 March–2 April 2009; Volume 5, pp. 656–660.
6. Kondo, T.; Ueno, J.; Takao, S. Hybrid Multi-layered GMDH-type Neural Network Using Principal Component Regression Analysis and its Application to Medical Image Diagnosis of Liver Cancer. *Procedia Comput. Sci.* **2013**, *22*, 172–181. doi:10.1016/j.procs.2013.09.093. [CrossRef]
7. Gandhamal, A.; Talbar, S.; Gajre, S.; Hani, A.F.M.; Kumar, D. Local gray level S-curve transformation: A generalized contrast enhancement technique for medical images. *Comput. Biol. Med.* **2017**, *83*, 120–133. doi:10.1016/j.compbiomed.2017.03.001. [CrossRef] [PubMed]
8. Huang, S.M.; Liu, H.L.; Li, M.L. Improved temperature imaging of focused ultrasound thermal therapy using a sigmoid model based cross-correlation algorithm. In Proceedings of the 2012 IEEE International Ultrasonics Symposium, Dresden, Germany, 7–10 October 2012; pp. 2766–2768.
9. Shao, H.; Zhang, Y.; Xian, M.; Cheng, H.; Xu, F.; Ding, J. A saliency model for automated tumor detection in breast ultrasound images. In Proceedings of the 2015 IEEE International Conference on Image Processing—ICIP, Quebec City, QC, Canada, 27–30 September 2015; pp. 1424–1428.
10. Almajalid, R.; Shan, J.; Du, Y.; Zhang, M. Development of a Deep-Learning-Based Method for Breast Ultrasound Image Segmentation. In Proceedings of the 2018 17th IEEE International Conference on Machine Learning and Applications (ICMLA), Orlando, FL, USA, 17–20 December 2018; pp. 1103–1108.
11. Luo, X.; Zhang, Z.; Zhang, C.; Wu, X. Multi-focus image fusion using HOSVD and edge intensity. *J. Vis. Commun. Image Represent.* **2017**, *45*, 46–61. doi:10.1016/j.jvcir.2017.02.006.

[CrossRef]
12. Zhang, X.; Li, X.; Feng, Y. A new multifocus image fusion based on spectrum comparison. *Signal Process.* **2016**, *123*, 127–142. doi:10.1016/j.sigpro.2016.01.006. [CrossRef]
13. Luo, X.; Zhang, Z.; Wu, X. Image Fusion Using Region Segmentation and Sigmoid Function. In Proceedings of the 2014 22nd International Conference on Pattern Recognition, Stockholm, Sweden, 24–28 August 2014; pp. 1049–1054.
14. Jung, C.; Cao, L.; Liu, H.; Kim, J. Visual comfort enhancement in stereoscopic 3D images using saliency-adaptive nonlinear disparity mapping. *Displays* **2015**, *40*, 17–23. doi:10.1016/j.displa.2015.05.006. [CrossRef]
15. Tsallis, C.; Mendes, R.; Plastino, A. The role of constraints within generalized nonextensive statistics. *Phys. A Stat. Mech. Its Appl.* **1998**, *261*, 534–554. doi:10.1016/S0378-4371(98)00437-3. [CrossRef]
16. Tsallis, C. Nonextensive Statistics: Theoretical, Experimental and Computational Evidences and Connections. *Braz. J. Phys.* **1999**, *29*, 1–35. [CrossRef]
17. Borges, E.P. Irreversibilidade, Desordem e Incerteza: Tres Visoes da Generalizacao do Conceito de Entropia. *Rev. Bras. De Ensino De Fis.* **1999**, *21*, 453.
18. Anastasiadis, A. Special Issue: Tsallis Entropy. *Entropy* **2012**, *14*, 174–176. [CrossRef]
19. Zang, W.; Wang, Z.; Jiang, D.; Liu, X.; Jiang, Z. Classification of MRI Brain Images Using DNA Genetic Algorithms Optimized Tsallis Entropy and Support Vector Machine. *Entropy* **2018**, *20*, 964. [CrossRef]
20. Albuquerque, M.P.; Albuquerque, M.P.; Esquef, I.A.; Mello, A.R.G. Image thresholding using Tsallis entropy. *Pattern Recognit. Lett.* **2004**, *25*, 1059–1065. [CrossRef]
21. Rodrigues, P.S.; Wachs-Lopes, G.A.; Erdmann, H.R.; Ribeiro, M.P.; Giraldi, G.A. Improving a firefly meta-heuristic for multilevel image segmentation using Tsallis entropy. *Pattern Anal. Appl.* **2015**, *20*, 1–20. [CrossRef]
22. Rodrigues, P.S.; Chang, R.F.; Giraldi, G.A.; Suri, J.S. Non-extensive entropy for cad systems of breast cancer images. In Proceedings of 19th Brazilian Symposium on Computer Graphics and Image Processing, Manaus, Brazil, 8–11 October 2006; pp. 121–128.
23. Sparavigna, A.C. On the Role of Tsallis Entropy in Image Processing. *Int. Sci. Res. J.* **2015**, *1*, 16–24. [CrossRef]
24. Sezgin, M.; Sankur, B. Survey over image thresholding techniques and quantitative performance evaluation. *J. Electron. Imaging* **2004**, *13*, 146–168.
25. Ronneberger, O.; Fischer, P.; Brox, T. U-Net: Convolutional Networks for Biomedical Image Segmentation. In *Medical Image Computing and Computer-Assisted Intervention (MICCAI)*; LNCS; Springer: Cham, Switzerland, 2015; Volume 9351, pp. 234–241.
26. Tsallis, C. What are the numbers that experiments provide. *Quim. Nova* **1994**, *17*, 468–471.
27. Rodrigues, P.S.; Giraldi, G.A.; Provenzano, M.; Faria, M.D.; Chang, R.F.; Suri, J.S. A new methodology based on q-entropy for breast lesion classification in 3-D ultrasound images. In Proceedings of the 28th International Conference of the IEEE Engineering in Medicine and Biology Society—EMBS'06, New York, NY, USA, 30 August–3 September 2006; pp. 1048–1051.
28. Rodrigues, P.S.; Giraldi, G.A. Computing the q-index for Tsallis Nonextensive Image Segmentation. In Proceedings of the SIBIGRAPI 2009, Rio de Janeiro, Brazil, 11–15 October 2009; pp. 232–237.
29. Rodrigues, P.S.; Giraldi, G.A. Improving the non-extensive medical image segmentation based on Tsallis entropy. *Pattern Anal. Appl.* **2011**, *14*, 369–379. [CrossRef]
30. Gonzalez, R.; Woods, R. *Processamento Digital de Imagens*, 3rd ed.; Pearson Prentice Hall: Upper Saddle River, NJ, USA, 2010.
31. Serra, J. (Ed.) *Image Analysis and Mathematical Morphology*; Academic: New York, NY, USA, 1982.
32. Chen, Y. An allometric scaling relation based on logistic growth of cities. *Chaos Solitons Fractals* **2014**, *65*, 65–77. doi:10.1016/j.chaos.2014.04.017. [CrossRef]
33. Chen, Y.G. Logistic Models of Fractal Dimension Growth of Urban Morphology. *Fractals* **2018**, *26*. [CrossRef]
34. Jain, A.K. *Fundamentals of Digital Image Processing*; Prentice-Hall, Inc.: Upper Saddle River, NJ, USA, 1989.
35. Giraldi, G.A.; Rodrigues, P.S.S. Theoretical Elements in Fourier Analysis of q-Gaussian Functions. *Theor. Appl. Inform.* **2016**, *27*, 16–44.
36. Borges, E.P.; Tsallis, C.; Miranda, J.G.V.; Andrade, R.F.S. Mother wavelet functions generalized through q-exponentials. *J. Phys. A Math. Gen.* **2004**, *37*, 9125.

[CrossRef]

37. Johal, R.S. q calculus and entropy in nonextensive statistical physics. *Phys. Rev. E* **1998**, *58*, 4147–4151. [CrossRef]
38. Wachs-Lopes, G.; Horvath, M.; Giraldi, G.; Rodrigues, P. A strategy based on non-extensive statistics to improve frame-matching algorithms under large viewpoint changes. *Signal Process. Image Commun.* **2019**, *75*, 44–54. [CrossRef]

entropy

Article

Large-Scale Person Re-Identification Based on Deep Hash Learning

Xian-Qin Ma, Chong-Chong Yu *, Xiu-Xin Chen and Lan Zhou

Beijing Key Laboratory of Big Data Technology for Food Safety, Beijing Technology and Business University, Beijing 100048, China; 18211128839@163.com (X.-Q.M.); chenxx1979@126.com (X.-X.C.); zhou0210x@163.com (L.Z.)

* Correspondence: chongzhy@vip.sina.com; Tel.: +86-139-1111-9035

Received: 31 March 2019; Accepted: 28 April 2019; Published: 30 April 2019

Abstract: Person re-identification in the image processing domain has been a challenging research topic due to the influence of pedestrian posture, background, lighting, and other factors. In this paper, the method of harsh learning is applied in person re-identification, and we propose a person re-identification method based on deep hash learning. By improving the conventional method, the method proposed in this paper uses an easy-to-optimize shallow convolutional neural network to learn the inherent implicit relationship of the image and then extracts the deep features of the image. Then, a hash layer with three-step calculation is incorporated in the fully connected layer of the network. The hash function is learned and mapped into a hash code through the connection between the network layers. The generation of the hash code satisfies the requirements that minimize the error of the sum of quantization loss and Softmax regression cross-entropy loss, which achieve the end-to-end generation of hash code in the network. After obtaining the hash code through the network, the distance between the pedestrian image hash code to be retrieved and the pedestrian image hash code library is calculated to implement the person re-identification. Experiments conducted on multiple standard datasets show that our deep hashing network achieves the comparable performances and outperforms other hashing methods with large margins on Rank-1 and mAP value identification rates in pedestrian re-identification. Besides, our method is predominant in the efficiency of training and retrieval in contrast to other pedestrian re-identification algorithms.

Keywords: person re-identification; image analysis; hash layer; quantization loss; Hamming distance; cross-entropy loss

1. Introduction

Pedestrian re-identification is the task of searching for and finding the same pedestrian captured by different cameras or the same camera at different times. In recent years, due to its extremely important applications in video monitoring, human–computer interaction, and other fields, this research has been a hot topic in the field of computer vision. However, affected by many factors, such as variable shooting angles, surrounding circumstances, and pedestrian behaviors, pedestrian re-identification still faces tremendous challenges.

Traditional pedestrian re-identification research [1,2] mostly uses the picture-concerned operator to extract feature information, such as global features, local features, pedestrian image color, texture, edge, shape, and other characteristics of information for later re-identification. In recent research on measurement learning methods for nearest neighbor classification, Weinberger and Saul [3] proposed the large margin nearest neighbor (LMNN) method, which sets the boundary for the target neighbor and punishment for the non-matching points of the intrusion boundary. To avoid the problems of over-fitting in LMNN, Davis et al. [4] proposed the information theoretic metric learning (ITML) method, which balances satisfying the given similarity constraint and ensures that the measure is close to the initial distance function.

In recent years, methods based on approximate nearest neighbors [5] to retrieve large-scale image datasets have been proposed. Among them, hash learning is a new approximate nearest neighbor re-identification method that represents an image as a string of fixed length and makes similar samples have similar binary coding [6], which has excellent performance in large-scale image re-identification. Image hash re-identification methods generally consist of image feature extraction to obtain an image feature code and hash mapping of that feature code.

In the traditional hash re-identification method, when extracting the low-dimensional compact features of an image, the image color, texture, shape, gradient, and other underlying feature information can be attained by manipulating the picture with the designed descriptor operator. When applied in complex scenarios, this method often fails to cover all cases well. The data-independent hash method [7], unsupervised hash learning method [8–13], and supervised learning algorithm [14,15] mainly use related feature operators to extract features before hashing to retrieve images.

The method based on deep learning can obtain useful features from the data with better generalization, semantic feature extraction, and strong expression ability. In 2009, using the non-linear expression of the deep learning model, Salakhutdinov and Hinton [16] first proposed the hash learning method based on a deep learning algorithm, of which the input feature remains artificial. In 2014, Xia el al. [17] proposed the convolutional neural network hashing (CNNH) model, which divides the hash learning into two stages. In the first stage, the image-paired similarity matrix based on LMNN is decomposed into binary vectors to complete the binary coding of image data and to be used as the training tag of the second stage. In the second stage, the binary coding of the convolutional neural network (CNN) model fitting image is constructed, and the performance of the model is improved by the classification loss function. However, the image representation learned in this method does not react to the update of binary hash code. In 2015, on the basis of weighting the Hamming distance, Zhang et al. [18] proposed the Deep Regularized Similarity Comparison Hashing (DRSCH). In 2016, Peng and Li [19] achieved hash code mapping by introducing the hash function into the final layer of the deep learning network. The output of the final layer in the network is the hash code. Meanwhile, classification loss is considered, but the number of categories in the dataset is too high, which leads to a significant increase in the dimension of the hash code.

Pedestrian re-identification is a branch of image re-identification, but some mature methods in image re-identification have not been applied in the field of pedestrian re-identification. Inspired by the deep hash learning algorithm, this paper proposes convolutional feature network person hash learning (CFNPHL) for pedestrian re-identification, which could achieve the learning of hashing code in pedestrian images. By calculating the distance between the hashing codes of pedestrian images, the re-identification of large-scale pedestrian image data is realized. The pedestrian hash learning algorithm of deep convolutional feature network proposed in this paper is different from other hash learning algorithms. The main contributions of this paper are as follows:

(1) The hash layer is applied to the fully connected layer in the proposed algorithm, which generates a hash code to learn the pedestrian images in an end-to-end approach. Additionally, for the integrity of the information, the quantized loss function is utilized in the procedure of generating the hash code in the hash layer.

(2) In order to make the hash code from the same pedestrian obtained by the network be the same or similar and that from a different pedestrian be of great discrepancy, the Softmax layer is adopted after the hash layer, which utilizes the Softmax cross-entropy loss to realize the higher constraint on the generated hash code of the network.

The proposed deep convolution network can extract feature information from a pedestrian image. Then, after utilizing the connection between fully connected layers to learn the hash function and mapping the feature information into the corresponding hash code, the network constructs a classifier layer for the pedestrian image. These efforts could make this network suitable for any huge pedestrian re-identification datasets, instead of limited datasets.

2. Related Work

In recent years, deep learning has made unprecedented achievements in various research fields. With stronger ability to extract image feature information, deep convolutional network replaces the traditional method of extracting pedestrian feature information by using manually designed image feature expression operators [20–26] in the research on pedestrian recognition. Among the existing pedestrian re-identification learning methods, they are mainly divided into three categories: Representation Learning-, Metric Learning-, and Generative Adversarial Networks (GAN)-based method.

The pedestrian re-identification method based on Representation Learning is a common method [26–30], which mainly uses the CNN to automatically extract image feature information according to task requirements after training, and then classifies or verifies pedestrians according to extracted feature information to achieve the purpose of re-recognition. Metric Learning is the most widely used method in the field of image retrieval. In the pedestrian re-identification, the similarity between the same pedestrian and different peers is mainly recognized. In the process of network training, the distance between the same pedestrian images (positive sample pairs) is minimized, while the distance between different pedestrian images (negative sample pairs) is maximized. A commonly used method of measuring learning loss consists of contrastive loss [31], triplet loss [32–34] and quadruplet loss [35]. The GAN-based method [36–39] is proposed to address the problem that the current dataset of pedestrian recognition is small.

By generating new pedestrian images, the result of pedestrian recognition can be improved. The current pedestrian re-identification methods utilize images for pedestrian retrieval. However, in practice, the storage of each image needs to take up some space, and the retrieval of the image reduces the speed of recognition. Therefore, a commonly used hash retrieval method in the image retrieval filed was adopted in this paper. The length of hash code can be set manually. Usually, the number of hash codes will be much smaller than the size of the original picture, which will save a lot of storage space and also improve the speed of pedestrian recognition.

3. CFNPHL Methods

The network structure of the CFNPHL model proposed in this paper is shown in Figure 1. The input data of this model are pedestrian images and label information. First, the convolutional feature network in the model is used to learn the feature information of pedestrian images. Then, the feature information of the pedestrian image is input into the full-connection layer of the network and the hash layer is introduced into the full-connection layer, so that the hash function can be learned and mapped into a hash code. Finally, in the process of training, the quantization loss of hash layer, and Softmax loss of classifier are updated to optimize network structure parameters.

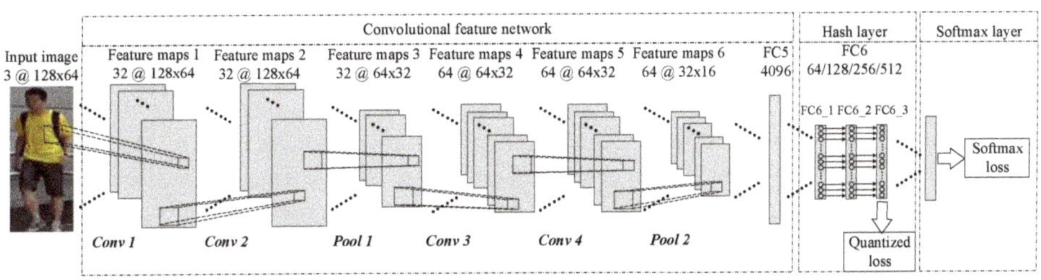

Figure 1. The network structure of the CFNPHL model.

As shown in Figure 1, the CFNPHL model we proposed is composed of three parts: convolutional feature network, hash layer and Softmax layer. In Figure 1, we take the 3-channel RGB color picture with a size of 128 × 64 as the input of the network for example. In the layer of convolution feature network, convolution and pooling are carried out, where *con* $]\varphi(]\varphi = 1, \cdots, 4)$ represents the convolution for four times and *pool* $\varepsilon(\varepsilon = 1, 2)$ indicates max pooling for two times. Different feature maps could be obtained after operations. $C@M \times N$ indicates the size of the feature map is M × N with C dimension. The specific parameters and relevant processes of operation in this part will be introduced in detail in Section 3.1. The second part is the hash layer. The main purpose of this part is to convert the 4096-dimensional data of FC5 layer into the hash code of specific length, in which FC_1, FC_2, and FC_3 operate hash function mapping, Sigmoid function, and threshold, respectively. Moreover, it would be introduced in Section 3.2 specifically. The third part is the Softmax layer, which mainly differentiates the information of the same pedestrian and different pedestrians according to the category. This part is introduced in Section 3.3.

3.1. Convolutional Feature Network

The function of the convolutional feature network is to learn the feature representation information of the pedestrian image. When the pedestrian image is input into the network, the image feature information of the pedestrian can be obtained through the convolutional feature network.

It is significant that the network could quickly extract the deep pedestrian feature information. Additionally, the network should be easy to be optimized. Therefore, the convolutional feature network in this paper adopts a structure of four CNN layers, two pooling layers and one full-connection layer, which is shown in Figure 1. The proposed hash layer does not constrain the network structure of convolution feature extraction, which means that the hash layer could be applied to other networks such as AlexNet or GoogLeNet. For example, if the input image size is 128 × 26, the parameter settings of the convolutional feature network are shown in Table 1. The dimension of pedestrian feature information extracted in the Pool2 layer of the convolutional feature network is relatively high. If it is directly connected with the FC5 of the full-connection layer, it can easily lead to over-fitting of the network. Thus, the dropout layer is introduced after the Pool2 layer, of which the purpose is to make the output of some nodes set to 0 with a certain probability in the training process. Pedestrian feature information obtained by the dropout layer is connected to the FC5 layer with a dimension of 4096, and the FC5 layer can obtain the representative information of pedestrian features extracted by the convolutional feature network.

Table 1. Convolutional feature network parameter setting.

Layer	Output Size	Parameter Setting
Conv1	128 × 64 × 32	3 × 3, 32, pad = 0
Conv2	128 × 64 × 32	3 × 3, 32, pad = 0
Pool1	64 × 32 × 32	2 × 2, max pool, stride = 2
Conv3	64 × 32 × 64	3 × 3, 64, pad = 0
Conv4	64 × 32 × 64	3 × 3, 64, pad = 0
Pool2	32 × 16 × 64	2 × 2, max pool, stride = 2
FC5	4096	4096

The convolutional feature network in this paper has a shallow structural hierarchy. In the process of training, the network has a fast convergence speed and a short training time, and the network weight parameters are easily optimized. It can extract pedestrian image feature information with a strong expressive ability.

3.2. *Hash Layer of the Network*

Instead of the traditional algorithm to build the corresponding hash function mapping the feature information into the hash code, the mapping relationship between hash layers was introduced in this paper to represent the procedure of hashing code from the hash function.

With the goals of efficient learning of the hash function and mapping into the hash code corresponding to the pedestrian image, this paper introduces a hash layer between FC5 and FC7 in the full-collection layer, of which the structure is depicted in Figure 1. By optimizing the weight of the network connection between FC5 of the full connection layer and FC6_1 of the hash layer, the pedestrian feature information used for FC5 layer learning $t_i(i = 1, 2, \cdots, 4096)$ is mapped into the hash function of the hash layer FC6_1 value $x_j(j = 1, 2, \cdots, n)$. The activation function used in the convolutional feature network is the rectified linear unit (ReLU), which is defined below as Equation (1):

$$f(x) = \max(0, x) \tag{1}$$

The output of the hash layer ranges from $[0, x)$ instead of a binary numeric string. Therefore, a constraint should be made to the hash layer FC6_1 so that it could generate a binary hash code.

Then, the FC6_2 layer is introduced into the hash layer, which could use the Sigmoid function to activate the value from the FC6_1 layer so that the output range of the FC6_2 layer is [40,41]. The Sigmoid function is defined below as Equation (2):

$$y = \frac{1}{1 + e^{-x}} \tag{2}$$

where y denotes the output of the FC6_2 layer. Therefore, the value of FC6_2 is not a binary numeric string. To further obtain binary numeric strings, the FC6_3 layer is introduced into the hash layer, which quantizes the output of the FC6_2 layer. The quantized function is defined below as Equation (3):

$$H(x) = \begin{cases} 0, & y \leq K \\ 1, & y > K \end{cases} \tag{3}$$

where K denotes a threshold. After Equation (3) is processed, the output of the FC6_3 layer is a binary hash code.

3.3. *Loss Function of Network*

In person re-identification, the deep networks often suffer from the over-fitting problem. It is difficult to solve the problem of overfitting by adjusting the hyperparameter of the network. In addition, changing the structure of the network can reduce the complexity of the network to a certain extent, but it is difficult to fundamentally solve the problem of overfitting in the problem of pedestrian re-identification. In this paper, in order to prevent the occurrence of network overfitting problem, we first design a shallow CNN on the network structure to extract the feature information of pedestrian images. Besides, two different loss functions are adopted to overcome the overfitting problem, namely using the measure loss function and the cross-entropy loss function to constrain the generated hash code value.

In the process of training the deep CNN, the weight parameters of the network need to be reversely adjusted by constantly reducing the error loss value of the network, of which the purpose is to achieve the optimal network. The error loss function in the CFNPHL model proposed in this paper consists of the following two parts.

3.3.1. Quantized Loss Function of the FC6_3 layer in the Hash Layer.

The quantized loss function of the FC6_3 layer in the hash layer indicates the information loss during the process that restricts the output of the FC6_3 layer to obtain the binary hash code.

For the purpose of minimizing the information loss, quantized loss is adopted, which is defined below as Equation (4):

$$Loss_q = \|y - H(x)\|_2^2 \tag{4}$$

where y denotes the output of the FC6_2 layer and $H(x)$ is mentioned in Equation (3). Motivated by quantized loss, the processed output of the FC6_2 layer in the hash layer will gradually approach 0 or 1 in the training process, which would result in less loss of information in the FC6_3 layer.

3.3.2. The Softmax Cross-Entropy Loss Function

The Softmax classifier is used in the full-connection layer of the network to enable the model to learn the characteristic information that distinguishes different pedestrians, so that the characteristic codes of the same pedestrian are similar, and the characteristic codes of different pedestrians are not similar.

In the logical regression of the Softmax classifier, a tagged sample is used as the dataset: $\{(x^{(1)}, y^{(1)}), (x^{(2)}, y^{(2)}), \cdots, (x^{(m)}, y^{(m)})\}$, where $x^{(i)} \in R^{n+1}$ denotes a feature vector with n+1 dimensions. In the multi-classification, the value of y can be taken as k, where k is the number of classification categories, denoted as $y^{(i)} \in \{1, 2, \cdots, k\}$. The mission of the Softmax classifier is to approximate the probability that the input vector belongs to j. The function is defined below as Equation (5):

$$h_k(x^{(i)}) = \begin{bmatrix} p(y^{(i)} = 1|x^{(i)}; \theta_1|) \\ p(y^{(i)} = 2|x^{(i)}; \theta_2|) \\ \vdots \\ p(y^{(i)} = k|x^{(i)}; \theta_k|) \end{bmatrix} = \frac{1}{\sum_{j=1}^{k} e^{\theta_j^T x^{(i)}}} \begin{bmatrix} e^{\theta_1^T x^{(i)}} \\ e^{\theta_2^T x^{(i)}} \\ \vdots \\ e^{\theta_k^T x^{(i)}} \end{bmatrix} \tag{5}$$

where $h_k(x^{(i)})$ indicates the output of the kth Softmax layer based on the image of pedestrian $x^{(i)} \in R^{n+1}$, $\theta_1, \theta_2, \cdots, \theta_k \in R^{n+1}$ denotes the parameters of the model, and T represents the number of iterations in network training. Besides, the final result of Equation (5) is the probability that the input $x^{(i)} \in R^{n+1}$ is classified into $y^{(i)} = j(k = 1, 2, \cdots, k)$ category.

$$Loss_c(\theta) = -\frac{1}{m}\left[\sum_{i=1}^{m}\sum_{j=1}^{k} 1\{y^{(i)} = j\} \log \frac{e^{\theta_j^T x^{(i)}}}{\sum_{l=1}^{k} e^{\theta_l^T x^{(i)}}}\right] \tag{6}$$

where $Loss_c(\theta)$ denotes the cross entropy loss value, m is the number of training samples, k represents the number of pedestrian types and $1\{y^{(i)} = j\}$ is the characteristic function. If $y^{(i)} = j$ is true, the function is 1, otherwise not.

The total loss of CFNPHL is the sum of the quantized loss value and cross-entropy loss value. Consequently, the total loss is defined below as Equation (7):

$$loss = Loss_q + Loss_c \tag{7}$$

where $loss$ denotes the total loss, $Loss_q$ denotes the quantized loss mentioned in Equation (4), and $Loss_c$ denotes the Softmax cross-entropy loss mentioned in Equation (6). In the process of training, the network weight is reversely adjusted through the loss value so that the network gradually reaches the optimal value and loss gradually approaches zero.

3.4. Distance Re-Identification of Hash Code in a Pedestrian Image

In image re-identification, the Euclidean distance, Markov distance, Cosine distance, and Hamming distance are commonly used to measure image similarity. The pedestrian re-identification method used in this paper first converts the pedestrian image into the hash code, and then measures the hash code. In the information theory, the Hamming distance represents the number of different characters in the corresponding position of two equal-length strings. The Hamming distance is often considered to measure the equal-length hash codes generated in image retrieval methods [17,18]. In this paper, the hash code generated by the pedestrian image is used to measure the Hamming distance to realize pedestrian re-identification. The Hamming distance is expressed as $H(p_x, p_y)$, where p_x denotes the hash code to be identified and p_y denotes the generated pedestrian hash code library.

The purpose of training CFNPHL model proposed in this paper is to make the Hamming distance of the hash code obtained by the same pedestrian closer and that of the hash code obtained by different pedestrians farther. A simple Hamming distance calculation example is provided in Figure 2 below, in which $p_{1,1}$ and $p_{1,2}$ are from the same pedestrian of which hash codes are 10,101,100 and 10,101,101, respectively, while $p_{2,1}$ is from a different pedestrian of which the hash code is 11,000,010. Supposing $p_{1,2}$ as the pedestrian image to be identified, the Hamming distance is calculated between the image to be identified and the image from the hash code library. Additionally, the result that $H(p_{1,2}, p_{1,1}) = 1, H(p_{1,2}, p_{2,1}) = 6$ could be obtained, which indicates the Hamming distance could measure the image similarity using the generated hash code. Eventually, the mission of pedestrian re-identification is completed by sorting the Hamming distance.

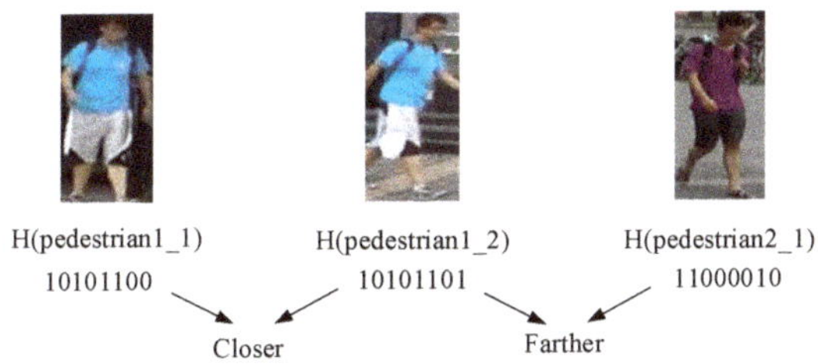

Figure 2. Hamming distance map.

4. Experiments

The computer used in this experiment is configured as follows: Ubuntu16.04 + caffe 1.0 + Tensorflow1.2 + MATLAB 2016b; CPU: Intel i7 8700k; RAM: 16G ×2; GPU: NVIDIA GTX1080Ti.

4.1. Datasets

The experimental dataset consists of three kinds of pedestrian datasets: CUHK02 [40], Market-1501 [41], and DukeMTMC [42]. Brief descriptions of each dataset are as follows:

CUHK02 dataset [40]: A total of 7264 images of 1816 people were collected from 5 different outdoor cameras. Each person has four images, and the collected images are 160 × 60 in size. Two images of each person were taken at different times under each camera.

Market-1501 dataset [41]: This dataset was collected on the campus of Tsinghua University, and the images came from six different cameras, one of which was low-pixel. The image sizes are 64 × 128. Images were automatically detected and segmented by the detector, including some detection errors (close to the actual use). There are 751 people in the training data and 750 people are in the test set. Therefore, in the training set, each person has 17.2 pieces of training data on average.

DukeMTMC dataset [42]: The dataset was collected at Duke University, and images were taken from eight different cameras. This dataset provides a training set and a test set. The training set contains 16,522 images, the test set contains 17,661 images, and the query set contains 2228 images. There are 702 people in the training data, with an average of 23.5 pieces of training data for each person. The size of the image is not fixed, making it the largest pedestrian recognition dataset at present.

4.2. Settings

In this paper, two evaluation indexes are used to evaluate the performance of the hash method in the pedestrian re-identification dataset:

(1) Mean average precision (mAP) [43]

$$AP = \frac{1}{\sum_{i=1}^{m} r_i} \sum_{i=1}^{m} r_i \left(\frac{\sum_{j=1}^{i} r_j}{i}\right) \tag{8}$$

$$mAP = \frac{1}{M} \sum_{p=1}^{M} AP(p) \tag{9}$$

A pedestrian image is picked from the query set, and then m images with the highest similarity are selected from the candidate database. Therefore, in Equation (8), AP represents a probability that the query image is in the m images. If the image of pedestrian to be re-identified is consistent with the ith image in m images, r_i is set 1, otherwise 0. In Equation (9), mAP is the mean of AP for M images in the query set.

(2) Cumulative match characteristic (CMC) [44]

$$CMC(R) = \frac{1}{N} \sum_{q=1}^{N} \begin{cases} 1, & r_q \leq R \\ 0, & r_q > R \end{cases} \tag{10}$$

In Equation (10), with N pedestrians in the query set and N times of query and rank, the result of the target pedestrian in each query could be represented in $r = (r_1, r_2, \cdots, r_M)$.

In this paper, Rank-1, Rank-5, Rank-10 and Rank-20 are selected from the evaluation CMC metric. Among the CMC performance evaluation metrics, the Rank-1 is the most commonly used one.

To test and compare the advantages of the proposed method, the following three comparative experiments were carried out across the above four datasets.

(1) Contrast experiment with other hash algorithms

The hash algorithms compared in this paper include five unsupervised hashing methods, including Locality-Sensitive Hashing (LSH) [7], Spectral Hashing (SH) [9], Principal Component Analysis Hashing (PCAH) [10], Spherical Hashing (SPH) [12], Sparse Embedding and Least Variance Encoding (SELVE) [13] and three supervised hashing methods, including Binary Reconstructive Embeddings (BRE) [14], Minimal Loss Hashing (MLH) [15], and Principal Component Analysis directions and random matrix R (PCA-RR) [8]. The experimental results of these eight methods were obtained according to the parameter setting scheme provided by the corresponding authors.

(2) Influence of different hash code dimensions on CMC.

(3) Ignoring the quantized loss function proposed in this paper, we compared the CMC metric on various datasets.

(4) Pose invariant embedding (PIE) [26] versus the method proposed in this paper: The comparative experiment analysis on pedestrian re-identification was conducted on the dataset Market-1501. Besides, the relevant parameters of the experiment were set according to the parameters provided by the author.

For the above four comparative experiments, the settings of the three datasets used in this paper are as follows. In the CUHK02 dataset, of which the testing set contains 1816 images, one image of each person was randomly selected as the test dataset. For the unsupervised hash methods, the remaining images were considered as training sets. In the supervised hash method and the method proposed in this paper, two images were randomly selected from the unchosen/remaining images, with 3832 images as the training set and the rest as the test set.

For the dataset provided by Market-1501, each pedestrian image was randomly selected from the data as the testing set. In this paper, in the supervised hash method and the proposed method, the remaining images were used as training sets. In the unsupervised hash method, 80% of the remaining data were selected as the training set and the rest were chosen as the verification set.

For the DukeMTMC dataset, the image size was uniformly set to 200 × 80. In the supervised hash method and the method proposed in this paper, the training set, testing set and verification set were all provided by the original data. In the unsupervised hash method, the training set and testing set were unified as the training set, and the verification set remained unchanged.

The traditional hash algorithm involves two steps: extracting the characteristic information of the image by using the manually designed image analysis operator and generating the hash code from extracted feature information by establishing the corresponding hash learning algorithm. In the comparative experiment of traditional hash learning algorithm, instead of the traditional operator for feature extraction, it is the feature information of a 4096-dimension image extracted by pre-trained AlexNet as the input. Then, the hash algorithm of comparative experiment is utilized to generate the hash code. Eventually, the Hamming distance between hash codes is calculated for pedestrian re-identification.

4.3. Results and Analysis

4.3.1. Comparison Across Diverse Hash Methods

The result shown in Figure 3 is the mAP value on three data points under the four kinds of hash code dimensions of 64 bits, 128 bits, 256 bits, and 512 bits. When the dimension of the hash code is 256 bits, Table 2 shows Rank-1 and mAP across the three datasets.

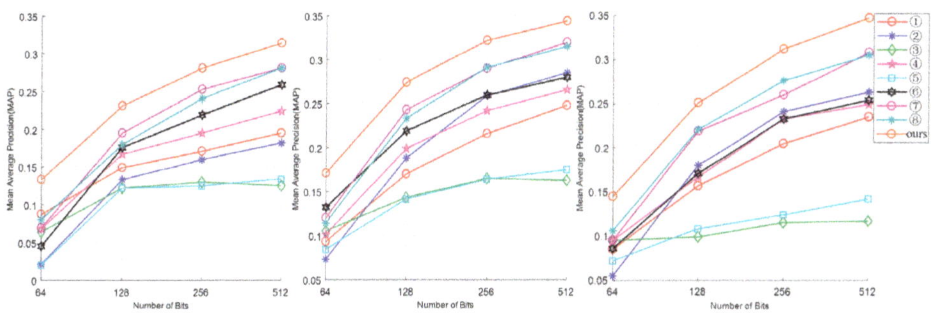

(a) (b) (c)

Figure 3. mAP values for different hash code dimensions of the three datasets: (**a**) mAP results of the CUHK02 dataset; (**b**) mAP results of the Market-1501 dataset; (**c**) mAP results of the DukeMTMC dataset.

Table 2. Comparative evaluation.

Method	CUHK02 [40]		Market-1501 [41]		DukeMTMC [42]	
	Rank-1	mAP	Rank-1	mAP	Rank-1	mAP
LSH [7]	20.3	17.1	23.5	21.6	23.3	20.5
PCA-RR [8]	26.8	24.1	34.3	29.5	29.5	27.6
SH [9]	18.1	16	28.6	25.9	27.9	24.1
PCAH [10]	14.3	13	18.9	16.5	15.4	11.5
SPH [12]	21.4	19.5	28.4	24.2	26.3	23.3
SELVE [13]	15.7	12.5	19.2	16.4	16.1	12.4
BRE [14]	23.4	21.9	30.4	26.1	25.4	23.4
MLH [15]	27.7	25.3	35.1	29.3	28.8	26.0
Our	**29.3**	**27.5**	**38.1**	**34.4**	**31.5**	**30.2**

As shown in Figure 3, eight sets of comparative experiments are indicated by the numbers ①–⑧, which represent LSH [7], SH [9], PCAH [10], SPH [12], SELVE [13], BRE [14], MLH [15], and PCA-RR [8], respectively. Additionally, 'ours' represents the proposed method.

It can be inferred from Figure 3 that our experimental results are significantly better than other comparative experimental methods. In the eight-group comparative experiment on pedestrian re-identification, the experimental results of MLH [15] and PCA-RR [8] with a supervised hashing learning method are better than those with an unsupervised hashing learning method. In addition, the experimental results of PCAH [10] and SELVE [13] are poor, and there is no significant improvement with the ascent of hash code dimension.

The following points can be gathered from the results:

1) Compared with other hash re-identification algorithms and the traditional hash algorithm (PCA-RR), the proposed method is equipped with a higher mAP value when pedestrian re-identification is performed across the three datasets. In addition, when the dimension of the hash code comes to 256 bits, the mAP values increase by 4%, 3.1%, and 3.5% for the respective datasets.

2) The mAP and Rank-1 are important evaluation metrics in the field of pedestrian re-identification. Conditioning on 256 bits of the hash code, Table 2 shows the result of both evaluations on three datasets. As can be inferred in Table 2, our method performs better than the traditional hash method in perspective of both evaluation metrics. It proves that our method can better learn the corresponding hash function according to the characteristics of the dataset and map it into the hash cod, which could improve the accuracy of pedestrian re-identification.

As can be seen from Figure 3 and Table 2, the mAP values and Rank-1 values obtained by PCAH and SELVE methods on the three datasets are relatively low. We hypothesize that a great deal of pedestrian feature information is lost when the algorithm processes the high-dimensional data, which leads to the hash code from the same pedestrian differing in similarity resulting in a low mAP value. It also could be seen from the result that supervised hash learning algorithms, MLH and PCA-RR, have greater mAP values than unsupervised learning algorithms, PCAH and SELVE. Besides, the results of the other algorithms like LSH, SH, SPH, and BRE are somewhat different. Our deep hashing network achieves the comparable performances and outperforms other hashing methods with large margins on Rank-1 and mAP value identification rates.

4.3.2. Influence of Variant Dimension of Hash Code on CMC and mAP

The effects of different dimensions of the hash layer on the CMC value and mAP is shown in Table 3 below.

Table 3. Effects of hash codes of different dimensions on CMC values and mAP.

Dataset	Hash Code Dimension	CMC				mAP
		Rank-1	Rank-5	Rank-10	Rank-20	
CUHK02 [40]	**64**	14.2	18.1	25.4	31.9	13.3
	128	24.1	31.3	39.2	46.7	22.1
	256	29.3	34.4	44.3	51.8	27.5
	512	**33.1**	**40.2**	**47.4**	**57.2**	**31.4**
Market-1501 [41]	**64**	17.4	24.2	29.6	40.1	17.1
	128	27.2	31.1	36.4	47	25.9
	256	33.1	35.4	45.9	56.3	31.2
	512	**38.1**	**46.3**	**55.2**	**65.8**	**34.4**
DukeMTMC [42]	**64**	15.1	17.2	22.3	28.3	14.5
	128	25.5	30.3	39.1	42.5	24.1
	256	31.5	38.1	40.5	46.1	30.2
	512	**36.5**	**40.5**	**47.8**	**54.9**	**33.5**

As can be seen from Table 3, accompanied with the growth of the hash code's dimension, the CMC metric and mAP increase simultaneously, in which the Rank-1 metric is significantly growing, and the Rank-10 and Rank-20 metrics are in a smooth growth when the dimension of hash code comes to 256 or 512 bits. The reason why such a phenomenon occurs is that the description of the feature is more detailed and concrete as the dimension of the hash code increases, which means that the probability of finding the same target in the first search is higher. Nevertheless, as the hash code length increases exponentially, the re-identification speed of the hard disk storage space rises.

4.3.3. Ignorance of the Quantized Loss Function Proposed

To further investigate the influence of the quantized loss function on the result, the CMC metric and mAP, a comparative experiment was conducted to identify whether the quantized loss function was considered or not. Quantized loss was not considered in the training, only classification loss was used for model training, and the other experimental parameters were set in accordance with the settings considering quantized loss, in which the dimension of hash code was set to 256.

From the experimental results in Table 4, it can be concluded that across the three datasets, the Rank-1 metric values in CMC obtained by the model without considering the loss of measurement are reduced by 5.1%, 4.4%, and 4.2%, respectively. The loss measurement function proposed in this paper improves the re-identification accuracy of the model to some extent.

Table 4. Effects of quantitative loss function on CMC values of experimental results.

Dataset	Method	CMC				mAP
		Rank-1	Rank-5	Rank-10	Rank-20	
CUHK02 [40]	Our-	24.2	32.8	41.9	49.3	20.2
	Our	**29.3**	**34.4**	**44.3**	**51.8**	**27.5**
Market-1501 [41]	Our-	28.7	31.0	40.8	53.2	25.4
	Our	**33.1**	**35.4**	**45.9**	**56.3**	**31.2**
DukeMTMC [42]	Our-	27.3	34.9	0.371	0.413	24.3
	Our	**31.5**	**38.1**	**40.5**	**46.1**	**30.2**

4.3.4. Comparative Experiment Analysis of Pedestrian Re-Identification on the PIE Method

For the fact that Market-1501 is the common dataset between our method and PIE method [26], the comparative experiment and analysis were conducted merely on the dataset Market-1501.

There are many research methods for pedestrian re-identification at present, and the reason why we choose the PIE method could be concluded as follow. First, the PIE method is one of the classical and convincing methods to solve the problem of pedestrian re-identification. Second, the result of PIE method is satisfying, indicating that it was a dominant baseline in the pedestrian re-identification filed. The direct comparison experiment between PIE and ours can highlight the advantages of the proposed method. Therefore, we believe that comparison experiments using the PIE method can emphasize the contribution of this paper to some extent.

In this part of the experiment, the dimension of the hash layer was set as 512. The mAP values of the two pedestrian re-identification methods obtained through the experiment in the Market-1501 dataset and the results of Rank-1 in the CMC are shown in the Table 5.

Table 5. The mAP and Rank-1 on the Market-1501 [41] dataset.

Method	Rank-1	mAP
PIE [26]	**78.65**	**53.87**
Ours	38.1	34.4

It can be seen from the experimental results in Table 5 that the mAP value and Rank-1 value of our deep hash pedestrian re-identification algorithm are lower than the experimental results of PIE. The reason for such a situation could be concluded that in the procedure of mapping the pedestrian feature information into the hash code in the hash layer, the feature information of pedestrian gets compressed, which results in inevitable loss of feature information. While the PIE method is to recognize the feature information generated by deep CNN directly. Therefore, the PIE method performs better than our method in pedestrian re-identification.

In the network, the extracted pedestrian feature information changes from a high-dimensional to a low-dimensional hash code, which will inevitably cause the loss of pedestrian feature information. Therefore, the ability to maintain the important feature information is the key to improving the result of pedestrian re-identification after the transformation. As can be seen from Table 3, the use of hash codes with higher dimensions can improve the result of pedestrian re-identification to some extent, but the result rate of pedestrian recognition with higher dimensions becomes slower and the retrieval time becomes longer. Therefore, it is necessary to determine appropriate dimensions and improve the network's ability to map pedestrian feature information into the hash code by using the hash function. In this way, excessive pedestrian feature information can be maintained and prevented from being lost, so as to maximize the efficiency of pedestrian re-identification with the hash code.

However, during the experiment, the loss of our method was calculated by quantization loss and cross-entropy loss (Equation (7)). However, the loss of PIE method is the sum of Ori.img Softmax loss, PIE Softmax loss, and PoseBox Softmax loss. It was found that the convergence speed of loss values in the proposed CFNPHL model is faster than that of PIE, as shown in Figure 4.

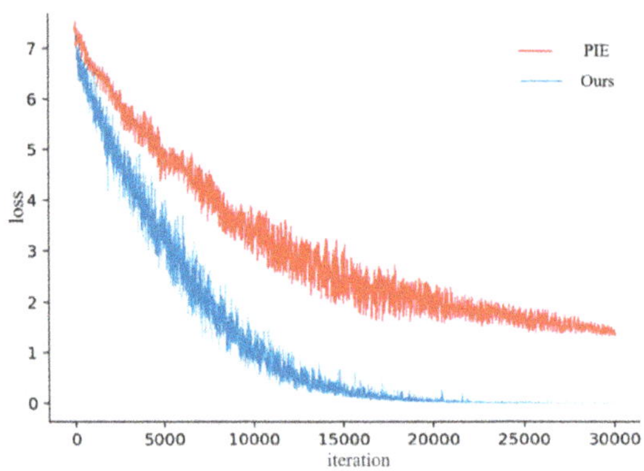

Figure 4. The loss value of CFNPHL and PIE.

It can be seen from Figure 4 that loss of CFNPHL model converges rapidly, and after 23,000 iterations, the loss value is close to 0. Compared with the CFNPHL, the PIE does not converge until 30,000 iterations are completed but the loss value is not close to 0.

Besides, as shown in Figure 5, the comparison of train time costs of 30,000 iterations in the same circumstance for both the models is conducted. PIE takes 12 h to converge while our proposed CFNPHL merely takes 7 h to converge. It could be inferred that the CFNPHL is trained more efficiently.

With the Market-1501 dataset for validation, our proposed method exceeds the PIE method in the speed of pedestrian re-identification. Moreover, the specific test time cost is shown in Table 6. Due to the transformation of pedestrian feature information extracted by the deep convolution network into the fixed length hash code, the dimension of data is immensely reduced. Therefore, our method achieves better performance than the PIE method.

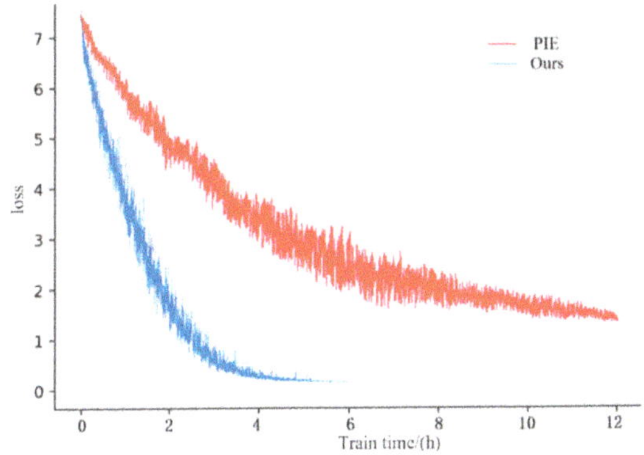

Figure 5. The train time costs of 30,000 iterations for CFNPHL and PIE.

Table 6. Test time of CFNPHL and PIE on the Market-1501 dataset.

Method	Test Time (min)			
PIE [26]	25.3			
Ours	64 bits 5.4	128 bits 7.1	256 bits 11.5	512 bits 17.7

As can be seen in Table 6, the PIE model spends much more time in testing than the CFNPHL model proposed in this paper. It can also be seen from the time taken by the different hash code dimensions of our model that the higher the hash code dimension, the longer the time taken.

5. Conclusions

Equipped with the deep neural network that has an excellent capacity to extract features and the hash layer that maps the pedestrian feature into a binary hash code, the CFNPHL model proposed in this paper utilizes quantized loss and the Softmax cross-entropy loss to achieve end-to-end training. This method applies the hash algorithm to the pedestrian re-identification domain for research, and results show that the proposed deep hash re-identification algorithm has a higher re-identification accuracy than other hash pedestrian re-identification algorithms. The experimental results show that compared with the PIE method, our method performs better in training time, convergence speed and retrieval time, but in the pedestrian recognition results (including Rank-1 and mAP), our method still needs to improve. Therefore, numerous endeavors could be devoted to researching hash algorithms in pedestrian re-identification.

Author Contributions: C.-C.Y. conceived and designed the experiments; X.-Q.M., X.-X.C. and L.Z. carried out the experiments and collected the date under the supervision of C.-C.Y.; X.-Q.M. analyzed the data and wrote the paper; C.-C.Y. and X.-X.C. reviewed and edited the manuscript.

Funding: This work was supported by the National Key Research & Development Program of China (project No. 2018YFC0807900, No. 2018YFC0807903), and the Beijing Municipal Education Commission Research Plan (project No. KM201510011010).

Acknowledgments: We would like to express thanks to the anonymous reviewers for their invaluable comments and suggestions.

Conflicts of Interest: The authors declare no conflict of interest.

References

1. Liu, C.; Chen, C.L.; Chen, C.L.; Lin, X.G. *Person re-identification: What Features Are Important? European Conference on Computer Vision*; Springer: Berlin/Heidelberg, Germany, 2012; pp. 391–401.

2. Gray, D.; Brennan, S.; Tao, H. Evaluating appearance models for recognition, reacquisition, and tracking. In Proceedings of the IEEE International Workshop on Performance Evaluation for Tracking and Surveillance, Rio de Janeiro, Brizil, 20 September 2007.
3. Weinberger, K.Q.; Saul, L.K. Distance Metric Learning for Large Margin Nearest Neighbor Classification. *J. Mach. Learn. Res.* **2009**, *10*, 207–244.
4. Davis, J.V.; Kulis, B.; Jain, P.; Sra, S.; Dhillon, I. Information-theoretic metric learning. In Proceedings of the 24th International Conference on Machine Learning, Corvalis, OR, USA, 20–26 June 2007; pp. 209–216.
5. Wang, J.; Kumar, S.; Chang, S.-F. Semi-supervised hashing for scalable image re-identification. In Proceedings of the IEEE Conference on Computer Vision and Pattern Recognition, San Francisco, CA, USA, 13–18 June 2010; pp. 3424–3431.
6. Chen, S.Z.; Guo, C.C.; Lai, J.H. Deep Ranking for Person Re-identification via Joint Representation Learning. *IEEE Trans. Image Process.* **2015**, *25*, 2353–2367. [CrossRef] [PubMed]
7. Datarm, M.; Immorlicam, N.; Indyk, P.; Mirroknim, V.S. Locality-sensitive hashing scheme based on p-stable distributions. In Proceedings of the Twentieth Annual Symposium on Computational Geometry, New York, NY, USA, 8–11 June 2004; pp. 253–262.
8. Gong, Y.; Lazebnik, S. Iterative quantization: A procrustean approach to learning binary codes. *IEEE Trans. Pattern Anal. Mach. Intell.* **2011**, *35*, 2916–2929. [CrossRef] [PubMed]
9. Weissm, Y.; Torralbam, A.; Fergus, R. Spectral hashing. In Proceedings of the 21st International Conference on Neural Information Processing Systems, Vancouver, BC, Canada, 8–10December 2008.
10. Yum, X.; Zhangm, S.; Lium, B.; Linm, Z.; Metaxasm, D. Large-Scale Medical Image Search via Unsupervised PCA Hashing. In Proceedings of the Computer Vision & Pattern Recognition Workshops, Portland, OR, USA, 23–28 June 2013.
11. Jin, Z.; Li, C.; Lin, Y.; Cai, D. Density Sensitive Hashing. *IEEE Trans. Cybern.* **2013**, *44*, 1362–1371. [CrossRef]
12. Heo, J.P.; Lee, Y.; He, J.; Chang, S.; Yoon, S. Spherical hashing. In Proceedings of the CVPR, IEEE Computer Society Conference on Computer Vision and Pattern Recognition. IEEE Computer Society Conference on Computer Vision and Pattern Recognition, Providence, RI, USA, 16–21 June 2012.
13. Zhu, X.; Zhang, L.; Huang, Z. A Sparse Embedding and Least Variance Encoding Approach to Hashing. *IEEE Trans. Image Process.* **2014**, *23*, 3737–3750. [CrossRef] [PubMed]
14. Kulis, B.; Darrell, T. Learning to Hash with Binary Reconstructive Embeddings. In Proceedings of the NIPS'09 Proceedings of the 22nd International Conference on Neural Information Processing Systems, Vancouver, BC, Canada, 7–10 December 2009.
15. Norouzi, M.E.; Fleet, D.J. Minimal Loss Hashing for Compact Binary Codes. In Proceedings of the 28th International Conference on Machine Learning, ICML 2011, Bellevue, WA, USA, 28 June– 2 July 2011.
16. Salakhutdinov, R.; Hinton, G. Semantic hashing. *Inter. J. Approx. Reason.* **2009**, *50*, 969–978. [CrossRef]
17. Xia, R.; Pan, Y.; Lai, H. Supervised hashing for image re-identification via image representation learning. In Proceedings of the AAAI Conference on Artificial Intelligence, Québec, Canada, 27–31 July 2014.
18. Zhang, R.; Lin, L.; Zhang, R.; Zuo, W.; Zhang, L. Bit-Scalable Deep Hashing with Regularized Similarity Learning for Image Re-identification and Person Re-Identification. *IEEE Trans. Image Process.* **2015**, *24*, 4766–4779. [CrossRef] [PubMed]
19. Peng, T.; Li, F. Image Re-identification Based on Deep Convolutional Neural Networks and Binary Hashing Learning. *J. Electron. Inform. Technol.* **2016**, *38*, 2068–2075.
20. Bazzani, L.; Cristani, M.; Murino, V. Symmetry-driven accumulation of local features for human characterization and re-identification. *Comput. Vis. Image Und.* **2013**, *117*, 130–144. [CrossRef]
21. Liao, S.; Hu, Y.; Zhu, X.; Li, S. Person re-identification by Local Maximal Occurrence representation and metric learning. In Proceedings of the 2015 IEEE Conference on Computer Vision and Pattern Recognition (CVPR), Boston, MA, USA, 7–12 June 2015.
22. Lisanti, G.; Karaman, S.; Masi, I. Multichannel-Kernel Canonical Correlation Analysis for Cross-View Person Reidentification. *ACM Trans. Multimedia Comput. Communi. Appli.* **2017**, *13*, 1–19. [CrossRef]
23. Matsukawa, T.; Okabe, T.; Suzuki, E.; Sato, Y. Hierarchical Gaussian Descriptor for Person Re-identification. In Proceedings of the 2016 IEEE Conference on Computer Vision and Pattern Recognition (CVPR), Las Vegas, NV, USA, 27–30 June 2016.
24. Yang, Y.; Yang, J.; Yan, J.; Liao, S.; Yi, D.; Li, S. Salient Color Names for Person Re-identification. *Lect. Note. Comput. Sci.* **2014**, *8689*, 536–551.

25. Zhao, R.; Ouyang, W.; Wang, X. Unsupervised Salience Learning for Person Re-identification. In Proceedings of the 2013 IEEE Conference on Computer Vision and Pattern Recognition (CVPR), Portland, OR, USA, 23–28 June 2013.
26. Zheng, L.; Huang, Y.; Lu, H.; Yang, Y. Pose Invariant Embedding for Deep Person Re-identification. *arXiv* **2017**, arXiv:1701.07732. [CrossRef] [PubMed]
27. Geng, M.; Wang, Y.; Xiang, T.; Tian, Y. Deep Transfer Learning for Person Re-identification. *arXiv* **2016**, arXiv:1611.05244.
28. Lin, Y.; Zheng, L.; Zheng, Z.; Wu, Y.; Yang, Y. Improving Person Re-identification by Attribute and Identity Learning. *arXiv* **2017**, arXiv:1703.07220.
29. Zheng, L.; Yang, Y.; Hauptmann, A.G. Person Re-identification: Past, Present and Future. *arXiv* **2016**, arXiv:1610.02984.
30. Matsukawa, T.; Suzuki, E. Person re-identification using CNN features learned from combination of attributes. In Proceedings of the 2016 23rd International Conference on Pattern Recognition (ICPR), Cancun, Mexico, 4–8 December 2016.
31. Varior, R.R.; Haloi, M.; Wang, G. Gated Siamese Convolutional Neural Network Architecture for Human Re-Identification. In *European Conference on Computer Vision*; Springer: Cham, Switzerland, 2016.
32. Liu, H.; Feng, J.; Qi, M.; Jiang, J.; Yan, S. End-to-End Comparative Attention Networks for Person Re-identification. *IEEE Trans. Image Process.* **2017**. [CrossRef]
33. Hermans, A.; Beyer, L.; Leibe, B. In Defense of the Triplet Loss for Person Re-Identification. *arXiv* **2017**, arXiv:1703.07737.
34. Cheng, D.; Gong, Y.; Zhou, S.; Wang, J.; Zheng, N. Person re-identification by multi-channel parts-based CNN with improved triplet loss function. In Proceedings of the 2016 IEEE Conference on Computer Vision and Pattern Recognition (CVPR), Las Vegas, NV, USA, 27–30 June 201.
35. Chen, W.; Chen, X.; Zhang, J.; Huang, K. Beyond triplet loss: A deep quadruplet network for person re-identification. *arXiv* **2017**, arXiv:1704.01719.
36. Zheng, Z.; Zheng, L.; Yang, Y. Unlabeled Samples Generated by GAN Improve the Person Re-identification Baseline in vitro. *arXiv* **2017**, arXiv:1701.07717.
37. Zhong, Z.; Zheng, L.; Zheng, Z.; Li, S.; Yang, Y. Camera Style Adaptation for Person Re-identification. *arXiv* **2018**, arXiv:1711.10295.
38. Wei, L.; Zhang, S.; Gao, W.; Tian, Q. Person Transfer GAN to Bridge Domain Gap for Person Re-Identification. *arXiv* **2017**, arXiv:1711.08565.
39. Qian, X.; Fu, Y.; Xiang, T.; Wang, W.; Wu, Y.; Jiang, Y.; Xue, X. Pose-Normalized Image Generation for Person Re-identification. *arXiv* **2017**, arXiv:1712.02225.
40. Li, W.; Wang, X. Locally Aligned Feature Transforms across Views. In Proceedings of the IEEE Conference on Computer Vision and Pattern Recognition, Portland, OR, USA, 23–28 June 2013; pp. 3594–3601.
41. Zheng, L.; Shen, L.; Tian, L.; Wang, S.; Wang, J.; Tian, Q. Scalable Person Re-identification: A Benchmark. In Proceedings of the IEEE International Conference on Computer Vision, Santiago, Chile, 7–13 December 2015; pp. 1116–1124.
42. Ristani, E.; Solera, F.; Zou, R.; Cucchiara, R.; Tomasi, C. Performance Measures and a Data Set for Multi-Target, Multi-Camera Tracking. *arXiv* **2016**, arXiv:1609.01775.
43. Turpin, A.; Scholer, F. User performance versus precision measures for simple search tasks. In Proceedings of the SIGIR '06 Proceedings of the 29th Annual International ACM SIGIR Conference on Research and Development in Information Retrieval, Seattle, WA, USA, 6–11 August 2006; pp. 11–18.
44. Liu, C. POP: Person Re-identification Post-rank Optimisation. In Proceedings of the IEEE International Conference on Computer Vision, Sydney, NSW, Australia, 1–8 December 2013.

MDPI
St. Alban-Anlage 66
4052 Basel
Switzerland
Tel. +41 61 683 77 34
Fax +41 61 302 89 18
www.mdpi.com

Entropy Editorial Office
E-mail: entropy@mdpi.com
www.mdpi.com/journal/entropy

www.ingramcontent.com/pod-product-compliance
Lightning Source LLC
Chambersburg PA
CBHW051703210326
41597CB00032B/5356
9783039210923